Texts and Monographs in Physics

Series Editors:

R. Balian, Gif-sur-Yvette, France
W. Beiglböck, Heidelberg, Germany
H. Grosse, Wien, Austria
W. Thirring, Wien, Austria

Yakov M. Shnir

Magnetic Monopoles

Springer

Dr. Yakov M. Shnir
Institute of Physics
Carl von Ossietzky University Oldenburg
26111 Oldenburg
Germany
E-mail: shnir@theorie.physik.uni-oldenburg.de

Library of Congress Control Number: 2005930438

ISBN-10 3-540-25277-0 Springer Berlin Heidelberg New York
ISBN-13 978-3-540-25277-1 Springer Berlin Heidelberg New York

This work is subject to copyright. All rights are reserved, whether the whole or part of the material is concerned, specifically the rights of translation, reprinting, reuse of illustrations, recitation, broadcasting, reproduction on microfilm or in any other way, and storage in data banks. Duplication of this publication or parts thereof is permitted only under the provisions of the German Copyright Law of September 9, 1965, in its current version, and permission for use must always be obtained from Springer. Violations are liable for prosecution under the German Copyright Law.

Springer is a part of Springer Science+Business Media
springeronline.com
© Springer-Verlag Berlin Heidelberg 2005
Printed in The Netherlands

The use of general descriptive names, registered names, trademarks, etc. in this publication does not imply, even in the absence of a specific statement, that such names are exempt from the relevant protective laws and regulations and therefore free for general use.

Typesetting: by the authors and TechBooks using a Springer LATEX macro package
Cover design: *design & production* GmbH, Heidelberg

Printed on acid-free paper   SPIN: 11398127   55/TechBooks   5 4 3 2 1 0

To Marina with love

Preface

> *"One would be surprised if Nature had made no use of it."*
>
> P.A.M. Dirac

According to some dictionaries, one meaning of the notion of "beauty" is "symmetry". Probably, beauty is not entirely "in the eye of the beholder". It seems to be related to the symmetry of the object. From a physical viewpoint, this definition is very attractive: it allows us to describe a central concept of theoretical physics over the last two centuries as being a quest for higher symmetry of Nature. The more symmetric the theory, the more beautiful it looks.

Unfortunately, our imperfect (at least at low-energy scale) world is full of nasty broken symmetries. This has impelled physicists to try to understand how this happens. In some cases, it is possible to reveal the mechanism of violation and how the symmetry may be recovered; then our picture of Nature becomes a bit more beautiful.

One of the problems of the broken symmetry that we see is that, while there are electric charges in our world, their counterparts, *magnetic monopoles*, have not been found. Thus, in the absence of the monopoles, the symmetry between electric and magnetic quantities is lost. Can this symmetry be regained?

In the history of theoretical physics, the hypothesis about the possible existence of a magnetic monopole has no analogy. There is no other purely theoretical construction that has managed not only to survive, without any experimental evidence, in the course of more than a century, but has also remained the focus of intensive research by generations of physicists.

Over the past 25 years the theory of magnetic monopoles has surprisingly become closely connected with many actual directions of theoretical physics. This includes the problem of confinement in Quantum Chromodynamics, the problem of proton decay, astrophysics and evolution of the early Universe, and the supersymmetrical extension of the Standard Model, to name just a few. It seems plausible that the answer to the question: "Why do magnetic monopoles not exist?" is a key to understanding the very foundations of Nature. Furthermore, the mathematical problem of construction and investigation of the exact multimonopole configurations is at the frontier of the most fascinating directions of modern field theory and differential geometry. The techniques developed in this area of theoretical physics find many other applications and have become very important mathematical tools.

The theory of monopoles seems to be tailor-made for demonstrating beautiful interplay between mathematics and physics. Therefore, I believe that an introduction to the basic ideas and techniques that are related to the description and construction of monopoles may be useful to physicists and mathematicians interested in the modern developments in this direction. Moreover, there is a second aspect of the monopoles. These objects arise in many different contexts running through all levels of modern theoretical physics, from classical mechanics and electrodynamics to multidimensional branes. This provides an alternative point of view on the subject, which may be of interest to readers.

My original motivation was to provide a comprehensive review on the monopole that would capture the current status of the problem, something which could be entitled *"Everything you always wanted to know about the monopole but did not have time to ask"*. However, it soon became clear that such a project was too ambitious. An estimate of the related literature approaches 6000 papers. The original paper by Dirac [200] has been quoted more than 1000 times and the citation index of the papers by 't Hooft and Polyakov [270, 428] is approaching 1400.

I have therefore tried to give a restricted introduction to the classical and quantum field theory of monopoles, a more or less compact review, which could give a "bird's eye view" on the entire set of problems connected with the field theoretical aspects of the monopole.

The book is divided into three parts. This approach reproduces in some sense that used by S. Coleman in his famous lectures [43]; that is, I start the discussion with a simple classical consideration of a monopole as seen at large distances and then go on to its internal structure.

In Part I, the monopole is considered "from afar", at the large distances where pure electrodynamical description works well. In the first chapter, I review some features of the classical interaction between a static monopole and an electric charge. The quantum mechanical consideration in terms of the Dirac potential is described in Chapter 2. Next, in Chapter 3 the notions of topology, which are closely related to the theory of monopole, are described. Chapter 4 is devoted to the generalization of QED, which includes the monopoles. Part II forms the core of the book. There I discuss the theory of non-Abelian monopoles, construction of the multimonopole solutions, and some applications. In Chapter 5 the famous 't Hooft–Polyakov solution, the simplest specimen of the monopole family, is discussed. This is the first step inside the monopole core. I review the basic properties of the classical non-Abelian monopoles, which arise in spontaneously broken $SU(2)$ gauge theory, and the relation that exists between the magnetic charge of the configuration and the topological charge. The Bogomol'nyi–Prasad–Sommerfield (BPS) monopole appears here for the first time as a particular analytic solution with vanishing potential. Here I also give a brief account of the gauge zero mode and comment on its relation to the electric charge. Chapter 6 contains a survey of the classical multimonopoles, both in the BPS limit and beyond. A powerful formalism for investigation of the low-energy dynamics of the BPS monopoles is the moduli space approach, which arises from consideration of the monopole collective coordinates. In Chapter 7 some of the results related to the quantum field theory of the $SU(2)$ monopoles are reviewed.

Next, in Chapter 8 the consideration is extended to a more general class of $SU(3)$ theories containing different limits of symmetry breaking. It turns out that the multimonopole configurations are natural in a model with the gauge group of higher rank. Here I discuss fundamental and composite monopoles and consider the limiting situation of the massless states.

Chapter 9 contains a brief survey of the role that the monopoles may play in the phenomenon of confinement. I discuss here the compact lattice electrodynamics, formalism of Abelian projection in gluodynamics and the Polyakov solution of confinement in the 2+1-dimensional Georgi–Glashow model. In Chapter 10 the original Yang–Mills–Higgs system is extended by inclusion of fermions. Here I consider the details of the monopole–fermion interaction, especially the role of the fermionic zero modes of the Dirac equation. In this context, I briefly describe the current status of the Rubakov–Callan effect.

The last part of the book reveals the intersection of many lines of the previous discussion. Indeed, the spectrum of states of $N = 2$ supersymmetric (SUSY) Yang–Mills theory includes the monopoles. There the arguments of duality become well-founded and the BPS mass bound arises in a new context. Moreover, the geometrical moduli space approach, which was originally developed to describe the dynamics of BPS monopoles, turns out to be a key element of the Seiberg–Witten solution of the low-energy $N = 2$ SUSY Yang–Mills theory. Chapter 11 is an introductory account of supersymmetry. Construction of the $N = 2$ $SU(2)$ supersymmetric monopoles is described in Chapter 12 and the Seiberg–Witten solution is presented in Chapter 13. Evidently, this is a separate topic, which has been intensively discussed in recent years. However, the very structure of the book does not make it possible to avoid such a discussion. The reader will definitely find this topic well presented elsewhere.

Let us mention some omissions. An obvious gap is the current experimental situation. I do not venture to discuss the numerous experiments directed to the search for a monopole. This must be the subject of a separate survey. I would like to point the reader to the very good reviews [47, 48, 50]. However, the most important thing we know from experiment is that there are probably no monopoles around.

I do not consider the astrophysical aspects of monopoles, the problem of relic monopoles, or other related directions. I do not discuss some by-product topics like, for example, the conception of the Berry phase. Neither do I consider some specific mathematical problems of the Abelian theory of monopoles (e.g., singularities and regularization). In considering construction of the BPS multimonopoles, I have made no attempt to discuss one of the approaches that is related to the application of the inverse scattering method to the linearized Bogomol'nyi equation. Instead, the discussion concentrates on the modern development due to the Nahm technique and twistor approach. I would like to draw attention to the recent excellent monograph by N. Manton and P. Sutcliffe, "Topological Solitons" [54], which provides the reader with a solid framework of modern classical theory of solitons, not only monopoles, in a very general context.

Because of the restricted size of the book, I do not consider the very interesting properties of gravitating monopoles, which are solutions of the Einstein–Yang–Mills–Higgs theory. I pay more attention to the general properties of the non-Abelian monopoles, namely, to their topological nature. Coupling with gravity yields a number of classical solutions that are not presented in flat space, so that the related discussion becomes rather involved. Another omission is the Kaluza–Klein monopole and, more generally, the analysis of multidimensional theories. For more rigor and broader discussion I refer the reader to the original publications.

Though extensive, the list of references at the end of the book cannot be considered an exhaustive bibliography on monopoles. I apologize to those authors whose contributions are not mentioned here.

The work on this project coincided with a period of serious personal turmoil. I am grateful to all my friends and colleagues who supported me. I am deeply indebted to Ana Achucarro, Emil Akhmedov, Alexander Andrianov, Dmitri Antonov, Jürgen Baacke, Pierre van Baal, Askhat Gazizov, Dmitri Diakonov, Conor Houghton, Iosif Khriplovich, Viktor Kim, Valerij Kiselev, Ken Konishi, Boris Krippa, Steffen Krusch, Dieter Maison, Stephane Nonnenmacher, Alexander Pankov, Murray Peshkin, Victor Petrov, Lutz Polley, Mikhail Polikarpov, Maxim Polyakov, Kirill Samokhin, Ruedi Seiler, Andrei Smilga, Joe Sucher, Paul Sutcliffe, Tigran Tchrakian, Arthur Tregubovich, Andreas Wipf, and Wojtek Zakrzewski for many useful discussions, critical interest and remarks. I am very thankful to L.M. Tomilchik and E.A. Tolkachev, who were my teachers and advisors, for their valuable support, encouragement, and guidance. They awakened my interest in the monopole problem.

Many of the ideas discussed here are due to Nick Manton, who played a very important role in my understanding of the monopoles, both through his papers and in private discussions. He commands my deepest personal respect and gratitude. The year I spent in Cambridge in his group strongly influenced my life.

This book originates from work in collaboration with Per Osland which, unfortunately, was not completed. Without his support and encouragement I would never have started to work on this extended project. A draft version of the first five chapters was prepared in collaboration with him during my stays at the Institute of Physics, University of Bergen.

I am deeply indebted to Burkhard Kleihaus and Jutta Kunz for collaboration and help in numerous ways. The support I received in Oldenburg has been invaluable.

My special thanks go to Milutin Blagojević, Maxim Chernodub, Adriano Di Giacomo, Fridrich W. Hehl, and Valentine Zakharov for reading a preliminary version of several chapters and providing many helpful comments, suggestions, and remarks.

I would like to acknowledge the hospitality I received at the Service de Physique Théorique, CEA-Saclay, the Max-Planck-Institut für Physik (Werner-Heisenberg-Institut), München, and the Abdus Salam International Center for Theoretical Physics, Trieste, where some parts of this work were carried out. A substantial part of the work on the manuscript was done in 1999–2002 at the Institute of Theoretical Physics, University of Cologne. Some chapters of the book are elaborations of lectures given on several occasions.

Oldenburg, *Yakov Shnir*
June 2005

Contents

Part I Dirac Monopole

1 Magnetic Monopole in Classical Theory 3
   1.1 Non-Relativistic Scattering on a Magnetic Charge 3
   1.2 Non-Relativistic Scattering on a Dyon..................... 10
   1.3 Vector Potential of a Monopole Field 12
   1.4 Transformations of the String 15
   1.5 Dynamical Symmetries of the Charge-Monopole System 20
   1.6 Dual Invariance of Classical Electrodynamics............... 22

**2 The Electron–Monopole System:
Quantum-Mechanical Interaction** 27
   2.1 Charge Quantization Condition 27
   2.2 Spin-Statistics Theorem in a Monopole Theory 31
   2.3 Charge-Monopole System: Quantum-Mechanical Description . 33
      2.3.1 The Generalized Spherical Harmonics 34
      2.3.2 Solving the Radial Schrödinger Equation 37
   2.4 Non-Relativistic Scattering on a Monopole:
Quantum Mechanical Description......................... 40
   2.5 Charge-Monopole System: Spin in the Pauli Approximation.. 42
      2.5.1 Dynamical Supersymmery
of the Electron-Monopole System................... 44
      2.5.2 Generalized Spinor Harmonics: $j \geq \mu + 1/2$ 46
      2.5.3 Generalized Spinor Harmonics: $j = \mu - 1/2$ 48
      2.5.4 Solving the Radial Pauli Equation 49
   2.6 Charge-Monopole System: Solving the Dirac Equation 53
      2.6.1 Zero Modes and Witten Effect 55
      2.6.2 Charge Quantization Condition
and the Group $SL(2,\mathbb{Z})$ 61

3 Topological Roots of the Abelian Monopole............... 67
   3.1 Abelian Wu–Yang Monopole 67
   3.2 Differential Geometry and Topology 70
      3.2.1 Notions of Topology 70
      3.2.2 Notions of Differential Geometry 81

	3.2.3	Maxwell Electrodynamics and Differential Forms	89
3.3	Wu–Yang Monopole and the Fiber-Bundle Topology		93
	3.3.1	Fiber Bundles..	93
	3.3.2	Principal Bundle and Connection.....................	97
	3.3.3	Wu–Yang Monopole Bundle	102
	3.3.4	Hopf Bundle...	103

4 Abelian Monopole: Relativistic Quantum Theory 109

- 4.1 Two Types of Charges 110
- 4.2 Two-Potential Formulation of Electrodynamics 112
  - 4.2.1 Energy-Momentum Tensor and Angular Momentum .. 115
- 4.3 Canonical Quantization 118
  - 4.3.1 Relativistic Invariance of Two-Charge Electrodynamics 121
- 4.4 Renormalization of QED with a Magnetic Charge 125
- 4.5 Vacuum Polarization by a Dyon Field 128
- 4.6 Effective Lagrangian of QED with a Magnetic Charge 132

Part II Monopole in Non-Abelian Gauge Theories

5 't Hooft–Polyakov Monopole............................... 141

- 5.1 $SU(2)$ Georgi–Glashow Model and the Vacuum Structure.... 141
  - 5.1.1 Non-Abelian Wu–Yang Monopole.................. 141
  - 5.1.2 Georgi–Glashow Model............................ 143
  - 5.1.3 Topological Classification of the Solutions 146
  - 5.1.4 Definition of Magnetic Charge 148
  - 5.1.5 't Hooft–Polyakov Ansatz 151
  - 5.1.6 Singular Gauge Transformations and the Connection between 't Hooft–Polyakov and Dirac monopoles 154
  - 5.1.7 Dyons 155
- 5.2 The Bogomol'nyi Limit................................... 157
  - 5.2.1 Gauge Zero Mode and the Electric Dyon Charge 161
- 5.3 Topological Classification of Non-Abelian Monopoles 163
  - 5.3.1 $SO(3)$ vs $SU(2)$ 163
  - 5.3.2 Magnetic Charge and the Topology of the Gauge Group 165
  - 5.3.3 Equivalence of Topological and Magnetic Charge 166
  - 5.3.4 Topology of the Dyon Sector....................... 168
- 5.4 The θ Term and the Witten Effect Again 170

Contents

6 Multimonopole Configurations 173
  6.1 Multimonopoles Configurations
      and Singular Gauge Transformations 174
      6.1.1 Singular $SU(2)$ Monopole with Charge $g = ng_0$ 174
      6.1.2 Magnetic Dipole 176
  6.2 Rebbi–Rossi Multimonopoles, Chains
      of Monopoles and Closed Vortices 178
  6.3 Interaction of Magnetic Monopoles 192
      6.3.1 Monopole in External Magnetic Field 192
      6.3.2 The Interaction Energy of Monopoles 194
      6.3.3 Classical Interaction
           of Two Widely Separated Dyons 197
  6.4 The n-Monopole Configuration in the BPS Limit 201
      6.4.1 BPS Multimonopoles: A Bird's Eye View 201
      6.4.2 Projective Spaces and Twistor Methods 203
      6.4.3 The n-Monopole Twistor Construction 206
      6.4.4 Hitchin Approach and the Spectral Curve 214
      6.4.5 Nahm Equations 217
      6.4.6 Solution of the Nahm Equations 220
      6.4.7 The Nahm Data and Spectral Curve 223
  6.5 Moduli Space and Low-Energy Multimonopoles Dynamics ... 227
      6.5.1 Zero Modes Lagrangian and the Moduli Space Metric . 227
      6.5.2 Metric on the Space $\mathcal{M}_2$ 232
      6.5.3 Low-Energy Scattering of Two Monopoles 236

7 $SU(2)$ Monopole in Quantum Theory 241
  7.1 Field Fluctuations on Monopole Background 241
      7.1.1 Generalized Angular Momentum and the Spectrum
           of Fluctuations 245
      7.1.2 Quantum Correction to the Mass of a Monopole 250
  7.2 Non-Abelian Monopole: Quasiclassical Quantization 254
      7.2.1 Collective Coordinates and Constraints 254
      7.2.2 Quantum Mechanics on the Moduli Space 258
      7.2.3 Evaluation of the Generating Functional 263
  7.3 $g\bar{g}$ Pair Creation in an External Magnetic Field 267
      7.3.1 Dynamics of Non-Abelian Monopole
           in Weak External Field 267
      7.3.2 Metastable Vacuum Decay and Monopole Pair
           Creation in an External Field 269

8 Monopoles Beyond $SU(2)$ Group 275
  8.1 $SU(N)$ Monopoles 276
      8.1.1 Generalization of the Charge Quantization Condition . 276
      8.1.2 Towards Higher Rank Gauge Groups 277
      8.1.3 Montonen–Olive Conjecture 279

8.1.4 Cartan–Weyl Basis and the Simple Roots............ 282
 8.1.5 $SU(3)$ Cartan Algebra 284
 8.1.6 $SU(3)$ Monopoles 287
8.2 Massive and Massless Monopoles 301
 8.2.1 Pathologies of Non-Abelian Gauge Transformations... 303
8.3 $SU(3)$ Monopole Moduli Space........................... 306
 8.3.1 $SU(3)$ Monopoles: Nahm Equations 314

9 Monopoles and the Problem of Confinement.............. 319
9.1 Quark Confinement in QCD 319
 9.1.1 Dual Superconductor 324
9.2 Monopoles in the Lattice QCD 328
 9.2.1 Compact QED and Lattice Monopoles 331
 9.2.2 Lattice Duality.................................. 334
9.3 Abelian Projection 339
 9.3.1 "Monopoles" from Abelian Projection.............. 339
 9.3.2 Maximal Abelian Gauge.......................... 346
9.4 Polyakov Solution of Confinement
 in the $d = 3$ Georgi–Glashow Model...................... 349
 9.4.1 Dilute Gas of Monopoles in the $d = 3$
 Georgi–Glashow Model............................ 349
 9.4.2 Wilson Loop Operator in $d = 3$
 Georgi–Glashow Model............................ 355

10 Rubakov–Callan Effect 359
10.1 Dirac Hamiltonian
 on the Non-Abelian Monopole Background 359
 10.1.1 Fermionic Zero Modes 363
 10.1.2 Zero Modes and the Index Theorem 367
 10.1.3 S-Wave Fermion Scattering on a Monopole 373
10.2 Anomalous Non-Conservation of the Fermion Number 378
 10.2.1 Axial Anomaly and the Vacuum Structure........... 378
 10.2.2 Effective Action of Massless Fermions 379
 10.2.3 Properties of the Anomalous Fermion Condensate 385
 10.2.4 Properties of Other Condensates 388
10.3 Monopole-Fermion Scattering
 in the Bosonisation Technique........................... 390
 10.3.1 Vertex Operator and Bosonization
 of the Free Model 391
 10.3.2 Monopole Catalysis of the Proton Decay 397
 10.3.3 Monopole Catalysis of the Proton Decay:
 Semiclassical Model............................... 400

Part III Supersymmetric Monopoles

11 Supersymmetric Yang-Mills Theories 407
  11.1 What is Supersymmetry? 407
    11.1.1 Poincaré Group and Algebra of Generators 408
    11.1.2 Algebra of Generators of Supersymmetry 412
  11.2 Representations of SUSY Algebra 415
    11.2.1 $N = 1$ Massive Multiplets 415
    11.2.2 $N = 1$ Massless Multiplets 417
    11.2.3 $N = 2$ Extended SUSY 418
  11.3 Local Representations of SUSY 420
    11.3.1 $N = 1$ Superspace 420
    11.3.2 $N = 1$ Superfields 424
    11.3.3 Non-Abelian Multiplets 428
  11.4 $N = 1$ SUSY Lagrangians 429

12 Magnetic Monopoles in the $N = 2$ Supersymmetric Yang–Mills Theory 437
  12.1 $N = 2$ Supersymmetric Lagrangian 437
    12.1.1 Praise of Beauty of $N = 2$ SUSY Yang–Mills 441
  12.2 $N = 2$ Supersymmetric $SU(2)$ Magnetic Monopoles 443
    12.2.1 Construction of $N = 2$ Supersymmetric $SU(2)$ Monopoles 443
  12.3 Central Charges in the $N = 2$ SUSY Yang–Mills 446
  12.4 Fermionic Zero Modes in Supersymmetric Theory 449
  12.5 Low Energy Dynamics of Supersymmetric Monopoles 451
  12.6 $N = 2$ Supersymmetric Monopoles beyond $SU(2)$ 453
    12.6.1 $SU(3)$ $N = 2$ Supersymmetric Monopoles 458

13 Seiberg–Witten Solution of $N = 2$ SUSY Yang–Mills Theory 465
  13.1 Moduli Space 466
    13.1.1 Moduli Space and its Parameterization 466
    13.1.2 Quantum Moduli Space of $N = 2$ SUSY Yang–Mills Theory 472
  13.2 Global Parametrization of the Quantum Moduli Space 478
    13.2.1 Transformation of Duality for $N = 2$ Low-Energy Effective Theory 478
    13.2.2 BPS Bound Reexamined 483
  13.3 Seiberg–Witten Explicit Solution 485
    13.3.1 Monodromies on the Moduli Space 485
    13.3.2 Solution of the Monodromy Problem 492
    13.3.3 Confinement and the Monopole Condensation 496
  13.4 Concluding Remarks 498

A	**Representations of $SU(2)$**	501
B	**Quaternions**	505
C	**$SU(2)$ Transformations of the Monopole Potential**	509

References 513

Index 529

Part I

Dirac Monopole

1 Magnetic Monopole in Classical Theory

1.1 Non-Relativistic Scattering on a Magnetic Charge

One could set up a naive definition of a monopole as being just a point-like particle with a magnetic charge instead of an electric one. Then almost all non-trivial features caused by its presence would manifest themselves in the process of interaction between a monopole and "normal" electrically charged particles. One can see these features already on the level of classical mechanics by comparing the electric-charge-monopole scattering and the standard Coulomb problem. Historically that problem was first considered by H. Poincaré in the context of interaction of an electron beam and the pole of a very long and very thin magnet already more than a century ago, in 1896 [425]. This work could be considered a first brick in the foundation of the modern history of the monopole. Nevertheless, one should say that for a long time before H. Poincaré's work, the question about the possible existence of a single magnetic pole was raised many times.[1]

In this section, we will consider the classical non-relativistic motion of a charge in an external field. That is why it would be correct to define a magnetic charge g as a source of a static Coulomb-like magnetic field

$$\mathbf{B} = g \frac{\mathbf{r}}{r^3} \,. \tag{1.1}$$

Then the equation of motion of an electrically charged particle e in such a field is

$$m \frac{d^2 \mathbf{r}}{dt^2} = e \left[\mathbf{v} \times \mathbf{B} \right] = \frac{eg}{r^3} \left[\frac{d\mathbf{r}}{dt} \times \mathbf{r} \right], \tag{1.2}$$

where a static monopole is situated at the origin and the vector $\mathbf{r}$ defines the position of the electric charge (see Fig. 1.1). For the sake of simplicity we will use units such that the speed of light c is equal to 1 and in this section consider only positive values of both electric and magnetic charges.

---

[1] A very detailed description of the "stone age history" of the monopole problem is given in [35], where the genesis of it has been traced up to the notes by Petrus Pelegrinius, written at the Crusades in 1269! We will not go into this fascinating story.

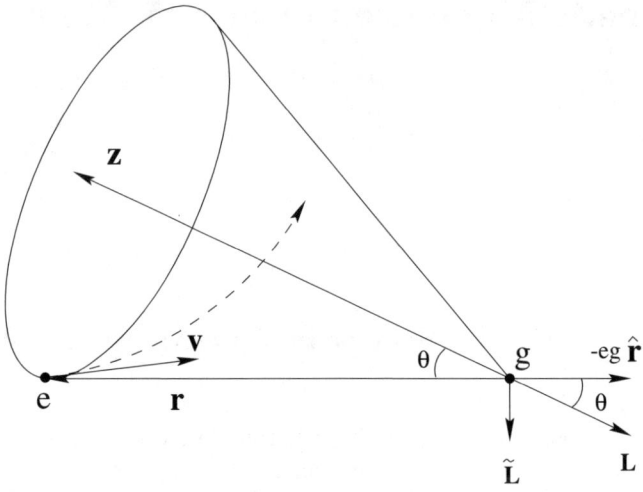

Fig. 1.1. Motion of an electric charge in the monopole field

One could obtain the corresponding integrals of motion just by making use of (1.2). Scalar multiplication of (1.2) by a vector of velocity **v** gives:

$$\frac{1}{2}\frac{d}{dt}\left(mv^2\right) = 0, \tag{1.3}$$

so that the kinetic energy of an electric charge in a monopole field is a constant:

$$E = \frac{mv^2}{2} = \text{const.}, \tag{1.4}$$

as is the absolute value v of the velocity vector.

On the other hand, the scalar product of the equation of motion (1.2) and the radius vector **r** gives:

$$\mathbf{r} \cdot \frac{d^2\mathbf{r}}{dt^2} \equiv \frac{1}{2}\frac{d^2}{dt^2}r^2 - v^2 = 0.$$

Taking into account the conservation of energy (1.3), one can write

$$r = \sqrt{v^2 t^2 + b^2}, \tag{1.5}$$

and therefore $\mathbf{r} \cdot (d\mathbf{r}/dt) = \mathbf{r} \cdot \mathbf{v} = v^2 t$. Thus, there is no closed orbit in the charge-monopole system: the electric charge is falling down from infinitely far away onto the monopole, approaching a minimal distance b and reflected back to infinity (so-called "magnetic mirror" effect).

A very special feature of such a motion is that the conserved angular momentum is different from the ordinary case. Indeed, one can see that the absolute value of the vector of ordinary angular momentum

$$\widetilde{\mathbf{L}} = \mathbf{r} \times m\mathbf{v} \tag{1.6}$$

is conserved, because the cross product of $\mathbf{r}$ and (1.2) is

$$\frac{d}{dt}[\mathbf{r} \times m\mathbf{v}] \equiv \frac{d\widetilde{\mathbf{L}}}{dt} = \frac{eg}{mr^3}\left[\widetilde{\mathbf{L}} \times \mathbf{r}\right]. \tag{1.7}$$

Scalar multiplication of this equation with the vector $\widetilde{\mathbf{L}}$ gives

$$\frac{d}{dt}|\widetilde{\mathbf{L}}| = 0, \tag{1.8}$$

and, because the absolute value of the velocity vector is a constant, one can write

$$\widetilde{L} \equiv |\widetilde{\mathbf{L}}| = mvb. \tag{1.9}$$

The very important difference from the ordinary Coulomb problem is that now the direction of the vector of angular momentum is not a constant, because from (1.7) it follows that

$$\frac{d}{dt}\left(\widetilde{\mathbf{L}} - eg\frac{\mathbf{r}}{r}\right) = \frac{d\mathbf{L}}{dt} = 0, \tag{1.10}$$

where the generalized angular momentum is an integral of motion:

$$\mathbf{L} = [\mathbf{r} \times m\mathbf{v}] - eg\frac{\mathbf{r}}{r} = \widetilde{\mathbf{L}} - eg\hat{\mathbf{r}}. \tag{1.11}$$

Let $\hat{\mathbf{r}}$ be a unit vector in the direction of $\mathbf{r}$. Taking into account (1.9) one can write (see Fig. 1.1)

$$L^2 \equiv \mathbf{L}^2 = \widetilde{\mathbf{L}}^2 + e^2g^2 = (mvb)^2 + (eg)^2. \tag{1.12}$$

As was demonstrated by J.J. Thompson already in 1904 [13, 500], the appearance of an additional term in the definition of the angular momentum (1.11) originates from a non-trivial field contribution. Indeed, since a static monopole is placed at the origin, its magnetic field is given by (1.1). Then the classical angular momentum of the electric field of a point-like electric charge, whose position is defined by its radius vector $\mathbf{r}$, and the magnetic field of a monopole is a volume integral involving the Poynting vector

$$\widetilde{\mathbf{L}}_{eg} = \frac{1}{4\pi}\int d^3r'\,[\mathbf{r}' \times (\mathbf{E} \times \mathbf{B})] = -\frac{g}{4\pi}\int d^3r'(\nabla' \cdot \mathbf{E})\,\hat{\mathbf{r}}' = -eg\hat{\mathbf{r}}, \tag{1.13}$$

where we perform the integration by parts, take into account that the fields vanish asymptotically and invoke the Maxwell equation

$$(\nabla' \cdot \mathbf{E}) = 4\pi e\,\delta^{(3)}(\mathbf{r} - \mathbf{r}').$$

At first sight, this conclusion looks rather paradoxical. Indeed, according to (1.13) even a static charge-monopole system has a non-zero angular momentum.

Notice that this formula could easily be generalized to the case of a pair of *dyons*, dual charged particles having both electric and magnetic charges, (e_1, g_1) and (e_2, g_2), respectively [549]. Let one of the dyons be placed at the origin and the position of the other one be given by the vector $\mathbf{r}$. Then the fields are

$$\mathbf{E} = e_1 \frac{\mathbf{r}}{r^3} + \mathbf{E}(e_2), \qquad \mathbf{B} = g_1 \frac{\mathbf{r}}{r^3} + \mathbf{B}(g_2),$$

and by analogy with (1.13) one has

$$\begin{aligned}\widetilde{\mathbf{L}}_{dd} &= \frac{1}{4\pi} \int d^3 r' \, [\mathbf{r}' \times (\mathbf{E} \times \mathbf{B})] \\ &= \frac{1}{4\pi} \int d^3 r' \left(\left[\mathbf{r}' \times \left(e_1 \frac{\mathbf{r}'}{r'^3} \times \mathbf{B}(g_2) \right) \right] + \left[\mathbf{r}' \times \left(\mathbf{E}(e_2) \times g_1 \frac{\mathbf{r}'}{r'^3} \right) \right] \right) \\ &= \frac{e_1}{4\pi} \int d^3 r' [\nabla' \cdot \mathbf{B}(g_2)] \hat{\mathbf{r}}' - \frac{g_1}{4\pi} \int d^3 r' [\nabla' \cdot \mathbf{E}(e_2)] \hat{\mathbf{r}}' \\ &= (e_1 g_2 - g_1 e_2) \hat{\mathbf{r}}. \end{aligned} \qquad (1.14)$$

Later we will come back to the definition of the generalized angular momentum by making use of standard variational procedure. Here we would like only to note that the conservation of the magnitude of the velocity together with the constant modulus of the angular momentum vector means that the impact parameter of the scattering problem coincides with the minimal separation b between the monopole and the electric charge. Also note that the energy of a charge in a monopole field (1.4) can be written as

$$E = \frac{m\dot{r}^2}{2} + \frac{L^2 - (eg)^2}{2mr^2} = \text{const.}, \qquad (1.15)$$

where we make use of the definition (1.12).

Thus, unlike the standard problem of charge scattering in a Coulomb field, now the trajectory does not lie in the plane of scattering that is orthogonal to the vector $\widetilde{\mathbf{L}}$. To define the character of the motion note that

$$|\mathbf{L} \cdot \hat{\mathbf{r}}| = eg = \text{const.}, \qquad (1.16)$$

i.e., the angle between the vectors $\mathbf{L}$ and $\mathbf{r}$ is a constant and the electric charge is moving on the surface of a cone whose axis is directed along $-\mathbf{L}$ with the cone angle θ, which can be defined using simple geometry (see Fig. 1.2) as

$$\cot \theta = \frac{eg}{|\widetilde{\mathbf{L}}|} = \frac{eg}{mvb}, \qquad (1.17)$$

or

1.1 Non-Relativistic Scattering on a Magnetic Charge

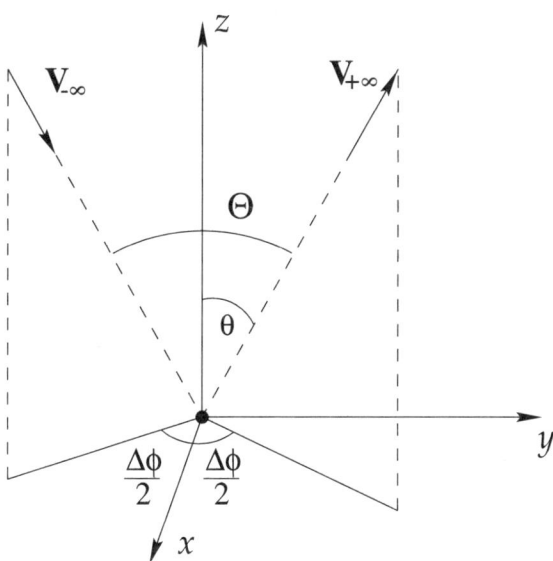

Fig. 1.2. Geometry of scattering of an electron by a monopole

$$\sin\theta = \frac{\tilde{L}}{L} = \frac{mvb}{\sqrt{(mvb)^2 + (eg)^2}}, \quad \cos\theta = \frac{eg}{L} = \frac{eg}{\sqrt{(mvb)^2 + (eg)^2}}. \quad (1.18)$$

Thus, the motion becomes planar only in the limit $g \to 0$, or $\theta = \pi$, which corresponds to the degeneration of the cone.

In the same way the ordinary vector of angular momentum $\tilde{\mathbf{L}}$ is precessing on the surface of a cone with a different cone angle but the same axis, because

$$\mathbf{L} \cdot \tilde{\mathbf{L}} = \tilde{\mathbf{L}}^2 = (mvb)^2 = \text{const}.$$

As was noted already by H. Poincaré [425], the existence of the integrals of motion (1.11) and (1.7) links the system of interacting electric and magnetic charges with a simple mechanical analog, a spherical top. One can understand it as a rotating disk with a thin rod of variable length as an axis of rotation. The charge and the monopole are sitting at the opposite ends of the rod.

Finally, the cross product of $\mathbf{L}$ (1.11) and the radius vector $\mathbf{r}$, together with (1.5), yields

$$\mathbf{v} = \frac{d\mathbf{r}}{dt} = \frac{1}{mr^2}[\mathbf{L} \times \mathbf{r}] + \frac{v^2 t}{r}\hat{\mathbf{r}} = \frac{1}{mr^2}[\mathbf{L} \times \mathbf{r}] + \frac{v}{\sqrt{1+(b/vt)^2}}\hat{\mathbf{r}}$$
$$= [\boldsymbol{\omega} \times \mathbf{r}] + v_r\hat{\mathbf{r}}, \quad (1.19)$$

where the angular and radial components of the velocity vector are

$$\boldsymbol{\omega} = \frac{\mathbf{L}}{mr^2}, \quad v_r = \frac{v}{\sqrt{1+(b/vt)^2}}. \quad (1.20)$$

Hence, asymptotically

$$\omega\big|_{t=\pm\infty} = 0, \qquad v_r\big|_{t=\pm\infty} = v.$$

At the turning point of the path, where the distance between the charge and the monopole is minimal

$$\omega\big|_{t=0} = \frac{\sqrt{(mvb)^2 + (eg)^2}}{mb^2}, \qquad v_r\big|_{t=0} = 0.$$

Thus, because the angular velocity is defined as $\omega = d\varphi/dt$, the azimuthal angle φ as a function of time can be obtained by simple integration[2]

$$\varphi(t) = \frac{1}{\sin\theta}\arctan\frac{vt}{b}, \tag{1.21}$$

where we made use of (1.5) and fix the boundary condition to $\varphi = 0$ at $t = 0$. Furthermore, θ is given by (1.18).

Since asymptotically

$$\hat{\mathbf{v}}\big|_{t=\pm\infty} = \left(\pm\sin\theta\cos\frac{\Delta\varphi}{2},\; \sin\theta\sin\frac{\Delta\varphi}{2},\; \pm\cos\theta\right),$$

where $\Delta\varphi = \varphi(\infty) - \varphi(-\infty) = \pi/\sin\theta$ (see Fig. 1.2), we can now calculate the angle of scattering on a monopole

$$\cos\Theta = \hat{\mathbf{v}}\big|_{t=-\infty}\cdot\hat{\mathbf{v}}\big|_{t=+\infty} = 2\sin^2\theta\sin^2\left(\frac{\pi}{2\sin\theta}\right) - 1, \tag{1.22}$$

or

$$\cos\left(\frac{\Theta}{2}\right) = \sin\theta\left|\sin\left(\frac{\pi}{2\sin\theta}\right)\right|, \tag{1.23}$$

where θ is a function of the impact parameter b, (1.18).

Unlike the standard problem of scattering in a Coulomb field, the angle of scattering Θ is not a monotonous function of the impact parameter b [462]. The dependence $\Theta(b)$ is depicted in Fig. 1.3, where the impact parameter b is rescaled in units of the parameter eg/mv. That is why, in order to calculate the effective cross-section, one has to take into account the contributions from all values of the impact parameter (or, equivalently, from all values of the cone angles θ_i), leading to scattering into the surface element $d\sigma$:

$$\frac{d\sigma}{d\Omega} = \left|\frac{b\,db}{d(\cos\Theta)}\right| = \sum_{\theta_i}\left(\frac{eg}{mv}\right)^2\frac{1}{2\cos^4\theta}\left|\frac{\sin 2\theta\, d\theta}{\sin\Theta\, d\Theta}\right|. \tag{1.24}$$

Here we made use of (1.17).

---

[2] Remember, v is constant, but $\dot{r}$ is not.

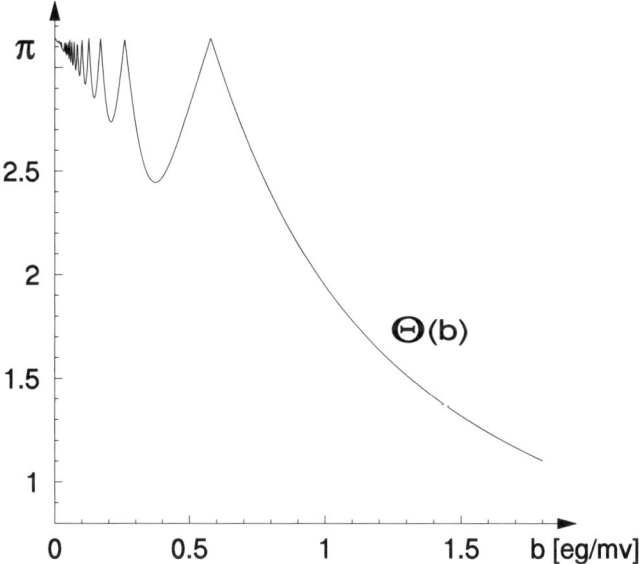

Fig. 1.3. Dependence of the scattering angle Θ on the impact parameter b

One can see that the effective cross-section of an electric charge (1.24) on the monopole is singular if $\sin \Theta = 0$ or $d\Theta/d\theta = 0$. In the scattering theory these two situations are referred to as the glory and rainbow respectively [462]. The first case corresponds to the back scattering, where[3] $\Theta = \pi$, while the cone is not degenerated, i.e., $\theta \neq \pi$. The formula (1.22) allows us to define corresponding "critical" values of the cone angles [131, 462]:[4]

$$\sin \theta_n = \frac{1}{2n}, \quad n = 1, 2, 3 \ldots \quad (1.25)$$

or $\theta_1 = 0.5236$, $\theta_2 = 0.2527$, $\theta_3 = 0.1674\ldots$

The rainbow scattering corresponds to cone angles θ_r being the solutions of the transcendental equation

$$\tan\left(\frac{\pi}{2 \sin \theta_r}\right) = \frac{\pi}{2 \sin \theta_r}. \quad (1.26)$$

These angles are $\theta_I = 0.3571$, $\theta_{II} = 0.2048$, $\theta_{III} = 0.1446\ldots$ Note that in both situations of glory and rainbow scattering the singularities of the cross-section are integrable and the total cross-section for scattering on a monopole is well defined. Note that such singularities are absent for small-angle scattering, defined by the condition $\Theta \approx \pi - 2\theta = 2eg/mvb \ll 1$. In such a case the differential cross-section is

---

[3] The case $\Theta \to 0$, or $\theta \to \pi/2$, would correspond to $eg \to 0$.
[4] Other authors use the complementary angle $\pi/2 - \theta$.

$$\frac{d\sigma}{d\Omega} = \frac{1}{\Theta^4}\left(\frac{2eg}{mv}\right)^2, \qquad (1.27)$$

which is evidently analogous to the Rutherford formula.

1.2 Non-Relativistic Scattering on a Dyon

Let us generalize the results of the previous section to the case of classical non-relativistic scattering of an electrically charged particle on a static dyon having both electric (Q) and magnetic (g) charges[5]. For simplicity we restrict our consideration to the case of an attractive electrostatic potential, i.e., suppose that $V = eQ/r$, where $eQ < 0$. A qualitative analysis suggests that unlike the charge-monopole scattering, described above, there are closed trajectories in such a system. Indeed, let us consider the corresponding equation of motion (cf. (1.2))

$$m\frac{d^2\mathbf{r}}{dt^2} = eQ\frac{\mathbf{r}}{r^3} - \frac{eg}{r^3}\left[\mathbf{r} \times \frac{d\mathbf{r}}{dt}\right]. \qquad (1.28)$$

Obviously, the generalized angular momentum $\mathbf{L}$ given by (1.11) is still an integral of motion. Also, the projection of the total angular momentum onto the radial direction $L_r = |\mathbf{L} \cdot \hat{\mathbf{r}}| = eg$, as well as the magnitude of the orbital angular momentum $\widetilde{L} = mbv_0$, where v_0 is the initial velocity of the electric charge given at an infinitely large distance from the scattering center, are conserved. Thus the motion is restricted to the same surface of a cone with a cone angle $\cot\theta = eg/mbv_0$, as it was in the case of charge-monopole scattering. The difference is that now the magnitude of the velocity is no longer an integral of motion, because unlike (1.4) the total energy conserved is now

$$E = \frac{m\mathbf{v}^2}{2} + \frac{eQ}{r} = \frac{m\dot{r}^2}{2} + \frac{\widetilde{L}^2}{2mr^2} + \frac{eQ}{r} = \text{const.} \qquad (1.29)$$

Here, one of the basic features of the interaction between a monopole and an electrically charged particle manifests itself: if the radial part of the Hamiltonian is determined by a Coulomb interaction, then the interaction of a charge and a monopole is described by its angular part. Indeed, we have seen that the magnitude of the radius vector of a charge moving in a magnetic Coulomb field depends on time just as in the case of free motion (see (1.5)). Hence, in the system of reference, which rotates with the angular velocity $\omega(t)$

---

[5] The problem of charge motion in a monopole (dyon) field was probably considered first by S.A. Boguslavsky [128], who also derived an expression for a vector potential of a monopole field a decade before the celebrated paper by P.A.M. Dirac [200]. The author is grateful to E.A. Tolkachev and L.M. Tomilchik for kindly informing him about that undeservedly forgotten paper [497]. Other references include [114, 388].

1.2 Non-Relativistic Scattering on a Dyon

(cf. (1.20)), the equation of motion is trivial: in the rotating plane orthogonal to the vector $\tilde{\mathbf{L}}$ the electric charge is moving with a constant velocity v along a straight line. Just in the same way, in the case of motion in a dyon field, the time dependence of the magnitude of the radius vector of the electric charge, is the same as for the ordinary interaction of two electric charges e and Q. Indeed, from (1.29) follows that

$$\dot{r} = \sqrt{\frac{2}{m}\left(E + \frac{|eQ|}{r} - \frac{\tilde{L}^2}{2mr^2}\right)}. \tag{1.30}$$

For a bound motion $E < 0$ and according to the standard procedure (see for example [18]) we can write

$$t = \sqrt{\frac{m}{2|E|}} \int \frac{r dr}{\sqrt{-r^2 + (|eQ|r)/|E| - \tilde{L}^2/(2m|E|)}}, \tag{1.31}$$

where the constant of integration can be chosen to fix the parameters $t_0 = 0$ and $r_0 = d$. The latter denotes the minimal distance between the charge e and the dyon, which unlike the problem of charge-monopole scattering is no longer equal to the impact parameter.

The elementary integration of (1.31) allows us to find the parametric dependence of coordinates on time. Putting

$$a = \frac{|eQ|}{2|E|}, \quad b = \frac{\tilde{L}}{\sqrt{2m|E|}}, \quad \varepsilon = \sqrt{1 - \frac{b^2}{a^2}} = \sqrt{1 - \frac{2|E|\tilde{L}^2}{me^2Q^2}}, \tag{1.32}$$

the integral (1.31) can be rewritten as

$$t = \sqrt{\frac{m}{2|E|}} \int \frac{r dr}{\sqrt{a^2 - b^2 - (r-a)^2}} = \sqrt{\frac{m}{2|E|}} \int \frac{r dr}{\sqrt{a^2\varepsilon^2 - (r-a)^2}},$$

which gives the parametric equation

$$t = \sqrt{\frac{ma^2}{2|E|}}(\xi - \varepsilon \sin \xi), \quad r = a(1 - \varepsilon \cos \xi). \tag{1.33}$$

An azimuthal angle φ as a function of time could be defined in the same way. Because the angular velocity of a charge in a dyon field is given by (1.20), the elementary integration of this relation gives

$$\varphi(t) = \frac{|\mathbf{L}|}{\sqrt{2m}} \int \frac{dr}{r^2 \sqrt{-|E| + (|eQ|)/r - \tilde{L}^2/(2mr^2)}}, \tag{1.34}$$

or

$$\cos(\varphi \sin \theta) = \frac{-r + \dfrac{\widetilde{L}^2}{m|eQ|}}{\varepsilon r}. \tag{1.35}$$

This means that a charged particle in a dyon field is moving along an ellipse with semi-axes a and b and eccentricity ε. However, unlike the classic Kepler problem of the relative motion of two electric charges, the ellipse itself is precessing on the conic surface with the cone angle θ and the precession angle is $\nu = 2\pi(L/\widetilde{L} - 1)$ per each radial period [128, 356].

If $Q > 0$, the motion is infinite and the trajectories are hyperbolic orbits. The scattering angle can easily be written by analogy with (1.22) as [462]:

$$\cos \frac{\Theta}{2} = \sin \theta \left| \sin \left(\frac{\chi}{\sin \theta} \right) \right|, \tag{1.36}$$

where

$$\chi = \varphi \sin \theta \big|_{r=\infty} = \arctan \left(\frac{g v_0}{Q} \tan \theta \right). \tag{1.37}$$

The cross-section can be calculated by making use of the same formula (1.24). The difference consists of another form of dependence of the angle of scattering on the boundary conditions, i.e., on the cone angle. The details of the calculation are given in [462].

1.3 Vector Potential of a Monopole Field

All the above formulae describing the classical non-relativistic dynamics of an electric charge in a monopole (or dyon) field, were obtained by consideration of a monopole as a static external source of a Coulomb-like magnetic field. Nevertheless, a consistent consideration, taking into account further quantization of the theory, requires a generalization of the standard Lagrangian description of the system of interacting charges of two types: electric and magnetic.

At a first glance, such a generalization is trivial. As in conventional electrodynamics, one has to introduce the Lagrangian of an electric charge in an external field as

$$L = \frac{1}{2} m \dot{\mathbf{r}}^2 + e \dot{\mathbf{r}} \cdot \mathbf{A}, \tag{1.38}$$

which gives the required equation of motion (1.2), while the second term here is the Lagrangian of minimal interaction between electric and magnetic charges. However, according to the standard definition, the vector potential of the magnetic field $\mathbf{A}$ must satisfy the relation

$$\mathbf{B} = g \frac{\mathbf{r}}{r^3} = \boldsymbol{\nabla} \times \mathbf{A}. \tag{1.39}$$

It would seem that is not problematic to integrate this expression to define the function $\mathbf{A}(\mathbf{r})$. However, there is a problem, since at the same time the

1.3 Vector Potential of a Monopole Field

second pair of Maxwell equations demands the magnetic charge to be the source of such a field, i.e.,

$$\nabla \cdot \mathbf{B} = 4\pi g\, \delta^{(3)}(\mathbf{r})\,, \tag{1.40}$$

which is in contradiction with the condition (1.39), which requires $\nabla \cdot \mathbf{B} = 0$.

Let us analyze the situation. Because the magnetic field $\mathbf{B}$ is spherically symmetric, the corresponding vector-potential could be written as

$$\mathbf{A}(\mathbf{r}) = A(\theta)\nabla\varphi\,, \tag{1.41}$$

where φ is an azimuthal angle in the spherical coordinates and $A(\theta)$ is a function of the polar angle only. One can easily see that since in spherical coordinates $\hat{\mathbf{e}}_\varphi = -\hat{\mathbf{e}}_x \sin\varphi + \hat{\mathbf{e}}_y \cos\varphi$, the choice $A(\theta) = -g(1+\cos\theta)$ yields, after a straightforward calculation,

$$\nabla\varphi = \left(-\frac{\sin\varphi}{r\sin\theta},\ \frac{\cos\varphi}{r\sin\theta},\ 0\right),$$

$$\mathbf{A}(\mathbf{r}) = \left(g\frac{1+\cos\theta}{r\sin\theta}\sin\varphi,\ -g\frac{1+\cos\theta}{r\sin\theta}\cos\varphi,\ 0\right). \tag{1.42}$$

One can rewrite this expression in a covariant form as

$$\mathbf{A}(\mathbf{r}) = \frac{g}{r}\frac{[\mathbf{r}\times\mathbf{n}]}{r-(\mathbf{r}\cdot\mathbf{n})}\,, \tag{1.43}$$

where the unit vector $\mathbf{n}$ is directed along the z-axis: $\mathbf{n} = (0,0,1)$. This is the celebrated *Dirac potential* [200].

At first sight that is just the potential we need, because after a simple calculation we have, for example,

$$B_x = -\partial_z A_y = \partial_z \left(\frac{gx}{r(r-z)}\right) = g\frac{x}{r^3}\,, \tag{1.44}$$

that is

$$\mathbf{B}(\mathbf{r}) = [\nabla\times\mathbf{A}] = g\frac{\mathbf{r}}{r^3}\,.$$

At the same time, the radial character of the magnetic field means that

$$\mathbf{A}\cdot\mathbf{r} = 0,\qquad \frac{\partial}{\partial r}\big|[\mathbf{r}\times\mathbf{A}]\big| = g\frac{\partial}{\partial r}\left[\frac{1+\cos\theta}{\sin\theta}\right] = 0\,.$$

However, one should be careful with such a calculation, because the vector potential $\mathbf{A}$ of (1.43) that we made use of is singular along the line $\theta = 0$, although it is regular[6] along the direction $\theta = \pi$. This means that the

---

[6] Two angles of the spherical coordinates parameterize a sphere S^2. Since the potential (1.43) has no singularities on its south hemisphere one could introduce an additional index to label it: $\mathbf{A} \to \mathbf{A}^S$.

straightforward calculation of the magnetic field above is *not* correct along the semi-infinite line of singularity. In the vicinity of it we have $\mathbf{A} \sim -2g\boldsymbol{\nabla}\varphi$.

Let us note that such a potential is typical for a singular string of magnetic flux along the positive semi-axis z. Indeed, generally speaking, the potential (1.41) can be written as

$$\mathbf{A}(\mathbf{r}) = -g(1+\cos\theta)\boldsymbol{\nabla}\varphi = (1+\cos\theta)\,\frac{i}{e}U^{-1}\boldsymbol{\nabla}U\,, \tag{1.45}$$

where $U = e^{-ieg\varphi}$. Thus, the Dirac potential is the pure gauge transformation, which is complemented by the polar-angle-depending factor. However this pure gauge is singular. That is why we have to be careful and try to use instead of (1.43) a regularized form of the potential [131]

$$\mathbf{A}_R(\mathbf{r},\varepsilon) = \frac{g}{R}\frac{[\mathbf{r}\times\mathbf{n}]}{R-(\mathbf{r}\cdot\mathbf{n})}\,, \tag{1.46}$$

where $R = \sqrt{r^2 + \varepsilon^2} = \sqrt{x^2 + y^2 + z^2 + \varepsilon^2}$. Thus, the regularized magnetic field is

$$\mathbf{B}_R(\mathbf{r},\varepsilon) = g\frac{\mathbf{r}}{R^3} - g\varepsilon^2\left(\frac{\mathbf{n}}{R^3[R-(\mathbf{r}\cdot\mathbf{n})]} + \frac{\mathbf{n}}{R^2[R-(\mathbf{r}\cdot\mathbf{n})]^2}\right)\,. \tag{1.47}$$

To lift the regularization we go to the limit $\varepsilon^2 \to 0$, which implies

$$\mathbf{B}_R(\mathbf{r},\varepsilon) \stackrel{\varepsilon^2\to 0}{\sim} g\frac{\mathbf{r}}{r^3} - 2g\varepsilon^2\mathbf{n}\theta(z)\left(\frac{1}{r^2(x^2+y^2+\varepsilon^2)} + \frac{2}{(x^2+y^2+\varepsilon^2)^2}\right)\,. \tag{1.48}$$

It is easy to see that the singular terms differ from zero only on the positive infinite semi-axis z. To compute the magnetic field flux in that direction we evaluate the integral over the infinitesimal element of the surface, orthogonal to this axis. Then only the second term in the parentheses contribute and the result is

$$\mathbf{B}(\mathbf{r}) = \mathbf{B}_g + \mathbf{B}_{sing} = g\frac{\mathbf{r}}{r^3} - 4g\pi\mathbf{n}\theta(z)\delta(x)\delta(y)\,. \tag{1.49}$$

Therefore, in addition to the expected Coulomb field we obtain a singular flux of the magnetic field $\mathbf{B}_{sing}$. The very important point is that this extra piece resolves the above-mentioned paradox with the Maxwell equations, because the flux of the string field exactly cancels the contribution from the Coulomb-like part and a total flux of the fields through the closed surface with a monopole inside is (see Fig. 1.4):

$$\Phi_{tot} = \oint d\boldsymbol{\sigma}\mathbf{B} = g\left(\oint d\boldsymbol{\sigma}\frac{\mathbf{r}}{r^3} - 4\pi\oint d\boldsymbol{\sigma}\mathbf{n}\,\theta(z)\delta(x)\delta(y)\right)$$
$$= 4g\pi - 4g\pi = 0\,. \tag{1.50}$$

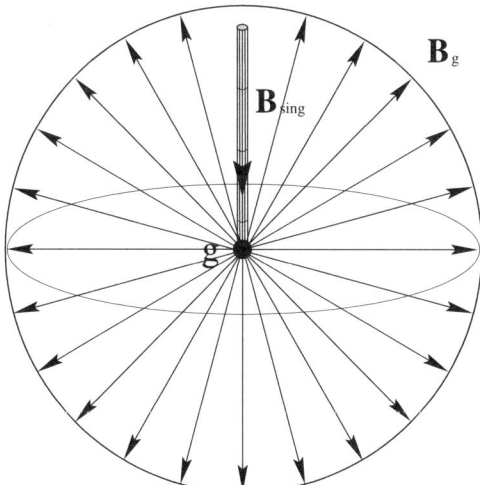

Fig. 1.4. Magnetic field of the singular Dirac potential

Thus, the potential (1.43) corresponds not to a single isolated magnetic pole, but rather to a semi-infinite and infinitely thin solenoid. Of course, the position of such a solenoid is not fixed and it can have any form.

If we choose the Coulomb gauge by taking the gauge condition $\boldsymbol{\nabla} \cdot \mathbf{A} = 0$, for an arbitrary curvilinear string we can write

$$\boldsymbol{\nabla} \times [\mathbf{B}_g + \mathbf{B}_{sing}] = \boldsymbol{\nabla} \times \left[-g \boldsymbol{\nabla} \left(\frac{1}{r} \right) + \mathbf{B}_{sing} \right] = \boldsymbol{\nabla} \times \mathbf{B}_{sing}$$
$$= \boldsymbol{\nabla} \times \boldsymbol{\nabla} \times \mathbf{A} = -\boldsymbol{\nabla}^2 \mathbf{A}. \qquad (1.51)$$

Therefore we can express the Dirac potential as

$$\mathbf{A} = g \int d^3 x' \left[\boldsymbol{\nabla} \frac{1}{|r - r'|} \times \mathbf{B}_{sing}(\mathbf{r}) \right] = g \int \frac{(\mathbf{r} - \mathbf{r}') \times d\mathbf{r}'}{|\mathbf{r} - \mathbf{r}'|^3}, \qquad (1.52)$$

where the integral has to be taken along the string from the position of the monopole to infinity [41, 225, 305]. The physical meaning of this remarkable representation of the Dirac potential derived by P. Jordan is that it can be viewed as the integral sum of the vector potentials of the infinitesimal magnetic dipoles $g d\mathbf{r}'$ located along the string [305]. To check the correctness of this picture note that for a straight string the singular field $\mathbf{B}_{sing}$ is defined by relation (1.49) and we can easily prove that the substitution of this expression in (1.52) yields the Dirac potential (1.43).

1.4 Transformations of the String

As we saw in the previous section, the singular potential (1.43) leads to the appearance of an extra field of a string (1.49), directed along the singularity, in addition to the spherically symmetric Coulomb magnetic field. That is not what we expected. In order to get rid of this singular piece, we can set up a condition making it unobservable and therefore unphysical. A first step to secure this on the classical level is to demand that all the possible string configurations and their positions have to be physically equivalent. In fact we have to describe transformations from one string configuration to another and discuss under which conditions they are identical.

Recall that an electromagnetic potential is defined up to a $U(1)$ gauge transformation, $U(\mathbf{r}) = \exp\{ie\lambda(\mathbf{r})\}$:

$$\mathbf{A} \to \mathbf{A}' = \mathbf{A} - \frac{i}{e} U^{-1}\nabla U = \mathbf{A} + \nabla\lambda(\mathbf{r}). \qquad (1.53)$$

Both potentials $\mathbf{A}'$ and $\mathbf{A}$ correspond to the same magnetic field. Usually the gauge function $\lambda(\mathbf{r})$ is taken to be single-valued. This is not a necessary condition and we may consider any multivalued gauge function $\lambda(\mathbf{r})$ on an equal footing. However, the gauge transformation itself is single-valued. On the other hand, the gradient of the gauge function $\nabla\lambda(\mathbf{r})$ is single-valued almost everywhere in space with the exception that it becomes singular along the line connecting the different sheets of the multivalued gauge function $\lambda(\mathbf{r})$. Thus, such a gauge function is a source of additional singular terms in the transformed vector-potential $\mathbf{A}'$ as well as in the corresponding field $\mathbf{B}'$. Indeed, let us consider how the magnetic flux through a closed surface σ changes under the gauge transformation (1.53)

$$\Delta\Phi = \int_\sigma d^2S\,\hat{\mathbf{n}}_S \cdot (\mathbf{B}' - \mathbf{B}) = \int_\sigma d^2S\,\hat{\mathbf{n}}_S \cdot [\nabla \times (\nabla\lambda)] = \oint d\mathbf{l}\cdot\nabla\lambda.$$

Certainly, the flux is a constant only if the gauge function satisfies the condition $\lambda(\varphi + 2\pi) = \lambda(\varphi)$, otherwise, $\Delta\Phi \neq 0$.

This feature turns out to be a key point in the analysis of the magnetic monopole potential. Let us consider a singular gauge transformation $U = \exp\{2ieg\varphi\}$ resulting in the transformation[7] of the potential $\mathbf{A}^S$ (1.43)

$$\mathbf{A}^S \to \mathbf{A}^S - \frac{i}{e}e^{-2ieg\varphi}\nabla e^{2ieg\varphi} = -\frac{g}{r}\frac{1+\cos\theta}{\sin\theta}\hat{\mathbf{e}}_\varphi + \frac{2g}{r\sin\theta}\hat{\mathbf{e}}_\varphi$$

$$= \frac{g}{r}\frac{1-\cos\theta}{\sin\theta}\hat{\mathbf{e}}_\varphi \equiv \mathbf{A}^N. \qquad (1.54)$$

The gauge transformation is given by the function

$$\lambda(\mathbf{r}) = 2g\varphi = 2g\arctan(y/x).$$

---

[7] Note that, strictly speaking, such transformations have to be defined as distributions. There are plenty of paradoxes and incorrect conclusions in the monopole theory that originated from naive treatments of singular expressions like (1.54).

This results in the appearance of an additional singular magnetic field $\Phi_{\text{str}} = 4\pi g$ along the z-axis. The physical picture of this transformation looks like the addition of the extra flux, generated by the gauge transformation (1.54), and the flux of the monopole string field along the positive semi-axis z. The sum is obviously a semi-infinite string field along the negative semi-axis z that corresponds to the singularity of the transformed vector-potential $\mathbf{A}^N$. Indeed, this potential is singular along the line $\theta = \pi$, but regular in the opposite direction $\theta = 0$.

Thus, the gauge transformation (1.54) acts as a rotation of the line of singularity of the monopole by an angle of π. More generally, this means that the position of the string is defined up to such a transformation. Therefore its field is not physical.

Indeed, one can define a general gauge transformation, rotating the line of singularity from a position given by a unit vector $\mathbf{n}$ to the new direction along the vector $\mathbf{n}'$ [49, 225]:

$$\begin{aligned}\mathbf{A}(\mathbf{r}) &\to \mathbf{A}(\mathbf{r}) + \boldsymbol{\nabla}\lambda(\theta,\varphi) = \\ &- g(1+\cos\theta)\boldsymbol{\nabla}\varphi + \boldsymbol{\nabla}\lambda(\theta,\varphi) = -g(1+\cos\theta')\boldsymbol{\nabla}\varphi',\end{aligned} \quad (1.55)$$

where we make use of (1.45). Thus, the angles θ' and φ' define the direction of the new line of singularity $\mathbf{n}'$ and the gauge function $\lambda_{n,n'}(\mathbf{r})$ yields the rotation of the string

$$\mathbf{A}_n \longrightarrow \mathbf{A}_{n'} = \mathbf{A}_n + \boldsymbol{\nabla}\lambda_{n,n'}(\mathbf{r}). \quad (1.56)$$

Let us consider a surface σ spanned on the strings of directions $\mathbf{n}$ and $\mathbf{n}'$ (see Fig. 1.5). Then relation (1.52) yields

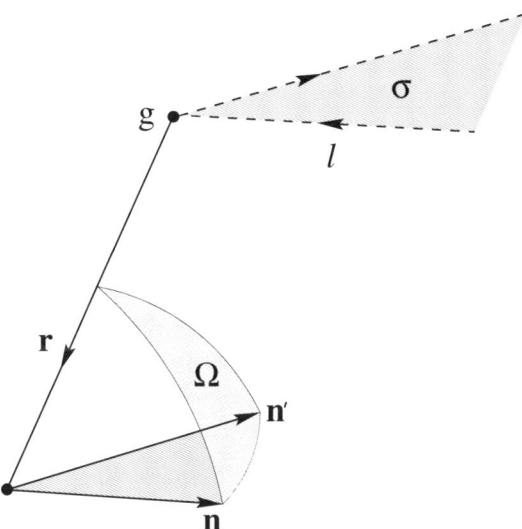

Fig. 1.5. Geometrical interpretation of the string displacement from $\mathbf{n}$ to $\mathbf{n}'$

18   1 Magnetic Monopole in Classical Theory

$$\mathbf{A}_{n'} - \mathbf{A}_n = g \int_l \frac{(\mathbf{r} - \mathbf{r}') \times d\mathbf{r}'}{|\mathbf{r} - \mathbf{r}'|^3}, \qquad (1.57)$$

where the integral has to be taken along the contour l.

The open strings can be closed at infinity since the contribution of the infinite separated singular magnetic flux along this segment of l is vanishing. Then we get

$$\nabla \lambda_{n,n'}(\mathbf{r}) = g \nabla \Omega_{nn'}(\mathbf{r}), \qquad (1.58)$$

where $\Omega_{nn'}(\mathbf{r})$ is a solid angle under which the surface σ is seen from the point $\mathbf{r}$ (see Fig. 1.5). This gives a geometrical interpretation of the parameter of the gauge transformation which rotates the string.

The function $\Omega_{nn'}(\mathbf{r})$ has a jump on the surface σ if $\mathbf{n}' = \mathbf{n}$ because then it becomes degenerated. In this case we have $\Omega_{n,-n} = 2\varphi + \varphi_0$, which corresponds to the gauge transformation (1.54) above. We can see that this discontinuity of the function $\Omega_{nn'}(\mathbf{r})$ can be canceled if we redefine the gauge function to be [41]

$$\nabla \lambda_{n,n'}(\mathbf{r}) = g \nabla \Omega_{nn'}(\mathbf{r}) - 4\pi g \int_\sigma d\sigma \delta^{(3)}(\mathbf{r} - \mathbf{r}'). \qquad (1.59)$$

Then the integral over the δ-function in (1.59) exactly cancels the jump 4π of the solid angle $\Omega_{nn'}(\mathbf{r})$ on the plane $\{\mathbf{n}, -\mathbf{n}\}$. Thus, the gauge transformation

$$U = \exp\{ie\lambda_{n,n'}(\mathbf{r})\} \qquad (1.60)$$

is regular everywhere but in the direction of the "old" and "new" strings $\mathbf{n}, \mathbf{n}'$.

Note that such gauge transformations are able not only to rotate the string but also to split it. For example, the gauge transformation $U = \exp\{ieg\varphi\}$ of the potential $\mathbf{A}^S$ leads to

$$\mathbf{A}^S \to \mathbf{A}^S - \frac{i}{e} e^{-ieg\varphi} \nabla e^{ieg\varphi} = -\frac{g}{r} \frac{1 + \cos\theta}{\sin\theta} \hat{\mathbf{e}}_\varphi + \frac{g}{r \sin\theta} \hat{\mathbf{e}}_\varphi$$

$$= -\frac{g}{r} \frac{\cos\theta}{\sin\theta} \hat{\mathbf{e}}_\varphi \equiv \mathbf{A}^{Sch}, \qquad (1.61)$$

that is the so-called *Schwinger potential* [459]. Schwinger insisted on such a form of the monopole potential because he thought it more appropriate to relativistic field theory.

This potential has a singularity along the entire infinite z-axis. Indeed, the magnetic flux generated by the singular gauge transformation annihilates only half of the original flux along the positive semi-axis z and at the same time is equal to half the "normal" string flux along the negative z direction. Thus one could think of the Schwinger potential as the sum of two "half-Dirac" potentials:

$$\mathbf{A}^{\text{Sch}} = -\frac{g}{r}\frac{\cos\theta}{\sin\theta}\hat{\mathbf{e}}_\varphi = \frac{g}{2r}\left(\frac{1-\cos\theta}{\sin\theta} - \frac{1+\cos\theta}{\sin\theta}\right)\hat{\mathbf{e}}_\varphi = \frac{1}{2}\mathbf{A}^N + \frac{1}{2}\mathbf{A}^S. \tag{1.62}$$

Another, rather funny example of this kind, is the almost unknown *Banderet* monopole potential [110]. Again, we can obtain it starting from the Dirac potential (1.43) and performing the gauge transformation $U = \exp\{ieA(\theta)\varphi\}$. Here we make use of the function $A(\theta) = -g(1+\cos\theta)$ introduced above (1.42). A simple calculation gives

$$\mathbf{A}^N \to \mathbf{A}^N - \frac{i}{e}e^{-ieA(\theta)\varphi}\boldsymbol{\nabla}e^{ieA(\theta)\varphi} \tag{1.63}$$

$$= \frac{g}{r}\frac{1-\cos\theta}{\sin\theta}\hat{\mathbf{e}}_\varphi - g\boldsymbol{\nabla}\left[(1+\cos\theta)\varphi\right] = -\frac{g}{r}\varphi\sin\theta\,\hat{\mathbf{e}}_\varphi \equiv \mathbf{A}^B.$$

It is easy to see that such a potential is singular on the entire half-plane $x \geq 0, y \geq 0$ and the corresponding magnetic field is (compare with (1.49))

$$\mathbf{B}(\mathbf{r}) = g\frac{\mathbf{r}}{r^3} + 2g\pi\mathbf{r}\theta(x)\delta(y)\frac{x}{r^3}. \tag{1.64}$$

Thus, the magnetic flux spreads out on the radial directions from the origin of coordinates where the monopole is situated, and comes back from infinity along this semi-plane. Again, the total flux is zero.

Finally, we note that the presence of the string in the monopole theory requires modification of all space-time transformations, complementing usual translations, rotations and reflections by gauge transformations like (1.53) [134, 301]. The latter are effectively moving the string in space. For example, extended in such a way, rotational symmetry of the theory means that it has to be invariant with respect to both ordinary $O(3)$ rotation $\mathbf{r} \to O\mathbf{r}$ and gauge transformation (1.53). In the same way, the discrete transformations are modified. As a simple example, we consider modification of the operation of space reflection [225, 494]. Indeed, the Dirac potential (1.43) is no longer an eigenfunction of the operator of reflection $P: \mathbf{r} \to -\mathbf{r}$, because

$$P:\ \mathbf{A}(\mathbf{r}) = \mathbf{A}(-\mathbf{r}) = -\frac{g}{r}\frac{[\mathbf{r}\times\mathbf{n}]}{r+(\mathbf{r}\cdot\mathbf{n})} \neq \pm\mathbf{A}(\mathbf{r}), \tag{1.65}$$

since $\mathbf{n}$ is fixed. However, a combination of the reflection and the gauge transformation $U = e^{2ieg\varphi}$ of (1.54), rotating the singularity line by an angle of π gives

$$\mathcal{P}:\ \mathbf{A}(\mathbf{r}) = P:\mathbf{A}(\mathbf{r}) - \frac{i}{e}U^{-1}\boldsymbol{\nabla}U = \mathbf{A}(\mathbf{r}). \tag{1.66}$$

Thus, with respect to the operator of generalized space reflection $\mathcal{P}$ the potential $\mathbf{A}(\mathbf{r})$ (1.43) transforms as a pseudo-vector. This means that the presence of a monopole leads to a series of effects connected with violation of parity [438, 466]. Indeed, unlike conventional electrodynamics, the spherically symmetric magnetic field of a monopole $\mathbf{B} = g\mathbf{r}/r^3$ is a vector rather than a pseudo-vector.[8]

1.5 Dynamical Symmetries of the Charge-Monopole System

The Lagrangian (1.38) governing the dynamics of charge-monopole interaction is hiding a lot of non-trivial features behind its simple form. One of them is a hidden dynamical symmetry of this interaction [300].

Let us recall that an action invariance with respect to some transformation is defined up to a total time derivative. For example, an infinitesimal transformation of the coordinates $\mathbf{r} \to \mathbf{r} + \delta\mathbf{r}$ leads to a variation of the form

$$\delta L = (m\mathbf{v} \cdot \delta\mathbf{v}) + \delta\,(e\mathbf{v} \cdot \mathbf{A})$$
$$= -m\left(\frac{d\mathbf{v}}{dt}\right) \cdot \delta\mathbf{r} + e\frac{d}{dt}(\delta\mathbf{r} \cdot \mathbf{A}) + e\,(\mathbf{v} \cdot [\mathbf{B} \times \delta\mathbf{r}]) \,. \qquad (1.67)$$

Here, the second term is a total time derivative and is not connected with the restrictions on the properties of the symmetry. The latter are fixed by the third (gauge invariant) term. Of course, the total time derivative does not affect the equation of motion and can usually be dropped. In monopole theory this operation is not so harmless because it is exactly the boundary terms in the action, like the second term in (1.67), that are responsible for the topological properties of the system. In what follows we shall clarify this point.

Note that the relation (1.67) easily allows us to obtain the expression (1.11) for the conserved angular momentum of the system. Since under a standard $O(3)$ rotation around an arbitrary axis $\boldsymbol{\omega}$, the radius-vector transforms as

$$\delta\mathbf{r} = [\boldsymbol{\omega} \times \mathbf{r}] \,,$$

and the magnetic field of a monopole $\mathbf{B}$ is spherically symmetric, we can write

$$[\mathbf{B} \times \delta\mathbf{r}] = [\mathbf{B} \times [\boldsymbol{\omega} \times \mathbf{r}]] = g\boldsymbol{\nabla}\,(\hat{\mathbf{r}} \cdot \boldsymbol{\omega}) \,.$$

Therefore, the last term in the variation of the action (1.67) takes the form $e\,(\mathbf{v} \cdot [\mathbf{B} \times \delta\mathbf{r}]) = eg(d\hat{\mathbf{r}}/dt)\boldsymbol{\omega}$. Thus, according to the Noether theorem, the invariance of the action (1.38) with respect to spacial rotations leads to the conservation of the generalized angular momentum (1.11)

$$\mathbf{L} = [\mathbf{r} \times m\mathbf{v}] - eg\frac{\mathbf{r}}{r} \,. \qquad (1.68)$$

---

[8] We do not go into the old discussion about the situation with an axial magnetic charge. The curious reader could look at the paper [117] and references therein.

1.5 Dynamical Symmetries of the Charge-Monopole System

It is remarkable that the extra term here originates from the part of the Lagrangian (1.38) that describes the charge-monopole interaction.

Another integral of motion is the standard Hamiltonian. To derive it we have to consider the variation $\delta \mathbf{r} = \mathbf{v}\delta t$ that follows from a simple shift of the time variable $t \to t + \delta t$. Then the last term in the variation of the action (1.67) vanishes and we obtain

$$H = \frac{mv^2}{2}. \tag{1.69}$$

Taking into account the form of the Lagrangian (1.38) describing the dynamics of an electric charge in a monopole field, the canonical momentum is

$$\mathbf{P} = \frac{\delta L}{\delta \dot{\mathbf{r}}} = m\dot{\mathbf{r}} + e\mathbf{A}. \tag{1.70}$$

Thus we can write the Hamiltonian in the form

$$H = \mathbf{P}\dot{\mathbf{r}} - L = \frac{1}{2m}(\mathbf{P} - e\mathbf{A})^2 \equiv \frac{\pi^2}{2m}. \tag{1.71}$$

It is easy to check that the Poisson bracket of the Hamiltonian (1.71) and the generalized angular momentum (1.11) vanish [49].

So far, the difference from ordinary electrodynamics was not very dramatic and all the integrals of motion have an analogue there. However, this is not the end of the story, because there are some other additional symmetries of the charge-monopole system, which form the so-called group of *dynamical symmetry*. Following the work by A. Barut and his collaborators [116] R. Jackiw noted [300] that the variation of the kinetic energy of a charge in a monopole field is a total time derivative also under transformation of dilatation $\delta \mathbf{r} = \mathbf{v}t - \mathbf{r}/2$ and the special conformal transformation $\delta \mathbf{r} = \mathbf{v}t^2 - \mathbf{r}t$. The generators of these transformations

$$D = Ht - \frac{m}{4}\left[(\mathbf{r} \cdot \mathbf{v}) + (\mathbf{v} \cdot \mathbf{r})\right], \qquad K = -Ht^2 + 2Dt + \frac{m\mathbf{r}^2}{2} \tag{1.72}$$

form, together with the Hamiltonian, the algebra of the conformal group $SO(2,1)$:

$$[H, D] = iH, \qquad [D, K] = iK, \qquad [H, K] = 2iD, \tag{1.73}$$

which is the group of dynamical symmetry of the non-relativistic charge-monopole system.

Thus, the non-relativistic Lagrangian of the system is invariant under the group of transformation $O(3) \times SO(2,1)$. However, the Casimir operator of the $SO(2,1)$ subgroup is not independent of the $O(3)$ generators, because [116, 300]

$$\mathcal{J}^2 = \frac{1}{2}(KH + HK) - D^2 = \frac{1}{4}\left(\mathbf{L}^2 - e^2 g^2\right), \tag{1.74}$$

and the eigenvalues of the $SO(2,1)$ Casimir operator are determined by eigenvalues of the angular momentum operator.

Note that, generally speaking, any property of invariance of a theory with respect to a transformation involving a time dependence does not lead to the existence of a new integral of motion, but introduces some constraints on the configuration space of the classical system. A well-known example is ordinary electrodynamics, where Gauss law appears as a constraint on physical configurations. The situation we considered in the present section is analogous.

1.6 Dual Invariance of Classical Electrodynamics

The very idea of the possible existence of monopoles is closely connected with the notion of *duality* in classical electrodynamics. As was noted a long time ago by O. Heaviside [263], the free equations of the electromagnetic field

$$\boldsymbol{\nabla} \cdot \mathbf{E} = 0, \qquad \boldsymbol{\nabla} \cdot \mathbf{B} = 0,$$
$$\boldsymbol{\nabla} \times \mathbf{E} + \frac{\partial \mathbf{B}}{\partial t} = 0, \qquad \boldsymbol{\nabla} \times \mathbf{B} - \frac{\partial \mathbf{E}}{\partial t} = 0, \qquad (1.75)$$

possess a very remarkable invariance with respect to the transformations

$$D: \begin{cases} \mathbf{E} \to \mathbf{E} \cos\theta - \mathbf{B} \sin\theta, \\ \mathbf{B} \to \mathbf{E} \sin\theta + \mathbf{B} \cos\theta. \end{cases} \qquad (1.76)$$

This $O(2)$ symmetry that is parameterized by an arbitrary angle θ is called *dual*. In compact covariant notation, the free Maxwell equations are written as

$$\partial_\mu F^{\mu\nu} = 0, \qquad \partial_\mu \widetilde{F}^{\mu\nu} = 0, \qquad (1.77)$$

where the electromagnetic field strength tensor and its dual are

$$F_{\mu\nu} = \partial_\mu A_\nu - \partial_\nu A_\mu,$$
$$\widetilde{F}_{\mu\nu} = \frac{1}{2} \varepsilon_{\mu\nu\rho\sigma} F^{\rho\sigma} = \varepsilon_{\mu\nu\rho\sigma} \partial^\rho A^\sigma, \qquad (1.78)$$

where $\varepsilon_{\mu\nu\rho\sigma}$ is the totally antisymmetric pseudo-tensor, $\varepsilon_{0123} = 1$. Then the dual transformations of the field strength tensor are

$$D: \begin{cases} F^{\mu\nu} \to F^{\mu\nu} \cos\theta - \widetilde{F}^{\mu\nu} \sin\theta, \\ \widetilde{F}^{\mu\nu} \to F^{\mu\nu} \sin\theta + \widetilde{F}^{\mu\nu} \cos\theta. \end{cases} \qquad (1.79)$$

By composing the electric and magnetic fields into a complex vector $\mathbf{F} = \mathbf{E} + i\mathbf{B}$, the dual transformations can be written in a compact form

$$D: \quad \mathbf{E} + i\mathbf{B} \to e^{i\theta} (\mathbf{E} + i\mathbf{B}). \qquad (1.80)$$

1.6 Dual Invariance of Classical Electrodynamics

In particular, a special choice of the parameter of the dual transformation $\theta = -\pi/2$ leads to an interchange of electric and magnetic fields

$$D: \quad \mathbf{E} \to \mathbf{B}; \quad \mathbf{B} \to -\mathbf{E}. \tag{1.81}$$

Thus, if the theory enjoys such a symmetry, the separation of the fields into electric and magnetic ones is a matter of convention.

We observe that the energy and the momentum of the electromagnetic field

$$\mathcal{E} = \frac{1}{2}|\mathbf{F}|^2 = \frac{1}{2}\left(\mathbf{E}^2 + \mathbf{B}^2\right),$$

$$\mathcal{P} = \frac{1}{2i}[\mathbf{F}^* \times \mathbf{F}] = \mathbf{E} \times \mathbf{B}, \tag{1.82}$$

remain invariant under dual transformations (1.76). However, a simple calculation tells us that the Lagrangian of the free electromagnetic field

$$L_0 = -\frac{1}{4e^2}F_{\mu\nu}F^{\mu\nu} = -\frac{1}{2e^2}\left(E^2 - B^2\right) \tag{1.83}$$

transforms as

$$L_0 = \to -\frac{1}{4e^2}F_{\mu\nu}F^{\mu\nu}\cos 2\theta - \frac{1}{4e^2}F_{\mu\nu}\widetilde{F}^{\mu\nu}\sin 2\theta$$

$$= -\frac{1}{2e^2}\cos 2\theta\left(\mathbf{E}^2 - \mathbf{B}^2\right) - \frac{1}{2e^2}\sin 2\theta\left(\mathbf{E} \cdot \mathbf{B}\right). \tag{1.84}$$

In other words, the real and imaginary parts of the complex quantity

$$\frac{1}{2e^2}\left(\mathbf{E} + i\mathbf{B}\right)^2 = \frac{1}{2e^2}\left[(\mathbf{E}^2 - \mathbf{B}^2) + 2i(\mathbf{E} \cdot \mathbf{B})\right] \tag{1.85}$$

are mixed under the dual rotations (1.76). Nevertheless, the Abelian electrodynamics is effectively invariant with respect to such a transformation because

$$F_{\mu\nu}\widetilde{F}^{\mu\nu} = 2\partial_\mu\left(A_\nu\widetilde{F}^{\mu\nu}\right).$$

Thus, the dual transformations produce a total divergence and multiplication by a constant in addition to the Lagrangian. Certainly, it does not affect the dynamical equations of the fields.

The physical meaning of the dual invariance of the free electromagnetic theory can be clarified if we consider the infinitesimal form of the transformation (1.76). Then the Lagrangian transforms as

$$L_0 \to L_0 - \partial_\mu\left(A_\nu\widetilde{F}^{\mu\nu}\right)\delta\theta = L_0 - \partial_\mu D^\mu \delta\theta, \tag{1.86}$$

where we introduce a *dual current* $D^\mu = A_\nu\widetilde{F}^{\mu\nu}$. Thus, the dual symmetry of the free Maxwell theory implies the current conservation

$$\partial_\mu D^\mu \equiv \partial_\mu(A_\nu \widetilde{F}^{\mu\nu}) = 0. \tag{1.87}$$

Let us consider a particular case of the field of a plane electromagnetic wave. Then $F_{\mu\nu}\widetilde{F}^{\mu\nu} = 0$ and the dual current is automatically conserved. The corresponding charge, the generator of an infinitesimal dual transformation, is $D = \int d^3x D_0 = \int d^3x (\mathbf{A}\cdot \mathbf{B})$. Since in the case of a plane wave $\mathbf{B} = [\mathbf{k}\times \mathbf{E}]$, where the wave vector $\mathbf{k}$ is a unit vector in a propagation direction, we have

$$D = \int d^3x \left([\mathbf{k}\times \mathbf{E}]\cdot \mathbf{A}\right) = (\mathbf{S}\cdot \mathbf{k}). \tag{1.88}$$

Here we used the definition of a vector of spin of the electromagnetic field $\mathbf{S} = \int d^3x\,[\mathbf{E}\times \mathbf{A}]$. Thus, in such a case the dual charge is identical to the helicity of the plane wave.

Note that there is a general connection between the dual current D^μ (1.87) and the spin characteristic of the electromagnetic field. Indeed, the spin tensor is defined as $S_{\nu\rho\sigma} = F_{\nu\rho}A_\sigma - F_{\nu\sigma}A_\rho$ and the dual current is just the product $D_\mu = (1/4)\varepsilon_{\mu\nu\rho\sigma}S^{\nu\rho\sigma} = \widetilde{F}_{\mu\nu}A^\nu$.

There is a different formulation of the idea of electromagnetic duality connected with the path integral formulation of the theory. The point is that it is possible to change the variable of the functional integration by incorporation of the Bianchi identity $\partial^\mu \widetilde{F}_{\mu\nu} \equiv 0$ into the Lagrangian (1.83). Introducing a Lagrange multiplier vector field $\widetilde{A}_\mu$, we can now write the path integral over the fields $F_{\mu\nu}$ and $\widetilde{A}_\mu$, rather than over the potentials A_μ:

$$Z \sim \int \mathcal{D}A \exp\left[-\frac{1}{4e^2}\int d^4x\, F_{\mu\nu}F^{\mu\nu}\right] \tag{1.89}$$

$$\simeq \int \mathcal{D}F\mathcal{D}\widetilde{A}\exp\left[-\frac{1}{4e^2}\int d^4x\, F_{\mu\nu}F^{\mu\nu} + \frac{1}{2}\int d^4x\, \varepsilon_{\mu\nu\rho\sigma}\widetilde{A}^\mu\partial^\nu F^{\rho\sigma}\right].$$

The functional integral (1.89) over $F_{\mu\nu}$ is Gaussian. Indeed, taking into account that $\int d^4x\,\varepsilon_{\mu\nu\rho\sigma}\widetilde{A}^\mu\partial^\nu F^{\rho\sigma} = \int d^4x\,\varepsilon_{\mu\nu\rho\sigma}(\partial^\mu \widetilde{A}^\nu)F^{\rho\sigma}$, we obtain

$$Z \sim \int \mathcal{D}\widetilde{A}\exp\left[e^2\int d^4x\, \widetilde{F}_{\mu\nu}\widetilde{F}^{\mu\nu}\right] \quad \text{or} \quad L^{dual} = e^2\widetilde{F}_{\mu\nu}\widetilde{F}^{\mu\nu},$$

with the definition of the dual electromagnetic field strength via the Lagrange vector multiplier: $\widetilde{F}_{\mu\nu} = \partial_\mu \widetilde{A}_\nu - \partial_\nu \widetilde{A}_\mu$. Thus, the latter could be identified with the *potential* dual to A_μ. It is worth noting that the form of the dual transformed Lagrangian L^{dual} in terms of $\widetilde{F}_{\mu\nu}$ coincides with the original Lagrangian L_0 (1.83) up to a transformation of the electromagnetic coupling constant $e^2 \to e_D^2 = -4/e^2$.

Indeed, if we do not restrict our consideration to the case of the free electromagnetic field, the duality transformation could be considered a general rotation of the electric and magnetic quantities into each other. Then the generalized system of Maxwell equations with sources could be written in the form

$$\nabla \cdot (\mathbf{E} + i\mathbf{B}) = \rho_e + i\rho_g, \qquad \nabla \times (\mathbf{E} + i\mathbf{B}) - i\frac{\partial}{\partial t}(\mathbf{E} + i\mathbf{B}) = \mathbf{j}_e + i\mathbf{j}_g, \quad (1.90)$$

which is obviously invariant under simultaneous transformations of the fields (1.80) and charges

$$D: \quad e + ig \to e^{i\theta}(e + ig) \qquad (1.91)$$

(here we restrict consideration to the case of point-like particles with electric charge e and magnetic charge g). Note that it is not a genuine two-charge electrodynamics, because the experimental observables would not be these charges separately, but their dual-invariant combination

$$q = \sqrt{e^2 + g^2}. \qquad (1.92)$$

Indeed, the dual rotation by the angle $\theta = \arctan(g/e)$ would transform the system (1.90) into the standard form with an effective charge q.

2 The Electron–Monopole System: Quantum-Mechanical Interaction

2.1 Charge Quantization Condition

In Sect. (1.4) we already considered some properties of the singular gauge transformations of the vector-potential $\mathbf{A}$. However, the potential of the electromagnetic field not only defines the fields but, according to the formulation of the gauge theory, gives the form of the interaction between an electric charge and the electromagnetic field. In quantum mechanics this interaction is described by the covariant derivative acting on the wave function of a particle:

$$D\psi(\mathbf{r}) \equiv [\boldsymbol{\nabla} - ie\mathbf{A}(\mathbf{r})]\,\psi(\mathbf{r})\,. \tag{2.1}$$

Recall that under the gauge transformations of the potential (1.53), both the wave function and the covariant derivative transform in the same way:

$$\psi(\mathbf{r}) \to U\psi(\mathbf{r}) = e^{ie\lambda(\mathbf{r})}\psi(\mathbf{r})\,, \quad D\psi(\mathbf{r}) \to UD\psi(\mathbf{r}) = e^{ie\lambda(\mathbf{r})}D\psi(\mathbf{r})\,. \tag{2.2}$$

We already noted that the gauge function $\lambda(\mathbf{r})$ can in principle be a periodic function. However, for this point the gauge degrees of freedom do not matter on the classical level: the equation of motion of a classical charged particle in an external electromagnetic field remains invariant under the transformation (1.53). In quantum mechanics the situation is different since the fundamental quantity is now the action of the system and the corresponding path integral that defines the transition amplitudes and canonical variables. Therefore, under the gauge transformations (1.53), the Lagrangian (1.38), and the action as well, change as

$$L \to L + e\dot{\mathbf{r}}\,\boldsymbol{\nabla}\lambda(\mathbf{r}) = L + \frac{d}{dt}\left[e\lambda(\mathbf{r})\right]\,,$$

$$S = \int_0^T dt L \to S + e\lambda(\mathbf{r})\Big|_0^T\,. \tag{2.3}$$

Since the transition amplitude $\sim e^{iS}$ must be a gauge invariant quantity, the gauge function $\lambda(\mathbf{r})$ is no longer an arbitrary function of the coordinates, because any changes of the quantum-mechanical action must leave the transition amplitude unchanged. This is possible if $\delta S = e\delta\lambda(\mathbf{r}) = 2\pi n, \ n \in \mathbb{Z}$. The consequence is the quantization of the parameters of the quantum-mechanical system.

To see it, let us consider the change of the action under the gauge transformation (1.53) given by $U = \exp\{2ieg\varphi\}$ and suppose that the path is closed, that is $\varphi \to \varphi + 2\pi$. Then the system comes back to the initial position, but the action can pick up an additional phase factor. Indeed, if we set $\mathbf{r}(T) = \mathbf{r}(0)$, the interaction term can be written as an integral along the closed contour l:

$$S_{\text{int}} = e \int_0^T \dot{\mathbf{r}} \cdot \mathbf{A} = e \oint_l d\mathbf{x} \cdot \mathbf{A}. \tag{2.4}$$

Under the gauge transformation (1.53) this term changes to

$$\delta S_{\text{int}} = 2eg\delta\varphi = 4\pi eg. \tag{2.5}$$

Obviously, the quantum-mechanical transition amplitude remains invariant if $\delta S_{\text{int}} = 4\pi eg = 2\pi n$, $n \in \mathbb{Z}$, that is if Dirac's *charge quantization condition* [200] is satisfied:

$$eg = \frac{n}{2}. \tag{2.6}$$

In the following we will repeatedly return to this simple and nice relation, which is one of the most attractive aspects of the monopole theory. The suggestion by Dirac was that a monopole provides a beautiful explanation to the problem of quantization of electric charge: it is well known that all charged particles have electric charges that are proportional to the minimal charge of an electron.[1] Then, if there is a monopole somewhere in the universe, even one such object placed anywhere would be enough to explain the quantization of electric charges according to (2.6).

Note that this situation has nothing in common with the standard approach to quantization. In the latter case, the quantized parameter comes from the discrete part of the spectrum of eigenvalues of a Hermitian operator. In the former the situation is completely different since the quantized parameter – the product of electric and magnetic charges – is *not* an eigenvalue of any quantum-mechanical operator. The reason is that the charge quantization condition has a topological origin. In the next chapter we shall discuss this mechanism in more detail.

The charge quantization condition can be derived in many different ways. Historically the first one is connected with the analysis of the Schrödinger equation describing a non-relativistic particle of mass M in an external field:

$$H\psi(\mathbf{r}) = -\frac{1}{2M}(\boldsymbol{\nabla} - ie\mathbf{A})^2\psi(\mathbf{r}) = E\psi(\mathbf{r}), \tag{2.7}$$

where, according to the canonical quantization procedure, the momentum $\mathbf{P}$ in the classical Hamiltonian (1.71) is replaced by the quantum-mechanical operator $-i\boldsymbol{\nabla}$. Then the covariant derivative is $D_\mu = \partial_\mu - ieA_\mu$.

---
[1] There were no fundamental quarks in 1931!

Dirac noted [200] that the term of interaction of a charged particle and an external electromagnetic field can be written in the form (2.4). Thus, the wave function of this particle can be represented as

$$\psi(\mathbf{r}) = \psi_0(\mathbf{r}) e^{ie \int_0^{\mathbf{r}} d\mathbf{x}' \cdot \mathbf{A}(\mathbf{x}')}, \qquad (2.8)$$

where $\psi_0(\mathbf{r})$ is a wave function satisfying the free Schrödinger equation. The point is that the wave function must be a continuous function of $\mathbf{r}$, but the phase of ψ can be discontinuous in some point. The only condition is that the change of the phase factor $\oint_l d\mathbf{x} \cdot \mathbf{A}$ with a single turn along the closed path l surrounding such a point must be a multiple of 2π:

$$e \oint_l d\mathbf{x} \cdot \mathbf{A} = 2\pi n, \quad n \in \mathbb{Z}. \qquad (2.9)$$

On the other hand, this phase has an obvious physical interpretation: this is the flux of a magnetic field through the surface σ with the boundary l. If this boundary encircles the line of singularity, which corresponds to the transition from one sheet of the phase factor to another, we can write

$$e \oint_l d\mathbf{x} \cdot \mathbf{A} = e \int_\sigma d\sigma \cdot \mathbf{B}_{\text{sing}} = 4\pi eg, \qquad (2.10)$$

and the wave function is single valued if the charge quantization condition (2.6) is satisfied.

Perhaps the most simple way to derive this condition was suggested as early as 1936 by Saha [451]. He noted that the standard quantization of the operator of generalized angular momentum (cf. (1.68))

$$\mathbf{L} = [\mathbf{r} \times \boldsymbol{\pi}] - eg\hat{\mathbf{r}} = [\mathbf{r} \times (\mathbf{p} - e\mathbf{A})] - eg\hat{\mathbf{r}} = \widetilde{\mathbf{L}} - e[\mathbf{r} \times \mathbf{A}] - eg\,\hat{\mathbf{r}}, \qquad (2.11)$$

immediately leads to the charge quantization condition. Indeed, we can demand that the components of the angular momentum operator satisfy the standard commutation relations. Then its eigenvalues must be either integer or half-integer, that is of the form $n/2$, where $n \in \mathbb{Z}$. If we suppose that the orbital angular momentum $\mathbf{r} \times \boldsymbol{\pi}$ has integer eigenvalues as usual, then the additional term in (2.11) must have half-integer eigenvalues, which requires $eg = n/2$. This is exactly the charge quantization condition (2.6).

In order to obtain a more strict derivation of this result, we need to define in a consistent way algebra of the components of the angular momentum operator (2.11). A set of canonical commutation relations take the form

$$[x_i, x_j] = 0, \quad [x_i, \pi_j] = i\delta_{ij}, \quad [\pi_i, \pi_j] = ie\varepsilon_{ijk}[\boldsymbol{\nabla} \times \mathbf{A}]_k,$$

where $\pi_i = -i\partial_i - eA_i$. Of course, if the potential $\mathbf{A}$ would be a non-singular function, it would not be a big problem to prove that the generalized angular momentum commutes with the Hamiltonian operator. However, we are

30   2 The Electron–Monopole System: Quantum-Mechanical Interaction

working with a singular potential, which requires some care. At first sight it looks like there are some problems, since a naive calculation of the contribution of the additional singular magnetic flux, given by (1.49), gives an anomalous commutation relation that violates rotational invariance of the theory [41,547]:

$$[L_i, L_j] = i\varepsilon_{ijk} L_k + ie\varepsilon_{ijk} x_k (\mathbf{r} \cdot \mathbf{B}_{\text{sing}}), \qquad (2.12)$$

$$[L^i, H] = \frac{ie}{2M} \left(\pi^i (\mathbf{r} \cdot \mathbf{B}_{\text{sing}}) + (\mathbf{r} \cdot \mathbf{B}_{\text{sing}}) \pi^i - (\pi \cdot \mathbf{r}) B^i_{\text{sing}} - x^k B^i_{\text{sing}} \pi^k \right).$$

However, the potential $\mathbf{A}$ *is* a singular function and therefore both the Hamiltonian H and the generalized angular momentum operator $\mathbf{L}$ are singular operators. The product of such operators must be regularized in some way already on the quantum-mechanical level, for example, by the point splitting method [457]. In this approach we can write the regularized operators [41,547]

$$H = \lim_{\epsilon \to 0} \frac{3}{m\epsilon^2} \left\{ 1 - \exp[-i(\mathbf{p} \cdot \boldsymbol{\epsilon})] E \exp[-i(\mathbf{p} \cdot \boldsymbol{\epsilon})] \right\},$$

$$L_i = \lim_{\epsilon \to 0} \varepsilon_{ijk} x_j \frac{\epsilon_k}{i\epsilon^2} \left\{ 1 - \exp[-i(\mathbf{p} \cdot \boldsymbol{\epsilon})] E \exp[-i(\mathbf{p} \cdot \boldsymbol{\epsilon})] \right\}, \qquad (2.13)$$

with

$$E \equiv \exp \left\{ ie \int_{\mathbf{r}-\boldsymbol{\epsilon}/2}^{\mathbf{r}+\boldsymbol{\epsilon}/2} \mathbf{A} \cdot d\boldsymbol{\tau} \right\}. \qquad (2.14)$$

Indeed, if the potential $\mathbf{A}$ is a regular function, by expanding the exponents in (2.13) and averaging over all directions of the parameter $\boldsymbol{\epsilon}$, we immediately recover the standard Hamiltonian operator and the generalized angular momentum operator. However, for the Dirac potential these relations must be considered as a definition. Then, making use of the the regularized operators (2.13) yields, instead of (2.12), the standard commutation relations

$$[L_i, L_j] = i\varepsilon_{ijk} L_k, \qquad [L_i, H] = 0. \qquad (2.15)$$

Finally, let us note that the Dirac quantization condition (2.6) can also be obtained if we set a requirement of "invisibility" of the singularity of the vector potential. This situation can be considered as an analog of the Aharonov–Bohm effect [61]. Indeed, in both situations the wave function of an electron, which scatters in the field of an infinitely long and thin solenoid, picks up a phase factor $\exp\{e\Phi/2\pi\}$, where Φ is the flux of the magnetic field. The only difference is that in the monopole case the singular flux along the string is fixed: $\Phi_{\text{str}} = 4g\pi$. Thus, the phase factor is unobservable if the charge quantization condition $eg = n/2$ is satisfied.

To sum up, the Dirac charge quantization condition (2.6) is a fundamental element of any model of a monopole. It is an interesting fact that this condition can be proved in a large number of ways. Actually, the charge quantization is related to the topological origin of the monopole. We shall give a more thorough discussion of these interrelated problems in the next chapter. Now let us consider some other features of the interaction between a monopole and a "normal" electric charge.

2.2 Spin-Statistics Theorem in a Monopole Theory

The theory of monopoles is rich in paradoxes, both real and fictional. One of these paradoxical effects is a generalization of the spin-statistics theorem in a theory with a monopole [245]. Here we draw on the lectures by S. Coleman [43].

We have already noted that an extra angular momentum $\mathbf{T} - eg\hat{\mathbf{r}}$ appears in the charge-monopole system. The charge quantization condition (2.6) means that it can take both integer and half-integer values. Therefore, the system of a charge and a monopole can possess a half-integer angular momentum, i.e., a bound system of two bosons behaves like a fermion.

In order to understand this effect better, let us consider the Hamiltonian of the Schrödinger equation (2.7), which describes the behavior of the electric charge in the monopole field,

$$H_e = -\frac{1}{2M_e} \left[\boldsymbol{\nabla}_e - ieg\mathbf{A}(\mathbf{r}_e - \mathbf{r}_g) \right]^2 . \tag{2.16}$$

In the same way, making use of the discrete form of the dual transformations (1.91), $e \to g$; $g \to -e$, we can formally write the Hamiltonian operator that describes the dynamics of the monopole g in the external field of a charge e:

$$H_g = -\frac{1}{2M_g} \left[\boldsymbol{\nabla}_g + ieg\widetilde{\mathbf{A}}(\mathbf{r}_g - \mathbf{r}_e) \right]^2 , \tag{2.17}$$

where $\widetilde{\mathbf{A}}(\mathbf{r})$ is the potential dual to $\mathbf{A}(\mathbf{r})$.

Note that the charge-monopole system is translationally invariant. Then the explicit form of this dual potential can be recovered from conservation of the total momentum. Indeed, the classical equations of motion following from (2.16) and (2.17) are

$$M_e \mathbf{v}_e = \mathbf{P}_e - eg\mathbf{A}(\mathbf{r}_e - \mathbf{r}_g), \qquad M_g \mathbf{v}_g = \mathbf{P}_g + eg\widetilde{\mathbf{A}}(\mathbf{r}_g - \mathbf{r}_e).$$

These equations are compatible with the conservation of momentum of the whole system, $\mathbf{P}_e + \mathbf{P}_g = 0$ only if $\mathbf{A}(\mathbf{r}) = \widetilde{\mathbf{A}}(-\mathbf{r})$. Therefore, these potentials are connected by the gauge transformation:

$$\mathbf{A}(\mathbf{r}) \to \mathbf{A}(\mathbf{r}) + \boldsymbol{\nabla}\lambda(\mathbf{r}) = \mathbf{A}(-\mathbf{r}) = \widetilde{\mathbf{A}}(\mathbf{r}),$$

and we can write the Hamiltonian operator (2.17) in the form

$$H_g = -\frac{1}{2M_g}[\boldsymbol{\nabla}_g + ieg\mathbf{A}(\mathbf{r}_e - \mathbf{r}_g)]^2 \ .$$

Let us consider a system of two identical dyons with electrical and magnetic charges e and g at $\mathbf{r}_1$ and $\mathbf{r}_2$, respectively. Obviously, a permutation of these particles cannot change any physical observable. Thus, the only thing that can happen is that the wave function of the system picks up an additional phase factor $e^{i\pi\alpha}$. The effect of two consecutive interchanges is the same as that of no interchange. Thus $e^{2i\pi\alpha} = 1$, i.e., $e^{i\pi\alpha} = 1$ (Bose–Einstein statistics) or $e^{i\pi\alpha} = -1$ (Fermi–Dirac statistics).

In the case under consideration it would be natural to expect that the dyons are bosons, since a dual rotation by the angle $\theta = \text{arctg}\,(g/e)$ transforms this system to a pair of identical effective charges $q = \sqrt{e^2 + g^2}$. Indeed, the total Hamiltonian of the system of two interacting dyons (2.16) and (2.17) is

$$\begin{aligned} H &= H_1 + H_2 \\ &= -\frac{1}{2M}[\boldsymbol{\nabla}_1 + ieg\mathbf{A}(\mathbf{r}_1 - \mathbf{r}_2) - ieg\mathbf{A}(\mathbf{r}_2 - \mathbf{r}_1)]^2 \\ &\quad - \frac{1}{2M}[\boldsymbol{\nabla}_2 + ieg\mathbf{A}(\mathbf{r}_2 - \mathbf{r}_1) - ieg\mathbf{A}(\mathbf{r}_1 - \mathbf{r}_2)]^2 \\ &\quad + V(e^2) + V(g^2), \end{aligned} \quad (2.18)$$

where

$$V(e^2) = \frac{e^2}{|\mathbf{r}_1 - \mathbf{r}_2|}, \qquad V(g^2) = \frac{g^2}{|\mathbf{r}_1 - \mathbf{r}_2|}. \quad (2.19)$$

We have already noted that two Dirac potentials $\mathbf{A}(\mathbf{r})$ and $\mathbf{A}(-\mathbf{r})$ are connected by the gauge transformation $U = e^{2ieg\varphi}$ (cf. (1.65)) and therefore

$$\mathbf{A}(\mathbf{r}_1 - \mathbf{r}_2) - \mathbf{A}(\mathbf{r}_2 - \mathbf{r}_1) = \frac{i}{e} U^{-1} \boldsymbol{\nabla} U \ .$$

Thus, the terms of interaction in (2.18) can be eliminated by such a transformation that rotates this expression to the standard Hamiltonian of two effective charges with Coulomb interaction

$$H = H_1 + H_2 = -\frac{1}{2M}\left[(\boldsymbol{\nabla}_1)^2 + (\boldsymbol{\nabla}_2)^2\right] + V(q^2) \ .$$

However, this transformation also gives a phase factor to the the wave function of the system:

$$\psi(\mathbf{r}) \to U\psi(\mathbf{r}) = e^{2ieg\varphi}\psi(\mathbf{r}) \ .$$

The interchange of the two dyons corresponds to the rotation $\varphi \to \varphi + \pi$ and the wave function of the whole system is symmetric with respect to permutation if the product

$$eg = \mu \quad (2.20)$$

is an integer and it is anti-symmetric if μ is a half-integer. Thus, the standard spin-statistics theorem is fulfilled, but a system of two identical dyons can satisfy either Bose–Einstein or Fermi–Dirac statistics.

This surprising result inspired some authors [113, 168, 460] to construct phenomenological composite models of hadrons not with quarks but with dyons as fundamental constituents. Of course, from a modern point of view these constructions look rather naive. Nevertheless, in some way they anticipated a very interesting development of the modern theory of strong interactions, which is connected with the possible role of topologically non-trivial configurations. Later, in Chap. 9, we discuss some models of the monopole-related mechanism of confinement in quantum chromodynamics in more detail.

2.3 Charge-Monopole System: Quantum-Mechanical Description

Let us analyze the solutions of the Schrödinger equation that describes the quantum-mechanical motion of a charged particle of mass M in the monopole external field. Since the operator of the generalized angular momentum (2.11) commutes with the Hamiltonian that enters the Schrödinger equation of (2.7):

$$H = -\frac{1}{2Mr^2}\left\{\frac{\partial}{\partial r}\left(r^2\frac{\partial}{\partial r}\right) + \mathbf{L}^2 - \mu^2\right\}, \quad (2.21)$$

they will have common eigenfunctions. Therefore, in a spherical coordinate system we can separate the variables and the eigenfunctions of the Hamiltonian operator (2.21) are of the form [214, 291, 486]

$$\Psi(\mathbf{r}) = F_{k\tilde{\ell}}(r)Y_{\mu lm}(\theta, \varphi), \quad (2.22)$$

and

$$\mathbf{L}^2 Y_{\mu lm}(\theta, \varphi) = \lambda Y_{\mu lm}(\theta, \varphi), \quad L_3 Y_{\mu lm}(\theta, \varphi) = m Y_{\mu lm}(\theta, \varphi). \quad (2.23)$$

Choosing the following form of the Dirac potential (see (1.54))

$$\mathbf{A} = \frac{g}{r}\frac{1-\cos\theta}{\sin\theta}\hat{\mathbf{e}}_\varphi = \frac{g}{r}\frac{[\mathbf{n}\times\mathbf{r}]}{r+(\mathbf{r}\cdot\mathbf{n})},$$

we can write the operator $\mathbf{L}$ of (2.11) as[2]

$$\mathbf{L} = \widetilde{\mathbf{L}} - \mu\frac{\hat{\mathbf{r}}+\hat{\mathbf{z}}}{1+\cos\theta} = \frac{1}{\sin\theta}\left(i\frac{\partial}{\partial\varphi} + \mu(1-\cos\theta)\right)\hat{\mathbf{e}}_\theta - i\frac{\partial}{\partial\theta}\hat{\mathbf{e}}_\varphi - \mu\hat{\mathbf{e}}_r, \quad (2.24)$$

---
[2] To simplify our considerations we suppose that all quantum numbers are positive. In particular, we set $\mu = eg > 0$.

or

$$\mathbf{L}^2 = -\frac{1}{\sin^2\theta}\left[\sin\theta\frac{\partial}{\partial\theta}\left(\sin\theta\frac{\partial}{\partial\theta}\right) + \left(\frac{\partial}{\partial\varphi} - i\mu(1-\cos\theta)\right)^2\right] + \mu^2,$$

$$L_3 = -i\frac{\partial}{\partial\varphi} - \mu, \qquad (2.25)$$

where we use the notation $\mu = eg$.

The explicit form of the operator L_3 suggests that the dependence of the wave functions on the azimuthal angle φ must be of the form $Y_{\mu lm}(\theta, \varphi) = P(\theta)e^{i(\mu+m)\varphi}$ where $m \in \mathbb{Z}$. However, in this case the commutation relations for the components of the generalized angular momentum are satisfied if $\mu = n/2$, that is, if the charge quantization condition (2.6) is fulfilled.

2.3.1 The Generalized Spherical Harmonics

We proceed by first finding[3] the eigenfunctions of the operator of the generalized angular momentum:

$$\mathbf{L}^2 Y_{\mu lm}(\theta, \varphi)$$
$$= -\left[(1-x^2)\frac{\partial^2}{\partial x^2} - 2x\frac{\partial}{\partial x} - \frac{1}{1-x^2}\left(i\frac{\partial}{\partial\varphi} + \mu(1-x)\right)^2 - \mu^2\right]Y_{\mu lm}(\theta, \varphi)$$
$$= \lambda Y_{\mu lm}(\theta, \varphi), \qquad (2.26)$$

where $x = \cos\theta$. Taking into consideration the second of the equations (2.25), we see that the dependence on the variable x is given by the equation

$$\left\{-(1-x^2)\frac{d^2}{dx^2} + 2x\frac{d}{dx} + \frac{(m+\mu x)^2}{1-x^2} + \mu^2\right\}P(x) = \lambda P(x). \qquad (2.27)$$

The procedure is rather standard. Separating the singularities $x = \pm 1$ we seek a solution of this equation in the form

$$P(x) = (1-x)^{-\frac{\mu+m}{2}}(1+x)^{-\frac{\mu-m}{2}}F(x). \qquad (2.28)$$

Substituting this into (2.27), we get

$$(1-x^2)\frac{dF^2}{dx^2} + 2[m + (\mu-1)x]\frac{dF}{dx} + (\mu - \mu^2 + \lambda)F = 0. \qquad (2.29)$$

Introducing the new variable $z = (1+x)/2$ this can be recognized as the standard hypergeometric equation

---

[3] This problem was solved in 1931 by Tamm [486], who was directly informed by Dirac about his work [200]. In our discussion we follow the approach of [291].

2.3 Charge-Monopole System: Quantum-Mechanical Description

$$z(1-z)F'' + \{m - \mu + 1 + 2z(\mu - 1)\}F' + (\mu - \mu^2 + \lambda)F$$
$$= z(1-z)F'' + \{c - (a+b+1)z\}F' - abF = 0, \qquad (2.30)$$

with the solution[4] $_2F_1(a,b;c;z)$; here obviously $c = m - \mu + 1$, $ab = \mu^2 - \mu - \lambda$, and $a + b + 1 = 2(1 - \mu)$.

Recall that the hypergeometric function is finite if it is a polynomial with a finite number of terms. This is satisfied if the parameter a or b is a negative integer. For the sake of definiteness, let us fix $a = -n$, $n = 0, 1, 2, \ldots$ Then the eigenvalues λ of the operator $\mathbf{L}^2$ must satisfy the relation

$$\lambda = n(n+1) - 2\mu n + \mu(\mu - 1),$$

or, introducing the variable $l = n - \mu$,

$$\lambda = l(l+1).$$

Thus, up to a normalization factor, a particular solution of the eigenvalue (2.25) can be written as

$$Y_{\mu l m}(\theta, \varphi) = (1-x)^{-\frac{\mu+m}{2}} (1+x)^{-\frac{\mu-m}{2}} {}_2F_1(-n, n+1-2\mu; m+1-\mu; z) e^{i(\mu+m)\varphi}.$$

For a general solution of (2.25) it is more convenient to use another representation, which is given by the *Jacobi polynomials* [2]

$$P_n^{(\alpha,\beta)}(x) = \frac{(-1)^n}{2^n n!}(1-x)^{-\alpha}(1+x)^{-\beta}\frac{d^n}{dx^n}\left((1-x)^{\alpha+n}(1+x)^{\beta+n}\right),$$

related to the hypergeometric function $_2F_1(-n,b;c;z)$ above. Then the eigenfunctions of the operator $\mathbf{L}^2$, the so-called *generalized spherical harmonics* are

$$Y_{\mu l m}(\theta, \varphi) = N(1-x)^{-(\mu+m)/2}(1+x)^{-(\mu-m)/2} P_{l+m}^{(-\mu-m,-\mu+m)}(x) e^{i(\mu+m)\varphi},$$
$$(2.31)$$

where the normalization factor N is

$$N = 2^m \left(\frac{(2l+1)(l-m)!(l+m)!}{4\pi(l-\mu)!(l+\mu)!}\right)^{1/2}.$$

The first few normalized generalized spherical harmonics for the minimal value of $\mu = 1/2$ are for the north hemisphere [530]

$$Y_{\frac{1}{2}\frac{1}{2}\frac{1}{2}}(\theta,\varphi) = -\frac{1}{\sqrt{2\pi}} \sin\frac{\theta}{2} e^{i\varphi},$$

---

[4] For the sake of simplicity we do not write here the second independent solution of the hypergeometric equation. However, we cannot neglect it because this solution is necessary to construct the complete set of eigenfunctions of the operator of the generalized angular momentum $\mathbf{L}$. The definition (2.31) below includes both solutions.

$$Y_{\frac{1}{2}\frac{1}{2}-\frac{1}{2}}(\theta,\varphi) = \frac{1}{\sqrt{2\pi}}\cos\frac{\theta}{2},$$

$$Y_{\frac{1}{2}\frac{3}{2}\frac{1}{2}}(\theta,\varphi) = -\frac{1}{\sqrt{4\pi}}\sin\frac{\theta}{2}(1+3\cos\theta)e^{i\varphi},$$

$$Y_{\frac{1}{2}\frac{3}{2}-\frac{1}{2}}(\theta,\varphi) = -\frac{1}{\sqrt{4\pi}}\cos\frac{\theta}{2}(1-3\cos\theta),$$

$$Y_{\frac{1}{2}\frac{3}{2}\frac{3}{2}}(\theta,\varphi) = \sqrt{\frac{3}{4\pi}}\cos\frac{\theta}{2}(1-\cos\theta)e^{2i\varphi},$$

$$Y_{\frac{1}{2}\frac{3}{2}-\frac{3}{2}}(\theta,\varphi) = \sqrt{\frac{3}{4\pi}}\sin\frac{\theta}{2}(1+\cos\theta)e^{-i\varphi}.$$

Obviously, in the case $\mu = 0$ these functions reduce to standard spherical harmonics, for example:

$$Y_{010}(\theta,\varphi) = \sqrt{\frac{3}{4\pi}}\cos\theta, \qquad Y_{01\pm1}(\theta,\varphi) = \mp\sqrt{\frac{3}{8\pi}}\sin\theta e^{\pm i\varphi}. \qquad (2.32)$$

As in the case of the standard spherical functions we can introduce two Hermitian-conjugated raising and lowering generators

$$L_{\pm} = L_1 \pm iL_2 = e^{\pm i\varphi}\left\{\pm\frac{\partial}{\partial\theta} + i\frac{\cos\theta}{\sin\theta}\frac{\partial}{\partial\varphi} - \mu\frac{\sin\theta}{1+\cos\theta}\right\}. \qquad (2.33)$$

Making use of the algebra of the components of the angular momentum operator (2.11), we can write the relation

$$L_{\pm}L_{\mp} = L^2 - L_3^2 \pm L_3.$$

Therefore

$$L_{\pm}Y_{\mu lm}(\theta,\varphi) = \sqrt{(l(l+1) - m(m\pm1)}Y_{\mu lm\pm1}(\theta,\varphi). \qquad (2.34)$$

We already noted that the components of the generalized angular momentum $\mathbf{L}$ satisfy standard commutation relations. Therefore, the system has spherical symmetry and eigenfunctions of the operator $\mathbf{L}^2$ correspond to an irreducible representation of the rotation group $SO(3)$. In other words, they are connected with the standard Wigner functions [10] as (see Appendix A)

$$Y_{\mu lm}(\theta,\varphi) = D^l_{\mu m}(-\varphi,\theta,\varphi) = <l,\mu\mid e^{-i\varphi L_3}e^{i\theta L_2}e^{i\varphi L_3}\mid l,m>. \qquad (2.35)$$

Here, the quantum numbers l and μ must simultaneously be integer or half-integer and the relations

$$l = \mu, \mu+1, \mu+2\ldots, \qquad -l \leq m \leq l$$

2.3 Charge-Monopole System: Quantum-Mechanical Description

must hold. Since $l \geq \mu$, the centrifugal potential of the Schrödinger equation is always repulsive. Thus, there is no bound state in the spectrum of a monopole and a spinless charged particle. This corresponds to the absence of closed trajectories in the classical problem of charge-monopole scattering. On the other hand, if we consider not a monopole but a dyon, then there are bound states in the spectrum caused by the normal Coulomb interaction. This system is an analogue of the hydrogen atom. Sometimes it is called a *dyogen atom* [129,130] or *dyonium* [115]. This simple model allows us to demonstrate non-trivial quantum mechanical properties caused by the monopole presence in a very clear form [356, 466].

2.3.2 Solving the Radial Schrödinger Equation

As was noted by Coleman [43], the effect of the magnetic monopole in the Schrödinger equation is very simple: it just modifies the centrifugal potential. Indeed, the radial function satisfies the equation

$$-\frac{1}{2M}\left[\frac{d^2}{dr^2} + \frac{2}{r}\frac{d}{dr} - \frac{l(l+1) - \mu^2}{r^2}\right] F_{k\tilde{\ell}}(r) = EF_{k\tilde{\ell}}(r), \quad (2.36)$$

which is solved by spherical Bessel functions of the order

$$\tilde{\ell} = \sqrt{\left(l + \frac{1}{2}\right)^2 - \mu^2} - \frac{1}{2}, \quad (2.37)$$

namely

$$F_{k\tilde{\ell}}(r) = \sqrt{\frac{k}{r}} J_{\tilde{\ell}+1/2}(kr) = \frac{1}{k}\sqrt{\frac{2}{\pi}} j_{\tilde{\ell}}, \quad k = \sqrt{2ME}, \quad (2.38)$$

with $E > 0$. Making the identification $\tilde{\ell}(\tilde{\ell}+1) = l(l+1) - \mu^2$, we can see that formally this is precisely the radial wave function of the standard Schrödinger equation that describes a state with (non-integer) angular momentum $\tilde{\ell}$.

Note that asymptotically, as $r \to \infty$, the behavior of the radial function is

$$F_{k\tilde{\ell}}(r) \xrightarrow[r \to \infty]{} \frac{1}{r}\sin\left(kr - \frac{\pi\tilde{\ell}}{2}\right). \quad (2.39)$$

If we consider not a monopole but a dyon, the angular dependence of the eigenfunctions is still given by the generalized spherical harmonics $Y_{\mu lm}(\theta, \varphi)$ of (2.31), which are eigenfunctions of the operator of angular momentum **L**. Then the radial function will satisfy the equation

$$\left[\frac{d^2}{dr^2} + \frac{2}{r}\frac{d}{dr} - \frac{l(l+1) - \mu^2}{r^2} + \frac{2MeQ}{r}\right] F(r) = -2MEF(r), \quad (2.40)$$

where for the sake of definiteness we fix the electric charge of the dyon Q to be positive and the scalar potential of the Coulomb interaction is $V(r) = -eQ/r < 0$. Knowledge of the asymptotic behavior of the radial wave function allows us to search for a regular solution of this equation in the form

$$F(r) = e^{-kr} r^{\tilde{\ell}} \rho(r), \qquad \text{where} \quad k^2 = -2ME > 0.$$

Substituting this ansatz into (2.40) we obtain

$$r\frac{d^2\rho(r)}{dr^2} + (2\tilde{\ell} + 2 - 2kr)\frac{d\rho(r)}{dr} + (2MeQ - 2k\tilde{\ell} - 2k)\rho(r)$$
$$+ \frac{\tilde{\ell}(\tilde{\ell}+1) - l(l+1) + \mu^2}{r}\rho(r) = 0. \tag{2.41}$$

The solution is regular at $r = 0$ if $\tilde{\ell}$ satisfies (2.37). Obviously, the solution of (2.41) is given by the confluent hypergeometric function $\rho(r) = {}_1F_1(-N; 2\tilde{\ell} + 2; 2kr)$, where $N = 0, 1, 2, \ldots$ is a radial quantum number. Furthermore, it follows from (2.41) that

$$N = \frac{MeQ}{k} - \tilde{\ell} - 1, \quad \text{and therefore} \quad E = -\frac{M(eQ)^2}{2(N + \tilde{\ell} + 1)^2}, \tag{2.42}$$

which defines the spectrum of bound states of an electron in a dyon field. Note that the ground state has quantum numbers $l = 1/2, N = 0, \mu = 1/2$, that is, it is doubly degenerated ($m = \pm 1/2$).

To sum up, the solution of the radial Schrödinger equation that describes an electron in a dyon field is

$$F_{N\tilde{\ell}}(r) = C(kr)^{\tilde{\ell}} e^{-kr} {}_1F_1(-N; 2\tilde{\ell} + 2; 2kr), \tag{2.43}$$

where the normalization factor is

$$C = \frac{2^{\tilde{\ell}+1}[MeQ\,\Gamma(N + 2\tilde{\ell} + 2)]^{\frac{1}{2}}}{(N!)^{\frac{1}{2}}(N + \tilde{\ell} + 1)\,\Gamma(2\tilde{\ell} + 2)}.$$

Obviously, in the limiting case $Q = 0$, (2.38) can be recovered from the expression (2.43) since in this case $\tilde{\ell} = -N - 1$. Then the hypergeometric function reduces to

$${}_1F_1(\tilde{\ell} + 1; 2\tilde{\ell} + 2; 2kr) = \Gamma(\tilde{\ell} + 3/2) e^{(\tilde{\ell}+1/2)\pi/2} \left(\frac{kr}{2}\right)^{-\tilde{\ell}+1/2} e^{kr} J_{\tilde{\ell}+1/2}(kr).$$

Finally, let us note that the relativistic description of this system of a dyon and a scalar particle in terms of a solution of the Klein–Gordon equation is practically identical to the present one. Indeed, the equation

2.3 Charge-Monopole System: Quantum-Mechanical Description

$$\left[(\partial_\mu - ieA_\mu)^2 + M^2\right]\phi(\mathbf{r},t) = 0, \tag{2.44}$$

where the 4-potential of a dyon is $A_\mu = (Q/r, \mathbf{A})$, can be written in a form almost identical to the non-relativistic Schrödinger equation, (2.31) and (2.43) up to the replacement $2ME \to k^2 = \omega^2 - M^2$. In Chap. 4 we will need the exact solutions of the relativistic equation. Let us therefore discuss it here in more detail [5].

We define the wave functions $\phi_{\mu klm}(\mathbf{r},t)$ describing both bound states and continuum states. Clearly, the time dependence factorizes, and we can separate the radial and angular variables, the latter being determined by the generalized spherical harmonics (2.31):

$$\phi_{\mu klm}(\mathbf{r},t) = F_{k\tilde{\ell}}(r) Y_{\mu lm}(\theta,\varphi) e^{-i\omega t}. \tag{2.45}$$

For the sake of convenience, we use wave functions that satisfy the normalization condition

$$\int_V d^3x |F_{k\tilde{\ell}}(r) Y_{\mu lm}(\theta,\varphi)|^2 = \frac{1}{2\omega}.$$

Then the radial functions are solutions of the equation

$$\left(\frac{d^2}{dr^2} + \frac{2}{r}\frac{d}{dr} - k^2 + 2\omega\frac{eQ}{r} - \frac{e^2Q^2 + l(l+1) - \mu^2}{r^2}\right) F_{k\tilde{\ell}}(r) = 0, \tag{2.46}$$

which can be recognized as a slightly modified hypergeometric equation of the non-relativistic radial problem (2.41). The solutions corresponding to the continuum are

$$F_{k\tilde{\ell}}(r) = C(kr)^{\tilde{\ell}} r^{-1} e^{-ikr} {}_1F_1\left(\tilde{\ell} + 1 - ieQ\frac{\omega}{k}; 2\tilde{\ell} + 2; 2ikr\right), \tag{2.47}$$

where $k^2 = \omega^2 - \mathbf{k}^2$, the parameter $\tilde{\ell}$ is defined as (cf. (2.37))

$$\tilde{\ell} = \sqrt{\left(l + \frac{1}{2}\right)^2 - \mu^2 + e^2 Q^2} - \frac{1}{2}, \tag{2.48}$$

and the normalization constant is

$$C = \frac{2^{\tilde{\ell}+1/2}}{\sqrt{\pi\omega}} \frac{\Gamma(\tilde{\ell} + 2 + ieQ\frac{\omega}{k})}{(2\tilde{\ell}+1)} \exp\left(-\frac{\pi eQ\omega}{2k}\right).$$

As for the solutions corresponding to the discrete part of the spectrum, the radial equation is solved by

$$F_{k\tilde{\ell}}(r) = G(kr)^{\tilde{\ell}} e^{-kr} {}_1F_1(-N; 2\tilde{\ell} + 2; 2kr), \tag{2.49}$$

where $N = 0, 1, 2, \ldots$ is the radial quantum number and G is a normalization constant. The energy spectrum of the relativistic particle in the dyon field is

$$\omega = M\sqrt{1 - \frac{e^2 Q^2}{e^2 Q^2 + (N + \widetilde{\ell} + 1)^2}}\,. \tag{2.50}$$

Thus, these solutions are not very different from the solutions of the Schrödinger equation discussed above. Much more interesting are the properties of the spectrum of a spin-1/2 particle in a dyon field since the addition of spin and the extra angular momentum due to the magnetic charge can drastically change the spectrum of the bound states. However, before we consider this effect in detail, let us analyze the quantum mechanical problem of the scattering of a spinless particle on a monopole, in order to establish the correspondence with the classical problem of scattering considered in Chap. 1.

2.4 Non-Relativistic Scattering on a Monopole: Quantum Mechanical Description

The solutions of the Schrödinger equation for a charged particle in a monopole field, which we discussed above, allow us to solve the quantum mechanical problem of scattering by the magnetic charge [110, 131, 462], i.e., to define the scattering wave function and calculate the cross-section. Recall that the standard approach is to write a general solution of the Schrödinger equation $\psi(\mathbf{r})$ describing the outgoing wave, as an expansion in partial waves, which correspond to states with a fixed angular momentum l and projection m. The initial state is taken to be a plane wave $\psi(\mathbf{r}) \sim e^{ikz}$ propagating in the z direction.

Note that the principal difference from the standard situation is that even if the particle is infinitely far away from the monopole, the projection of the angular momentum on the z-axis does not vanish. Indeed, the minimal value of the angular momentum is restricted to be $l = \mu$. Then, according to (2.25) the eigenvalue of the component L_3 is μ. Thus, the appropriate angular dependence of the asymptotic states must be given by the generalized spherical harmonic $Y_{\mu l \mu}(\theta, \varphi)$ containing the phase factor $e^{2i\mu\varphi}$ (the modified plane wave [131]).

Now, let us apply the standard expansion of the plane wave in spherical Bessel functions $j_{\widetilde{\ell}}(kr)$:

$$e^{ikr\cos\theta} = \pi \sum_{\widetilde{\ell}=0}^{\infty} \sqrt{(2\widetilde{\ell}+1)}\, i^{\widetilde{\ell}} \sqrt{\frac{2}{kr}} J_{\widetilde{\ell}+1/2}(kr) Y_{0\widetilde{\ell}0}(\theta, \varphi)\,. \tag{2.51}$$

However, the angular dependence is defined by the generalized spherical functions $Y_{\mu l \mu}(\theta, \varphi)$ and the summation over angular momentum l begins from $l = \mu$:

2.4 Non-Relativistic Scattering on a Monopole

$$e^{ikr\cos\theta} = \pi \sum_{l=\mu}^{\infty} \sqrt{(2l+1)}\, i^{l-\mu} \sqrt{\frac{2}{kr}} J_{\tilde{\ell}+1/2}(kr) Y_{\mu l \mu}(\theta,\varphi)$$

$$= e^{-i\pi\mu} \sum_{l=\mu}^{\infty} (2l+1) e^{i\pi l} j_{\tilde{\ell}}(kr) Y_{\mu l \mu}(\theta,\varphi). \tag{2.52}$$

Then the scattered wave function of the particle in a monopole field is [131]:

$$\psi(\mathbf{r}) = 2\sqrt{\pi}\, e^{-i\pi\mu} \sum_{l=\mu}^{\infty} \sqrt{(2l+1)}\, e^{i\pi l} e^{i\delta_{\tilde{\ell}}} j_{\tilde{\ell}}(kr) Y_{\mu l \mu}(\theta,\varphi). \tag{2.53}$$

The phase shift $e^{i\delta_\ell}$ in this expansion can be calculated from the condition that the outgoing wave (2.53) must correspond to the asymptotic form of the radial function (2.39), namely

$$\sqrt{\frac{\pi}{2kr}} J_{\tilde{\ell}+1/2}(kr) \sim \frac{1}{r} \sin\left(kr - \frac{\pi\tilde{\ell}}{2}\right) \sim \frac{1}{r}\left(e^{i(kr - \pi\tilde{\ell}/2)} - e^{-i(kr - \pi\tilde{\ell}/2)}\right). \tag{2.54}$$

Therefore we obtain $\delta_{\tilde{\ell}} = -\pi\tilde{\ell}/2$.

From (2.54) we can see that the scattered wave function is singular at the forward direction where $r \to \infty$ and $\theta = 0$. These limits are not uniform and cannot be interchanged. To analyze the structure of this singularity it is convenient to separate the outgoing wave into two parts as [131]

$$\psi(\mathbf{r}) = e^{-i\pi\mu} \{\psi_I(\mathbf{r}) + \psi_{II}(\mathbf{r})\}, \tag{2.55}$$

where

$$\psi_I(\mathbf{r}) = N(\mu) \sum_{l=\mu}^{\infty} \sqrt{(2l+1)}\, e^{i\pi l/2} j_l(kr) Y_{\mu l \mu}(\theta,\varphi),$$

$$\psi_{II}(\mathbf{r}) = N(\mu) \sum_{l=\mu}^{\infty} \sqrt{(2l+1)}\, e^{i\pi l} \left[e^{-i\pi\tilde{\ell}/2} j_{\tilde{\ell}}(kr) - e^{i\pi l/2} j_l(kr)\right] Y_{\mu l \mu}(\theta,\varphi). \tag{2.56}$$

where $N(\mu) = 2\sqrt{\pi}\, e^{-i\pi\mu}$.

The first term has a strong singularity for forward scattering, while the second one is less singular. Moreover, $\psi_I(\mathbf{r})$ has a closed representation in terms of the confluent hypergeometric function [131]:

$$\psi_I(\mathbf{r}) = e^{-i\pi\mu/2} \frac{\Gamma(\mu+1)}{\Gamma(2\mu+1)} e^{ikr} \left[kr(1-\cos\theta)\right]^\mu$$

$$\times {}_1F_1\left(\mu+1, 2\mu+1, -2ikr\sin^2\frac{\theta}{2}\right) e^{2i\mu\varphi}, \tag{2.57}$$

because the function (2.57) is an eigenfunction of the Hamiltonian (2.21) and can be expressed via an infinite sum, as in (2.56).

Small-angle scattering corresponds to the situation when $\theta \ll 1$, but $kr\sin^2\frac{\theta}{2} \to \infty$. Moreover, since $r \gg 1/k\theta^2$, the expression (2.57) reduces to the form

$$\psi_I(\mathbf{r})e^{-2i\mu\varphi} \simeq e^{ikr\cos\theta} - \frac{i\mu e^{-i\pi\mu}}{2kr\sin^2(\theta/2)}e^{ikr}. \tag{2.58}$$

This is recognized as a superposition of the initial plane wave propagating in the z direction and an outgoing spherical wave with an amplitude

$$|f(\theta)| = \left|\frac{\mu}{2k\sin^2(\theta/2)}\right|. \tag{2.59}$$

In the semi-classical limit where $k = Mv$, this amplitude corresponds exactly to the classical differential cross-section of the scattering at small angles (1.27):

$$\frac{d\sigma}{d\Omega} = |f(\theta)|^2 = \frac{1}{\theta^4}\left(\frac{2\mu}{Mv}\right)^2.$$

Since the incoming plane wave, the initial state of the scattering problem, is already separated from the expansion (2.53), the sum $\psi_{II}(\mathbf{r})$ consists only of corrections to the small angle scattering amplitude. Indeed, making use of the asymptotic behavior of the Bessel function (2.54), we can write

$$\psi_{II}(\mathbf{r}) \simeq -i\sqrt{\pi}\,\frac{e^{-i\pi\mu}}{k}\sum_{l=\mu}^{\infty}\sqrt{(2l+1)}\left[e^{i\pi(l-\tilde{\ell})} - 1\right]Y_{\mu l\mu}(\theta,\varphi)\,\frac{e^{ikr}}{r}.$$

Finally, collecting all the contributions to the scattering amplitude, we obtain

$$f(\theta) = \frac{e^{i\pi\mu}}{2ik}\left(\frac{\mu}{\sin^2(\theta/2)} + 2\sqrt{\pi}\sum_{l=\mu}^{\infty}\sqrt{(2l+1)}\left[e^{i\pi(l-\tilde{\ell})} - 1\right]Y_{\mu l\mu}(\theta,\varphi)\right),$$

which defines the quantum-mechanical amplitude for scattering by a monopole. Further numerical calculations and a discussion of the results can be found in the papers [462] and [131].

2.5 Charge-Monopole System: Spin in the Pauli Approximation

Very interesting features of the interaction between a monopole and a charged spin-1/2 particle (an electron) can be seen already on the level of the non-relativistic Pauli equation, which generalizes (2.7):

2.5 Charge-Monopole System: Spin in the Pauli Approximation

$$H\psi(\mathbf{r}) = -\frac{1}{2M}\left[\boldsymbol{\sigma}\cdot(\nabla - ie\mathbf{A})\right]^2 \psi(\mathbf{r}) = E\psi(\mathbf{r}). \tag{2.60}$$

This problem was considered by Malkus already in 1951 [360]. However, a more detailed analysis of the interaction of a monopole and an electron, which consistently takes into consideration all symmetry properties of the eigenfunctions of the Dirac Hamiltonian, was carried out many years later, in 1977 [306]. Here we will follow this work.

Let us first note that unlike the analogous spinless problem we discussed above, the Hamiltonian (2.60) now commutes with the operator of generalized angular momentum

$$\mathbf{J} = \mathbf{L} + \frac{1}{2}\boldsymbol{\sigma} = [\mathbf{r}\times(\mathbf{p} - e\mathbf{A})] - eg\hat{\mathbf{r}} + \frac{1}{2}\boldsymbol{\sigma}, \tag{2.61}$$

which includes the standard orbital angular momentum, $\mathbf{r}\times\boldsymbol{\pi}$, the extra orbital momentum $\mathbf{T} = -\mu\hat{\mathbf{r}}$ that appears in the charge-monopole system, and the spin operator $\mathbf{S} = \frac{1}{2}\boldsymbol{\sigma}$. Note that

$$\mathbf{J}^2 = \mathbf{L}^2 + (\boldsymbol{\sigma}\cdot\mathbf{L}) + \frac{3}{4}, \tag{2.62}$$

where $\mathbf{L}^2$ is defined by the expression (2.25).

This definition actually leads to a rather serious problem in the theory of the Abelian monopole. On a qualitative level this can be seen from the following argument. Let us consider a minimal value of the parameter $\mu = 1/2$ that is consistent with the charge quantization condition. Then the ground state is a spherically symmetric s-wave with zero angular momentum $\mathbf{J} = 0$, while the orbital angular momentum is zero and $\mathbf{S} + \mathbf{T} = 0$. The latter condition means that the spin angular momentum $\mathbf{S}$ has the same length as the extra angular momentum $\mathbf{T}$, but these vectors are antiparallel.

The subtle point here is that the direction of the vector $\mathbf{T}$ is given by the unit vector $\hat{\mathbf{r}}$ from the monopole to the charge. Therefore, if the electron somehow manages to go through the core of the monopole, this component of the angular momentum must invert its sign: $\mathbf{T} \to -\mathbf{T}$. However, the total angular momentum is conserved, which means that the spin of an electron falling down onto the center must also change its sign in order to compensate for the inversion of $\mathbf{T}$. However, if we consider a Dirac, or Pauli equation, which describes a massless particle interacting with a monopole, the helicity is a conserved quantum number labeling the states. Therefore, the Hamiltonian of the system of a massless spin-1/2 charged particle and a monopole is not self-adjoint at the origin for the s-wave states. Thus, the theory becomes pathological [351, 422].

A possible way to save the situation is to suppose that there is something unknown inside the monopole core. When an electron enters this "black box", some process of non-electrodynamical nature there would lead, for example, to the conjugation of the electron charge: $e \to -e$. Then there in no need to consider spin-flip scattering of the electron on the monopole, the amplitude of which is ill-defined in the s-wave.

This naive picture reflects in some respects the situation taking place in the scattering of a charged particle on the non-Abelian 't Hooft–Polyakov monopole. We shall discuss this problem in more detail below (see Chap. 10). Now let us continue the analysis of the properties of (2.60).

2.5.1 Dynamical Supersymmery of the Electron-Monopole System

In Sect. 1.5 we briefly discussed "hidden" dynamical $SO(2,1)$ symmetry of the non-relativistic Schrödinger equation, which describes the charge-monopole system. It is remarkable that the addition of spin-1/2 makes this system invariant with respect to the transformations of a dynamical conformal supergroup $OSp(1,2)$ [269].

Let us consider the classical counterpart of the Pauli Hamiltonian (2.60). Indeed, the spin-1/2 degrees of freedom can be represented via three-dimensional anticommuting Grassmann variables ξ_k, $\{\xi_i, \xi_j\} = \delta_{ij}$, as

$$S_i = -\frac{1}{2}\varepsilon_{ijk}\xi_j\xi_k \ .$$

This definition leads to $[S_i, S_j] = i\varepsilon_{ijk}S_k$, so the usual algebra of the spin operator is satisfied. The irreducible two-dimensional representation of the Clifford algebra of the variables ξ is given by the Pauli matrices: $\xi_i = \sigma_i/\sqrt{2}$. Thus, $S_i = \sigma_i/2$.

The anticommuting variables can be treated on the same footing as other "normal" coordinates r_i, that is $\boldsymbol{\xi}$ can be considered as a vector under the rotation group. Then we can complement the classical Hamiltonian of the charge-monopole system (1.71) by the terms that describe the dynamics of the Grassmann variables and write the extended Lagrangian of the classical non-relativistic spin-1/2 particle of mass M in the external field of a monopole as

$$L = \frac{M}{2}\dot{\mathbf{r}}^2 + \frac{i}{2}(\boldsymbol{\xi}\cdot\dot{\boldsymbol{\xi}}) + e\mathbf{A}\cdot\dot{\mathbf{r}} - \frac{\mu}{2Mr^2}\varepsilon_{ijk}\hat{r}_i\xi_j\xi_k \ . \tag{2.63}$$

Then the expression for generalized angular momentum (2.61) becomes

$$J_i = \varepsilon_{ijk}Mr_j\dot{r}_k - \mu\hat{r}_i - \frac{1}{2}\varepsilon_{ijk}\xi_j\xi_k \ .$$

Clearly, the Lagrangian (2.63) transforms as a scalar under the spatial rotations.

The Pauli Hamiltonian (2.60) can be derived from the Lagrangian of (2.63). Indeed, the canonical momenta conjugate to r_i and ξ_i are

$$P_i^{(r)} = M\dot{r}_i + eA_I \ , \qquad P_i^{(\xi)} = \frac{i}{2}\xi_i \ . \tag{2.64}$$

2.5 Charge-Monopole System: Spin in the Pauli Approximation

Then the generalized Legendre transformation

$$H = P_i^{(r)} \dot{r}_i + \dot{\xi}_i P_i^{(\xi)} - L,$$

allows us to recover the Pauli Hamiltonian (2.60). Note that the Hamiltonian equation of motion of the Grassmannian variables ξ becomes:

$$\dot{\xi}_i = \frac{i\mu}{Mr^2}\varepsilon_{ijk}r_j\xi_k,$$

which describes the classical spin precession in the monopole field.

As before, the system remains invariant with respect to the transformations of dilatation, which are generated by the charge D, and the special conformal transformations, which are generated by the charge K (see (1.72)). Together with the Hamiltonian H they form the conformal group $SO(2,1)$. However, in addition to these symmetries, the system (2.63) possesses a dynamical *supersymmetry* because the (time-dependent) supertransformations of the form

$$Q: \quad r_i \to r_i + \frac{i\epsilon}{\sqrt{m}}\xi_I, \qquad \xi_i \to \xi_i - \epsilon\sqrt{m}\dot{r}_i,$$

$$S: \quad r_i \to r_i + \frac{i\eta}{\sqrt{m}}t\xi_I, \qquad \xi_i \to \xi_i - \eta\sqrt{m}(t\dot{r}_i - r_i), \qquad (2.65)$$

where ϵ and η are Grassmannian transformation parameters, change the Lagrangian (2.63) by a total time derivative. The corresponding supercharges can easily be calculated using the Noether theorem [269]:

$$Q = \sqrt{m}\dot{r}_i\xi_i, \qquad S = -tQ + \sqrt{m}r_i\xi_i. \qquad (2.66)$$

They complement the generators of the conformal group (1.73). Note that all the charges commute with the operator of generalized angular momentum $\mathbf{J}$. Thus, the graded algebra of the complete set of generators of the dynamical group of symmetry becomes

$$[H, D] = iH, \qquad [D, K] = iK, \qquad [H, K] = 2iD,$$
$$[H, S] = -iQ, \qquad [K, Q] = iS, \qquad [K, S] = 0,$$
$$[H, Q] = 0, \qquad [D, Q] = -\frac{i}{2}Q, \qquad [D, S] = \frac{i}{2}S, \qquad (2.67)$$
$$\{Q, Q\} = 2H, \qquad \{Q, S\} = -2D, \qquad \{S, S\} = 2K.$$

This is the superalgebra $OSp(1,1)$, which extends the group of dynamical symmetry $SO(2,1)$ we discussed in Chap. 1.

The quadratic Casimir operator of the supergroup $OSp(1,1)$ is (cf. (1.74)) [269]

$$J^2 = \frac{1}{4}\left(i[Q,S] - \frac{1}{2}\right)^2 \equiv \frac{1}{4}(C)^2, \tag{2.68}$$

where the operator C in terms of the dynamical variables is

$$C = \sigma \cdot (\mathbf{J} + \mu\hat{\mathbf{r}}) - \frac{1}{2}. \tag{2.69}$$

Hence, if the eigenvalues of the operator of generalized angular momentum $\mathbf{J}^2$ are denoted by $j(j+1)$, the eigenvalues of the Casimir operator $\mathcal{J}^2$ are $\tilde{\ell}^2/4$ where

$$\tilde{\ell} = \sqrt{(j+1/2)^2 - \mu^2}. \tag{2.70}$$

Thus, because the eigenstates of the commuting operators $\mathbf{J}^2$, J_3 and C transform under some irreducible representation of the supergroup $OSp(1,1)$, we can determine the spectrum of the Pauli equation in the presence of a magnetic monopole in a very simple and elegant way [269]. However, we shall use another, more traditional formalism, and find the spectrum of the monopole-spin-1/2 particle system by applying "brute force", i.e., by solving the eigenvalue problem directly.

2.5.2 Generalized Spinor Harmonics: $j \geq \mu + 1/2$

Recall that the spectrum of eigenvalues of the operator of generalized angular momentum (1.11) of a spinless charge-monopole system starts from the minimal value $l = \mu$. Introduction of the spin angular momentum according to the standard rule of addition of angular momenta means that the total angular momentum $\mathbf{J}$ (see (2.61)), which is made up of three parts, has eigenvalues $j = l \pm 1/2$. Thus, the spectrum of eigenvalues of j can start either from the minimal value $\mu - 1/2$ or from the minimal value $\mu + 1/2$. These situations have to be discussed separately.

Let us start with the second case, that is, we set $j \geq \mu + 1/2$. Since we can write the Hamiltonian as

$$-\frac{1}{2M}[\sigma \cdot (\nabla - ie\mathbf{A})]^2 = -\frac{1}{2M}\left[(\nabla - ie\mathbf{A})^2 + e(\sigma \cdot \mathbf{B})\right]$$

$$= -\frac{1}{2M}\left[(\nabla - ie\mathbf{A})^2 + \mu\frac{(\sigma \cdot \hat{\mathbf{r}})}{r^2}\right] \tag{2.71}$$

$$= -\frac{1}{2Mr^2}\left[\frac{\partial}{\partial r}\left(r^2\frac{\partial}{\partial r}\right) + \mathbf{L}^2 - \mu^2 + \mu(\sigma \cdot \hat{\mathbf{r}})\right],$$

the angular and radial parts of the Pauli Hamiltonian can be separated again: $\psi(\mathbf{r}) = R(r)\Omega_{\mu jm}(\theta, \varphi)$. Let us first solve the angular equation, that is, define an explicit form of the spinors $\Omega_{\mu jm}(\theta, \varphi)$. Specific to this problem is that the states of the spin-1/2 particle in a monopole field, apart from the conventional quantum numbers l, m, are also labeled by eigenvalues of the operator K (cf. (2.69))

2.5 Charge-Monopole System: Spin in the Pauli Approximation

$$K = (\boldsymbol{\sigma} \cdot [\mathbf{r} \times \boldsymbol{\pi}]) = \boldsymbol{\sigma} \cdot (\mathbf{L} + \mu \hat{\mathbf{r}}),$$

which commutes with the Pauli Hamiltonian, as well as with the operators $\mathbf{J}^2$, $\mathbf{J}_3$, $\mathbf{L}^2$, and $(\boldsymbol{\sigma} \cdot \mathbf{L})$. Also, the helicity operator $\boldsymbol{\sigma} \cdot \boldsymbol{\pi} = \boldsymbol{\sigma} \cdot (-i\nabla - e\mathbf{A})$ is conserved [246].

The operator K is obviously a generalization of the parity operator of the spinors. Indeed, the angular part of the Pauli Hamiltonian can be written as

$$\frac{1}{2Mr^2}[\mathbf{L}^2 - \mu^2 + \mu(\boldsymbol{\sigma} \cdot \hat{\mathbf{r}})] = \frac{1}{2Mr^2}[(\boldsymbol{\sigma} \cdot [\mathbf{r} \times \boldsymbol{\pi}])^2 + (\boldsymbol{\sigma} \cdot [\mathbf{r} \times \boldsymbol{\pi}])]$$
$$= \frac{1}{2Mr^2}(K^2 + K). \qquad (2.72)$$

By analogy with the case of the conventional Coulomb problem, the eigenfunctions of the angular Hamiltonian can be separated into two types that correspond to the values of angular momentum $j = l \pm 1/2$. Thus, the eigenfunctions of the operators $\mathbf{J}^2$ with eigenvalues $j(j+1)$ and the operators $\mathbf{J}_3$ and $\mathbf{L}^2$ as well, are two-component spinorial angular harmonics [306]

$$\Phi^{(1)}_{\mu jm}(\theta, \varphi) = \begin{pmatrix} \sqrt{\frac{j+m}{2j}} Y_{\mu,j-1/2,m-1/2}(\theta,\varphi) \\ \sqrt{\frac{j-m}{2j}} Y_{\mu,j-1/2,m+1/2}(\theta,\varphi) \end{pmatrix},$$

$$\Phi^{(2)}_{\mu jm}(\theta, \varphi) = \begin{pmatrix} -\sqrt{\frac{j-m+1}{2j+2}} Y_{\mu,j+1/2,m-1/2}(\theta,\varphi) \\ \sqrt{\frac{j+m+1}{2j+2}} Y_{\mu,j+1/2,m+1/2}(\theta,\varphi) \end{pmatrix}, \qquad (2.73)$$

where $Y_{\mu l m}(\theta,\varphi)$ are the monopole harmonics (2.31) and the coefficients are defined from the standard rules of addition of angular momenta. These harmonics define a complete orthonormal set. The range of eigenvalues of j is $j - 1/2 = l \geq \mu$ for the states $\Phi^{(1)}_{\mu jm}(\theta,\varphi)$ and $j + 1/2 = l \geq \mu$ for the states $\Phi^{(2)}_{\mu jm}(\theta,\varphi)$, respectively.

However, the harmonics (2.73) are not eigenfunctions of the operator $K = \boldsymbol{\sigma} \cdot (\mathbf{L} + \mu \hat{\mathbf{r}})$. Although we see that

$$(\boldsymbol{\sigma} \cdot \mathbf{L}) \Phi^{(1)}_{\mu jm}(\theta,\varphi) = \left(\mathbf{J}^2 - \mathbf{L}^2 - \frac{3}{4}\right) \Phi^{(1)}_{\mu jm}(\theta,\varphi) = (j - \frac{1}{2}) \Phi^{(1)}_{\mu jm}(\theta,\varphi),$$

$$(\boldsymbol{\sigma} \cdot \mathbf{L}) \Phi^{(2)}_{\mu jm}(\theta,\varphi) = \left(\mathbf{J}^2 - \mathbf{L}^2 - \frac{3}{4}\right) \Phi^{(2)}_{\mu jm}(\theta,\varphi) = (-j - \frac{3}{2}) \Phi^{(2)}_{\mu jm}(\theta,\varphi),$$

the operator $(\boldsymbol{\sigma} \cdot \hat{\mathbf{r}})$ still mixes the spinors $\Phi^{(1)}_{\mu jm}(\theta,\varphi)$ and $\Phi^{(2)}_{\mu jm}(\theta,\varphi)$, for example

$$(\boldsymbol{\sigma} \cdot \hat{\mathbf{r}}) \Phi^{(1)}_{\mu jm} = \sqrt{\frac{4\pi}{3}} \begin{pmatrix} Y_{010} & \sqrt{2}Y_{01-1} \\ -\sqrt{2}Y_{011} & Y_{010} \end{pmatrix} \begin{pmatrix} \sqrt{\frac{j+m}{2j}} Y_{\mu,j-1/2,m-1/2} \\ \sqrt{\frac{j-m}{2j}} Y_{\mu,j-1/2,m+1/2} \end{pmatrix}.$$

Taking into account the rules of addition of generalized spherical harmonics [532], we can see that

$$(\boldsymbol{\sigma} \cdot \hat{\mathbf{r}})\Phi^{(1)}_{\mu j m} = A\Phi^{(1)}_{\mu j m} + B\Phi^{(2)}_{\mu j m}$$
$$(\boldsymbol{\sigma} \cdot \hat{\mathbf{r}})\Phi^{(2)}_{\mu j m} = B\Phi^{(1)}_{\mu j m} - A\Phi^{(2)}_{\mu j m}, \quad (2.74)$$

where the coefficients A and B are [306]

$$A = -\frac{\mu}{j+1/2}, \qquad B = -\frac{\sqrt{(j+1/2)^2 - \mu^2}}{j+1/2}. \quad (2.75)$$

Therefore, the eigenfunctions of the operator K are of two types [306]

$$\Omega^{(1)}_{\mu j m} = \frac{1}{2}\left(\sqrt{1 + \frac{\mu}{j+1/2}} + \sqrt{1 - \frac{\mu}{j+1/2}}\right)\Phi^{(1)}_{\mu j m}$$
$$- \frac{1}{2}\left(\sqrt{1 + \frac{\mu}{j+1/2}} - \sqrt{1 - \frac{\mu}{j+1/2}}\right)\Phi^{(2)}_{\mu j m},$$
$$\Omega^{(2)}_{\mu l m} = \frac{1}{2}\left(\sqrt{1 + \frac{\mu}{j+1/2}} - \sqrt{1 - \frac{\mu}{j+1/2}}\right)\Phi^{(1)}_{\mu j m}$$
$$+ \frac{1}{2}\left(\sqrt{1 + \frac{\mu}{j+1/2}} + \sqrt{1 - \frac{\mu}{j+1/2}}\right)\Phi^{(2)}_{\mu j m}, \quad (2.76)$$

which satisfy

$$(\boldsymbol{\sigma} \cdot \hat{\mathbf{r}})\Omega^{(1)}_{\mu j m} = -\Omega^{(2)}_{\mu j m}, \qquad (\boldsymbol{\sigma} \cdot \hat{\mathbf{r}})\Omega^{(2)}_{\mu j m} = -\Omega^{(1)}_{\mu j m}. \quad (2.77)$$

Making use of the relations (2.77) and (2.74), after some algebra we obtain the eigenvalues of the operator K:

$$K\Omega^{(1)}_{\mu j m} = \boldsymbol{\sigma} \cdot (\mathbf{L} + \mu\hat{\mathbf{r}})\Omega^{(1)}_{\mu j m} = (-1 + \widetilde{\ell})\Omega^{(1)}_{\mu j m},$$
$$K\Omega^{(2)}_{\mu j m} = \boldsymbol{\sigma} \cdot (\mathbf{L} + \mu\hat{\mathbf{r}})\Omega^{(2)}_{\mu j m} = (-1 - \widetilde{\ell})\Omega^{(2)}_{\mu j m}, \quad (2.78)$$

where by analogy with the spinless problem we introduce the quantum number $\widetilde{\ell} = \sqrt{(j+1/2)^2 - \mu^2}$ of (2.70). Recall that $(\widetilde{\ell}/2)^2$ is the eigenvalue of the Casimir operator (2.68).

Similar to the standard Coulomb problem of a spin-1/2 particle, we see that this quantum number corresponds to the parity. Its values label the eigenstates $\Omega^{(1)}_{\mu j m}$ and $\Omega^{(2)}_{\mu j m}$ according to the eigenvalues of the "parity" operator K.

2.5.3 Generalized Spinor Harmonics: $j = \mu - 1/2$

We have already noted that there is one more type of spinor eigenfunctions that corresponds to the values $j = \mu - 1/2$ (referred to as the "third type").

2.5 Charge-Monopole System: Spin in the Pauli Approximation

For the minimal value of the magnetic charge, the charge quantization condition yields $\mu = 1/2$. This state is a spherically symmetric state with a magnitude of the angular momentum $j = 0$, namely an s-wave. Later we will see that for the 't Hooft–Polyakov non-Abelian monopole only the *minimal* value of the magnetic charge is stable with respect to quantum corrections. This makes the states of the third type especially interesting.

Unlike the consideration above, now there is only one angular spinor $\Omega^{(3)}_{\mu,\mu-1/2,m} \equiv \Phi^{(2)}_{\mu,\mu-1/2,m}$, and

$$K\Omega^{(3)}_{\mu,\mu-1/2,m} = -\Omega^{(3)}_{\mu,\mu-1/2,m}, \quad (\boldsymbol{\sigma}\cdot\hat{\mathbf{r}})\Omega^{(3)}_{\mu,\mu-1/2,m} = \Omega^{(3)}_{\mu,\mu-1/2,m}, \quad (2.79)$$

where

$$\Omega^{(3)}_{\mu,\mu-1/2,m} = \begin{pmatrix} -\sqrt{\frac{\mu-m+1/2}{2\mu+1}} Y_{\mu\mu m-1/2}(\theta,\varphi) \\ \sqrt{\frac{\mu+m+1/2}{2\mu+1}} Y_{\mu\mu m+1/2}(\theta,\varphi) \end{pmatrix}. \quad (2.80)$$

With all this information at hand we can now turn to the solution of the radial Pauli equation.

2.5.4 Solving the Radial Pauli Equation

As in the case of the spinless Schrödinger equation, the interaction with the monopole only changes the centrifugal potential of the radial equation. However, this modification now depends on the angular spinor. Indeed, separating the variables in the Pauli equation (2.60), and using the relations (2.72) and (2.78), we obtain for states of the first type

$$-\frac{1}{2M}\left[\frac{d^2}{dr^2} + \frac{2}{r}\frac{d}{dr} - \frac{\tilde{\ell}(\tilde{\ell}-1)}{r^2}\right]R^{(1)}_{k\tilde{\ell}}(r) = ER^{(1)}_{k\tilde{\ell}}(r). \quad (2.81)$$

A regular solution of this equation (up to a normalization factor) is given by the modified Bessel function of the order $\tilde{\ell} - 1/2$ (cf. (2.38)):

$$R^{(1)}_{k\tilde{\ell}}(r) = \sqrt{\frac{k}{r}} J_{\tilde{\ell}-1/2}(kr), \quad k = \sqrt{2ME}. \quad (2.82)$$

Correspondingly, for states of the second type, the radial equation is written as

$$-\frac{1}{2M}\left[\frac{d^2}{dr^2} + \frac{2}{r}\frac{d}{dr} - \frac{\tilde{\ell}(\tilde{\ell}+1)}{r^2}\right]R^{(2)}_{k\tilde{\ell}}(r) = ER^{(2)}_{k\tilde{\ell}}(r), \quad (2.83)$$

with the solution

$$R^{(2)}_{k\tilde{\ell}}(r) = \sqrt{\frac{k}{r}} J_{\tilde{\ell}+1/2}(kr), \quad k = \sqrt{2ME}. \quad (2.84)$$

Obviously, close to the origin the wave function of the first type behaves as $R^{(1)}(r) \sim ar^{-\tilde{\ell}} + br^{\tilde{\ell}-1}$, where a and b are some coefficients. Since for both types of wave functions we have $\tilde{\ell} > 1$, these coefficients must vanish at the origin, $r = 0$, where the monopole is placed. On the classical level this corresponds to the effect of the "magnetic mirror" in the scattering of a charge by the monopole, which we discussed in Chap 1.

The situation changes drastically when we consider states of the third type. Here the centrifugal barrier vanishes and the solution of the radial equation

$$-\frac{1}{2M}\left[\frac{d^2}{dr^2} + \frac{2}{r}\frac{d}{dr}\right] R^{(3)}_{kj}(r) = E R^{(3)}_{kj}(r), \qquad (2.85)$$

are now the familiar spherical waves

$$R^{(3)}_{k,\mu-1/2}(r) = \frac{1}{\sqrt{\pi}} \frac{e^{\pm ikr}}{r}, \qquad k = \sqrt{2ME}. \qquad (2.86)$$

Independent of any kind of boundary conditions that could be imposed on the wave functions at the origin, such solutions must behave as $\sim 1/r$. Therefore, the Hamiltonian operator is not compatible with a smooth boundary condition at the origin for states of the third type. The reason is that the Pauli Hamiltonian (2.60) is not self-adjoint over the complete space of eigenfunctions, that is

$$\left(\psi^{(3)}, H\psi^{(3)}\right) - \left(H\psi^{(3)}, \psi^{(3)}\right) \neq 0, \qquad (2.87)$$

where we used the standard notation

$$(\psi_1, \psi_2) \equiv \int d^3x\, \psi_1^\dagger \psi_2. \qquad (2.88)$$

We shall discuss this condition in more detail below, when we consider the general case of the Dirac equation in the monopole background field. Note only that this situation corresponds to an electron falling down on the monopole, which we discussed in Chap. 1.

From the point of view of the dynamical supersymmetry of the system, the difference between states of the first two types and states of the third type is that the states with $j \geq \mu + 1/2$ transform under representation of the supergroup $OSp(1,1)$, as we discussed above. However, the eigenvalue of the Casimir operator (2.68) for states with $j = \mu - 1/2$ vanishes, supercharges Q and S of (2.66) are no longer self-adjoint and only the $SO(2,1)$ subgroup remains as a group of the dynamical symmetry of states of the third type [269].

One could suppose that an additional, non-electromagnetic interaction of an electron and a monopole could save the situation. Indeed, one can include into the Pauli Hamiltonian (2.60) an extra term [306]

2.5 Charge-Monopole System: Spin in the Pauli Approximation

$$H_{\text{extra}} = \kappa\mu \frac{(\boldsymbol{\sigma} \cdot \hat{\mathbf{r}})}{2Mr^2}, \tag{2.89}$$

which describes the interaction between the extra angular momentum $\mathbf{T} = -\mu\hat{\mathbf{r}}$ and an anomalous magnetic moment of magnitude κ (often taken to be infinitesimal). On the quantum mechanical level this term leads to drastic changes in the behavior of the wave functions at the origin. It was shown that the eigenfunctions of this modified Hamiltonian are regular at $r = 0$ [306].

Indeed, the radial equation for wave functions of the third type now becomes

$$-\frac{1}{2M}\left[\frac{d^2}{dr^2} + \frac{2}{r}\frac{d}{dr} + \frac{\kappa\mu}{r^2}\right] R_{k\tilde{\ell}}^{(3)}(r) = E R_{k\tilde{\ell}}^{(3)}(r). \tag{2.90}$$

Upon substituting $R_{k\tilde{\ell}}^{(3)}(r) = (1/r)U(r)$, (2.90) simplifies to

$$\left(\frac{d^2}{dr^2} + \frac{\kappa\mu}{r^2} - 2ME\right) U(r) = 0.$$

Obviously, for any $\kappa \neq 0$ this equation has a solution that is compatible with the boundary condition $U(r) = 0$ at $r = 0$.

However, the inclusion of an extra magnetic moment has another effect. For each possible value of the angular momentum j there is now a non-degenerate bound state with *zero* energy in the spectrum of the modified Hamiltonian (so-called *zero modes*). Moreover, even the limiting situation $\kappa \to 0$ still yields a third type bound state with $E = 0$. The meaning of these zero energy states is discussed in Sect. 2.6.1.

An alternative is to modify the Pauli Hamiltonian (2.60) by the introduction of a term $H_Q = -eQ/r$ describing a Coulomb interaction between the electron and a dyon charge Q at the origin [129, 130]. Thus, we consider the bound charge-dyon system again, now taking into account the electron spin. The angular dependence of the wave functions is still given by the three types of monopole harmonics described above. The radial equations are modified and (2.81) is replaced by

$$-\frac{1}{2M}\left[\frac{d^2}{dr^2} + \frac{2}{r}\frac{d}{dr} - \frac{\tilde{\ell}(\tilde{\ell}-1)}{r^2} + \frac{2MeQ}{r}\right] R_{k\tilde{\ell}}^{(1)}(r) = E R_{k\tilde{\ell}}^{(1)}(r), \tag{2.91}$$

where the parameter $\tilde{\ell}$ is defined by (2.70): $\tilde{\ell} = \sqrt{(j+1/2)^2 - \mu^2}$.

Just as in the spinless case, the regular solution of this equation can be found upon substituting

$$R_{k\tilde{\ell}}^{(1)}(r) = e^{-kr} r^{\tilde{\ell}-1} \rho^{(1)}(r), \quad \text{where } k^2 = -2ME,$$

into (2.91), which becomes

$$r\frac{d^2\rho^{(1)}(r)}{dr^2} + (2\tilde{\ell} - 2kr)\frac{d\rho^{(1)}(r)}{dr} + (2MeQ - 2k\tilde{\ell})\rho^{(1)}(r) = 0. \tag{2.92}$$

Thus, the regular solution of (2.92) is a confluent hypergeometric function, $\rho^{(1)}(r) = {}_1F_1(-N; 2\widetilde{\ell}; 2kr)$, which is now a polynomial of degree N, where $N = 0, 1, 2, \ldots$ is a radial quantum number that yields the quantization of the spectrum of bound states: $N = \frac{MeQ}{k} - \widetilde{\ell}$. Then the energy spectrum becomes

$$E^{(1)} = -\frac{M(eQ)^2}{2(N+\widetilde{\ell})^2}. \tag{2.93}$$

For completeness, let us write the corresponding normalization constant of the first type of radial wave functions:

$$N_1 = \frac{2^{\widetilde{\ell}}}{\sqrt{N!}} \frac{(MeQ)^{\widetilde{\ell}+1/2}}{(N+l)^{\widetilde{\ell}+1}} \frac{\sqrt{\Gamma(N+2\widetilde{\ell})}}{\Gamma(2\widetilde{\ell})}. \tag{2.94}$$

By complete analogy, the radial equation for the second type of wave function

$$-\frac{1}{2M}\left[\frac{d^2}{dr^2} + \frac{2}{r}\frac{d}{dr} - \frac{\widetilde{\ell}(\widetilde{\ell}+1)}{r^2} + \frac{2MeQ}{r}\right] R^{(2)}_{k\widetilde{\ell}}(r) = E R^{(2)}_{k\widetilde{\ell}}(r), \tag{2.95}$$

has a regular normalizable solution

$$R^{(2)}_{k\widetilde{\ell}}(r) = N_2 e^{-kr} r^{\widetilde{\ell}} {}_1F_1(1-N; 2\widetilde{\ell}+2; 2kr), \tag{2.96}$$

which gives the energy spectrum of the states of the second type:

$$E^{(2)} = -\frac{M(eQ)^2}{2(N+2+\widetilde{\ell})^2}. \tag{2.97}$$

Here the normalization constant is

$$N_2 = \frac{2^{\widetilde{\ell}+1}}{\sqrt{(N-1)!}} \frac{(MeQ)^{\widetilde{\ell}+3/2}}{(N+\widetilde{\ell}+1)^{\widetilde{\ell}+2}} \frac{\sqrt{\Gamma(N+2\widetilde{\ell}+2)}}{\Gamma(2\widetilde{\ell}+2)}. \tag{2.98}$$

Note that states of the first and second type with radial quantum numbers N and $N-1$, respectively, are degenerated in energy.

As for states of the third type, we have to fix $j = \mu - 1/2$, that is $\widetilde{\ell} = 0$. Therefore, the radial equation for these states has an especially simple form:

$$-\frac{1}{2M}\left[\frac{d^2}{dr^2} + \frac{2}{r}\frac{d}{dr} + \frac{2MeQ}{r}\right] R^{(3)}_{k\widetilde{\ell}}(r) = E R^{(3)}_{k\widetilde{\ell}}(r), \tag{2.99}$$

which is precisely the form of the spherically symmetric Schrödinger equation for a spinless particle in a Coulomb field. This corresponds to the mutual cancellation of the extra angular momentum and spin for the s-wave state that we discussed above. The solution of (2.99) is

$$R^{(3)}_{k\tilde{\ell}}(r) = N_3 e^{-kr} {}_1F_1(-N; 2; 2kr). \qquad (2.100)$$

and the energy spectrum is given by

$$E^{(3)} = -\frac{M(eQ)^2}{2(N+1)^2}. \qquad (2.101)$$

The normalization constant is

$$N_3 = 2\left(\frac{MeQ}{N+1}\right)^{3/2}. \qquad (2.102)$$

One can easily see that the ground state of an electron in the dyon external field is of the third type: $\psi^{(3)}(\mathbf{r}) = R^{(3)}_{k,1/2}(r)\Omega^{(3)}_{1/2,0,0}$ with quantum numbers $\mu = 1/2$, $N = 0$, $l = 0$ and the energy of the ground state $E_0 = -M(eQ)^2/2$. Note that this wave function does not vanish for $r \to 0$, while in this case we can make use of the approximation ${}_1F_1(-N, 2, 2kr) \sim (2kr)^{-1}$ and therefore

$$R^{(3)}_{k\tilde{\ell}}(r) \underset{r\to 0}{\longrightarrow} \frac{MeQ}{r} e^{-kr}.$$

2.6 Charge-Monopole System: Solving the Dirac Equation

The non-relativistic analysis of the spin-1/2 particle-monopole system can easily be generalized to the case of the Dirac equation

$$H\psi(\mathbf{r}) \equiv [-i\boldsymbol{\alpha} \cdot (\nabla - ie\mathbf{A}) + \beta M]\psi(\mathbf{r}) = E\psi(\mathbf{r}), \qquad (2.103)$$

which describes a relativistic electron in the external field of a point-like monopole. Here, α and β are Dirac matrices in the representation

$$\alpha_i = \gamma_0\gamma_i = I \otimes \sigma_i = \begin{pmatrix} 0 & \sigma_i \\ \sigma_i & 0 \end{pmatrix}, \quad \beta = \gamma_0 = \sigma_3 \otimes I = \begin{pmatrix} 1 & 0 \\ 0 & -1 \end{pmatrix}. \qquad (2.104)$$

Equation (2.103) was considered by Harish-Chandra in 1948 [254]. As in the non-relativistic case, the Dirac Hamiltonian commutes with the operator (2.61) of the generalized angular momentum $\mathbf{J} = \mathbf{L} + \mathbf{S} - \mu\hat{\mathbf{r}}$. This makes it possible to apply, without any significant change, the symmetry properties discussed above.

Let us find the eigenfunctions of the Hamiltonian operator decomposed as

$$H = \begin{pmatrix} M & -i\boldsymbol{\sigma} \cdot (\nabla - ie\mathbf{A}) \\ -i\boldsymbol{\sigma} \cdot (\nabla - ie\mathbf{A}) & -M \end{pmatrix}. \qquad (2.105)$$

Again, we must distinguish the cases $j \geq \mu + 1/2$ and $j = \mu - 1/2$.

Simple algebra of the Clebsch–Gordan coefficients supplemented by (2.77) gives the electron wave functions in terms of two-component angular harmonics of the first and second type

$$\psi^{(1)}(\mathbf{r}) = \frac{1}{r}\begin{pmatrix} F(r)\Omega^{(1)}_{\mu jm} \\ iG(r)\Omega^{(2)}_{\mu jm} \end{pmatrix}, \qquad \psi^{(2)}(\mathbf{r}) = \frac{1}{r}\begin{pmatrix} F(r)\Omega^{(2)}_{\mu jm} \\ iG(r)\Omega^{(1)}_{\mu jm} \end{pmatrix}. \qquad (2.106)$$

States of the third type are constructed from the harmonics $\Omega^{(3)}_{\mu jm}$ [306]:

$$\psi^{(3)}(\mathbf{r}) = \frac{1}{r}\begin{pmatrix} F(r)\Omega^{(3)}_{\mu jm} \\ iG(r)\Omega^{(3)}_{\mu jm} \end{pmatrix}. \qquad (2.107)$$

The helicity operator $(\boldsymbol{\sigma}\cdot\boldsymbol{\pi}) = \boldsymbol{\sigma}\cdot(\nabla - ie\mathbf{A})$ acts on these states as [306,307]:

$$(\boldsymbol{\sigma}\cdot\boldsymbol{\pi})F(r)\Omega^{(1)}_{\mu jm} = \left(-\frac{d}{dr} - \frac{1}{r} + \frac{\tilde{\ell}}{r}\right)F(r)\Omega^{(2)}_{\mu jm},$$

$$(\boldsymbol{\sigma}\cdot\boldsymbol{\pi})F(r)\Omega^{(2)}_{\mu jm} = \left(-\frac{d}{dr} - \frac{1}{r} - \frac{\tilde{\ell}}{r}\right)F(r)\Omega^{(1)}_{\mu jm},$$

$$(\boldsymbol{\sigma}\cdot\boldsymbol{\pi})F(r)\Omega^{(3)}_{\mu jm} = \left(\frac{d}{dr} + \frac{1}{r}\right)F(r)\Omega^{(3)}_{\mu jm}, \qquad (2.108)$$

where the eigenvalues $\tilde{\ell}$ of (2.70) of the Casimir operator are again used. Therefore for these states the radial equations for the function $F(r)$ and $G(r)$ are

Type 1: $\left(\dfrac{d}{dr} - \dfrac{\tilde{\ell}}{r}\right)F(r) = (M+E)\,G(r),$

$\left(\dfrac{d}{dr} + \dfrac{\tilde{\ell}}{r}\right)G(r) = (M-E)\,F(r),$

Type 2: $\left(\dfrac{d}{dr} + \dfrac{\tilde{\ell}}{r}\right)F(r) = (M+E)\,G(r),$

$\left(\dfrac{d}{dr} - \dfrac{\tilde{\ell}}{r}\right)G(r) = (M-E)\,F(r),$

Type 3: $\dfrac{dG(r)}{dr} = (E-M)\,F(r),$

$\dfrac{dF(r)}{dr} = -(E+M)\,G(r). \qquad (2.109)$

The solutions of the radial equations (2.109) for states of the first and second type (up to the normalization constant) are, respectively,

2.6 Charge-Monopole System: Solving the Dirac Equation 55

Type 1: $\quad G(r) = \sqrt{\dfrac{r}{k}} J_{\tilde{\ell}+1/2}(kr), \quad F(r) = \dfrac{\sqrt{kr}}{E-M} J_{\tilde{\ell}-1/2}$,

Type 2: $\quad F(r) = \sqrt{\dfrac{r}{k}} J_{\tilde{\ell}+1/2}(kr), \quad G(r) = \dfrac{\sqrt{kr}}{E+M} J_{\tilde{\ell}-1/2}$,

where $k = \sqrt{E^2 - M^2} > 0$.

The radial equations (2.109) for the states of the third type can easily be solved in terms of elementary functions and the conventionally normalized solutions are

Type 3: $\quad F(r) = \dfrac{1}{k}\sqrt{\dfrac{2}{\pi}} \sin(kr+\delta), \quad G(r) = -\dfrac{1}{E+M}\sqrt{\dfrac{2}{\pi}} \cos(kr+\delta)$,
(2.110)

or

Type 3: $\quad F(r) = \dfrac{1}{E-M}\sqrt{\dfrac{2}{\pi}} \cos(kr+\delta), \quad G(r) = \dfrac{1}{k}\sqrt{\dfrac{2}{\pi}} \sin(kr+\delta)$.
(2.111)

where the phase shift δ is an arbitrary parameter. Therefore the wave functions of the third type are

$$\psi_1^{(3)}(\mathbf{r}) = \dfrac{1}{kr}\sqrt{\dfrac{2}{\pi}} \begin{pmatrix} \sin(kr+\delta)\Omega_{\mu jm}^{(3)} \\ -\dfrac{ik}{E+M}\cos(kr+\delta)\Omega_{\mu jm}^{(3)} \end{pmatrix} \equiv \dfrac{1}{kr}\sqrt{\dfrac{2}{\pi}} \chi_1(r)\Omega_{\mu jm}^{(3)},$$

$$\psi_2^{(3)}(\mathbf{r}) = \dfrac{1}{kr}\sqrt{\dfrac{2}{\pi}} \begin{pmatrix} \dfrac{k}{E-M}\cos(kr+\delta)\Omega_{\mu jm}^{(3)} \\ i\sin(kr+\delta)\Omega_{\mu jm}^{(3)} \end{pmatrix} \equiv \dfrac{1}{kr}\sqrt{\dfrac{2}{\pi}} \chi_2(r)\Omega_{\mu jm}^{(3)}.$$
(2.112)

Obviously, the solutions of the first and second type satisfy the conventional boundary conditions at the origin: $F^{(1,2)}(0) = G^{(1,2)}(0) = 0$. However, as in the case of the Pauli equation, spherically symmetric states of the third type, the lowest of which is the ground state of the electron-monopole system, require special treatment. Indeed, there is also a normalizable solution of the radial equation for these wave functions, which corresponds to a complex energy.

Moreover, whatever boundary conditions we would impose, these wave functions behave at the origin as $\sim 1/r$. In fact, the boundary condition at $r = 0$ is not determined by the structure of the Hamiltonian. As mentioned previously, the problem is that the Hamiltonian operator is not self-adjoint over the space of the eigenfunctions it admits. A detailed analysis of this problem was carried out by Yamagishi in 1983 [533].

2.6.1 Zero Modes and Witten Effect

Note that the Hamiltonian operator (2.105) of the functions of the third type acting on the spinor $\chi(r)$ can be written as [246]:

2 The Electron–Monopole System: Quantum-Mechanical Interaction

$$H_0 = \begin{pmatrix} M & -i\frac{d}{dr} \\ -i\frac{d}{dr} & -M \end{pmatrix} = -i\gamma_5 \frac{d}{dr} + M\beta, \quad (2.113)$$

with

$$H_0\chi(r) \equiv H_0 \begin{pmatrix} F(r) \\ iG(r) \end{pmatrix} = E\chi(r).$$

Let us analyze whether there is a possibility to modify the problem in order to obtain a well-defined self-adjoint Hamiltonian for these states. The reduced Hamiltonian (2.113) is defined over a semi-infinite line $0 \leq r < \infty$. However, it is not Hermitian in one point on this line, $r = 0$.

A. Goldhaber pointed out [246] that, by making use of the Weyl–von Neumann theory of self-adjoint operators, it is possible to construct a self-adjoint extension of H_0. The idea is to impose some non-trivial self-consistent boundary conditions. The results turn out to be very interesting, giving rise to unexpected physical consequences [533]. In particular, the helicity operator $(\boldsymbol{\sigma} \cdot \boldsymbol{\pi})$ will not be Hermitian and it is no longer conserved.

The condition that the Hamiltonian (2.113) be self-adjoint on states of the third type can be written as the vanishing of the expression (cf. (2.87)):

$$\Delta = (\chi, H_0\chi) - (H_0\chi, \chi) = i\chi^\dagger(0)\gamma_5\chi(0) \quad (2.114)$$
$$= i\left[\chi_+^\dagger(0)\chi_+(0) - \chi_-^\dagger(0)\chi_-(0)\right] = -[F^*(0)G(0) - G^*(0)F(0)] = 0,$$

where $\chi_\pm = \frac{1}{2}(1 \pm \gamma_5)\chi$ are eigenfunctions of the operator γ_5 with positive and negative eigenvalues, respectively. Here we make use of the explicit form of the Hamiltonian H given by (2.113) and assume that the normalizable radial functions $\chi(r)$ decrease faster at spatial infinity than $1/r$.

Therefore, the boundary condition on the radial function, that we are seeking, must connect the states $\chi_-(0)$ and $\chi_+(0)$ with opposite chirality at the point $r = 0$. The most general condition of this form can be written as

$$\chi_-(0) = e^{i\theta}\chi_+(0), \quad (2.115)$$

where θ is an arbitrary angular parameter. In a different form the condition for Δ to vanish is

$$\frac{F^*(0)}{G^*(0)} = \frac{F(0)}{G(0)}. \quad (2.116)$$

In other words, the ratio $F(0)/G(0)$ is an arbitrary real number that can be parametrized by an angular parameter θ as[5] [155, 533]:

$$F(0) = G(0)\tan\left(\frac{\theta}{2} + \frac{\pi}{4}\right). \quad (2.117)$$

---

[5] Here, in contrast to (2.110), we choose the value of the phase shift δ to be $-\theta/2 - \pi/4$.

2.6 Charge-Monopole System: Solving the Dirac Equation

Therefore, at the origin the wave functions of the third type (2.112) must behave as

$$\chi(0) \sim \begin{pmatrix} \sin\left(\frac{\theta}{2} + \frac{\pi}{4}\right) \\ -\frac{ik}{E+M} \cos\left(\frac{\theta}{2} + \frac{\pi}{4}\right) \end{pmatrix}. \tag{2.118}$$

It is well-known that the appearance of the phase angle θ in the boundary conditions drastically affects the spectrum of the Hamiltonian. Obviously, there is a one-parameter family of self-adjoint Hamiltonians H_θ, which are specified by this condition.

For simplicity, let us consider a massless spin-1/2 particle in the monopole field. Imposing the boundary conditions (2.118) on the solutions $\chi(r)$ given by (2.110), one can see that the appearance of the strange angular parameter θ there can be explained as a chiral rotation of the initial wave functions (2.110) by the angle θ [533]:

$$\chi(r) \to \chi_\theta(r) \sim e^{i\theta\gamma_5/2} \begin{pmatrix} \sin\left(kr + \frac{\pi}{4}\right) \\ -i\cos\left(kr + \frac{\pi}{4}\right) \end{pmatrix} = \begin{pmatrix} \sin\left(kr + \frac{\theta}{2} + \frac{\pi}{4}\right) \\ -i\cos\left(kr + \frac{\theta}{2} + \frac{\pi}{4}\right) \end{pmatrix}. \tag{2.119}$$

The physical meaning of this modification is obvious if we decompose these wave functions into superpositions of in- and out-states, which in our case are just usual plane waves propagating in both directions, namely:

$$\chi(r) = e^{i\theta/2} \frac{1+i}{2\sqrt{2}} \begin{pmatrix} 1 \\ 1 \end{pmatrix} e^{ikr} - e^{-i\theta/2} \frac{1-i}{2\sqrt{2}} \begin{pmatrix} 1 \\ -1 \end{pmatrix} e^{-ikr}. \tag{2.120}$$

Thus, according to the discussion above, these states correspond to the changing of helicity when a particle passes through the origin and the phase shift is given by $e^{i\theta}$ (cf. (2.115)).

The result of this modification of the boundary conditions is that although the Hamiltonian (2.113) formally commutes with the operator γ_5, its eigenfunctions now depend on an arbitrary phase θ, which looks like a CP violating parameter. Indeed, for states of the third type, the CP inversion is defined as $CP: \chi(r) \to \gamma_5 \chi^*(r)$.

However, in the massless case, the model is invariant under chiral rotations $\chi \to e^{i\gamma_5 \theta'} \chi$, which shift the value of this parameter as $\theta \to \theta + \theta'$, and in particular allows us just to set it to zero. Therefore the physical observables are independent on the value of θ (in the absence of a chiral anomaly).

The situation is different in the massive case. Then the eigenfunctions of the Hamiltonian operator (2.113) that satisfy the boundary conditions (2.118) and correspond to the states of the continuum with positive and negative energy are [533]

$E = \sqrt{k^2 + M^2}$:

$$\chi_\theta(r) = \frac{k}{\sqrt{E(E - M\sin\theta)}} \left[\chi_1(r) \cos\left(\frac{\theta}{2} + \frac{\pi}{4}\right) + i\chi_2(r) \sin\left(\frac{\theta}{2} + \frac{\pi}{4}\right) \right],$$

$E = -\sqrt{k^2 + M^2}$:

$$\xi_\theta(r) = \frac{k}{\sqrt{|E|(|E|+M\sin\theta)}}\left[\chi_1(r)\cos\left(\frac{\theta}{2}+\frac{\pi}{4}\right) + i\chi_2(r)\sin\left(\frac{\theta}{2}+\frac{\pi}{4}\right)\right],$$
(2.121)

where the functions $\chi_1(r)$ and $\chi_2(r)$ are defined as in (2.112) above.

If $\cos\theta < 0$, there is a one-parameter family of bound states in the spectrum of the Hamiltonian (2.113), which corresponds to (2.118) [246, 533]:

$$\chi_\theta \sim \begin{pmatrix} \sin\left(\frac{\theta}{2}+\frac{\pi}{4}\right) \\ -i\cos\left(\frac{\theta}{2}+\frac{\pi}{4}\right) \end{pmatrix} e^{-kr}.$$
(2.122)

Its energy depends on the value of the parameter θ as $E = M\sin\theta$, $k = M|\cos\theta|$. In particular, if we set $\theta = 0$, the energy is equal to zero. This is the celebrated *zero mode* of the Dirac operator whose appearance is connected with the *index theorem* [78, 156]. Without going into detail, let us note that this mode is part of the complete set of eigenfunctions of the Hamiltonian and cannot be neglected.

Note that unlike the situation with the massless case, the eigenfunctions (2.121) obviously violate CP symmetry of the theory, since chiral rotations of the wave functions no longer leave the Hamiltonian invariant. If we still demand the theory to remain CP invariant, the value of θ must be fixed to $\theta = 0$ or $\theta = \pi$. However, the physical content of these cases is different. The point is that the existence of the fermionic zero mode on the monopole background transforms a monopole into a dyon!

To prove this, let us first note that the problem is effectively reduced to a two-dimensional system. Indeed, the states of the third type are spherically symmetric and only radial and time coordinates matter. Second, we recall that the θ-angle in the boundary conditions (2.121) arises after a chiral transformation of the wave functions

$$\chi \to \chi_\theta = e^{i\theta Q_5/2}\chi = \exp\left(\frac{i\theta}{2}\int dr\, J_5^0(r)\right)\chi,$$
(2.123)

where

$$J_5^\mu(r) = \chi^\dagger(r)\gamma_0\gamma_5\gamma^\mu\chi(r),$$

is the chiral (or axial) current, with

$$Q_5 = \int_0^\infty dr\, J_5^0(r),$$

being a generator of the chiral rotations.

One can now use the anomalous commutation relations between the operators of density of electric and chiral charges (see, e.g. [45])

2.6 Charge-Monopole System: Solving the Dirac Equation

$$[J_0(x), J_5^0(y)] = i\frac{e^2}{2\pi^2}\mathbf{B}\cdot\nabla\delta(x-y), \qquad (2.124)$$

where $J_0 = e\chi^\dagger\chi$ and $\mathbf{B}$ is the Coulomb magnetic field of the monopole. Thus, the operator of electric charge density under chiral rotations transforms as

$$e^{-i\theta Q_5/2}J_0(0)e^{i\theta Q_5/2} = J_0(0) + \frac{i\theta}{2}\int dy\,[J_0(0), J_5^0(y)]$$

$$= J_0(0) + \frac{e^2\theta}{4\pi^2}(\nabla\cdot\mathbf{B}) = J_0(0) + \frac{e^2 g\theta}{\pi}\delta(0).$$

The physical meaning of this result is rather obvious: the vacuum expectation value of the operator of electric charge on the monopole background is not equal to zero:

$$\langle Q\rangle = \frac{e\theta}{2\pi}n. \qquad (2.125)$$

Here, we use the charge quantization condition (2.6). In other words, in quantum field theory, the interaction with a fermion transforms a monopole into a dyon having an arbitrary electric charge characterized by the θ-angle. This is the so-called *Witten effect* [524]. Note that since the energy of this state is equal to zero, this effect can be treated as a Grassmannian deformation of the monopole configuration itself. This alternative description of the fermionic zero modes is especially important when we consider the non-Abelian generalization of the fermion-monopole system.

Generation of the electric charge due to the fermionic zero mode presence can Also be proved by a direct calculation of the vacuum expectation value of the electric charge operator [533]:

$$\langle\theta|Q|\theta\rangle = \frac{e}{2}\langle\theta|\,[\psi^\dagger,\psi]\,|\theta\rangle, \qquad (2.126)$$

where $|\theta\rangle$ corresponds to the vacuum state in the presence of a monopole and the wave function of the fermion is expanded as:

$$\psi = \frac{1}{\pi}\int_0^\infty dk\sum_m\left(a\chi_\theta(r)\Omega^{(3)}_{\mu lm} + b^\dagger\xi_\theta(r)\Omega^{(3)}_{\mu lm}\right)$$

$$+ \text{ contributions of states with higher }l, \qquad (2.127)$$

where a and $b^\dagger$ are annihilation and creation operators, with χ_θ and ξ_θ the spinors of (2.121). Substitution of the solutions (2.121) into this expansion yields[6] the expression for the density of the electric charge [533]

---

[6] Note that due to CP invariance, the contribution of the continuum states with quantum numbers $j > \mu + 1/2$ and positive energy is exactly compensated by the contribution of the negative continuum and only states of the third type with $j = \mu - 1/2$ give a non-trivial contribution.

$$Q(\mathbf{r}) = \langle\theta|Q|\theta\rangle = \frac{e}{2}\langle\theta|[\psi^\dagger(\mathbf{r}),\psi(\mathbf{r})] \tag{2.128}$$

$$= -\frac{e^2 gM\sin\theta}{2\pi^2 r^2}\int_M^\infty dk \frac{k}{\sqrt{k^2-M^2}(k+M\cos\theta)} e^{-2kr}.$$

One can see that for $r \to 0$, the charge density behaves as

$$Q(r) \sim \frac{eM\sin\theta}{4\pi^2 r^2}\ln(Mr),$$

while at $r \to \infty$ it vanishes exponentially.

Now, integrating over space we obtain the total charge

$$Q = \frac{e^2 gM\sin\theta}{\pi}\int_M^\infty dk \frac{k}{\sqrt{k^2-M^2}(k+M\cos\theta)} = \frac{e\theta}{2\pi}n, \tag{2.129}$$

where the integral relation

$$\int_0^\infty dx \frac{1}{\cosh^2 x - \sin^2(\theta/2)} = \frac{\theta}{\sin\theta}$$

is used. Thus, the charge Q exactly corresponds to (2.125) and we come to the conclusion that the Witten effect originates from a non-trivial contribution of the Dirac sea on the monopole background.

Finally note that we could argue by analogy with QCD that the physical value of the θ-angle must be set equal to zero for some reason of non-electrodynamical nature. In this case, we encounter the problem of the self-adjointness of the Hamiltonian. The solution was sketched above when we described the modification of the model by the introduction of an infinitesimal anomalous magnetic moment as a regulator [306]. Then the equations (2.109) for the "dangerous" radial wave function of the third type are replaced with (cf. (2.90))

$$\frac{dG(r)}{dr} = \left(M - E - \frac{\kappa\mu}{r^2}\right)F(r), \quad \frac{dF(r)}{dr} = \left(M + E - \frac{\kappa\mu}{r^2}\right)G(r), \tag{2.130}$$

which in the special case $E = 0$ can be explicitly solved:

$$F(r) = G(r) = \frac{1}{\sqrt{2}}\exp\{-\frac{\kappa\mu}{2Mr} - Mr\}.$$

This yields the regular $\theta = 0$ solution. A very detailed study of the electron-monopole spectrum was done in a series of paper in the 1980s [411–416, 532].

2.6.2 Charge Quantization Condition and the Group $SL(2,\mathbb{Z})$

Undoubtedly, the appearance of an arbitrary, non-quantizable electric charge (2.125) reflects some hidden picture of unknown processes that are taking place deep inside the monopole core. The situation becomes more transparent if instead of electrodynamics we consider an extended model of unification with a non-Abelian gauge group. We shall discuss these problems later in Chap. 10. Now let us consider the Witten formula for the electric charge of the monopole (2.125) from a somewhat different point of view.

Clear evidence of how little interest the problem of the monopole elicited for decades is the fact that the rather obvious generalization of the Dirac charge quantization condition (2.6) to the case of dyons was suggested by Schwinger [461] and Zwanziger [548] 37 years after the pioneering paper by Dirac [200]. This generalization can easily be obtained in the same way as the quantization condition for the angular momentum of the electromagnetic field of a static $e-g$ pair (1.14). For a pair of dyons having point-like electric and magnetic charges (e_1, g_1) (e_2, g_2), we obtain:

$$e_1 g_2 - e_2 g_1 = n, \quad n \in \mathbb{Z}. \qquad (2.131)$$

Note that unlike the Dirac charge quantization condition (2.6), this formula is dual invariant.

For many years following Dirac's celebrated work, the main argument in support of the monopole concept was the possibility of explaining quantization of the electric charge: since we know that there are only electrically charged particles around us, it would be enough to place a single monopole anywhere in the universe to provide this effect. Indeed, if the monopole has an arbitrary nonzero charge g, from (2.131) we see that all electric charges must be multiples of the elementary electric charge $e_0 = 1/g$:

$$e = \frac{n}{g} = n e_0. \qquad (2.132)$$

A decade later, Witten [524] noted that the generalization of the charge quantization condition (2.131) also leads to another, non-trivial consequence. Indeed, it is rather obvious that the magnetic charge is also quantizable, since for two states, one of which has a minimal electric charge $(e_0, 0)$ and the other has an arbitrary magnetic charge as well as an electric charge (e, g), the quantization condition (2.131) yields

$$e_0 g = n, \quad \text{or} \quad g = \frac{n}{e_0} = \frac{n}{n_0} g_0. \qquad (2.133)$$

Thus, there is a minimal magnetic charge

$$g_0 = \frac{n_0}{e_0},$$

where n_0 is a positive number that depends on the particular choice of the model. Clearly, the Dirac charge quantization condition (2.6) corresponds to $n_0 = 1/2$. (In some particular case, it is useful to take $n_0 = 2\pi$).

Now let us consider two dyons with the same minimal magnetic charge g_0, but different electric charges e_1 and e_2, respectively. Then the Schwinger–Zwanziger quantization condition (2.131) becomes

$$e_1 - e_2 = \frac{n}{g_0} = \frac{n}{n_0} e_0. \tag{2.134}$$

At the same time, for the state with a purely electric charge $e = e_1 - e_2$, this condition gives

$$e_1 - e_2 = m e_0, \quad \text{where } m \in \mathbb{Z}. \tag{2.135}$$

Thus, the integer n must be a multiple of n_0. This means that the possible electric charges $e_1, e_2, \ldots e_i$, of a dyon with a minimal magnetic charge g_0, must satisfy the general relation

$$e_i = e_0 \left(n_i + \frac{\theta}{2\pi} \right), \quad \text{where } n_i \in \mathbb{Z} \text{ and } \theta \text{ is an angular parameter.} \tag{2.136}$$

Note that a shift of this parameter by 2π corresponds to the change $n_i \to n_i + 1$.

Finally, let us apply the the Schwinger–Zwanziger quantization condition (2.131) to the case of two dyons with charges (e_1, mg_0) and (e_2, mg_0), respectively:

$$e_1 - e_2 = \frac{n}{mg_0} = \frac{n}{mn_0} e_0 = p e_0, \tag{2.137}$$

where p, m, n are integers. Here, the last equality is obtained by exploiting the charge quantization condition for a pure electric state $(e_1 - e_2, 0)$. Thus, the possible values of the dyon electric charges e_1, e_2 are given by the formula

$$e_i = e_0 \left(n_i + f_m \frac{\theta}{2\pi} \right), \tag{2.138}$$

where f_m is a number depending on the value of the magnetic charge of the dyon only.

One can define this number by making use of the Schwinger–Zwanziger quantization condition (2.131) for two states having charges (e_1, mg_0) and (e_2, g_0), respectively:

$$(e_1 - me_2) g_0 = n = n_0 \left(n_1 - mn_2 + \frac{\theta}{2\pi}(f_m - mf_1) \right).$$

This equation requires $f_m = mf_1 = m$, since we have seen that $f_1 = 1$ (see expression (2.136) above).

2.6 Charge-Monopole System: Solving the Dirac Equation

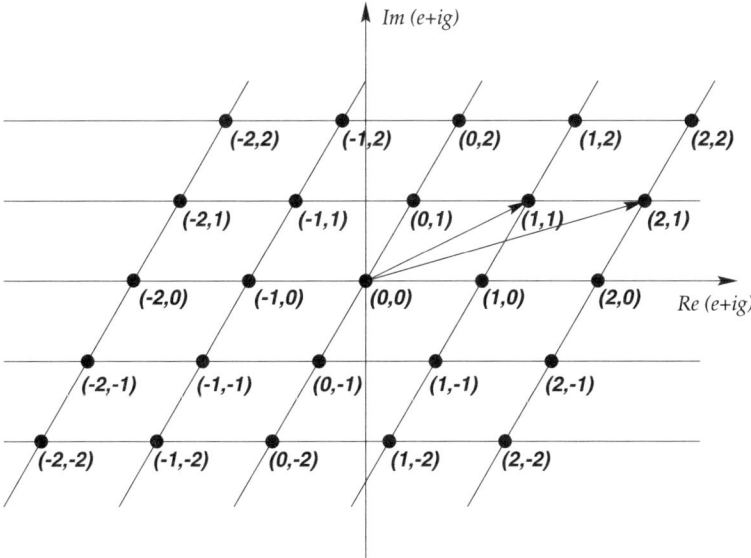

Fig. 2.1. Lattice of electric and magnetic charges in the complex plane $e + ig$

Therefore, we have come to the conclusion that the electric and magnetic charges of the dyon, represented as a complex vector $e + ig$ (cf. (1.91)), are quantized as

$$e + ig = e_0 \left(n + m\frac{\theta}{2\pi} \right) + im\frac{n_0}{e_0} = e_0(m\tau + n), \qquad (2.139)$$

where a new complex parameter is introduced:

$$\tau = \frac{\theta}{2\pi} + \frac{in_0}{e_0^2}. \qquad (2.140)$$

This formulation shows that it is not the charges e and g of a dyon that have a fundamental meaning, but the "magnetic" and "electric" quantum numbers m and n. The charge quantization condition tells us that these charges are not arbitrary, but must correspond to the sites of a discrete charge lattice with periods e_0 and $e_0\tau$ in the complex plane $e + ig$ as is shown in Fig. 2.1. Any other value of these charges would be forbidden. Therefore, the original, continuous $SO(2)$ dual symmetry is broken down to a new discrete group of transformations of the quantum numbers m and n [472]. A site on this lattice can be represented by a vector from the origin $(0,0)$ to this point. The basis vectors of this lattice are chosen to be from the origin to the points $(0,1)$ and $(1,0)$. However, this choice is not unique, since any "primitive vector" can be taken as a vector of an alternative basis[7]. Obviously, there is an infinite number of primitive vectors on the lattice. Figure 2.1 shows that, for example, all points with $m = \pm 1$ and arbitrary integer n correspond to such vectors. If $m = \pm 2$, every second lattice site gives a primitive vector, etc. More generally, any pair of non-collinear primitive vectors e'_0, $e'_0 \tau'$ form an alternative basis on the charge lattice and

$$e'_0 \tau' = ae_0 \tau + be_0, \qquad e'_0 = ce_0 \tau + de_0, \qquad (2.141)$$

where the parameters $a, b, c, d \in \mathbb{Z}$. On the other hand, the vectors e_0, $e_0 \tau$, which form the old basis for this lattice, can be written via a linear combination of e'_0, $e'_0 \tau'$, i.e., the determinant of the matrix must be equal to ± 1:

$$\det \begin{pmatrix} a & b \\ c & d \end{pmatrix} = \pm 1. \qquad (2.142)$$

We may fix the sign to be positive, then the transformations (2.141) form the group $SL(2, \mathbb{Z})$, whose quotient by its center $\mathbb{Z}_2$ is called the *modular group* or the *Möbius* group. The need for the quotient $\mathbb{Z}_2$ is caused by the invariance with respect to the simultaneous change of sign for all parameters of the transformations (2.141).

The action of the modular group on the parameter τ of (2.140) yields (cf. (2.141)):

$$\tau \longrightarrow \tau' = \frac{a\tau + b}{c\tau + d}, \qquad (2.143)$$

which forms the modular group and preserves the sign of the imaginary part of θ. This transformation relates the parameters e_0 and θ for different choices of basis. We shall discuss a nice geometrical meaning of this transformation in the next chapter.

Recall that the modular group $SL(2, \mathbb{Z})$ is generated by two elements

$$T: \quad \tau \to \tau + 1, \qquad S: \quad \tau \to -\frac{1}{\tau}. \qquad (2.144)$$

Obviously, T generates a shift of the θ-angle according to $\theta \to \theta + 2\pi$, and S, for the particular case of $\theta = 0$, corresponds to the transformation of electromagnetic duality $e \to -1/e$ discussed above[8].

Finally, we briefly consider the action of the modular group $SL(2, \mathbb{Z})$ on the quantum numbers m, n. As a result of this transformation, we obtain a new state

---

[7] A vector from the origin O to a point of the lattice site A is a primitive vector if the line OA crosses no other lattice site.
[8] For historical reasons this transformation is referred to as S-duality; although the complex compact parameter τ was used for the first time in the model on the lattice [160], in the string theory its analog is a dynamical field variable usually labeled as S [216].

2.6 Charge-Monopole System: Solving the Dirac Equation

$$\begin{pmatrix} m \\ n \end{pmatrix} \longrightarrow \begin{pmatrix} m' \\ n' \end{pmatrix} = \begin{pmatrix} a & b \\ c & d \end{pmatrix} \begin{pmatrix} m \\ n \end{pmatrix}. \qquad (2.145)$$

Thus, the transformations of the modular group not only replace one set of basis vectors for the charge lattice by another, but transforms a state with some quantum numbers m, n into another state with different values of the electric and magnetic charges. For example, transformations of the modular group connect a state with a pure electric charge (that is the state corresponding to a primitive vector $(0, 1)$) and having the quantum numbers $m = 0$, $n = 1$ and all other primitive vectors for a charge lattice:

$$\begin{pmatrix} a & b \\ c & d \end{pmatrix} \begin{pmatrix} 0 \\ 1 \end{pmatrix} = \begin{pmatrix} b \\ d \end{pmatrix}. \qquad (2.146)$$

Note that if $b = 1$, then there are dyon states with unit magnetic charge and an arbitrary non-quantized electric charge: $\begin{pmatrix} 1 \\ d \end{pmatrix}$. Later we will consider such dyons in more detail.

3 Topological Roots of the Abelian Monopole

3.1 Abelian Wu–Yang Monopole

Recall from Chap. 1 that the monopole vector potential cannot be smoothly defined everywhere in space. At the same time the electromagnetic field strength tensor is defined globally and we can expect that the singularity of the vector potential has no physical meaning[1] and that there is a mathematical description that does not involve non-physical singular expressions of any kind.

Indeed, the Dirac potential can be considered a distribution rather than a standard function of coordinates. Applying this formalism, we can prove that in this case all anomalous contributions to the physical observables vanish. One example is the calculation of the commutation relations (2.15) by making use of the point-splitting method. However such calculations are rather involved.

There is another possibility of constructing a non-singular theory of the Abelian monopole, which was discovered by T.T. Wu and C.N. Yang in 1975 [529]. The great interest of this theoretical construction is due to the fact that it touches the very foundations of field theory and provides a new insight into the connection between topology and physics. This elegant topological description is given in terms of differential geometry, which perfectly mirrors the underlying physical situation.

Before going into this formalism in more detail, let us briefly describe the original description of an Abelian monopole which is due to Wu and Yang. The basic observation is that the direction of the monopole string is defined up to a gauge transformation. Then a singularity-free description can be constructed, if we give up the traditional parametrization of the space $\mathbb{R}^3$ surrounding the monopole, by a single set of coordinates. Instead let us divide $\mathbb{R}^3/\{0\}$ into two slightly overlapping hemispheres, say the north hemisphere R^N and the south one R^S. The intersection, i.e., the "equator", is a region $R^N \cap R^S$ and the entire space surrounding the monopole now consists of two parts, each being parameterized by a separate set of coordinates (see

---

[1] However there is obviously a physical singularity at the origin $\{0\}$, where the monopole is placed. In the following, we shall suppose this point to be removed, that is, we consider the space $\mathbb{R}^3$ as removing just one point: $\mathbb{R}^3/\{0\}$.

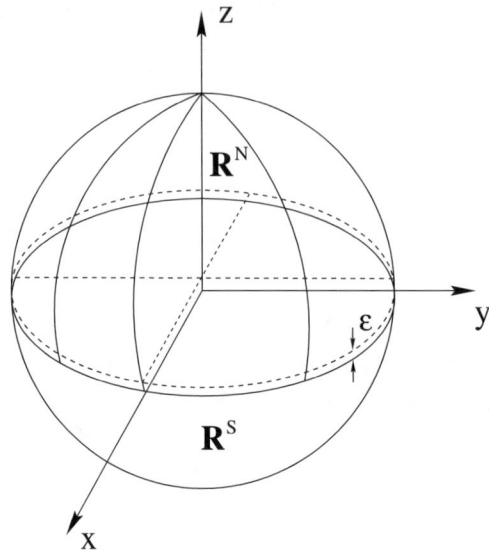

Fig. 3.1. Definition of two hemispheres R^N and R^S over the space $\mathbb{R}^3/\{0\}$

Fig. 3.1). Then, we can write two potentials $\mathbf{A}^N$ and $\mathbf{A}^S$ (cf. (1.54)), which are singularity-free everywhere in the domains of their definition:

$$\begin{cases} \mathbf{A}^N = g\dfrac{1-\cos\theta}{r\sin\theta}\hat{\mathbf{e}}_\varphi \implies 0 \leq \theta < \tfrac{\pi}{2} + \tfrac{\varepsilon}{2} : R^N \\ \mathbf{A}^S = -g\dfrac{1+\cos\theta}{r\sin\theta}\hat{\mathbf{e}}_\varphi \implies \tfrac{\pi}{2} - \tfrac{\varepsilon}{2} < \theta \leq \pi : R^S \end{cases} \quad (3.1)$$

In the intersection region $R^N \cap R^S$, both potentials are well-defined and there is a gauge transformation (1.54) connecting them:

$$\mathbf{A}^S \to \mathbf{A}^S - \frac{i}{e}e^{-2ieg\varphi}\nabla e^{2ieg\varphi} \quad (3.2)$$

$$= -\frac{g}{r}\frac{1+\cos\theta}{\sin\theta}\hat{\mathbf{e}}_\varphi + \frac{2g}{r\sin\theta}\hat{\mathbf{e}}_\varphi = \frac{g}{r}\frac{1-\cos\theta}{\sin\theta}\hat{\mathbf{e}}_\varphi \equiv \mathbf{A}^N.$$

In particular, this definition provides us with one more elegant derivation of the Dirac charge quantization condition. Let us consider a closed path l lying entirely in the overlap region. If a charged particle passes along this loop, then according to the charge-monopole interaction term in the Lagrangian (1.38), the corresponding wave function picks up a phase factor

$$e\int_0^T dt\,\dot{\mathbf{r}}\cdot\mathbf{A}(\mathbf{r}) = e\oint_l d\mathbf{r}\cdot\mathbf{A}(\mathbf{r}). \quad (3.3)$$

3.1 Abelian Wu–Yang Monopole

This is actually a piece of the action of a particle in a monopole field that describes the interaction, and under the vector **A** here we can understand either form of the potential: $\mathbf{A}^N$ or $\mathbf{A}^S$ of (3.1).

However, it appears the effect of the interaction is different for these Potentials, since within the domains of their definition they are regular, and we can apply Stokes theorem to write

$$e \oint_l d\mathbf{r} \cdot \mathbf{A}^N(\mathbf{r}) = e \int_{R^N} d\mathbf{s} \cdot [\nabla \times \mathbf{A}^N] = e \int_{R^N} d\mathbf{s} \cdot \mathbf{B},$$

$$e \oint_l d\mathbf{r} \cdot \mathbf{A}^S(\mathbf{r}) = -e \int_{R^S} d\mathbf{s} \cdot [\nabla \times \mathbf{A}^S] = -e \int_{R^S} d\mathbf{s} \cdot \mathbf{B}, \quad (3.4)$$

where the minus sign is due to the opposite orientations of the elements $d\mathbf{s}$ of the surfaces. Thus, in the overlap region the action is defined up to a term

$$\Delta S = e \int_{R^N \cup R^S} d\mathbf{s} \cdot \mathbf{B} = e \int_V d^3r \, \nabla \cdot \mathbf{B} = 4\pi e g, \quad (3.5)$$

that must not affect a physical observable of any kind. Here we applied the Gauss theorem to transform the integral over the surface **s** into a volume integral and made use of Maxwell's equation for the magnetic field generated by the monopole.

Let us recall that in quantum theory, the Lagrangian can not to be invariant under the gauge transformation in general, unlike the corresponding path integral. In other words, a physical amplitude is defined by the exponent of the action $\sim \exp\{iS\}$, which remains invariant if the change of the action is a multiple of 2π, that is, we again come to the charge quantization condition (2.6):

$$\Delta S = 4\pi e g = 2\pi n, \quad eg = \frac{n}{2}, \quad n \in \mathbb{Z}.$$

We can come to the same conclusion if we note that the wave function of a particle in a monopole field depends on the gauge. In the overlap region $R^N \cap R^S$, both forms of the potential $\mathbf{A}^N$ and $\mathbf{A}^S$ are regular and therefore the corresponding wave functions are connected there via the gauge transformation (1.54)

$$\psi^S = U\psi^N = e^{2ieg\varphi}\psi^N. \quad (3.6)$$

Here, each of the wave functions ψ^N and ψ^S must be single-valued in the hemispheres R^N and R^S, respectively. Let us again consider the closed path l in the overlap region. Then the azimuthal angle φ increases from 0 to 2π and we have[2]

$$\varphi^S(0) = \varphi^N(0), \quad \varphi^S(2\pi) = e^{4\pi ieg}\varphi^N(2\pi). \quad (3.7)$$

---
[2] We shall discuss adequate mathematical language of fiber-bundles and sections below.

Therefore, the wave functions are single-valued only if the phase factor is a multiple of 2π. Again, we arrive at the Dirac charge quantization condition (2.6).

3.2 Differential Geometry and Topology

In this section we would like to discuss briefly some of the mathematical notions used in this book. We would like to point out from the very beginning that we shall not provide mathematical rigor, preferring instead to present the ideas in practical use. We refer the reader wanting more information about the language and theorems of modern differential geometry, topology and fiber bundles to the books [7, 9, 17, 21, 22, 28], the review [207] and references therein.

3.2.1 Notions of Topology

Manifolds

We begin our discussion by going back to the fundamental concept of a *manifold*. Simply speaking, manifolds are generalizations of Euclidean spaces and the basic property of a manifold is that locally (i.e., in the vicinity of a point in the manifold) they look like a Euclidean space $\mathbb{R}^n$. In the following we shall discuss three kinds of manifold, having increasingly more sophisticated mathematical structures: topological manifolds, differentiable manifolds and complex manifolds.

The simplest is the *topological manifold*. This is a structure necessary to define the notion of *continuity*. In fairly general terms, a topological manifold is defined as a structure on a set of points X, which locally looks like a piece of $\mathbb{R}^n$. We can consider a collection U, finite or infinite, of open subsets $\{U_i\}$ of the set of points X. The collection of subsets defines a *topology* on X if the subsets $\{U_i\}$ are closed under finite intersections and arbitrary unions, and if the empty set $\{0\}$ and X itself are also included in U. Then X, or more precisely the pair (X, U) is called a *topological space*.

This definition allows us to consider a continuous map of one topological space onto another, $\phi : X \to Y$. The notion of continuity here means that the function ϕ must be continuous. This means that we can define an inverse map $\phi^{-1}(V_j)$ of an open set V_j in the space Y and such an inverse map shall be an open set in the space X. Obviously, for a standard Euclidean space $\mathbb{R}^n$ this agrees with the usual $\epsilon - \delta$ calculus.

The topological space X is a topological manifold if it can be covered with the family of open sets $\{U_i\}$, such that for each subset we can find a continuous map to Euclidean space $\phi_i : U_i \to \mathbb{R}^n$ with a continuous inverse map ϕ_i^{-1}. The pair (U_i, ϕ_i) is known as a *chart* on X, since the vector field ϕ_i provides natural local coordinates for points of U_i. Note that two different

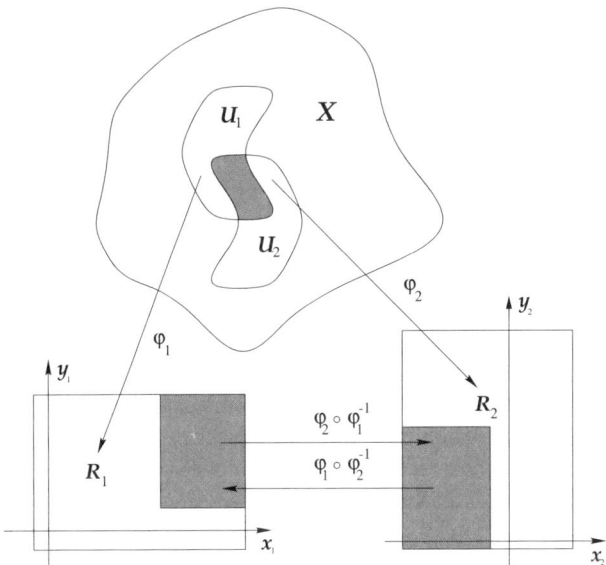

Fig. 3.2. Charts and transition functions of the topological manifold X

maps ϕ_1, ϕ_2 can overlap in some region. The two charts are compatible if the overlap maps $\phi_1 \circ \phi_2^{-1}$ and $\phi_2 \circ \phi_1^{-1}$ are continuous (see Fig. 3.2) and the set of compatible charts $\{U_i, \phi_i\}$ covering the topological space X (i.e., $X = \cup_i U_i$) is an *atlas*. The topological manifold X is *compact* if every collection of sets U_i that covers X has a finite sub-cover.

Using the definition of topological space we can define a *differentiable manifold*, which has a somewhat more refined structure. Here the notion of differentiability appears. This additional structure is required since the map $f : X \to \mathbb{R}$ can be analyzed in terms of its coordinate representation in region U_i, $f \circ \phi_i^{-1} : \phi_i(U_i) \to \mathbb{R}$. However, a function in a coordinate representation can be differentiated using standard multi-variable calculus.

Now let us consider the overlap region $U_i \cap U_j$, where we can use two different coordinates in $\mathbb{R}^n$ and we have two different representations $f_i = f \circ \phi_i^{-1}$ and $f_j = f \circ \phi_j^{-1}$. Obviously, both coordinate representations should be equivalent. More explicitly

$$f_i = f \circ \phi_i^{-1} = f \circ \phi_j^{-1} \circ (\phi_j \circ \phi_i^{-1}), \qquad (3.8)$$

where we introduce the composite mapping $(\phi_j \circ \phi_i^{-1}) : \mathbb{R}^n \to \mathbb{R}^n$, which is called the *transition function*. Thus, the additional structure that appears in comparison to the definition of a topological manifold, is the restriction on the transition functions to be infinitely differentiable in the ordinary sense (so-called $\mathbb{C}^\infty$-functions).

To complete this part of our discussion we note that we are speaking about a *complex manifold* if we replace the real space $\mathbb{R}^n$ by a complex space $\mathbb{C}^n$.

3 Topological Roots of the Abelian Monopole

Just as a differentiable manifold is connected with the notion of differentiable functions, a complex manifold is connected with the notion of *holomorphic functions* $f : X \to \mathbb{C}$. The argument is completely analogous to the reasoning above, but the condition on the transition functions $(\phi_j \circ \phi_i^{-1})$ now has to satisfy the Cauchy–Riemann equations.

In a given path on any even-dimensional manifold we can introduce local complex coordinates and the only restriction is that the transition functions from one region to another, expressed in terms of local complex coordinates, must be holomorphic maps. It should be stressed that the property of holomorphism is actually very restrictive, since the statement that a function analytically depends on a set of complex variables is much stronger than the statement that a function depends on a double set of real parameters. Therefore, many of the properties of a complex manifold are closely connected with the property of holomorphism.

Let us consider a few standard examples of manifolds and maps that ultimately match the physical situation of a magnetic monopole. The most simple example is a map of a one-sphere $X = S^1$ onto a line $Y = \mathbb{R}^1$ (see Fig. 3.3).

Let us prove that a circle, which is a real differentiable manifold, is locally Euclidean. Since S^1 is a subspace of $\mathbb{R}^2$ we can parametrize it by two coordinates (x_1, x_2): $S^1 = \{(x_1, x_2) \in \mathbb{R}^2 : x_1^2 + x_2^2 = 1\}$. Let the point $N = (0, 1)$ be the north pole and the point $S = (0, -1)$ be the south pole of S^1. Now we can introduce two open sets (coordinate patches) in $\mathbb{R}^2$ by extraction from the circle of either the north or south pole:

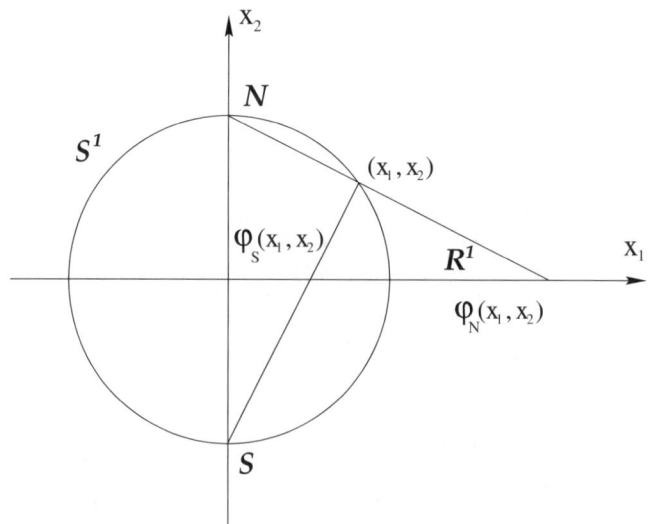

Fig. 3.3. Stereographic projection $S^1 \to \mathbb{R}^1$

3.2 Differential Geometry and Topology

$$U_N = \mathbb{R}^2/\{S\}, \qquad U_S = \mathbb{R}^2/\{N\},$$

and define two maps by

$$\phi_N(x_1, x_2) : U_N \to \mathbb{R}^1, \quad \text{where} \quad \phi_N(x_1, x_2) = \frac{x_1}{1 - x_2},$$

and

$$\phi_S(x_1, x_2) : U_S \to \mathbb{R}^1, \quad \text{where} \quad \phi_S(x_1, x_2) = \frac{x_1}{1 + x_2}.$$

Thus, we introduce a local Euclidean coordinate for a point (x_1, x_2). Actually this is the well-known *stereographic projection* from a circle. Moreover, by introducing a local coordinate y on the line $\mathbb{R}^1$, we can write the inverse continuous transformations

$$\phi_N^{-1}(y) : \mathbb{R}^1 \to U_N, \quad \text{where} \quad \phi_N^{-1}(y) = \left(\frac{2y}{1+y^2}, \frac{1-y^2}{1+y^2}\right), \qquad (3.9)$$

and

$$\phi_S^{-1}(y) : \mathbb{R}^1 \to U_S, \quad \text{where} \quad \phi_S^{-1}(y) = \left(\frac{2y}{y^2+1}, \frac{y^2-1}{y^2+1}\right). \qquad (3.10)$$

Note that now the patch intersection region is the whole circle except for two points: $U_N \cap U_S = S^1/\{N, S\}$ and there we have $\phi_N(U_N \cap U_S) = \phi_S(U_N \cap U_S) = \mathbb{R}^1/\{0\}$. The transition functions which map a local coordinate in one patch to those of another patch are therefore simple

$$\phi_S \circ \phi_N^{-1} = \phi_N \circ \phi_S^{-1} = \frac{1}{y}. \qquad (3.11)$$

Furthermore, we referred to the circle S^1 as a subspace of two-dimensional real space $\mathbb{R}^2$. The reason is that it can be promoted to the complex plane $\mathbb{C}$, if we introduce the operation of complex multiplication. Then S^1 is identified with the set of complex numbers $U = e^{i\alpha}$ of modulus one, which is closed under the complex multiplication $U(\alpha_1)U(\alpha_2) = U(\alpha_1 + \alpha_2)$ and inversion $U^{-1}(\alpha) = U(-\alpha)$. Thus, the complexification of the topological space directly leads to the Abelian group of transformations. Let us note that we can write a local complex coordinate as $z = x + iy$. Then, complex multiplication yields

$$z_1 z_2 = (x_1 x_2 - y_1 y_2, x_1 y_2 + x_2 y_1), \quad z^{-1} = \left(\frac{x}{x^2+y^2}, -\frac{y}{x^2+y^2}\right),$$

when we write a complex number as the pair $z = (x, y)$.

A straightforward generalization of the first example is a map of the two-sphere

$$S^2 = \{(x_1, x_2, x_3) \in \mathbb{R}^3, \ x_1^2 + x_2^2 + x_3^2 = 1\},$$

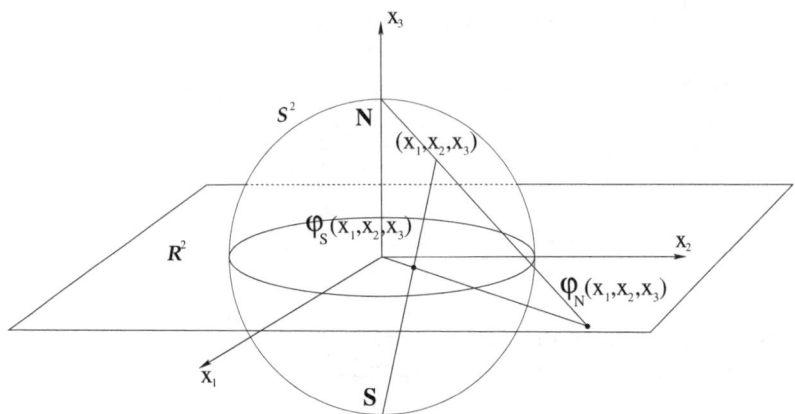

Fig. 3.4. Stereographic projection $S^2 \to \mathbb{R}^2$

onto a plane $\mathbb{R}^2$ (see Fig. 3.4). Again, it is most convenient to introduce a couple of two-dimensional patches by means of a stereographic projection from the north $N = (0,0,1)$ and south $S = (0,0,-1)$ poles, respectively. By analogy with the S^1 sphere considered above, we can write the maps

$$\phi_N(x_1, x_2, x_3) : U_N \to \mathbb{R}^2, \quad \text{where} \quad \phi_N(x_1, x_2, x_3) = \left(\frac{x_1}{1-x_3}, \frac{x_2}{1-x_3} \right),$$

and

$$\phi_S(x_1, x_2, x_3) : U_S \to \mathbb{R}^2, \quad \text{where} \quad \phi_S(x_1, x_2, x_3) = \left(\frac{x_1}{1+x_3}, \frac{x_2}{1+x_3} \right),$$

which defines the local Euclidean coordinates in the patches. Thus, geometrically, the map ϕ_N is the intersection with the horizontal x_1, x_2 plane of the straight line joining the points N and (x_1, x_2, x_3) (see Fig. 3.4), while the map ϕ_S is the intersection of the line joining the pole S and the point (x_1, x_2, x_3) with the same plane. A simple calculation yields the inverse transformation taking a point (y_1, y_2) from the plane $\mathbb{R}^2$ onto the sphere S^2

$$\begin{aligned}\phi_N^{-1}(y_1, y_2) &= \left(\frac{2y_1}{1+y_1^2+y_2^2}, \frac{2y_2}{1+y_1^2+y_2^2}, \frac{1-y_1^2-y_2^2}{1+y_1^2+y_2^2} \right) \\ &= \left(\frac{z+\bar{z}}{1+z\bar{z}}, i\frac{\bar{z}-z}{1+z\bar{z}}, \frac{1-z\bar{z}}{1+z\bar{z}} \right),\end{aligned} \quad (3.12)$$

and

$$\begin{aligned}\phi_S^{-1}(y_1, y_2) &= \left(\frac{2y_1}{1+y_1^2+y_2^2}, \frac{2y_2}{1+y_1^2+y_2^2}, \frac{y_1^2+y_2^2-1}{y_1^2+y_2^2+1} \right) \\ &= \left(\frac{z+\bar{z}}{1+z\bar{z}}, i\frac{\bar{z}-z}{1+z\bar{z}}, \frac{z\bar{z}-1}{z\bar{z}+1} \right),\end{aligned} \quad (3.13)$$

3.2 Differential Geometry and Topology

where we define a local complex coordinate $z = y_1 + iy_2$, $\bar{z} = y_1 - iy_2$ in two patches. Once again, this identifies a real two-dimensional space $\mathbb{R}^2$ with the complex plane $\mathbb{C}$. The transition functions for any point $z \neq (0,0)$ are

$$\phi_S \circ \phi_N^{-1} = \phi_N \circ \phi_S^{-1} = \left(\frac{y_1}{y_1^2 + y_2^2}, \frac{y_2}{y_1^2 + y_2^2}\right) = \frac{1}{\bar{z}},$$

and, like S^1, the sphere S^2 is equivalent to an extended complex manifold, a plane with a point at infinity $\mathbb{C}^* = \mathbb{C} \cup \{\infty\}$. The difference is that there is no natural way to introduce a group structure on the two-sphere.

Our last example is a real two-torus $T^2 = S^1 \times S^1$, which is a Cartesian product of two circles. We can introduce two charts $(U_i^{(1)}, \phi_i^{(1)})$ and $(U_i^{(2)}, \phi_i^{(2)})$ for each of these circles in exactly the same manner as before. Then, the collection of sets that cover the torus is defined by the product of the charts

$$(U_i^{(1)}, \phi_i^{(1)}) \times (U_i^{(2)}, \phi_i^{(2)}) = (U_i^{(1)} \times U_i^{(2)}, \phi_i^{(1)} \times \phi_i^{(2)}).$$

Moreover, since each of the circles S^1 is a set of complex numbers of modulus one, which is closed under complex multiplication, we have a simple parametrization of T^2

$$T^2 = \left\{ e^{i\alpha_1} e^{i\alpha_2} = e^{i(\alpha_1 + \alpha_2)} : \alpha_1, \alpha_2 \in \mathbb{R} \right\}.$$

A remarkable property of the two-torus T^2 is that it can be described in terms of an elliptic curve. Since in the following we shall use this language to describe multimonopole configurations, let us consider this notion in some detail. A complex (elliptic) curve of genus one is defined by the cubic equation

$$y^2 = x^3 + \alpha x + \beta, \tag{3.14}$$

where $\alpha, \beta \in \mathbb{C}$. The coordinates x and y can be thought as a meromorphic two-periodic function $W(z)$ of the complex argument z and its derivative dW/dZ, respectively. This allows us to set a correspondence between the algebraic equation (3.14) and the differential equation

$$\left(\frac{dW}{dz}\right)^2 = 4W^3 - \alpha W - \beta. \tag{3.15}$$

A general solution of this equation can be written in terms of the *Weierstrass function* [12], which is periodic on the torus obtained by the compactification of the complex plane:

$$W(z) = \frac{1}{z^2} + \sum_{m,n} \left\{ \frac{1}{(z - 2m\omega_1 - 2n\omega_2)^2} - \frac{1}{(2m\omega_1 + 2n\omega_2)^2} \right\}.$$

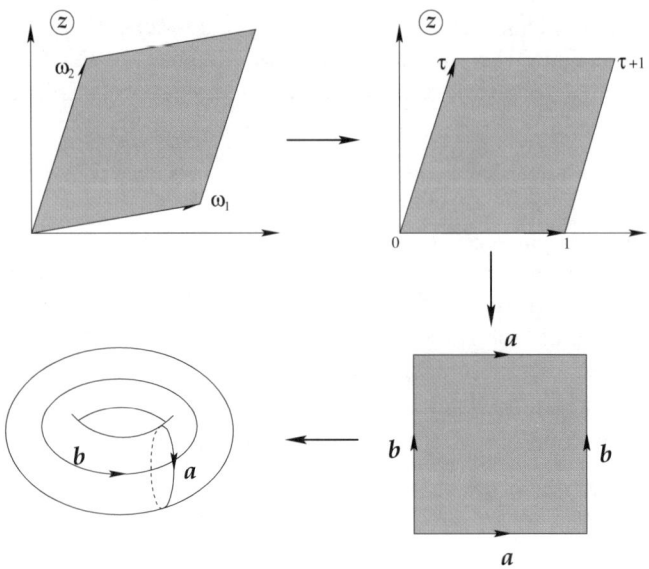

Fig. 3.5. Equivalence between the spaces $\mathbb{C}/\Omega$ and a two-torus T^2

Here $2\omega_1, 2\omega_1$ are two periods and we suppose that the parameters of the differential equation α and β are functions of the semi-periods ω_1 and ω_2. The asymptotic behavior of the Weierstrass function $W(z)$ is $\sim z^2$.

The corresponding compactification of the complex plane on a torus can be constructed by a quotient $\mathbb{C}/\Omega$ of a complex plane by a lattice $\Omega(\omega_1, \omega_2) = \{n\omega_1 + m\omega_2 | n, m \in \mathbb{Z}\}$, as illustrated in Fig. 3.5. The lattice divides the complex space $\mathbb{C}$ into parallelograms that share boundaries. The *orbit* of a point $z \in \mathbb{C}$ is a map $z \to z + n\omega_1 + m\omega_2$. Now we can identify the opposite boundaries since they belong to the same orbit that provides a two-torus T^2. Two points in the plane ω_1, ω_2 generate the entire lattice and are called a basis of it. Note that the choice of a basis is not unique. It is defined up to a transformation of the complex Möbius group $SL(2, \mathbb{C})$

$$\begin{pmatrix} \omega_1 \\ \omega_2 \end{pmatrix} \longrightarrow \begin{pmatrix} \omega_1' \\ \omega_2' \end{pmatrix} = \begin{pmatrix} d & c \\ b & a \end{pmatrix} \begin{pmatrix} \omega_1 \\ \omega_2 \end{pmatrix}. \tag{3.16}$$

Here a, b, c, d are complex parameters and we fix the determinant of the matrix of transformation $ad - bc$ equal to unity. Thus, the complex Möbius group $SL(2, \mathbb{C})$ is parametrized by six real parameters. Note that this complex finite transformation generalizes the corresponding transformations of (2.145) with three real parameters that arise in the context of the dual transformations on the charge lattice.

The transformation (3.16) relates any torus generated by a lattice $\Omega(\omega_1, \omega_2)$ to that of a torus generated by a fundamental lattice $\Omega(1, \tau)$, where τ is an element of the upper complex half-plane defined as

$$\tau = \frac{\omega_2}{\omega_1}, \qquad \text{Im } \tau > 0. \tag{3.17}$$

Then the different choices of a basis for the lattice Ω are related by fractional linear transformations and all such bases correspond to the same elliptic curve:

$$\tau \longrightarrow \frac{a\tau + b}{c\tau + d}. \tag{3.18}$$

Homeomorphism, Diffeomorphism and Biholomorphism

Let us consider two different manifolds X and Y. The practical question is whether they are identical or different from each other. That is exactly the main aim of topology: to classify spaces with certain properties. Actually, the equivalence between the manifolds X and Y can be established by comparing the structures we considered above. If X and Y are topological manifolds we can identify them if they satisfy the same notion of continuity. In a more formal way, we shall call them *homeomorphic* to each other if there is a one-to-one map $\phi : X \to Y$ having an inverse $\phi^{-1} : Y = X$ such that both ϕ and ϕ^{-1} are continuous. Such a map is also called a *bijection*. We may understand the homeomorphism as a simple distortion of one topological space into another. The important consequence of the equivalence between a map and its inverse is the property of *transitivity*. Indeed, if X is homeomorphic to Y and Y is homeomorphic to T, then the composition of these two maps makes X homeomorphic to T. Thus we can divide all topological manifolds into equivalence classes.

If we consider X and Y to be differentiable manifolds, in order to be identical they must not only be homeomorphic, but the map ϕ and its inverse ϕ^{-1} must be continuously differentiable. If such a map exists, then X and Y are called *diffeomorphic* and ϕ is a diffeomorphism. In the same way, we can establish an equivalence between two complex manifolds. If there is a holomorphic map $\phi : X \to Y$ and its inverse $\phi^{-1} : Y \to X$ is also a holomorphic map, we can say that these two complex manifolds are isomorphic. Such a map is called *biholomorphism*. Actually, the structure of the complex manifolds can be rather complicated, since there are some homeomorphic but not diffeomorphic manifolds. Moreover, there are also complex manifolds X and Y that are diffeomorphic but not biholomorphic, that is they are equivalent as differentiable manifolds but not equivalent as complex manifolds. A famous example of such manifolds are two tori T^2 generated by the lattices

$$\Omega(\omega_1, \omega_2) = ((1,0), (0,1)), \qquad \Omega'(\omega_1', \omega_2') = ((1,0), (0,2)).$$

Now we can introduce local coordinates (α_1, α_2) and (α_1', α_2') on $X = T^2(\alpha_1, \alpha_2)$ and $Y = T^2(\alpha_1', \alpha_2')$. Obviously, as differentiable manifolds these tori are equivalent, since there is a diffeomorphism

$$\phi : X \to Y, \qquad (\alpha_1', \alpha_2') = \phi(\alpha_1, \alpha_2) = (\alpha_1, 2\alpha_2).$$

However, we can parameterize these tori by local complex coordinates $z = \alpha_1 + i\alpha_2$ and $w = \alpha_1' + i\alpha_2'$, respectively. Then the map

$$w = \phi(z, \bar z) = \frac{3z}{2} - \frac{\bar z}{2}$$

is not a holomorphic function of z. Two tori would have the same complex structure if the value of the parameter $\tau = \omega_2/\omega_1$ were the same for both X and Y.

Homotopy and Homotopy Groups

The notion of homeomorphism allows us to separate the topological spaces according to equivalence classes and it is now necessary to understand how to characterize any particular equivalence class. We can thus decide whether two topological manifolds are the same or different.

Now we can consider a "one-way" map $\phi_1(x) : X \to Y$, which has no inverse. If there is another map $\phi_2(x) : X \to Y$ and the function $\phi_1(x)$ can be continuously deformed into $\phi_2(x)$, the map $\phi_1(x)$ is considered *homotopic* to $\phi_2(x)$. More precisely[3] $\phi_1(x) \sim \phi_2(x)$ if there is a continuous family of functions $f(x, t)$ parameterized over the product space $X \times [0, 1]$:

$$f : X \times [0, 1] \to Y,$$

where $f(x, 0) = \phi_1(x)$ and $f(x, 1) = \phi_2(x)$. Thus, when a parameter $t \in [0, 1]$ varies from 0 to 1, the map $\phi_1(x)$ is deformed into $\phi_2(x)$. The family of functions $f(x, t)$ is called a *homotopy*. Obviously, the homotopy is an equivalence relation and it divides the space of continuous maps from X to Y into equivalence classes.

The idea of classifying topological spaces is to compare spaces Y that are different in the sense of homotopy with the same "reference space" X. A standard choice is to set $X = S^n$, since the n-sphere has very simple topological properties. Two topological spaces X and Y are said to be homotopically equivalent, if there exist continuous maps f, g that satisfy

$$f : X \to Y; \qquad g : Y \to X,$$

and $f \circ g \sim I_Y$ (an identity on the space Y), and $g \circ f \sim I_X$ (an identity on the space X). The advantage of this scheme is that the equivalence classes reveal a group structure.

However, before we consider these structures, let us give an example of homotopic spaces. From the point of view of topology, there is no difference between the space $\mathbb{R}^n$ with the origin extracted, $X = \mathbb{R}^n/\{0\}$, and the sphere $Y = S^{n-1}$, that is $\mathbb{R}^n/\{0\} \sim S^{n-1}$ (see Fig. 3.6). Note that this homotopy is very important for our considerations since the space of a theory with an Abelian monopole is exactly $\mathbb{R}^3/\{0\} \sim S^2$. To prove this homotopy, let

---

[3] We shall use the standard symbol $\sim$ to denote homotopic maps.

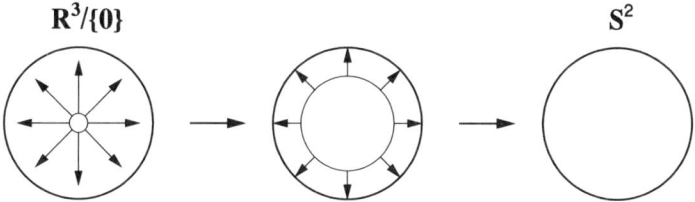

Fig. 3.6. Homotopic map $\mathbb{R}^3/\{0\} \to S^2$

us construct two maps $f : \mathbb{R}^n/\{0\} \to S^{n-1}$ via $x \to \hat{x} = x/|x|$ and $g : S^{n-1} \to \mathbb{R}^n/\{0\}$ via $x \to \{x, |x| = 1\}$. Then there is a continuous map $F(x,t) : X \times [0,1] \to X$, where $F(x,t) = (1-t)x + tx/|x|$ with $F(x,0) = x$ and $F(x,1) = g \circ f = x/|x|$. Therefore, $g \circ f \sim I_X$ and $f \circ g \sim I_Y$, which completes the proof of homotopy.

There is an important example of a homotopic map $\phi : S^1 \to S^1$ that could clarify the criteria to characterize any particular equivalence class. Let X be a unit circle in $\mathbb{R}^2$. It can be parameterized by an angle $\theta \in [0, 2\pi]$, where the points $\theta = 0$ and $\theta = 2\pi$ are identified. Let us consider the map $\phi : X \to Y$ where $Y = U(\alpha) \sim S^1$ is a set of unimodular complex numbers $U = e^{i\alpha}$, that is, the group space of $U(1)$. To classify such mappings of a circle onto a circle, let us consider the continuous function

$$\phi(\theta) = \exp\{i(n\theta + \delta)\}, \qquad \phi(\theta) : S^1 \to S^1, \tag{3.19}$$

whereby a particular value of θ is mapped into $e^{i\alpha}$ as $e^{i(n\theta+\delta)}$. This mapping generates a *homotopy class* for different δ and fixed integer n. Indeed, the map $\phi_0(\theta) = e^{i(n\theta+\delta_0)}$ is homotopic to $\phi_1(\theta) = e^{i(n\theta+\delta_1)}$, since there is a continuous family of functions

$$F(\theta, t) : X \times [0,1] \to Y,$$

where the homotopy is $F(\theta, t) = \exp\{i(n\theta + (1-t)\delta_0 + t\delta_1)\}$ with $F(\theta, 0) = \phi_0(\theta)$ and $F(\theta, 1) = \phi_1(\theta)$.

Clearly, the function $\phi(\theta)$ maps a circle in the space $\mathbb{R}^2$ onto the group space of $U(1)$. However, when θ varies from 0 to 2π, that is when we complete one turn around the circle S^1 in $\mathbb{R}^2$, the circle in the group space is covered n times. We refer to n as the *winding number*, which is a characteristic of the homotopy class.

Indeed, for a given mapping $\phi(\theta) : S^1 \to S^1$, the winding number can be expressed as

$$n = \frac{i}{2\pi} \int_0^{2\pi} d\theta \, \phi \frac{\partial}{\partial \theta} \phi^{-1}. \tag{3.20}$$

We shall see that this integer exactly corresponds to the magnetic charge of an Abelian monopole, which can be promoted to a topological charge.

The power of homotopy is that this notion reveals a group structure on the topological spaces. Indeed, the notion of continuity can be used to define classes of *simply connected* manifolds. These have to satisfy that: (i) any two points of the given manifold can be connected by a continuous curve and (ii) any closed curve (a loop) can be shrunk continuously to a point. If only the first requirement is fulfilled, then the manifold is called *linearly connected*. The circle S^1 is an example of a *linearly*, but not *simply* connected manifold. If a function f is single-valued in some region of a simply connected manifold, it can be continued to the whole space along paths connecting the points of this region with any other point. The condition for f to be single-valued requires that for any two points x_0 and x, continuations of $f(x_0)$ along any path connecting these two points must give the same result, $f(x)$. In particular, the continuation along any closed curve going through x_0 must lead to the initial value $f(x_0)$. This is automatically true if any two paths connecting x_0 and x can be deformed into each other, i.e., all paths are topologically equivalent (i.e., homotopic).

If we introduce a parameter $t \in I = [0, 1]$ along the loop in a topological manifold X, it can be formally defined as a continuous map $\gamma(t) : I \to X$, such that $\gamma(0) = \gamma(1) = x_0 \in X$. The point x_0 is called a *base point* of the loop $\gamma(t)$. Now we can call two loops $\alpha(t)$ and $\beta(s)$ based at the same point equivalent or homotopic if one loop may be continuously deformed into the other, that is if there is a continuous map $H(t, s) : [0, 1] \times [0, 1] \to X$, $s, t \in [0, 1]$ such that $H(0, s) = H(1, s) = x_0$, $H(t, 1) = \beta(t), H(t, 0) = \alpha(t)$. The map H is called a homotopy between the loops α and β and we can consider the equivalence class $[\alpha]$ of all loops homotopic to α.

It is important that we are also able to define multiplication of such mappings. Indeed, a product of two loops can be defined as

$$\alpha \circ \beta = \begin{cases} \alpha(2t) & \text{if } 0 \leq t \leq 1/2 \\ \beta(2t - 1) & \text{if } 1/2 \leq t \leq 1 \end{cases} \sim \gamma(t).$$

There is an inverse loop $\alpha^{-1}(t) = \alpha(1-t)$ (the loop in the opposite direction to α) and the unit element $I(t) = x_0$ (see Fig. 3.7).

Note that the product of a loop with its inverse is not the identity and therefore loops themselves do not admit a group structure. However, the complete set of group axioms are satisfied by the equivalence classes of loops $[\alpha]$, since the product of a loop and its inverse is a loop that can be shrunk continuously to the base point x_0, that is $\alpha \circ \alpha^{-1} \sim I$. Such a group is called the *first homotopy group* $\pi_1(X)$ (also called the *fundamental group*) of the topological manifold X.

The first homotopy group is called nontrivial if it consists of more than one element. In this case, there are contours that cannot be continuously shrunk to a point. Thus a non-simply connected space is precisely the space

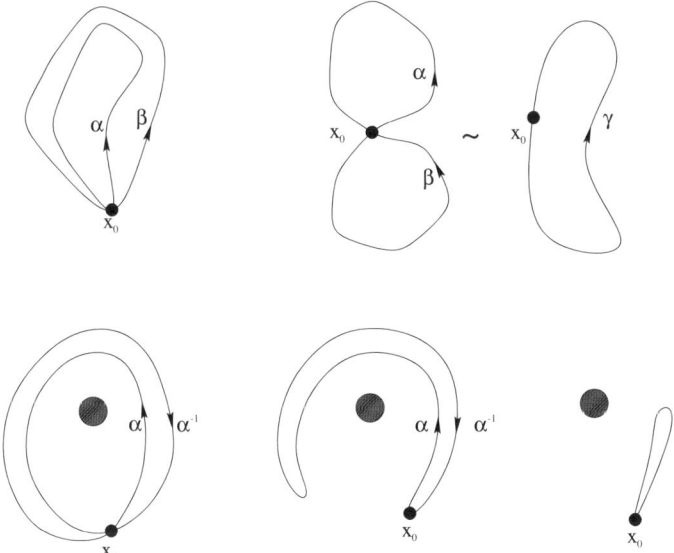

Fig. 3.7. Products of the loops $\alpha(t) \circ \beta(t)$ and $\alpha \circ \alpha^{-1} \sim I$

with the nontrivial first homotopy group. In the case of the mapping of a circle S^1 onto S^1, we have (cf. (3.20))

$$\pi_1(S^1) = \mathbb{Z}.$$

If we consider the mapping of a loop into the sphere S^2, the fundamental group $\pi_1(S^2)$ is trivial because the sphere is simply connected and any loop there can be continuously deformed to a point. This is also the case for the n-sphere, $\pi_1(S^n) = 0$ for $n \geq 2$.

A straightforward generalization of 1-loops to the n-dimensional case provides the higher rank homotopy groups. We can define an n-loop in X as a continuous map of a sphere S^n into a topological manifold X. The corresponding homotopy group is $\pi_n(X)$.

3.2.2 Notions of Differential Geometry

So far we have discussed rather general topological properties of the manifolds and not geometry. To refine our discussion and bring in geometry we have to add an additional structure on the differentiable manifold, a geometric structure. Such a manifold equipped with a geometric structure is called a *Riemannian manifold*.

Let X be an n-dimensional differentiable manifold. To introduce a geometric structure on X, we must consider a *vector field* V on it. Fairly generally, such a vector field can be introduced as a map that establishes a correspondence between each point p on X and a vector $v(p)$, namely $v : X \to V; p \mapsto v(p)$. Obviously this defines a tangent space $T_p(X)$ that yields closest linear approximation to X at the point p. More precisely, we can define $T_p(X)$ in a patch with local coordinates x^i, if we introduce a basis of linearly independent partial derivative operators of translation in the direction x_i

$$\left\{ \frac{\partial}{\partial x^1}\bigg|_p, \ldots, \frac{\partial}{\partial x^n}\bigg|_p \right\}. \tag{3.21}$$

Thus, the tangent space has the same dimension as X. Then, the "velocity" vector $v \in T_p(X)$ is $v = v^i(\partial/\partial x^i)|_p$ and the space of all velocity vectors forms the vector field $V = T_p(X)$ on X.

The next step is to define a dual vector space V^* consisting of all real linear maps on V. An element of this space ω is called a *one-form*. It is a linear map $\omega : V \to \mathbb{R}^n$, $v(p) \mapsto \omega(v)$ that sets an element of V into correspondence with a real number. For example, the differential of a function $df = (\partial/\partial x^i)f(x)dx^i$ is such an element of $V^* = T_p^*(X)$. Therefore, a convenient basis for V^* is dual to (3.21):

$$\left\{ dx^1\bigg|_p, \ldots, dx^n\bigg|_p \right\}, \tag{3.22}$$

where the space dual to $T_p(X)$ is defined as $T_p^*(X)$. The inner product defines a duality between the spaces:

$$\left(dx^i, \frac{\partial}{\partial x^j} \right) = \delta^i_j.$$

In such a basis an element of the space $T_p^*(X)$, a one-form ω, can be written as

$$\omega = \omega_i dx^i \in T_p^*(X). \tag{3.23}$$

The components ω_i are dual to the components of the vector v in $T_p(X)$ and the action of a one-form on a vector is defined via the inner product:

$$(\omega, v) = \left(\omega_i dx^i, v^j \frac{\partial}{\partial x^j} \right) = \omega_i v^i. \tag{3.24}$$

We shall see that the underlying geometry of monopoles requires us to consider a complex differentiable manifold rather than a real one. Thus we briefly describe the extension of the formalism to complex manifolds.

Obviously, a complex manifold has an underlying real analytic structure. Thus, if X has a complex dimension $d = n/2$, the r-form on X can be defined in exact analogy with the real case. The complexified tangent space $T_p(X^{\mathbb{C}})$ is defined as a space of complex vectors $v = v^i(\partial/\partial x^i)|_p \in T_p(X^{\mathbb{C}})$, where the components v^i are now complex numbers. In analogy with the expression (3.21), we can introduce a basis in the tangent space. However, it is convenient to rearrange the basis vectors on $T_p(X^{\mathbb{C}})$ to be linear combinations

3.2 Differential Geometry and Topology 83

$$\left\{ \frac{1}{2}\left[\left.\frac{\partial}{\partial x^1}\right|_p + i\frac{\partial}{\partial x^{d+1}}\right], \ldots, \frac{1}{2}\left[\left.\frac{\partial}{\partial x^d}\right|_p + i\frac{\partial}{\partial x^{2d}}\right], \right.$$
$$\left. \frac{1}{2}\left[\left.\frac{\partial}{\partial x^1}\right|_p - i\frac{\partial}{\partial x^{d+1}}\right], \ldots, \frac{1}{2}\left[\left.\frac{\partial}{\partial x^d}\right|_p - i\frac{\partial}{\partial x^{2d}}\right] \right\}, \quad (3.25)$$

or, in terms of complex coordinates $z^k = x^k + ix^{d+k}$, $\bar{z} = x^k - ix^{d+k}$, $k = 1, 2, \ldots, d$, the basis on $T_p(X^{\mathbb{C}})$ is

$$\left\{ \left.\frac{\partial}{\partial z^1}\right|_p, \ldots, \left.\frac{\partial}{\partial z^d}\right|_p, \left.\frac{\partial}{\partial \bar{z}^1}\right|_p, \ldots, \left.\frac{\partial}{\partial \bar{z}^n}\right|_p \right\}. \quad (3.26)$$

A standard convention is to label the tensor indices, which correspond to the conjugated coordinates $\bar{z}$, as "bar"-indices, e.g., $\bar{i}$. Then a complex vector, which is tangent to the complex manifold $X^{\mathbb{C}}$ at the point P can be decomposed as

$$v = v^i \left.\frac{\partial}{\partial z^i}\right|_p + v^{\bar{i}} \left.\frac{\partial}{\partial \bar{z}^i}\right|_p,$$

that is $v = (v^i, v^{\bar{i}})$.

We may also decompose the holomorphic and anti-holomorphic directions in the complex tangent space

$$T_p(X^{\mathbb{C}}) = T_p(X^{(1,0)}) + T_p(X^{(0,1)}),$$

where the holomorphic part $T_p(X^{(1,0)})$ and the anti-holomorphic part $T_p(X^{(0,1)})$ have the bases

$$\left\{ \left.\frac{\partial}{\partial z^1}\right|_p, \ldots, \left.\frac{\partial}{\partial z^d}\right|_p \right\}, \quad \text{and} \quad \left\{ \left.\frac{\partial}{\partial \bar{z}^1}\right|_p, \ldots, \left.\frac{\partial}{\partial \bar{z}^d}\right|_p \right\},$$

respectively. Both holomorphic and antiholomorphic tangent spaces have the complex dimension d.

Again, we can define the space $T_p^*(X^{\mathbb{C}})$ dual to $T_p(X^{\mathbb{C}})$ with the basis

$$\left\{ \left.dz^1\right|_p, \ldots, \left.dz^d\right|_p, \left.d\bar{z}^1\right|_p, \ldots, \left.d\bar{z}^d\right|_p \right\}, \quad (3.27)$$

where $dz^k = dx^k + idx^{d+k}$, $d\bar{z} = dx^k - idx^{d+k}$, $k = 1, 2, \ldots, d$. This space can also be decomposed into holomorphic and anti-holomorphic subspaces.

Note that the consideration of the tangent space $T_p(X)$ allows us to define a new type of manifold that is more general than the notion of the complex manifold. This is a so-called *almost complex manifold*, which is related with a complex structure not on the manifold X itself, but on the tangent space $T_p(X)$. If at each point p of a differentiable manifold, a complex structure is defined in the space $T_p(X)$, the manifold has an almost complex structure and is itself called an almost complex manifold. One may define an *almost complex structure* as an isomorphism of the tangent space $I : T_p(X) \to T_p(X)$, such that a second-rank tensor $I_{ij}(x)$ with real components satisfies the relation

$$I_{ij}I^{jk} = -\delta_i^k \, .$$

If the torsion on the space $X^{\mathbb{C}}$ is vanishing, that is I is covariantly constant, the tensor I is called a *complex structure* and $X^{\mathbb{C}}$ is reduced to be a complex manifold with holomorphic transition functions in all charts.

Furthermore, one may define an almost quaternionic structure on a manifold as a set of three linearly independent almost complex structures $I^{(n)}$, $n = 1, 2, 3$ that are covariantly constant and satisfy the quaternionic algebra

$$I^{(m)}I^{(n)} = -\delta_{mn} + \varepsilon_{mnk}I_{(k)} \, .$$

The manifold equipped with an almost quaternionic structure is called a *quaternionic manifold*. We shall see that the geometry of monopoles is closely related with properties of this space.

Differential Forms

The advantage of differential forms is that their notation is independent of the particular choice of coordinate system. This provides a very clear and simple description of the geometrical structure of the manifold.

We can generalize the notion of differential forms introduced above. A one-form is a real-valued linear map acting on $T_p(X)$. Now we can define an r-tensor, which is a real-valued multi-linear map on the space $T_p(X) \otimes \cdots \otimes T_p(X)$ (with r factors of $T_p(X)$). Furthermore, a tensor of type (r, q) can be defined as a real multi-linear map acting on the product space $T_p(X) \otimes \cdots \otimes T_p(X) \otimes T_p^*(X) \otimes \cdots \otimes T_p^*(X)$, where there are r factors of $T_p(X)$ and q factors of $T_p^*(X)$.

We saw that the dx^i form a basis for the one-form on X. Analogously, a basis for a two-form can be constructed as a bilinear map

$$dx_i \otimes dx_j : T_p(X) \otimes T_p(X) \to \mathbb{R}^2 \, ,$$

according to

$$dx_i \otimes dx_j \left(\frac{\partial}{\partial x^k}, \frac{\partial}{\partial x^l} \right) = \delta_k^i \delta_l^j \, .$$

Let us now introduce the antisymmetric *wedge* product

$$dx^i \wedge dx^j = \left(dx^i \otimes dx^j - dx^j \otimes dx^i \right) \, . \tag{3.28}$$

By definition, we have $dx^i \wedge dx^j = -dx^j \wedge dx^i$ and $dx^i \wedge dx^i = 0$. It is easy to show that the wedge products $dx^i \wedge dx^j$, $i < j$ are linearly independent and therefore form a basis. The generalization to higher wedge products being totally antisymmetric tensor products yields differential forms on an n-dimensional manifold X

$$
\begin{aligned}
0 - \text{form} \quad & \omega = \omega(x)\,, \\
1 - \text{form} \quad & \omega = \omega_i(x)dx^i\,, \\
2 - \text{form} \quad & \omega = \frac{1}{2!}\omega_{ij}(x)dx^i \wedge dx^j\,, \\
\ldots \quad & \ldots \\
r-\text{form} \quad & \omega = \frac{1}{r!}\omega_{i_1\ldots i_r}(x)dx^{i_1} \wedge dx^{i_2} \wedge \ldots dx^{i_r}\,.
\end{aligned}
\tag{3.29}
$$

The wedge product vanishes if the rank r of the totally antisymmetric tensor $\omega_{i_1\ldots i_r}(x)$ exceeds the dimension n of the manifold X.

Now let us note that the set of all r-forms $\Lambda^r(V)$ on the n-dimensional tangent space $T_p(X)$ itself forms a vector space of dimension

$$
\dim \Lambda^r(V) = \binom{n}{r} = \frac{n!}{r!(n-r)!}\,.
$$

There is a natural differential operation that transforms an r-form on X onto an $(r+1)$-form on X. This is the map

$$
d : \Lambda^r(V) \to \Lambda^{r+1}(V)\,,
$$

called *exterior differentiation*. Explicitly, in local coordinates this map is given by

$$
d : \omega \mapsto d\omega = \frac{1}{r!}\frac{\partial \omega_{i_1,\ldots i_r}}{\partial x^k} dx^k \wedge dx^{i_1} \wedge \cdots \wedge dx^{i_r}\,.
\tag{3.30}
$$

We can see that because of the antisymmetry of the wedge product, this operation is *niltopent*, that is $d^2\omega = 0$ for any ω. If there is an r-form ω such that $d\omega = 0$, then this form is called *closed*. Now two possibilities arise: either ω is *exact*, which means that it can be written as $\omega = d\alpha$ for an $(r-1)$-form α, or ω has no such representation. The *Poincaré lemma* states that any closed form can be locally expressed as an exact form. We shall see that this statement is directly related to the monopole in $\mathbb{R}^3/\{0\}$.

The vector spaces $\Lambda^r(V)$ and $\Lambda^{n-r}(V)$ obviously have the same dimension. There is an isomorphism between these spaces known as the *Hodge star operation* $*$

$$
\Lambda^r \xrightarrow{*} \Lambda^{n-r}\,.
\tag{3.31}
$$

To write an explicit form for this operation, we must further develop our discussion of the geometric structure of a differentiable manifold X. We can define an operation that is inverse to the exterior derivative of an r-form ω. It is the integration on a Riemannian manifold X. A nice property of the differential forms is that they automatically give a measure of integration on X. Let us consider an n-form

$$\Omega = dx^1 \wedge dx^2 \wedge \cdots \wedge dx^n \, .$$

Then a volume integral over a function $f(x_1, \ldots x_n)$ of local coordinates in a coordinate patch U_i becomes:

$$\int_{U_i} f \, \Omega = \int_{R_i} f(x_1, \ldots x_n) \, dx^1 dx^2, \ldots, dx^n \, , \qquad (3.32)$$

that is, the n-form Ω is a volume element of the n-dimensional space. Such an integral can be computed for any particular patch, and for an open set of patches covering X we can define a volume integral over the whole manifold.

In addition to the two-form, which is defined as an antisymmetric bilinear map, we can consider a *symmetric* positive bilinear map

$$g : T_p(X) \otimes T_p(X) \to \mathbb{R}^2 \, , \qquad (3.33)$$

which is called a *metric* on the real differentiable manifold X. In local coordinates it can be written as $g = g_{ij} dx^i \otimes dx^j$, where $g_{ij} = g_{ji}$. This allows us to use the notion of distance on X by measuring lengths of tangent vectors at the point $p(x)$ in the usual way: $ds^2 = g_{ij} dx^i dx^j$, where ds is the distance between the points x and $x+dx$. It is known that any Riemannian manifold is a metric space with the positive definite metric tensor g_{ij}. For a given metric on X we can calculate the Christoffel symbols

$$\Gamma^i_{jk} = \frac{g_{il}}{2} \left(\frac{\partial g_{lk}}{\partial x^j} + \frac{\partial g_{lj}}{\partial x^k} - \frac{\partial g_{jk}}{\partial x^l} \right) \, ,$$

which defines a unique connection one-form $\Gamma^i_k = \Gamma^i_{jk} dx^j$.

The notion of the metric on the manifold X allows us to define so-called *isometries*. Let $T_p(X)$ and $T'_p(X)$ be two tangent spaces on X with metrics g and g', respectively. The isometry I is a linear mapping $I : T_p(X) \to T'_p(X)$; $p \to I(p)$ that does not change the metric. For example, let us consider two flat Euclidean vector spaces $\mathbb{R}^4$. The corresponding isometries are just the transformations of the rotation group $O(4)$. Isometries of the space with a Minkovski metric are obviously transformations of the Lorentz group $O(3,1)$.

Furthermore, let us consider the parallel transport of a vector around a contractible closed loop on an n-dimensional Riemannian manifold using the Christoffel symbols. The transported vector is related to the original vector by some $SO(n)$ rotations. The matrices of this rotation form a group that is referred to as the *local holonomy group*.

Now we can define the Hodge star operation $*$ of (3.31), which maps an r-form ω on an n-dimensional differentiable manifold with metric g onto an $(n-r)$-form $*\omega$:

$$\omega \mapsto *\omega = \frac{1}{r!(n-r)!} \varepsilon_{i_1 \ldots i_n} \sqrt{|\det g|} g^{i_1 j_1} \ldots g^{i_r j_r} \omega_{j_1 \ldots j_r} dx^{i_{r+1}} \wedge \cdots \wedge dx^{i_n} \, ,$$

$$(3.34)$$

where $\varepsilon_{i_1...i_n}$ is antisymmetric in all indices, with $\varepsilon_{0123...n} = +1$. This operation formally defines the symmetric inner product of two r-forms $\alpha, \beta \in \Lambda^r(V)$ by

$$(\alpha, \beta) = \int_X \alpha \wedge *\beta \in \mathbb{R}.$$

If the manifold X is Riemannian, then the operation d has an adjoint $d^\dagger$. This is the map

$$d^\dagger : \Lambda^r(V) \to \Lambda^{r-1}(V), \tag{3.35}$$

which is defined by $d^\dagger : \omega \mapsto d^\dagger \omega = (-1)^{nr+n+1+s} * d * \omega$, where s is the signature of the metric. Then we can introduce the Laplacian

$$\Delta = dd^\dagger + d^\dagger d = (d + d^\dagger)^2. \tag{3.36}$$

If this operator annihilates a form ω, $\Delta \omega = 0$, such a form is called *harmonic*.

In the following we shall use the *Stokes theorem*, which we give here without proof. Let X be an n-dimensional compact manifold with a nonempty boundary ∂X. Then for an $(r-1)$-form ω, we have the identity

$$\int_X d\omega = \int_{\partial X} \omega. \tag{3.37}$$

Finally, we can simply consider a map between two differentiable manifolds

$$\phi : X \to Y, \qquad x \mapsto \phi(x).$$

The function ϕ here also induces a map of the corresponding tangent spaces, $\phi_* : T_p(X) \to T_{\phi p}(Y)$. More precisely, if $f(x)$ is a function on X with the corresponding function $f(\phi(x))$ on Y, then ϕ_* acts on the vector $v \in T_p(X)$ as

$$(\phi_* v) f = v f(\phi(x)).$$

This operation ϕ_* has a counterpart ϕ^* that maps forms from the dual spaces $T_p^*(X)$ to $T_{\phi p}^*(Y)$ when ϕ acts on the differentiable manifolds themselves. However, the forms are dual to vectors and therefore while ϕ_* transfers a vector from X to Y, the map ϕ^* transforms a form from Y to X. More precisely, if ω is a 1-form on $T_{\phi p}^*(Y)$, then $\phi^* \omega$ is a one-form on $T_p^*(X)$, which can be defined from the equation

$$(\phi^* \omega, v) = (\omega, \phi_* v), \qquad v \in T_p(X),$$

where the inner product is defined according to (3.24). The map $\phi^* \omega$ is called the *pullback*.

Complex Differential Geometry

A further generalization is to define differential forms on a complex manifold $X^{\mathbb{C}}$ of complex dimension $n/2$. Since we have already introduced a basis (3.27) of the dual complex vector space $T_p^*(X^{\mathbb{C}})$, a complex r-form ω can be written as a linear combination of

$$dz^{i_1} \wedge \cdots \wedge dz^{i_p} \wedge d\bar{z}^{\bar{j}_1} \wedge \cdots \wedge d\bar{z}^{\bar{j}_q},$$

where $r = p + q$. Thus, ω can be written as

$$\omega = \omega_{r,\bar{0}} + \omega_{r-1,\bar{1}} + \cdots + \omega_{0,\bar{r}},$$

where each summand, $\omega_{p,\bar{q}}$, is labeled by the number p of holomorphic differentials and the number q of anti-holomorphic differentials it contains:

$$\omega_{p,\bar{q}} = \frac{1}{p!}\frac{1}{q!} \sum_{i_1,\bar{j}_1}^{i_p,\bar{j}_q} \omega_{i_1\ldots i_p \bar{j}_1\ldots \bar{j}_q} dz^{i_1} \wedge \cdots \wedge dz^{i_p} \wedge d\bar{z}^{\bar{j}_1} \wedge \cdots \wedge d\bar{z}^{\bar{j}_q}.$$

This defines the so-called (p,q)-form on $T_p^*(X^{\mathbb{C}})$ at the point p on $X^{\mathbb{C}}$.

Then the differential operator d naturally decomposes into a sum of two operators of exterior differentiation in the holomorphic and antiholomorphic directions: $d = \partial + \bar{\partial}$, where

$$\partial : \omega_{p,\bar{q}} \to \omega_{p+1,\bar{q}}, \qquad \bar{\partial} : \omega_{p,\bar{q}} \to \omega_{p,\bar{q}+1},$$

and $\partial\bar{\partial} = -\bar{\partial}\partial$, $\partial^2 = \bar{\partial}^2 = 0$.

Finally, we can define a metric on $X^{\mathbb{C}}$ as a map

$$g : T_p(X^{\mathbb{C}}) \otimes T_p(X^{\mathbb{C}}) \to \mathbb{C}. \tag{3.38}$$

This metric includes holomorphic, anti-holomorphic and mixed components

$$g_{ij} = g\left(\frac{\partial}{\partial z^i}, \frac{\partial}{\partial z^j}\right), \quad g_{i\bar{j}} = g\left(\frac{\partial}{\partial z^i}, \frac{\partial}{\partial \bar{z}^{\bar{j}}}\right), \quad g_{\bar{i}\bar{j}} = g\left(\frac{\partial}{\partial \bar{z}^{\bar{i}}}, \frac{\partial}{\partial \bar{z}^{\bar{j}}}\right),$$

such that $g_{ij} = g_{ji}$, $g_{i\bar{j}} = g_{\bar{j}i}$, $g_{\bar{i}\bar{j}} = \bar{g}_{ij}$, $\bar{g}_{i\bar{j}} = g_{\bar{i}j}$

There is a particular case of the metric on a complex manifold. If the functions $g_{ij} = g_{\bar{i}\bar{j}} = 0$, the metric is called *Hermitian*:

$$g = g_{i\bar{j}} dz^i \otimes d\bar{z}^{\bar{j}} + g_{\bar{i}j} d\bar{z}^{\bar{i}} \otimes dz^j.$$

For such a metric we can build a complex differential $(1,1)$-form on $T_p^*(X^{\mathbb{C}})$:

$$K = ig_{i\bar{j}} \, dz^i \wedge d\bar{z}^{\bar{j}},$$

and if this form is closed, that is

$$dK = i(\partial + \bar{\partial})g_{i\bar{j}}dz^i \wedge d\bar{z}^{\bar{j}} = 0 \, ,$$

then the form K is called a *Kähler form* and the complex manifold $X^{\mathbb{C}}$ itself is called a *Kähler manifold*. The geometrical properties of such a manifold turn out to be extremely simple, since the fact that $dK = 0$ implies that locally we can write the Hermitian metric as a second derivative of some scalar function $\mathcal{F}$, the *Kähler potential*:

$$g_{i\bar{j}} = \frac{\partial^2 \mathcal{F}}{\partial z^i \partial \bar{z}^{\bar{j}}} \, . \tag{3.39}$$

Indeed, the form of the metric on $X^{\mathbb{C}}$ allows us to calculate the standard Levi-Civita connection Γ^i_{jk} in complex coordinates. Now, the condition that the metric be of Kähler form leaves as the only non-vanishing components of Γ^i_{jk} only those indices that are all holomorphic or anti-holomorphic:

$$\Gamma^i_{km} = g^{i\bar{j}}\frac{\partial g_{m\bar{j}}}{\partial z^k} \, , \qquad \Gamma^{\bar{i}}_{\bar{k}\bar{m}} = g^{\bar{i}j}\frac{\partial g_{\bar{m}j}}{\partial \bar{z}^{\bar{j}}} \, . \tag{3.40}$$

However, the connection is responsible for parallel transport along a contour on the complex manifold $X^{\mathbb{C}}$. The form of the Christoffel symbols (3.40) implies that the decomposition of any tangent vector v onto the holomorphic and antiholomorphic components remains the same after a parallel transport, that is, the group of holonomy of a Kähler manifold with real dimension $2n$ is reduced to the unitary group $U(n)$.

Furthermore, we can consider the hyper-Kähler manifold equipped with an almost quaternionic structure. This manifold possesses a Riemannian metric that is Kählerian with respect to three almost complex structures on the tangent space. It turns out that such a quaternionic manifold is directly related to the geometry of monopoles. We shall discuss these topics in Chap. 6. For a hyper-Kähler manifold with real dimension $4n$, the holonomy group of the metric is reduced from $SO(4n)$ to $SP(2n)$ and any hyper-Kähler manifold is Ricci-flat. Actually, a four-dimensional Euclidean hyper-Kähler manifold is always characterized by the self-dual Riemann curvature. In turn, it is equivalent to the conditions of the Ricci-flatness and the Kähler form of the metric together. In other words, in Kähler geometry the Ricci-flatness means that the metric determinant is a constant.

3.2.3 Maxwell Electrodynamics and Differential Forms

Recall that a familiar formulation of Maxwell electrodynamics in vacuum is given in the component notation of the electromagnetic field strength tensor and the 4-current $j_\mu = (\rho, \mathbf{j})$:

$$\partial_\mu F^{\mu\nu} = j_\nu \, , \qquad \partial_\mu \widetilde{F}^{\mu\nu} = 0 \, , \tag{3.41}$$

where

$$F_{\mu\nu} = \begin{pmatrix} 0 & E_1 & E_2 & E_3 \\ -E_1 & 0 & B_3 & -B_2 \\ -E_2 & -B_3 & 0 & B_1 \\ -E_3 & B_2 & -B_1 & 0 \end{pmatrix}, \quad \tilde{F}_{\mu\nu} = \begin{pmatrix} 0 & B_1 & B_2 & B_3 \\ -B_1 & 0 & -E_3 & E_2 \\ -B_2 & E_3 & 0 & -E_1 \\ -B_3 & -E_2 & E_1 & 0 \end{pmatrix}, \quad (3.42)$$

and $\tilde{F}_{\mu\nu} = \frac{1}{2}\varepsilon_{\mu\nu\rho\sigma}F^{\rho\sigma}$, with the convention $\varepsilon_{0123} = +1$. Obviously, the temporal and spatial components of these equations give precisely the familiar Maxwell equations for the electric and magnetic fields **E** and **B**, respectively.

This beautiful formulation obviously reflects the relativistic covariance of the theory and the underlying symmetries. However, from a modern point of view, the most concise and adequate formulation of the gauge field theory, a simplest example of which is Abelian electrodynamics, needs extensive use of the language of differential geometry and topology. This is especially the case of the magnetic monopole, which naturally arises in this picture.

Let us apply this rather abstract mathematical formalism to the Maxwell electrodynamics. First, the theory is defined on the differentiable four-dimensional Minkowski space M^4, the space $\mathbb{R}^4$ with the pseudo-Riemannian metric $g_{\mu\nu} = \text{diag}\,(-1,1,1,1)$, $\mu,\nu = 0,1,2,3$.

We can introduce local coordinates $\{x_0 = ct, x_1, x_2, x_3\}$ in a local patch on M^4. Then, a convenient basis of the dual tangent space $T_p^*(M^4)$ is given by one-forms of coordinates $\{dx^0, dx^i\}$ and we can define a one-form $A = A_\mu(x)dx^\mu = A_0(x)dx^0 + A_i(x)dx^i$. This is nothing but the potential of the electromagnetic field. Indeed, let us consider a two-form F generated from A by the action of the differential operator d as

$$F = dA = \frac{1}{2}F_{\mu\nu}dx^\mu \wedge dx^\nu. \quad (3.43)$$

In components we find then the standard expression for the electromagnetic field strength tensor via the vector-potential $F_{\mu\nu} = \partial_\mu A_\nu - \partial_\nu A_\mu$. Now, note that the form F is closed, $dF = d^2 A = 0$. This is precisely the homogeneous Maxwell equation in terms of differential forms. Indeed, the basis for a two-form F is given by $\{dx^\mu \wedge dx^\nu\} \equiv \{dx^0 \wedge dx^i, dx^i \wedge dx^j\}$ and we can decompose it as[4]

$$F = E \wedge dx^0 + B = E_i dx^i \wedge dx^0 + \frac{1}{2}B_{ij}dx^i \wedge dx^j, \quad (3.44)$$

where we separated the one-form E and the two-form B. This relation can be rewritten in a more symmetrical form as we note that the Hodge $*$-operation provides the relations for a dual basis of a three-form $*A = A_\mu * dx^\mu = \tilde{A}_{\mu\nu\rho}dx^\mu dx^\nu dx^\rho$ dual to A:

$$*dx^0 = -dx^1 \wedge dx^2 \wedge dx^3, \quad *dx^1 = -dx^0 \wedge dx^2 \wedge dx^3,$$

---

[4] In this subsection, Greek indices label components of Minkowski space, whereas Latin indices label three-space.

$$*dx^2 = dx^0 \wedge dx^1 \wedge dx^3, \quad *dx^3 = -dx^0 \wedge dx^1 \wedge dx^2, \quad (3.45)$$

and for the two-form $*F = \frac{1}{2}F_{\mu\nu} * dx^\mu \wedge dx^\nu = \frac{1}{2}\widetilde{F}_{\mu\nu}dx^\mu \wedge dx^\nu$:

$$*dx^0 \wedge dx^i = -\frac{1}{2}\varepsilon_{ijk}dx^j \wedge dx^k, \quad *dx^i \wedge dx^j = \varepsilon_{ijk}dx^0 \wedge dx^k, \quad (3.46)$$

dual to F. Then we can write

$$F = E_i dx^i \wedge dx^0 + \frac{1}{2}\varepsilon_{ijk}B^{jk} * dx^i \wedge dx^0 = E_i dx^i \wedge dx^0 + B_i * dx^i \wedge dx^0, \quad (3.47)$$

where we define the vector $B_i = \frac{1}{2}\varepsilon_{ijk}B^{jk}$. Clearly, this decomposition corresponds to the electric and magnetic components of the field strength tensor (3.43).

Using the relations (3.46), we see that the Hodge $*$-operation acts on the form (3.47) as[5]

$$*F = \frac{1}{2}F_{\mu\nu} * dx^\mu \wedge dx^\nu = E_i * dx^i \wedge dx^0 - B_i dx^i \wedge dx^0$$
$$= B_i dx^0 \wedge dx^i + \frac{1}{2}E^i \varepsilon_{ijk} dx^j \wedge dx^k. \quad (3.48)$$

This corresponds to the transformation of duality of (1.81):

$$E_i \to \widetilde{E}_i = -B_i, \quad B_i \to \widetilde{B}_i = E_i, \quad (3.49)$$

and clarifies the geometrical meaning of electromagnetic duality as the Hodge $*$-operation: $F \to *F$.

Now, let us note that the exterior derivative acts on the components of the two-form F defined by (3.44) as

$$dB = \frac{1}{2}\frac{\partial B_{jk}}{\partial x_i} dx^i \wedge dx^j \wedge dx^k + \frac{1}{2}\frac{\partial B_{jk}}{\partial x_0} dx^0 \wedge dx^j \wedge dx^k$$
$$= \frac{1}{2}\varepsilon^{ijk}\frac{\partial B_{jk}}{\partial x_i} dx^1 \wedge dx^2 \wedge dx^3 + \frac{1}{2}\frac{\partial B_{jk}}{\partial x_0} dx^0 \wedge dx^j \wedge dx^k$$
$$= \frac{\partial B_i}{\partial x_i} dx^1 \wedge dx^2 \wedge dx^3 + \frac{1}{2}\frac{\partial B_{jk}}{\partial x_0} dx^0 \wedge dx^j \wedge dx^k, \quad (3.50)$$
$$dE \wedge dx^0 = \frac{1}{2}\left(\frac{\partial E_i}{\partial x_j} - \frac{\partial E_j}{\partial x_i}\right) dx^0 \wedge dx^j \wedge dx^i,$$

and therefore

$$dF = \frac{1}{2}(\partial_k E_i - \partial_i E_k - \varepsilon_{jik}\partial_0 B_j) dx^0 \wedge dx^i \wedge dx^k + \partial_i B_i \, dx^1 \wedge dx^2 \wedge dx^3, \quad (3.51)$$

and

---
[5] Recall that in Minkowski space $* * dx^i \wedge dx^j = -dx^i \wedge dx^j$.

$$d*F = \frac{1}{2}\left(\partial_i B_k - \partial_k B_i - \varepsilon_{jik}\partial_0 E_j\right) dx^0 \wedge dx^i \wedge dx^k - \partial_i E_i \, dx^1 \wedge dx^2 \wedge dx^3\,.$$
(3.52)

Further, we consider the relativistic current three-form, which is defined as

$$J = \frac{1}{3!}J_{\mu\nu\lambda}dx^\mu \wedge dx^\nu \wedge dx^\lambda = \rho - j \wedge dx_0\,,\qquad(3.53)$$

where $\rho = \frac{1}{3!}\rho_{ijk}\,dx^i \wedge dx^j \wedge dx^k$ and $j = \frac{1}{2}j_{ij}dx^i \wedge dx^j$ are the three-form of charge density and the two-form of current, respectively. We can see that the standard four-vector of current corresponds to the components of the dual one-form $j = *J = j_0 dx^0 + j_k dx^k$. Thus, the system of Maxwell equations can be written compactly as

$$dF = 0\,,\qquad d*F = J\,,\qquad(3.54)$$

or, using the definition of the adjoint derivative $d^\dagger$ of (3.35)

$$dF = 0\,,\qquad d^\dagger F = *d*F = *J = j\,.$$

Obviously, current conservation emerges if we apply the $d^\dagger$ operation to the second Maxwell equation: $d^\dagger d^\dagger F = 0 = d^\dagger j$, and then

$$d^\dagger j = *d * j_\mu dx^\mu = \partial_\alpha j^\alpha * \frac{1}{4!}\varepsilon_{\mu\nu\rho\lambda}dx^\mu \wedge dx^\nu \wedge dx^\rho \wedge dx^\lambda = 0\,.\qquad(3.55)$$

Also, the Lagrangian of the Abelian electrodynamics can be written in an extremely compact way in this notation. Geometrically it is a four-form on M^4 constructed from the forms A, F and J and all other forms that can be obtained from them by the operations $*$, d and $d^\dagger$. The simplest possible combination is

$$L = dA \wedge *dA - A \wedge J = \frac{1}{2}F \wedge *F + A \wedge *j\,.$$

It can be also supplemented by a topological term $F \wedge F = d(A \wedge dA)$, which is an exact form. The variation of this function with respect to A yields:

$$\frac{\partial L}{\partial A} = -J\,,\qquad \frac{\partial L}{\partial dA} = *dA\,,$$

and the Euler–Lagrange equation becomes $d*dA - J = 0$, or just $d*F = J$ as above. However, the other Maxwell equation, which would connect the field form F with a monopole current, does not follow from the variation approach. It is a direct consequence of the Poincaré lemma for a closed form $F = dA$.

Obviously, the one-form A is not unique and we have a freedom of $U(1)$ gauge Transformations, which act as $A \to A + d\lambda$ and leaves the two-form $F = dA$ unchanged. The gauge function λ may be taken, for example, to set the Lorentz gauge

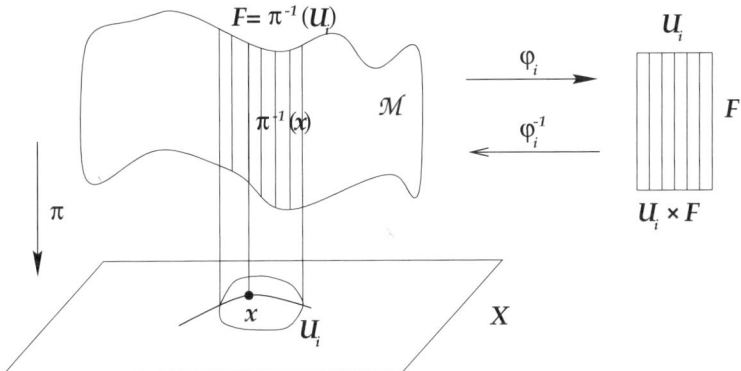

Fig. 3.8. Construction of the fiber bundle $\mathcal{M}$

$$d^\dagger A' = d^\dagger(A + d\lambda) = d^\dagger A + (d^\dagger d + d d^\dagger)\lambda = d^\dagger A + \Delta\lambda = 0\,,$$

since the 0-form λ satisfies $d^\dagger \lambda = 0$. The gauge function can then be found as a solution of the Poisson equation $\Delta\lambda = -d^\dagger A$. Finally, the invariance of the action with respect to the gauge transformation is obviously related to the conservation of the current j, since $L[A + d\lambda] - L[A] = d\lambda \wedge *j = -\lambda d * j$.

Before we address the question of how a monopole could be incorporated into this picture, we need to clarify one very important point. Much of the discussion above has no direct connection with the gauge invariance of the theory. To incorporate it into the geometrical discussion above, we must extend our discussion of manifolds and construct so-called *fiber bundles* on M^4.

3.3 Wu–Yang Monopole and the Fiber-Bundle Topology

It is very remarkable that the highly formal and abstract language of differential geometry and fiber bundles is perfectly tailor-made to describe the underlying topology of monopoles. More surprising may be the fact that the fundamental paper by H. Hopf [278], where the corresponding purely mathematical construction was introduced, was published exactly in the same year as the celebrated work by Dirac [200]. However, the beautiful relationships between these two classical works were only discovered more than 40 years later, in the mid-1970s. The obvious reason for that is that the language and notions of topology were not familiar to physicists at that time. The ideas and methods of topological calculus invaded physics first in the seventies and subsequently became commonly used.

3.3.1 Fiber Bundles

When we discussed simple examples of topological manifolds above, we briefly mentioned the possibility of constructing a topological manifold as a global

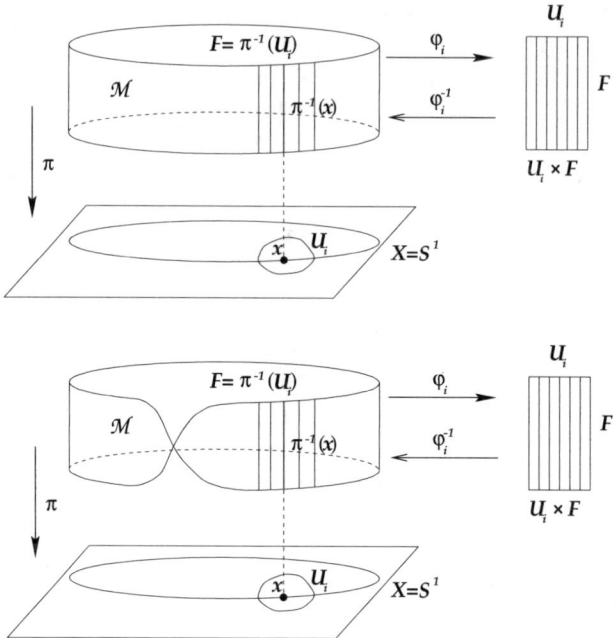

Fig. 3.9. Two fiber bundles over the base S^1 with the fiber $F = [0, 1]$ with identical local structures but different transition functions

product of two other spaces. The two-torus $T^2 = S^1 \times S^1$ is such a product space. Roughly speaking, we can define a fiber bundle as a space $\mathcal{M}$ that locally, but not globally, is a product of two spaces X and F.

To make the notion of this structure more precise, we shall introduce some basic definitions (see, e.g., [17, 207]).

Let $\mathcal{M}$ be a topological manifold. The idea of the fiber bundle is to divide $\mathcal{M}$ into sets of subspaces, the *fibers*, each of which is homeomorphic to F. Then we require that there exists a map $\pi : \mathcal{M} \to X$, where X is called the *base* space. Thus all the points of a fiber F are set into correspondence to a single point $x \in X$ and such a fiber F is homeomorphic to the inverse image $\pi^{-1}(x)$. This map is sometimes called the *projection* (see Fig. 3.8).

The notion of a fiber bundle arises when at each point of a given manifold X, we somehow define another manifold, a fiber F. One example is the linear tangent space $F = T_p(X)$ of the differentiable manifold we considered above. Such a bundle is called the *tangent bundle*. The fiber at the point p is a vector space of dimension n and $\mathcal{M}$ is called a vector bundle.

Note that, in general, it is not enough to define the base space X and the fiber F to describe the fiber bundle $\mathcal{M}$. Let us give the standard example of the Möbius strip versus a cylinder (see Fig. 3.9). There are two different fiber bundles on $X = S^1$ with the same fiber, a line segment $F = [0, 1]$. Clearly

3.3 Wu–Yang Monopole and the Fiber-Bundle Topology

the local structure of both spaces is identical, while the global topology is different.

In order to consider the global properties of $\mathcal{M}$, we must cover the base space X with a collection of subsets $\{U_i\}$. Then there is a homeomorphism $\phi_i : \pi^{-1}(U_i) \to U_i \times F$, such that $(\pi \circ \phi^{-1})(x, f) = x$ with $x \in U_i$, $f \in F$. Thus, in each given chart (U_i, ϕ_i) a fiber bundle $\mathcal{M}$ over X can be represented as a product $U_i \times F$.

Furthermore, such a structure, even though it is local, can provide information about the global topological properties of the space $\mathcal{M}$. The point is that to make our description complete, we must define the transition functions from one chart (U_i, ϕ_i) to another chart (U_j, ϕ_j). The *structure group* G of homeomorphisms of the fiber F arises when we consider such a transition from one local patch to another. Indeed, let us consider a continuous homeomorphic map in the overlap region $U_i \cap U_j$

$$g_{ij} \equiv \phi_i \circ \phi_j^{-1} : (U_i \cap U_j) \times F \to (U_i \cap U_j) \times F. \quad (3.56)$$

The transition function $g_{ij} = \phi_i \circ \phi_j^{-1}$ is a homeomorphism of the fiber F. It gives information as to how the fibers are glued together in the overlap region[6]. Then the set of all these homeomorphisms $\{g_{ij}\}$ for all charts on X form the structure group G of the fiber bundle $\mathcal{M}$.

A simple and instructive example is again the Möbius strip. Let us reconstruct the whole fiber bundle using all ingredients at our disposal. The base space, the circle S^1, can be parameterized by an azimuthal angle θ. Now we can cover the base by two subsets U_i and U_j, as shown in Fig. 3.10:

$$U_i = \{\theta : \; -\varepsilon < \theta < \pi + \varepsilon\}, \qquad U_j = \{\theta : \; \pi - \varepsilon < \theta < 2\pi + \varepsilon\}, \quad (3.57)$$

and take the fiber F to be a line segment in $\mathbb{R}^1$ parameterized by a local coordinate $t \in [-1, 1]$. Then the fiber bundle consists of two patches $U_i \times F$ and $U_j \times F$ with coordinates (θ, t_i) and (θ, t_j), respectively. The transition functions are defined in two overlap regions $U_i \cap R_j = I_R \cup I_L$, $I_R \in [-\varepsilon, \varepsilon]$, $I_L \in [\pi - \varepsilon, \pi + \varepsilon]$ (see Fig. 3.10), and there connect the local coordinate in the fibers: $t_i = g_{ij} t_j$. The structure group $G = \mathbb{Z}_2 = \{1, -1\}$ consists of two elements and we can choose

$$\begin{aligned} t_i = g_{ij} t_j, & \quad g_{ij} = 1 \quad \text{if } \theta \in I_R, \\ t_i = g_{ij} t_j, & \quad g_{ij} = -1 \quad \text{if } \theta \in I_L, \end{aligned} \quad (3.58)$$

---

[6] When we compare the definition of a fiber bundle $\mathcal{M}$ with the definition of a differentiable manifold above, we can see the similarities between these notions. Indeed, the transition functions g_{ij} and the maps (3.8) that we used to change the coordinates on the manifold are defined in the same way. Furthermore, as a real differentiable manifold is locally isomorphic to $\mathbb{R}^n$, the fiber bundle $\mathcal{M}$ is locally a direct product of two spaces $X \times F$. There is no miracle since the differentiable manifold is a base space for the tangent bundle T_p with the structure group $GL(n, \mathbb{R})$.

96   3 Topological Roots of the Abelian Monopole

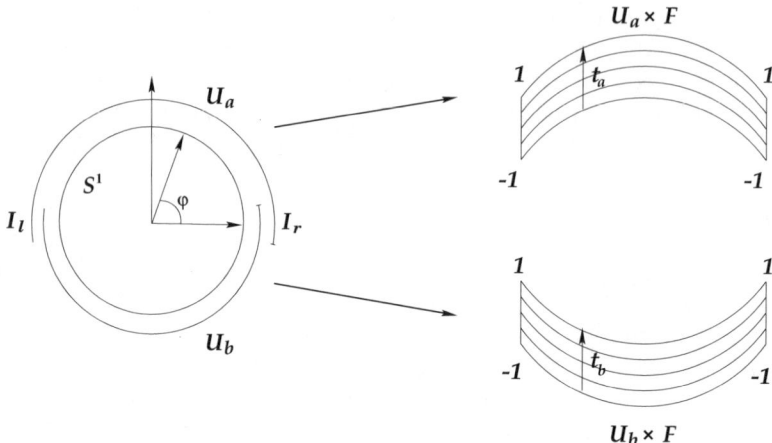

Fig. 3.10. Two local patches of the fiber bundle with the base S^1 and fiber $F = [0, 1]$

to get the twist for the Möbius strip. If we choose $g_{ij} = 1$ in both overlap regions, the structure group is trivial, $G = \{1\}$, and we have a trivial bundle, the cylinder $S^1 \times F$.

Thus, the local topology in each chart is trivial, but the global topology hidden behind the transition functions might be quite different due to the twists of the neighboring fibers. If all the transition functions are trivial, $\{g_{ij}\} = 1$ for all charts, we come back to the direct product of two spaces defined globally: $\mathcal{M} = X \times F$. Such a bundle is called *trivial*.

An important property of the transition functions is that they are defined up to some unitary transformation. Indeed, let us consider two bundles $\mathcal{M}$ and $\mathcal{M}'$ with the same base X, fiber F (but different maps), and structure group G. Let the charts (U_i, ϕ_i), (U_i, ψ_i) define the coordinates in $\mathcal{M}$ and $\mathcal{M}'$, respectively. We require that the homeomorphism $h_i \equiv \phi_i \circ \psi_i^{-1}$, $h_i : U_i \times F \to U_i \times F$ belongs to the structure group G. Then the transition functions g_{ij} and g'_{ij} are related by

$$g'_{ij} = \psi_i \circ \psi_j^{-1} = h_i^{-1} \circ g_{ij} \circ h_j \,. \tag{3.59}$$

Since we require $h_i \in G$, the two fiber bundles $\mathcal{M}$ and $\mathcal{M}'$ are topologically equivalent. Thus we have to deal not with a single Bundle, but with an *equivalence class* of bundles. The example of the cylinder and the Möbius strip discussed above gives two different equivalence classes. The former are bundles twisted an even number of times and the latter are bundles twisted an odd number of times.

An important consequence of the definition (3.59) is that we can define under which condition a fiber bundle $\mathcal{M}$ is trivial. If a transition function can be "split" as a product

$$g_{ij} = h_i \circ h_j^{-1}, \tag{3.60}$$

then, according to (3.59), we have $g'_{ij} = 1$ and such a bundle is trivial.

3.3.2 Principal Bundle and Connection

Let us introduce some more definitions necessary to establish a correspondence with our further physical discussion. The projection π can be supplemented by a dual operation, namely the *section* of a bundle $\mathcal{M}$. This is a map of an element of the base X to $\mathcal{M}$:

$$s : U_i \to \pi^{-1}(U_i), \quad \text{where} \quad \pi \circ s(x) = x, \quad x \in X. \tag{3.61}$$

In the following, we shall restrict our consideration to the so-called *principal bundles*. We can construct such a bundle by choosing the fiber to be identical to the structure group: $F = G$. The reason is that a gauge theory has a natural structure of the principal bundle with the gauge group as a fiber. In the following we shall consider only principal bundles.

As we saw above, the transition functions are elements of a bundle containing information about the global topology. However, knowledge of the topology is not enough to provide a complete description of a topological space $\mathcal{M}$ and we must define an additional characteristic to define its "shape".

Actually, when we discussed the geometrical properties of a differentiable manifold, which is a particular case of the fiber bundle construction, we already introduced such a differential structure, the metric and the associated Levi-Civita connection Γ^i_{jk}. This is the structure that arises when we consider the parallel transport of a vector of the tangent space and, roughly speaking, it tells us about the geometry of a manifold.

This construction can be generalized for an arbitrary non-trivial bundle $\mathcal{M}$, when the base is no longer locally homeomorphic to Euclidean space. Moreover, we can define a connection associated with the general notion of parallel transport without any reference to a metric. Generally, it contains information as to how a path in the base space maps onto the corresponding collection of fibers. Such a structure can be introduced if we consider a path between two different points x and x' in the base space X. They can be "lifted" up to the points p and p' in the bundle $\mathcal{M}$ and then we can compare two tangent spaces $T_p(\mathcal{M})$ and $T_{p'}(\mathcal{M})$. Since locally the bundle has the structure of a direct product $U_i \times F$, where U_i is a subset in the base X, a tangent vector $v \in T_p(\mathcal{M})$ can be decomposed into two components: one along the fiber ("vertical" direction) and one along the base ("horizontal" direction), respectively.

Now we note that the vertical component of a vector v points along a single fiber and this direction is fixed by the action of the structure group G. The horizontal direction is certainly not unique and there are many ways to choose it. However, the transport in the horizontal direction is exactly what we need, since it maps a path in $\mathcal{M}$ from a point p to another point p' onto a path from one fiber to another.

The *connection* is defined as an assignment to each point of the fiber $p \in \mathcal{M}$, a horizontal subspace $H_p(\mathcal{M})$ of the tangent space $T_p(\mathcal{M}) = V_p(\mathcal{M}) + H_p(\mathcal{M})$. Since the structure group G is responsible for transport along the fiber via the right multiplication

$$p \mapsto p' = pg, \qquad g \in G,$$

the horizontal subspaces $H_p(\mathcal{M})$ and $H_{pg}(\mathcal{M})$ on the same fiber are linked together by a differential map.

From the point of view of physics, the subjects of primary interest are principal bundles with a Lie group G as a fiber. If the field takes values in some Lie algebra $\mathbf{g}$ of G, then the connection can be introduced as a one-form ω on $\mathcal{M}$ also taking values in $\mathbf{g}$. This gives rise to the gauge potential as it is known in physics.

Let us take a patch U_i of the n-dimensional base space parameterized by local coordinates x_μ and consider there a Lie algebra-valued one-form $A = A_\mu^a T^a dx^\mu$, where T^a are the generators of a Lie group satisfying the algebra $[T^a, T^b] = f_{abc} T^c$. The local *section* on U_i is $s_i : U_i \to \pi^{-1}(U_i)$. Given s there is a map s^*, which pulls the form on the bundle back to a form on U_i. Then the pullback $s^*\omega$ of the connection ω down to the base is exactly the form A.

Indeed, we can define a canonical 1-form $\omega_0 = g^{-1}dg$ on the group G, which is an isomorphism between each point of the vertical tangent space $V_p(\mathcal{M})$ and the Lie algebra $\mathbf{g}$. Here the operation d is the exterior derivative on the bundle and $g \in G$ is a local coordinate in the bundle.

Then the connection is defined as a horizontal projection $\pi: T_p(\mathcal{M}) \to H_p(\mathcal{M})$ mapped onto a Lie algebra: $\omega_0 : H_p(\mathcal{M}) \to \mathbf{g}$. Thus, the superposition of $\omega_0 \circ \pi$ is a one-form

$$\omega = \omega_0 + g^{-1}\pi^* A g = g^{-1}dg + g^{-1}\pi^* A g, \qquad (3.62)$$

and we can check that ω formally satisfies the definition of the connection above, and the pullback within a local section s (cf. (3.61)) yields $s^*\omega = A$.

Note that the connection ω is uniquely defined on the bundle, since the decomposition of the tangent space $T_p(\mathcal{M})$ onto vertical and horizontal subspaces is unique. Therefore the potential A must satisfy some restrictions when we pull back the connection down to different patches U_i and U_j:

$$g_i^{-1}dg_i + g_i^{-1}\pi^* A(x_i) g_i = g_j^{-1}dg_j + g_j^{-1}\pi^* A(x_j) g_j = \omega.$$

In the overlap region $U_i \cap U_j$ the transition functions g_{ij} of (3.56), here denoted U_{ij}, connect the local coordinates according to $g_j = U_{ji} g_i$ and therefore the gauge potential transforms when we change the fiber coordinates according to

3.3 Wu–Yang Monopole and the Fiber-Bundle Topology

$$A_j = U_{ij}^{-1} A_i U_{ji} + U_{ij}^{-1} dU_{ji}. \tag{3.63}$$

This relation is called the *affine transformation* of the form A. In the context of gauge theory it is known as the gauge transformation of the vector-potential A_μ.

The notion of curvature of a differentiable manifold can also be generalized when we consider a fiber. To do so we have to define the operation of covariant exterior differentiation D of the connection one-form ω. We can do it in a rather intuitive way by saying that such a derivative $D\omega$ must transform under the action of the group G in the same way as the form ω itself. For example, if ω is invariant with respect to a transformation of the group G, $\omega' = \omega$, then the covariant derivative must also be invariant: $(D\omega)' = D\omega$. Then we can identify the covariant derivative and the ordinary exterior derivative d above. If a one-form transforms as a vector, $\omega' = g\omega, g \in G$, then the covariant derivative must transform as $(D\omega)' = g(D\omega)$ and we can write

$$D\omega = d\omega + A \wedge \omega, \tag{3.64}$$

where the one-form A obeys the affine transformation law (3.63). In the same way, covariant derivatives of quantities transforming as tensors with respect to the group action could be constructed.

The exterior covariant derivative must produce a form of one degree higher than ω and obey a certain group transformation property. If we consider a vector-potential one-form A transforming according to (3.63), its covariant derivative defined as in (3.64) transforms as

$$F \equiv DA = dA + A \wedge A, \quad (DA)' = g^{-1}(DA)g = g^{-1} F g. \tag{3.65}$$

The curvature two-form Ω on the bundle $\mathcal{M}$ is defined as the covariant derivative of the connection one-form: $\Omega = D\omega$. The pullback $s^*\Omega$ down to a local patch U_i on the base is a two-form $F = DA$ that locally represents the gauge field strength tensor. Again, the local changes of sections $s' = gs$ in the principal bundle induce compatibility conditions in the overlap region $U_i \cap U_j$ (cf. (3.63))

$$F_j = U_{ij}^{-1} F_i U_{ij},$$

which is known as a gauge transformation of the field strength tensor.

The marvelous property of the curvature form is that it provides a topological classification of bundles. We have seen above that given a base X and a fiber $F = G$, they can be assembled into different bundles, like in the case of the cylinder versus the Möbius strip. Similarly to the classification of the homotopic maps above, which were given in terms of homotopy classes, we can produce a classification of principal bundles in terms of *characteristic classes*. The basic element of this classification is the observation that a closed but not exact r-form ω on a differentiable manifold is equivalent to the forms $\omega + d\alpha$, where α is $(r-1)$-form. To define the equivalence classes of the differential forms, we have to consider the space of all closed forms $Z^r = \{\omega \mid d\omega = 0\}$.

100   3 Topological Roots of the Abelian Monopole

These have the subspace of exact forms $B^r = \{\omega \mid \omega = d\alpha\}$, which must be excluded. Then the elements $[\omega]$ of the coset space $H^r = Z^r/B^r$ represent the equivalence classes of differential forms. Clearly, these classes are defined in the same manner as the equivalence classes $[\alpha]$ of a topological manifold discussed above. This analogy can be expanded since the equivalence classes $[\omega]$ also have a group structure. The corresponding group is called the *DeRham cohomology group*.

Recall that each of the homotopy classes is characterized by an integer, the winding number n (see (3.20)). A similar characteristic can be introduced to characterize the equivalence classes of the differential forms. Let us consider an integral over a closed differential form ω' on a manifold X with an empty boundary $\partial X = 0$. Then the Stokes theorem (3.37) yields

$$\int_X \omega' = \int_X (\omega + d\alpha) = \int_X \omega + \int_{\partial X} \alpha = \int_X \omega. \tag{3.66}$$

Thus, properly normalized, such an integral can be used as a characteristic of the cohomology class.

Let us return to the closed two-form of a curvature. An equivalence class is formed by the different connection one-forms, which corresponds to the same curvature. Now we can introduce a quantity that is an invariant of the Lie algebra $\mathfrak{g}$ in the fiber, a *characteristic polynomial*

$$\det(1 + \Omega) = 1 + \mathrm{tr}\,\Omega + \frac{1}{2}\left[(\mathrm{tr}\,\Omega)^2 - \mathrm{tr}\,\Omega^2\right] + \dots$$

This polynomial allows us to define a form on the base X, which is called the *Chern form*

$$\det\left(1 + \frac{\lambda}{2\pi}F\right) = \sum_{k=0}^{\infty} \lambda^k\, c_k, \tag{3.67}$$

where F is a Lie-algebra valued curvature two-form and the coefficient $1/2\pi$ is introduced in order to have a convenient normalization. The coefficients of the expansion c_k are invariant polylinear closed forms on X.

Let us write the explicit expressions for the first few of these forms:

$$c_1 = \frac{1}{2\pi}\mathrm{tr}\,F, \qquad c_2 = \frac{1}{8\pi^2}\left[\mathrm{tr}\,F \wedge \mathrm{tr}\,F - \mathrm{tr}(F \wedge F)\right], \tag{3.68}$$
$$c_3 = \frac{1}{48\pi^3}\left[\mathrm{tr}\,F \wedge \mathrm{tr}\,F \wedge \mathrm{tr}\,F - 3\mathrm{tr}(F \wedge F) \wedge \mathrm{tr}\,F + 2\mathrm{tr}(F \wedge F \wedge F)\right].$$

The coefficients of these forms c_k, which originate from the factor $1/2\pi$ in the definition of the characteristic polynomial (3.67), provide the normalization of the integrals over compact space without boundary:

$$\int_X c_k = n, \quad \text{with} \quad n \in \mathbb{Z}. \tag{3.69}$$

3.3 Wu–Yang Monopole and the Fiber-Bundle Topology

The so-called *Gauss–Bonnet formula* relates these integers with the Euler characteristic of the compact space X.

The integrals over the forms c_k on compact even-dimensional manifolds define the Chern characteristic classes[7]. These integers are precisely the topological characteristics of the bundle, according to the Gauss–Bonnet theorem.

Let us consider two examples. The monopole bundle that we shall discuss below has the structure group $U(1)$ over the base space S^2. Then we have two Chern classes: $c_0 = 1$ (trivial bundle) and $c_1 = F/2\pi$, the integral over which must be an integer associated with the magnetic charge.

The second example concerns the Chern classes for the principal bundle over S^4 with group $SU(2)$. Then the curvature takes values in the $su(2)$ algebra and can be written as $F = \frac{1}{2}\sigma^a F^a$, where the σ^a are Pauli matrices,

$$\sigma_1 = \begin{pmatrix} 0 & 1 \\ 1 & 0 \end{pmatrix}, \qquad \sigma_2 = \begin{pmatrix} 0 & -i \\ i & 0 \end{pmatrix}, \qquad \sigma_3 = \begin{pmatrix} 1 & 0 \\ 0 & -1 \end{pmatrix}. \tag{3.70}$$

Therefore tr $F = 0$, we have $c_0 = 1$, $c_1 = 0$, and the bundle is characterized by the *second Chern class*[8]

$$c_2 = \frac{1}{32\pi^2} F^a \wedge F^a. \tag{3.71}$$

This is known as the *instanton bundle*.

We stop here our discussion of the basic ingredients of differential geometry. The connection of the bundle and its curvature is exactly the notion that provides a bridge between gauge theory and geometry. If we restrict our consideration to the case of the Abelian theory, the structure group must be taken as $G = U(1)$, as the base space we choose four-dimensional Minkowski space M^4, and the fiber is the group space of $g = \{e^{i\alpha}\} \in U(1)$, that is the circle S^1 parameterized by the group coordinate α.

If a bundle admits a global section, the base can be covered by just one chart. Such a trivial bundle can be characterized by the zero Chern class c_0 and has a global structure as a direct product of the base X and the Lie group G. In general, the triviality of the bundle depends on the contractibility of the base. For example, if we take the base to be just $\mathbb{R}^3$, it can be covered by a single chart and therefore both the potential and the field strength tensor are defined globally in this case. This is exactly the case of ordinary Maxwell electrodynamics, where the field strength two-form F is closed and exact: $DF = 0$ and $f = dA$, where A is continuous and differentiable on the whole base manifold. Thus, the topology is trivial and there is no place for a monopole.

---

[7] Strictly speaking, for the principal bundles with the structure group $O(n)$ they are called *Pontryagin classes*, while the Chern classes arise if we are dealing with the structure group $U(n)$.

[8] For the sake of brevity, we do not discuss the sign arising from the orientation of the manifold.

3.3.3 Wu–Yang Monopole Bundle

Let us change the base space of Maxwell electrodynamics a little. Namely, we extract just one point, the origin $\{0\}$, from $\mathbb{R}^3$. This minor change affects the situation in a very drastic way. The point is that topologically, the space $\mathbb{R}^3/\{0\}$ is homotopic to the two-sphere S^2 of unit radius. It can be parameterized by a polar angle $\theta \in [0, \pi[$ and an azimuthal angle $\varphi \in [0, 2\pi[$. Now we can define a basis of the tangent space $T_p(S^2)$ as components of the gradient $\{\partial/\partial\theta, \partial/\partial\varphi\}$ and the basis of the dual tangent space of 1-forms as $\{d\theta, d\varphi\}$. Thus the exterior derivative (cf. (3.30)) is

$$d = d\varphi \frac{\partial}{\partial \varphi} + d\theta \frac{\partial}{\partial \theta}.$$

Now we can cover the base space by two hemispheres R^N and R^S (see Fig. 3.1 at the beginning of this chapter), which are our patches with local coordinates on the base. However in our discussion, it would be more convenient to take the limit $\varepsilon \to 0$, i.e., to consider the equatorial circle S^1 as the overlap region.

Again, the elements of the Abelian gauge group $g = \{e^{i\alpha}\} \in U(1)$ form a fiber of the principal bundle over $\mathbb{R}^3/\{0\}$. The group parameter α is a cyclic coordinate along the fiber and we have two charts $R^N \times S^1$ and $R^S \times S^1$ with bundle coordinates $\{\theta, \varphi, \alpha_N\}$ and $\{\theta, \varphi, \alpha_S\}$, respectively. The pull back of the connection one-form to these local patches yields the potential one-form A (cf. (1.41)):

$$A = A(\theta) \wedge dr = \begin{cases} A^N = \frac{n}{2}(1 - \cos\theta) d\varphi & \text{on } R^N \in S^2 \\ A^S = -\frac{n}{2}(1 + \cos\theta) d\varphi & \text{on } R^S \in S^2, \end{cases} \quad (3.72)$$

and the gauge field strength two-form is

$$F = dA = \frac{n}{2} \sin\theta\, d\theta \wedge d\varphi. \quad (3.73)$$

Now if we take into account that the basis of common spherical coordinates on S^2 is given by two unit vectors $\hat{\mathbf{e}}_\varphi = \sin\theta\, d\varphi$, $\hat{\mathbf{e}}_\theta = d\theta$, we see that this two-form F corresponds to the radial magnetic field of a monopole. As before, the form F is closed, $dF = d^2A = 0$, but now it is not exact since the exterior derivatives dA^N and dA^S are defined only locally in the patches R^N and R^S, respectively. However, in the overlap region $R^N \cap R^S$ the potentials are connected via the gauge transformation $A^N = A^S + nd\varphi$ and we can define transition functions there. They are elements of the structure group $U \in U(1)$ that connects the fiber coordinates as $e^{i\alpha_N} = Ue^{i\alpha_S}$. We can identify the overlap region with the equator of the sphere S^2. There $\theta = \pi/2$ and U are functions of the azimuthal angle φ only. They can be written as $U = e^{in\varphi}$, where n is an integer.

3.3 Wu–Yang Monopole and the Fiber-Bundle Topology

The wave function of a charged particle in a monopole field cannot be defined globally either. In each of the two patches we have a single-valued function ψ^N and ψ^S, respectively (cf. (3.7)). They are sections of the associated bundle, and are related in the overlap region as $\psi^N = U\psi^S$.

The topological meaning of the transition functions of the monopole bundle is that they define a map from the equatorial circle S^1 of the base manifold S^2 onto the structure group $U(1)$. We have seen above (see (3.19) and the corresponding discussion) that these maps can be classified according to the first homotopy group $\pi_1(U(1)) = \mathbb{Z}$ and that the integer n is a winding number of the corresponding homotopy class.

On the other hand, we mentioned above that the topology of the monopole bundle can also be characterized by the *first Chern number*

$$c_1 = \frac{1}{2\pi}\int_{S^2} F = \frac{1}{2\pi}\left(\int_{R^S} dA^S + \int_{R^N} dA^N\right)$$

$$= \frac{1}{2\pi}\int_{S^1}(A^N - A^S) = \frac{1}{2\pi}\int_0^{2\pi} d\varphi \cdot n = n, \qquad (3.74)$$

where we used the Stokes theorem (3.37). Thus, the magnetic charge of an Abelian monopole in this description coincides with the first Chern number c_1. If $n = 0$ the bundle is trivial, i.e., it has the global form $S^2 \times S^1$. The case of unit topological charge $n = 1$ is much more interesting. This is the case of the *Hopf bundle*, which we discuss below.

3.3.4 Hopf Bundle

Recall that there are three coordinates that parameterize the monopole bundle: two angles θ and φ on the sphere S^2 and the fiber coordinate $\alpha \in S^1$. However, the set of 3 coordinates can be used to parameterize a unit sphere S^3.

Let us consider the geometry of such a space. It is customary to work with Cartesian coordinates on S^3, which can be introduced as

$$x_1 = \cos\frac{\theta}{2}\cos\alpha, \qquad x_3 = \sin\frac{\theta}{2}\cos(\varphi + \alpha)$$

$$x_2 = \cos\frac{\theta}{2}\sin\alpha, \qquad x_4 = \sin\frac{\theta}{2}\sin(\varphi + \alpha). \qquad (3.75)$$

Clearly, $x_\mu x^\mu = 1$. Then we can consider the tangent space to S^3 with the basis $\{\partial/\partial x_\mu\}$ and define a metric on the sphere S^3. It is non-diagonal in terms of the angular variables above:

$$ds^2 = g_{\mu\nu}dx^\mu dx^\nu = \frac{1}{4}d\theta^2 + d\alpha^2 + \sin^2\frac{\theta}{2}d\varphi^2 + 2\sin^2\frac{\theta}{2}d\varphi d\alpha. \qquad (3.76)$$

3 Topological Roots of the Abelian Monopole

Therefore, it is convenient to introduce the orthogonal coordinates θ, α, $\xi = \varphi + \alpha$, which diagonalize the metric:

$$ds^2 = \frac{1}{4}d\theta^2 + \cos^2\frac{\theta}{2}d\alpha^2 + \sin^2\frac{\theta}{2}d\xi^2. \tag{3.77}$$

The Hopf bundle is a map $S^3(\theta, \varphi, \alpha) \to S^2(\theta, \varphi)$, which can be constructed if we introduce two complex coordinates [69, 278]

$$z_1 = x_1 + ix_2 = \cos\frac{\theta}{2}e^{i\alpha}, \qquad z_2 = x_3 + ix_4 = \sin\frac{\theta}{2}e^{i(\varphi+\alpha)}, \tag{3.78}$$

and compose these coordinates into a two-component spinor

$$z = \begin{pmatrix} z_1 \\ z_2 \end{pmatrix} = \begin{pmatrix} \cos\frac{\theta}{2}e^{i\alpha} \\ \sin\frac{\theta}{2}e^{i(\varphi+\alpha)} \end{pmatrix}, \qquad z^\dagger z = 1. \tag{3.79}$$

The advantage of this construction is that now we can define a non-singular monopole connection on S^3. Indeed, let us consider the connection one-form on S^3, which can be constructed from the spinor coordinates (3.79) (see the related discussion p. 98):

$$\omega = -iz^\dagger dz = x_1 dx_2 - x_2 dx_1 + x_3 dx_4 - x_4 dx_3$$
$$= d\alpha + \sin^2\frac{\theta}{2}d\varphi = d\alpha + \frac{1}{2}(1 - \cos\theta)d\varphi, \tag{3.80}$$

where we have used the parameterization (3.75) and take into account that $d(x_\mu x^\mu) = 0$. If we use the orthogonal coordinates, the connection 1-form becomes

$$\omega = \cos^2\frac{\theta}{2}d\alpha + \sin^2\frac{\theta}{2}d\xi = \frac{1}{2}\omega_\theta d\theta + \cos\frac{\theta}{2}\omega_\alpha d\alpha + \sin\frac{\theta}{2}\omega_\xi d\xi, \tag{3.81}$$

that is, we found non-singular components of the connection on S^3

$$\omega_\theta = 0, \qquad \omega_\alpha = \cos\frac{\theta}{2}, \qquad \omega_\xi = \sin\frac{\theta}{2}. \tag{3.82}$$

This is a particular realization of the connection form on the bundle (3.62) that we discussed above. The curvature two-form Ω is then

$$\Omega = d\omega = -idz^\dagger \wedge dz = 2(dx_1 \wedge dx_2 + dx_3 \wedge dx_4)$$
$$= \frac{1}{2}\sin\theta d\theta \wedge d\varphi. \tag{3.83}$$

This form is exact and closed on S^3 and obviously corresponds to (3.73) for the particular case $n = 1$. Obviously it is nothing but half of the volume form of the sphere S^2, which is parameterized by the two coordinates θ and φ. Therefore,

3.3 Wu–Yang Monopole and the Fiber-Bundle Topology

$$\int_{S^2} \Omega = 2\pi.$$

Now let us go down to the base space S^2. The map $S^3 \to S^2$ (so-called *Hopf fibration*) is given by the following transformation (*Hopf map*)

$$n_k = z^\dagger \sigma_k z, \tag{3.84}$$

where the Pauli matrices σ^a are defined according to (3.70). Indeed, substituting (3.75) into (3.84), we obtain

$$\begin{aligned} n_1 &= 2(x_1 x_3 + x_2 x_4) = \sin\theta \cos\varphi, \\ n_2 &= 2(x_1 x_4 - x_2 x_3) = \sin\theta \sin\varphi, \\ n_3 &= (x_1^2 + x_2^2 - x_3^2 - x_4^2) = \cos\theta. \end{aligned} \tag{3.85}$$

Obviously, the components of the unit vector n are identical to the Cartesian coordinates on the sphere S^2. The magic of the transformation (3.84) is that it completely removes the dependence on the third coordinate on S^3, the cyclic variable α. In other words, the transformation (3.84) maps a circle S^1 onto a point on S^2, and the Hopf bundle has the structure group $S^1 = U(1)$.

The section of the fiber bundle can be taken if we fix a particular value of the angle α. Let us take the local patches to be two hemispheres R^N and R^S, as above. Then we must take a section in such a way that the metric on S^3 (3.76) reduces to the metric of the 2-sphere $ds^2 = d\theta^2 + \sin^2\theta d\varphi^2$ in the north ($\theta/2 \to \theta$) and south ($\pi/2 - \theta/2 \to \theta$) hemispheres of S^2 respectively. From (3.76) it is clear that this reduction to S^2 is provided by the restriction on α to be equal to 0 and $-\varphi$, respectively. However, if we take $\alpha = 0$ or $\alpha = -\varphi$, the globally defined connection one-form (3.80) is reduced to the local connections (3.72) above:

$$A = \begin{cases} A^N = \tfrac{1}{2}(1 - \cos\theta) d\varphi & \text{on } R^N \in S^2, \\ A^S = -\tfrac{1}{2}(1 + \cos\theta) d\varphi & \text{on } R^S \in S^2. \end{cases} \tag{3.86}$$

The transition functions of the bundle are defined in the overlap region $R^N \cap R^S$ as a map $S^1 \to S^1$ to be given by (cf. (3.59))

$$\begin{aligned} g_{NS} &= \phi_N \circ \phi_S^{-1} = \frac{z_2/|z_2|}{z_1/|z_1|} = e^{i\varphi}, \\ g_{SN} &= \phi_S \circ \phi_N^{-1} = \frac{z_1/|z_1|}{z_2/|z_2|} = e^{-i\varphi}. \end{aligned} \tag{3.87}$$

We can look at this map from a different point of view. Note that the sphere S^3 coincides with the group manifold of the group $SU(2)$. Thus we reformulate the Hopf map (3.84) in terms of elements of this group, the matrices $U(\theta, \varphi, \alpha) \in SU(2)$ (we use a slightly different parameterization in Appendix A, where general properties of the $SU(2)$ group matrices are described)

$$U(\theta, \varphi, \alpha) = \begin{pmatrix} z_1^* & z_2^* \\ -z_2 & z_1 \end{pmatrix} = \begin{pmatrix} \cos\frac{\theta}{2} e^{-i\alpha} & \sin\frac{\theta}{2} e^{-i(\varphi+\alpha)} \\ -\sin\frac{\theta}{2} e^{i(\varphi+\alpha)} & \cos\frac{\theta}{2} e^{i\alpha} \end{pmatrix}. \quad (3.88)$$

It is straightforward to see now that the relation $z^\dagger z = |z_1|^2 + |z_2|^2 = 1$ corresponds to the unitarity of these matrices: $U^\dagger U = UU^\dagger = 1$. Then the Hopf fibration from S^3 to S^2 is given by

$$\sigma_k \cdot n_k = U^{-1} \sigma_3 U, \quad (3.89)$$

which is an analog of (3.84).

On the other hand, this is a rotation in the group space, which takes a spherically symmetric "hedgehog" $\sigma_k \cdot n_k$ to the third axis in the group space. In the following, we shall analyze such a transformation, which actually relates non-Abelian and Dirac monopoles, in more detail. Here, we only note that the rotation to the third axis (3.89) is defined up to a left $U(1)$ multiplication:

$$U \longrightarrow gU = e^{i\sigma_3\alpha} U, \quad g = e^{i\sigma_3\alpha} \in U(1)$$
$$\sigma_k \cdot n_k \longrightarrow U^{-1} e^{-i\sigma_3\alpha} \sigma_3 e^{i\sigma_3\alpha} U = U^{-1} \sigma_3 U = \sigma_k \cdot n_k, \quad (3.90)$$

which changes the phase of the two-component spinor z (3.79). Obviously this is a transformation of the gauge group $U(1)$. Thus, the Hopf map identifies each gauge orbit in S^3 with a single point on S^2 and we have constructed a quotient space $SU(2)/U(1)$.

According to (3.80) the Lagrangian of interaction between a charged particle and a monopole (1.38) in terms of the geometrical variables z can be written as [4, 105]:

$$L_{\text{int}} = inz\frac{dz^\dagger}{dt} = i\frac{n}{2}\text{tr}\left(\sigma_3 U \frac{d}{dt} U^{-1}\right), \quad (3.91)$$

where the integer n specifies the magnetic charge according to (3.73). This form of the Lagrangian is known as the Wess–Zumino term. It is particularly convenient for analyzing the underlying topological properties of the model.

Recall that the gauge group acts as $z \to ze^{i\alpha}$, which corresponds to the transformation (3.90) above. Then the Lagrangian of interaction (3.91) transforms as

$$L_{\text{int}} \longrightarrow L_{\text{int}} + n\dot\alpha, \quad (3.92)$$

which corresponds to (2.3). The extra term causes no trouble, since the charge quantization condition is automatically satisfied with n being an integer.

To conclude this section, we note that the above discussion of topology, which is basically aimed at establishing the basis for understanding the mathematical properties of monopoles, provides a new understanding of the relations between topology, geometry and physics. Actually, the message is that an Abelian monopole could exist only if the geometry of our Universe were different from the standard Euclidean description. Then the sad negative conclusion, which comes from all possible experiments aimed at discovering monopoles, could be considered as an argument in support of the "triviality" of the observable part of the Universe.

However the situation is more subtle, because so far we do not pay much attention to the consistent relativistic treatment of the theory. We shall address this question in the next chapter.

4 Abelian Monopole: Relativistic Quantum Theory

There are two completely distinct periods in the long history of the monopole problem. Soon after the pioneering work by Dirac [200], the interest in this field of research faded away although some enthusiasts continued working on it. Actually, for more than 40 years the monopole problem was considered to be rather esoteric, beyond mainstream theoretical physics of that time. Such an attitude was caused in part by negative results of all the experiments searching for monopoles, and it was reinforced by the cumbersome character of the theoretical constructions that were connected with the corresponding generalization of quantum electrodynamics. As a matter of fact, this problem still remains unsolved and within an Abelian theory the monopole looks like a stranger. Nevertheless, some important results were obtained during this period. Certainly, the second "monopole-related" paper by P. Dirac [201] is an example of this. In 1948 Dirac returned to a re-investigation of this problem, considering a relativistic generalization of quantum theory including two different charges, electric and magnetic ones. His approach was based on the Hamiltonian formulation of generalized Maxwell electrodynamics.

There are a few alternatives to the Dirac method to construct relativistic two-charge electrodynamics (QEMD). A very interesting development is connected with the mathematically refined approach by J. Schwinger [459, 461], who developed a non-local Hamiltonian formalism and introduced corresponding commutation relations. An alternative approach aimed at avoiding the difficulties connected with using the non-physical string variables was employed by T.M. Yan [537]. The guiding idea of this work is connected with a reformulation of the Schwinger approach on a classical level and making use of a non-local action functional. This allows us to apply path integral techniques to quantize such a non-local model [541].

A somewhat more radical approach originated from work by S. Mandelstam [361] and by N. Cabibbo and E. Ferrari [148]. Within this framework, the dynamical variables are not vector potentials, but rather the electromagnetic field strength tensor and its dual, which are considered to be functionals of trajectories of charged particles. It was shown that the corresponding non-local action can be used to construct a variational formulation of the two-charge electrodynamics [439].

From a modern point of view, all these constructions are subjects of rather historical interest, although some of the ideas have been implemented in different contexts[1]. Moreover, it has been shown [41,135] that all the different formulations of generalized electrodynamics with monopoles are in practice equivalent. Thus, we briefly give an account of the argumentation by P. Dirac [201] and then consider the formalism developed by D. Zwanziger and his collaborators in the papers [136,137,548–550]. For further details, we refer the interested reader to the excellent review by M. Blagojević and P. Senjanović [41], which was intensively used throughout this chapter. Here we only note that the most self-consistent mathematical description can be constructed within the framework of the Wu–Yang formalism [529], which has no problem in treating singularities by its very definition.

4.1 Two Types of Charges

A starting point in the construction of a generalized form of Abelian electrodynamics is a postulated system [201] of dynamical Lorentz–Maxwell equations, which describe in a covariant form a closed system of interacting electromagnetic fields and charges:

$$\partial_\mu F^{\mu\nu} = j^{(e)\nu}, \qquad \partial_\mu \widetilde{F}^{\mu\nu} = j^{(g)\nu}, \qquad (4.1)$$

$$m_e \frac{du_\mu^{(e)}}{d\tau^{(e)}} = eF_{\mu\nu}u^{(e)\nu}, \qquad m_g \frac{du_\mu^{(g)}}{d\tau^{(g)}} = g\widetilde{F}_{\mu\nu}u^{(g)\nu},$$

where $u_\mu^{(i)} = dx_\mu^{(i)}/d\tau^{(i)}$ is the relativistic 4-velocity of an electric ($i = e$) or a magnetic ($i = g$) charge, whose trajectory $x_\mu(\tau^{(i)})$ is parameterized by the proper time $\tau^{(i)}$. The relativistic four-currents of point-like charges are defined as

$$j_\mu^{(e)} = e \int dx_\mu^{(e)} \delta^{(4)}(x - x^{(e)}) = e \int u_\mu^{(e)} \delta^{(4)}(x - x^{(e)}) d\tau^{(e)},$$

$$j_\mu^{(g)} = g \int dx_\mu^{(g)} \delta^{(4)}(x - x^{(g)}) = g \int u_\mu^{(g)} \delta^{(4)}(x - x^{(g)}) d\tau^{(g)}, \qquad (4.2)$$

with both currents conserved:

$$\partial_\mu j^{\mu(e)} = \partial_\mu j^{\mu(g)} = 0.$$

These postulated equations look very symmetric and elegant. The real fun begins when we try to derive an action, whose stationary point corresponds to the system (4.1), or attempt to represent the electromagnetic energy-momentum tensor via potentials that would be necessary in order

---

[1] This is especially the case of the dynamics of the Dirac string, which seems to be the first non-local object considered in quantum theory.

to apply the canonical quantization scheme. Indeed, the standard definition $F_{\mu\nu} = \partial_\mu A_\nu - \partial_\nu A_\mu$ is not compatible with the field equations of the dual field strength tensor, since $\partial^\mu \widetilde{F}_{\mu\nu} = \varepsilon_{\mu\nu\rho\sigma}\partial^\mu\partial^\rho A^\sigma \equiv 0$. Therefore, this definition must be modified, at least in one point on the surface surrounding the magnetic charge.

In such a simple case, this modification actually leads to the appearance of the singular string that is already familiar to us. In 3+1-dimensional space it is lying on a two-dimensional sheet $y_\mu(\tau,\sigma)$ that is parameterized by temporal and spatial coordinates τ and σ, respectively. Since this string is attached to a magnetic charge, $y_\mu(\tau,0) = x_\mu^{(g)}(\tau)$. Thus, it is convenient to define the dynamical variable of the string $n_\mu(\tau,\sigma)$ as

$$y_\mu(\tau,\sigma) = n_\mu(\tau,\sigma) + x_\mu^{(g)}(\tau),$$

with the boundary condition $n_\mu(\tau,0) = 0$.

Thus, the Dirac approach [201] is to define the field tensor as

$$F_{\mu\nu} = \partial_\mu A_\nu - \partial_\nu A_\mu - \widetilde{G}_{\mu\nu},$$

where the singular part $\widetilde{G}_{\mu\nu}$ (the dual of $G_{\mu\nu}$) is associated with the field of the string and is required to satisfy the relation $\partial^\mu G_{\mu\nu} = j_\nu^{(g)}$. Making use of the magnetic current definition (4.2), we can write this tensor as

$$G_{\mu\nu}(x,y) = g \int d\tau d\sigma \left(\frac{\partial y_\mu}{\partial \tau}\frac{\partial y_\nu}{\partial \sigma} - \frac{\partial y_\mu}{\partial \sigma}\frac{\partial y_\nu}{\partial \tau} \right) \delta(x - y(\tau,\sigma)). \quad (4.3)$$

Evidently, this definition of the field tensor solves the problem of obtaining an action functional, the stationary point of which is given by the system of equations (4.1):

$$S = -\frac{1}{4} \int d^4x F_{\mu\nu}^2 + \int d^4x A^\mu j_\mu^{(e)} + m_e \int d\tau^{(e)} + m_g \int d\tau^{(g)}. \quad (4.4)$$

Indeed, variation of this action with respect to the potential A_μ, as usual gives the first field equation $\partial_\mu F^{\mu\nu} = j^{(e)\nu}$, and the second field equation for the dual field strength tensor appears as a constraint imposed by the definition (4.3). Moreover, the variation of the action (4.4) with respect to the trajectories of the magnetic charges leads to the equation of motion

$$m_g \frac{du_\mu^{(g)}}{d\tau^{(g)}} = g\widetilde{F}_{\mu\nu} u^{(g)\nu},$$

as anticipated.

However, the variation of the action (4.4) with respect to the electric charge trajectories gives

$$m_e \frac{du_\mu^{(e)}}{d\tau^{(e)}} = e \left(\partial_\mu A_\nu - \partial_\nu A_\mu \right)\big|_{x=x^{(e)}} u^{(e)\nu}, \quad (4.5)$$

which is not exactly the equation of motion of an electric charge that was postulated in the system (4.1). The form of the Lorentz force acting on the electric charge can be reproduced only if $G_{\mu\nu}(x^{(e)}) = 0$. According to the definition (4.3), this condition can be satisfied if $x_\mu^{(e)}(\tau) \ne y_\mu(\tau,\sigma)$, that is, the trajectory of the electric charge must not touch the string. This restriction is known as *"Dirac's veto"* [201].

Note that "Dirac's veto" is not an external constraint imposed on the dynamics of the system, but it follows from the variational formalism. It can be obtained after variation of the action (4.4) with respect to the string variable n_μ, which yields

$$\frac{\partial F_{\mu\nu}}{\partial y_\mu} = 0. \tag{4.6}$$

Since the field strength tensor obeys the equation $\partial_\mu F^{\mu\nu} = j^{(e)\nu}$, this relation means that the current of an electric charge vanishes on the string: $j_\nu^{(e)}(y) = 0$. Thus, a sufficient condition for self-consistency of the system of equations, which corresponds to the stationary point of the action (4.4), can be achieved only if "Dirac's veto is fulfilled". On the other hand, this system does not contain dynamical equations for the string itself. This corresponds to the non-physical nature of the string variables; recall that in Chap. 1 we saw that the position of the singular string is defined up to a gauge transformation of the electromagnetic potential $A_\mu(x)$.

In principle, the Dirac string can be attached not to a magnetic, but to an electric charge. This mutual interchange between electric and magnetic characteristics reflects the dual symmetry of the system of dynamical Maxwell–Lorentz equations (4.1). However, this symmetry, which is obvious at the level of the field variables, becomes subtle when we go to the potentials of the electromagnetic field. Indeed, the interaction term $L_{\text{int}} = A^\mu j_\mu^{(e)}$ in the Lagrangian (4.4) describes the minimal interaction between an electric charge and the electromagnetic field A_μ. Recovering the dual invariance at this level is possible within the framework of the two-potential formalism suggested in the 1960s in the papers [148, 548, 550].

4.2 Two-Potential Formulation of Electrodynamics

We can try to construct a naive dual-invariant Lagrangian formulation of Abelian electrodynamics with two types of charges, by making use of a non-standard definition of the electromagnetic field strength tensor, which shall include not only the 4-vector potential A_μ, but also a pseudovector $\widetilde{A}_\mu$ [148]:

$$\begin{aligned} F_{\mu\nu} &= \partial_\mu A_\nu - \partial_\nu A_\mu - \varepsilon_{\mu\nu\rho\sigma}\partial^\rho \widetilde{A}^\sigma, \\ \widetilde{F}_{\mu\nu} &= \partial_\mu \widetilde{A}_\nu - \partial_\nu \widetilde{A}_\mu + \varepsilon_{\mu\nu\rho\sigma}\partial^\rho A^\sigma. \end{aligned} \tag{4.7}$$

This obviously allows us to introduce dual rotations of the potentials

4.2 Two-Potential Formulation of Electrodynamics

$$D: \begin{cases} A_\mu \to A_\mu \cos\theta - \widetilde{A}_\mu \sin\theta, \\ \widetilde{A}_\mu \to A_\mu \sin\theta + \widetilde{A}_\mu \cos\theta, \end{cases} \quad (4.8)$$

which are compatible with (1.76). However, the number of physical degrees of freedom of the electromagnetic field does not change, i.e., the potentials A_μ and $\widetilde{A}_\mu$ are not independent dynamical variables, but must be connected via an additional constraint. In the simplest possible case it may be written as

$$\partial_\mu A_\nu - \partial_\nu A_\mu + \varepsilon_{\mu\nu\rho\sigma} \partial^\rho \widetilde{A}^\sigma = 0,$$

which actually eliminates one of the potentials completely. We can come to same result in a somewhat more elegant way if we note that the condition

$$e \widetilde{A}_\mu - g A_\mu = 0$$

allows complete elimination of the potential $\widetilde{A}_\mu$ after the dual transformation (4.8) with the parameter $\theta = -\arctan(g/e)$. Thus, effectively, this transformation reduces the system of equations (4.1) to the case of standard one-charge electrodynamics [60, 380, 493]. Therefore, we consider another way to construct a local two-potential Lagrangian formulation of electrodynamics developed by D. Zwanziger [548, 550].

First, note that a general solution of the equation of the dual field strength $\partial_\mu \widetilde{F}^{\mu\nu} = j^{(g)\nu}$ formally can be written as

$$F_{\mu\nu} = \partial_\mu A_\nu - \partial_\nu A_\mu + \frac{1}{2} \varepsilon_{\mu\nu\rho\sigma} (n^\alpha \cdot \partial_\alpha)^{-1} n^\rho j^{(g)\sigma}, \quad (4.9)$$

where n_α is an arbitrary fixed unit four-vector, which can be chosen, for example, to be $n_\alpha = (0, \mathbf{n})$, and the integral operator $(n^\alpha \cdot \partial_\alpha)^{-1}$ is defined as a resolvent of the equation

$$(n^\alpha \cdot \partial_\alpha) \left(n^\beta \cdot \partial_\beta\right)^{-1} (x) = \delta^{(4)}(x).$$

Clearly, the physical observables shall be independent of the vector n_α and variation of the direction of n_α corresponds to the gauge degrees of freedom of the theory. One can visualize this vector as the direction in space along which all Dirac strings are directed.

Obviously, the two local terms in the expression (4.9) correspond to the solution of the homogeneous Maxwell equation in vacuum. The non-local term corresponds to a particular solution in the presence of an external current of magnetic charge $j^{(g)\mu}$, where the vector n_α defines the position of the singularity line. When this vector n_α vanishes, one automatically recovers the standard formulation.

A particular representation of the operator $(n^\alpha \partial_\alpha)^{-1}$ can, for example, be taken as [41, 550]

$$(\mathbf{n} \cdot \boldsymbol{\nabla})^{-1}(x) = [a\theta(\mathbf{n} \cdot \mathbf{x}) - (1-a)\theta(-\mathbf{n} \cdot \mathbf{x})] \prod_{k=1}^{2} \delta(\boldsymbol{\tau}^{(k)} \cdot \mathbf{x}), \quad (4.10)$$

where $\tau^{(k)}$ are two unit vectors, which are orthogonal to $\mathbf{n}$ and to each other, and the parameter a defines the type of singularity. Choosing $a = 0$, we arrive at the semi-infinite Dirac string, while the case $a = 1/2$ corresponds to the infinite Schwinger string in the expression (1.61)[2].

Furthermore, a general solution of the first field equation $\partial_\mu F^{\mu\nu} = j^{(e)\nu}$ can be formally written as

$$\widetilde{F}_{\mu\nu} = \partial_\mu \widetilde{A}_\nu - \partial_\nu \widetilde{A}_\mu - \frac{1}{2}\varepsilon_{\mu\nu\rho\sigma}(n^\alpha \cdot \partial_\alpha)^{-1} n^\rho j^{(e)\sigma}, \quad (4.11)$$

where the pseudovector $\widetilde{A}_\mu$ is the second potential. Obviously, the potentials A_μ and $\widetilde{A}_\mu$ are independent dynamical variables, but they are connected by some non-local relation. The difference from the naive two-potential formulation discussed above is that now the dual transformations could rotate the system (4.1) to standard electrodynamics only if $u_\mu^{(e)} = u_\mu^{(g)}$, that is if we consider a dyon having both electric and magnetic charges.

To complete the definition of the two-potential electrodynamics, note that the non-local terms in the definition of the field strength tensor, (4.9) and (4.11), can be excluded. Indeed, an arbitrary antisymmetric second rank tensor can be written as

$$F_{\mu\nu} = \tfrac{1}{2}\hat{n}^\alpha \left[\hat{n}_\mu F_{\alpha\nu} - \hat{n}_\nu F_{\alpha\mu} - \varepsilon_{\mu\nu\rho\sigma}\hat{n}^\rho \widetilde{F}_\alpha{}^\sigma \right]. \quad (4.12)$$

Since the formal definitions (4.9) and (4.11) lead to the relations

$$\hat{n}^\mu F_{\mu\nu} = \hat{n}^\mu \left[\partial_\mu A_\nu - \partial_\nu A_\mu \right],$$
$$\hat{n}^\mu \widetilde{F}_{\mu\nu} = \hat{n}^\mu \left[\partial_\mu \widetilde{A}_\nu - \partial_\nu \widetilde{A}_\mu \right], \quad (4.13)$$

the decomposition (4.12) immediately provides us with a local two-potential form of the electromagnetic field strength tensor [41, 550]

$$F_{\mu\nu} = \hat{n}^\alpha \left[\hat{n}_\mu (\partial_\alpha A_\nu - \partial_\nu A_\alpha) - \tfrac{1}{2}\varepsilon_{\mu\nu\rho\sigma}\hat{n}^\rho \left(\partial_\alpha \widetilde{A}^\sigma - \partial^\sigma \widetilde{A}_\alpha \right) \right],$$
$$\widetilde{F}_{\mu\nu} = \hat{n}^\alpha \left[\hat{n}_\mu \left(\partial_\alpha \widetilde{A}_\nu - \partial_\nu \widetilde{A}_\alpha \right) + \tfrac{1}{2}\varepsilon_{\mu\nu\rho\sigma}\hat{n}^\rho \left(\partial_\alpha A^\sigma - \partial^\sigma A_\alpha \right) \right]. \quad (4.14)$$

Substitution of the relations (4.14) into the field equations leads to a rather cumbersome system of coupled equations

---

[2] An alternative representation is

$$(\mathbf{n} \cdot \boldsymbol{\nabla})^{-1}(x) = \frac{1}{2}\left[\frac{1}{(n_i \partial_i) + i\varepsilon} + \frac{1}{(n_i \partial_i) - i\varepsilon} \right].$$

A four-dimensional, useful generalization of (4.10), is given as

$$(n^\alpha \partial_\alpha)^{-1}(x) = \tfrac{1}{2}\int_0^\infty ds\, [\delta(x + sn) - \delta(x - sn)].$$

This formula will be used in the following calculations.

4.2 Two-Potential Formulation of Electrodynamics

$$(\hat{n}^\alpha \partial_\alpha)(\hat{n}^\beta \partial_\beta) A_\mu - (\hat{n}^\alpha \partial_\alpha)(\partial_\mu \hat{n}^\beta A_\beta) - \hat{n}_\mu (\hat{n}^\alpha \partial_\alpha)(\hat{n}^\beta A_\beta)$$
$$+ \hat{n}_\mu \Box (\hat{n}^\alpha A_\alpha) - (\hat{n}^\alpha \partial_\alpha)\varepsilon_{\mu\nu\rho\sigma} \hat{n}^\nu \partial^\rho \widetilde{A}^\sigma = j_\mu^{(e)},$$

$$(\hat{n}^\alpha \partial_\alpha)(\hat{n}^\beta \partial_\beta) \widetilde{A}_\mu - (\hat{n}^\alpha \partial_\alpha)(\partial_\mu \hat{n}^\beta \widetilde{A}_\beta) - \hat{n}_\mu (\hat{n}^\alpha \partial_\alpha)(\hat{n}^\beta \widetilde{A}_\beta)$$
$$+ \hat{n}_\mu \Box (\hat{n}^\alpha \widetilde{A}_\alpha) + (\hat{n}^\alpha \partial_\alpha)\varepsilon_{\mu\nu\rho\sigma} \hat{n}^\nu \partial^\rho A^\sigma = j_\mu^{(g)}, \qquad (4.15)$$

which generalize the well-known relation $\Box A_\mu - \partial_\mu \partial^\nu A_\nu = j_\mu$. Fixing a gauge, we can simplify these awkward equations (4.15) a lot.

Now we can finally write a local and dual invariant Lagrangian, the variation of which with respect to the dynamical variables leads to the system of equations (4.15):

$$L = L_{\text{em}} + L_{\text{int}} + L_m$$
$$= -\frac{1}{2}\left[\hat{n}^\mu (\partial_\mu A_\nu - \partial_\nu A_\mu)\right]^2 - \frac{1}{2}\left[\hat{n}^\mu \left(\partial_\mu \widetilde{A}_\nu - \partial_\nu \widetilde{A}_\mu\right)\right]^2$$
$$-\frac{1}{4}\varepsilon_{\mu\nu\rho\sigma}\hat{n}^\nu \partial^\rho \widetilde{A}^\sigma \hat{n}^\alpha \partial_\alpha A^\mu + \frac{1}{4}\varepsilon_{\mu\nu\rho\sigma}\hat{n}^\nu \partial^\rho A^\sigma \hat{n}^\alpha \partial_\alpha \widetilde{A}^\mu$$
$$- j_\mu^{(g)} \widetilde{A}^\mu - j_\mu^{(e)} A^\mu + L_m, \qquad (4.16)$$

where $L_{\text{int}} = -j_\mu^{(g)} \widetilde{A}^\mu - j_\mu^{(e)} A^\mu$ is an interaction Lagrangian and L_m is a Lagrangian of the matter field, which, for example, in the case of spinor particles is

$$L_m = \bar{\psi}\left(i\gamma^\mu \partial_\mu - m\right)\psi, \qquad (4.17)$$

and the currents of electric and magnetic particles are

$$j_\mu^{(e)} = e\bar{\psi}_e \gamma_\mu \psi_e, \qquad j_\mu^{(g)} = g\bar{\psi}_g \gamma_\mu \psi_g,$$

respectively. For a dyon of a mass m, a covariant derivative D_μ, which includes both potentials, appears in the dynamical equation for the matter field:

$$(i\gamma^\mu D_\mu - m)\psi \equiv \left[i\gamma^\mu \left(\partial_\mu + ieA_\mu + ig\widetilde{A}_\mu\right) - m\right]\psi = 0. \qquad (4.18)$$

In Sect. 4.6 we shall consider in more detail some properties of this equation, which will be used in an analysis of quantum effects in the monopole background field.

Finally, we would like to note that the kinetic part L_{em} of the Lagrangian (4.16) does not have the common form $L_{\text{em}} \sim F_{\mu\nu}^2$. The action (4.16), which is local, but depends on the vector n_μ, can be re-written in terms of the field variables. It turns out to be equal to the non-local Schwinger description [135]. In other words, this is a way to establish a correspondence between the Schwinger and Zwanziger formulations.

4.2.1 Energy-Momentum Tensor and Angular Momentum

The Lagrangian (4.16) can be used to derive an expression for a conserved canonical energy-momentum tensor [550]. Following the standard procedure, let us consider an arbitrary variation of the Lagrangian (4.16)

116   4 Abelian Monopole: Relativistic Quantum Theory

$$\delta L = \partial_\mu \left(\frac{\partial L}{\partial(\partial_\mu A_\nu)} \delta A_\nu \right) + \partial_\mu \left(\frac{\partial L}{\partial(\partial_\mu \tilde{A}_\nu)} \delta \tilde{A}_\nu \right) + \partial_\mu \left(\frac{\partial L}{\partial(\partial_\mu \psi)} \delta \psi \right). \quad (4.19)$$

Under infinitesimal displacements the Lagrangian density changes as $\delta L = a^\mu \partial_\mu L$, where the 4-vector a_μ is a parameter of the transformation. In the same way

$$\delta A_\mu = a^\nu \partial_\nu A_\mu, \quad \delta \tilde{A}_\mu = a^\nu \partial_\nu \tilde{A}_\mu, \quad \delta \psi = a^\nu \partial_\nu \psi,$$

and therefore

$$a^\mu \partial_\mu L = a^\alpha \partial_\mu \left(\frac{\partial L}{\partial(\partial_\mu A_\nu)} \partial_\alpha A_\nu + \frac{\partial L}{\partial(\partial_\mu \tilde{A}_\nu)} \partial_\alpha \tilde{A}_\nu + \frac{\partial L}{\partial(\partial_\mu \psi)} \partial_\alpha \psi \right), \quad (4.20)$$

that is the conserved canonical energy-momentum tensor is

$$\partial_\mu T^{\mu\nu} = 0, \quad (4.21)$$

$$T^{\mu\nu} = \frac{\partial L}{\partial(\partial_\mu A_\lambda)} \partial^\nu A^\lambda + \frac{\partial L}{\partial(\partial_\mu \tilde{A}_\lambda)} \partial^\nu \tilde{A}^\lambda + \frac{\partial L}{\partial(\partial_\mu \psi)} \partial^\nu \psi - \eta^{\mu\nu} L,$$

and the model is obviously translational invariant.

The situation with rotational, and with relativistic invariance in general, is a bit more complicated. The problem is that the Lagrangian (4.16) includes a four-vector n_α, which under transformations $\Lambda_{\alpha\beta}$ of the Lorentz group transforms as $n_\alpha \to n'_\alpha = \Lambda_{\alpha\beta} n^\beta$, or, in infinitesimal form

$$n_\alpha \to n_\alpha + \omega_{\alpha\beta} n^\beta, \quad (4.22)$$

where $\omega_{\alpha\beta}$ is an antisymmetric tensor. The coordinates transform in the same way, that is $\delta x_\mu = \omega_{\mu\nu} x^\nu$, and therefore the transformations of the fields are

$$A_\mu(x_\nu) \to A_\mu(x_\nu) + x_\alpha \omega^{\alpha\beta} \partial_\beta A_\mu(x_\nu) + \omega_{\mu\alpha} A^\alpha(x_\nu)$$
$$\equiv A_\mu(x_\nu) + \frac{1}{2} \omega_{\alpha\beta} m^{\alpha\beta}_{(A)} A_\mu(x_\nu),$$

$$\tilde{A}_\mu(x_\nu) \to \tilde{A}_\mu(x_\nu) + x_\alpha \omega^{\alpha\beta} \partial_\beta \tilde{A}_\mu(x_\nu) + \omega_{\mu\alpha} \tilde{A}^\alpha(x_\nu)$$
$$\equiv \tilde{A}_\mu(x_\nu) + \frac{1}{2} \omega_{\alpha\beta} m^{\alpha\beta}_{(\tilde{A})} \tilde{A}_\mu(x_\nu),$$

$$\psi(x_\nu) \to \psi(x_\nu) + x_\alpha \omega^{\alpha\beta} \partial_\beta \psi(x_\nu) + \frac{1}{4} \gamma^\mu \omega_{\mu\nu} \gamma^\nu \psi(x_\nu)$$
$$\equiv \psi(x_\nu) + \frac{1}{2} \omega_{\alpha\beta} m^{\alpha\beta}_{(\psi)} \psi(x_\nu). \quad (4.23)$$

For the sake of brevity, let us introduce an abbreviation $\phi^a(x) = (A_\mu, \tilde{A}_\mu, \psi)$. Then, the variation of the Lagrangian density (4.16), which is given as $L(x) = L(\phi^a(x), \partial \phi^a(x), n)$, can be written as

4.2 Two-Potential Formulation of Electrodynamics

$$\delta L = \partial_\mu \sum_a \left(\frac{\partial L}{\partial(\partial_\mu \phi^a)} \delta \phi^a \right), \qquad (4.24)$$

that is, for the infinitesimal Lorentz transformations (4.23) we have

$$x_\mu \omega^{\mu\nu} \partial_\nu L = \frac{1}{2} \partial_\mu \sum_a \left(\frac{\partial L}{\partial(\partial_\mu \phi^a)} \omega_{\alpha\beta} m_a^{\alpha\beta} \phi^a \right) + n_\mu \omega^{\mu\nu} \frac{\partial L}{\partial n_\nu}. \qquad (4.25)$$

Now we can define the tensor of angular momentum by analogy with ordinary electrodynamics as:

$$\begin{aligned} M^{\mu\nu\rho} &= \sum_a \frac{\partial}{\partial(\partial_\rho \phi^a)} m_a^{\mu\nu} \phi^a - (x^\mu \eta^{\nu\rho} + x^\nu \eta^{\mu\rho}) L \\ &= x^\mu T^{\rho\nu} - x^\nu T^{\rho\mu} + S^{\mu\nu\rho}, \end{aligned} \qquad (4.26)$$

where we make use of the definition of the energy-momentum tensor $T^{\mu\nu}$ (4.21) and define the usual tensor of an intrinsic angular momentum

$$\omega_{\mu\nu} S^{\mu\nu\rho} = \frac{\partial L}{\partial(\partial_\rho A^\alpha)} \omega^{\alpha\beta} A_\beta + \frac{\partial L}{\partial(\partial_\rho \widetilde{A}^\alpha)} \omega^{\alpha\beta} \widetilde{A}_\beta + \frac{1}{4} \frac{\partial L}{\partial(\partial_\rho \psi)} \gamma_\alpha \omega^{\alpha\beta} \gamma_\beta \psi.$$

The relation (4.25) then yields

$$\partial_\rho M^{\mu\nu\rho} = n^\mu \frac{\partial L}{\partial n_\nu} - n^\nu \frac{\partial L}{\partial n_\mu}. \qquad (4.27)$$

Thus, at a first glance, such an energy momentum tensor which we introduced in the canonical way, is not conserved. Obviously, the underlying reason is that the Lagrangian density (4.16) depends on a fixed vector n_μ, that is the rotational invariance of the model is violated.

Let us recall that we already encountered a problem of the same nature when we considered the algebra of the operators of non-relativistic angular momentum (2.12). We have shown there that the correct treatment of the corresponding singularity leads to cancellation of the anomalous terms (2.15). We could come to the same conclusion, if we take into account that the spatial rotations in electrodynamics are defined up to a gauge transformation that effectively restores rotational invariance of the theory [135]. Later we shall return to the general proof of relativistic invariance of two-charge electrodynamics.

Let us consider the question as to under which conditions the anomalous right-hand side of the expression (4.27) could vanish. Exploiting the relation (4.16) we can rewrite (4.27) in the form

$$\begin{aligned} & n^\mu \frac{\partial L}{\partial n_\nu} - n^\nu \frac{\partial L}{\partial n_\mu} \\ & = (-n^\mu n^\alpha \varepsilon_{\alpha\beta\rho\nu} + n^\nu n^\alpha \varepsilon_{\alpha\beta\rho\mu}) (n^\sigma \cdot \partial_\sigma)^{-1} j_{(e)}^\beta (n^\tau \cdot \partial_\tau)^{-1} j_{(g)}^\rho, \end{aligned}$$

where we make use of the general solutions (4.9) and (4.11) of the field equations.

In the simplest case of a couple of dyons with charges (e_1, g_1) and (e_2, g_2), respectively, this relation is proportional to the combination of charges

$$\mu = e_1 g_2 - e_2 g_1,$$

which obviously vanishes if $e_1/g_1 = e_2/g_2$, or if in the general case of i particles $e_i/g_i = $ constant [550]. However, we have seen that when this condition is satisfied, the dual rotation of the variables allows us to transform the model to standard electrodynamics of effective charges $q_i = \sqrt{e_i^2 + g_i^2}$ [60, 380, 493]. Thus, the model has no room for genuine monopoles, and we will not consider this situation further.

The definition of the angular momentum tensor (4.26) that we obtained above, by simple analogy with standard electrodynamics makes it possible to write the symmetrized energy-momentum tensor[3] [550]:

$$\begin{aligned}\theta^{\mu\nu} &= \frac{1}{2}\left(F^{\mu\nu}F_{\mu\nu} + \widetilde{F}^{\mu\nu}\widetilde{F}_{\mu\nu}\right) \\ &+ \frac{1}{2}\left[\bar\psi\gamma^\mu\left(\partial_\nu + ieA_\nu + ig\widetilde{A}_\nu\right) + \gamma^\nu\left(\partial_\mu + ieA_\mu + ig\widetilde{A}_\mu\right)\right]\psi \\ &- n^\mu n^\alpha \varepsilon_{\alpha\beta\rho\nu}(n^\sigma \cdot \partial_\sigma)^{-1} j_{(e)}^\beta (n^\tau \cdot \partial_\tau)^{-1} j_{(g)}^\rho. \end{aligned} \qquad (4.28)$$

The first two terms here exactly correspond to the naive energy-momentum tensor that could be obtained by a straightforward dual rotation of the standard electrodynamics of a massless spinor particle. The last non-local term is due to the anomalous part of the expression (4.27).

4.3 Canonical Quantization

In order to construct a self-consistent canonical quantization scheme of two-charge electrodynamics, it is necessary to identify the physical degrees of freedom of the electromagnetic field. The problem is that the two-potential dual invariant reformulation of electrodynamics is possible only at the cost of formal invariance of the model with respect to the extended gauge group $U(1)_{(e)} \times U(1)_{(g)}$, that is

$$\begin{aligned} A_\mu &\to A_\mu - \frac{i}{e} U_{(e)}^{-1} \partial_\mu U_{(e)} = A_\mu + \partial_\mu \lambda_{(e)}, \\ \widetilde{A}_\mu &\to \widetilde{A}_\mu - \frac{i}{g} U_{(g)}^{-1} \partial_\mu U_{(g)} = \widetilde{A}_\mu + \partial_\mu \lambda_{(g)}, \\ \psi &\to U_{(e)} U_{(g)} \psi = \exp\{ie\lambda_{(e)} + ig\lambda_{(g)}\}\psi. \end{aligned} \qquad (4.29)$$

---

[3] Here we again consider spinor matter fields.

Therefore, even fixing a "double" Lorentz gauge $\partial^\mu A_\mu = \partial^\mu \widetilde{A}_\mu = 0$ we would still have four transversal degrees of freedom of the electromagnetic field, twice as many as we need.

To tackle this problem, D. Zwanziger [550] noted that there is an important difference between standard electrodynamics, where a potential A_μ satisfies a second-order wave equation, and the two-potential formulation. In the latter approach, the number of physical degrees of freedom of the electromagnetic field does not change if the potentials A_μ and $\widetilde{A}_\mu$ satisfy not the equations (4.15), but two other equations of first-order. The gauge condition must reduce equations (4.15) to these two first-order equations.

Indeed, were it not for terms $\Box(\hat{n}^\alpha A_\alpha)$ and $\Box(\hat{n}^\alpha \widetilde{A}_\alpha)$, the left-hand sides of (4.15) would factorize into two differential operators of first-order. Therefore, the natural choice is to fix the generalized axial gauge

$$\hat{n}^\alpha A_\alpha = \hat{n}^\alpha \widetilde{A}_\alpha = 0, \qquad (4.30)$$

which solves the problem of separating the physical degrees of freedom of the electromagnetic field in the two-potential formalism. Indeed, the projection of the equations (4.15) onto a unit time-like four-vector τ_μ, which is orthogonal to n_μ, yields

$$(\hat{n}^\alpha \partial_\alpha) \left[(\hat{n}^\beta \partial_\beta)(\tau^\mu A_\mu) - \tau^\mu \varepsilon_{\mu\nu\rho\sigma} \hat{n}^\nu \partial^\rho) \widetilde{A}^\sigma \right] = \tau^\mu j_\mu^{(e)},$$
$$(\hat{n}^\alpha \partial_\alpha) \left[(\hat{n}^\beta \partial_\beta)(\tau^\mu \widetilde{A}_\mu) + \tau^\mu \varepsilon_{\mu\nu\rho\sigma} \hat{n}^\nu \partial^\rho) A^\sigma \right] = \tau^\mu j_\mu^{(g)}, \qquad (4.31)$$

where we take into account the gauge condition (4.30). Since the expressions (4.31) contain only derivatives in directions orthogonal to the vector τ_μ, there are only two degrees of freedom left.

D. Zwanziger used a similar approach in order to construct a scheme of canonical quantization of two-potential electrodynamics. He suggested to supplement the Lagrangian density (4.16) by a gauge-fixing term

$$L_g = \frac{1}{2} \left\{ [\partial_\mu(\hat{n}^\nu A_\nu)]^2 + [\partial_\mu(\hat{n}^\nu \widetilde{A}_\nu)]^2 \right\}, \qquad (4.32)$$

that is the Lagrangian of the electromagnetic field can be written as

$$L_{em} + L_g = -\frac{1}{4}\varepsilon_{\mu\nu\rho\sigma}\hat{n}^\nu \partial^\rho \widetilde{A}^\sigma \hat{n}^\alpha \partial^\alpha A_\mu + \frac{1}{4}\varepsilon_{\mu\nu\rho\sigma}\hat{n}^\nu \partial^\rho A^\sigma \hat{n}^\alpha \partial_\alpha \widetilde{A}^\mu$$
$$- \frac{1}{2}[(\hat{n}^\nu \partial_\nu A_\mu)^2 - 2(\hat{n}^\nu \partial_\nu A_\mu)(\partial^\mu \hat{n}^\rho A_\rho) \qquad (4.33)$$
$$+ (\hat{n}^\nu \partial_\nu \widetilde{A}_\mu)^2 - 2(\hat{n}^\nu \partial_\nu \widetilde{A}_\mu)(\partial^\mu \hat{n}^\rho \widetilde{A}_\rho)].$$

It is easy to see that the variation of L_g with respect to the potentials gives the equation

$$\Box \left(\hat{n}^\nu A_\nu + \hat{n}^\nu \widetilde{A}_\nu \right) = 0 ,$$

which obviously plays the role of the axial gauge (4.30).

Now the complete Lagrangian of the system of interacting fields and charges $L = L_{em} + L_{int} + L_m + L_g$ provides the factorized dynamical equations (cf. (4.15)):

$$(\hat{n}^\alpha \partial_\alpha)\left[(\hat{n}^\nu \partial_\nu)A_\mu - \partial_\mu(\hat{n}^\nu A_\nu) - \hat{n}_\mu(\partial^\nu A_\nu) - \varepsilon_{\mu\nu\rho\sigma}\hat{n}^\nu \partial^\rho \widetilde{A}^\sigma \right] = j^{(e)}_\mu ,$$

$$(\hat{n}^\alpha \partial_\alpha)\left[(\hat{n}^\nu \partial_\nu)\widetilde{A}_\mu - \partial_\mu(\hat{n}^\nu \widetilde{A}_\nu) - \hat{n}_\mu(\partial^\nu \widetilde{A}_\nu) + \varepsilon_{\mu\nu\rho\sigma}\hat{n}^\nu \partial^\rho A^\sigma \right] = j^{(g)}_\mu .$$

(4.34)

Let us consider the quantization scheme of a free electromagnetic field without sources. Then the system (4.34) can be written in compact form as

$$(\hat{n}^\alpha \partial_\alpha)\left[(\hat{n}^\nu \partial_\nu)V^a_\mu - \partial_\mu(\hat{n}^\nu V_\nu)^a - \hat{n}_\mu(\partial^\nu V_\nu)^a - \varepsilon_{ab}\varepsilon_{\mu\nu\rho\sigma}\hat{n}^\nu \partial^\rho (V^\sigma)^b \right] = 0 ,$$

(4.35)

where we introduced a generalized matrix potential $(V_\mu)^a \equiv (A_\mu, \widetilde{A}_\mu)$, $a = 1, 2$. Thus we have two equations of first-order in time derivatives for the 8 independent variables A_μ and $\widetilde{A}_\mu$.

We now have to identify the pairs of canonically conjugate variables. To do this, we write the Lagrangian in the standard form

$$L = \sum \pi_a \dot{\phi}_a - H ,\qquad (4.36)$$

where the multiplet ϕ^a now contains the fields $A_\mu, \widetilde{A}_\mu$. Since the corresponding Hamiltonian is [550]

$$H = \frac{1}{2}\Big\{[\boldsymbol{\nabla} \times \mathbf{A}]^2 + [\boldsymbol{\nabla} \times \widetilde{\mathbf{A}}]^2 \qquad (4.37)$$

$$- \left(\hat{\mathbf{n}} \cdot \boldsymbol{\nabla} A_0 - \hat{\mathbf{n}} \cdot [\boldsymbol{\nabla} \times \widetilde{\mathbf{A}}]\right)^2 + \left(\hat{\mathbf{n}} \cdot \boldsymbol{\nabla} \widetilde{A}_0 + \hat{\mathbf{n}} \cdot [\boldsymbol{\nabla} \times \mathbf{A}]\right)^2$$

$$+ (\boldsymbol{\nabla}\hat{\mathbf{n}} \cdot \mathbf{A})^2 + (\boldsymbol{\nabla}\hat{\mathbf{n}} \cdot \widetilde{\mathbf{A}})^2 \Big\} ,$$

a straight comparison with (4.33) yields

$$\phi_1 = A_1, \quad \phi_2 = \widetilde{A}_1, \quad \phi_3 = A_3, \quad \phi_4 = \widetilde{A}_3, \qquad (4.38)$$
$$\widetilde{A}_2 = -\partial_3^{-1}\pi_1, \quad A_2 = \partial_3^{-1}\pi_2, \quad A_0 = \partial_3^{-1}\pi_3, \quad \widetilde{A}_0 = \partial_3^{-1}\pi_4,$$

where, for the sake of simplicity, we choose the vector $\mathbf{n}$ directed along the third axis.

Thus the equal-time canonical commutation relations

$$[\phi_a(\mathbf{x}), \phi_b(\mathbf{y})] = [\pi_a(\mathbf{x}), \pi_b(\mathbf{y})] = 0, \qquad (4.39)$$
$$[\pi_a(\mathbf{x}), \phi_b(\mathbf{y})] = -i\delta_{ab}\delta(\mathbf{x} - \mathbf{y}),$$

provide non-local commutation relations between the potentials [550]:

$$[A_\mu(\mathbf{x}), A_\nu(\mathbf{y})] = \left[\widetilde{A}_\mu(\mathbf{x}), \widetilde{A}_\nu(\mathbf{y})\right] = -i(\delta_\mu^0 \hat{n}_\nu + \delta_\nu^0 \hat{n}_\mu)(\hat{n}^\alpha \partial_\alpha)^{-1}(\mathbf{x} - \mathbf{y}),$$
$$\left[A_\mu(\mathbf{x}), \widetilde{A}_\nu(\mathbf{y})\right] = i\varepsilon_{\mu\nu\rho 0}\hat{n}^\rho(\hat{n}^\alpha \partial_\alpha)^{-1}(\mathbf{x} - \mathbf{y}). \qquad (4.40)$$

Both electric and magnetic fields satisfy the standard electrodynamical relations

$$[E_m(\mathbf{x}), E_n(\mathbf{y})] = [B_m(\mathbf{x}), B_n(\mathbf{y})] = 0,$$
$$[B_m(\mathbf{x}), B_n(\mathbf{y})] = i\varepsilon_{mnk}\nabla_k \delta(\mathbf{x} - \mathbf{y}), \qquad (4.41)$$

without anomalous contributions.

Let us recall that although there are eight canonically conjugated variables, the pairs of canonical coordinates $(\hat{n}^\mu A_\mu)$, $(\hat{n}^\mu \widetilde{A}_\mu)$ and conjugated momenta must satisfy the free equations. Therefore there are actually only four dynamical variables over the configuration space, i.e., only two physical degrees of freedom left.

4.3.1 Relativistic Invariance of Two-Charge Electrodynamics

Obviously, the price paid for using such a perverted two-potential formulation in quantum electrodynamics instead of the standard quantization scheme, is that the commutation relations (4.40) depend on a fixed vector **n**. Moreover, we have seen that in two-charge electrodynamics, the standard tensor of angular momentum is not conserved. Thus, generally speaking, we have to prove that such a generalization is compatible with the relativistic invariance of the theory.

Let us recall that according to the Schwinger formulation [458], the condition of Lorenz-invariance is

$$[\theta^{00}(x), \theta^{00}(y)] = -i\left(T^{0k}(x) + T^{k0}(y)\right)\partial_k \delta^{(4)}(x - y). \qquad (4.42)$$

At first glance, this is not satisfied. Indeed, if we choose $n_0 = 0$, the contribution of the last term in (4.28) to the naive commutation relations, which are defined without a proper point-splitting regularization, would be (cf. (2.13)) [41]

$$[\theta^{00}(x), \theta^{00}(y)] = -i\left(T^{0k}(x) + T^{k0}(y)\right)\partial_k \delta(x - y) \qquad (4.43)$$
$$+ \varepsilon_{ijk} j_{(e)}^i(x) n^j (n^\sigma \cdot \partial_\sigma)^{-1}(x, y) j_{(g)}^k(y)$$

$$-\varepsilon_{ijk}j^i_{(e)}(y)n^j\,(n^\sigma\cdot\partial_\sigma)^{-1}(y,x)j^k_{(g)}(x)\,.$$

However, we already noted that in the monopole theory, spatial rotations are defined up to a complementary unitary transformation that rotates the vector **n** in space [301]. The details of this mechanism of restoration of the effective rotational invariance of a non-relativistic quantum theory were discussed in Chap. 1 (p. 19). With minor changes, the same arguments can also be used to prove the effective relativistic invariance of the model with monopoles. It was shown by Schwinger [459] that the commutation relations of the type (4.42) are satisfied in two-charge quantum electrodynamics.

Moreover, the charge quantization condition again arises from the phase factor of the corresponding Abelian gauge transformation. Here we will not discuss the argument by Schwinger in more detail. We shall turn instead to more general and elegant considerations in terms of the path integral formulation of the theory, as was suggested in [136, 137]. This formalism is especially helpful, because it easily allows us to write a formal expression for propagators in complete analogy with standard QED. Hence, we can try to make use of the common language of Feynman diagrams.

Following [41, 136], let us consider a simplified version of a scalar model, extending the Lagrangian (4.16)

$$L = L_{\text{em}}(A, \widetilde{A}, n) + D_\mu\phi^* D^\mu\phi - m^2\phi^*\phi\,, \quad (4.44)$$

where ϕ is a complex scalar field representing a dyon, $L_{\text{em}}(A, \widetilde{A}, n)$ is the Zwanziger action of free fields appearing in (4.16), and the covariant derivative is $D_\mu\phi = (\partial_\mu + ieA_\mu + ig\widetilde{A}_\mu)\phi$.

We can write the partition function of the model as

$$Z(\eta_\mu, \xi_\mu) = \int \mathcal{D}A\mathcal{D}\widetilde{A}\mathcal{D}\phi^*\mathcal{D}\phi \exp\left\{i\int d^4x\left[L + j^\mu_{(e)}\eta_\mu + j^\mu_{(g)}\xi_\mu\right]\right\}\,. \quad (4.45)$$

Here the currents are

$$j^{(e)}_\mu = \frac{ie}{2}\left[\phi^*\partial_\mu\phi - (\partial_\mu\phi^*)\phi\right] - 2e(eA_\mu + g\widetilde{A}_\mu)\phi^*\phi\,, \quad (4.46)$$

$$j^{(g)}_\mu = \frac{ie}{2}\left[\phi^*\partial_\mu\phi - (\partial_\mu\phi^*)\phi\right] - 2g(eA_\mu + g\widetilde{A}_\mu)\phi^*\phi\,,$$

and $\eta_\mu(x)$ and $\xi_\mu(x)$ are the external sources.

Let us recall that relativistic invariance of two-charge quantum electrodynamics means that the corresponding functional integral must be independent on the direction of the vector n_μ. To demonstrate this, let us first evaluate the Gaussian integral over the scalar fields in (4.45):

$$\begin{aligned}Z(\eta_\mu,\xi_\mu) = &\int \mathcal{D}A\mathcal{D}\widetilde{A}\exp\{i\int d^4x\ L_{\text{em}}(A,\widetilde{A},n)\}\\ &\times \exp\left\{-\text{Tr}\ \ln[(p_\mu - J_\mu)^2 - m^2]/(p^2_\mu - m^2)\right\}\,,\end{aligned} \quad (4.47)$$

where
$$J_\mu \equiv e(A_\mu + \eta_\mu) + g(\widetilde{A}_\mu + \xi_\mu), \qquad (4.48)$$

and the momentum operator is as usual defined as $p_\mu = -i\partial_\mu$. Now we can apply the Fock–Schwinger proper time formalism. The operator identity

$$\ln \frac{A}{B} = -\int \frac{d\tau}{\tau} \left[\exp\frac{iA\tau}{2} - \exp\frac{iB\tau}{2}\right]$$

allows us to rewrite the exponent in (4.47) as[4]

$$\mathrm{Tr}\ln[(p_\mu - J_\mu)^2 - m^2] \qquad (4.49)$$
$$p_\mu^2 - m^2$$
$$= -\int \frac{d\tau}{\tau} \int d^4x \langle x | \exp\left\{\frac{i\tau}{2}[(p_\mu - J_\mu)^2 - m^2]\right\} - \exp\left\{\frac{i\tau}{2}(p_\mu^2 - m^2)\right\} | x \rangle.$$

Gaussian integration of the second exponent in this expression gives a trivial result that only affects the normalization factor of the path integral. To evaluate the remaining functional integral over the first exponent in (4.49), we make use of the well-known Feynman relation between the matrix element appearing in (4.49) and the path integral over an auxiliary trajectory $z(\tau)$:

$$U(\tau, x, y) = \langle x | \exp\left\{\frac{i\tau}{2}[(p_\mu - J_\mu)^2 - m^2]\right\} | y \rangle$$
$$= \int_{\substack{z(0)=x \\ z(\tau)=y}} \mathcal{D}z \exp\left\{-i\int_0^\tau d\tau' \left[\frac{1}{2}\dot{z}_\mu^2(\tau') - J_\mu(z)\dot{z}^\mu(\tau') + \frac{1}{2}m^2\right]\right\}. \qquad (4.50)$$

Thus [136, 137]

$$Z(\eta_\mu, \xi_\mu) = \int \mathcal{D}A \mathcal{D}\widetilde{A} \exp[i\int d^4x \, L_{\mathrm{em}}(A, \widetilde{A}, n)]$$
$$\times \exp\left\{\int \frac{d\tau}{\tau} \exp\left(-\frac{im^2\tau}{2}\right)\right\} \int \mathcal{D}z f(z) \exp\left\{i\int_0^\tau d\tau' \, J_\mu(z)\dot{z}^\mu\right\}, \qquad (4.51)$$

where $f(z) = \exp\left\{-\frac{i}{2}\int_0^\tau d\tau' \dot{z}_\mu^2(\tau')\right\}$ and the paths of integration are closed: $z(0) = z(\tau)$.

We can now expand the last exponent in (4.51) involving the external currents J_μ into a series. Then the partition function can be written as

---

[4] Recall that here the trace operator is a functional trace, that is we must evaluate the corresponding matrix element of the operator involved in a complete basis $\langle x|, |x\rangle$ and then sum over all states. In the case under consideration, we must integrate over the momentum of free-particle states having the form of plane waves.

124   4 Abelian Monopole: Relativistic Quantum Theory

$$Z(\eta_\mu, \xi_\mu) = \sum_{k=0}^{\infty} Z_n^{(k)}, \qquad (4.52)$$

where

$$Z_n^{(k)} = \frac{1}{k!} \int \mathcal{D}A \mathcal{D}\widetilde{A} \exp[i \int d^4x \, L_{\text{em}}(A, \widetilde{A}, n)]$$
$$\times \prod_{i=1}^{k} \left[\frac{d\tau_i}{\tau_i} \int \mathcal{D}z(\tau_i) f(z(\tau_i)) \right] \sum_\rho C_\rho \exp\left\{ -\frac{i}{2} m^2 \tau_i \right\}$$
$$\times \exp\left\{ i \int d^4x \left[j_\mu^{(e)}(A^\mu + \eta^\mu) j_\nu^{(g)}(\widetilde{A}^\nu + \xi^\nu) \right]_\rho \right\}. \qquad (4.53)$$

Here C_ρ is a combinatoric coefficient and the classical currents of particles moving along closed trajectories $z_i(\tau_i)$ are (cf. (4.2))

$$j_\mu^{(e)} = \sum_i e_i \oint dz_\mu^i \delta^{(4)}(x - z_i),$$
$$j_\mu^{(g)} = \sum_i g_i \oint dz_\mu^i \delta^{(4)}(x - z_i). \qquad (4.54)$$

The dependence of the partition function on the vector n_μ is now hidden in the Lagrangian $L_{\text{em}}(A, \widetilde{A}, n)$. Consequently, integration of each of the terms of the expansion (4.53) over the field variables A_μ and $\widetilde{A}_\mu$ yields

$$\int \mathcal{D}A \mathcal{D}\widetilde{A} \exp\left\{ i \int d^4x \, L_{\text{em}}(A, \widetilde{A}, n) \right\}$$
$$\times \exp\left\{ i \int d^4x \left[j_\mu^{(e)}(A^\mu + \eta^\mu) j_\nu^{(g)}(\widetilde{A}^\nu + \xi^\nu) \right] \right\}$$
$$= \exp\left[-\frac{1}{2} \int \int \left(j_\mu^{(e)} D_{AA}^{\mu\nu} j_\nu^{(e)} + j_\mu^{(g)} D_{\widetilde{A}A}^{\mu\nu} j_\nu^{(e)} \right. \right.$$
$$\left. \left. + j_\mu^{(e)} D_{A\widetilde{A}}^{\mu\nu} j_\nu^{(g)} + j_\mu^{(g)} D_{\widetilde{A}\widetilde{A}}^{\mu\nu} j_\nu^{(g)} \right) \right]. \qquad (4.55)$$

Here we define the propagators of free fields A_μ and $\widetilde{A}_\mu$. In momentum representation they can be written as [436, 512, 547, 550]:

$$D_{AA}^{\mu\nu}(k) = D_{\widetilde{A}\widetilde{A}}^{\mu\nu}(k) = i \left[-\eta^{\mu\nu} + \frac{k^\mu n^\nu - n^\mu k^\nu}{n \cdot k} \right] \frac{1}{k^2 + i\varepsilon},$$
$$D_{A\widetilde{A}}^{\mu\nu}(k) = D_{\widetilde{A}A}^{\mu\nu}(k) = i\varepsilon^{\mu\nu\rho\sigma} \frac{n_\rho k_\sigma}{n \cdot k} \frac{1}{k^2 + i\varepsilon}. \qquad (4.56)$$

The propagator of the scalar field has a standard form and vertices of the interaction between the scalar (dyon) and electromagnetic fields are multiplied by e and g. However, the charge quantization condition means that the weak coupling regime with respect to the electric charge corresponds to the strong coupling regime with respect to the magnetic charge, and vice versa. Therefore the diagram techniques for expansion over the magnetic coupling inevitably becomes formal, since there is no perturbation theory with respect to the charge g. This is a pathology that, within QED with a magnetic charge, makes the standard renormalization scheme completely meaningless. Nevertheless, there have been some attempts to calculate amplitudes of elementary quantum-electrodynamical processes with monopoles by making use of the propagators (4.56). The simplest example is the evaluation of the quantum amplitude of scattering of an electron by a magnetic charge [436].

Let us now return to the proof of relativistic invariance of the functional integral (4.51). To simplify our consideration, we shall consider again the interaction between two point-like dyons. Then the explicit form of the Green functions (4.56) suggests that the n_μ-dependent part of the partition function has the form

$$\exp S_n = \exp\left\{-i(e_1 g_2 - e_2 g_1) \oint_{\Gamma_1} dz_1^\mu \oint_{\Gamma_2} dz_2^\nu \varepsilon_{\mu\nu\rho\sigma} \frac{n^\rho \partial^\sigma}{n \cdot \partial} \frac{1}{(z_1 - z_2)^2 + i\varepsilon}\right\},$$
(4.57)

where the contours of integration, Γ_i, $i = 1, 2$, are closed trajectories of the particles. Exploiting Stokes theorem we can transform the first integral over the contour Γ_1 into an integral over the surface Σ_1:

$$\exp S_n = \exp\left\{i(e_1 g_2 - e_2 g_1) \int_{\Sigma_1} dS^{\rho\sigma} \varepsilon_{\mu\nu\rho\sigma}\right.$$
(4.58)
$$\left.\times \oint_{\Gamma_2} dz_2^\mu \left[\partial^\nu \left(\frac{1}{(z_1 - z_2)^2 + i\varepsilon}\right) + in_\nu (n \cdot \partial)^{-1} \delta(z_1 - z_2)\right]\right\}.$$

The first term in the exponent is obviously n_μ-independent. Now we can make use of the integral representation (4.10) of the operator $(n \cdot \partial)^{-1}$ to see that the second term can be interpreted as the number of intersections between the closed world line Γ_2 of the second dyon and an oriented three-surface Σ_1 spanned by the path of the first dyon. Obviously, this is an integer for all trajectories, except those completely in the surface Σ_1. The latter belong to a manifold of measure zero in functional space. Thus, the partition function of two-charge electrodynamics (4.52) is n_μ-independent if the Schwinger–Zwanziger charge quantization condition (2.131) is fulfilled. Thus, under this condition the model remains relativistically invariant.

4.4 Renormalization of QED with a Magnetic Charge

By analogy with the standard QED, we can try to renormalize the generalized model with two types of charges. However, we immediately run into

4 Abelian Monopole: Relativistic Quantum Theory

trouble, since the charge quantization condition means that one of the couplings inevitably must be strong, so we cannot apply the usual perturbative expansion in electric and magnetic couplings simultaneously. Therefore the standard procedure of renormalization in two-charge quantum electrodynamics becomes just formal and questionable.

Let us illustrate this statement, which is common to all evaluations of amplitudes of different processes in QED with monopoles, with the well-known problem of vacuum polarization [43, 135, 419, 459, 520]. It is obvious that the fundamental question is: how are the electric and magnetic coupling constants modified by quantum corrections, i.e., what are the β-functions corresponding to electric and magnetic couplings? It is clear that the the pole of the photon propagator (4.56), despite its rather unusual structure, corresponds to a single photon that can be emitted by an electric or magnetic source. However, the loop corrections to the photon propagator now include both usual electron diagrams and contributions from virtual pairs of monopoles and dyons.

The procedure of renormalization of electric and magnetic charges was the subject of intensive discussions for a long time (see, for example, the reviews [41,45]). According to the conclusion reached by Schwinger [459], and supported in [191, 419], both electric and magnetic charges are renormalized in a similar way, that is

$$e_r^2 = Z_e e_0^2, \qquad g_r^2 = Z_g g_0^2, \qquad Z_e = Z_g < 1. \qquad (4.59)$$

By contrast, Coleman [43] and other authors [158, 242, 496] concluded that

$$Z_e = Z_g^{-1}, \qquad (4.60)$$

that is, when the energy increases, the effect of vacuum fluctuations lead not only to the standard electric charge screening, but simultaneously to the effect of anti-screening of the magnetic charge.

In essence, the argument by Schwinger was as follows. For the sake of simplicity, let us consider scalar electrodynamics. There are photon propagators of three different types. The explicit form of the propagator $D_{AB}^{\mu\nu}(k)$ corresponds to a photon that is emitted at a vertex of one type and absorbed at a vertex of a different type. Its contribution to the self-energy of the scalar matter field and the corresponding vertex functions are equal to zero.

At the same time, there is a non-vanishing contribution of the type $\langle 0|BB|0\rangle$, which comes from a photon that is emitted and absorbed by magnetic vertices. The latter is identical to the contribution of the standard photon propagator $\langle 0|AA|0\rangle$ up to the substitution $e^2 \to g^2$. Therefore, the vertices and propagators of the matter fields are renormalized just as in normal QED with an effective charge $q^2 = e^2 + g^2$ and the only non-trivial effects are connected with vacuum polarization due to dyons.

On the one-loop level, the renormalized photon propagator has the form

$$\tilde{D}_{\mu\nu} = D_{\mu\nu} + D_{\mu\rho}\Pi^{\rho\lambda}D_{\lambda\nu}, \qquad (4.61)$$

4.4 Renormalization of QED with a Magnetic Charge

where the vacuum polarization tensor includes the contributions:

$$\begin{aligned}
\Pi^{(ee)}_{\mu\nu}(x) &= \langle 0|j^{(e)}_\mu(x)j^{(e)}_\nu(0)|0\rangle\,, \\
\Pi^{(gg)}_{\mu\nu}(x) &= \langle 0|j^{(g)}_\mu(x)j^{(g)}_\nu(0)|0\rangle\,, \\
\Pi^{(eg)}_{\mu\nu}(x) &= \langle 0|j^{(e)}_\mu(x)j^{(g)}_\nu(0)|0\rangle\,, \\
\Pi^{(ge)}_{\mu\nu}(x) &= \langle 0|j^{(g)}_\mu(x)j^{(e)}_\nu(0)|0\rangle\,.
\end{aligned} \quad (4.62)$$

As in the case of standard QED, the condition of conservation of electric and magnetic currents fix the form of the Fourier transform of the polarization tensor to be

$$\Pi_{\mu\nu}(k^2) = -i(\eta_{\mu\nu}k^2 - k_\mu k_\nu)\Pi(k^2, m^2)\,, \quad (4.63)$$

where m is the mass of the virtual particle. Thus, substituting (4.56) into (4.61), and keeping only gauge invariant terms, we can write

$$\begin{aligned}
\widetilde{D}^{(ee)}_{\mu\nu} &= -\frac{i\eta_{\mu\nu}}{k^2}\left(1 - \Pi^{(ee)} - \Pi^{(gg)}\right) - i\frac{T_{\mu\nu}}{(n\cdot k)^2}\Pi^{(gg)}\,, \\
\widetilde{D}^{(gg)}_{\mu\nu} &= -\frac{i\eta_{\mu\nu}}{k^2}\left(1 - \Pi^{(ee)} - \Pi^{(gg)}\right) - i\frac{T_{\mu\nu}}{(n\cdot k)^2}\Pi^{(ee)}\,, \\
\widetilde{D}^{(eg)}_{\mu\nu} &= -\frac{i}{k^2}S_{\mu\nu}\left(1 - \Pi^{(ee)} - \Pi^{(gg)}\right) + i\frac{T_{\mu\nu}}{(n\cdot k)^2}\Pi^{(eg)}\,, \\
\widetilde{D}^{(ge)}_{\mu\nu} &= -\frac{i}{k^2}S_{\mu\nu}\left(1 - \Pi^{(ee)} - \Pi^{(gg)}\right) + i\frac{T_{\mu\nu}}{(n\cdot k)^2}\Pi^{(ge)}\,,
\end{aligned} \quad (4.64)$$

where

$$S_{\mu\nu} = \varepsilon_{\mu\nu\rho\sigma}\frac{n^\rho k^\sigma}{(n\cdot k)}\,, \qquad T_{\mu\nu} = n^2\eta_{\mu\nu} - n_\mu n_\nu\,, \quad (4.65)$$

$$\Pi^{(ee)} = e^2\Pi(k^2);\quad \Pi^{(gg)} = g^2\Pi(k^2)\,,\quad \Pi^{(eg)} = \Pi^{(ge)} = eg\Pi(k^2)\,,$$

and $\Pi(k^2)$ is the standard reduced one-loop vacuum polarization tensor. We shall write its explicit form below.

Note that, as we approach the photon pole, $k^2 \to 0$, the terms in (4.64) that are proportional to $T_{\mu\nu}$ become negligible compared to the first terms. Let us define the renormalization constant

$$Z = 1 - \Pi^{(ee)}(0) - \Pi^{(gg)}(0) = 1 - (e^2 + g^2)\Pi(0)\,. \quad (4.66)$$

Then the photon propagators (4.64) at zero momentum transfer become

$$\begin{aligned}
\widetilde{D}^{(ee)}_{\mu\nu} &\approx \widetilde{D}^{(gg)}_{\mu\nu} \approx -\frac{i\eta_{\mu\nu}}{k^2}Z\,, \\
\widetilde{D}^{(eg)}_{\mu\nu} &\approx -\widetilde{D}^{(ge)}_{\mu\nu} \approx -\frac{iS_{\mu\nu}}{k^2}Z\,,
\end{aligned} \quad (4.67)$$

that is, the renormalized charges can be defined according to (4.59). The conclusion by Schwinger about the similar character of the renormalization of both electric and magnetic charges follows from (4.67). Moreover, due to the charge quantization Condition, the constant Z must be a rational number smaller than one [459].

This result is rather general, since it follows directly from the form of the photon propagator (4.56) and the definition of the renormalized charge, (4.59). However, let us note that if the relations (4.59) are fulfilled, the charge quantization condition will not be independent of the renormalization. That would be a puzzle and cause trouble. On the other hand, if the conclusion reached by Coleman (4.60) and others is correct, the quantization condition does not depend on the renormalization point, i.e., the electric and magnetic β-functions run in opposite ways.

Let us try to understand the difference between the two approaches presented in [191, 419, 459] and [43, 158, 242, 496], respectively. The Schwinger approach indirectly relies on the treatment of both electric and magnetic charges on the same footing. Coleman and others argue that, according to the experimental data and predictions of non-Abelian gauge theories, the monopole must be much heavier than its electric counterpart and therefore its contribution to the vacuum polarization amplitude in the low-energy limit would be strongly suppressed.

Note that even within the framework of pure QED there still remains an unsolved question: can one treat the quantum effects caused by monopoles or dyons in the same manner, because the magnetic coupling is strong: $eg \sim 1$? Calucci and Iengo [158] attempt to solve this problem by making use of the lattice formulation of the theory. We will not go into details of this consideration [157, 159], which is based on the loop representation of the partition function of two-charge electrodynamics. Note just that it looks like the reason for the disagreement between the relations (4.59) and (4.60) is connected with the treatment of the n-dependent term in the action [41]. In the Calucci–Iengo representation [158] the charge quantization condition removes this term in just the same way as was done above when we proved the relativistic invariance of the model. Then the relation (4.60) is recovered.

4.5 Vacuum Polarization by a Dyon Field

As we have seen above, there is a serious inconsistency in QEMD when it is formally constructed according to the standard prescriptions. The charge quantization condition must be satisfied for the theory to be self-consistent but, at the same time, this condition means that we cannot apply the perturbative expansion in electric and magnetic couplings simultaneously.

However, there still remains the possibility of using the loop expansion of the partition function, which relies on the smallness of the Planck constant. In this way, we can obtain some results even without making use of the standard Feynman diagram technique but, for example, by applying a rather old-fashioned approach that is based on the exact solution of the relativistic

4.5 Vacuum Polarization by a Dyon Field

wave equations in external fields [242]. Indeed, in this case, the one-loop correction to the energy of the electromagnetic field can be calculated by "brute force", that is, by direct summation of the corresponding one-particle modes over all quantum numbers n (see, e.g., [31]):

$$\Delta \mathcal{E} = \frac{1}{2V} \sum_n \left(\omega_n^2 |\phi_n|^2 + |\nabla \phi_n - ieA_k \phi_n|^2 + m^2 |\phi_n|^2 + e^2 A_0^2 |\phi_n|^2 \right), \quad (4.68)$$

where ϕ_n are the known solutions (2.47) of the scalar relativistic wave equation for a spinless charged particle in an external dyon field [5] and the scalar potential of a dyon is $A_0 = Q/r$.

Let us recall that these solutions describe bound states, as well as the continuum (see page 38). Since we suppose that the mass of the monopole or dyon is very large in comparison with the mass of the charged particle, in the evaluation of the effects of the vacuum polarization at the one-loop level, we can restrict ourselves to the consideration of the contribution that is caused by electrically charged virtual particles.

Let us substitute the solutions (2.47) and (2.49) into the expression (4.68). Note that the spherical symmetry of the configuration allows us to calculate the sum over the magnetic quantum number m by exploiting the relations [531]:

$$\sum_{m=-j}^{j} |Y_{\mu l m}(\theta, \varphi)|^2 = \frac{2j+1}{4\pi}, \quad (4.69)$$

$$\sum_{m=-j}^{j} |\nabla Y_{\mu l m}(\theta, \varphi) - ieA_k Y_{\mu l m}(\theta, \varphi)|^2 = \frac{2j+1}{4\pi} [j(j+1) - \mu^2].$$

Now we can split the expression (4.68) into two parts corresponding to the contributions of the bound states and the continuum, respectively. Since the explicit form of the radial wave functions of the discrete part of the spectrum (2.49) shows that their contribution to the sum (4.68) is exponentially suppressed, we can neglect those terms and consider only the sum over the continuum states. Invoking (4.69), we can roughly estimate the zero-point energy density as

$$\Delta \mathcal{E} = \frac{1}{4\pi^2} \sum_{j=\mu}^{\infty} (2j+1) \int_0^\infty \frac{dk}{\omega} \left((\omega^2 + m^2) |F_{kj}(r)|^2 \right. \quad (4.70)$$

$$\left. + \frac{e^2 Q^2 + j(j+1) - \mu^2}{r^2} |F_{kj}(r)|^2 + |\partial_r F_{kj}(r)|^2 \right),$$

where $F_{kj}(r)$ are the radial functions (2.49). Note that in the limiting case $Q = 0$, this relation coincides with the one-loop correction to the energy density of a charged particle in a monopole magnetic field obtained in [242].

Using the Watson–Sommerfeld integral formula

$$\sum_{n=0}^{\infty} F(n) = \oint_C dy \frac{F(y)}{1 - \exp(2\pi i y)}$$

$$= \int_{-1/2}^{\infty} dy F(y) + 2 \int_0^{\infty} \frac{dx}{1 + \exp(2\pi x)} \operatorname{Im} F(-1/2 + ix), \quad (4.71)$$

we can perform the summation over the quantum number j:

$$\Delta \mathcal{E} = \sum_{j=\mu}^{\infty} (2j+1) F(l(j), Q) = 2 \int_0^{\infty} dl (l + 1/2) F(l, Q) \quad (4.72)$$

$$+ 4 \int_0^{\infty} \frac{dx}{1 + \exp(2\pi x)} \operatorname{Re}\left[(x - i\mu) F\left(-\tfrac{1}{2} + \sqrt{2i\mu x - x^2 - e^2 Q^2}\right)\right],$$

where we have introduced the abbreviation

$$F(l, Q) = \frac{1}{4\pi^2} \int_0^{\infty} \frac{dk}{\omega} \left((\omega^2 + m^2) |F_{kj}(r)|^2 \right.$$

$$\left. + \frac{e^2 Q^2 + j(j+1) - \mu^2}{r^2} |F_{kj}(r)|^2 + |\partial_r F_{kj}(r)|^2 \right). \quad (4.73)$$

The next step is to consider the asymptotic behavior of the radial functions of (4.70)

$$F_{kj}(r) r \overset{r \to \infty}{\approx} \frac{1}{r} (G \cos \theta + F \sin \theta),$$

where $\theta = kr - (eQ\omega/k) \ln(2kr) - \tfrac{1}{2} \pi l + \delta$ and the coefficients G, F are defined by the expansion

$$G = \frac{j(j+1) - \mu^2}{2kr} - \frac{e^2 Q^2}{2kr} \left(1 - \frac{\omega^2}{k^2}\right)$$

$$- \left(\frac{eQ}{4(kr)^2} - \frac{eQ(j(j+1) - \mu^2)}{2(kr)^2} + \frac{e^3 Q^3}{2(kr)^2} \left(1 - \frac{\omega^2}{k^2}\right) \right) \frac{\omega}{k} + \ldots,$$

$$F = 1 + \frac{eQ}{2kr} \frac{\omega}{k} + \frac{j(j+1) - \mu^2}{4(kr)^2} - \frac{(j(j+1) - \mu^2)^2}{8(kr)^2} + \frac{5}{8} \frac{e^2 Q^2}{(kr)^2} \frac{\omega^2}{k^2}$$

$$- \frac{e^2 Q^2}{4(kr)^2} + \frac{e^2 Q^2 (j(j+1) - \mu^2)}{4(kr)^2} \left(1 - \frac{\omega^2}{k^2}\right) - \frac{e^4 Q^4}{8(kr)^2} \left(1 - \frac{\omega^2}{k^2}\right)^2$$

$$+ \ldots \quad (4.74)$$

The recipe of regularization of the divergent integral (4.73), is to subtract the analogous vacuum expression for vanishing charges: $Q = 0, g = 0$. Then,

4.5 Vacuum Polarization by a Dyon Field

substituting the asymptotic expansions (4.74) into (4.72), after some algebra we obtain, to leading order in $1/kr$

$$\Delta \mathcal{E} = \frac{1}{8\pi^2 r^4} \sum_{j=\mu}^{\infty} (2j+1)\left[j(j+1) - \mu^2 + 2e^2 Q^2\right]\left(\ln\frac{2\Lambda}{m} - 1\right), \quad (4.75)$$

having omitted all the Q and g independent terms and performing an integration over the variable $x = kr$. The last step is to sum over the quantum number j. This can be performed by again making use of the Watson–Sommerfeld formula:

$$\Delta \mathcal{E} \approx \frac{(\ln\frac{2\Lambda}{m} - 1)}{2\pi^2 r^4} \int_0^\infty \frac{dx}{1 + \exp(2\pi x)}$$
$$\times \mathrm{Im}\left[(ix + \mu)\left((ix+\mu)^2 + 2e^2 Q^2 - \frac{1}{4} - \mu^2\right)\right]. \quad (4.76)$$

Keeping only the charge-dependent terms, we find

$$\Delta \mathcal{E} \approx \frac{\alpha}{6\pi}\left(\ln\frac{2\Lambda}{m} - 1\right)\frac{E^2}{2} + \frac{\alpha}{6\pi}\left(\ln\frac{2\Lambda}{m} - 1\right)\frac{B^2}{2}, \quad (4.77)$$

where $\alpha = e^2/4\pi$. The final step of our calculation is to include the terms corresponding to the energy density of the electromagnetic field in the tree approximation. Then, we obtain

$$\mathcal{E} = \frac{E^2}{2}\left(1 + \frac{\alpha}{6\pi}\ln\frac{2\Lambda}{m} - \frac{\alpha}{6\pi}\right) + \frac{B^2}{2}\left(1 + \frac{\alpha}{6\pi}\ln\frac{2\Lambda}{m} - \frac{\alpha}{6\pi}\right). \quad (4.78)$$

Thus, the effect of vacuum polarization in an external dyon field is that both electric and magnetic fields are multiplied by the same divergent coefficient. Formally, this is in agreement with the Schwinger conclusion (4.59) that we discussed above.

As in the case of usual QED, the logarithmic divergence in (4.78) can be eliminated by a standard renormalization of the external electric and magnetic fields E, B and the electric charge of the scalar field e:

$$e_R^2 = e^2 Z, \qquad E_R^2 = E^2 Z^{-1}, \qquad B_R^2 = B^2 Z^{-1}, \quad (4.79)$$

where the renormalization factor

$$Z = 1 - \frac{\alpha}{6\pi}\ln\frac{2\Lambda}{M}$$

corresponds to the usual renormalization constant of scalar QED and its generalization (4.66) considered above. Thus, the electric charge of a dyon is screened by the vacuum polarization effects as it should be.

Let us define the "dielectric" (ϵ) and "magnetic" (μ) vacuum permittivities [242] as

$$\epsilon = \mu^{-1} = \left(1 + \frac{\alpha}{6\pi} \ln \frac{2\Lambda}{m} - \frac{\alpha}{6\pi}\right).$$

Then we can write

$$\mathcal{E} = \epsilon \frac{E^2}{2} + \mu^{-1} \frac{B^2}{2}. \tag{4.80}$$

We can also introduce renormalized quantities

$$\epsilon_R = \mu_R^{-1} = 1 - \frac{\alpha}{6\pi}, \tag{4.81}$$

which define the finite corrections to the energy density $\mathcal{E}$ of the dyon electromagnetic field originating from the vacuum polarization. The regularized expression for $\mathcal{E}$, taking into account the one-loop correction to the first order in α takes the form:

$$\mathcal{E} = \frac{1}{2}\left(\epsilon_R E_R^2 + \mu_R^{-1} B_R^2\right). \tag{4.82}$$

The physical interpretation of this result is that the contribution of the one-loop corrections to the energy density of the electric and magnetic fields of the dyon leads to a shielding of the electric charge of the dyon and, taking into account the definition (4.81), to the effect of antishielding of the magnetic charge g.

4.6 Effective Lagrangian of QED with a Magnetic Charge

In the previous section, we considered the vacuum polarization induced by electron-positron virtual pairs in a dyon external field. Now, let us consider what in some sense is the opposite problem, i.e., we shall investigate how one-loop corrections connected with a virtual pair of dyons could modify the vacuum energy. As is well-known (see, for example, [1]) in the case of a weak, almost homogeneous electromagnetic field, such a correction can be evaluated again without using the perturbative expansion in a coupling constant. Actually, our goal is to write an effective Lagrangian of QEMD [337], which is a generalization of the known form of the Euler–Heisenberg non-linear Lagrangian.

Let us consider the simple case of weak, constant, parallel electric and magnetic fields **E** and **B**. We impose the conditions $eE/m^2 \ll 1$ and $eB/m^2 \ll 1$, such that the creation of particles is not possible. As in the previous section, the one-loop correction can be calculated by summing the one-particle modes — the solutions of the Dirac equation in the external electromagnetic field — over all quantum numbers[5].

---

[5] Here, in order to make the comparison with ordinary QED more transparent, we will consider not scalar but spinor excitations.

4.6 Effective Lagrangian of QED with a Magnetic Charge

Let us recall the basic elements of such calculations. For example, if there is just a magnetic field, $\mathbf{B} = (0, 0, B)$, the corresponding equation of motion of an electron is

$$\left[i\gamma^\mu (\partial_\mu + ieA_\mu) - m^{(e)}\right]\psi = 0, \tag{4.83}$$

where the electromagnetic potential is $A^\mu = (0, -By, 0, 0)$. The solution to this equation gives the energy levels of an electron in a magnetic field [1,5]

$$\varepsilon_n = \sqrt{m^2 + eB(2n - 1 + s) + k^2}, \tag{4.84}$$

where $n = 0, 1, 2 \ldots$ and $s = \pm 1$. Furthermore, k is the electron momentum along the field. In this case, the correction to the Lagrangian is [1,6]

$$\begin{aligned}\Delta L_B &= \frac{eB}{2\pi^2} \int_0^\infty dk \left[(m^2 + k^2)^{1/2} + 2\sum_{n=1}^\infty (m^2 + 2eBn + k^2)^{1/2}\right] \\ &= -\frac{1}{8\pi^2} \int_0^\infty \frac{ds}{s^3} e^{-m^2 s} \left[(esB)\coth(esB) - 1 - \frac{1}{3}e^2 s^2 B^2\right],\end{aligned} \tag{4.85}$$

where the terms independent of the external field $\mathbf{B}$ have been dropped and a standard renormalization of the electron charge has been made.

It is known [5] that if we consider simultaneously homogeneous magnetic and electric fields, then (4.83), as well as its classical analogue, can be separated into two uncoupled equations, each in two variables. Indeed, in this case we can take $A^\mu = (Ez, -By, 0, 0)$, and the interaction of an electron with the fields $\mathbf{E}$ and $\mathbf{B}$ is determined independently. For such a configuration of electromagnetic fields, the correction to the Lagrangian is [1,6]

$$\Delta L = \frac{eB}{2\pi^2} \sum_{n=1}^\infty \int_0^\infty dk\, \varepsilon_n^{(E)}(k). \tag{4.86}$$

Here, $\varepsilon_n^{(E)}$ is the correction to the energy of an electron (4.84) in the combined external magnetic and electric fields, which to the first order is proportional to $e^2 E^2$. Thus the total Lagrangian is $L = L_0 + \Delta L$, where $L_0 = (\mathbf{E}^2 - \mathbf{B}^2)/2$ is just the Lagrangian of the free electromagnetic field in the tree approximation, and can be written as

$$L = \left(1 + \frac{\alpha}{3\pi} \int_0^\infty \frac{ds}{s} e^{-m^2 s}\right) \frac{\mathbf{E}^2 - \mathbf{B}^2}{2} + \Delta L'. \tag{4.87}$$

The logarithmic divergence can again be removed by the standard renormalization of the external fields and the electron charge (4.79). The only difference from the case of scalar electrodynamics that we considered above is that the renormalization constant now is

$$Z^{-1} = 1 + (\alpha/3\pi) \int_0^\infty (ds/s) e^{-m^2 s}.$$

Thus, the finite part of the correction to the Lagrangian, $\Delta L'$, can be written in terms of physical quantities as (see, e.g., [1])

$$\Delta L' = -\frac{1}{8\pi^2} \int_0^\infty \frac{ds}{s^3} e^{-m^2 s} \left[(esE)(esB) \cot(esE) \coth(esB) - 1 \right], \quad (4.88)$$

which in the limit $E = 0$ reduces to the renormalized form of (4.85).

Since we consider the case of weak electromagnetic fields, we can expand the expression (4.88) in the parameters $eE/m^2 \ll 1$, $eB/m^2 \ll 1$, which yields the well-known Euler–Heisenberg correction [264]:

$$\Delta L' \approx \frac{e^4}{360\pi^2 m^4} \left[(\mathbf{B}^2 - \mathbf{E}^2)^2 + 7(\mathbf{B} \cdot \mathbf{E})^2 \right]. \quad (4.89)$$

Let us consider how the situation changes, if we allow for virtual pair creation of dyons in the external electromagnetic field. We recall that the dynamics of a spinor particle in the external field of a dyon is described by the (4.18)

$$\left[i\gamma^\mu \left(\partial_\mu + ieA_\mu + ig\widetilde{A}_\mu \right) - M \right] \psi = 0,$$

where M is the dyon mass.

Here the notations are slightly different from the two-potential formulation considered above. The potential A_μ and its dual $\widetilde{A}_\mu$ are defined by $F_{\mu\nu} = \partial_\mu A_\nu - \partial_\nu A_\mu = \varepsilon_{\mu\nu\rho\sigma} \partial^\rho \widetilde{A}^\sigma$ where $F_{\mu\nu}$ is the electromagnetic field strength tensor.[6]

The potentials in the case of constant parallel electric and magnetic fields can be expressed as

$$A_\mu = (Ez, -By, 0, 0), \qquad \widetilde{A}_\mu = (Bz, Ey, 0, 0). \quad (4.90)$$

It is easily seen that the solution to the equation of motion for a dyon in an external electromagnetic field can be obtained from the solution to the equation for an electron by the dual transformation

$$eE \to QE + gB, \qquad eB \to QB - gE. \quad (4.91)$$

Using this substitution we obtain the following expression for the quantum correction to the Lagrangian, due to the vacuum polarization caused by dyons:

---

[6] This definition is consistent only if $\Box A_\mu = \Box \widetilde{A}_\mu = 0$, i.e., for constant electromagnetic fields or for free electromagnetic waves.

4.6 Effective Lagrangian of QED with a Magnetic Charge

$$L = \left(1 + \frac{Q^2}{12\pi^2}\int_0^\infty \frac{ds}{s}e^{-M^2 s} - \frac{g^2}{12\pi^2}\int_0^\infty \frac{ds}{s}e^{-M^2 s}\right)\frac{\mathbf{E}^2 - \mathbf{B}^2}{2} + \Delta L', \quad (4.92)$$

where a total derivative has been dropped.

The procedure of renormalization of electric and magnetic charges and the fields can be done in complete analogy with the problem of vacuum polarization by a dyon field that we considered above. The difference is that now we have to take into account both electric and magnetic charges of a virtual particle in a loop. Thus, we can introduce two different renormalization factors [41]:

$$Z_e^{-1} = 1 + \frac{Q^2}{12\pi}\int_0^\infty \frac{ds}{s}e^{-M^2 s}, \quad Z_g^{-1} = 1 - \frac{g^2}{12\pi}\int_0^\infty \frac{ds}{s}e^{-M^2 s}, \quad (4.93)$$

which corresponds to the decomposition $Z = Z_e Z_g^{-1}$ of the renormalization constant of (4.66). In this case, the fields and charges are renormalized as

$$E_R^2 = Z_e^{-1} Z_g E^2, \qquad B_R^2 = Z_e^{-1} Z_g B^2,$$
$$e_R^2 = Z_e Z_g^{-1} e^2, \qquad g_R^2 = Z_e^{-1} Z_g g^2. \quad (4.94)$$

Considering now the case of weak electromagnetic fields by analogy with (4.89), we can write the finite part of the renormalized correction to the Lagrangian $\Delta L'$ as

$$\Delta L' = \frac{1}{360\pi^2 M^4}\{[(Q^2 - g^2)^2 + 7Q^2 g^2](\mathbf{B}^2 - \mathbf{E}^2)^2$$
$$+ [16Q^2 g^2 + 7(Q^2 - g^2)^2](\mathbf{B}\cdot\mathbf{E})^2 + 6Qg(Q^2 - g^2)(\mathbf{B}\cdot\mathbf{E})(\mathbf{B}^2 - \mathbf{E}^2)\}. \quad (4.95)$$

Let us recall that this expression describes nonlinear corrections to the Maxwell equations that correspond to photon-photon interactions. The principal difference between the formula (4.95) and the standard Euler–Heisenberg effective Lagrangian (4.89) consists in the appearance of P and T non-invariant terms proportional to $(\mathbf{BE})(\mathbf{B}^2 - \mathbf{E}^2)$. It should, however, be noted that this term is invariant under charge conjugation C, since then both Q and g would change sign [438].

Let us make one more remark on the loop corrections. If we would consider the corrections connected with the contribution of the virtual dyons only (and not the electrons), then the dual invariance of the model would mean that the charges Q and g would not separately have any meaning, the physics would be determined by the effective charge $\sqrt{Q^2 + g^2}$. However, in the case under consideration, we are dealing with two effects simultaneously, that is, we have two different contributions from vacuum polarization by electron-positron and dyon pairs, respectively. Then it is not possible to reformulate the theory

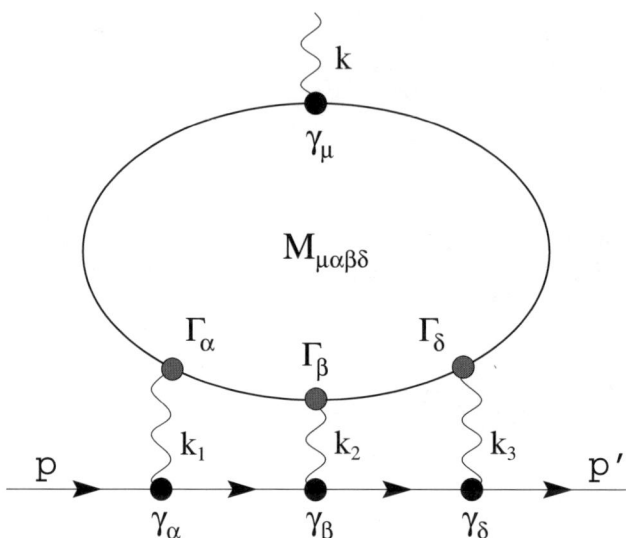

Fig. 4.1. T-violating contribution to the electron-photon vertex due to the dyon loop

in terms of just one effective charge by means of a dual transformation. Moreover, the Dirac charge quantization condition connects just the electric charge of the electron and the magnetic charge of the dyon: $eg = n/2$, whereas the electric dyon charge Q is not quantized. Its quantization becomes natural in the quantum field theory, which we shall consider later.

We can further simplify the expression (4.95), taking into account the difference between the monopole and electron masses. Indeed, it is widely believed, based both on experimental bounds and predictions of non-Abelian theories, that the dyon mass would be large. Thus, for $M \gg m$, we can write the dominant contributions of (4.89) and (4.95) as

$$\Delta L' \approx \frac{e^4}{360\pi^2 m^4}\left[(\mathbf{B}^2 - \mathbf{E}^2)^2 + 7(\mathbf{B}\cdot\mathbf{E})^2\right] + \frac{Qg(Q^2 - g^2)}{60\pi^2 M^4}(\mathbf{B}\cdot\mathbf{E})(\mathbf{B}^2 - \mathbf{E}^2), \quad (4.96)$$

where the P and T invariant terms contributing to vacuum polarization by dyons have been dropped because they are suppressed by factors M^{-4}.

Finally, let us make two remarks. First, we note that the expression (4.96) yields the matrix element for low-energy photon-photon scattering, which contains P and T non-invariant terms [337]. This results in interference between two one-loop diagrams corresponding to loops with dyons and those with simply electrically charged particles, causing an asymmetry between the processes of photon splitting $\gamma \to 2\gamma$ and photon coalescence $2\gamma \to \gamma$ [336]. The asymmetry is linear in the product of the dyon charges, and proportional to the fourth power of the electron to dyon mass ratio m/M.

4.6 Effective Lagrangian of QED with a Magnetic Charge

Second, the violation of T invariance caused by the contribution of the photon-photon scattering sub-diagram to the sixth-order radiative corrections to the electron-photon vertex (see Fig. 4.1) could induce an electric dipole moment of the electron. The corresponding estimate yields [337]

$$d_e \sim \frac{e^2 Qg(Q^2 - g^2)}{(4\pi^2)^3} \frac{m}{M^2}. \tag{4.97}$$

This estimate can be used to obtain a new bound on the dyon mass. Indeed, recent experimental progress in the search for an electron electric dipole moment [120,174,302] gives a rather strict upper limit: $d_e < 9 \times 10^{-28} e$ cm. If we suppose that $Q \sim e$, then from (4.97) we obtain $M \geq 2 \times 10^6 \, m \approx 10^3$ GeV. This coincides with the bound obtained from the experimental data of the e^+e^- collider LEP by De Rújula [449] for virtual monopoles. Thus, we can definitely conclude that a monopole (dyon) is not a natural object in QED and its mass belongs at least to the electroweak scale.

Part II

Monopole in Non-Abelian Gauge Theories

5 't Hooft–Polyakov Monopole

5.1 $SU(2)$ Georgi–Glashow Model and the Vacuum Structure

5.1.1 Non-Abelian Wu–Yang Monopole

We have seen in the previous chapters that a magnetic monopole could be introduced into the Abelian electrodynamics on the classical level, if the vector potential is not defined globally or if there are singular objects in the theory. However, even in this case, the quantum theory of the monopole is full of contradictions that one can hardly avoid within the framework of an Abelian model.

The situation changes drastically if we take into account that the Abelian electrodynamics is part of a unified model, i.e., the generator of the electromagnetic $U(1)$ subgroup is embedded into a non-Abelian gauge group of higher rank. Indeed, let us consider the simplest possible version of such an embedding, the $SU(2)$ Yang–Mills model. Suppose that the Dirac potential (1.43) with the string directed along the positive direction of the z-axis is associated with the Cartan subgroup of $SU(2)$. In other words, it is embedded into $SU(2)$ as[1] $A_\mu = A_\mu^{\text{Dirac}} \sigma_3/2$. Thus, the components of the non-Abelian potential $A_\mu = A_\mu^a \sigma^a$ are

$$A_\mu^1 = 0, \qquad A_\mu^2 = 0,$$
$$A_r^3 = 0, \qquad A_\theta^3 = 0, \qquad A_\varphi^3 = -\frac{g}{r}\frac{1+\cos\theta}{\sin\theta}. \qquad (5.1)$$

As before, we can also write $A_\mu^3 = -g(1+\cos\theta)\partial_\mu \varphi$, where the four-vector

$$\partial_\mu \varphi = \frac{1}{r\sin\theta}(0, -\sin\varphi, \cos\varphi, 0)$$

is singular along the entire z-axis (see the related discussion of (1.41) in Chap. 1).

---

[1] We have adopted a notation in which Greek indices should always be understood to run from 1 to 4 and Roman indices from 1 to 3.

Let us recall that under gauge transformations the non-Abelian vector potential transforms as

$$A_\mu \to A'_\mu = U A_\mu U^{-1} + \frac{i}{e} U \partial_k U^{-1}, \tag{5.2}$$

where the matrix of $SU(2)$ gauge transformations is defined by (3.88). The group parameters θ and φ in the standard parameterization are identified with the polar and azimuthal angles on the sphere S^2. The third angle parameter α is defined[2] on the sphere S^3.

Let us suppose $\alpha = 0$ and consider the gauge transformations rotating the unit vector on the sphere S^2 to the third axis in isospace:

$$U(\theta, \phi) = e^{i\sigma_3 \frac{\varphi}{2}} e^{i\sigma_2 \frac{\theta}{2}} e^{-i\sigma_3 \frac{\varphi}{2}} = \begin{pmatrix} \cos\frac{\theta}{2} & -\sin\frac{\theta}{2} e^{-i\varphi} \\ \sin\frac{\theta}{2} e^{i\varphi} & \cos\frac{\theta}{2} \end{pmatrix}. \tag{5.3}$$

Obviously, the range of the rotation angle φ varies from the value 4π for $\theta = 0$ to 0 for $\theta = \pi$; recall that in the general case of the $SU(N)$ group, the unit element is equal to a rotation by the angle $2\pi N$.

After some straightforward, but lengthy calculations, the gauge transformation (5.2) gives [76, 86, 270]

$$A'_n = A^a_n \frac{\sigma^a}{2} = \varepsilon_{amn} \frac{r_m}{r^2} \frac{\sigma^a}{2}. \tag{5.4}$$

This potential is obviously regular everywhere but the origin, that is the string singularity of the original expression (5.1) disappears due to some miracle! One can understand the mechanism of that miracle by taking into account that the gauge transformation (5.3) itself is singular. Indeed, the third isotopic component of the affine part of the gauge transformation

$$\left[-\frac{i}{e} U \partial_k U^{-1}\right]_3 = \frac{1}{e}(1 + \cos\theta)\partial_k \varphi,$$

possesses a singularity of the same type as the original string singularity of the embedded Dirac potential (5.1). The exact cancellation of both singularities is possible if the charge quantization condition (2.6) is satisfied.

The gauge transformation of the potential (5.1) leads us to the expression (5.4), which is just the classical solution of the pure $SU(2)$ Yang–Mills theory discovered by Wu and Yang in 1969 [528]. They also mentioned that such a solution is connected with the magnetic monopole, because the substitution of the Wu–Yang potential (5.4) into the definition of the non-Abelian field strength tensor

$$F_{\mu\nu} = \partial_\mu A_\nu - \partial_\nu A_\mu + ie[A_\mu, A_\nu] = \frac{1}{2} F^a_{\mu\nu} \sigma^a, \tag{5.5}$$

---

[2] We shall discuss the space of parameters of the gauge group $SU(2)$ in Sect. 5.3.1 below.

yields
$$F^a_{mn} = \varepsilon_{mnk}\frac{r^a r^k}{er^4} \sim \frac{1}{r^2}.\quad (5.6)$$

This suggests that such a solution could be a source of a non-Abelian Coulomb magnetic field. However, if we try naively to calculate the color magnetic charge that corresponds to this field, the integral over the color magnetic field on the surface of an infinite spatial sphere vanishes

$$\int dS_k B^a_k = \int dS_k \frac{r^a r^k}{er^4} = 0.$$

The problem could be solved, if for some reason there would be a special direction $\hat{r}_k$ in the spatial asymptotic and the non-Abelian field strength tensor could be projected onto this direction. On the other hand, the singularity at the origin of the Wu–Yang solution leads to some trouble with the definition of the energy of the configuration. We will see that the coupling of the gauge field with the Higgs field and related mechanism of spontaneous symmetry breaking shall cure both these problems.

5.1.2 Georgi–Glashow Model

The modern era of the monopole theory started in 1974, when 't Hooft and Polyakov independently discovered monopole solutions of the $SO(3)$ Georgi–Glashow model [270, 428]. The essence of this break-through is that while a Dirac monopole *could be* incorporated in an Abelian theory, some non-Abelian models, like that of Georgi and Glashow, *inevitably contain* monopole-like solutions.

For many years, starting from the pioneering paper by Dirac, the most serious argument to support the monopole concept, apart from his emotional belief that "one would be surprised if Nature had made no use of it" [200], was the possible explanation of the quantization of the electric charge. However, as time went on and the idea of grand unification emerged, it seemed that the latter argument had lost some power.

Indeed, the modern point of view is that the operator of electric charge is the generator of a $U(1)$ group. The charge quantization condition arises in models of unification if the electromagnetic subgroup is embedded into a semi-simple non-Abelian gauge group of higher rank. In this case, the electric charge generator forms nontrivial commutation relations with all other generators of the gauge group. Therefore, the electric charge quantization today is considered as an argument in support of the unification approach.

However, it turns out that both the "old" and "new" explanations of the electric charge quantization are just two sides of the same problem, because it was realized that any model of unification with an electromagnetic $U(1)$ subgroup embedded into a semi-simple gauge group, which becomes spontaneously broken by the Higgs mechanism, possesses monopole-like solutions!

An example of such a model is the well-known Georgi–Glashow model [234] with the classical Lagrangian density, which describes coupled gauge and Higgs fields:

$$L = -\frac{1}{2}\mathrm{Tr}\, F_{\mu\nu}F^{\mu\nu} + \mathrm{Tr}\, D_\mu\phi D^\mu\phi - V(\phi)$$
$$= -\frac{1}{4}F^a_{\mu\nu}F^{a\mu\nu} + \frac{1}{2}(D^\mu\phi^a)(D_\mu\phi^a) - V(\phi)\,. \quad (5.7)$$

Here, $F_{\mu\nu} = F^a_{\mu\nu}T^a$, $\phi = \phi^a T^a$ and we use standard normalization of the generators of the gauge group: $\mathrm{Tr}\,(T^a T^b) = \frac{1}{2}\delta_{ab}$, $a, b = 1, 2, 3$, which for the gauge group $G = SU(2)$ satisfy the algebra

$$[T^a, T^b] = i\varepsilon_{abc}T^c\,. \quad (5.8)$$

The generators[3] could be taken in the fundamental representation, $T^a = \frac{1}{2}\sigma^a$ or in the adjoint representation, $(T^a)_{bc} = -i\varepsilon_{abc}$. In this chapter, we will choose the adjoint representation, but in some cases we will use convenient matrix notations. Also note that there is a typical "birth-mark" that allows us to distinguish a mathematician from a theoretician: whereas, for some reasons, the former prefers to work with a field in anti-Hermitian representation of a group, the latter definitely likes Hermitian objects. For the $SU(2)$ group, for example, an anti-Hermitian basis may be taken as $T^a = -\frac{i}{2}\sigma^a$. Then the structure constants of the $su(2)$ algebra are real. Here we shall use the "physical" Hermitian basis.

The covariant derivative is defined as

$$D_\mu = \partial_\mu + ieA_\mu\,, \quad (5.9)$$

which yields

$$D_\mu\phi = \partial_\mu\phi + ie[A_\mu, \phi]\,, \quad \text{or} \quad D_\mu\phi^a = \partial_\mu\phi^a - e\varepsilon_{abc}A^b_\mu\phi^c\,, \quad (5.10)$$

and the potential of the scalar fields is taken to be

$$V(\phi) = \frac{\lambda}{4}(\phi^a\phi^a - v^2)^2\,, \quad (5.11)$$

where e and λ are gauge and scalar coupling constants, respectively. The field strength tensor is

$$F^a_{\mu\nu} = \partial_\mu A^a_\nu - \partial_\nu A^a_\mu - e\varepsilon_{abc}A^b_\mu A^c_\nu\,, \quad (5.12)$$

---

[3] Note that the $SO(3)$ group is locally isomorphic to the simply connected covering group $SU(2)$. However, in the monopole theory the global difference between these two groups is very important, because the topological properties of the corresponding group spaces are different. We discuss this situation in Sect. 5.3.1.

5.1 $SU(2)$ Georgi–Glashow Model and the Vacuum Structure

or in matrix form

$$F_{\mu\nu} = \partial_\mu A_\nu - \partial_\nu A_\mu + ie[A_\mu, A_\nu] \equiv \frac{1}{ie}[D_\mu, D_\nu]. \tag{5.13}$$

The field equations corresponding to the Lagrangian (5.7) are

$$D_\nu F^{a\mu\nu} = -e\varepsilon_{abc}\phi^b D^\mu \phi^c, \qquad D_\mu D^\mu \phi^a = -\lambda\phi^a(\phi^b\phi^b - v^2). \tag{5.14}$$

Moreover, the Bianchi identities $D^\nu \tilde{F}^a_{\mu\nu} \equiv 0$ for the dual non-Abelian field strength tensor generalize the second pair of the field equations.

The symmetric stress-energy tensor $T_{\mu\nu}$, which follows from the Lagrangian (5.7) and the field equations (5.14), is

$$\begin{aligned}T_{\mu\nu} &= F^a_{\mu\rho} F^{a\rho}_\nu + (D_\mu \phi^a)(D_\nu \phi^a) - g_{\mu\nu} L \\ &= F^a_{\mu\alpha} F^{\nu\alpha a} + D_\mu \phi^a D_\nu \phi^a - \frac{1}{2} g_{\mu\nu} D_\alpha \phi^a D^\alpha \phi^a - \frac{1}{4} g_{\mu\nu} F^a_{\alpha\beta} F^{\alpha\beta a} \\ &\quad - g_{\mu\nu} \frac{\lambda}{4}(\phi^2 - v^2), \end{aligned} \tag{5.15}$$

and is conserved by virtue of the field equations:

$$\partial_\mu T^{\mu\nu} = 0.$$

From (5.15) we can easily obtain the static Hamiltonian, aka the total energy of the system:

$$\begin{aligned}E &= \int d^3x\, T_{00} = \int d^3x \left[\frac{1}{4} F^a_{\mu\nu} F^{\mu\nu a} + \frac{1}{2}(D_\mu \phi^a)(D^\mu \phi^a) + \frac{\lambda}{4}(\phi^a\phi^a - v^2)^2\right] \\ &= \int d^3x \frac{1}{2}[E^a_n E^a_n + B^a_n B^a_n + (D_n\phi^a)(D_n\phi^a)] + V(\phi), \end{aligned} \tag{5.16}$$

where

$$E^a_n \equiv F^a_{0n} \quad \text{and} \quad B^a_n \equiv \frac{1}{2}\varepsilon_{nmk} F^a_{mk}, \tag{5.17}$$

are "color" electric and magnetic fields. We see that the energy is minimal if the following conditions are satisfied:

$$\phi^a\phi^a = v^2, \qquad F^a_{mn} = 0, \qquad D_n\phi^a = 0. \tag{5.18}$$

These conditions define the Higgs vacuum of the system and the constant v is the scalar field vacuum expectation value.

The perturbative spectrum of the model can be found from analyzing small fluctuations around the vacuum. Let us suppose that the system under consideration is static, $E^a_n = 0$. Then the energy of the vacuum is equal to zero. Furthermore, let us consider a fluctuation χ of the scalar field ϕ around the trivial vacuum $|\phi| = v$, where only the third isotopic component of the Higgs field is non-vanishing:

$$\phi = (0, 0, v + \chi). \tag{5.19}$$

Substitution of the expansion (5.19) into the Lagrangian (5.7) yields, up to terms of the second-order

$$(D_n\phi^a)(D_n\phi^a) \approx (\partial_n\chi^a)(\partial_n\chi^a) + e^2v^2\left[\left(A_n^1\right)^2 + \left(A_n^2\right)^2\right], \tag{5.20}$$

and

$$V(\phi) \approx \frac{\lambda}{2}v^2\chi^2. \tag{5.21}$$

Thus, the vacuum average of the scalar field is non-vanishing and the model describes spontaneous symmetry breaking. Further analysis given in Chap. 7 shows that the perturbative spectrum consists of a massless photon A_μ^3 corresponding to the unbroken $U(1)$ electromagnetic subgroup, massive vector fields $A_\mu^\pm = (1/\sqrt{2})(A_\mu^1 \pm A_\mu^2)$ with mass $m_v = ev$, and neutral scalars having a mass $m_s = v\sqrt{2\lambda}$.

We will discuss some properties of such fluctuations later. Here we only note that the electric charge of the massive vector bosons $A^\pm$ is given by the unbroken $U(1)$ subgroup. In general, this is a subgroup H of the gauge group G, the action of which leaves the Higgs vacuum invariant. Obviously, this is a little group of the rotation in isospace about the direction given by the vector ϕ^a. The generator of it, $(\phi^a T^a)/v$, must be identified with the operator of electric charge Q. Thus, the expression for the covariant derivative (5.10) can be written in the form

$$D_\mu = \partial_\mu + ieA_\mu^a T^a = \partial_\mu + iQA_\mu^{\text{em}}, \tag{5.22}$$

which allows us to define the "electromagnetic projection" of the gauge potential

$$A_\mu^{\text{em}} = \frac{1}{v}A_\mu^a\phi^a, \quad Q = \frac{e}{v}\phi^a T^a. \tag{5.23}$$

Taking into account the definition of the generators T^a of the gauge group, we can easily see that the minimal allowed eigenvalues of the electric charge operator are now $q = \pm e/2$.

5.1.3 Topological Classification of the Solutions

> "The fox knows many things,
> but the hedgehog knows one big thing"
> Archiolus

The spectrum of possible solutions of the Georgi–Glashow model is much richer than one would naively expect. There are stable soliton-like static solutions of the complicated system of field equations (5.14) having a finite energy density on the spatial asymptotic. An adequate description of these objects needs the topological methods we discussed in Chap. 3.

5.1 SU(2) Georgi–Glashow Model and the Vacuum Structure 147

Indeed, the very definition (5.18) forces the classical vacuum of the Georgi–Glashow model to be degenerated. The condition $V(\phi) = 0$ means that $|\phi| = v$, i.e., the set of vacuum values of the Higgs field forms a sphere S^2_{vac} of radius v in $d = 3$ isotopic space. All the points on this sphere are equivalent because there is a well-defined $SU(2)$ gauge transformation that connects them.

The solutions of the classical field equations map the vacuum manifold $\mathcal{M} = S^2_{\text{vac}}$ onto the boundary of 3-dimensional space, which is also a sphere S^2. These maps are charactered by a *winding number* $n = 0, \pm 1, \pm 2 \ldots$, which is the number of times S^2_{vac} is covered by a single turn around the spatial boundary S^2. The crucial point is that the solutions having a finite energy on the spatial asymptotic could be separated into different classes according to the behavior of the field ϕ^a. The trivial case is that the isotopic orientation of the fields does not depend on the spatial coordinates and asymptotically the scalar field tends to the limit

$$\phi^a = (0, 0, v). \tag{5.24}$$

This situation corresponds to the winding number $n = 0$.

We can also consider another type of solutions with the property that the direction of isovector and isoscalar fields in isospace are functions of the spatial coordinates. One could suppose that since the absolute minimum of the energy corresponds to the trivial vacuum, such configurations would be unstable. However, their stability will be secured by the topology: if we try to deform the fields continuously to the trivial vacuum (5.24), the energy functional would tend to infinity. In other words, all the different topological sectors are separated by infinite barriers.

To construct the solutions corresponding to the non-trivial minimum of the energy functional (5.16), we again consider the scalar field on the spatial asymptotic $r \to \infty$, taking values on the vacuum manifold $|\phi| = v$. However, we suppose that the isovector of the scalar field is now directed in the isotopic space along the direction of the radius vector on the spatial asymptotic[4]

$$\phi^a \xrightarrow[r \to \infty]{} \frac{vr^a}{r}. \tag{5.25}$$

This asymptotic behavior obviously mixes the spatial and isotopic indices and defines a single mapping of the vacuum $\mathcal{M}$ onto the spatial asymptotic. A single turn around the boundary S^2 leads to a single closed path on the sphere S^2_{vac} and the winding number of such a mapping is $n = 1$.

As was mentioned by 't Hooft [270], the configurations that are characterized by different winding numbers cannot be continuously deformed into each other. Indeed, we have seen that the gauge transformation (5.3) of the form $U = e^{i(\sigma_k \hat{r}_k)\theta/2}$ rotates the isovector to the third axis. However, if we

---

[4] In the pioneering paper by Polyakov [428] this solution was coined a "hedgehog".

try to "comb the hedgehog", that is, to rotate the scalar field everywhere in space to a given direction (so-called *unitary* or *Abelian gauge*), the singularity of the gauge transformation on the south pole does not allow us to do it globally. Thus, there is no well-defined global gauge transformation that connects the configurations (5.24) and (5.25) and this singularity results in the infinite barrier separating them.

5.1.4 Definition of Magnetic Charge

The condition of vanishing covariant derivative of the scalar field on the spatial asymptotic (5.18) together with the choice of the nontrivial hedgehog configuration implies that at $r \to \infty$

$$\partial_n \left(\frac{r^a}{r}\right) - e\varepsilon_{abc} A_n^b \frac{r^c}{r} = 0 \,. \tag{5.26}$$

The simple transformation

$$\partial_n \left(\frac{r^a}{r}\right) = \frac{r^2 \delta_{an} - r_a r_n}{r^3} = \frac{1}{r}\left(\delta_{an}\delta_{ck} - \delta_{ak}\delta_{nc}\right) \frac{r_c r_k}{r^2} = -\varepsilon_{abc}\varepsilon_{bnk} \frac{r_c r_k}{r^3} \,,$$

then provides an asymptotic form of the gauge potential

$$A_k^a(r) \xrightarrow[r\to\infty]{} \frac{1}{e}\varepsilon_{ank}\frac{r_n}{r^2} \,. \tag{5.27}$$

This corresponds to the non-Abelian magnetic field

$$B_n^a \xrightarrow[r\to\infty]{} \frac{r_a r_n}{er^4} \,. \tag{5.28}$$

Therefore, the boundary conditions (5.25) and (5.27) are compatible with the existence of a long-range gauge field associated with an Abelian subgroup that is unbroken in the vacuum. Since this field falls off like $1/r^2$, which characterizes the Coulomb-like field of a point charge, and since the electric components of the field strength tensor (5.12) vanish, this regular field configuration can be identified with a monopole.

To prove it, we first have to define the electromagnetic field strength tensor. Recall that the unbroken electromagnetic subgroup $U(1)$ is associated with rotations about the direction of the isovector ϕ. Thus, it would be rather natural to introduce the electromagnetic potential as a projection of the $SU(2)$ gauge potential A_μ^a onto that direction, see (5.23). Furthermore, as was mentioned in the paper [177], a general solution of the equation $D_\mu \phi^a = 0$, for $\phi^a \phi^a = v^2$ can be written as

$$A_\mu^a = \frac{1}{v^2 e}\varepsilon_{abc}\phi^b \partial_\mu \phi^c + \frac{1}{v}\phi^a \Lambda_\mu \,, \tag{5.29}$$

5.1 SU(2) Georgi–Glashow Model and the Vacuum Structure

where Λ_μ is an arbitrary four-vector. It can be identified with the electromagnetic potential because (5.29) yields for $\phi^a \phi^a = v^2$:

$$\frac{\phi^a}{v} A_\mu^a = \Lambda_\mu \equiv A_\mu^{\text{em}}.$$

Inserting (5.29) into the definition of the field strength tensor (5.12) yields

$$F_{\mu\nu}^a = F_{\mu\nu} \frac{\phi^a}{v}, \quad \text{where} \quad F_{\mu\nu} = \partial_\mu A_\nu - \partial_\nu A_\mu + \frac{1}{v^3 e} \varepsilon_{abc} \phi^a \partial_\mu \phi^b \partial_\nu \phi^c. \tag{5.30}$$

This gauge-invariant definition of the electromagnetic field strength tensor $F_{\mu\nu}$, given in [49], is close to the original definition of the 't Hooft tensor [270]:

$$\mathcal{F}_{\mu\nu} = \text{Tr}\left\{\hat\phi F_{\mu\nu} - \frac{i}{2e}\hat\phi D_\mu \hat\phi D_\nu \hat\phi\right\} = \hat\phi^a F_{\mu\nu}^a + \frac{1}{e}\varepsilon_{abc}\hat\phi^a D_\mu \hat\phi^b D_\nu \hat\phi^c, \tag{5.31}$$

where $\hat\phi^a = \phi^a/|\phi|$ is a normalized Higgs field. Obviously, both definitions coincide on the spatial boundary. The difference is that the 't Hooft tensor (5.31) is singular at the zeros of the Higgs field, while (5.30) is regular everywhere. These zeros, as we will see, are associated with positions of the monopoles.

Note that the definition of an electromagnetic field strength tensor is always somewhat arbitrary in a non-Abelian gauge theory, for example, one can also consider $\mathcal{F}_{\mu\nu} = \hat\phi^a F_{\mu\nu}^a$ [49, 367].

It is rather obvious that in the topologically trivial sector (5.24), the last term in (5.30) vanishes and then we have

$$F_{\mu\nu} = \partial_\mu A_\nu - \partial_\nu A_\mu.$$

This is precisely the case of standard Maxwell electrodynamics. Of course, in this sector there is no place for a monopole, because the Bianchi identities are satisfied: $\partial^\mu \widetilde{F}_{\mu\nu} \equiv 0$.

However, for the configuration with non-trivial boundary conditions (5.25) and (5.27), the Higgs field also gives a non-vanishing contribution to the electromagnetic field strength tensor (5.30). Then, the second pair of Maxwell equations becomes modified:

$$\partial^\mu \widetilde{F}_{\mu\nu} = k_\nu. \tag{5.32}$$

Note that if the electromagnetic potential A_μ^{em} is regular, the magnetic, or topological current k_μ is expressed via the scalar field alone

$$k_\mu = \frac{1}{2}\varepsilon_{\mu\nu\rho\sigma}\partial^\nu F^{\rho\sigma} = \frac{1}{2v^3 e}\varepsilon_{\mu\nu\rho\sigma}\varepsilon_{abc}\partial^\nu \phi^a \partial^\rho \phi^b \partial^\sigma \phi^c. \tag{5.33}$$

From the first glance this current is independent of any property of the gauge field. It is conserved by its very definition:

$$\partial_\mu k^\mu \equiv 0, \qquad (5.34)$$

unlike a Noether current that is conserved because of some symmetry of the model.

Now we can justify the definition of magnetic charge [76]. According to (5.33)

$$\begin{aligned} g &= \int d^3x\, k_0 = \frac{1}{2ev^3} \int d^3x\, \varepsilon_{abc}\varepsilon_{mnk}\, \partial_m \left(\phi^a \partial_n \phi^b \partial_k \phi^c\right) \\ &= -\frac{1}{2ev^3} \int d^2 S_n\, \varepsilon_{abc}\varepsilon_{mnk}\, \phi^a \partial_n \phi^b \partial_k \phi^c, \end{aligned} \qquad (5.35)$$

where the last integral is taken over the surface of the sphere S^2 on the spatial asymptotic. One can parameterize it by local coordinates ξ_α, $\alpha = 1, 2$. Then we can write

$$\partial_n \phi^a = \frac{\partial \xi^\alpha}{\partial r^n}\frac{\partial \phi^a}{\partial \xi^\alpha}, \qquad d^2 S_n = \frac{1}{2}\,\varepsilon_{nmk}\frac{\partial r^m}{\partial \xi^\alpha}\frac{\partial r^k}{\partial \xi^\beta}\varepsilon_{\alpha\beta}d^2\xi. \qquad (5.36)$$

After some simple algebra we arrive at

$$g = \frac{1}{2ev^3}\int d^2\xi\, \varepsilon_{\alpha\beta}\varepsilon_{abc}\, \phi^a \partial_\alpha \phi^b \partial_\beta \phi^c = \frac{1}{e}\int d^2\xi\,\sqrt{\mathbf{g}}, \qquad (5.37)$$

where $\mathbf{g} = \det(\partial_\alpha \hat\phi^a \partial_\beta \hat\phi^a)$ is the determinant of the metric tensor on the S^2_{vac} sphere in isospace. The magnetic charge is proportional to an integer n, which mathematicians refer to as the Brouwer degree. The geometrical interpretation of this integer is clear: it is the number of times the isovector ϕ^a covers the sphere S^2_{vac}, while r^a covers the sphere S^2 on the spatial asymptotic once. Thus [76]:

$$g = \frac{4\pi n}{e}, \qquad n \in \mathbb{Z}, \qquad (5.38)$$

where the factor 4π is due to integration over the unit sphere. This is the non-Abelian analog of the Dirac charge quantization condition (2.6).

Another remark about the definition of the magnetic charge is that the Brouwer degree and homotopic classification are equivalent to the Poincaré–Hopf index [76]. The latter is defined as a mapping of a sphere S^2 surrounding an isolated point r_0, where the scalar field vanishes, i.e., $\phi(r_0) = 0$, onto a sphere of unit radius S^2_ϕ. In other words, the magnetic charge of an arbitrary field configuration can be defined as a sum of the Poincaré–Hopf indices i of non-degenerated zeros $r_0^{(k)}$ of the Higgs field:

$$g = \frac{4\pi}{e}\sum_k i(r_0^{(k)}). \qquad (5.39)$$

Indeed, if we consider a scalar field that is constant everywhere in space and satisfies the boundary condition (5.24), it has no zero at all. Thus, the

5.1 $SU(2)$ Georgi–Glashow Model and the Vacuum Structure

Poincaré–Hopf index aka magnetic charge is equal to zero. However, in the case of the hedgehog configuration, which satisfies the boundary condition (5.25)

$$\phi^a = r^a h(r), \tag{5.40}$$

where $h(r)$ is a smooth function having no zeros, there is a single zero at the origin. Thus, $i(0) = 1$ and this is a configuration of unit magnetic charge.

This approach allows us to identify monopoles according to the positions of zeros of the Higgs field. Such an identification is very useful from the point of view of constructing multimonopole solutions, which we will consider in the following chapter. However, first we have to find a solution of the field equations (5.14), that would satisfy the boundary conditions (5.25) and (5.27).

5.1.5 't Hooft–Polyakov Ansatz

We showed that, asymptotically, the monopole field configuration must satisfy the conditions (5.25) and (5.27). Now, we try to define the structure functions that form the radial shape of the monopole. As usual, this problem can be simplified, if we take into account the constraints following from the symmetries of the configuration.

Note that we consider static fields. This condition leaves only rotational $SO(3)$ symmetry from the original Poincaré invariance of the Lagrangian (5.7). Therefore, the full invariance group of the system is $SO(3) \times SO(3)$, the product of spatial and isotopic rotations. Moreover, the non-trivial asymptotic of the Higgs field (5.25) corresponds to the symmetry with respect to the transformation from the diagonal $SO(3)$, subgroup which mixes spatial and group rotations. Thus, one can make use of the ansatz [270, 428]:

$$\phi^a = \frac{r^a}{er^2} H(\xi), \quad A_n^a = \varepsilon_{amn} \frac{r^m}{er^2}[1 - K(\xi)], \quad A_0^a = 0, \tag{5.41}$$

where $H(\xi)$ and $K(\xi)$ are functions of the dimensionless variable $\xi = ver$. The explicit forms of these shape functions of the scalar and gauge field can be found from the field equations. However, it would be much more convenient to make use of the condition that the monopole solution corresponds to a local minimum of the energy functional. Substituting the ansatz (5.41) back into (5.16), we have

$$E = \frac{4\pi v}{e} \int_0^\infty \frac{d\xi}{\xi^2} \left[\xi^2 \left(\frac{dK}{d\xi}\right)^2 + \frac{1}{2}\left(\xi \frac{dH}{d\xi} - H\right)^2 \right. \tag{5.42}$$

$$\left. + \frac{1}{2}\left(K^2 - 1\right)^2 + K^2 H^2 + \frac{\lambda}{4e^2}\left(H^2 - \xi^2\right)^2 \right].$$

Variations of this functional with respect to the functions H and K yields

$$\xi^2 \frac{d^2 K}{d\xi^2} = KH^2 + K(K^2 - 1), \qquad \xi^2 \frac{d^2 H}{d\xi^2} = 2K^2 H + \frac{\lambda}{e^2} H(H^2 - \xi^2). \quad (5.43)$$

The functions K and H must satisfy the following boundary conditions:

$$\begin{aligned} K(\xi) &\to 1, & H(\xi) &\to 0 & \text{as} \quad \xi &\to 0, \\ K(\xi) &\to 0, & H(\xi) &\to \xi & \text{as} \quad \xi &\to \infty, \end{aligned} \quad (5.44)$$

which correspond to the asymptotics (5.25) and (5.27). Indeed, the substitution of the ansatz (5.41) into the expressions for the covariant derivative of the scalar field and the non-Abelian magnetic field yields

$$D_n \phi^a = \frac{\delta_{an}}{er^2} KH + \frac{r^a r^n}{er^4} \left(\xi \frac{dH}{d\xi} - H - KH \right) \xrightarrow[r \to \infty]{} 0,$$

$$B_n^a = \frac{r_n r^a}{er^4} \left(1 - K^2 + \xi \frac{dK}{d\xi} \right) - \frac{\delta_{an}}{er^2} \xi \frac{dK}{d\xi} \xrightarrow[r \to \infty]{} \frac{r_n r^a}{er^4}. \quad (5.45)$$

Evidently, in the Higgs vacuum the vector potential of the gauge field takes the form of the Wu–Yang potential (5.4). However, unlike this potential, the configuration (5.41) is regular everywhere and corresponds to finite energy both at the origin and at the spatial boundary.

Let us note that in the Higgs vacuum, $D_n \phi^a = 0$ and the electromagnetic field strength is $F_{\mu\nu} = \phi^a F_{\mu\nu}^a / v$. Clearly, the magnetic charge could be calculated as an integral over the surface of the sphere S^2 on spatial infinity (compare with (5.35)):

$$g = \frac{1}{v} \int d^2 S_n B_n = \frac{1}{v} \int d^2 S_n B_n^a \phi^a = \frac{1}{v} \int d^3 x B_n^a D_n \phi^a, \quad (5.46)$$

where we make use of the Bianchi identity for the tensor of non-Abelian magnetic field $D_n B_n^a = 0$. Substituting the ansatz (5.41), we obtain

$$\begin{aligned} g &= \frac{4\pi}{e} \int_0^\infty \frac{d\xi}{\xi^2} \left\{ (K^2 - 1)(H - \xi H') - 2\xi K' KH \right\} \\ &= \frac{4\pi}{e} \int_0^\infty d\xi \frac{d}{d\xi} \left\{ \frac{1 - K^2}{\xi} \right\} = \frac{4\pi}{e}. \end{aligned} \quad (5.47)$$

Again, we see that the boundary conditions (5.44) correspond to a monopole of unit magnetic charge.

Unfortunately, the system of non-linear coupled differential equations (5.43) in general has no analytical solution. The only known exception is the very special case $\lambda = 0$ [127, 171, 431]. This is the so-called *Bogomol'nyi–Prasad–Sommerfield (BPS) limit*, which deserves a special consideration. However, before we come to this limit, we have to describe the general properties of the non-Abelian monopole.

5.1 SU(2) Georgi–Glashow Model and the Vacuum Structure

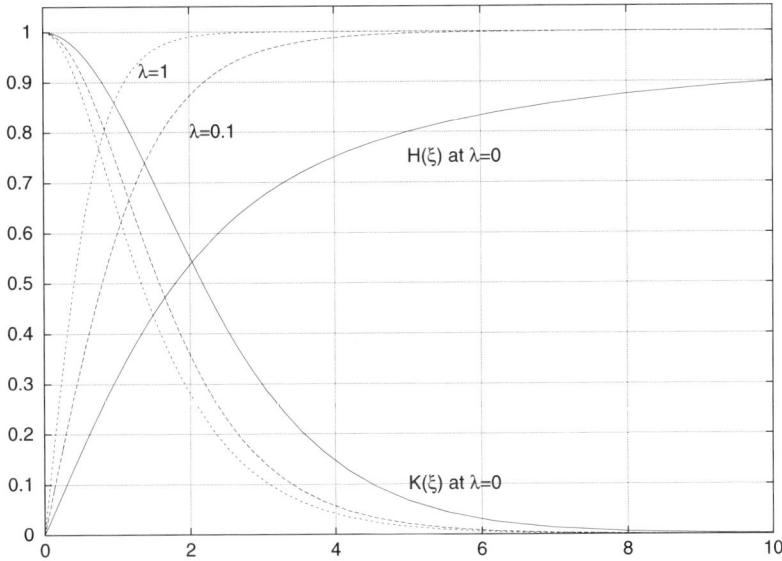

Fig. 5.1. The profile functions $K(\xi)$ and $H(\xi)/\xi$ are shown for the 't Hooft–Polyakov monopole at $\lambda = 0$, (BPS limit) $\lambda = 0.1$ and $\lambda = 1$

Numerical solutions of the system (5.43) were discussed in the papers [85, 315]. It turns out that the shape functions $H(\xi)$ and $K(\xi)$ approach rather fast to the asymptotic values (see Fig. 5.1).

Thus, there is a Higgs vacuum outside of some region of the order of the characteristic scale R_c, which is called the *core* of the monopole. One could estimate this size by simple arguments [55]. The total energy of the monopole configuration consists of two components: the energy of the Abelian magnetic field outside the core and the energy of the scalar field inside the core:

$$E = E_{\text{mag}} + E_s \sim 4\pi g^2 R_c^{-1} + 4\pi v^2 R_c \sim \frac{4\pi}{e^2}\left(R_c^{-1} + m_v^2 R_c\right).$$

This sum is minimal if $R_c \sim m_v^{-1}$. In other words, inside the core at distances shorter than the wavelength of the vector boson $m_v^{-1} \sim (ve)^{-1}$, the original $SU(2)$ symmetry is restored. However, outside the core this symmetry is spontaneously broken down to the Abelian electromagnetic subgroup. In this sense there is no difference between the 't Hooft–Polyakov and the Dirac monopole outside the core [49].

Numerical calculations [85, 315] show that the mass of the monopole, which in the classical static case is identical to its energy, depends on the scalar coupling constant as

$$M = \frac{4\pi v}{e} f\left(\frac{\lambda}{e^2}\right). \tag{5.48}$$

The smooth function $f(\lambda/e^2)$ is a monotonically increasing function, interpolating between the limits [315]

$$f(0) = 1, \qquad f(\infty) = 1.787. \qquad (5.49)$$

The reason why the mass becomes independent of the values of the coupling constant λ for $\lambda \gg 1$ is that in this limit the scalar field approaches the asymptotic form faster than the vector field. The correction to the monopole mass connected with the scalar field is of the order $\Delta E \sim M(m_v/m_s)$, which is negligible for large values of m_s. A high precision numerical analysis of the monopole mass was performed recently [221].

5.1.6 Singular Gauge Transformations and the Connection between 't Hooft–Polyakov and Dirac Monopoles

We showed above that in the Higgs vacuum, the gauge symmetry is spontaneously broken down to the Abelian subgroup. This means that outside the monopole core, there is no difference between the 't Hooft–Polyakov and the Dirac monopoles. At a first glance, there is a disagreement because even though the magnetic charge in each case has a topological root, the origin seems to be different. Indeed, in the former case it looks like the topological current (5.33) and the topological charge associated with it are connected with the mapping of the scalar field onto the spatial asymptotic and could be defined in a formal way even in the absence of the gauge field A_μ^a. In contrast to this, in the case of the Dirac monopole, the topological properties of the gauge field completely defined the Abelian magnetic charge that is associated with the first Chern class.

This puzzle could be solved, if we note that the gauge transformation of type (5.3) could change the homotopy class of the configuration[5] [76]. Indeed, we have already mentioned that the singular gauge transformations (5.3) could rotate the "hedgehog-like" Higgs field (5.25) on the spatial asymptotic S^2 to the unitary gauge:

$$\hat{\phi}^a = (\sin\theta\cos\varphi, \sin\theta\sin\varphi, \cos\theta) \to U^{-1}\hat{\phi}^a U = (0, 0, 1). \qquad (5.50)$$

This transformation is well-defined everywhere on the sphere S^2, except for the south pole $\theta = \pi$. This is exactly the singularity that corresponds to the Dirac string in the Abelian theory [76]. Indeed, in the Higgs vacuum the vector potential has the form of the Wu–Yang potential (5.4). As we have shown, if we perform the gauge transformation (5.3), the singularity of the term $U^{-1}\partial_n U$ exhibits the Dirac string:

---

[5] However, it cannot, of course, change the topology!

5.1 $SU(2)$ Georgi–Glashow Model and the Vacuum Structure

$$A_n = A_n^a \frac{\sigma^a}{2} = \varepsilon_{amn} \frac{r_m}{r^2} \frac{\sigma^a}{2}$$

$$\rightarrow U^{-1} A_n U - \frac{i}{e} U^{-1} \partial_n U = \frac{1}{2r} \frac{1-\cos\theta}{\sin\theta} \hat{\varphi}_n \sigma_3. \qquad (5.51)$$

In this Abelian gauge, the monopole potential is just the celebrated Dirac potential embedded into the $SU(2)$ group.

Note that the singular character of the gauge transformation (5.3) requires us to be very careful in the calculations. It is a good idea to make use of a regularized expression. Such a procedure (see Appendix A) demonstrates that the magnetic field of the non-Abelian monopole is purely Coulomb-like and has no string-like singularity, neither in the Abelian nor in the "hedgehog" gauges [49, 131].

It was noted in [76, 86] that the gauge-invariant electromagnetic field strength tensor (5.30)

$$F_{\mu\nu} = \partial_\mu A_\nu - \partial_\nu A_\mu + \frac{1}{v^3 e} \varepsilon_{abc} \phi^a \partial_\mu \phi^b \partial_\nu \phi^c,$$

is composed of two parts that are not separately invariant with respect to the gauge transformations (5.3). These two parts are connected with the contributions of the gauge and the Higgs field, respectively. Thus, the singular gauge transformation changes the homotopy class of the configuration; the Brouwer degree of the mapping is equal to 0 in the Abelian gauge, whereas in the "hedgehog" gauge it is equal to 1. Of course, the topological (alias magnetic) charge does not change, because the singularity of the gauge transformation (5.3) corresponds to a non-trivial contribution to the electromagnetic field strength. The difference is that in the unitary gauge, the contribution of the scalar field to the magnetic charge completely vanishes. In other words, the gauge transformation (5.3) has "transferred" the magnetic charge from the scalar to the gauge field. Below, we justify this conclusion by demonstrating the equivalence between the topological properties of the Higgs and the gauge fields.

5.1.7 Dyons

Note that the solution given by the 't Hooft–Polyakov ansatz (5.41) corresponds to the condition $A_0^a = 0$. One could consider a more general case, where this time component of the vector potential is not equal to zero, but is also a function of the spatial coordinates [304]:

$$A_0^a = \frac{r^a}{er^2} J(r). \qquad (5.52)$$

This field configuration corresponds to the non-Abelian *dyon*, which has both magnetic and electric charges. Indeed, by analogy with (5.46), the electric charge of the system of the fields can be defined as

$$q = \frac{1}{v} \int dS_n E_n = \frac{1}{v} \int dS_n E_n^a \phi^a = \frac{1}{v} \int d^3x E_n^a D_n \phi^a . \quad (5.53)$$

Here, we invoked the field equations (5.14), according to which $D_n E_n^a = 0$, and made use of the relation $\varepsilon_{abc} \phi^b D_0 \phi^c = 0$, which is valid for the ansatz under consideration. The magnetic charge of the dyon is as before given by the formula (5.46). However, on the classical level there is no reason for the electric charge (5.53), unlike the magnetic charge, to be quantized.

The system of differential equations for the radial shape functions of the dyon is different from (5.43):

$$\xi^2 \frac{d^2 K}{d\xi^2} = K(H^2 - J^2) + K(K^2 - 1),$$

$$\xi^2 \frac{d^2 H}{d\xi^2} = 2K^2 H + \frac{\lambda}{e^2} H(H^2 - \xi^2),$$

$$\xi^2 \frac{d^2 J}{d\xi^2} = 2K^2 J. \quad (5.54)$$

This system can be solved numerically. Note that the asymptotic behavior of the profile function $J(r)$ is very similar to that of the scalar field (see (5.44)):

$$J(r) \to 0, \quad \text{as} \quad r \to 0, \qquad J(r) \to Cr \quad \text{as} \quad r \to \infty. \quad (5.55)$$

The arbitrary constant C is connected with the electric charge of the dyon (5.53) [304]. The charge vanishes if $C = 0$.

Indeed, substituting the ansatz (5.41) into the integral (5.53), by analogy with (5.47) we obtain

$$q = \frac{4\pi}{e} \int_0^\infty \frac{d\xi}{\xi^2} \left\{ 2JHK^2 + \xi^2 J'H' + JH - \xi(J'H + H'J) \right\}$$

$$= \frac{4\pi}{e} \int_0^\infty d\xi \frac{d}{d\xi} \left\{ \xi H \frac{d}{d\xi} \left(\frac{JH}{\xi} \right) \right\} = \frac{4\pi C}{e} = Cg, \quad (5.56)$$

where the magnetic charge of the dyon g is as before given by the formula (5.46). However, on the classical level there is no reason for the electric charge (5.53), unlike the magnetic charge, to be quantized and the constant C in (5.56) remains an arbitrary parameter.

Finally, we note that the time component of the vector-potential (5.52) is in isospace parallel to the direction of the Higgs field. Moreover, one can consider it as an additional triplet of the scalar fields. This is the so-called Julia–Zee correspondence $\phi^a \rightleftharpoons A_0^a$.

5.2 The Bogomol'nyi Limit

Let us try to find a lower bound on the monopole mass. We start from the energy of the static monopole configuration (5.16) and write it in a general form as [49, 171]

$$E = \int d^3x \left\{ \frac{1}{2} [E_n^a E_n^a + B_n^a B_n^a + (D_n \phi^a)(D_n \phi^a)] + V(\phi) \right\}$$

$$= \frac{1}{2} \int d^3x \, (E_n^a - D_n \phi^a \sin \alpha)^2 + \frac{1}{2} \int d^3x \, (B_n^a - D_n \phi^a \cos \alpha)^2$$

$$+ \sin \alpha \int d^3x \, E_n^a D_n \phi^a + \cos \alpha \int d^3x \, B_n^a D_n \phi^a + \int d^3x \, V(\phi). \quad (5.57)$$

Here, α is an arbitrary real angular parameter. Obviously, the minimum of the energy corresponds to the situation when the potential of the scalar field vanishes and the the following equations hold [127, 171, 431]:

$$E_n^a = D_n \phi^a \sin \alpha, \qquad B_n^a = D_n \phi^a \cos \alpha. \quad (5.58)$$

These are the *Bogomol'nyi–Prasad–Sommerfield (BPS) equations*.

Note that the electric and magnetic charges of the configuration are given by (5.53) and (5.46), respectively. Thus, we can write

$$E = M \geq v(q \sin \alpha + g \cos \alpha). \quad (5.59)$$

The energy, as a function of α, has an extreme at $\tan \alpha = q/g$, which provides us with a lower bound on the dyon mass

$$M \geq v\sqrt{q^2 + g^2} = v|q + ig|. \quad (5.60)$$

This condition, known as the *Bogomol'nyi bound*, will be a key to what follows.

The situation is simple to analyze if the electric charge q vanishes. We now substitute the 't Hooft–Polyakov ansatz (5.41) into the BPS equation:

$$B_n^a = D_n \phi^a. \quad (5.61)$$

The result is the system of coupled differential equations of first-order

$$\xi \frac{dK}{d\xi} = -KH, \qquad \xi \frac{dH}{d\xi} = H + (1 - K^2), \quad (5.62)$$

which have an analytical solution in terms of elementary functions:

$$K = \frac{\xi}{\sinh \xi}, \qquad H = \xi \coth \xi - 1. \quad (5.63)$$

Note, that the solution to the first-order BPS equation (5.61) automatically satisfies the system of field equations of the second-order, (5.14). Of course, the Euler–Lagrange equations (5.14) could have other solutions that correspond to some local minima of the action, but the vacuum state with minimal energy in the given topological sector must be just the Bogomol'nyi state.

As was mentioned in [507], the BPS equation together with the Bianchi identity means that $D_n D_n \phi^a = 0$, which precisely corresponds to the field (5.14). Therefore, the condition

$$D_n \phi^a D_n \phi^a = (\partial_n \phi^a)(\partial_n \phi^a) + \phi^a (\partial_n \partial_n \phi^a) = \frac{1}{2} \partial_n \partial_n (\phi^a \phi^a)$$

holds. The energy of the monopole configuration in the BPS limit is independent on the properties of the gauge field and completely defined by the Higgs field alone [58, 507]:

$$E = \frac{1}{2} \int d^3 x \, \partial_n \partial_n (\phi^a \phi^a) = \frac{4\pi v}{e} \int_0^\infty d\xi \frac{d}{d\xi} \left[\xi H \frac{d}{d\xi} \left(\frac{H}{\xi} \right) \right]$$

$$= \frac{4\pi v}{e} \left(\coth \xi - \frac{1}{\xi} \right) \left(1 - \frac{\xi^2}{\sinh^2 \xi} \right) \bigg|_0^\infty = \frac{4\pi v}{e} .$$

Making use of the Julia–Zee correspondence, it is easy to see that in the BPS Limit, the dyon solution could be constructed by a simple rotation of the pure monopole solution (5.63) [49, 127, 431]:

$$H \to H = \frac{1}{\cos \alpha} (\xi \coth \xi - 1) , \quad J = \tan \alpha \, (\xi \coth \xi - 1) , \qquad (5.64)$$

while the gauge field profile function remains unchanged:

$$K = \frac{\xi}{\sinh \xi} .$$

This rotation affects the vacuum expectation value of the scalar field as $v \to v/\cos \alpha$.

Clearly, the parameter α is related to the electric charge of a dyon. Indeed, according to the definition of the angle α above, this electric charge is

$$q = g \tan \alpha .$$

Comparing this equation to (5.56), we can identify the constant that appears in the latter relation as $C = \tan \alpha$. Recall that the state with $q = 0$ corresponds to the absolute minimum of the energy. Below, we discuss the connection between the generation of electric charge of a dyon and the so-called gauge zero mode.

In comparison with the 't Hooft–Polyakov solution, the behavior of the Higgs field of the monopole in the BPS limit has changed drastically. As we can see from (5.63), alongside with the exponentially decaying component it also obtains a long-distance Coulomb tail

$$\phi^a \to v\hat{r}^a - \frac{r^a}{er^2} \quad \text{as} \quad r \to \infty. \tag{5.65}$$

The reason for this is that in the limit $V(\phi) = 0$, the scalar field becomes massless. Because an interaction, which is mediated by a massless scalar field, always leads to attraction, the picture of the interaction between the monopoles is very different in the BPS limit, as compared with the naive picture based on electromagnetic interaction. As we will see in Chap. 6, this long-range monopole-monopole interaction is composed of two parts originating from the long-range scalar force and the standard electromagnetic interaction, which could be either attractive or repulsive [366]. Mutual compensation of both contributions leaves the pair of BPS monopoles static but the monopole and anti-monopole would interact with double strength.

Many of the remarkable properties of the BPS equation (5.61) are connected with its property of integrability. As was pointed out by Manton [367], integrability of the BPS system is connected with a one-to-one correspondence between the system of BPS equations and the reduced equations of self-duality of the pure Euclidean Yang–Mills theory. Indeed, the Julia–Zee correspondence means that

$$D_n \phi^a \rightleftharpoons D_n A_0^a \equiv F_{0n}^a,$$
$$B_n^a = D_n \phi^a \rightleftharpoons \widetilde{F}_{0n}^a = F_{0n}^a. \tag{5.66}$$

Therefore, if we suppose that all the fields are static, the Euclidean equations of self-duality $F_{\mu\nu}^a = \widetilde{F}_{\mu\nu}^a$ reduce to the equations (5.61) and the monopole solutions in the Bogomol'nyi limit could be considered as a special class of self-dual fields.

Of course, it would not be quite correct to make a direct identification between these fields and instantons, because the instanton configuration could be independent from Euclidean time only in the limit of infinite action. Nevertheless, this analogy opens a way to apply in the $d = 3+1$ monopole theory the same very powerful methods of algebraic and differential geometry that were used to construct multi-instanton solutions of the self-duality equations in $d = 4$ [39]. In particular, in the case of the BPS monopole, the solution of the self-duality equations could be constructed on the ansatz of Corrigan and Fairlie [179]. We will discuss these very exciting topics in Chap. 6.

The analogy between the Euclidean Yang–Mills theory and the BPS equations can be traced up to the solutions. It was shown [443, 444] that the solutions of these equations are exactly equal to an infinite chain of instantons directed along the Euclidean time axis t in $d = 4$. It has also been shown by Manton [367] that such a multi-instanton configuration can be written in the 't Hooft ansatz with the help of a superpotential $\rho(r,t)$ as:

$$A_n^a = \varepsilon_{anm}\partial_m \ln \rho + \delta_{an}\partial_0 \ln \rho, \qquad A_0^a = -\partial_a \ln \rho, \tag{5.67}$$

where the sum over the infinite number of instantons is performed in the superpotential:

5 't Hooft–Polyakov Monopole

$$\rho = \sum_{n=-\infty}^{n=\infty} \frac{1}{\xi^2 + (\tau - 2\pi n)^2}, \quad \text{where} \quad \xi = ver, \quad \tau = vet.$$

Here, the distance between the neighboring instantons is equal to 2π in units of τ and the size of the instanton is equal to one in units of ξ.

Rossi noted [443] that this sum over instantons could be calculated analytically. Indeed, the superpotential can be decomposed into two sums over Matsubara frequencies $\omega_n = 2\pi n$, which are well-known from statistical physics:

$$\rho = \frac{1}{2\xi} \left\{ \sum_{n=-\infty}^{n=\infty} \frac{1}{\xi + i\tau - 2i\pi n} + \sum_{n=-\infty}^{n=\infty} \frac{1}{\xi - i\tau + 2i\pi n} \right\}. \tag{5.68}$$

Introducing the complex variable $z = \xi + i\tau$, we can write

$$\rho = \frac{1}{2\xi} \left\{ \sum_{n=-\infty}^{n=\infty} \frac{1}{z - i\omega_n} + \sum_{n=-\infty}^{n=\infty} \frac{1}{z^* + i\omega_n} \right\} = \frac{1}{2\xi} \left\{ \coth \frac{z}{2} + \coth \frac{z^*}{2} \right\}$$

$$= \frac{1}{2\xi} \frac{\sinh \xi}{\cosh \xi - \cos \tau}. \tag{5.69}$$

Substitution of this result into the potential (5.67) corresponds to the "dyon in the 't Hooft gauge". This solution is periodic in time. However, the time-dependent periodic gauge transformation of the form [367]

$$U = \exp \left\{ \frac{i}{2v} \hat{r}^a \sigma^a \omega \right\}, \quad \text{where} \quad \tan \omega = \frac{\sin \tau \sinh \xi}{\cosh \xi \cos \tau - 1}, \tag{5.70}$$

transforms the infinite chain of instantons (5.67) into the form:

$$A_n^a = \varepsilon_{anm} \frac{r^m}{er^2} \left(1 - \frac{\xi}{\sinh \xi} \right), \quad A_0^a = v\hat{r}^a \left(\coth \xi - \frac{1}{\xi} \right). \tag{5.71}$$

This is exactly the monopole solution of the BPS equation (5.63), but with the time component of the gauge potential replacing the scalar field. This is the so-called "dyon in the Rossi gauge". Thus, the Julia–Zee correspondence establishes an exact relation between a single BPS monopole and an infinite instanton chain.

As mentioned above, the action of the infinite number of instantons is divergent:

$$S = \sum_n S_1 = \sum_n \frac{8\pi^2 n}{e^2} \to \infty. \tag{5.72}$$

However, the mass of the monopole being defined as an action per unit of Euclidean time is, of course, finite [443]:

$$\frac{dS}{dt} = \frac{8\pi^2}{e^2} \frac{ve}{2\pi} = \frac{4\pi v}{e} \equiv M. \tag{5.73}$$

To sum up, the BPS monopole is equivalent to an infinite chain of instantons having identical orientation in isospace and separated by an interval $\tau_0 = 2\pi$. An alternative configuration is a chain of correlated instanton–anti-instanton pairs, which corresponds to an infinite monopole loop.

5.2.1 Gauge Zero Mode and the Electric Dyon Charge

In the BPS limit the Julia–Zee dyonic solutions have a very interesting interpretation [238, 513]. First we note that for the *static* ansatz (5.41), (5.52) and the choice $A_0 = 0$, the kinetic energy of the configuration

$$T = \int d^3x \, \text{Tr} \left(E_n E_n + D_0 \phi D_0 \phi \right), \quad (5.74)$$

is equal to zero. Moreover, in this case the Gauss law

$$D_n E_n - ie\left[\phi, D_0\phi\right] = 0, \quad (5.75)$$

can be satisfied trivially, with $E_n = D_0\phi = 0$.

Let us now consider time-dependent fields $A_n(\mathbf{r}, t)$, $\phi^a(\mathbf{r}, t)$, but suppose that their time-dependence arises as a result of a gauge transformation of the original static configuration:

$$A_n(\mathbf{r}, t) = U(\mathbf{r}, t) A_n(\mathbf{r}, 0) U^{-1}(\mathbf{r}, t) - \frac{i}{e} U(\mathbf{r}, t) \partial_n U^{-1}(\mathbf{r}, t). \quad (5.76)$$

Here, $U(\mathbf{r}, t) = e^{ie\omega t}$ with $\omega(\mathbf{r})$ a parameter of the transformation. If the time interval δt is very small, we can expand

$$U(\mathbf{r}, \delta t) \approx 1 + ie\omega \delta t + \dots \quad (5.77)$$

Now it follows from (5.76) that

$$A_n(\mathbf{r}, \delta t) \approx A_n(\mathbf{r}) + (ie[\omega, A_n(\mathbf{r})] - \partial_n \omega) \delta t, \quad (5.78)$$

and we have

$$\partial_0 A_n = ie[\omega, A_n(\mathbf{r})] - \partial_n \omega = -D_n \omega. \quad (5.79)$$

In a similar way we obtain for the time-dependence of the scalar field:

$$\partial_0 \phi = ie[\omega, \phi]. \quad (5.80)$$

These gauge transformations simultaneously affect the time component of the gauge potential, which for the monopole configuration (5.41), (5.52) is a pure gauge:

$$A_0(\mathbf{r}, t) = -\frac{i}{e} U(\mathbf{r}, t) \partial_0 U^{-1}(\mathbf{r}, t) = -\omega. \quad (5.81)$$

Since the gauge transformations (5.79) and (5.80) do not change the potential energy of the configuration, the parameter ω can be identified with the *gauge zero mode*. This is one of four collective coordinates (they are also called *moduli*) of the one-monopole configuration [513]. The other three specify the position of the monopole in space. Their appearance reflects an obvious breaking of translational invariance of the original Lagrangian (5.7) by the monopole configuration: the position of the monopole in $\mathbb{R}^3$ can be chosen arbitrarily.

However, defined in this way, the gauge zero mode is not physical, since the gauge transformations (5.79) and (5.80) do not affect the non-Abelian electric field:

$$E_n^a = \partial_0 A_n - D_n A_0 = -D_n \omega + D_n \omega \equiv 0,$$
$$D_0 \phi = \partial_0 \phi + ie[A_0, \phi] = ie[\omega, \phi] - ie[\omega, \phi] \equiv 0. \qquad (5.82)$$

Thus, as before, the Gauss law is satisfied trivially and the kinetic energy of the monopole (5.74) is still equal to zero.

Now let us suppose that the time-dependence of the fields again appears as a result of the gauge transformation (5.79) and (5.80), but that the corresponding gauge zero mode $(\partial_0 A_n, \partial_0 \phi)$ satisfies the *background gauge condition*:

$$D_n(\partial_0 A_n) - ie[\phi, (\partial_0 \phi)] = 0. \qquad (5.83)$$

Then the Gauss law (5.75) is satisfied, if $A_0 = 0$ and there is a non-trivial solution of the equations (5.79), (5.80) and (5.83) [238], where ω is proportional to ϕ and an additional time dependence is allowed:

$$\omega = \dot{\Upsilon}(t)\phi,$$

which corresponds to the gauge transformation

$$U(\mathbf{r}, t) = \exp\{ie\Upsilon(t)\phi(\mathbf{r})\} \approx 1 + ie\dot{\Upsilon}\phi\delta t. \qquad (5.84)$$

Here $\Upsilon(t)$ is an arbitrary function of time. Indeed, in this case we have $\partial_0 A_n = \dot{\Upsilon} D_n \phi$ and $\partial_0 \phi = 0$, and, since in the Bogomol'nyi limit $D_n D_n \phi = 0$, the background gauge condition (5.83) is satisfied by the ansatz (5.84). However, this solution corresponds to the generation of a non-Abelian electric field

$$E_n = \partial_0 A_n = \dot{\Upsilon}(t) D_n \phi = \dot{\Upsilon}(t) B_n, \qquad D_0 \phi = 0, \qquad (5.85)$$

so the kinetic energy of the monopole (5.74) is no longer zero:

$$T = \tfrac{1}{2}\dot{\Upsilon}^2 \int d^3x \, D_n\phi^a D_n\phi^a$$
$$= \tfrac{1}{2}\dot{\Upsilon}^2 \int d^3x \, B_n^a B_n^a = 2\pi v g \dot{\Upsilon}^2 = \tfrac{1}{2} M \dot{\Upsilon}^2, \qquad (5.86)$$

where we make use of the definition of the magnetic charge (5.46) and take into account that the mass of the BPS monopole is

$$M = \frac{4\pi v}{e}.$$

Since the potential energy of the configuration is time-independent, the gauge transformations (5.79) and (5.80), supplemented with the condition $A_0 = 0$, define a physical collective coordinate $\Upsilon(t)$, that is a gauge zero mode. Its excitation corresponds to the generation of an electric charge $Q = \dot{\Upsilon} g$. Thus, such a gauge-induced time-dependence of the fields transforms the monopole into a dyon.

Note that this collective coordinate is an angular variable, which is defined on a circle S^1. Indeed, the points $\Upsilon = 2\pi n$, $n \in \mathbb{Z}$ correspond to the same gauge transformation $U(\mathbf{r}, t)$, which is unity on the spatial asymptotic [238]. However, the points $\Upsilon = 0$ and, for example $\Upsilon = 2\pi$, correspond to different topological classes. We discuss this situation in more detail below, in Sect. 5.3.4.

To sum up, the one-monopole configuration in the BPS limit could be characterized by four zero modes (moduli) that form the so-called *moduli space* $\mathcal{M}_1$. It is clear from the discussion above that $\mathcal{M}_1 = \mathbb{R}^3 \times S^1$.

Note that we can come back to the Julia–Zee description of a dyon configuration just by inverting the discussion above: we could start from a system of time-dependent fields and apply the gauge transformations (5.79) and (5.80) to compensate for that dependence. The price we would have to pay, would be the appearance of a non-zero time component of the gauge potential A_0. This corresponds to the static ansatz (5.52).

5.3 Topological Classification of Non-Abelian Monopoles

5.3.1 $SO(3)$ vs $SU(2)$

It is very important for our consideration to note that there is an essential difference between $SO(3)$ and its simply connected covering group $SU(2)$, even though the corresponding Lie algebras are identical. The group $SO(3)$ is locally isomorphic to the group $SU(2)$, but there is a global topological difference between them. Since in the monopole theory this feature plays a very important role, we shall discuss it in some detail.

First, we show that the group manifold of $SU(2)$ has the topology of the hyper-sphere S^3. Let us parameterize it by three angular parameters θ, φ, α. The interpretation of first two parameters is rather obvious: the azimuthal angle $0 \leq \varphi \leq 2\pi$, such that $\varphi = 2\pi$ is equivalent to $\varphi = 0$, and the polar angle $0 \leq \theta \leq \pi$, such that the values $\theta = 0$ and $\theta = \pi$ correspond to two different points, each of which is the same for all values of φ. These points are just the north and south poles of the conventional sphere S^2, which is described by these two parameters.

The third angular parameter is more subtle. It is defined over the interval $0 \leq \alpha \leq \pi$. This can be visualized if one imagines subsequent sections of

S^3 (with unit radius) as three-dimensional hyperplanes. The sections are, of course, spheres S^2 with radius $\sim \sin \alpha$.

We must now show that the group manifold of $SU(2)$ is S^3. The transformations of the $SU(2)$ group correspond to the rotations of spinors. Let us recall that the fundamental representation of $SU(2)$ is given by the set of 2×2 Pauli matrices $\boldsymbol{\sigma} = (\sigma_1, \sigma_2, \sigma_3)$. They can be used to parameterize the standard transformation of a rotation in three-dimensional space, which is given by an element of $SU(2)$:

$$U(\mathbf{n}) = \exp\left(\frac{i\omega}{2}\mathbf{n}\cdot\boldsymbol{\sigma}\right), \tag{5.87}$$

where $\mathbf{n}$ is a unit vector that fixes the direction of the axis of rotation, and ω is the rotation angle about this axis. An expansion of the exponent yields

$$U(\mathbf{n}) = \cos\frac{\omega}{2} + i\,(\mathbf{n}\cdot\boldsymbol{\sigma})\sin\frac{\omega}{2}. \tag{5.88}$$

Obviously, the rotation angle $\omega = 4\pi$ corresponds to the identity transformation. However, in order to parameterize all the transformations generated by (5.87), it would be enough to take the angle of rotation ω in the interval $[0, 2\pi]$, because the rotation about $\mathbf{n}$ by an angle $2\pi+\delta$ is exactly the rotation about the reflected vector $-\mathbf{n}$ by the angle $-\delta$.

Thus, the direction of the vector $\mathbf{n}$ is given by two angles θ and ϕ, which parameterize a sphere S^2 of unit radius. The angle of rotation ω could be identified with the third parameter α on the hypersphere S^3. Its boundary points 0 and 2π must be put into correspondence to the north and the south poles of S^3, respectively.

One can use almost the same parameterization in the case of the group $SO(3)$, which corresponds to the rotations of three-dimensional vectors. The only difference is that in this case two rotations about a unit vector $\mathbf{n}$ by angles ω and $2\pi + \omega$ are identical, because a rotation by the angle 2π is an identity transformation. Thus, the group manifold of $SO(3)$ is still a sphere S^3, but all its antipodal points are identified.

There is a difference between the group manifolds of $SU(2)$ and $SO(3)$. Any closed contour in $SU(2)$ can be continuously deformed to a point, i.e., the first homotopy group of $SU(2)$ is trivial: $\pi_1(SU(2)) = 0$. However, alongside contours of that type, which correspond to the winding number 0, there are closed paths in $SO(3)$ that begin and end at the antipodal points of S^3. These paths cannot be continuously deformed to a point and they have a winding number of 1. Thus, the first homotopy group of $SO(3)$ is $\pi_1(SO(3)) = \mathbb{Z}_2$, where $\mathbb{Z}_2$ is the additive group of integers consisting of two elements [0,1]. Of course, two subsequent loops in $SO(3)$, both having winding number 1, are equal to the trivial path. Therefore, for the $SO(3)$ group, the inverse element of the homotopy group (see p. 80) is equal to the identity: $0+1 = 1+0 = 1$ and $1+1 = 0$.

Thus, the covering group $SU(2)$ has the center $\mathbb{Z}_2$, which is a discrete subgroup consisting of two elements $[-1,1]$ commuting with all elements of $SU(2)$. Mathematicians refer to this as an isomorphism between the quotient $SU(2)/\mathbb{Z}_2$ and $SO(3)$.

5.3.2 Magnetic Charge and the Topology of the Gauge Group

The Wu–Yang formalism, which was described in Chap. 3, provides a straightforward explanation of the topological roots of an Abelian monopole. Recall that within this framework, a magnetic charge has been identified with the first Chern class c_1 or the winding number. This is the number of times the gauge group is covered, while the equator S^1 of the boundary of Euclidean three-dimensional space, S^2, is covered once.

This description can be generalized to the case of an arbitrary non-Abelian gauge group H [43, 55, 355]. Let us consider the gauge field taking values in the Lie algebra of H. As before, let us consider this field on the spatial asymptotic S^2. In order to construct a non-trivial bundle over the base S^2, let us cover this sphere by two hemispheres R^N and R^S, and introduce gauge potentials A_μ^N and A_μ^S, which are non-singular in the respective hemispheres. A single-valued gauge transformation U relates the two potentials in the overlap region. Since U is a function of an angular coordinate on the sphere (for example, the azimuthal angle φ), this is a mapping of a loop S^1 (the equator) from the spatial asymptotic into a closed path in the group manifold H.

A particular non-trivial example is $H = SO(3)$. As we saw above, $\pi_1(SO(3)) = \mathbb{Z}_2$, i.e., the winding number of a closed path in the $SO(3)$ group manifold can only have two possible values: $n = 0$ and $n = 1$. This means that there are only two stable topological sectors in an $SO(3)$ gauge theory: a monopole with unit charge and a trivial sector without monopoles. This rather surprising conclusion can be proved by a direct analysis of the stability of the generalized 't Hooft–Polyakov configuration (5.41) [84, 138]. The result of these calculations shows that for the $SO(3)$ gauge model only the spherically symmetric configuration with unit charge is stable with respect to fluctuations of the fields.

In the general case, an arbitrary Lie group H has a simply connected covering group $\bar{H}$, i.e., $\pi_1(\bar{H}) = 0$. In the case of $SO(3)$ considered here, the covering group is $SU(2)$. Recall that the group H is isomorphic to the factor group $\bar{H}/K$, where K is the center of $\bar{H}$. All contours on the group manifold H, which begin and end at the identity element of H, correspond to contours on $\bar{H}$, which begin at the identity and end at an element of the center K. Thus, for an arbitrary Lie group the first homotopy group is

$$\pi_1(H) = \pi_1(\bar{H}/K) = K. \tag{5.89}$$

In the Abelian case we have $H = U(1)$ with the additive covering group of real numbers $\bar{H} = R$. Then, the center is the group of integer numbers $K = \mathbb{Z}$. We saw above that the center of $H = SO(3)$ is $K = \mathbb{Z}_2$. In general, for any simple Lie group $K = \mathbb{Z}_N$, where N is an integer.

Once again, we note that this definition of topological charge in terms of the homotopy group is connected with the asymptotic behavior of the fields at spatial infinity. However, as noted by Coleman [43], if the gauge field is non-singular and the gauge transformation U is an element of the group H, then the winding number is a constant independent of the radius of the sphere S^2. Thus, the magnetic charge is not connected with the behavior of the fields at spatial infinity, but rather resides on a point-singularity at the origin (Wu–Yang Abelian monopole), or there is a monopole core where gauge fields other than H are excited ('t Hooft–Polyakov monopole). In the former case, the singularity of the gauge field is a guarantee of monopole stability. On the other hand, if the configuration is regular everywhere in space, nothing can prevent the monopole from decaying [43, 55, 138].

5.3.3 Equivalence of Topological and Magnetic Charge

At a first glance, the description of the previous section has nothing in common with the topological definition of the magnetic charge of the 't Hooft–Polyakov monopole (5.35) that we discussed above. Indeed, the latter was connected with the asymptotic behavior of the Higgs field, while the winding number of the gauge field is entirely associated with the first homotopy group $\pi_1(H)$. However, both definitions are identical.

Let us consider an arbitrary generalization of the Georgi–Glashow model with the gauge group G spontaneously broken down to a subgroup H. An example could be $G = SU(3)$ and $H = U(1) \times U(1)$. The potential of the Higgs fields $V(\phi)$ has a minimum at ϕ_0, which is invariant under the action of H. Furthermore, it would be convenient to assume that the gauge group G is compact and simply connected, i.e., $\pi_1(G) = 0$.

As before, in order to construct topologically non-trivial configurations with finite energy, we consider the field configuration with Higgs field ϕ approaching its vacuum value at spatial infinity. The vacuum manifold $\mathcal{M}$ is defined as the set of values of Φ that minimize the potential $V(\phi)$ [49]:

$$\mathcal{M} = \{\phi : V(\phi) = 0\}.$$

As the fields $\mathbf{A}_\mu$ and ϕ are supposed to be transformed under an irreducible representation of the gauge group G, the vacuum manifold $\mathcal{M}$ is determined by the structure of the group manifold. As we have seen above, there are transformations in the group that leave a fixed point of $\mathcal{M}$ unchanged. Such transformations form a subgroup H in G and $\mathcal{M}$ is topologically equivalent to the right coset space of H in G:

$$\mathcal{M} = G/H = \{\phi : \phi = U\phi_0;\ U \in H\}.$$

5.3 Topological Classification of Non-Abelian Monopoles

Topologically non-trivial configurations correspond to the mapping from the spatial boundary S^2 into the vacuum $\mathcal{M} = G/H$, which cannot be continuously deformed to the trivial mapping.

The situation is more complicated than it was in the case of $G = SO(3)$. There we had $H = U(1)$ and $\mathcal{M} = SO(3)/U(1) \cong S^2$. Thus, the topological classification of the solutions was given by the mapping of the spatial asymptotic S^2 into the vacuum manifold S^2. Now we need to analyze the more general situation.

A mapping from S^2 to the vacuum falls into homotopy classes with a natural group structure. The corresponding group $\pi_2(G/H)$ is the second homotopy group of G/H. There is a group homomorphism from $\pi_2(G/H)$ to $\pi_1(H)$. To see this [55], let us consider a sphere S^2 parameterized by the two angles θ and φ. Each point on this sphere has to be mapped into the vacuum G/H. As before, we suppose that the sphere S^2 consists of two hemispheres with an overlap region at the equator. Then we can consider the smooth gauge transformations $U^N(\theta, \varphi)$ and $U^S(\theta, \varphi)$, which rotate the Higgs field to the vacuum configuration ϕ_0:

$$\begin{cases} U^N(\theta,\varphi)\phi(\theta,\varphi) = \phi_0 \longrightarrow 0 \leq \theta \leq \frac{\pi}{2} & : R^N \\ U^S(\theta,\varphi)\phi(\theta,\varphi) = \phi_0 \longrightarrow \frac{\pi}{2} \leq \theta \leq \pi & : R^S \end{cases} \quad (5.90)$$

The overlap region is the equatorial circle $\theta = \pi/2$, where the quotient of the transition function is defined:

$$U^N\left(\theta = \frac{\pi}{2}, \varphi\right) U^{S-1}\left(\theta = \frac{\pi}{2}, \varphi\right) \equiv U(\varphi).$$

Defined in this way, the gauge transformation $U(\varphi)$ leaves the Higgs vacuum invariant and is therefore an element of the unbroken subgroup: $U(\varphi) \in H$. Thus, this construction sets a correspondence between each mapping from S^2 into G/H and a mapping of the equatorial circle on S^2 into a closed path in H. Since the composition of the mapping from spatial infinity into G/H corresponds to the composition of loops in H, this mapping has a group structure, or in other words, such a correspondence defines a group homomorphism from $\pi_2(G/H)$ to $\pi_1(H)$.

Moreover, if we consider the transition function that defines a trivial mapping with winding number zero, the loop can be continuously deformed to a point. This mapping corresponds to the unit element of H. In the topologically trivial sector there is a smooth global gauge transformation that rotates ϕ to ϕ_0. Since we suppose that G is compact, $\pi_2(G) = 0$, and this transformation can be continuously deformed to a trivial gauge transformation, which is the unit element of G/H. Therefore, the homomorphism takes the unit elements of $\pi_2(G/H)$ into the unit element of $\pi_1(H)$.

This consideration indicates that both mappings are equivalent. More generally, one can prove that there is a natural isomorphism[6]

$$\pi_2(G/H) = \pi_1(H). \tag{5.91}$$

Thus, the topological classification of mappings from the spatial asymptotic sphere S^2 into the vacuum manifold G/H is equivalent to the topological classification of paths in H. We saw that elements of the first homotopy group $\pi_1(H)$ correspond to the topological charge of the configuration. On the other hand, the elements of the second homotopy group $\pi_2(G/H)$ are identified with the magnetic charge. Thus, the meaning of the relation (5.91) is that both these charges are identical.

5.3.4 Topology of the Dyon Sector

In the discussion above, we considered configurations with only magnetic charge. Let us consider now the topological properties of the dyon sector. We already mentioned above that if we start from the Julia–Zee ansatz (5.41) and (5.52), the time component of the vector potential can be eliminated by a time-dependent gauge transformation whose parameter is the component $A_0^a(r)$ itself [167, 417]:

$$\begin{aligned} U(\mathbf{r},t) &= \exp\{ieA_0 t\} = \exp\{ieA_0^a T^a t\} \\ &= \exp\left\{i\frac{r^a T^a}{r^2}J(r)t\right\} \xrightarrow[r\to\infty]{} \exp\{iCt(\hat{r}^a T^a)\}. \end{aligned} \tag{5.92}$$

Clearly, the identical asymptotic behavior of the fields ϕ and A_0 on the spatial infinity makes this transformation similar to that of (5.84).

Note that even eliminating the temporal component of the vector potential, this gauge transformation cannot change the electric field of the dyon since the spatial components of the vector potential are no longer static (cf. (5.76)):

$$A_n(\mathbf{r},t) = U(\mathbf{r},t)A_n(\mathbf{r},0)U^{-1}(\mathbf{r},t) - \frac{i}{e}U(\mathbf{r},t)\partial_n U^{-1}(\mathbf{r},t). \tag{5.93}$$

Because the matrix of the $SU(2)$ gauge transformation generated by $T^a = \sigma^a/2$ can be written as

$$U(\mathbf{r},t) = e^{iCt(\hat{r}^a T^a)} = \cos(Ct) + i(\hat{r}^a T^a)\sin(Ct),$$

Equation (5.92) means that the gauge field $A_\mu(\mathbf{r},t)$, up to a gauge transformation, is on the spatial asymptotic periodic in time with period $T = 2\pi/C$:

---

[6] We draw our discussion from the review by Preskill [55]. For a mathematically more rigorous discussion, see [52].

$$A_\mu(\mathbf{r}, t+T) = U(\mathbf{r})A_\mu(\mathbf{r},t)U^{-1}(\mathbf{r}) - \frac{i}{e}U(\mathbf{r})\partial_\mu U^{-1}(\mathbf{r}), \qquad (5.94)$$

where

$$U(\mathbf{r}) \equiv U(\mathbf{r}, T) = \exp\left\{2i\pi(\hat{r}^a T^a)\right\}. \qquad (5.95)$$

Thus, we supplement the spatial boundary S^2 with the circle S^1, which corresponds to the cyclic collective coordinate discussed in Sect. (5.2.1).

The transformation (5.95) is obviously a mapping of the sphere $S^3 = S^2 \times S^1$ into the $SO(3)$ group manifold S^3. This mapping can be characterized by the so-called *second Chern class* or *Pontryagin index* (see the definition of the Chern classes above, (3.68)):

$$c_2 = \frac{e^2}{8\pi^2}\int d^4x \, \mathrm{Tr} F_{\mu\nu}\widetilde{F}^{\mu\nu} = \frac{e^2}{4\pi^2}\int_0^T dt \int d^3x \, \partial^\mu K_\mu, \qquad (5.96)$$

where we make use of the definition of the topological current

$$K_\mu = \varepsilon^{\mu\nu\rho\sigma}\left(A_\nu^a \partial_\rho A_\sigma^a - \frac{e}{3}\varepsilon^{abc}A_\nu^a A_\rho^b A_\sigma^c\right),$$

and take into account that $\mathrm{Tr}(T^a T^b) = \frac{1}{2}\delta_{ab}$, $\mathrm{Tr}(T^a T^b T^c) = i\varepsilon_{abc}$. Therefore

$$\partial^\mu K_\mu = 2\varepsilon_{\mu\nu\rho\sigma}\partial^\mu \mathrm{Tr}\left(A^\nu \partial^\rho A^\sigma + \frac{ie}{6}A^\nu A^\rho A^\sigma\right) = \frac{1}{2}\mathrm{Tr}F_{\mu\nu}\widetilde{F}^{\mu\nu}. \qquad (5.97)$$

The topological index (5.96) is a gauge invariant quantity. If we suppose that $A_0 = 0$, this index can be represented in the form [167, 417]

$$c_2 = w(T) - w(0), \qquad (5.98)$$

where

$$w(t) = \frac{e^2}{4\pi^2}\int d^3x K_0(t) = \frac{e^2}{4\pi^2}\int d^3x \varepsilon_{mnk} \, \mathrm{Tr}\left(A_m \partial_n A_k + \frac{ie}{6}A_m A_n A_k\right).$$

This is the winding number of the gauge transformation (5.95). Note that this index of the dyon configuration is very similar to the topological charge of an instanton solution of pure Yang–Mills theory. This similarity has very deep roots and has been discussed above.

The specifics of the situation in the dyon sector is that the winding number (5.96) is not a new independent topological characteristic, but is identical to the magnetic charge (5.35) [167, 417]. To see this[7], let us consider the infinitesimal gauge transformations (5.92) on the spatial asymptotic S^2

$$U(\mathbf{r}, t) = \exp\{i\omega\delta t\} \equiv \exp\{iC(\hat{r}^a T^a)\delta t\} \approx 1 + i\omega\delta t + \ldots \qquad (5.99)$$

---

[7] Here we reproduce a simplified version of the proof given in the paper [167].

The relation (5.79) means that[8]

$$\frac{\partial A_n}{\partial t} = \frac{1}{e} D_n \omega. \tag{5.100}$$

Let us note that the Pontryagin index (5.96) can be represented as an integral over non-Abelian electric and magnetic fields:

$$c_2 = \frac{e^2}{8\pi^2} \int d^4x \, \mathrm{Tr} F_{\mu\nu} \widetilde{F}^{\mu\nu} = \frac{e^2}{4\pi^2} \int_0^T dt \int d^3x \, \mathrm{Tr}(B_n E_n) \tag{5.101}$$

$$= -\frac{e^2}{4\pi^2} \int_0^T dt \int d^3x \, \mathrm{Tr}\left(B_n \frac{\partial A_n}{\partial t}\right) - \frac{e^2}{4\pi^2} \int_0^T dt \int d^3x \, \mathrm{Tr}\left(B_n (D_n A_0)\right).$$

The last term here vanishes upon integration by parts, since the Bianchi identity gives $D_n B_n = 0$, and the fields B_n and A_n decay asymptotically in such a way that the surface term vanishes. Thus, making use of the definition (5.79), we have

$$c_2 = \frac{e}{4\pi^2} \int_0^T dt \int d^3x \, \mathrm{Tr}\left(B_n D_n \omega\right) = \frac{e}{4\pi^2} \int_0^T dt \int dS_n \, \mathrm{Tr}(B_n \omega). \tag{5.102}$$

Invoking the definition of ω, (5.99), into this formula and using $T = 2\pi/C$, we have

$$c_2 = \frac{e}{4\pi^2} \frac{2\pi}{C} \frac{C}{2} \int dS_n B_n^a \hat{r}^a = \frac{e}{4\pi} \int dS_n B_n^a \hat{r}^a = \frac{eg}{4\pi}. \tag{5.103}$$

Now we can compare this result with (5.37), which relates the magnetic charge of the monopole to the Brouwer degree, the topological characteristic of the Higgs field:

$$c_2 = \frac{eg}{4\pi} = n, \quad n \in \mathbb{Z}.$$

Thus, the Pontryagin index is not an additional topological characteristic of the dyon configuration; it is equal to the Brouwer degree, the integer that appears in the charge quantization condition.

5.4 The θ Term and the Witten Effect Again

Despite being equivalent to the magnetic charge, the possibility of labeling the dyon configuration by a Pontryagin index (5.96) leads to very interesting

---
[8] Recall that in our discussion of the corresponding gauge zero modes, we put $U = \exp\{i e \omega t\}$.

5.4 The θ Term and the Witten Effect Again

consequences. Its existence means that the Lagrangian of the Georgi–Glashow model could be supplemented by a so-called θ-*term*:

$$L_\theta = -\frac{\theta e^2}{32\pi^2} F^a_{\mu\nu} \widetilde{F}^{a\mu\nu}. \tag{5.104}$$

Here θ is an arbitrary parameter. It is easy to see that the inclusion of such a term leads to the appearance of a non-quantized electric charge of the configuration. This would be in correspondence with the discussion above and (2.129). A very simple explanation of this effect on the electrodynamical level was given by Coleman [43]. In the pure Abelian case, the θ term can be written as

$$L_\theta = \frac{\theta e^2}{8\pi^2} \mathbf{E} \cdot \mathbf{B}.$$

Now let us suppose that the fields E_n and B_n can be represented as a composition of the background classical field of a static monopole and a quantum fluctuation a_μ around it:

$$E_n = \partial_n a_0, \qquad B_n = \varepsilon_{nmk} \partial_m a_k + \frac{g}{4\pi} \frac{r_n}{r^3}. \tag{5.105}$$

Thus, the θ-term becomes

$$\int d^3x \, L_\theta = \frac{\theta e^2}{8\pi^2} \int d^3x \, \partial_n a_0 \left(\varepsilon_{nmk} \partial_m a_k + \frac{g}{4\pi} \frac{r_n}{r^3} \right)$$
$$= -\frac{\theta e^2 g}{32\pi^3} \int d^3x \, a_0 \, \partial_n \left(\frac{r_n}{r^3} \right) = -\frac{\theta e^2 g}{8\pi^2} \int d^3x \, a_0 \, \delta^3(\mathbf{x}), \tag{5.106}$$

which has an interpretation as the interaction Lagrangian of a scalar potential a_0 and a static electric charge $q = -\theta e^2 g / 8\pi^2$ located at the origin. Using the charge quantization condition, we see that the electric charge of a monopole described by the θ-term is $q = -e\theta/2\pi$.

A deeper derivation of the same result could be obtained within a non-Abelian model by making use of the definition of the electromagnetic $U(1)$ subgroup [524]. As we have seen above, the electric charge of the monopole is generated by the $U(1)$ gauge transformations (5.79) and (5.84), which are constant on the spatial asymptotic. These transformations are just infinitesimal rotations $U = 1 + in^a T^a$ about the unit vector $n^a = \phi^a/v$, which leave the Higgs vacuum invariant. Under this transformation, the scalar field remains invariant but the gauge field changes as

$$A^a_\mu T^a \rightarrow U A^a_\mu T^a U^{-1} + \frac{i}{e} U \partial_\mu U^{-1} \approx A^a_\mu T^a + \frac{1}{ev} D_\mu \phi^a T^a, \tag{5.107}$$

that is

$$\delta A^a_\mu = \frac{1}{ev} D_\mu \phi^a.$$

The standard Noether theorem then allows us to define a generator of this transformation, which has to be identified with the electric charge operator:

$$n = \int d^3x \left(\frac{\delta L}{\delta \partial_0 A_\mu^a} \delta A_\mu^a + \frac{\delta L}{\delta \partial_0 \phi^a} \delta \phi^a \right). \tag{5.108}$$

Since $\delta \phi^a = 0$, only the variation of the gauge field contributes to this expression. Thus, taking into account the effect of the θ-term, we have:

$$n = \frac{1}{ev}\int d^3x\, D_n\phi^a E_n^a - \frac{\theta e}{8\pi^2 v}\int d^3x\, D_n\phi^a B_n^a = \frac{q}{e} - \frac{\theta eg}{8\pi^2}, \tag{5.109}$$

where the electric and magnetic charges of the configuration are (see (5.53) and (5.35), respectively):

$$q = \frac{1}{v}\int d^3x\, D_n\phi^a E_n^a; \quad g = \frac{1}{v}\int d^3x\, D_n\phi^a B_n^a.$$

Now we can use the condition of single-valuedness of the $U(1)$ gauge transformation. The rotation about the axis n^a by the angle 2π must produce an identity transformation, $\exp\{2\pi i n\} = 1$, which is possible if

$$q = en + \frac{\theta e^2 g}{8\pi^2} = en + \frac{e\theta}{2\pi}m. \tag{5.110}$$

Here we used the charge quantization condition $eg/4\pi = m$ again[9]. Thus, we arrived at the Witten formula (2.138), but this time we obtain it within the framework of a non-Abelian gauge theory.

Finally, we note that the complex parameter τ (2.140), which includes both the gauge coupling constant and the θ-angle introduced in the last section of Chap. 2, appears in a very natural way, if we rescale the gauge field as $A_\mu^a \to \frac{1}{e}A_\mu^a$ and write the complete Lagrangian of the Georgi–Glashow model as a sum of (5.7) and (5.104):

$$\begin{aligned}L &= -\frac{1}{4e^2}F_{\mu\nu}^a F^{a\mu\nu} - \frac{\theta}{32\pi^2}F_{\mu\nu}^a \widetilde{F}^{a\mu\nu} + \frac{1}{2}(D^\mu \phi^a)(D_\mu \phi^a) - V(\phi)\\ &= -\frac{1}{32\pi}\text{Im}\left(\frac{\theta}{2\pi} + \frac{4\pi i}{e^2}\right)\left(F_{\mu\nu}^a + i\widetilde{F}_{\mu\nu}^a\right)^2 + \frac{1}{2}(D^\mu \phi^a)(D_\mu \phi^a) - V(\phi).\end{aligned} \tag{5.111}$$

We shall use this very elegant formula in the following analysis of the remarkable properties of the supersymmetric extensions of the Georgi–Glashow model.

---

[9] In the Georgi–Glashow model, $m = 1$.

6 Multimonopole Configurations

So far, we have considered a single static monopole that has the topological charge $n = 1$. An obvious generalization would be a solution of the field equation with an arbitrary integer topological charge n. At this stage, we have to consider two possibilities: a single "fat" (possibly unstable!) monopole, having a charge $n > 1$, or a system of several monopoles with a total charge equal to n. Obviously, the second situation could be much more interesting, because in this case one would have to take into account the effects of interaction between the monopoles, their scattering and decay.

Over the last 20 years, the investigation of the exact multi-monopole configurations has definitely been at the crossing of the most fascinating directions of modern field theory and differential geometry. This kind of research may have caused more enthusiasm from the side of mathematicians, rather than physicists. The point is that the Bogomol'nyi equation may be treated as a three-dimensional reduction of the integrable self-duality equations. Thus, its solution is simpler than the investigation of the multi-instanton configurations that arise in Yang–Mills theory in $d = 4$.

The property of integrability makes it possible to find the whole set of multi-monopole solutions. This, however, requires sophisticated mathematical techniques, for example the Atiyah–Drinfeld–Hitchin–Manin (ADHM) construction modified by Nahm to the case of BPS monopoles, and other methods developed over the last years. Of course, any attempt to give a detailed description of this fast developing and very intriguing subject is outside the scope of the present book. We consider in this chapter only some elementary aspects of multimonopoles. The reader wishing to know more about this subject should consult the classic book by Atiyah and Hitchin [39] and the excellent presentation by Manton and Sutcliffe [54], which contains a very detailed discussion of recent developments. A comprehensive review of the mathematical aspects of this problem can be found in [52]. The very detailed review by Nahm in the collection [45] and the original papers [81,83,507] are essential reading. The construction of BPS multimonopoles and other topics are discussed in a very good review by Sutcliffe [58].

6.1 Multimonopole Configurations and Singular Gauge Transformations

6.1.1 Singular $SU(2)$ Monopole with Charge $g = ng_0$

As we attempt to construct in a non-Abelian theory a system of several (anti)monopoles, we run into a problem connected with the self-coupling of the gauge fields and non-linearity of the field equations. A simple superposition of the fields of two or more monopoles is no longer a stationary point of the action. However, as we saw in Chap. 5, in the Yang–Mills–Higgs theory, the gauge invariant electromagnetic field strength tensor (5.30) can be defined as

$$F_{\mu\nu} = \partial_\mu A_\nu - \partial_\nu A_\mu - \frac{1}{e}\varepsilon_{abc}\hat\phi^a\partial_\mu\hat\phi^b\partial_\nu\hat\phi^c \, , \tag{6.1}$$

where the Abelian potential is projected out as $A_\mu = A_\mu^a \hat\phi^a$. Thus, we can perform a singular gauge Transformation, which rotates this configuration to an *Abelian gauge*, where the scalar field is constant: $\tilde\phi^a = v\delta_{a3}$ and the gauge field has only one isotopic component, $\tilde A_k = A_k^3$. In such a gauge, the second term in the definition of the electromagnetic field strength tensor (6.1) transforms into the singular field of the Dirac string, and the magnetic charge is entirely associated with the topology of the gauge field.

Recall that gauge transformations, which connect the "hedgehog" and Abelian gauges, are singular and we have to be very careful in dealing with them. However, there is an obvious advantage in working in the Abelian gauge: here the gauge potentials are additive and the field equations are linear. That is why the authors of the paper [76] suggested to implement the following program in order to construct a multimonopole configuration: (i) start with an Abelian gauge, (ii) suppose that the gauge potential $\tilde A_k^3$ is a simple sum of a few singular Dirac monopoles embedded into the $SU(2)$ gauge group and then try to define a gauge transformation, which

1. removes the singularity of the potential $\tilde A_k^3$ in the string gauge;
2. provides proper asymptotic behavior of the fields in the Higgs vacuum: the scalar field must smoothly tend to the vacuum value $|\phi| = v$, while the gauge potential must vanish as $1/r$.

Let us try to implement this program to construct a possible generalization of the 't Hooft–Polyakov solution (5.41) to the case of non-minimal magnetic charge $g = ng_0 = 4\pi n/e$ [86, 223, 228]. First, consider a "fat" Dirac potential, given by the generalization of (1.43), and then embed it into the $SU(2)$ group. Thus, we start with the Abelian gauge where the electromagnetic subgroup corresponds to rotation about the third axis of isospace:

$$\tilde A_k(\mathbf{r}) = \frac{n}{er}\frac{1-\cos\theta}{\sin\theta}T_3\hat\varphi_k \, . \tag{6.2}$$

6.1 Multimonopoles Configurations and Singular Gauge Transformations

In the following discussion we will consider the Hermitian fundamental representation of the $SU(2)$ group, $T_a = \frac{1}{2}\sigma_a$ as before.

Now we can make use of an analogy with the gauge transformation (5.3) considered above and rotate the Dirac potential to the non-Abelian Wu–Yang potential (5.4):

$$A_k = \frac{1}{2}A_k^a \sigma^a = U^{-1}\tilde{A}_k U - \frac{i}{e}U^{-1}\partial_k U, \qquad \phi^a = U\tilde{\phi}^a U^{-1}, \qquad (6.3)$$

where

$$U(\theta, \phi) = e^{-i(\boldsymbol{\sigma}\cdot\hat{\boldsymbol{\varphi}}^{(n)})\theta/2} = \begin{pmatrix} \cos\frac{\theta}{2} & -\sin\frac{\theta}{2}e^{-in\varphi} \\ \sin\frac{\theta}{2}e^{in\varphi} & \cos\frac{\theta}{2} \end{pmatrix}. \qquad (6.4)$$

Here the n-fold rotation in azimuthal angle φ is needed to balance the singular part of the Abelian potential (6.2), $\hat{\boldsymbol{\varphi}}^{(n)} = -\hat{\mathbf{e}}_x \sin n\varphi + \hat{\mathbf{e}}_y \cos n\varphi$.

In order to describe the rotated potential in a compact form, we define the $su(2)$ matrices

$$\begin{aligned}
\tau_r^{(n)} &= (\hat{\mathbf{r}}^{(n)} \cdot \boldsymbol{\sigma}) = \sin\theta\cos n\varphi\,\sigma_1 + \sin\theta\sin n\varphi\,\sigma_2 + \cos\theta\,\sigma_3 , \\
\tau_\theta^{(n)} &= (\hat{\boldsymbol{\theta}}^{(n)} \cdot \boldsymbol{\sigma}) = \cos\theta\cos n\varphi\,\sigma_1 + \cos\theta\sin n\varphi\,\sigma_2 - \sin\theta\,\sigma_3 , \\
\tau_\varphi^{(n)} &= (\hat{\boldsymbol{\varphi}}^{(n)} \cdot \boldsymbol{\sigma}) = -\sin n\varphi\,\sigma_1 + \cos n\varphi\,\sigma_2 .
\end{aligned} \qquad (6.5)$$

In this notation we obtain

$$A_k = \frac{1}{2er}\left(\tau_\varphi^{(n)}\hat{\theta}_k - n\tau_\theta^{(n)}\hat{\varphi}_k\right), \qquad \phi^a = v\hat{r}_a^{(n)}. \qquad (6.6)$$

As one could expect, when rotated into the "hedgehog" gauge, this configuration is spherically symmetric and the corresponding magnetic field

$$B_k = n\frac{r_k}{er^3} \qquad (6.7)$$

is exactly the field of a static magnetic monopole with charge $g = 4\pi n/e$ at the origin. One can prove that for the configuration (6.6) the condition $D_n\phi^a = 0$ holds.

Note that the potential (6.6) is not a naive generalization of the Wu–Yang potential (5.4)

$$A_n = n\,\varepsilon_{amn}\frac{r_m}{r^2}\frac{\sigma^a}{2}, \qquad (6.8)$$

which was considered in [542]. The configuration (6.8) with $n = 2$ has some amusing properties: the corresponding field strength tensor vanishes identically since the commutator terms precisely cancel the derivative terms. Indeed, such a potential is a pure gauge:

$$A_n = iU^{-1}\partial_n U, \quad \text{where} \quad U = i\sigma^a \hat{r}^a ,$$

unlike the potential (6.6). Such a configuration may exist as an unstable deformation of the topologically trivial sector. Below we shall consider these deformations.

Recall that the Wu–Yang configuration (5.4), which was constructed via the gauge rotation of the embedded Dirac potential, is only the asymptotic limit of the 't Hooft–Polyakov solution (5.41) at $r \to \infty$. Unlike the former, the latter corresponds to finite energy of the configuration. It is easy to see that the configuration (6.6) is singular at the origin as well. A generalization of the 't Hooft–Polyakov solution that we are looking for must not only have proper asymptotic behavior of the fields, but it must also be regular at the origin.

One can try to exploit an analogy with the 't Hooft–Polyakov ansatz, i.e., modify the asymptotic form of the fields (6.6) by including shape functions $H(r)$ and $K(r)$, respectively [86, 223]:

$$A_k = \frac{K(r)}{2er}\left(\tau_\varphi^{(n)}\hat{\theta}_k - n\tau_\theta^{(n)}\hat{\varphi}_k\right), \qquad \phi^a = vH(r)\hat{r}_a^{(n)}. \tag{6.9}$$

However, substitution of this ansatz into the field equations of the Yang–Mills–Higgs system (5.14) for $|n| \geq 2$ leads to a contradiction with the assumption that a regular solution of the form (6.9) could exist. Thus, we need to introduce more profile functions to obtain a smooth non-spherically symmetric solution of the field equations. However, even that configuration will be unstable.

Here, we can see a manifestation of the very general Lubkin theorem [355] (see, for example, the discussion in the Coleman lectures [43] and Nahm review in [45]). According to this theorem, there is a unique spherically symmetric monopole in the $SU(2)$ Yang-Mills–Higgs theory with minimal magnetic charge. Both analytical and numerical calculations have proved this conclusion [84, 138]. Therefore, the configuration, which has the asymptotic form (6.9), is a saddle point of the energy functional and it decays into a system of a few separated single monopoles and antimonopoles with total charge $n = n_+ - n_-$.

6.1.2 Magnetic Dipole

Let us continue our attempts to construct multi-monopole configurations. We assume that the electromagnetic subgroup is associated with $U(1)$ rotations about the third isotopic component of the scalar field $\tilde{\phi}^a = a\delta_{a3}$. We first consider a magnetic dipole: a monopole-antimonopole pair located on the z axis at the points $(0, 0, \pm L)$ with both strings directed along the positive z axis as in Fig. 6.1. A simple addition of the corresponding two singular Dirac potentials yields

$$A_k(\mathbf{r}) = \frac{1}{2e}\left(\frac{1 - \cos\theta_1}{r_1 \sin\theta_1} - \frac{1 - \cos\theta_2}{r_2 \sin\theta_2}\right)\sigma_3\,\hat{\varphi}_k. \tag{6.10}$$

6.1 Multimonopoles Configurations and Singular Gauge Transformations

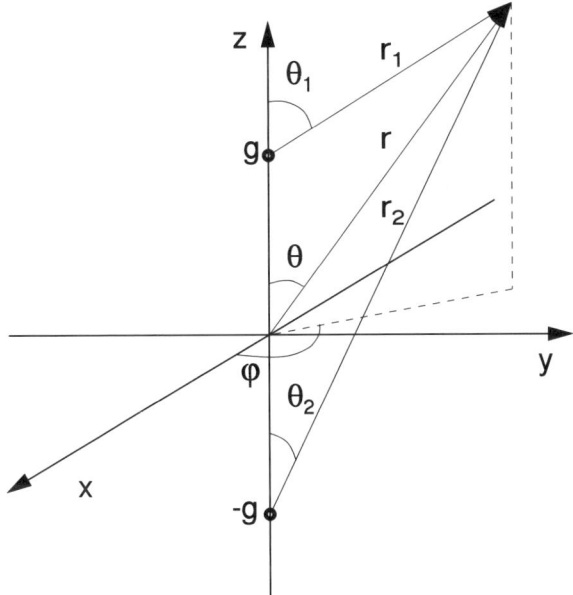

Fig. 6.1. Magnetic dipole configuration

It was noted by S. Coleman that a crucial feature of this construction is that the Dirac strings of both monopoles lie along the same axis [43]. This allows us to define a gauge transformation that removes the singularity of the potential (6.10) embedded into the $SU(2)$ group [76, 86]. Let us exploit the analogy with the transition of a monopole embedded into $SU(2)$, from the singular Dirac monopole in the Abelian gauge to the Wu–Yang non-Abelian monopole in the "hedgehog" gauge. We see that the gauge transformation that would remove the string singularity must rotate a unit isovector $\hat{n}$ associated with the direction of the string about the third isospace axis by an angle 4π. However, this vector now originates from the points $|z| = L$. Therefore, the gauge transformation that we are looking for must become an element of unity for $|z| > L$. A proper choice is [76, 86]

$$U(\theta_1 - \theta_2, \varphi) = e^{-\frac{i}{2}\varphi\sigma_3} e^{\frac{i}{2}(\theta_1-\theta_2)\sigma_2} e^{\frac{i}{2}\varphi\sigma_3}$$

$$= \begin{pmatrix} \cos\frac{\theta_1-\theta_2}{2} & -\sin\frac{\theta_1-\theta_2}{2}e^{-i\varphi} \\ \sin\frac{\theta_1-\theta_2}{2}e^{i\varphi} & \cos\frac{\theta_1-\theta_2}{2} \end{pmatrix}. \quad (6.11)$$

A simple calculation shows that the Higgs field after rotation into the "hedgehog" gauge is

$$\tilde{\phi}^a = U\phi^a U^{-1} = v[\sin(\theta_1 - \theta_2)\cos\varphi, \sin(\theta_1 - \theta_2)\sin\varphi, \cos(\theta_1 - \theta_2)]. \quad (6.12)$$

One can prove that the gauge transformation (6.11) removes the singularity of the embedded potential (6.10).

A generalization of this procedure also allows us to rotate into a non-singular gauge some other configurations: a monopole-antimonopole pair connected with a Dirac string, a monopole-monopole pair, or a system of a few monopoles lying along a line [86]. Moreover, it is possible to generalize this procedure to the case of an arbitrary gauge group, for example, an $SU(3)$ magnetic dipole was considered in [406]. The only restriction is that all the monopole strings must be directed along the same line; otherwise it is impossible to remove all the singularities of the multi-monopole potential by making use of a singular gauge transformation [43]. In other words, after rotation to a "hedgehog" gauge such a multimonopole configuration describes a system of a few monopoles having identical orientation in isotopic space. Evidently, such a system is not a solution of the field equations, it is generally unstable and that is not really the case of the arbitrary multimonopole system we are looking for.

Another inconsistency of the description above is that these expressions have a restricted domain of applicability.

Indeed, there was a hidden contradiction in our discussion above. Actually, so far we are dealing with point-like monopoles, because our configuration is just a generalization of the non-Abelian Wu–Yang potential. The regular 't Hooft–Polyakov solution coincides with it only asymptotically. Thus, there is still a question of the inner structure of the monopoles, or in other words, the problem of finding a solution that would make the Higgs field vanish at some points, which are associated with the positions of the monopoles.

The contradiction is that, on the one hand we suppose that each monopole is characterized by a topological charge connected with the spatial asymptotic of the scalar field. On the other hand, when we calculated the field of a magnetic dipole, we suppose that the monopoles are separated by a finite distance $2L$, and moreover, $L \ll r$. A proper approximation would, therefore, not be a magnetic dipole, but rather a monopole-anti-monopole pair separated by a distance that is very large compared to the core. Indeed, there is a smooth, finite energy magnetic dipole solution to the model (5.7), where two zeros of the Higgs field are relatively close to each other [327, 448].

6.2 Rebbi–Rossi Multimonopoles, Chains of Monopoles and Closed Vortices

Discussion of the magnetic dipole "from afar" yields some clue to the structure of the solution we sought for. The configuration space of Yang–Mills–Higgs theory consists of sectors characterized by the topological charge of the Higgs field. While the unit charge 't Hooft–Polyakov hedgehog solution (5.41) corresponds to a single covering of the vacuum manifold S^2_{vac} by a single

6.2 Rebbi–Rossi Multimonopoles, Chains of Monopoles   179

turn around the spatial boundary S^2, multimonopole configurations have to be characterized by an n-fold covering of the vacuum manifold. Recall that the 't Hooft–Polyakov solution is spherically symmetric. It was shown that $SU(2)$ monopoles with higher topological charge cannot be spherically symmetric [251] and possess at most axial symmetry [219, 434, 440, 507] or no rotational symmetry at all [58].

We shall see that the pattern of interaction between the monopoles is very different from naive picture of Coulomb long-range electromagnetic interaction of point-like charges. Indeed, there is such an attractive force between well separated monopole and antimonopole that, in the singular gauge, is mediated by the A^3 component of the vector field. However, this field is massless only outside of the monopole core. Furthermore, the massive vector bosons $A_\pm$ also mediate the short-range Yukawa interaction between the monopoles.

Taubes pointed out that the latter contribution to the net potential depends on the relative orientation of the monopoles in the group space that is parameterized by an angle δ. Thus, the potential of the gauge interaction between the wide separated monopole and anti-monopole depends on two parameters: the distance between the locations of the poles r and relative angle $\delta \in [-\pi, \pi]$:

$$V_{gauge} = -\frac{e^{-r}}{r} + \frac{2e^{-r}}{r} \cos \delta \qquad (6.13)$$

Thus, there is a saddle point configuration at $\delta = 0$ and we may conclude that there is some equilibrium distance r_0 at which the attractive contribution to the net potential energy of the monopole-antimonopole pair is compensated by the short-range Yukawa interaction mediated by the massive vector $A_\pm$ bosons. Furtermore, there is also a scalar interaction between the monopoles mediated by the Higgs boson, which is always attractive but remains short-ranged until the scalar coupling constant in not zero. In our consideration we have to take into account all these contributions.

Using infinite dimensional Morse theory, which relates the topology of a manifold to the number and types of critical points of a function defined on this manifold, Taubes proved that in the $SU(2)$ Yang-Mill–Higgs theory a smooth, finite energy magnetic dipole solution of the second-order field equations could exist [489]. In his consideration, the space of the field configurations and the energy functional are considered as the manifold and the function, respectively.

Recall that the functional space of the finite energy configuration if classified according to the homotopy classes. For a monopole-antimonopole pair the map $S^2 \to S^2$ has a degree zero, thus it is a deformation of the topologically trivial sector. A generator for this homotopy group is a non-contractible loop, which describes the creation of a monopole-antimonopole pair from the vacuum with relative orientation in the isospace $\delta = -\pi$, separation of the pair, rotation of the monopole by 2π and annihilation of the pair back into vacuum[1].

Minimization of the energy functional along such a loop yields an equilibrium state in the middle of the loop where the monopole is rotated by π and $\delta = 0$. Taubes argued that this corresponds to the vanishing of the potential of the interaction (6.13), because of the balance between the short-range Yukawa interactions and the long-range electromagnetic contribution. Such an axially symmetric configuration, with two zeros of the Higgs field located symmetrically on the positive and negative z-axis, corresponds to a saddlepoint of the energy functional, a monopole and antimonopole in static equilibrium. On the level of classical theory, this configuration cannot annihilate because the loop is not contractible. Numerical computation confirms that such a solution really exists [327, 448].

New classical axially symmetric solutions, which are associated with monopole-antimonopole systems in $SU(2)$ YMH theory, were discovered recently [328–330]. In these solutions, the Higgs field vanishes either at some set of discrete isolated points or at rings. The latter configurations correspond to the closed vortices while the former are (multi)monopole-antimonopole bound systems. There is also a third class of solutions, which corresponds to a single (multi)monopole bounded with a system of vortex rings centered around the symmetry axis. We review these configurations below.

Since the Higgs field takes values in $su(2)$ Lie algebra, we may consider a triplet of unit vectors

$$\hat{\mathbf{e}}_r^{(n,m)} = [\sin(m\theta)\cos(n\varphi), \sin(m\theta)\sin(n\varphi), \cos(m\theta)],$$
$$\hat{\mathbf{e}}_\theta^{(n,m)} = [\cos(m\theta)\cos(n\varphi), \cos(m\theta)\sin(n\varphi), -\sin(m\theta)], \quad (6.14)$$
$$\hat{\mathbf{e}}_\varphi^{(n)} = [-\sin(n\varphi), \cos(n\varphi), 0],$$

which describe both rotations in azimuthal angle and in polar angle.

Now we define the $su(2)$ matrices $\tau_r^{(n,m)}$, $\tau_\theta^{(n,m)}$, and $\tau_\varphi^{(n)}$ as a product of these vectors with the usual Pauli matrices $\tau^a = (\tau_x, \tau_y, \tau_z)$:

$$\tau_r^{(n,m)} = \sin(m\theta)\tau_\rho^{(n)} + \cos(m\theta)\tau_z,$$
$$\tau_\theta^{(n,m)} = \cos(m\theta)\tau_\rho^{(n)} - \sin(m\theta)\tau_z,$$
$$\tau_\varphi^{(n)} = -\sin(n\varphi)\tau_x + \cos(n\varphi)\tau_y,$$

where $\tau_\rho^{(n)} = \cos(n\varphi)\tau_x + \sin(n\varphi)\tau_y$ and $\rho = \sqrt{x^2 + y^2} = r\sin\theta$. This is a generalization of the basis (6.5).

We parametrize the gauge potential and the Higgs field by the static, purely magnetic Kleihaus–Kunz ansatz

---

[1] We shall discuss the process of creation of a monopole-antimonopole pair in a weak external magnetic field in Sect. 7.3.2.

6.2 Rebbi–Rossi Multimonopoles, Chains of Monopoles

$$A_\mu dx^\mu = \left(\frac{K_1}{r}dr + (1-K_2)d\theta\right)\frac{\tau_\varphi^{(n)}}{2e}$$
$$- n\sin\theta\left(K_3\frac{\tau_r^{(n,m)}}{2e} + (1-K_4)\frac{\tau_\theta^{(n,m)}}{2e}\right)d\varphi, \quad (6.15)$$

$$\phi = \Phi_1\tau_r^{(n,m)} + \Phi_2\tau_\theta^{(n,m)}, \quad (6.16)$$

which generalizes the spherically symmetric 't Hooft–Polyakov ansatz (5.41). The latter can be recovered if we impose the constraints $K_1 = K_3 = \Phi_2 = 0$, $K_2 = K_4 = K(\xi)$, $\Phi_1 = H(\xi)$.

We refer to the integers m and n in (6.14)–(6.16) as θ winding number and φ winding number, respectively. Indeed, as the unit vector (6.14) parameterized by the polar angle θ and azimuthal angle φ covers the sphere S_2 once, the fields defined by the ansatz (6.15) and (6.16) wind n and m times around the z-axis and ρ-axis, respectively.

There are six structure functions in the Kleihaus–Kunz ansatz; four for the gauge field (K_i, $i = 1\ldots 4$) and two for the scalar field (Φ_1, Φ_2). They depend on the coordinates r and θ only. Thus, the modulus of the scalar field is $|\phi| = \sqrt{\Phi_1^2 + \Phi_2^2}$ and the Higgs vacuum corresponds to the condition $\sqrt{\Phi_1^2 + \Phi_2^2} = v$.

The ansatz (6.15) and (6.16) is axially symmetric in the sense that a spatial rotation around the z-axis can be compensated by an Abelian gauge transformation $U = \exp\{i\omega(r,\theta)\tau_\varphi^{(n)}/2\}$, which leaves the ansatz form-invariant. However, the structure functions of the ansatz transform as [327]

$$K_1 \to K_1 - r\partial_r\omega, \quad K_2 \to K_2 + \partial_\theta\omega,$$

$$\left(K_3 + \frac{\cos(m\theta)}{\sin\theta}\right) \to \left(K_3 + \frac{\cos(m\theta)}{\sin\theta}\right)\cos\omega + \left(1 - K_4 - \frac{\sin(m\theta)}{\sin\theta}\right)\sin\omega,$$

$$\left(1 - K_4 - \frac{\sin(m\theta)}{\sin\theta}\right) \to -\left(K_3 + \frac{\cos(m\theta)}{\sin\theta}\right)\sin\omega + \left(1 - K_4 - \frac{\sin(m\theta)}{\sin\theta}\right)\cos\omega,$$

$$\Phi_1 \to \Phi_1\cos\omega + \Phi_2\sin\omega, \quad \Phi_2 \to -\Phi_1\sin\omega + \Phi_2\cos\omega. \quad (6.17)$$

To obtain a regular solution, we make use of the $U(1)$ gauge symmetry to fix the gauge [324]. We impose the condition

$$G_f = \frac{1}{r^2}(r\partial_r K_1 - \partial_\theta K_2) = 0.$$

The gauge fixing term $L_\eta = \eta G_f^2$ must be added to the Lagrangian (5.7).

With this ansatz the field strength tensor components become

$$F_{r\theta} = -\frac{1}{r}\left(\partial_\theta K_1 - r\partial_r K_2\right)\frac{\tau_\varphi^{(n)}}{2e},$$

$$F_{r\varphi} = -\frac{n}{r}\sin\theta\left[\left(K_1\frac{\sin(m\theta)}{\sin\theta} + K_1(K_4 - 1) - r\partial_r K_3\right)\frac{\tau_r^{(n,m)}}{2e}\right.$$
$$\left. + \left(K_1\frac{\cos(m\theta)}{\sin\theta} + K_1 K_3 + r\partial_r K_4\right)\frac{\tau_\theta^{(n,m)}}{2e}\right],$$

$$F_{\theta\varphi} = n\left[((1-K_2)\sin(m\theta) + (1-K_4)(K_2+n-1)\sin\theta - \partial_\theta[K_3\sin\theta])\frac{\tau_r^{(n,m)}}{2e}\right.$$
$$\left. + ((1-K_2)\cos(m\theta) - K_3(K_2+n-1)\sin\theta - \partial_\theta[(1-K_4)\sin\theta])\frac{\tau_\theta^{(n,m)}}{2e}\right],$$
$$(6.18)$$

and the components of the covariant derivative of the Higgs field become

$$D_r\phi = \frac{1}{r}\left([r\partial_r \Phi_1 + K_1\Phi_2]\tau_r^{(n,m)} + [r\partial_r \Phi_2 - K_1\Phi_1]\tau_\theta^{(n,m)}\right),$$

$$D_\theta\phi = [\partial_\theta \Phi_1 - K_2\Phi_2]\tau_r^{(n,m)} + [\partial_\theta \Phi_2 + K_2\Phi_1]\tau_\theta^{(n,m)},$$

$$D_\varphi\phi = n\sin\theta\left[\left(\Phi_1\frac{\sin(m\theta)}{\sin\theta} + \Phi_2\frac{\cos(m\theta)}{\sin\theta} + K_3\Phi_2 - (1-K_4)\Phi_1\right)\right]\tau_\varphi^{(n)}.$$
$$(6.19)$$

Variation of the Lagrangian (5.7) with respect to the profile functions yields a system of six second-order non-linear partial differential equations in the coordinates r and θ, which is rather cumbersome. Nevertheless, these equations can be solved numerically.

Boundary Conditions

To obtain regular solutions with finite energy density and correct asymptotic behavior, we impose the boundary conditions. Regularity at the origin requires

$$K_1(0,\theta) = 0, \quad K_2(0,\theta) = 1, \quad K_3(0,\theta) = 0, \quad K_4(0,\theta) = 1,$$
$$\sin(m\theta)\Phi_1(0,\theta) + \cos(m\theta)\Phi_2(0,\theta) = 0,$$
$$\partial_r\left[\cos(m\theta)\Phi_1(r,\theta) - \sin(m\theta)\Phi_2(r,\theta)\right]\big|_{r=0} = 0$$

that is $\Phi_\rho(0,\theta) = 0$ and $\partial_r \Phi_z(0,\theta) = 0$.

To obtain the boundary conditions at infinity, we require that solutions in the vacuum sector ($m = 2k$) tend to a gauge transformed trivial solution,

$$\phi \longrightarrow U\tau_z U^\dagger, \quad A_\mu \longrightarrow i\partial_\mu U U^\dagger,$$

and the solutions in the topological charge n sector ($m = 2k+1$) tend to

$$\phi \longrightarrow U\phi_\infty^{(1,n)} U^\dagger, \quad A_\mu \longrightarrow UA_{\mu\infty}^{(1,n)} U^\dagger + i\partial_\mu U U^\dagger,$$

where

$$\phi_\infty^{(1,n)} = v\tau_r^{(1,n)}, \quad A_{\mu\infty}^{(1,n)} dx^\mu = \frac{\tau_\varphi^{(n)}}{2e} d\theta - n\sin\theta \frac{\tau_\theta^{(1,n)}}{2e} d\varphi,$$

is the asymptotic solution of a charge n multimonopole, and $SU(2)$ matrix $U = \exp\{-ik\theta\tau_\varphi^{(n)}\}$, both for even and odd m. Consequently, solutions with even m have vanishing magnetic charge, whereas solutions with odd m possess magnetic charge n. Thus, if we suppose, for example, that $k = 1$, the boundary conditions at the spatial asymptotic yield the rotation of the configuration on the negative semi-axis z by π, with respect to the configuration placed on the positive semi-axis z. This corresponds to the Taubes conjecture for a magnetic dipole.

In terms of the functions K_i $i = 1\ldots 4$, Φ_1, Φ_2, these boundary conditions read

$$K_1 \to 0, \quad K_2 \to 1 - m, \tag{6.20}$$

$$K_3 \to \frac{\cos\theta - \cos(m\theta)}{\sin\theta} \ m \text{ odd}, \quad K_3 \to \frac{1 - \cos(m\theta)}{\sin\theta} \ m \text{ even}, \tag{6.21}$$

$$K_4 \to 1 - \frac{\sin(m\theta)}{\sin\theta}, \tag{6.22}$$

$$\Phi_1 \to 1, \quad \Phi_2 \to 0. \tag{6.23}$$

Note that the gauge transformation (6.17) allows us to tune these boundary conditions, e.g., the configuration with winding numbers $m = 2, n = 1$ and the boundary conditions (6.20)–(6.23) is identical to the configuration with winding numbers $m = 1, n = 1$, which satisfy

$$\begin{aligned} K_1 \to 0, \quad K_2 \to -1, \quad K_3 \to 0, \quad K_4 \to -1, \\ \Phi_1 \to \cos\theta, \quad \Phi_2 \to \sin\theta. \end{aligned} \tag{6.24}$$

This corresponds to the particular choice of the parameter $\omega = \theta$.

Regularity on the z-axis, finally, requires

$$K_1 = K_3 = \Phi_2 = 0, \quad \partial_\theta K_2 = \partial_\theta K_4 = \partial_\theta \Phi_1 = 0,$$

for $\theta = 0$ and $\theta = \pi$.

Subject to the above boundary conditions, we constructed numerical solutions with $1 \le m \le 6$, $1 \le n \le 6$ and several values of the Higgs selfcoupling constant λ [327–330].

Rebbi–Rossi Multimonopoles

Note that asymptotic behavior of the profile functions allows us to check the equivalence between the topological charge Q, which is defined as a winding number of the Higgs field, and the magnetic charge g. Indeed,

6 Multimonopole Configurations

$$Q = \frac{1}{2ev^3} \int_{S_2} d^2\xi \; \varepsilon_{\alpha\beta}\varepsilon_{abc}\phi^a \partial_\alpha\phi^b \partial_\beta\phi^c$$

$$= \frac{2nm}{2ev^3} \int d\theta d\varphi \, [\Phi_1 \sin(m\theta) + \Phi_2 \cos(m\theta)] = \frac{4\pi n}{2e} [1 - (-1)^m] ,$$

$$g = \frac{1}{v} \int \text{Tr}\,\varepsilon_{ijk} (F_{ij} D_k \phi) \, d^3r = \frac{2}{v} \int_{S_2} d\theta d\varphi \; \text{Tr}(F_{\theta\varphi}\phi)$$

$$= \frac{4\pi n}{2e} \int d\theta \, [m \sin(m\theta) - \partial_\theta(\sin\theta K_3)] = \frac{4\pi n}{2e} [1 - (-1)^m] ,$$

(6.25)

where we used the definitions (5.35) and (5.47), and substituted the asymptotic behavior of the profile functions (6.21), (6.22) and (6.23). Recall that for a spherically symmetric monopole, we have the charge quantization condition (5.38): $g = 4\pi n/e$.

Thus, the configurations given by the axially symmetric Kleihaus–Kunz ansatz (6.15) and (6.16) are either deformations of the topologically trivial sector (even winding number m), or deformations of the core of charge n multimonopoles with minimal winding number $m = 1$. The latter configuration corresponds to the finite λ extension of the Rebbi–Rossi BPS multimonopoles [440]. For these solutions all zeros of the Higgs field are superimposed at a single point. In Fig. 6.2. we show the energy density and the modulus of the Higgs field of configurations with topological charges $n = 2, 3, 4$ as a function of the coordinates $\rho = \sqrt{x^2 + y^2}$ and z. Here individual monopoles cannot be distinguished and the scalar field has a multiple zero at the origin. When $\lambda > 0$, the energy of these monopoles per unit charge is higher than the energy of n infinitely separated charge one monopoles [324]. Thus, this configuration is unstable and non-BPS monopoles repulse each other.

m-Chains

Zeros of the Higgs field can be separated if the θ winding number $m > 1$. Let us consider $n = 1$ configurations first. These m-chains possess m nodes of the Higgs field on the z-axis. Due to reflection symmetry, each node on the negative z-axis corresponds to a node on the positive z-axis. The nodes of the Higgs field $x_0(k)$ are associated with the location of the monopoles and antimonopoles (see Fig. 6.3). For odd m ($m = 2k + 1$) the Higgs field possesses k nodes on the positive z-axis and one node at the origin. The node at the origin corresponds to a monopole, if k is even and to an antimonopole, if k is odd. For even m ($m = 2k$), there is no node of the Higgs field at the origin and the topological charge of the configurations (6.25) is zero.

The $m = 1$ solution is the spherically symmetric 't Hooft–Polyakov monopole that we discussed in the Chap. 5. The $m = 3$ and $m = 5$ chains represent saddlepoints with unit topological charge. The $m = 2$ chain is identical to the monopole-antimonopole pair (the magnetic dipole) discussed in [327,448]. Indeed, asymptotic expansion of the profile function shows that only the gauge

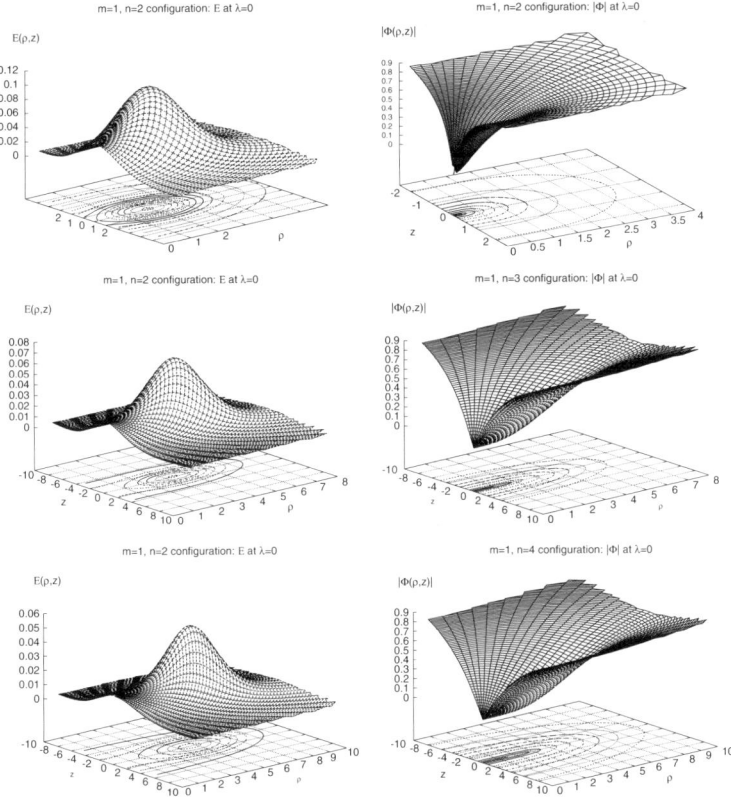

Fig. 6.2. The rescaled energy density $E(\rho, z)$ and the modulus of the Higgs field $|\phi(\rho, z)|$ are shown for the Rebbi–Rossi multimonopoles solutions $m = 1, n = 2, 3, 4$ at $\lambda = 0$

field function K_3 decays like $O(r^{-1})$, while other gauge functions decay exponentially [325, 327]. In the singular gauge, where the Higgs field is constant at infinity, $\phi = \sigma_3$, we obtain [328]

$$K_3 \to \frac{1 - \cos(m\theta)}{\sin \theta} + \frac{d}{r} \sin \theta$$

and the asymptotic gauge potential precisely corresponds to the field of a magnetic dipole:

$$A_\mu dx^\mu \to d \frac{\sin^2 \theta}{2er} \sigma_3 d\varphi \, .$$

However, this dipole moment originates not only from a distribution of the magnetic charge, as one may naively expect. The point is that for the axially symmetric configuration (6.15) and (6.16), the Abelian 't Hooft tensor

(5.31) corresponds not only to the topological (magnetic) current k_μ (5.33), but also to the electric current j^ν_{el}:

$$\partial_\mu \mathcal{F}^{\mu\nu} = 4\pi j^\nu_{\text{el}} . \tag{6.26}$$

Evaluation of the 't Hooft tensor (5.31) with the above ansatz yields

$$\mathcal{F}_{\theta\varphi} = \partial_\theta \mathcal{A}_\varphi , \quad \mathcal{F}_{\varphi r} = -\partial_r \mathcal{A}_\varphi , \quad \mathcal{F}_{r\theta} = 0 , \tag{6.27}$$

with the Abelian potential

$$\mathcal{A}_\varphi = \frac{n}{e} \left[-\hat{\Phi}_1 \left[K_3 \sin\theta + \cos(m\theta) \right] + \hat{\Phi}_2 \left[(K_4 - 1)\sin\theta + \sin(m\theta) \right] \right] , \tag{6.28}$$

and $\hat{\Phi}_1 = \Phi_1/\sqrt{\Phi_1^2 + \Phi_2^2}$, $\hat{\Phi}_2 = \Phi_2/\sqrt{\Phi_1^2 + \Phi_2^2}$. Evidently, for the 't Hooft–Polyakov monopole this potential is reduced to $e\mathcal{A}_\varphi = -n\hat{\Phi}_1 \cos\theta$.

As can be seen from (6.27)–(6.28), contour lines of the vector potential component $\mathcal{A}_\varphi$, correspond to the field lines of the Abelian magnetic field $\mathbf{B}$. Thus, the magnetic dipole moment can be obtained from [330],

$$\boldsymbol{\mu} = \int d^3x \left(\mathbf{r} \frac{k_0}{e} + \frac{1}{2} [\mathbf{r} \times \boldsymbol{j}_{\text{el}}] \right) , \tag{6.29}$$

where k_0/e and $\boldsymbol{j}_{\text{el}}$ are the magnetic charge density and the electric current density, respectively. Therefore, the physical picture of the source of the dipole moment is that it originates both from a distribution of magnetic charges and electric currents. Because of the axial symmetry of the configurations, $\boldsymbol{\mu} = \mu \mathbf{e}_z$ and the contribution of the electric current density to the magnetic moment is

$$\mu_{\text{current}} = \frac{1}{2} \int j_\varphi r^2 \sin\theta\, dr d\theta d\varphi$$

$$= \frac{1}{4} \int dr d\theta \left[r^2 \sin\theta\, \partial_r^2 \mathcal{A}_\varphi + \sin^2\theta\, \partial_\theta \frac{1}{\sin\theta} \partial_\theta \mathcal{A}_\varphi \right] .$$

In Fig. 6.3 we present[2] the dimensionless energy density and nodes of the Higgs field for the $n = 1$ solutions with θ winding number $m = 1, \ldots, 6$. The energy density of the m-chain possesses m maxima on the z-axis and decreases with increasing ρ. The locations of the maxima are close to the nodes of the Higgs field. For a given m the maxima are of similar magnitude, but their height decreases with increasing m. Increasing λ makes these maxima sharper and decreases the distance between the locations of the monopoles. We observe that for a given λ, the distances between the corresponding nodes increase with increasing m.

---

[2] I am grateful to Burkhard Kleihaus for his kind permission to reproduce this plot here [328].

6.2 Rebbi–Rossi Multimonopoles, Chains of Monopoles 187

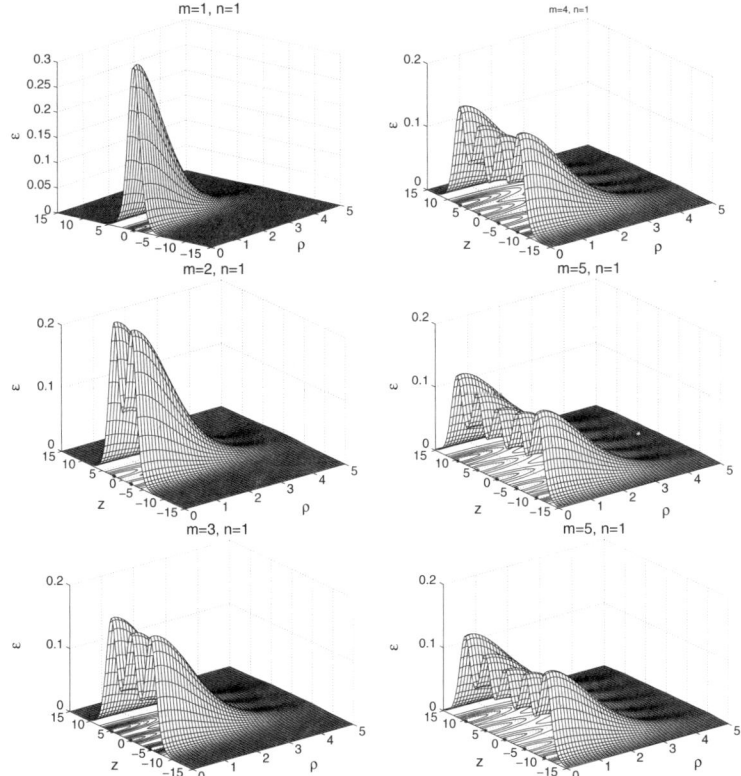

Fig. 6.3. The rescaled energy density $E(\rho, z)$ for the monopole-antimonopole chains with winding numbers $n = 1, m = 1, \ldots 6$ at $\lambda = 0$

We observe that the energy $E^{(m)}$ of an m-chain is always smaller than the energy of m single monopoles or antimonopoles (with infinite separation between them), i.e., $E^{(m)} < E_\infty = 4\pi m v/e = mM$, where $M = 4\pi v/e$ is a mass of a single BPS monopole. On the other hand, $E^{(m)}$ exceeds the minimal energy bound given by the Bogomol'nyi limit $E_{\min} = 0$ for even m, and $E_{\min} = M$ for odd m.

A linear dependence of the energy $E^{(m)}$ on m can be modelled by taking into account only the energy of m single (infinitely separated) monopoles and the next-neighbor interaction between monopoles and antimonopoles on the chain. Defining the interaction energy as the binding energy of the monopole-antimonopole pair,

$$\Delta E = 2M - E^{(2)} ,$$

we obtain as energy estimate for the m-chain

$$E^{(m)}_{\text{est}}/M = m + (m-1)\Delta E .$$

If $\lambda = 0$, we interpret the m-chains as equilibrium states of m non-BPS monopoles and antimonopoles. Indeed, these configurations are essentially *non-BPS solutions*.

To see this in another way, let us consider the limit $\lambda = 0$. Then the energy can be written in the form:

$$E = \int \left\{ \frac{1}{4}\text{Tr}\left((\varepsilon_{ijk}F_{ij} \pm D_k\phi)^2\right) \mp \frac{1}{2}\varepsilon_{ijk}\text{Tr}\left(F_{ij}D_k\phi\right) \right\} d^3x \ . \tag{6.30}$$

The second term is proportional to the topological charge and vanishes when m is even. The first term is just the integral of the square of the Bogomol'nyi equations. Thus, for even m the energy is a measure of the deviation of the solution from self-duality.

Recall that the physical reason for the existence of such a chain is a balance of the attractive and repulsive contributions to the net potential energy of the static configurations. However, we can also make use of the effective electromagnetic interaction associated with the Abelian 't Hooft tensor (5.31).

Let us consider the magnetic dipole. Indeed, the numerical results show that the separation between two nodes of the scalar field, that are identified with positions of the monopole and antimonopole, respectively, is not too large. Thus, both photon and scalar particles remain massive and, together with the vector bosons $A_\mu^\pm$, they all contribute to the short-range Yukawa-type interactions.

Therefore we may conclude that the saddle point configuration exists because of the balance of these short-range interactions. In the BPS limit such a solution is proved to be an equilibrium state [489]; it corresponds to a saddle point of the energy functional and there are negative modes among the fluctuations that do not posses the axial symmetry of the Kleihaus–Kunz ansatz (6.15) and (6.16).

However, the numerical results shows that in the limit $\lambda = 0$ the energy of interaction of the magnetic dipole is $E_{int} = 2M - E_0 = 0.3$, while the distance d between the location of the monopole and antimonopole is 4.23 [327]. Thus, the corresponding Coulomb energy, $E_{Coulomb} = 1/d \approx 0.24$, almost saturates the interaction energy and we may try to make use of an effective electromagnetic interaction to model the configuration. Indeed, the 't Hooft tensor (5.31) allows us to project all components of the gauge field onto the scalar field and we can consider the corresponding effective Abelian forces. Evidently, this interaction is mediated by a massive photon.

At a first glance, there is no place for a repulsive Abelian force: the electromagnetic interactions between a monopole and an anti-monopole is attractive. However, the structure of the solution shows that the $U(1)$ magnetic field, which is defined by the 't Hooft tensor (5.31), is generated both by the monopoles and by the electric current ring (cf. Fig. 6.4). There is the magnetic field $\mathbf{B}_j$ of the current loop, which stabilizes the configuration.

6.2 Rebbi–Rossi Multimonopoles, Chains of Monopoles

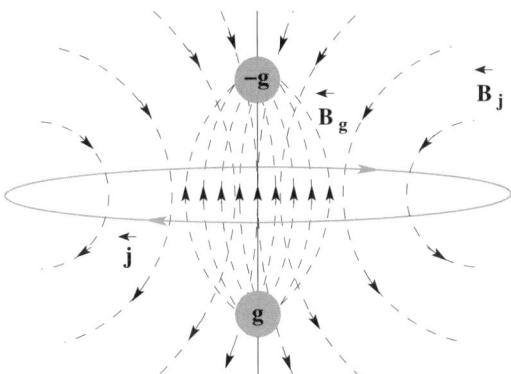

Fig. 6.4. Magnetic dipole: the Abelian magnetic fields of the charges and the current loop

This configuration can exist as an equilibrium electrodynamical state of two charges and a circular current.

Let us now consider chains consisting of multimonopoles with winding number $n = 2$ [329, 330, 421]. Identifying the locations of the Higgs zeros on the symmetry axis with the locations of the monopoles and antimonopoles, we observe that when each pole carries charge $n = 2$, the zeros form pairs, where the distance between the monopole and the antimonopole of a pair is less than the distance to the neighboring monopole or antimonopole, belonging to the next pair.

We observe furthermore, that the equilibrium distance of the monopole-antimonopole pair composed of $n = 2$ multimonopoles is smaller than the equilibrium distance of the monopole-antimonopole pair composed of single monopoles. Thus the higher attraction between the poles of a pair with charge $n = 2$ is balanced by the repulsion only at a smaller equilibrium distance. The difference from $n = 1$ chains is that for $n = 2$ multimonopoles, the maxima of the energy density no longer coincide with (double) zeros of the Higgs field. The latter are still placed on the z axis. One may understand the reason by considering the energy associated with the electromagnetic current, which has to be much stronger to balance the attraction between two monopoles of double charge. There is a current contribution that moves the maxima of the energy density to the ρ-plane.

Closed Vortices

The boundary conditions imposed at the origin on the Higgs field (6.16) means that $\Phi_\rho(0, \theta) = 0$ and

$$\phi(0, \theta) = \Phi_z(0, \theta)\tau_z^{(n,m)}.$$

Thus, the scalar field can either vanish there (for an odd θ winding number m), or be directed along the z-axis (for an even m). In the former case, there is a single n-monopole placed at the origin, whereas the latter configuration in the limit of very large scalar coupling approaches the vacuum expectation value, not only on the spacial boundary, but also in the vicinity of the origin. Then the solutions with different winding number n link the trivial configuration $\phi(0,\theta) = (0,0,\Phi_z(0,\theta))$ and its gauge rotated on the spatial infinity. For $n = 1,2$, these solutions are the monopole-antimonopole chains we discussed above.

The situation changes dramatically, if the φ winding number $n > 2$. The solutions of the second type, which appears in that case, are not multimonopole chains, but systems of vortex rings. They exist both in the BPS limit and beyond it, and represent either a system of closed vortices bounded with a single n-monopole placed at the origin, or without it [329, 330].

This situation is a bit surprising, because one could expect that, when the charge of poles is increasing further beyond $n = 2$, the similar chain solutions consisting of multimonopoles with winding number $n > 2$ should exist; the monopoles and antimonopoles of the pairs should approach each other further, settling at a still smaller equilibrium distance.

Constructing solutions with charge $n = 3$ in the BPS limit ($\lambda = 0$), however, we do not find any chains at all. Now there is no longer sufficient repulsion to balance the strong attraction between the 3-monopoles and 3-antimonopoles. Instead of chains, we now observe solutions with vortex rings, where the Higgs field vanishes on closed rings around the symmetry axis.

To better understand these findings, let us consider unphysical intermediate configurations, where we allow the φ winding number n to continuously vary between the physical integer values[3]. Beginning with the simplest such solution, the $m = 2$ solution, we observe that the zeros of the solution with the winding number n continue to approach each other when n is increased beyond 2, until they merge at the origin. Here the pole and antipole do not annihilate, however. We conclude that this is not allowed by the imposed symmetries and boundary conditions. Instead, the Higgs zero changes its character completely, when n is further increased. It turns into a ring with increasing radius for increasing n. The physical three-monopole-three-antimonopole solution in the BPS limit then has a single ring of zeros of the Higgs field and no point zeros.

Considering the dipole magnetic moment of the $m = 2$ solutions, we observe that it is (roughly) proportional to n. The pair of poles on the z-axis for $n = 2$ clearly gives rise to the magnetic dipole moment of a physical dipole. The ring of zeros also corresponds to a magnetic dipole field, which however, looks like the field of a ring of mathematical dipoles. This corresponds to the simple picture that the positive and negative charges have merged, but not

---

[3] An alternative is to include an electromagnetic interaction with an external magnetic field directed along z-axis.

annihilated, and have then spread out on a ring. Thus, we can identify the $m = 2$, $n = 3$ solution with a closed vortex configuration.

While the dipole moment of monopole-antimonopole chains with an equal number of monopoles and antimonopoles has its origin in the magnetic charges of the configuration [327, 328, 330], the dipole moment of the closed vortices is associated with loops of electric currents [330].

Other solutions with even θ winding number reside in the vacuum sector as well (cf. Fig. 6.5). For $m = 2k > 2$ solutions with zero scalar coupling, it is now clear how they evolve, when the φ winding number is increased beyond $n = 2$. Starting from k pairs of physical dipoles, the pairs merge and form k vortex rings, which carry the dipole strength of the solutions.

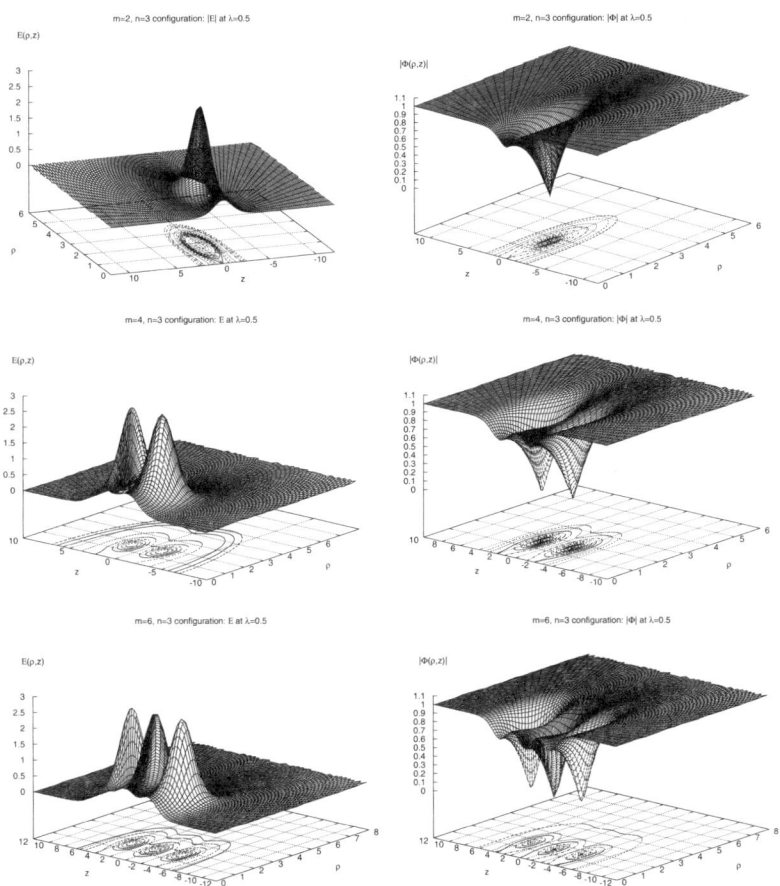

Fig. 6.5. The rescaled energy density $E(\rho, z)$ and the modulus of the Higgs field $|\phi(\rho, z)|$ are shown for the circular vortex solutions $m = 2, n = 3$ (*single vortex*), $m = 4, n = 3$ (*double vortex*) and $m = 6, n = 3$ (*triple vortex*) at $\lambda = 0.5$

The solutions with odd θ winding number have an isolated zero of the Higgs field at the origin, thus they reside in the topological sector with charge n. For $m = 4k + 1$ the situation is somewhat similar to the above. Here a single n-monopole remains at the origin, whereas all other zeros form pairs, which for $n > 2$ approach each other, merge and form $2k$ rings carrying dipole strength. Since, however, a dipole on the positive axis and its respective counterpart on the negative axis have opposite orientation, their contributions cancel in the total magnetic moment. Thus, the magnetic moment remains zero, as it must, because of the symmetry of the ansatz [328]. Non-zero scalar coupling does not change the situation, but the maxima of the energy density distribution are getting sharper.

For $m = 4k - 1$, on the other hand, the situation is more complicated. Let us consider the simplest case $m = 3$ in the limit $\lambda = 0$. Again, we consider unphysical configurations with winding number n continuously varying between the $n = 2$ chain solution and the $n = 3$ configuration. The difference from the case of an even m is that in the initial state there are 3 poles on the z-axis, which cannot form pairs, such that all zeros belong to a pair, symmetrically located around the origin, thus the dipole moment of that configuration is zero. For $m = 3$, we observe in the BPS limit that two vortices appear in the charge $n = 3$ solution, emerging from the upper and lower unpaired zero, respectively, carrying opposite dipole strength. Increasing the scalar coupling decreases the radius of the vortices, as well as the distance between then. However, both rings remain individual.

Once again, we are reminded that the axially symmetric multimonopole chains and vortices, which were discussed here, correspond to the saddle point of the action, not to an absolute minimum. This means that these configurations are unstable. The situation becomes different if we consider the system of monopoles, especially in the BPS limit. To see this we have to analyze the effects of the interaction between the monopoles in more detail.

6.3 Interaction of Magnetic Monopoles

6.3.1 Monopole in External Magnetic Field

Let us consider first the 't Hooft–Polyakov monopole interacting with an external weak magnetic field. We shall consider a monopole initially at rest, i.e., the time derivatives of the fields vanish. One could expect that, to leading order, the effect of this perturbation is that an initially static monopole would start to move with a small constant acceleration[4] w_k. Thus, up to second-order oscillation corrections, the configuration (5.41) obtains a time-dependence of the form [366]:

---

[4] One can treat this motion as an excitation of the translational zero modes of the scalar and vector fields of the system [314].

6.3 Interaction of Magnetic Monopoles

$$\phi^a(\mathbf{r},t) = \phi^a\left(\mathbf{r} - \mathbf{w}\frac{t^2}{2}\right) \approx \phi^a(\mathbf{r}) - \mathbf{w}\frac{t^2}{2}\nabla\phi^a(\mathbf{r}), \tag{6.31}$$

$$A_k(\mathbf{r},t) = A_k^a\left(\mathbf{r} - \mathbf{w}\frac{t^2}{2}\right)T^a = A_k^a(\mathbf{r})T^a - \mathbf{w}\frac{t^2}{2}\nabla A_k^a(\mathbf{r})T^a.$$

Obviously, such a perturbation leads to the generation of an electric field. In the instantaneous rest frame, we still have an initial 't Hooft–Polyakov configuration (5.41), where $A_0(\mathbf{r}) = 0$. The non-relativistic character of the motion then means that in the non-accelerating frame $A_0(\mathbf{r},t) = -w_k t A_k(\mathbf{r},t)$ [366].

Differentiation of the perturbed configuration (6.31) with respect to time gives

$$\partial_0 \phi^a = -w_k t \, \partial_k \phi^a, \qquad \partial_0 A_k = -w_m t \, \partial_m A_k, \tag{6.32}$$

and therefore the non-Abelian electric field is

$$\begin{aligned} F_{0k}^a &= \partial_0 A_k^a - \partial_k A_0^a - e\,\varepsilon_{abc}\, A_0^b A_k^c \\ &= -w_m t(\partial_m A_k^a - \partial_k A_m^a - e\,\varepsilon_{abc}\, A_m^b A_k^c) = -w_m t F_{mk}^a. \end{aligned} \tag{6.33}$$

Another differentiation of this relation yields, to first order in the perturbation $\boldsymbol{\omega}$:

$$D^0 F_{0k}^a = -w_m F_{km}^a. \tag{6.34}$$

This allows us to write the spatial part of the first of the Yang–Mills–Higgs field equations (5.14) as

$$(D_m - w_m) F_{mn} = -e\,\varepsilon_{abc}\,\phi^b D_n \phi^c. \tag{6.35}$$

Next, we note that the first of the relations (6.32) means that

$$D_0 \phi^a = -w_k t \, D_k \phi^a, \tag{6.36}$$

which after another differentiation with respect to time gives $w_k D_k \phi^a$. Thus, in the case of an accelerated monopole, the spatial part of the second of the field equations (5.14) takes the form

$$D_m (D_m - w_m) \phi^a = -\lambda(\phi^a - v^2)\phi^a. \tag{6.37}$$

The system of dynamical equations (6.35) and (6.37) that we obtained describes the dynamics of a 't Hooft–Polyakov monopole in an external homogeneous magnetic field.

We have already seen that the analysis of the monopole properties is very simple in the BPS limit. This is also the case of the monopole dynamics because one can easily prove that in the limit $\lambda = 0$, the field equations (6.35) and (6.37) are solved by the ansatz [366]

$$B_k^a = (D_k - w_k)\phi^a, \tag{6.38}$$

which is a rather straightforward generalization of the Bogomol'nyi equation (5.61). Taking into account that the scalar field of the monopole is defined as $\phi^a = vh(r)\hat{r}^a$, with asymptotic behavior of the shape function $h(r) \to 1$ as $r \to \infty$, we obtain from (6.38), using $B_k = B_k^a \hat{\phi}^a$:

$$B_k = v\partial_k h(r) - vh(r)w_k \ . \tag{6.39}$$

Thus, on the spatial asymptotic, the ansatz (6.38) corresponds to the superposition of the magnetic field of a monopole and a constant weak field $B_k^{\text{ext}} = vw_k$. Furthermore, the magnetic charge and the mass of the BPS monopole are just $g = 4\pi/e$ and $M = 4\pi v/e$, respectively. Thus, the external force on a monopole is given by the standard Newton force law: $F_k = -Mw_k = -gB^{\text{ext}}$.

There is another possible interpretation of the system (6.35) and (6.37). One can consider it as an original Yang–Mills–Higgs system (5.14), which is modified due to an external perturbation caused by a homogeneous magnetic field. This interaction can be described by an extra term

$$L_{\text{int}} = \frac{1}{2v}\varepsilon_{mnk}F_{mn}^a \phi^a B_k^{(\text{ext})} \ , \tag{6.40}$$

added to the Lagrangian of the Georgi–Glashow model (5.7) [314]. In the BPS limit, the equations of motion resulting from such a generalized Lagrangian are identical to the system (6.35) and (6.37).

Note that the effect of the interaction term (6.40), which is linear in the scalar field, is to lift the degeneration of the Higgs vacuum. Indeed, it can be considered as a correction to the Higgs potential

$$V(\phi) = \frac{\lambda}{4}(\phi^2 - v^2)^2 + \frac{1}{v}B_n^a \phi^a B_n^{(\text{ext})} \ , \tag{6.41}$$

and, therefore, there is a unique minimum of the potential on the spatial asymptotic at

$$\phi_{\text{min}}^a = v\hat{r}^a \left(1 + e\frac{\hat{r}_n}{r^2}\frac{B_n^{(\text{ext})}}{m_s^2 m_v^2}\right) \ , \tag{6.42}$$

where we made use of standard notations for the masses of the scalar and vector excitations ($m_s^2 = 2\lambda v^2$, $m_v^2 = e^2 v^2$). The true vacuum is unique, its location is given by the direction of the external magnetic field. Note that there is a certain similarity between the process of metastable vacuum decay (see, e.g., [505]) and the acceleration of a monopole in an external field.

6.3.2 The Interaction Energy of Monopoles

Now we consider the mechanism of interaction between two widely separated monopoles. As mentioned above, there is no analytic solution of the system of Yang–Mills–Higgs equations that would correspond to a system of two separated 't Hooft–Polyakov monopoles, and the monopole-antimonopole chain solutions that we described above, in the BPS limit corresponds to the static unstable equilibrium.

Let us consider a system of two monopoles. The simplest possible approach to estimate the energy of interaction is a straightforward calculation of the difference between the minimal value of the energy of the whole system of the fields and the sum of the masses of two identical individual monopoles:

$$E_{\text{int}} = \frac{1}{2}\int d^3x \, (B_k^a - D_k\phi^a)^2 + v\sum_{i=1,2} g_i + \int d^3x \, V(\phi) - 2M. \quad (6.43)$$

Here we used the form of the energy functional (5.57) suggested by Bogomol'nyi [127]. Evidently, in the limit of vanishing scalar coupling it reduces to (6.30).

This approach was applied to calculate the interaction energy of axially symmetric non-BPS Rebbi–Rossi multimonopoles [324]. It confirms that there is only repulsive phase in that system. In the paper [328], we evaluated the energy of interaction of the chain solutions. Then the separation parameter is of order of a few vacuum expectation values of the scalar field, thus the cores of monopoles overlap and, for example, the mass of the monopole-antimonopole pair is smaller than the energy of two widely separated monopoles.

If we consider well separated monopoles, there is some simplification. We may suppose that the monopole core has a radius that is much smaller than the distance between the monopoles. Moreover, outside of this core the covariant derivatives of the scalar field vanish and thus the gauge fields obey the free Yang–Mills equations. This approximation is a standard assumption in the analysis of monopole interactions.

A numerical study of the interaction between two well separated monopoles using the two-monopole ansatz (6.10) was given in [377]. The idea was to apply a variational approach, which reduces a trial configuration to the known Wu–Yang one-monopole solution in the neighborhood of each monopole. Their positions were defined as the zeros of the Higgs field. Further calculations of that type [389] included the effects of the deformations of the monopole core in the presence of other monopoles. A different approach was used in the work [243], where in order to estimate the energy of the interaction of a monopole-(anti)monopole pair, the conserved stress-energy tensor of such a system (5.15) was analyzed

$$T_{\mu\nu} = F_{\mu\alpha}^a F^{\nu\alpha a} + D_\mu\phi^a D_\nu\phi^a - \frac{1}{2}g_{\mu\nu}D_\alpha\phi^a D^\alpha\phi^a - \frac{1}{4}g_{\mu\nu}F_{\alpha\beta}^a F^{\alpha\beta a}$$
$$- g_{\mu\nu}\frac{\lambda}{4}\left(\phi^2 - v^2\right). \quad (6.44)$$

Again, we can consider as a simple example, a widely separated monopole-(anti)monopole pair centered on the z axis at the points $(0, 0, \pm L)$. The approach of the paper [243] was to consider two surfaces σ_1 and σ_2, surrounding each monopole, to make use of an analogy with the standard electrostatic problem of the interaction between two conducting spheres initially at rest. The radius of each sphere R corresponds to the core scale, thus, we suppose $R \ll L$. The solution of this electrostatic problem can be given by solving the static Maxwell equation with the proper boundary condition on the sphere and an asymptotic condition at $r \to \infty$. This allows us to compute the momentum transfer across the surface, which amounts to calculating the force on each sphere.

To define the force F_n on a monopole, let us write the equation of the stress-energy tensor conservation $\partial_\mu T^{\mu\nu} = 0$ in integral form:

$$\frac{d}{dt} \int_V d^3x \, P_n = \int_S dS_n \, T^{mn} \,, \tag{6.45}$$

where the four-momentum density is $P_n = T_{0n}$ and V is the space volume with boundary S. Thus, identification of the first monopole core with the volume surrounded by the sphere σ_1, together with a proper choice for the boundary conditions on this sphere, allows us to calculate an instant force exerted on the core of this monopole

$$F_n = \int_{\sigma_1} dS_n \, T^{mn} \,. \tag{6.46}$$

However, the conservation of the stress-energy tensor means that it would be enough to consider the flux through a plane xy between the two monopoles.

The result of both analytical [243,437] and variational [377,389,390] calculations confirm a rather surprising conclusion, first observed by Manton [366]: the character of the monopole interaction in the BPS limit changes drastically. While the energy of interaction between 't Hooft–Polyakov monopoles with finite coupling λ turns out to be a rather standard potential energy of charged particles exerting Coulomb-like magnetic fields, there is no interaction between two BPS monopoles at all, but the monopole-antimonopole pair attract each other with double strength. Nahm [390] presented a formal proof that the energy of interaction of two BPS monopoles decays faster than any inverse power of the distance between them. This conclusion is obvious, if we note that in the Bogomol'nyi limit (5.61)

$$B_k^a = \pm D_k \phi^a \,,$$

and $V(\phi) = 0$. In this case, the stress-energy tensor (6.44) vanishes and hence, a force on any spatial volume that we choose is just zero. Thus, the multi-monopole solutions of the Bogomol'nyi equation (5.61) are by construction in equilibrium. The reason for this unusual behavior is that the normal magnetostatic repulsion of the two monopoles is balanced by the long-range scalar interaction: in the BPS limit the quanta of the scalar field are also massless.

Indeed, we already noted that there is a crucial difference between the asymptotic behavior of the Higgs field in the non-BPS and the BPS cases: there is a long-range tail of the BPS monopole

$$\phi^a \to v\hat{r}^a - \frac{r^a}{er^2} \quad \text{as} \quad r \to \infty. \quad (6.47)$$

The result is that, in a system of two widely separated monopoles, the asymptotic value of the Higgs field in the region outside the core of the first monopole is distorted according to (6.47) due to the long-range scalar field of the other monopole: the mass of the first monopole will decrease and the size of its core is increased. In other words, the additional long-range force appears as a result of violation of the original scale invariance of the model in the BPS limit $\lambda \to 0$. A corresponding Goldstone particle, a dilaton, is connected with small fluctuations of the Higgs field

$$\phi^a = v\hat{r}^a e^D = v\hat{r}^a + v\hat{r}^a D + \dots, \quad (6.48)$$

where D is the dilaton field. Separation of the corresponding kinetic term of the dilaton action $L_D = -\frac{1}{2}v^2 \partial_\mu D \, \partial^\mu D$ allows us to establish an identity between the dilaton charge and the magnetic charge of the configuration

$$Q_D = v \int dS_n \partial_n D = \frac{4\pi}{e} = g = \frac{M}{v}. \quad (6.49)$$

Obviously, the mass of the monopole configuration is decreased as $\Delta M = \int d^3x L_D = -Q_D \phi$.

If the configuration has not only a magnetic but also an electric charge q, then the dilaton charge is defined by the dual invariant combination [369,384]

$$Q_D = \sqrt{g^2 + q^2}. \quad (6.50)$$

As a final remark, let us note that the choice of sign in (5.61) corresponds to the monopole (plus) or antimonopole (minus) configuration. This choice is fixed and cannot vary from one space region to another. This is why a solution of the Bogomol'nyi equation describes a static multi-monopole configuration that consists of only monopoles or only antimonopoles. There is another manifestation of the connection that exists between solutions of self-dual Yang–Mills equations and BPS monopoles: there is no interaction in a system of self-dual instantons [522].

6.3.3 Classical Interaction of Two Widely Separated Dyons

Now we can apply the discussion of the previous section to the case of the classical interaction between two dyons that are separated by a distance r. Let us suppose that they have identical magnetic charges g, but different electric charges q_1 and q_2. This problem was studied by Manton [369] (see

6 Multimonopole Configurations

also [92, 93]). Again, the situation is greatly simplified by the assumption of a large separation between them. Thus, we can neglect the inner structure and consider each dyon as a classical point-like particle. Since in the BPS limit the dyons possess both electric and magnetic charges, and the dilaton charge (6.50), as well, the total interaction of two static dyons is composed of electromagnetic repulsion or attraction, caused by the electric and magnetic charges, and attraction caused by the dilaton charges. Thus, the net Coulomb force is

$$\mathbf{F}_{12} = \frac{\mathbf{r}}{r^3}\left(g^2 + q_1 q_2 - \sqrt{g^2 + q_1^2}\sqrt{g^2 + q_2^2}\right). \quad (6.51)$$

An additional simplification comes from assuming that the electric charge of the dyon is much smaller than its magnetic charge. Then an expansion in q^2/g^2 yields

$$\mathbf{F}_{12} \approx -\tfrac{1}{2}(q_1 - q_2)^2 \frac{\mathbf{r}}{r^3}.$$

In this limit, there is no interaction between two dyons with identical electric charges. In general, only the relative electric charge of the system $Q = q_1 - q_2$ enters in the energy of interaction.

Now, let us consider a dyon moving with a velocity $\mathbf{v}_1$ in the background field of another dyon, which is placed at rest at the origin [369]. Since the electromagnetic part of the interaction is described by the Dirac potential (1.43), the canonical momentum of the first dyon, according to (2.16) and (2.17), is

$$\mathbf{P} = M\mathbf{v}_1 + q_1 \mathbf{A} + g\widetilde{\mathbf{A}}, \quad (6.52)$$

where the electromagnetic potentials corresponding to the fields of a static dyon are

$$\mathbf{A} = g\mathbf{a}, \qquad \widetilde{\mathbf{A}} = -q_2 \mathbf{a},$$

and we use the notation (see (1.41))

$$\mathbf{a} = (1 - \cos\theta)\boldsymbol{\nabla}\varphi. \quad (6.53)$$

The scalar potentials that correspond to the static dyon are simply

$$A_0 = \frac{q_2}{r}, \qquad \widetilde{A}_0 = \frac{g}{r}.$$

In addition, we have to take into account the scalar potential connected with the dilaton charge of the BPS dyon: $\phi = \sqrt{q_2^2 + g^2}/r$. As we have already mentioned above, the effect of this potential is to decrease the mass of the first dyon as[5]

$$M \to M - Q_D\phi = M - \frac{1}{r}\sqrt{q_1^2 + g^2}\sqrt{q_2^2 + g^2}.$$

---

[5] Note that a dyon is slightly heavier than a monopole: $M = M_0\sqrt{1 + q^2/g^2}$. However, in the case under consideration, $q \ll g$. Therefore, the difference is of second-order and can be neglected.

Collecting all this together, we arrive at the Lagrangian of the motion of the dyon in the external field of another static dyon:

$$L_1 = \left(-M + \phi\sqrt{q_1^2 + g^2}\right)\sqrt{1 - v_1^2} + \mathbf{v}_1(q_1\mathbf{A} + g\tilde{\mathbf{A}}) - q_1 A_0 - g\tilde{A}_0. \quad (6.54)$$

The next step is to incorporate the effect of motion of both dyons. It is well-known that, if the background field is generated by a moving source, the corresponding fields have to be written in the form of Lienard–Wiechert potentials [92, 369, 492]:

$$\mathbf{A} = g\mathbf{a} + q_2 \frac{\mathbf{v}_2}{\sqrt{r^2 - [\mathbf{r} \times \mathbf{v}_2]^2}},$$

$$\tilde{\mathbf{A}} = -q_2\mathbf{a} + g \frac{\mathbf{v}_2}{\sqrt{r^2 - [\mathbf{r} \times \mathbf{v}_2]^2}},$$

$$A_0 = \frac{q_2}{\sqrt{r^2 - [\mathbf{r} \times \mathbf{v}_2]^2}} + g(\mathbf{a}\cdot\mathbf{v}_2), \quad \tilde{A}_0 = \frac{g}{\sqrt{r^2 - [\mathbf{r} \times \mathbf{v}_2]^2}} - q_2(\mathbf{a}\cdot\mathbf{v}_2),$$

$$\phi = \frac{\sqrt{q_2^2 + g^2}}{\sqrt{r^2 - [\mathbf{r} \times \mathbf{v}_2]^2}}\sqrt{1 - v_2^2}. \quad (6.55)$$

In the a-dependent terms we made use of the non-relativistic character of the motion. Furthermore, in this case we can make the approximation $\sqrt{r^2 - [\mathbf{r} \times \mathbf{v}_2]^2} \approx r$. Substitution of the potentials (6.55) into the Lagrangian (6.54) yields, up to terms of order $q^2\mathbf{v}^2$ and $\mathbf{v}^4$,

$$L_1 = \frac{1}{2}M\mathbf{v}_1^2 - \frac{g^2}{2r}(\mathbf{v}_1 - \mathbf{v}_2)^2 + Qg(\mathbf{v}_1 - \mathbf{v}_2)\cdot\mathbf{a} + \frac{Q^2}{2r}. \quad (6.56)$$

It is important that, as we can see from the second term of this expression, the scalar and magnetic interactions depend on the relative velocity of the dyons in different ways. The third term here describes the minimal interaction between the relative charge Q and the magnetic charge g, while the last term is half the standard Coulomb energy of an electric charge Q. (The other half is associated with the other dyon.)

If we note that all the interaction terms remain the same in the case of the inverse problem of the dynamics of the second dyon in the background field of the first one, then the Lagrangian of the relative motion can be obtained by factorisation of the motion of the center of mass, $M(v_1 + v_2)^2/2$, from (6.56) [369]:

$$L = \left(\frac{M}{4} - \frac{g^2}{2r}\right)\dot{\mathbf{r}}\cdot\dot{\mathbf{r}} + Qg\,\dot{\mathbf{r}}\cdot\mathbf{a} + \frac{Q^2}{2r}, \quad (6.57)$$

where $\dot{\mathbf{r}} = (\mathbf{v}_1 - \mathbf{v}_2)$ is the relative velocity of the dyons.

Note that the total electric charge is conserved. The corresponding collective coordinate $q = q_1 + q_2$ can also be factored out with the motion of the center of mass.

6 Multimonopole Configurations

The equations of motion that follow from the Lagrangian (6.57) are

$$\left(\frac{M}{2} - \frac{g^2}{r}\right)\ddot{\mathbf{r}} = \frac{g^2}{r^3}\left\{\frac{1}{2}\mathbf{r}(\dot{\mathbf{r}}\cdot\dot{\mathbf{r}}) - (\mathbf{r}\cdot\dot{\mathbf{r}})\dot{\mathbf{r}}\right\} + \frac{Qg}{r^3}[\dot{\mathbf{r}}\times\mathbf{r}] - \frac{Q^2}{2r^3}\mathbf{r}. \tag{6.58}$$

Note that the dynamical equation does not change if we transform the Lagrangian of the relative motion (6.57) as

$$L = \frac{1}{4}\left(M - \frac{2g^2}{r}\right)\left(\dot{\mathbf{r}}\cdot\dot{\mathbf{r}} - \frac{Q^2}{g^2}\right) + Qg\dot{\mathbf{r}}\cdot\mathbf{a}, \tag{6.59}$$

where the constant term $MQ^2/4g^2$ is dropped out.

Obviously, for $g = 0$, the (6.58) is identical to the standard equation of motion of a charged particle in a Coulomb field. In the general case, (6.58) can be solved by making use of the corresponding integrals of motion. For example, the energy is composed of three terms: normal kinetic energy, a velocity-depending term originating from the difference between the dilatonic and magnetic interaction of the dyons, and the standard potential energy of interaction of the two charges:

$$E = \left(\frac{M}{2} - \frac{g^2}{r}\right)\dot{\mathbf{r}}^2 - \frac{Q^2}{r}. \tag{6.60}$$

The second integral of motion is the vector of angular momentum

$$\mathbf{L} = \left(\frac{1}{2} - \frac{g^2}{Mr}\right)\tilde{\mathbf{L}} - Qg\frac{\mathbf{r}}{r}, \tag{6.61}$$

where $\tilde{\mathbf{L}} = M[\mathbf{r}\times\dot{\mathbf{r}}]$ is the standard orbital angular momentum. These formulae are obviously generalizations of the expressions (1.4) and (1.11). Since the relation $\mathbf{L}\cdot\hat{\mathbf{r}} = -Qg = const.$ holds again, the same argumentation can be applied to show that the trajectory of the relative motion of the dyons lies on the surface of a cone. The motion becomes flat only if the magnetic charge vanishes.

We have already noted (see Sect. 5.2.1) that the electric charge of a static dyon is connected with its fourth cyclic collective coordinate Υ as $Q \sim \dot{\Upsilon}$. Excitation of this collective coordinate can be treated as the appearance of the kinetic energy (5.86). This analogy now can be generalized in the spirit of Kaluza–Klein theory [369]. Let us perform the Legendre transform of the Lagrangian (6.59)

$$L(\mathbf{r},\Upsilon) = L(\mathbf{r},Q) + gQ\dot{\Upsilon},$$

where

$$Q \equiv \frac{2g^3}{M - \frac{2g^2}{r}}\left(\dot{\Upsilon} + (\mathbf{a}\cdot\dot{\mathbf{r}})\right). \tag{6.62}$$

Then the Lagrangian (6.59) can be rewritten in the form

$$L = \frac{1}{4}\left(M - \frac{2g^2}{r}\right)\dot{\mathbf{r}}\cdot\dot{\mathbf{r}} + \frac{g^4}{M - \frac{2g^2}{r}}\left(\dot{\Upsilon} + (\mathbf{a}\cdot\dot{\mathbf{r}})\right)^2, \qquad (6.63)$$

which is structure of the Kaluza–Klein Lagrangian describing geodesic motion in the four-dimensional space $\mathcal{M}_0$ with one compact variable. Note that (6.63) does not depend explicitly on Υ. Thus, the corresponding equation of motion is just the conservation law of the relative electric charge Q, (6.62).

To sum up, the relative motion of well-separated BPS dyons is a geodesic motion in the space $\mathcal{M}_0$ governed by the Taub–NUT (Newman-Unti–Tamburino) metric

$$ds^2 = \left(1 - \frac{2g^2}{Mr}\right)d\mathbf{r}^2 + \frac{\left(\frac{2g^2}{M}\right)^2}{1 - \frac{2g^2}{Mr}}(d\Upsilon + \mathbf{a}\cdot d\mathbf{r})^2. \qquad (6.64)$$

This metric is well-known from general relativity; it was obtained as early as in 1951 (see [381]). The Taub–NUT metric corresponds to the spatially homogeneous solution of the Einstein equations in empty space. The length parameter of this metric is $2g^2/M$.

If the angular momentum (6.61) is quantized at integer values, the rotation group is $SO(3)$. Geometrically, this rotational invariance means that the conserved vector $\mathbf{L}$ implies a set of Killing vector fields on the space $\mathcal{M}_0$ that generate an $SO(3)$ symmetry, which is an *isometry* of the moduli space. If the charge quantization condition yields half-integer eigenvalues of $\mathbf{L}$, the rotational invariance induce an $SU(2)$ isometry of the space $\mathcal{M}_0$ [238, 346].

Finally, we recall that geodesic motion in a space with the Taub–NUT metric could be used only to describe the relative motion of widely separated dyons. A general description of the low-energy dynamics of BPS monopoles on the moduli space is given by the Atiyah–Hitchin metric, whose asymptotic form is the Taub–NUT metric. In the last section of this chapter we return to this approximation.

6.4 The n-Monopole Configuration in the BPS Limit

6.4.1 BPS Multimonopoles: A Bird's Eye View

The exact cancellation of the electromagnetic attraction and the dilaton repulsion in the two-monopole BPS system suggests the conjecture that there are multimonopole static solutions of the Bogomol'nyi equations. Since Bogomol'nyi found that the explicit spherically symmetric solution with $n = 1$ is unique [127], any possible multi-monopole configuration with $n > 1$ cannot have such a symmetry. Furthermore, it was shown [367] that even a configuration of n monopoles lying on a straight line [288, 289, 440] is unstable. Therefore, the structure of the configuration we would deal with is rather complicated and it is difficult to build up these solutions.

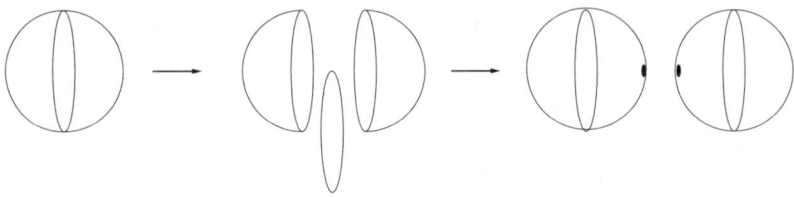

Fig. 6.6. Geometry of the transformation that allows us to construct two separated monopoles starting from a charge one spherically symmetric configuration

Naively, one can visualize the following geometric transformation that could help to construct a multi-monopole configuration starting from a given single spherically symmetric BPS monopole. Recall that magnetic charge is associated with the asymptotic behavior of the scalar field; the vacuum manifold of a monopole of unit topological charge is a sphere S^2_{vac}.

To construct a two-monopole configuration, we shall remove from this sphere the equatorial circle S^1, as shown in Fig. 6.6, and then identify all the points on the equators of the two hemispheres with the north and south poles of the two new spheres, respectively. Construction of an n-monopole configuration requires a simple iteration of this procedure[6].

However, an adequate mathematical description of the geometrical transformation that could solve the multimonopole problem is not so trivial and emerged only as the result of the work of many mathematicians over a decade. The problem is not to construct an explicit solution for an arbitrary n-monopole configuration, but to prove the regularity and completeness of the solution, i.e., to prove that all possible solutions are generated by this procedure.

To give some clue as to which methods we have to use, let us note that the naive picture above is closely connected with the mathematical apparatus of projective geometry. Indeed, making a standard stereographic projection of the sphere S^2 onto a plane, we see that removing the equator of the sphere corresponds to a cut on the projective plane, which then becomes isomorphic to a doubly covered Riemann surface. Recall also that identification of the antipodal points of a sphere S^n transform it into a real projective space $\mathbb{R}P^n$. Thus, it is no surprise that the powerful methods of twistor geometry were fully exploited to construct the multi-monopole configurations.

Generally speaking, to construct a general explicit n-monopole solution, one has to make use of the integrability of the Bogomol'nyi equation. There are three different approaches to this problem, which use:

1. twistor technique (the so-called Atiyah–Ward ansatz [79, 180, 181, 507]);

---

[6] With some imagination one can compare this picture with the well-known process of biological cell division...

2. the Atiyah–Drinfeld–Hitchin–Manin (ADHM) construction [80], which was modified by Nahm [393];

3. the inverse scattering method (Riemann–Hilbert problem), which was applied to the linearized Bogomol'nyi equation [217, 218, 220].

For a detailed description of the last approach, we refer the reader to the comprehensive review [220]. Here, we briefly outline only the first two directions, which are closely connected to the modern development of differential geometry.

6.4.2 Projective Spaces and Twistor Methods

The following discussion substantially uses the language of the twistor theory. Here we recall basic notions of this formalism. For more rigor and a broader presentation the reader should, for example, consult the book [30].

Let us extend our classification of spaces of Chap. 3. The general notion of *projective spaces* can be defined as a transformation of the n-dimensional real space $\mathbb{R}^n$ or the complex space $\mathbb{C}^n$ into equivalent $n - 1$-dimensional manifolds $\mathbb{R}P^{n-1}$ and $\mathbb{C}P^{n-1}$, respectively. These spaces are defined as the spaces of all unoriented lines L through the origin of $\mathbb{R}^n$ or $\mathbb{C}^n$, respectively.

The simple example is the *projective plane* $\mathbb{R}P^2$ of the lines through the origin of $\mathbb{R}^3$. Clearly, such a line L is defined by a point x with coordinates (x_1, x_2, x_3) and for any $a \neq 0$ a point (ax_1, ax_2, ax_3) corresponds to the same line. This is an equivalence class, which defines a point in the real projective space $\mathbb{R}P^2$; each such point is set into correspondence to a line in $\mathbb{R}^3$. A convenient choice is to consider points on the unit sphere S^2, which is defined by the equation $x_1^2 + x_2^2 + x_3^2 = 1$.

Now let us consider the *directed lines* $\vec{L}$ through the origin of $\mathbb{R}^3$. The difference is that such a line intersects the sphere S^2 only once, while an undirected line meets this unit sphere twice, in a pair of antipodal points. Thus, the projective space $\mathbb{R}P^2$ contains a set of pairs of these points. It can be constructed if we take only one of the hemispheres of the sphere S^2 and identify the antipodal points of the equator. Figure 6.7 illustrates this procedure (see also Fig. 6.6), which yields the disk in the equatorial plane of S^2 with antipodal points identified. Topologically it is the sphere S^2.

The projective space $\mathbb{R}P^2$ has the structure of the manifold that we discussed in Chap. 3. It can be covered by three local sets: (i) U_1, which includes the lines not lying in the x_2x_3-plane; (ii) U_2, which includes the lines not lying in the x_1x_3-plane; (iii) U_3, which includes the lines not lying in the x_1x_2-plane. In each patch we can introduce some local coordinates, for example, in the patch U_1 we have $u_1 = x_2/x_1$, $u_2 = x_3/x_1$, etc. The transition functions are defined in the overlap regions.

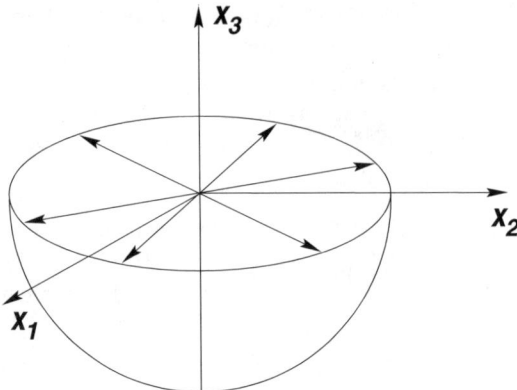

Fig. 6.7. Definition of the projective space $\mathbb{R}P^2$

In the same way, we can set a correspondence between the four-dimensional real space $\mathbb{R}^4$ and the three-dimensional projective space $\mathbb{R}P^3$. This situation is very interesting since the corresponding unit three-sphere S^3 with the antipodal points having been identified can be thought of as the group manifold of the group $SO(3)$. Thus, there is a diffeomorphism between the group manifold of $SO(3)$ and the real projective space $\mathbb{R}P^3$.

Similarly, we can define the *complex projective space* $\mathbb{C}P^{n-1}$ as a space of complex lines through the origin of $\mathbb{C}^n$. Local coordinates are now the complex numbers. For example, a complex projective line $\mathbb{C}P^1$ is defined as a set of equivalence classes $(z_1, z_2) \sim (\lambda z_1, \lambda z_2)$, where $\lambda \in \mathbb{C}$ and $|z_1|^2 + |z_2|^2 \neq 0$. Then there are two local patches U_1 and U_2, with the local coordinates $u = z_1/z_2$ for $z_2 \neq 0$, and $w = z_2/z_1$ for $z_1 \neq 0$. In the overlap region we have $u = 1/w$ and comparing this picture to (3.11), we conclude that the complex projective line $\mathbb{C}P^1$ is diffeomorphic to the sphere S^2.

In more general terms, a complex projective space $\mathbb{C}P^n$ can be obtained from the unit sphere S^{2n+1} via identification $z \sim ze^{i\alpha}$ and each point of $\mathbb{C}P^n$ corresponds to the circle $S^1 \sim e^{i\alpha}$ on the sphere S^{2n+1}. In particular, there is a map $S^3 \to \mathbb{C}P^1 \sim S^2$, which is exactly the Hopf map which we discussed in Chap. 3. Note that the projective space $\mathbb{C}P^n$ equipped with the Fubini–Study metric

$$g_{ij} = \frac{i}{2} \frac{\delta_{ij}(1+|w|^2) - w_i \bar{w}_j}{(1+|w|^2)^2},$$

written in the local coordinates $w_i = z_i/z_1$, $i = 1, 2, \ldots n$ for $z_1 \neq 0$, is an example of Kähler manifolds (recall the corresponding definition given in Chap. 3, p. 89).

This formalism can be applied to replace the physical four-dimensional space-time by a complex manifold of three-complex dimension, the projective twistor space $PT \equiv \mathbb{C}P^3$.

Let us explain these notions. The twistor description assumes that we are working in complexified Euclidean space, which is constructed by the correspondence between each point of x_μ in $\mathbb{R}^4$ and a complex quaternion

$$x = x_0 + ix_k\sigma_k = \begin{pmatrix} x_0 + ix_3 & ix_1 + x_2 \\ ix_1 - x_2 & x_0 - ix_3 \end{pmatrix} = \begin{pmatrix} z_1 & z_2 \\ -z_2^* & z_1^* \end{pmatrix}, \quad (6.65)$$

where we introduce two complex homogeneous coordinates $z_1 = x_0 + ix_3$ and $z_2 = x_2 + ix_1$ in $\mathbb{C}^2 \sim \mathbb{R}^4$. The notations (6.65) are especially convenient for analyzing self-dual systems because then the equations of self-duality $F_{\mu\nu} = \widetilde{F}_{\mu\nu}$ reduce to

$$\begin{aligned} F_{z_1 z_2} &= 0, & F_{z_1^* z_2^*} &= 0 \; ; \\ F_{z_1 z_1^*} &+ F_{z_2 z_2^*} = 0 \, . \end{aligned} \quad (6.66)$$

The advantage here is that in the complex coordinates z_i, the first pair of the equations (6.66) is integrable and we can see that the solution is a pure gauge. This is a key element of the following analysis. However, we need to introduce some more mathematical notions before going into details of the related discussion.

The space of quaternions $\mathbb{H}$ is isomorphic to $\mathbb{C}^2$, therefore $\mathbb{H}^2 \simeq \mathbb{C}^4$. Thus, each element θ of $\mathbb{C}P^3$, that is a complex line through the origin in $\mathbb{C}^4$, corresponds to a quaternionic line in $\mathbb{H}^2$. The latter is an element of the one-dimensional *quaternionic projective space* $\mathbb{H}P^1$, which is isomorphic to S^4.

A line in $\mathbb{C}P^3$ can be parameterized by four local homogeneous coordinates (z_1, z_2, z_3, z_4). In the twistor description, they are usually set into correspondence to a pair of 2-spinors λ and ρ as $(\lambda_0, \lambda_1, \rho^0, \rho^1)$. This defines a point $\theta = (\lambda \rho)$ in $\mathbb{C}P^3$, a 4-spinor formed from λ and ρ.

Let us consider a two-dimensional complex plane in $\mathbb{C}^4$. In terms of the local coordinates, it is defined by two orthogonal tangent vectors v_μ and u_μ. If, for any displacement in this plane, the tensor $\Omega_{\mu\nu} \equiv v_\mu u_\nu - u_\mu v_\nu$ is anti-self-dual, that is $\Omega_{\mu\nu} = -\widetilde{\Omega}_{\mu\nu}$, then this plane is called anti-self-dual (or β-plane [506]). If the tensor $\Omega_{\mu\nu}$ is self-dual, rather than anti-self-dual, then such a plane is called α-plane[7]. For such planes we shall use the notations $\alpha_{[\theta]}$ and $\beta_{[\theta]}$, respectively.

Now we observe that the components of a self-dual tensor $F_{\mu\nu}$ vanish when contracted with those of an anti-self-dual tensor $\Omega_{\mu\nu}$:

$$F_{\mu\nu}\Omega_{\mu\nu} = 0 \quad \text{if} \quad \Omega_{\mu\nu} = -\widetilde{\Omega}_{\mu\nu}, \quad F_{\mu\nu} = \widetilde{F}_{\mu\nu} \, .$$

In the twistor theory language, this result is usually referred to as the vanishing of the self-dual tensor $F_{\mu\nu}$ in the β-plane. Vice-versa, an anti-self-dual

---

[7] Clearly, a general plane in $\mathbb{C}^4$ is neither α- nor β-plane.

tensor vanishes in the α-plane. This is a very important element in the following discussion. The importance of these planes in $\mathbb{C}^4$ is that the set of all α-planes is isomorphic to complex projective space $\mathbb{C}P^3$, while the set of all β-planes is isomorphic to a dual complex projective space $\mathbb{C}P^{3*}$. Thus, this allows us to define the notion of duality in the context of the projective geometry.

6.4.3 The n-Monopole Twistor Construction

As pointed out by Ward, the multimonopole solutions of the Bogomol'nyi equations can be constructed by the use of the twistor methods (see, e.g., [30]). The basic elements of this approach were used to describe self-dual fields in Euclidean space $\mathbb{R}^4$. However, the analogy between the static Bogomol'nyi equations (5.61) in $\mathbb{R}^3$ and the self-duality equations in $\mathbb{R}^4$ [367] allows us to apply the twistor formalism to obtain n-monopole configurations.

Let us start from the formal description of the Ward twistor transform for a self-dual Yang–Mills field [79,506,539]. The idea is to exploit the correspondence between the self-dual gauge field and certain holomorphic (i.e., analytic in a complex variable) vector bundles over a standard three-dimensional twistor space $\mathbb{C}P^3$ [506]. The basic element of this construction is an observation that the components of the self-dual field strength tensor $F_{\mu\nu}$ vanish when restricted to any β-plane [506,539]. Therefore, the gauge potential on such a plane becomes a pure gauge

$$A_\mu = -iU^{-1}\partial_\mu U . \tag{6.67}$$

We shall see how this statement allows us to restore the gauge potential by knowledge of the element of the gauge group [180,506].

Moreover, both α-planes and β-planes have the property that their tangent vectors v_μ and u_μ are null vectors, that is $v_\mu v^\mu = u_\mu u^\mu = 0$. This means that any displacement in such a plane is zero and this allows us to define the α-plane from the equation

$$\alpha_{[\theta]} = x : \quad \lambda = x\rho , \tag{6.68}$$

where λ and ρ are two-spinors, which we defined above, and x is a matrix (6.65).

Let us explain this statement. Indeed, for any other point y this equation yields $\lambda = y\rho$. Thus, $(x - y)\rho = 0$ and, therefore, $\det(x - y) = 0$. Since $\det(x-y) = (x-y)_\mu(x-y)^\mu$, this implies $x_\mu = y_\mu$ and any translation in the plane defined according (6.68) is null, and this is a α-plane that is parameterized by the homogeneous coordinates $[\theta] = (x\rho, \rho)$ of $\mathbb{C}P^3$.

Similarly, we can consider the equation

$$\beta_{[\theta]} = x : \quad \omega = \bar{x}\pi , \tag{6.69}$$

where $\bar{x} = x_0 - ix_k\sigma_k$ and (ω, π) are coordinates of the dual complex projective space $\mathbb{C}P^{3*}$. Then the β-plane arises as a solution of the equation (6.69).

Note that the two points $\theta = (\lambda, \rho)$ and $\chi\theta = (\chi\lambda, \chi\rho)$ of $\mathbb{C}P^3$, where $\chi \in \mathbb{C}$ define the same α-plane. Thus, there is a direct correspondence between the points θ of the complex projective space $\mathbb{C}P^3$ and the points $[\theta]$ of the α-planes in $\mathbb{C}^4$ [506]. The same correspondence exists between the points $\theta = (\omega, \pi)$ of the dual complex projective space $\mathbb{C}P^{3*}$ and the points $[\theta]$ of the β-planes.

Let us explain the relation of the duality between the α-planes and the β-planes. A plane (x, y) passing through the origin in $\mathbb{C}^4$ is defined by set of lines that satisfy the equation

$$c_1 x + c_2 y = 0 \, ,$$

where $c_1, c_2 \in \mathbb{C}$. The pair of numbers (c_1, c_2) can be set into correspondence to any line, that is it can be thought of as a point of some dual space. This means that in the three-dimensional projective space $\mathbb{C}P^3$, the points θ are dual to the two-planes. If $\mathbb{C}P^3$ is the space of α-planes, then the β-planes are given by the points of the dual space $\mathbb{C}P^{3*}$ and vice-versa.

Thus, we have to analyse the properties of the space of β-planes, keeping in mind that a similar consideration can also be applied to the space of the α-planes. First, the β-plane is defined by (6.69), that is

$$\omega_1/\pi_1 = z_1^* - z_2\xi^{-1}, \quad \omega_2/\pi_1 = z_2^* + z_1\xi^{-1},$$
$$\omega_1/\pi_2 = z_1^*\xi - z_2, \quad \omega_2/\pi_2 = z_1 + z_2^*\xi \, , \qquad (6.70)$$

where $\xi = \pi_1/\pi_2$.

Second, we note that this space has a fibre bundle structure. Indeed, the initial gauge group $SU(2)$ is now replaced by its twistor analog $SL(2, C)$. An element of the latter defines the parallel transport of spinors along a path in the plane $\beta_{[\theta]}$:

$$\psi_{[\theta]}(x) = U_{[\theta]}(x, y)\psi_{[\theta]}(y), \quad U(x, y) = \mathcal{P} \exp\left\{i \int_x^y A_\mu dx^\mu\right\} \, , \qquad (6.71)$$

and both the points x, y and the path of integration lie entirely within $\beta_{[\theta]}$. This allows us to introduce the two-dimensional vector space $V_{[\theta]}$ of spinor fields over an β-plane. A set of these spaces forms a two-dimensional holomorphic vector bundle over $\mathbb{C}P^{3*}$ [506].

Note that this bundle is non-trivial. Indeed, we have to set up a correspondence between the coordinates on base $x_{[\theta]} \in \beta_{[\theta]}$ and the coordinates on the bundle $\psi_{[\theta]}(x)$ for each of spinors π, ω. However, it is not possible to choose the coordinates $x_{[\theta]}$ smoothly everywhere in the space $\mathbb{C}P^{3*}$. There are four homogeneous coordinates $(\pi_0, \pi_1, \omega^0, \omega^1)$ and we need four patches to cover

$\mathbb{C}P^{3*}$. For example, in the patch $\pi_0 \neq 0$, three complex coordinates are given by $(\pi_1/\pi_0; \omega^0/\pi_0; \omega^1/\pi_0)$. Since the four-fold intersection of all patches is not empty, we need six transition functions to specify the bundle.

Even if we were to consider only those four-spinors $[\theta]$ for which $\pi \neq 0$ (this corresponds to the reduction from $\mathbb{C}P^{3*}$ to $\mathbb{C}P^{1*}$), there are two singularities at the points $\pi_1 = 0$ and $\pi_2 = 0$. The way in which we have to treat such singularities is identical to the situation we confront in the case of the Abelian Wu–Yang monopole: one has to cover the space by two maps as [180, 506]

$$x^1_{[\theta]} = \begin{pmatrix} \omega_1/\pi_1 & 0 \\ \omega_2/\pi_1 & 0 \end{pmatrix}, \quad \pi_1 \neq 0, \quad x^2_{[\theta]} = \begin{pmatrix} 0 & \omega_1/\pi_2 \\ 0 & \omega_2/\pi_2 \end{pmatrix}, \quad \pi_2 \neq 0. \tag{6.72}$$

This bundle can be characterized by a holomorphic transition function $U_{[\theta]}(x^1_{[\theta]}, x^2_{[\theta]})$, which relates the coordinates in different regions

$$\psi_{[\theta]}(x^1_{[\theta]}) = U_{[\theta]}(x^1_{[\theta]}, x^2_{[\theta]})\psi_{[\theta]}(x^2_{[\theta]}) . \tag{6.73}$$

This is a 2×2 matrix of $SL(2, \mathbb{C})$.

Ward pointed out [506] that for any point $x_{[\theta]} \in \beta_{[\theta]}$, such a function can be written as

$$U(\omega, \pi) = U_{[\theta]}(x^1_{[\theta]}, x)U_{[\theta]}(x, x^2_{[\theta]}) .$$

Thus, one can "split" it as

$$U(x\pi, \pi) = U_{[\theta]}(x^1_{[\theta]}, x)U^{-1}_{[\theta]}(x^2_{[\theta]}, x) = U_1(x, \xi)U_2^{-1}(x, \xi) . \tag{6.74}$$

For a fixed $\xi = \pi_1/\pi_2$, the function $U_1(x, \xi)$ is analytic everywhere, but $\xi = 0$, while $U_2(x, \xi)$ is singular at $\xi = \infty$. Since the functions $U_1(x, \xi)$ and $U_2(x, \xi)$ are holomorphic in these regions, they can be expanded in the Laurent series: in positive degrees of ξ for $U_1(x, \xi)$ and in negative degrees of ξ for $U_2(x, \xi)$. Then (6.74), together with Liouville's theorem, determines the form of the function $U_1(x, \xi), U_2(x, \xi)$ up to $SL(2, \mathbb{C})$ gauge transformations

$$U_1(x, \xi) \to U_1(x, \xi)V(x) , \quad U_2(x, \xi) \to U_2(x, \xi)V(x) ,$$

which corresponds to the gauge transformations of the vector-potential A_μ. Moreover, since the transition function is defined by (6.74), the matrix of the gauge transformation $V(x) \in SL(2, \mathbb{C})$ is regular everywhere in $\mathbb{C}P^{1*}$, that is it does not contain an explicit dependence on ξ.

The problem of the splitting of the gauge function into two matrices $U_1(x, \xi)$ and $U_2(x, \xi)$ is known as the celebrated *Riemann–Hilbert problem*. A different formulation of this problem is to find multi-valued functions (sections) knowing the form of the given *monodromy* at the singularities $\xi = 0$ and $\xi = \infty$. This problem has a unique solution.

In the last chapter of this book we shall discuss the solution of the Riemann–Hilbert problem and the properties of the matrices of monodromy

6.4 The n-Monopole Configuration in the BPS Limit

of $SL(2,\mathbb{C})$ in more detail. For our consideration here it is enough to recall that the gauge potential A_μ is a pure gauge. Thus, by making use of the relation (6.67), one can regain it from the matrix $U_1(x,\xi), U_2(x,\xi)$ of $SL(2,\mathbb{C})$ [180]. Indeed, the self-dual connection $A_\mu(x)$ is given by $A(x) = -iU^{-1}dU = A_{ab}dx^{ab} = A_\mu dx^\mu$ and, because on the β-plane $\omega = x\pi$, we have $dx\pi = 0$. This implies that for the entries x_{ab} of the 2×2 matrix x, we have

$$dx_{a2} = -\xi dx_{a1}, \quad a,b = 1,2,$$

which in turn yields

$$\left(\frac{\partial}{\partial x_{a1}} - \xi \frac{\partial}{\partial x_{a2}}\right) U(\omega, \pi) = 0.$$

Substitution of the splitted function (6.74) then gives

$$A_{a1} - \xi A_{a2} = -iU_1^{-1}(x,\xi) \left(\frac{\partial}{\partial x_{a1}} - \xi \frac{\partial}{\partial x_{a2}}\right) U_1(x,\xi)$$
$$= -iU_2^{-1}(x,\xi) \left(\frac{\partial}{\partial x_{a1}} - \xi \frac{\partial}{\partial x_{a2}}\right) U_2(x,\xi). \quad (6.75)$$

This relation allowed Ward to argue that a gauge function $U(\omega, \pi)$, which can be splitted as in (6.74) into two functions that are holomorphic in the domains of their definitions, yields a self-dual gauge field [506]. Atiyah and Ward [79] and Corrigan et al. [180] showed that, at least in the case of the instanton self-dual fields, the patching matrix can be taken in the triangular form

$$U = \begin{pmatrix} \xi^n & \rho(x,\xi) \\ 0 & \xi^{-n} \end{pmatrix}, \quad (6.76)$$

where n is some positive integer and a function $\rho(x,\xi)$ satisfies certain properties. Actually this is only a function of ξ and [79]

$$\omega_1/\pi_1 = z_1^* - z_2\xi^{-1}, \quad \omega_2/\pi_2 = z_1 + z_2^*\xi.$$

The advantage of this form is that for such a matrix, the splitting can be done by a contour integration. Indeed, this limited dependence of the function $\rho(x,\xi)$, which appears in the right upper corner of the matrix (6.76), means that both the function $\rho(x,\xi)$ and the coefficients of its expansion into Laurent series

$$\Delta_p = \frac{1}{2\pi i} \oint_{|\xi|=1} \frac{d\xi}{\xi} \xi^p \rho(x,\xi), \quad (6.77)$$

satisfy the homogeneous four-dimensional Laplace equation

$$\partial^2 \rho = 0, \quad \partial^2 \Delta_q = 0. \quad (6.78)$$

Then the condition that a patching function may be splitted as in (6.74) is equivalent to the statement that for a banded matrix $D^{(n)}$ of dimension $n \times n$ with entries

$$D^{(n)}_{pq} = \Delta_{p+q-n-1}, \qquad 1 \leq p, q \leq n, \qquad (6.79)$$

we have

$$\det D^{(n)} \neq 0.$$

Then the gauge potential can be restored as[8]

$$A_i \equiv \frac{1}{2} A_i^a \sigma^a = -\frac{1}{2\Lambda_2} \begin{pmatrix} \eta^3_{ij} \partial_j \Lambda_2 & (\eta^1 - i\eta^2)_{ij} \partial_j \Lambda_1 \\ (\eta^1 + i\eta^2)_{ij} \partial_j \Lambda_3 & -\eta^3_{ij} \partial_j \Lambda_2 \end{pmatrix}, \qquad (6.80)$$

where $\eta^a_{ij} = \varepsilon_{oaij} + \delta_{ai}\delta_{0j} - \delta_{aj}\delta_{0i}$ is the 't Hooft tensor [272] and

$$\Lambda_1 = \left(D^{(n)-1}\right)_{11}, \quad \Lambda_2 = \left(D^{(n)-1}\right)_{1n}, \quad \Lambda_3 = \left(D^{(n)-1}\right)_{nn}$$

are components of the matrix (6.79). This is exactly the potential of the multi-instanton configuration in the Yang gauge [539].

Let us consider how this formalism can be applied to the case of the BPS multimonopole configurations. We already showed that the Julia–Zee correspondence $\phi \rightleftharpoons A_0$ allows us to consider the Bogomol'nyi equation as a reduced self-duality equation (see (5.66)). This analogy was used by Ward [507] to construct a vector bundle, which corresponds to the known spherically symmetric one-monopole configuration (5.71). An analogy with the multi-instanton case then can be used to generalize this construction to obtain another holomorphic vector bundle over $\mathbb{C}P^1$, which corresponds to the n-monopole configuration. This procedure was carried out in the paper [181] for a general multimonopole configuration.

The first difference from the case of the multi-instanton configuration considered above is that in a suitable gauge, the vector potential of the BPS monopole must be static: $\partial_0 A_\mu = 0$. This condition can be satisfied if one takes the patching function of the form [507]

$$U(x, \xi) \to U_0(\eta, \xi), \qquad (6.81)$$

where

$$\eta = \frac{i}{2}\left(\frac{\omega_2}{\pi_2} - \frac{\omega_1}{\pi_1}\right)\xi = \frac{x_1 + ix_2}{2}\xi^2 - x_3\xi - \frac{x_1 - ix_2}{2}. \qquad (6.82)$$

Thus, the temporal variable x_0 is excluded from the matrix $U_0(\eta, \xi)$ and the coefficients of the Laurent series (6.77), which satisfy the (6.78), become periodic in time:

$$\Delta_p(x) = e^{ix_0} \widetilde{\Delta}_p(\mathbf{x}). \qquad (6.83)$$

---

[8] Recall that we are using a "physical" Hermitian basis.

6.4 The n-Monopole Configuration in the BPS Limit

Another difference from the instanton problem is related to the asymptotic behavior of the Higgs field (5.65). For the BPS monopoles of charge n, its long-distance tail is $|\hat{\phi}| \sim 1 - n/r + O(r^{-2})$. Here the key component is the relation between the determinant D of the matrix $D^{(n)}$ (6.79) and the square of the length of the scalar field [432]:

$$|\hat{\phi}|^2 = 1 - \partial_i \partial_i \ln D. \tag{6.84}$$

Since the coefficients $\Delta_p(x)$ satisfy (6.78), we obtain:

$$\partial_i \partial_i \widetilde{\Delta}_p(x_i) = \widetilde{\Delta}_p(x_i). \tag{6.85}$$

Taking into account the relation (6.84), we see that on the spatial asymptotic

$$\widetilde{\Delta}_p(x_i) \sim \frac{e^r}{r} \delta_p(\theta, \phi). \tag{6.86}$$

The Higgs field has proper asymptotic behavior if this relation holds for all values $|p| \le n - 1$. Consequently, since the Laurent coefficients are time-periodic dependent, the function $\rho(\eta, \xi)$, which appears in the right upper corner of the Atiyah-Ward matrix, depends on the coordinate x_0 as $\rho \sim e^{ix_0}$ [181]. Thus we can guess that simple elimination of this time-oscillating exponent in the patching matrix and the following replacement of $\Delta \to \widetilde{\Delta}$ in (6.79) could yield the multimonopole solution of the Bogomol'nyi equations.

Indeed, let us consider the function of the form:

$$\rho(\mathbf{x}, \xi) = \frac{\xi e^{-ix_0}}{\eta} \left(e^{i\omega_2/\pi_2} - e^{i\omega_1/\pi_1} \right) = \frac{\xi}{\eta} \left(e^{(x_1+ix_2)\xi - x_3} - e^{(x_1-ix_2)\xi^{-1}+x_3} \right). \tag{6.87}$$

Then the explicit form of the matrix $\Delta_p(x)$ can be defined by evaluation of the contour integral (6.77) along the unit circle in the complex plane of ξ:

$$\Delta_0 = \frac{1}{2\pi i} \oint_{|\xi|=1} \frac{d\xi}{\xi} \rho(x, \xi) \tag{6.88}$$

$$= \frac{1}{\pi i} \oint_{|\xi|=1} d\xi \frac{e^{(x_1+ix_2)\xi - x_3} - e^{(x_1-ix_2)\xi^{-1}+x_3}}{(x_1 + ix_2)\left(\xi - \frac{r+x_3}{x_1+ix_2}\right)\left(\xi + \frac{r-x_3}{x_1+ix_2}\right)}.$$

Here the contour of integration encircles a simple pole of first order at $\xi_0 = \dfrac{r + x_3}{x_1 + ix_2}$, thus a simple calculation yields

$$\Delta_0 = \frac{1}{r}\left(e^r - e^{-r}\right) = \frac{2\sinh r}{r}. \tag{6.89}$$

Since in this case $D^{(1)} = \Delta_0$, by making use of (6.84), we have

$$|\hat{\phi}|^2 = 1 - \partial_i \partial_i \ln 2 \frac{\sinh r}{r} = \frac{(r \cosh r - \sinh r)^2}{(r \sinh r)^2} = \left(\coth r - \frac{1}{r} \right)^2, \quad (6.90)$$

which obviously corresponds to (5.71). The same conclusion can in principle be obtained from the direct calculation the gauge potential A_μ, which corresponds to the matrix (6.81), by analogy with the multi-instanton self-dual field. As one can expect in advance, this procedure yields the potential (5.67) of the infinite chain of instantons, which we already considered in Chap. 5. Recall that, up to a gauge transformation, this potential is equivalent to the one-monopole BPS solution (5.63). Thus, we can see that the patching matrix (6.81) really encoded all the information about the monopole field[9].

Note that the position of the monopole corresponds to the zero of the Higgs field. An obvious shift

$$\eta \to \eta + \frac{R_1 + iR_2}{2} \xi^2 - \frac{R_1 - iR_2}{2} - R_3 \xi ,$$

gives a monopole at the point with coordinates (R_1, R_2, R_3).

Ward was able to find, as guesswork, a generalization of this one-monopole solution, which corresponds to the static two-monopole configuration [507]. He suggested to substitute in the Atiyah–Ward transition matrix $U = \begin{pmatrix} \xi^2 & \rho(x, \xi) \\ 0 & \xi^{-2} \end{pmatrix}$, which generalizes (6.76), the function $\rho(x, \xi)$ of the following form:

$$\rho(x, \xi) = \frac{\xi^2}{\eta^2 + \frac{\pi^2}{4}\xi^2} \left(e^{(x_1 + ix_2)\xi - x_3} - e^{(x_1 - ix_2)\xi^{-1} + x_3} \right) . \quad (6.91)$$

Numerical analysis shows [58, 507] that the fields, which can be recovered from such a patching matrix, have proper asymptotic behavior and satisfy other requirements on the solutions of the BPS equations [507]. However, this configuration has not a spherical but an axial symmetry, thus, the surface of the constant energy density is a torus. The configuration (6.91) corresponds to the toroidal BPS monopole with double zero of the Higgs field at the origin (see Fig. 6.8). Actually, we have discussed this solution in Sect. 6.2.

Evidently this BPS configuration coincides with the charge two Rebbi–Rossi solution given by the axially symmetric ansatz (6.15) and (6.16) with winding numbers $m = 1$ and $n = 2$ in the limit $\lambda = 0$. This allows us to check the mathematically refined twistor approach.

---

[9] An interesting feature of the twistor approach is that it easily allows us to prove that some given patching matrices and algebraic curves really correspond to the multimonopoles. Much more difficult is the problem of finding of these matrices and curves themselves...

6.4 The n-Monopole Configuration in the BPS Limit

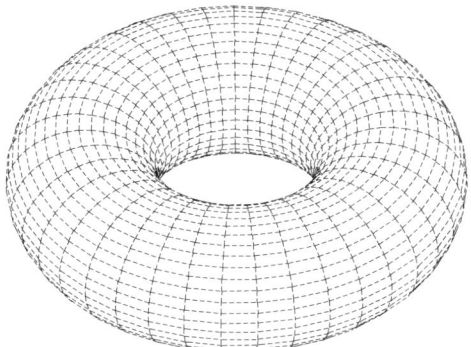

Fig. 6.8. Energy density surface for the BPS monopole of charge 2

Thus, the configuration (6.91) can be thought of as a superposition of two single $n = 1$ static BPS monopoles. However, as we know from consideration of the axially symmetric ansatz, unlike the fundamental $n = 1$ monopole, there is no analytical solution that would describe the coordinate dependence of the fields A_μ, ϕ of the two-monopole configuration in terms of elementary functions. As we shall see in the next section, the problem may be reduced to the calculation of the elliptic integrals [181, 507].

The discussion above shows that a basic element of the multimonopole construction is the function $\rho(x, \xi)$, which for an arbitrary n-monopole configuration can be taken as [181, 266]

$$\rho(x,\xi) = \frac{\xi^n}{S_n} \left(e^{(x_1+ix_2)\xi - x_3} - e^{(x_1-ix_2)\xi^{-1}+x_3} \right), \quad (6.92)$$

where S_n is a polynomial in n of degree η with coefficients that are polynomials in ξ. Thus, the location of the poles of the meromorphic function $\rho(x, \xi)$ is defined by the equation

$$S_n = \eta^n + a_1(\xi)\eta^{n-1} + \cdots + a_{n-i}(\xi)\eta^i + \cdots + a_{n-1}(\xi)\eta + a_n(\xi) = 0, \quad (6.93)$$

which defines the so-called *spectral curve* – a geometrical object that plays a very special role in all methods of construction of multimonopole configurations known today. Note that this general definition agrees with the particular cases (6.81) and (6.91). For example,

$$S_1 = \eta - \frac{x_1 + ix_2}{2}\xi^2 + \frac{x_1 - ix_2}{2} + x_3\xi = 0, \quad (6.94)$$

which corresponds to the set of all lines of the space R^3 directed from a point (x_1, x_2, x_3) to the point where a monopole is placed.

The Ward ansatz (6.91) for the axially symmetric two-monopole configuration with double zero of the scalar field as the origin, corresponds to the choice of the spectral curve

$$S_2 = \eta^2 + \frac{\pi^2}{4}\xi^2 . \tag{6.95}$$

Further generalization of this relation allows us to describe two well-separated one-monopoles [508]

$$S_2 = \eta^2 + \frac{K(k)^2}{4}(1-k^2)\left(\xi^4 - 2\frac{1+k^2}{1-k^2}\xi^2 + 1\right) = 0 , \tag{6.96}$$

where the parameter (the elliptic modulus) $k \in [0, 1]$ and

$$K(k) = \int_0^1 \frac{ds}{\sqrt{1-s^2}\sqrt{1-k^2 s^2}} \tag{6.97}$$

is a complete elliptic integral of the first kind. If its elliptic modulus $k = 0$, then $K = \pi/2$ and (6.96) obviously reduced to (6.95). This is the case of two coinciding one-monopoles.

Another limit is $k \to 1$, which corresponds to $K \to \infty$. Then the spectral function (6.96) can be factorized as

$$S_2 \to \left(\eta + \frac{K}{2}(1-\xi^2)\right)\left(\eta - \frac{K}{2}(1-\xi^2)\right) .$$

Comparison with (6.94) allows us to identify this as a product of two spectral curves of two widely separated one-monopoles located at the points $(0, 0, \pm K)$, where $K \to \infty$. Thus, the modulus of the elliptic integrals, taking values within the range $[0; 1]$, is actually a parameter of the separation between the monopoles.

The geometrical meaning of this construction becomes clearer in the Hitchin approach [51, 266], which we will briefly discuss below.

6.4.4 Hitchin Approach and the Spectral Curve

The approach to the construction of multimonopole configurations described above is related with dimensional reduction of Euclidean space $\mathbb{R}^4$ to $\mathbb{R}^3$ and a subsequent twistor transform of the self-duality equations. However, the dimensional reduction can be made directly at the twistor level. This technique was elaborated by N. Hitchin [51, 266, 267]. From the modern point of view, this construction seems to be in adequate correspondence with the underlying geometry of the multi-monopoles.

We have to make a reservation at this point. Nowadays, this branch of mathematical physics is developing rapidly and a lot of surprising results were obtained recently. For more rigor and broader discussion, we refer the reader to original publications, or to the recent review by P. Sutcliffe [58]. A very rewarding exposition of the modern development in this direction can be found in the excellent book by N. Manton and P. Sutcliffe [54].

6.4 The n-Monopole Configuration in the BPS Limit 215

The procedure suggested by N. Hitchin is to consider a geometrical description of the Atiyah–Ward twistor construction. To this end one can define a complex structure on the space of oriented lines (geodesics) in $\mathbb{R}^3$. The latter, which sometimes is referred to as *mini-twistor space*, can be identified with a two-dimensional manifold of the complex planes TP^1, the tangent bundle to the space of complex projective lines $\mathbb{C}P^1$ [51, 266]. Indeed, one can parameterize a complex plane in TP^1 by two coordinates (η, ξ). A tangent to this plane has the coordinate $\eta \, d/d\xi$ and, since the complex projective line $\mathbb{C}P^1$ is diffeomorphic to the sphere S^2, we actually have a tangent bundle over the usual Riemann sphere (see Fig. 6.9).

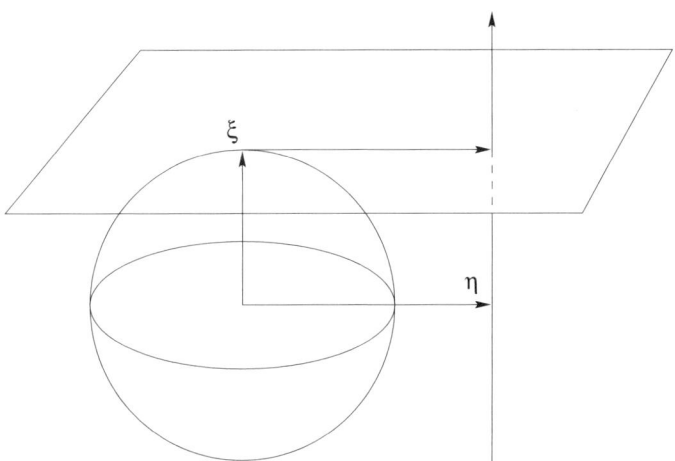

Fig. 6.9. Tangent bundle TP^1 as a space of directed lines in $\mathbb{R}^3$

The corresponding geodesic line is directed along ξ and pierces a perpendicular plane, which is tangent to the sphere S^2, at η. Thus, there is one-to-one correspondence between the points of TP^1 and the space of geodesics in $\mathbb{R}^3$.

A line bundle over TP^1 is defined as follows. Let us consider an operator of the covariant derivative D_l along an arbitrary line l on a distance r from the origin. One can set a kernel of the differential operator into correspondence to each geodesic

$$(D_l - \phi)v(r) = 0 \;, \tag{6.98}$$

where $v(r)$ is a complex doublet. This is the so-called *Hitchin equation*. In other words, this is a one-dimensional Dirac equation along the line l, whose independent solutions form a two-dimensional space. This is the vector bundle V over the space TP^1. Moreover, this bundle is holomorphic, if (D_l, ϕ) satisfy the Bogomol'nyi equation. Indeed, in this case the asymptotic behavior of the fields at $r \to \infty$ is known and the Hitchin equation becomes

$$\left[\frac{d}{dr} + \left(1 - \frac{n}{2r}\right)\begin{pmatrix}-1 & 0\\ 0 & 1\end{pmatrix}\right]v(r) = 0 \ . \tag{6.99}$$

This equation has a solution

$$v(r) \sim r^{n/2} e^{-r} \begin{pmatrix}0\\ 1\end{pmatrix}, \tag{6.100}$$

which belongs to the set of the solutions decaying on the asymptotic as $\sim e^{-r}$. This set forms a holomorphic sub-bundle $L^+ \subset V$. Another set of the solutions of the Hitchin equation (6.99), which are exponentially suppressed at the asymptotic $r \to -\infty$, form second holomorphic sub-bundle $L^- \subset V$. The curve on the complex plane TP^1 along which these sub-bundles coincide corresponds to the $L^2(-\infty, \infty)$ solutions of the Hitchin equations.

The line l, along which we differentiate in (6.99), is called a spectral line if the solutions of this equation are bounded in both directions. The set of all spectral lines defines an algebraic curve of genus $(n-1)^2$ in TP^1. This is precisely the polynomial S_n (6.93), which enters the definitions of the Atiyah–Ward matrix that we discussed above and whose zeros give a spectral curve.

The question is, does any such a curve corresponds to a multi-monopole configuration? The answer was given in the paper [267], where the corresponding set of restrictions was considered. First, a spectral curve S_n in TP^1 must be real to separate real structures in the complex space. This can be reformulated as a restriction on the solutions of the Hitchin equation (6.99), which should satisfy $L^+ = L^-$. To formulate the reality condition, Hitchin considered the transformation of the coordinates in the complex plane TP^1 of the form $(\eta, \xi) \to (-\bar{\eta}/\bar{\xi}^2, -1/\bar{\xi})$. Back to the space of the oriented lines in $\mathbb{R}^3$, this is equal to reversing the orientation of the line: if the Hitchin equation on an oriented line has $L^2(-\infty, \infty)$ solutions, it has $L^2(-\infty, \infty)$ solutions on the same line with opposite orientation. As a result, the coefficients of the spectral curve of the general form (6.93) must satisfy the condition [267]

$$a_i(\xi) = (-1)^i \xi^{2i} \overline{a_i(-1/\bar{\xi})} \ .$$

The general algebraic curve given by (6.93) has $(n+1)^2 - 1$ degrees of freedom. The reality condition results in $(n-1)^2$ constraints and the number of degrees of freedom of the multi-monopole configuration is $4n - 1$. These are collective coordinates of the n-monopole system.

6.4 The n-Monopole Configuration in the BPS Limit

Second, there is another restriction on the spectral curve connected with the condition of non-singularity of a multi-monopole configuration. Exactly this problem of singularity was the main source of trouble in the analysis of the Atiyah–Ward twistor construction. Ward [507] was able to prove that a two-monopole axial symmetric configuration is smooth everywhere. Further investigation of the spectral curve S_2 of (6.96) by Hurtubise [292] showed non-singularity of the corresponding general two-monopole configuration. However, a complete proof of the non-singular character of the twistor approach was obtained only when Nahm [393] invented another method for the construction of multimonopole configurations. It is related to the modification of the ADHM construction [80], the twistor transform for the solutions of the self-duality equations in $\mathbb{R}^4$.

6.4.5 Nahm Equations

Recall that the Julia–Zee correspondence allows us to identify the scalar field with the time component of the potential, thus the Bogomol'nyi equation may be considered as a self-duality equation in $\mathbb{R}^4$ with an additional constraint on the connection. All the remarkable properties of the Bogomol'nyi equation are connected, in one way or another, to the property of integrability, that is, the space of all possible solutions of this equation is exactly defined and all particular solutions can be represented in the form of integrals over some algebraic combination of elementary functions.

The ADHM construction was introduced to construct the solutions of the self-duality equations in $\mathbb{R}^4$, in terms of linear algebra in a vector space whose dimension is related to the topological charge of the instantons [80]. Since we already know that a single BPS monopole is identical to the infinite chain of instantons, we would expect that a modification of the ADHM construction to the case of monopoles can be done in an infinite dimensional vector space.

The starting point for the ADHM construction is the observation that an $SU(2)$-connection $A = A_\mu dx^\mu$ takes values in the corresponding Lie algebra and, therefore, can be represented also as taking values in the space of quaternions $\mathbb{H}$, that is, we can write by analogy with relation (6.65)

$$A = A_0 + iA_k \sigma_k = \begin{pmatrix} A_0 + iA_3 & iA_1 + A_2 \\ iA_1 - A_2 & A_0 - iA_3 \end{pmatrix},$$

and $\frac{1}{2}\text{tr } Adx = A_\mu dx^\mu$, where x is a complex quaternion that we set into correspondence to a point of $\mathbb{R}^4$. Since the space $\mathbb{H}$ is isomorphic to $\mathbb{C}^2$, this allows us to implement the twistor formalism to obtain multi-instanton solutions in $\mathbb{R}^4$. In this language, we have a bundle $\mathbb{R}^4 \times M$ with Euclidean base space $\mathbb{R}^4$ and M the Hermitian n-dimensional vector space, which is the space of solutions of the self-duality equations.

Originally, the ADHM construction was formulated in terms of an orthogonal basis in the space M, which is given by the quaternionic vectors v_i,

$i = 1, 2 \ldots n$. Then the self-dual connection is given by the projection from M to the subspace $M(x)$, which corresponds to a given point x of the base, and we can write

$$A_\mu(x) = \bar{v}_i \partial_\mu v_i \,.$$

Nahm pointed out [393] that there is a relation between the space M and the space of solutions of the Weyl equation

$$\mathcal{D}\psi \equiv \gamma_\mu D_\mu \psi = 0 \,, \tag{6.101}$$

where D_μ is a standard covariant derivative (5.10). Actually we are discussing the fermionic zero modes of the Dirac operator $\mathcal{D}$. Therefore, the dimension of the space of the square integrable solutions $\psi_i(x)$, that form an orthonormal basis,

$$\int \psi_i^\dagger(x) \psi_j(x) = \delta_{ij} \,, \tag{6.102}$$

can be defined via the Atiyah–Singer index theorem, which we shall discuss in Chap. 10. It is given by the index of the Dirac operator.

The homogeneous matrix function on $M(x)$ can now be introduced via the transformation

$$(M^\mu)_{ij} = -i \int d^4x \, \psi_i^\dagger \, x^\mu \, \psi_j \,,$$

which transforms the self-dual connection in $\mathbb{R}^4$ to the $n \times n$ quaternionic matrix form $M = M^\mu dq_\mu$, where q_μ are four auxiliary variables.

The observation by Nahm is that this approach can be generalized to describe multimonopoles. Formally, they can be written as pure self-dual gauge field configurations in $\mathbb{R}^4$, with the Higgs field ϕ replacing the temporal component of the gauge potential. However, monopoles are solutions with finite mass, not an action like multi-instantons. Thus, the space M for BPS monopoles becomes an infinite dimensional vector space with positive inner product (6.102), that is, a Hilbert space.

To take into account the invariance of the system with respect to the shift in the direction x_0, Nahm considered, instead of (6.101), the equation

$$\mathcal{D}\psi \equiv (\sigma_k D_k - \phi + s)\psi(\mathbf{x}, s) = 0 \,, \tag{6.103}$$

where s is a real constant.

The Nahm transform is a transition from the coordinates of the Euclidean space x_k to the square $n \times n$ Hermitian matrix functions

$$\left(T^k(s)\right)_{ij} = -i \int d^3x \, \psi_i^\dagger x^k \psi_j \,, \qquad \left(T^0(s)\right)_{ij} = \int d^3x \, \psi_i^\dagger \frac{\partial \psi_j}{\partial s} \,, \tag{6.104}$$

which are analytic on the interval $s \in [-1, 1]$, but have simple poles on its boundary $s = \pm 1$. One can now define the connection

$$T = T^k dp_k + T^4 ds,$$

where $p_k, k = 1, 2, 3$ are three dummy variables.

It is easy to see that, for the connection $T_\mu = (T^k, T^4)$ of the group $SU(n)$, the self-duality equations holds:

$$\frac{dT^1}{ds} - i[T^4, T^1] = -i[T^2, T^3],$$

$$\frac{dT^2}{ds} - i[T^4, T^2] = -i[T^3, T^1],$$

$$\frac{dT^3}{ds} - i[T^4, T^3] = -i[T^1, T^2]. \qquad (6.105)$$

Since the orthonormal basis ψ_i is defined up to a transformation of $SU(n)$

$$T^4 \to U^{-1}T^4 U + \frac{dU^{-1}}{ds}U, \qquad T^k \to U^{-1}T^k U,$$

a component T^4 can be set equal to zero and the system (6.105) is reduced to the *Nahm equation*

$$\frac{dT^k}{ds} = -\frac{i}{2}\varepsilon_{kij}[T^i, T^j]. \qquad (6.106)$$

Obviously, this is a dimension-one reduced version of the self-duality equations. In other words, the Nahm transform can be considered as a duality transformation that connects the gauge potential $A_\mu(x)$ and spatial coordinates $\mathbf{x}$ with the dual connection $T_\mu(s)$ and a $d = 1$ coordinate s.

Let us go back now to the Bogomol'nyi equation. This transition is given by the Weyl equation, with the connection T_k on a complex $2n$-dimensional vector $v(s)$:

$$\left[-\frac{\partial}{\partial s} + \frac{1}{2}(T_k + x_k) \otimes \sigma_k\right] v(s) = 0. \qquad (6.107)$$

Let the functions $v_i(s)$, $i = 1, 2 \ldots n$ form an orthonormal basis. Then one can define an inverse to the (6.104) transformation that allows us to restore the fields:

$$(A_k)_{ij} = -\frac{i}{2}\int ds\, v_i^\dagger \frac{\partial}{\partial x^k} v_j, \qquad (\phi)_{ij} = \frac{1}{2}\int ds\, v_i^\dagger s v_j. \qquad (6.108)$$

Obviously, the eigenvector $v(s)$ and the eigenfunction $\psi(x)$ of (6.103) are connected by the action of the covariant Laplace operator [392]

$$\left(\mathcal{D}^2 - (\phi + s)^2\right) v^\dagger(s) = 2i(2\pi)^{1/2}\sigma_2 \psi(x),$$

$$\left(-\frac{\partial^2}{\partial s^2} + \frac{1}{4}(T^k + ix^k)(T_k + ix_k)^\dagger\right) \psi(x) = 2i(2\pi)^{-1/2}\sigma_2 v(s).$$

$$(6.109)$$

Thus, the transformation between the functions ψ and v resembles the standard Fourier transform.

Note that there is a clear similarity between the Hitchin equation (6.99) and the (6.107). This is directly related to the fact that the restoration of the monopole fields from the Nahm data is a process opposite to the construction of the Nahm data T_k from a self-dual monopole connection by making use of the Hitchin equation (6.99) (for more details on this subject see the comprehensive papers [267, 392, 393]).

6.4.6 Solution of the Nahm Equations

An advantage of the Nahm construction, by its very definition, is the property of regularity. Now let us demonstrate that this, rather clumsy formalism, really describes multimonopole configurations.

For the simplest case $n = 1$, the Hermitian matrices T^k have a dimension 1×1, that is the Nahm data are a triplet of some real numbers. Therefore, the solution of the Nahm equations (6.106) can be written in the form $T^k = a^k/2$, where the numbers a^k correspond to the coordinates of the one-monopole. If the monopole is placed at the origin, the Nahm data are trivial, i.e., $T^k = 0$, and because of the spherical symmetry of the configuration, we can choose $x = (0, 0, r)$. Then (6.107) can be written in the simple form $(d/ds - r\sigma_3/2)v = 0$. Decomposing the two-component spinor v into the components (w_1, w_2), we arrive to the decoupled equations

$$-\frac{dw_1}{ds} + \frac{r}{2}w_1 = 0, \qquad \frac{dw_2}{ds} + \frac{r}{2}w_2 = 0. \tag{6.110}$$

It is easy to solve them:

$$w_1 = C_1 e^{rs/2}, \qquad w_2 = C_2 e^{-rs/2}. \tag{6.111}$$

Furthermore, the condition of orthonormality on the entire interval $s \in [-1, 1]$ yields

$$C_1^2 = 0, \qquad C_2^2 = \frac{r}{2\sinh r}, \quad \text{or} \quad C_1^2 = \frac{r}{2\sinh r}, \qquad C_2^2 = 0.$$

Thus, we can fix a basis of two-dimensional space of the solutions of (6.107) as

$$v_1 = \sqrt{\frac{r}{2\sinh r}} \begin{pmatrix} 0 \\ e^{rs/2} \end{pmatrix}, \qquad v_2 = \sqrt{\frac{r}{2\sinh r}} \begin{pmatrix} e^{-rs/2} \\ 0 \end{pmatrix}.$$

Choosing another basis corresponds to the gauge transformation of the field of a single monopole to some other, non-Abelian gauge.

Furthermore, the Higgs field can be recovered from the Nahm data According to (6.108):

6.4 The n-Monopole Configuration in the BPS Limit

$$\phi = \frac{1}{2} \begin{pmatrix} \int ds v_1^\dagger s v_1 & \int ds v_1^\dagger s v_2 \\ \int ds v_2^\dagger s v_1 & \int ds v_2^\dagger s v_2 \end{pmatrix} = \frac{r}{4 \sinh r} \begin{pmatrix} \int_{-1}^{1} ds\, s e^{rs} & 0 \\ 0 & \int_{-1}^{1} ds\, s e^{-rs} \end{pmatrix}$$

$$= \frac{1}{2}\left(\coth r - \frac{1}{r}\right)\sigma_3, \tag{6.112}$$

which, of course, coincides with the rescaled solution (5.71) of the Bogomol'nyi equations. In the same way, one can restore the gauge potential A_k. For example, using the expression (6.108), we obtain

$$A_r = -\frac{i}{2} \begin{pmatrix} \int ds v_1^\dagger \partial_r v_1 & \int ds v_1^\dagger \partial_r v_2 \\ \int ds v_2^\dagger \partial_r v_1 & \int ds v_2^\dagger \partial_r v_2 \end{pmatrix}$$

$$= \frac{ir}{4 \sinh r} \begin{pmatrix} \int_{-1}^{1} ds\left(\frac{1}{r} - \coth r + s\right)e^{rs} & 0 \\ 0 & \int_{-1}^{1} ds\left(\frac{1}{r} - \coth r - s\right)e^{-rs} \end{pmatrix}$$

$$= 0, \tag{6.113}$$

that is, the radial component of the gauge potential vanishes, as it should be for a spherically symmetric one-monopole configuration.

To describe a two-monopole configuration, it is convenient to use the ansatz for 2×2 Nahm matrices of the form $T^k = f^k(s)\sigma_k/2$ (there is no summation on k). By substituting this ansatz into (6.106), we obtain a system of equations on the set of the functions $f_k(s)$:

$$\frac{df_1}{ds} = f_2 f_3, \quad \frac{df_2}{ds} = f_3 f_1, \quad \frac{df_3}{ds} = f_1 f_2. \tag{6.114}$$

There is an obvious analogy with the well-known mechanical system of the Euler–Poinsot equations for a physical spinning top (see e.g. [3]):

$$\frac{dx_1}{dt} = (I_3 - I_2) x_2 x_3; \quad \frac{dx_2}{dt} = (I_1 - I_3) x_3 x_1; \quad \frac{dx_3}{dt} = (I_2 - I_1) x_1 x_2, \tag{6.115}$$

where $x(t)$ are time-dependent coordinates and I_k are components of the vector of moment of inertia. Obviously, (6.115) transforms to (6.114) by replacing $t \to is$ and choosing $I_1 = 1, I_2 = 0, I_3 = 2$ with the identification $f_1 \to \sqrt{2}x_1$; $f_2 \to \sqrt{2}x_2$; $f_3 \to ix_3$. Thus, the Nahm equations can be solved by a straightforward analogy with the classical Euler–Poinsot equations. In particular, this analogy means that there are two integrals of motion: the energy and the momentum

$$S = f_2^2 - f_1^2, \qquad T = 2f_3^2 - f_1^2 - f_2^2. \tag{6.116}$$

These two constraints define two surfaces in the flat Euclidean space $\mathbb{R}^3$ with coordinates f_k. The solutions of the Nahm equation (6.114) correspond to the curves along which these surfaces intersect. The formal difference between the classical equations of a spinning top (6.115) and Nahm equation (6.114) is that, in the former case, these surfaces are ellipsoids and their intersection gives a smooth curve. In the latter case, the surfaces are hyperboloids whose intersections give an equation with a solution with simple poles. Note that this is required by the boundary conditions on the Nahm dates.

Suppose that $f_1^2 \leq f_2^2 \leq f_3^2$. Then substituting the integrals of motion into the system (6.114), we obtain a general solution of the Nahm equations

$$f_1 = \pm \frac{D\,\text{cn}_k\,D(s+\tau)}{\text{sn}_k\,D(s+\tau)}\,, \qquad f_2 = \pm \frac{D\,\text{dn}_k\,D(s+\tau)}{\text{sn}_k\,D(s+\tau)}\,,$$

$$f_3 = \pm \frac{D}{\text{sn}_k\,D(s+\tau)}\,, \qquad (6.117)$$

where $D = \sqrt{S}$, and $\text{cn}_k(z)$, $\text{sn}_k(z)$ and $\text{dn}_k(z)$ are Jacobi elliptic functions with arguments z and spectral parameter k. They satisfy the equations

$$\frac{d\,\text{sn}_k\,z}{dz} = \text{cn}_k\,z\,\text{dn}_k\,z\,,$$

and the standard identities [8]

$$\text{sn}_k^2\,z + \text{cn}_k^2\,z = 1\,, \qquad k^2\,\text{sn}_k^2\,z + \text{dn}_k^2\,z = 1\,. \qquad (6.118)$$

In the limiting case, when the spectral parameter $k = 0$, the elliptic functions $\text{sn}_k z$ and $\text{cn}_k z$ reduce to the standard trigonometrical functions $\sin z$ and $\cos z$, respectively.

It is well-known that an elliptic function by its definition is a meromorphic two-periodic function of a complex argument [8]. They can be considered as a periodic functions on a torus obtained by a compactification of the complex plane, as we discussed in Chap. 3. Furthermore, the sum of residues in any of two periods of an elliptic function must be zero. Thus, there are two possibilities: either the elliptic function has two simple poles (Jacobi functions) or it has one double pole (Weierstrass function $W(z)$). Moreover, any elliptic function can be expressed as a rational function of $W(z)$ and its derivative (cf. discussion on page 76).

To complete the solution of the Nahm equations, we have to define τ and D in (6.117) in such a way that the boundary conditions are satisfied, that is they have two single poles at the boundaries $s = \pm 1$. The corresponding solutions are [147]

$$f_1 = -\frac{K\,\text{cn}_k\,(K(s+1))}{\text{sn}_k\,(K(s+1))}\,, \qquad f_2 = -\frac{K\,\text{dn}_k\,(K(s+1))}{\text{sn}_k\,(K(s+1))}\,,$$

$$f_3 = -\frac{K}{\text{sn}_k\,(K(s+1))}\,, \qquad (6.119)$$

6.4.7 The Nahm Data and Spectral Curve

Like the Bogomol'nyi equations, the Nahm equations are completely integrable. An important consequence is that they can be reformulated as a Lax equation with a spectral parameter [266]. Indeed, rewriting the Nahm equations (6.106) as

$$\frac{d(T_1 + iT_2)}{ds} = [T_3, T_1 + iT_2],$$

one can see that $\mathrm{tr}[(T_1 + iT_2)^n]$ is a constant for any values of n, that is the eigenvalues of the matrix $T_1 + iT_2$ are constants. However, that is exactly the property of the Lax equation: if $[d/ds + B, A] = 0$, then the eigenvalues of the matrix A are constants (this is so-called *isospectral evolution*).

However, in the combination $T_1 + iT_2$ only two Nahm matrices of three appear. In other words, in a three-dimensional space parameterized by the coordinates $1/4\ (T_1, T_2, T_3)$, a particular choice of this combination corresponds to an oriented line. The set of all possible orientations with a Lax equation along each direction gives a projective line P^1 parameterized by the inhomogeneous coordinate ξ. Thus, we have a family of Lax equations

$$\frac{d\Lambda}{ds} = [\Lambda, \Lambda_+], \tag{6.120}$$

where

$$\Lambda = (T_1 + iT_2) - 2T_3\xi + (T_1 - iT_2)\xi^2, \qquad \Lambda_+ = -T_3 + (T_1 - iT_2)\xi. \tag{6.121}$$

In a general case, we have to deal with the $n \times n$ Nahm matrices. Thus, there are n eigenvalues, which correspond to the n-fold branched covering of three-dimensional sphere formed by the set of all directions of P^1. If the eigenvalues Λ are constants, that is they are s-independent, there is an algebraic curve defined by equation

$$S = \det(\eta + \Lambda) \equiv \det(\eta + (T_1 + iT_2) - 2T_3\xi + (T_1 - iT_2)\xi^2) = 0. \tag{6.122}$$

Since it is a curve of eigenvalues, it is called a *spectral curve*. This is precisely the spectral curve which appears, in one way or another, in any description of multimonopoles. A proof of the equivalence between the spectral curves of the Hitchin approach (6.93) and the Nahm construction (6.122) is given in [267].

Let us turn back to the case of the spectral curve of a two-monopole configuration. Substituting the ansatz for the Nahm data into (6.122), we obtain

$$S_2(\eta, \xi) = \eta^2 + \frac{1}{4}\left((f_2^2 - f_1^2) + 2(f_1^2 + f_2^2 - 2f_3^2)\xi^2 + (f_2^2 - f_1^2)\xi^4\right) = 0, \tag{6.123}$$

that is, the coefficients of the polynomial S_2 are the constants (6.116). Substituting now the solutions (6.119) and making use of the identities on the elliptic functions (6.118), we obtain

$$S_2 = \eta^2 + \frac{1}{4}K(k)(1-k^2)\left(\xi^4 - 2\frac{1+k^2}{1-k^2}\xi^2 + 1\right) = 0. \tag{6.124}$$

This is precisely the spectral curve of the form (6.96).

Let us recall some geometrical notions that we shall need. Generally, a set of curves, which satisfy the (6.124), defines a Riemann surface that can be represented in terms of two sheets (complex plane) with four branch points $\xi_1, \xi_2, \xi_3, \xi_4$, which are solutions of the equation

$$\xi^4 - 2\frac{1+k^2}{1-k^2}\xi^2 + 1 = 0.$$

Indeed, (6.124) is quadratic in η and a loop around any branch point corresponds to the reflection $\eta \to -\eta$. Since η is supposed to be a single-valued function of ξ, the space parameterized by the coordinate ξ must be doubly covered by a complex plane. The branch points are connected through the cuts $[\xi_1, \xi_2]$ and $[\xi_3, \xi_4]$, and crossing a cut results in a transition from one sheet to another. This surface is shown in Fig. 6.10.

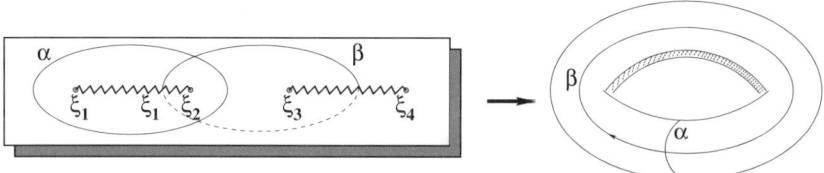

Fig. 6.10. Canonical circles of the torus and on the branched plane

It is known that this Riemann surface has the topology of a torus. It can be characterized by two circles α and β. The former goes around the cut connecting the branch points ξ_1 and ξ_2 and lies in one of the sheets, while the latter goes through the cut $[\xi_3, \xi_4]$, that is, it connects two sheets of the Riemann surface. The number of intersections of these contours is equal to one. The point at infinity is supposed to be added to each sheet and we obtain the torus $S^1 \times S^1$ with a canonical homology basis (α, β).

Since in the case of 2×2 Nahm data, the spectral curve is elliptic, the general solution of the Nahm equation can be obtained in terms of the elliptic functions. This is why the fields of a two-monopole configuration can

6.4 The n-Monopole Configuration in the BPS Limit

be written via elliptic integrals. In the general case of n-monopole configuration, the spectral curve has the form (6.93), which defines the relevant Riemann surface of genus $(n-1)^2$. Then the solution of the Nahm equation can be expressed through the theta-functions defined on this surface, and by introducing the required boundary conditions we could describe a general multimonopole configuration. However, in the general case, this procedure becomes a highly nontrivial problem, which has not been solved yet.

Some simplification ensues if we suppose that the configuration has an extra-symmetry. For example, we could expect that the n-monopole configuration is invariant with respect to the standard $SO(3)$ spatial rotations. Then we can look for a quotient spectral curve $S/SO(3)$, rather than for the original curve S [268]. The axially symmetric configurations constructed by Prasad and Rossi [433, 434] are an example of this kind. Other examples are the tetrahedral symmetrical three-monopole and the octahedral symmetrical cubic four-monopole configurations constructed by Hitchin, Manton and Murray in [268].

Without going into details, let us make a few remarks about the construction of multimonopole configurations. Recent impetuous development in this direction is connected with the rational map description and an idea about the discrete (platonic) symmetry of multimonopoles [54, 58, 283–285]. Numerical calculations allows us to construct different multi-monopole configurations in this way, some of them are depicted[10] the Fig. 6.11.

Briefly speaking, the rational map approach originates from the observation by Donaldson [204]. By making use of the Nahm transform, he proved that the n-monopole moduli space is diffeomorphic to the space of rational functions of degree n, which vanish at infinity, that is, to the space of rational maps $\mathbb{C}P^1 \to \mathbb{C}P^1$:

$$R(z) = \frac{P(z)}{Q(z)} = \frac{a_{n-1}z^{n-1} + \cdots + a_1 z + a_0}{z^n + b_{n-1}z^{n-1} + \cdots + b_1 z + b_0},$$

where $P(z)$ is a monic polynomial of degree n in a complex variable z and $Q(z)$ is a polynomial of a degree of less than n.

Hurtubise [293] explained the origin of this diffeomorphism by analysis of the Hitchin equation (6.98). Recall that there are two linearly independent solutions of this equation, which are defined along the line parameterized by a coordinate r. These solutions have the asymptotics $v_0 \sim e^{-r}$, $v_1 \sim e^r$, respectively. The idea is to fix a direction in $\mathbb{R}^3$ that gives a decomposition $\mathbb{R}^3 \cong \mathbb{C} \times \mathbb{R}$.

Let the complex plane $\mathbb{C}$ be parameterized by the coordinate $z = x_1 + ix_2$ and the direction $\mathbb{R}$ correspond to the x_3-axis. Then the basis of the independent solutions of the Hitchin equation (6.98) (v_0, v_1) asymptotically tends to the limits

---
[10] I am very grateful to Paul Sutcliffe for his kind permission to reproduce here the picture 6.11.

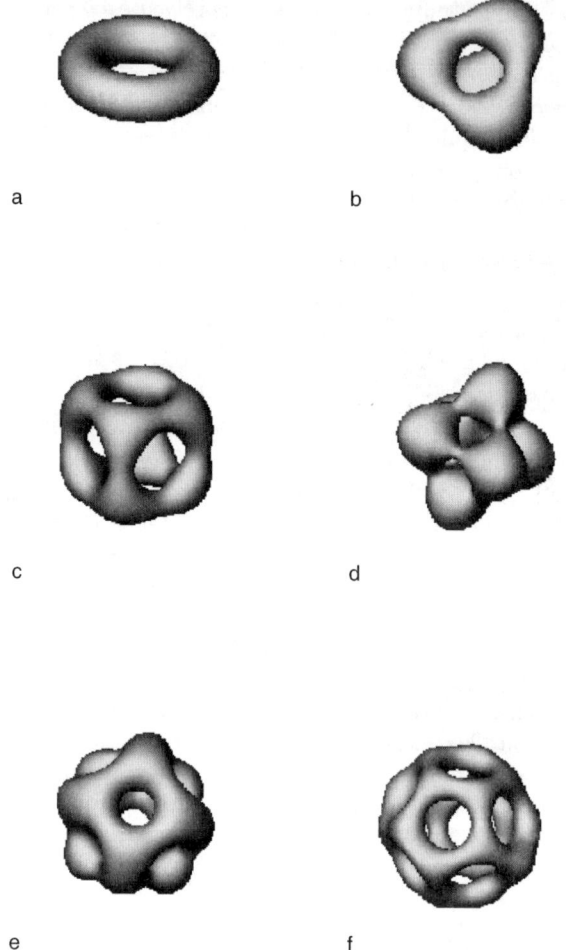

Fig. 6.11. Energy density surfaces for $SU(2)$ BPS monopoles of charges 2 to 7 constructed by rational map ansatz [58]

$$\lim_{x_3 \to \infty} v_0(x_3) x_3^{-n/2} e^{x_3} = e_0 \;, \qquad \lim_{x_3 \to \infty} v_1(x_3) x_3^{n/2} e^{-x_3} = e_1 \;,$$

where e_0, e_1 are two constants. This choice is not unique. One may consider another basis (v_0', v_1'), which corresponds to the scattering along another line, parallel to the x_3-axis:

$$v_0' = a(z) v_0 + b(z) v_1 \;, \qquad v_0 = a(z)' v_0' + b(z) v_1' \;,$$

where $a(z)$ and $b(z)$ are the parameters of the scattering. Then the rational map is defined as

$$R(z) = \frac{a(z)}{b(z)} \;.$$

Now recall that the line along which the Hitchin equation is considered, is a spectral line. Therefore, there is a correspondence $b(z) = S(z, 0) = 0$ and it can be shown [39] that
$$aa' = 1 + bb'.$$

A disadvantage of the Donaldson rational maps is that the choice of a direction in $\mathbb{R}^3$ violates the symmetry of a given multimonopole configuration. The decomposition $\mathbb{R}^3 \cong \mathbb{C} \times \mathbb{R}$ preserves the rotations in the plane $\mathbb{C}$, translations in the plane $\mathbb{C}$, translation in the direction $\mathbb{R}$ and constant gauge transformation [268]. These symmetries are *isometries*. An alternative rational map $\mathbb{C}P^1 \to \mathbb{C}P^1$, which preserves the full rotational symmetry, was suggested by Jarvis [303] (following a suggestion of M. Atiyah) in an analogy with the Donaldson construction. The only difference is that the Hitchin equation must be considered along each radial line from the origin to infinity.

The rational map approach is a very general construction. For example, surprisingly enough, it turns out that there is rather a close connection between the rational maps of the skyrmions and monopoles [54, 286]. In principle, one can directly recover the monopole field from the rational map. However, to do this, one must solve a non-linear differential equation, which is not much simpler than the original BPS equation. In this connection, Nahm's description, which reduces the problem to solution of a set of first-order ordinary differential equations appears more attractive.

So far we have discussed the properties of static multimonopoles. In the following section, we consider the low-energy dynamics of the monopoles by making use of the *moduli space* of the n-monopole configuration.

6.5 Moduli Space and Low-Energy Multimonopoles Dynamics

6.5.1 Zero Modes Lagrangian and the Moduli Space Metric

The nice property of integrability of the Bogomol'nyi equation, which made it possible to construct the multimonopole solution, works only for a static configuration. The complete, time-dependent field equations of the Yang–Mills–Higgs system are not completely integrable. Thus, it would be rather useless to apply the twistor transform or some other construction, which we briefly described above, in order to solve these equations, i.e., in order to find a complete solution describing the dynamical properties of the monopoles.

However, this problem can be solved by making some approximation. The guiding idea by Manton [368, 371] was that one can truncate the infinite-dimensional configuration space of the Yang–Mills–Higgs system (5.7) to a finite-dimensional Lagrangian dynamical system. In other words, the dynamics of an infinite number of degrees of freedom of the multi-monopole system can be reduced to that of a few collective coordinates of a soliton, or its *zero* modes, which include position coordinates and some internal coordinates as well. This description can be self-consistent only if we consider the low-energy dynamics. Then, at any fixed time, the multimonopole configuration can approximately be considered a static n-monopole solution of the Bogomol'nyi equation and the motion of the monopoles is connected with time-evolution of their collective coordinates only. In such a case, all information about the low-energy dynamics of the multi-monopoles is encoded in the kinetic part $\mathcal{T}$ of the action.

Let us write the Lagrangian of the system (5.7) as

$$L = \mathcal{T} - \mathcal{V} \tag{6.125}$$
$$= \int d^3x \, \text{Tr}\left(E_n E_n + D_0\phi D_0\phi\right) - \int d^3x \, \text{Tr}\left(B_n B_n + D_n\phi D_n\phi\right).$$

Recall that in the Bogomol'nyi limit, the potential energy of the system $\mathcal{V}$ is constant:

$$\mathcal{V} = \int d^3x \, \text{Tr}\left(B_n - D_n\phi\right)^2 + v\sum_i g_i = v\sum_i g_i.$$

If we set the time component of the gauge potential equal to zero, $A_0 = 0$, then the kinetic energy is

$$\mathcal{T} = \frac{1}{2}\int d^3x \left(\dot{A}_n^a \dot{A}_n^a + \dot{\phi}^a \dot{\phi}^a\right). \tag{6.126}$$

As we discussed in Sect. 5.2.1, for a static configuration it can be written in the form

$$\mathcal{T}_{gauge} = \frac{1}{2}M\dot{\Upsilon}^2, \tag{6.127}$$

where Υ is the gauge cyclic collective coordinate, the excitation of which corresponds to the generation of an electric charge. Now, we would like to excite another, spatial collective coordinate, that is to "push" the monopole. Obviously, the kinetic energy of the moving monopole would be higher than (6.127).

To define the corresponding correction to (6.127), let us first consider small translations of a single monopole in $\mathbb{R}^3$. For this purpose, we introduce three collective coordinates $X^k(t)$ and expand the fields in these perturbations:

$$A_n(X(t),x) \approx A_n(x) + \delta_k A_n(x) X^k(t) \equiv A_n(x) + a_n^{(k)}(x) X^k(t),$$
$$\phi(X(t),x) \approx \phi(x) + \delta_k\phi(x) X^k(t) \equiv \phi(x) + \chi^{(k)}(x) X^k(t). \tag{6.128}$$

Here, $a_n^{(k)}(x)$ and $\chi^{(k)}(x)$ are the relevant translational zero modes, excitations of which correspond to a small shift of the monopole in the direction X_k. Obviously,

$$\dot{A}_n = \dot{X}_k a_n^{(k)}(x), \qquad \dot{\phi} = \dot{X}_k \chi^{(k)}(x), \tag{6.129}$$

6.5 Moduli Space and Low-Energy Multimonopoles Dynamics

and then the kinetic energy is

$$T_{\text{trans}} = \dot{X}_k^2 \int d^3x \, \text{Tr}\left[a_n^{(k)} a_n^{(k)} + (\chi^{(k)})^2\right], \quad \text{for a given } k. \tag{6.130}$$

The explicit form of the translational zero modes can be determined by the condition that the perturbation must preserve the Bogomol'nyi bound. In other words, the "shifted" configuration $A_n^a + a_n^{a\,(l)}$, $\phi^a + \chi^{a\,(l)}$ must still satisfy the Bogomol'nyi equation $B_n^a = D_n \phi^a$. Therefore, the translational zero modes obey the linearized equation [67, 70, 385]

$$\varepsilon_{nmk} D_m a_k^{a\,(l)} = D_n \chi^{a\,(l)} - e\varepsilon_{abc} a_n^{b\,(l)} \phi^c. \tag{6.131}$$

Furthermore, recall that the time component of the gauge potential enters the Gauss law $D_n E_n - ie[\phi, D_0\phi] = 0$, which is imposed as a constraint on the time-independent physical fields. Expansion of it gives, in the Coulomb gauge, $A_0 = 0$:

$$D_n a_k^{a\,(l)} - e\varepsilon_{abc} \phi^b \chi^{c\,(l)} = 0. \tag{6.132}$$

In other words, the zero modes satisfy the background gauge condition. Actually, this condition ensures that the translational zero modes are orthogonal to all modes obtained by gauge transformation of the monopole configuration with a gauge function that vanishes at spatial infinity.

One can see that the normalizable zero modes are

$$a_n^{a\,(l)} = \partial_l A_n^a - D_n A_l^a = F_{ln}^a, \quad \chi^{a\,(l)} = \partial_l \phi^a - e\varepsilon_{abc} A_l^b \phi^c = D_l \phi^a, \tag{6.133}$$

i.e., they are standard infinitesimal translations $\partial_l A_n^a$, $\partial_l \phi^a$ in $\mathbb{R}^3$ that are supplemented by a gauge transformation with a special gauge function, the gauge potential itself [385]. Here, the condition of normalizability means that the translation in the corresponding direction is possible. Otherwise, one would need an infinite amount of energy to shift the monopole. Note that in the case of the BPS monopole, all zero modes are normalizable, this is not the case of an arbitrary multi-soliton configuration [511].

Substituting the zero modes (6.133) into the definition of the kinetic energy (6.130), we obtain

$$T_{\text{trans}} = \frac{\dot{X}_l^2}{2} \int d^3x \, (F_{ln}^a F_{ln}^a + D_l \phi^a D_l \phi^a) = \frac{1}{2} M \dot{X}_l^2, \tag{6.134}$$

and, taking into account the contribution of the gauge zero mode (6.127), we then find that

$$T = \frac{1}{2} M \left(\dot{X}_l^2 + \dot{\Upsilon}^2\right). \tag{6.135}$$

An obvious interpretation of this result is to consider it as the kinetic energy of a classical particle with a mass M moving in four-dimensional moduli space $\mathcal{M}_1 = \mathbb{R}^3 \times S^1$ (see the discussion on page 163). Here, the excitation of translational zero modes in $\mathbb{R}^3$ leads to the appearance of a non-zero momentum of the monopole and the motion on S^1 is connected with the generation of an electric charge.

The set of four collective coordinates of a single monopole $X_\alpha = (X_l, \Upsilon)$ defines the tangent vectors to the manifold $\mathcal{M}_1$, which naturally induce a metric structure on the moduli space. Normalizability of the zero modes means that there is a one-to-one correspondence between the zero modes and the coordinates on $\mathcal{M}_1$, i.e., the metric is finite.

The situation looks, of course, very simple in the case of a single monopole. Then the potential energy $\mathcal{V}$ is constant and the low-energy action of the moving monopole is

$$S = \frac{1}{2} M \int dt \, \dot{X}^\alpha \dot{X}^\beta g_{\alpha\beta}, \qquad (6.136)$$

where $g_{\alpha\beta} = \delta_{\alpha\beta}$. Thus, the metric on $\mathcal{M}_1$ is flat.

In the general case, an n-monopole configuration is characterized by $4n$ collective coordinates [490, 513]. The corresponding $4n$-dimensional moduli space $\mathcal{M}_n$ can asymptotically be decomposed into a product of n spaces $\mathcal{M}_1$, i.e., n individual monopoles that are widely separated from each other. However, in the interior region, the manifold $\mathcal{M}_n$ cannot be represented as a combination of these. This is the reason for the non-trivial behavior of the monopoles in head-on collisions [81].

The idea of Manton [368], developed in the works [82, 238], is that the classical dynamics of slowly moving multi-monopoles can be considered as geodesic motion in the moduli space. Here, an analogy with the motion in Euclidean space $\mathbb{R}^4$ of a classical particle in some potential profile would be helpful. It is obvious that the trajectories of particles correspond to the minima of the potential energy (the so-called "flat directions" or "valleys" of the potential). This restriction separates some subspace M of $\mathbb{R}^4$ and, up to the oscillations in transverse directions, the trajectory of a particle lies along a geodesic line in this subspace. This description can, in principle, be inverted; we can say that the metric on M defines the dynamics of the particle.

This analogy works also in the case of multimonopole low-energy dynamics. The difference is that the Euclidean space $\mathbb{R}^4$ is replaced by an infinite-dimensional configuration space and the condition on the energy to be minimal separates the moduli space $\mathcal{M}_n$. This space has a natural Riemannian metric, which asymptotically must be flat, as we noted above.

Another restriction on the metric on $\mathcal{M}_n$ is that it must be finite. For a metric on $\mathcal{M}_1$, this condition is fulfilled: all zero modes are normalizable and there are no restrictions on the motion of the monopole. Also, the metric on the moduli space must be complete, otherwise the topological charge would not be conserved.

Below, we describe a metric on $\mathcal{M}_2$ that satisfies all these conditions. However, first, let us make a few remarks on the most general properties of the

6.5 Moduli Space and Low-Energy Multimonopoles Dynamics

metric on the multi-monopole moduli space. This analysis was summarized in the book by Atiyah and Hitchin [39] (see also [82]).

Note that there is a very elegant way to construct a metric on $\mathcal{M}_n$. Indeed, let us consider the space $\mathcal{A}$ of finite energy configurations $A_\mu = (A_n, \phi)$. Here, we again adopt a notation in which the scalar field is treated as the fourth component of a 4-connection $A = A_\mu dx^\mu$ on $\mathbb{R}^4$, which is translationally invariant in the Euclidean time direction, i.e., the condition $\partial_4 A_\mu = 0$ is implemented.

Since the fields are defined up to the action of the gauge group G, the configuration space of the system is given by the quotient $\mathcal{A}/G$. Then the tangent vectors $\delta_\alpha A_\mu$ define tangent space $T_{\mathcal{A}/G}$ and a natural Riemannian metric on $\mathcal{A}/G$ can formally be written in terms of these vectors:

$$g_{\alpha\beta} = \int d^3x \, \mathrm{Tr} \left(\delta_\alpha A_\mu \delta_\beta A_\mu \right). \tag{6.137}$$

The n-monopole moduli space $\mathcal{M}_n$ is a subspace of the configuration space $\mathcal{A}/G$. It is parameterized by the collective coordinates X_α. However, there is a close relation between the zero modes and the tangent vectors $\delta_\alpha A_\mu$. Indeed, an arbitrary tangent vector to $\mathcal{M}_n$ can be written as $\dot{A}_\mu = \dot{X}^\alpha \delta_\alpha A_\mu$ and we can write the moduli space effective Lagrangian as

$$L = \frac{1}{2} g_{\alpha\beta} \dot{X}^\alpha \dot{X}^\beta. \tag{6.138}$$

An important observation is that the equations of the zero modes (6.131), together with Gauss law (6.132), have a quaternionic structure. Indeed, let us introduce the basis of the real four-dimensional space for the unit quaternions $\{e_\mu\} = (1, e_n)$, as in Appendix B. Then a vector $\delta_\alpha A = \delta_\alpha A_\mu e^\mu$, which is tangent to the moduli space, satisfies the equation (compare (B.4) in Appendix B):

$$D^* \delta_\alpha A = 0,$$

which in component notation exactly reproduces (6.132) and (6.131):

$$D_\mu \delta_\alpha A_\mu = 0,$$

$$D_\mu \delta_\alpha A_\nu - D_\nu \delta_\alpha A_\mu - \frac{1}{2} \varepsilon_{\mu\nu\rho\sigma} D_\rho \delta_\alpha A_\sigma = 0. \tag{6.139}$$

Clearly, the former equation is the condition of orthogonality of the tangent vector and the gauge orbits, while the latter is exactly the linearized Bogomol'nyi equation for a zero mode. Thus, the metric on the moduli space $\mathcal{M}_n$ is given by the restriction of the metric (6.137) to the subspace of the zero modes.

As was noted by Taubes [490], the moduli space $\mathcal{M}_n$ is by definition a *hyper-Kähler* manifold. This means its Riemannian metric is Kählerian[11] with

---

[11] Recall that a Kähler manifold is a complex manifold equipped with a non-singular, positive Hermitian metric, which can be locally written as a second derivative of some scalar function, see p. 89.

respect to three almost complex structures[12] $I^{(m)}$, $m = 1, 2, 3$, which satisfy the algebra of quaternions (see Appendix B). The almost complex structures are covariantly constant and obey the quaternionic algebra generating relations

$$I^{(m)} I^{(n)} = -\delta_{mn} + \varepsilon_{mnk} I_{(k)} \,. \tag{6.140}$$

In terms of the collective coordinates z_α on $\mathcal{M}_n$ they can be represented as matrices $(I^{(m)})_{\alpha\beta}$, which satisfy

$$-\varepsilon_{\alpha\beta\gamma\delta} = (I^{(m)})_{\alpha\beta}(I^{(m)})_{\gamma\delta} + (I^{(m)})_{\alpha\gamma}(I^{(m)})_{\delta\beta} + (I^{(m)})_{\alpha\delta}(I^{(m)})_{\beta\gamma} \,.$$

Indeed, the quaternionic structure of the zero mode equations (6.139) means that there are three covariantly constant tensors acting on the tangent bundle:

$$I_\alpha^{(m)\,\beta} \delta_\beta A = \delta_\alpha A \; e_m \,, \tag{6.141}$$

such that if $\dot{A}_\mu$ is a tangent vector to $\mathcal{M}_n$, then $I^{(m)} \dot{A}_\mu$ is also a vector of the tangent space. Thus, $I^{(m)}$ are precisely the three complex structures we defined above.

Recall that in Chap. 3 we briefly mentioned an important property: if the complex space $X^{\mathbb{C}}$ is a Kähler manifold with complex dimension n (that is, real dimension $2n$), the holonomy group of the metric is reduced to the unitary group $U(n)$. If $\mathcal{M}_n$ is a hyper-Kähler manifold with real dimension $4n$, the holonomy group of the metric on the moduli space $\mathcal{M}_n$ is reduced from $SO(4n)$ to $SP(2n)$. Actually, a hyper-Kähler manifold is always characterized by the self-dual Riemann curvature.

It is beyond the scope of this book to discuss these topics in more detail. Here, we only note that there is a remarkable connection between the equations of the zero modes, the quaternionic structures, which act on the tangent bundle of the moduli space, and the property of the multi-monopole moduli space being hyper-Kählerian, and refer the reader to [39].

6.5.2 Metric on the Space $\mathcal{M}_2$

We reduced the original Lagrangian of the Yang–Mills–Higgs system to (6.138) dealing with a finite number of degrees of freedom. Now the problem is to find the explicit form of the metric on the moduli space. There are three ways of doing this [519].

---

[12] Generally, an almost complex structure I on a manifold $\mathcal{M}$ is defined as an isomorphism of the tangent space $I : T_\mathcal{M} \to T_\mathcal{M}$ such that $I^2 = -\mathbb{I}$. Furthermore, by analogy with the standard differential geometry of the complex manifolds (cf. Chap. 3), these almost complex structures allow us to define complex differential forms on $\mathcal{M}$ for each tangent space. The metric is Kählerian with respect to these structures, if the corresponding two-forms are closed [39].

- The simplest method is just to substitute the exact solutions of the zero mode equation into the definition of the metric tensor (6.137). Actually, we already did this to obtain the flat metric of the one-monopole moduli space (6.135). Unfortunately, this approach does not work for an arbitrary n-monopole configuration, because we are not able to solve the BPS equations analytically. Thus, the explicit form of the n-monopole zero modes is, in general, not known.
- Another way was also considered above, when an asymptotic form of the metric on the two-monopole moduli space (6.64) was recovered from knowledge of the low-energy monopole dynamics [369]. However, this method works only for well-separated monopoles and can only give information about the asymptotic form of the metric.
- The restrictions imposed by the symmetries of the moduli space can in some particular cases completely determine the metric [39].

Let us briefly discuss the last approach, which is due to Atiyah and Hitchin. Recall that an arbitrary n-monopole configuration has $4n$ parameters. However, one can separate three parameters $(X_1, X_2, X_3) \in \mathbb{R}^3$, which correspond to the position of the center of the mass of the system, one angular parameter that specifies the global $U(1)$ phase angle on S^1, whose time-dependence determines the total electric charge of all monopoles. Thus, the moduli space can be factorized as [39, 81, 82]

$$\mathcal{M}_n = \mathbb{R}^3 \times \frac{S^1 \times \mathcal{M}_n^0}{\mathbb{Z}_n} . \qquad (6.142)$$

Here, the factor $\mathbb{Z}_n$ reflects that the monopoles cannot be distinguished. The metric on $\mathbb{R}^3 \times S^1$ is, as before, a flat one. Therefore, all non-trivial information about the low-energy dynamics of the monopoles is encoded in the $(4n-4)$-dimensional curved manifold $\mathcal{M}_n^0$. This is the most interesting part of the moduli space: the space of parameters describing relative positions and orientations of the monopoles, as well as their relative phases. Furthermore, because the metric on $\mathbb{R}^3 \times S^1$ is flat and $\mathcal{M}_n$ is a hyper-Kähler manifold, the metric on $\mathcal{M}_n^0$ is also hyper-Kähler.

Let us consider the case $n = 2$. Then the space $\mathcal{M}_2^0$ is a four-dimensional space. Since the holonomy group of a hyper-Kähler metric in $d = 4$ is $SU(2) \in SO(4)$, the hyper-Kähler manifold is just an Einstein self-dual space. Its Ricci tensor is proportional to the metric tensor, i.e., the scalar curvature of this space is zero. The isometry of this space is $SO(3)$: there are only rotations left from the complete symmetry group of Euclidean space when we separate the translations. Therefore, a proper parameterization of $\mathcal{M}_2^0$ can be given by a radial coordinate r and three Euler angles θ, φ and ψ (see Appendix A). The physical meaning of these parameters is that the radial coordinate determines the separation between the two monopoles, the angles θ and φ give the orientation of the axis joining the monopoles and ψ is the rotation angle about this axis [238]. This angle is associated with the relative electric charge of the monopoles.

The symmetry requirements on $\mathcal{M}_2^0$ are rather restrictive [39]. Indeed, an $SO(3)$-invariant metric on a self-dual Euclidean space, which is a four-dimensional hyper-Kähler manifold, has the unique form

$$ds^2 = f(r)^2 + a(r)^2 R_1^2 + b(r)^2 R_2^2 + c(r)^2 R_3^2, \qquad (6.143)$$

where R_n are the one-forms on $SO(3) = S^3/\mathbb{Z}_2$, whose definition and basic properties are described in Appendix A, and $f(r)$, $a(r)$, $b(r)$ and $c(r)$ are functions of the radial coordinate r.

Furthermore, the self-duality of the metric implies that these functions obey a set of first-order ordinary differential equations [237]:

$$\frac{2bc}{f}\frac{da}{dr} = b^2 + c^2 - a^2 - 2\lambda bc,$$

$$\frac{2ac}{f}\frac{db}{dr} = c^2 + a^2 - b^2 - 2\lambda ca,$$

$$\frac{2ab}{f}\frac{dc}{dr} = a^2 + b^2 - c^2 - 2\lambda ab, \qquad (6.144)$$

where $\lambda = 1$ or $\lambda = 0$.

Of course, the last case is the simplest one. This solution corresponds to the Eguchi–Hanson gravitational instanton [206]. The analysis of the system (6.144) in the case $\lambda = 1$ shows that there are only three solutions corresponding to the complete non-singular manifolds [39]:

- $a = b = c$: Flat metric on $\mathbb{R}^4$,
- $a = b \neq c$: Taub-NUT space,
- $a \neq b \neq c$: The Atiyah–Hitchin metric.

The first situation is trivial, because it implies that there is no interaction between the monopoles. Let us analyze the two remaining possibilities. First, note that the function $f(r)$ is defined up to a redefinition of the radial coordinate, that is, it can be chosen arbitrarily, assuming, for example, that $f = abc$ [39,81,82]. The second of these situations corresponds to the situation when two of three functions a, b, c coincide. As a result, the configuration will have an additional $SO(2)$ symmetry. This is the Taub-NUT metric [381] that already appears in the consideration of the low-energy dynamics of two dyons in Sect. 6.3.3. This metric is also an asymptotic limit of the third solution: the Atiyah–Hitchin metric discussed below, which asymptotically approaches the form $a \sim b$. The only difference is that in the last case, the function c has opposite sign as compared with the signs of a and b.

6.5 Moduli Space and Low-Energy Multimonopoles Dynamics

An alternative choice, $f = -b/r$ [238], taken together with the parameterization

$$r = 2K(\rho), \quad \text{with} \quad \rho = \sin\frac{\beta}{2},$$

where K is an elliptic integral (compare with (6.97))

$$K(\rho) = \int_0^1 \frac{ds}{\sqrt{1-\rho^2 s^2}}, \tag{6.145}$$

makes it possible to find the solution of the system (6.144):

$$bc = -r\sin\beta\frac{dr}{d\beta} - \frac{r^2}{2}(1+\cos\beta),$$

$$ca = -r\sin\beta\frac{dr}{d\beta}, \tag{6.146}$$

$$ab = -r\sin\beta\frac{dr}{d\beta} + \frac{r^2}{2}(1-\cos\beta).$$

The result of the numerical solutions of this set of equations can be found in [39, 238]. However, their asymptotic behavior can be determined analytically. For large monopole separation $r \to \infty$, the variable β tends to π and we can make use of the asymptotic expansion of the elliptic integral (6.145). The result, up to exponentially suppressed terms, is [39, 238]

$$a(r) \approx b(r) = r\sqrt{1-\frac{2}{r}} + O(e^{-r}), \quad c(r) = -2\frac{1}{\sqrt{1-\frac{2}{r}}} + O(e^{-2r}).$$
(6.147)

In the opposite limit $r \to \pi$ (which corresponds to $\beta \to 0$), one can use the approximate expansion of the elliptic integral $K(\rho) = \pi(1/2 + \rho^2/8 + \ldots)$. This yields

$$a(r) = 2(r-\pi)\left(1 - \frac{1}{4\pi}(r-\pi)\right) + \ldots$$

$$b(r) = \pi\left(1 + \frac{1}{2\pi}(r-\pi)\right) + \ldots \tag{6.148}$$

$$c(r) = -\pi\left(1 - \frac{1}{2\pi}(r-\pi)\right) + \ldots$$

Let us return now to the metric on the moduli space. Substituting the asymptotic (6.147) into the general formula (6.143), one finds the asymptotic metric on $\mathcal{M}_2^0$:

$$ds^2 = \left(1 - \frac{2}{r}\right)(dr^2 + r^2 d\theta^2 + r^2\sin^2\theta d\varphi^2) + \frac{4}{1-\frac{2}{r}}(d\psi + \cos\theta d\varphi)^2.$$
(6.149)

In this expression one can recognize, up to the obvious re-definitions, the classical Taub–NUT metric (6.64) that we have already encountered above. Thus, in this limit the Atiyah–Hitchin geometry describes two widely separated spherically symmetric monopoles, the inner structure of which is negligible (compare this with (6.96)). Since asymptotically $a(r) \approx b(r)$, an "accidental" $SO(2)$ symmetry appears. The physical content behind this symmetry is that the relative electric charge of well-separated dyons must be conserved, in addition to the total electric charge.

Let us consider the opposite limit $r = \pi$. Since the singularity of the metric (6.149) at $r = 2$ lies out of the range of r, which is $\pi \leq r \leq \infty$, this singularity has no physical meaning. However, if $r = \pi$, using (6.148) one can see that in this limit $a = 0$. This is a coordinate singularity, which corresponds to the collapse of three-dimensional orbits of $SO(3)$ to a two-dimensional sphere S^2 [39, 238]. In the theory of gravity, this singularity is called a "Bolt". The corresponding limit of the Atiyah–Hitchin metric on the "Bolt" describes the axially symmetric Ward configuration [507] with charge $n = 2$ and double zero of the Higgs field as the origin, which we discussed above.

To describe the geometry close to the "Bolt", Gibbons and Manton [238] make use of the general form of the Atiyah–Hitchin metric (6.143) as before, but introduce a new set of angular coordinates $\tilde{\psi}, \tilde{\theta}, \tilde{\varphi}$ on $SO(3)$, which parameterize the rotation matrices as

$$U(\tilde{\varphi}, \tilde{\theta}, \tilde{\psi}) = U_1(\tilde{\varphi})U_3(\tilde{\theta})U_1(\tilde{\psi}) = U(\varphi, \theta, \psi) = U_z(\varphi)U_y(\theta)U_z(\psi).$$

In terms of the right one-forms on the group $SO(3)$, which we describe in Appendix A, the metric close to the "Bolt" can be written as

$$ds^2 = dr^2 + (4r - \pi)^2 \left(d\tilde{\psi} + \cos\tilde{\theta}d\tilde{\varphi}\right)^2 + \pi^2 \left(d\tilde{\theta}^2 + \sin^2\tilde{\theta}d\tilde{\varphi}^2\right). \quad (6.150)$$

If we introduce the relative coordinate $\tilde{r} = r - \pi$, simple algebra gives

$$ds^2 = d\tilde{r}^2 + 4\tilde{r}^2 d\tilde{\psi}^2 + ds^2_{\text{Bolt}},$$

where the metric on the "Bolt" itself, that is on the sphere S^2, is

$$ds^2_{\text{Bolt}} = \pi^2 \left(d\tilde{\theta}^2 + \sin^2\tilde{\theta}d\tilde{\varphi}^2\right).$$

6.5.3 Low-Energy Scattering of Two Monopoles

The most interesting application of the Atiyah–Hitchin metric is connected with the problem of low-energy geodesic scattering of monopoles. Using (6.135) and (6.143), we can write the complete Lagrangian of a two-monopole system in terms of 8 collective coordinates. This Lagrangian is a sum of terms, each one proportional to the square of the velocity in configuration space:

6.5 Moduli Space and Low-Energy Multimonopoles Dynamics

$$L = M\left(\dot{X}_l^2 + \dot{Y}^2\right) + \frac{M}{4}\left[f(r)^2\dot{r}^2 + a(r)^2 l_1^2 + b(r)^2 l_2^2 + c(r)^2 l_3^2\right]. \quad (6.151)$$

Here, M is the mass of the monopole. The components of the vector of angular velocity l_n are defined via the rotation one-forms R_n on $SO(3)$ that are defined in Appendix A:

$$l_1 = -\sin\psi\dot{\theta} + \cos\psi\sin\theta\dot{\phi}, \quad l_2 = \cos\psi\dot{\theta} + \sin\psi\sin\theta\dot{\phi}, \quad l_3 = \dot{\psi} + \cos\theta\dot{\phi}. \quad (6.152)$$

As we saw in the case of the Nahm equations, (6.106), there is an obvious analogy between the monopole motion and the problem of rotation of a classical asymmetric rigid body. The difference is that this "top" is no longer a solid: in the case under consideration the analog of the principal momentum of inertia has the components

$$I_1 = a(r)^2 l_1, \quad I_2 = b(r)^2 l_2, \quad I_3 = c(r)^2 l_3, \quad (6.153)$$

which vary with monopole separation r, i.e., our "top" looks like a jelly changing its shape on the way. In the limit $r \to \infty$, this is a very long and thin rod with the monopoles at the ends (compare to the comment on page 7).

Indeed, the equations of motion corresponding to the Lagrangian (6.151) in the flat part of the moduli space $\mathbb{R}^3 \times S^1$ are trivial [238]

$$\ddot{X}_l = 0, \quad \ddot{Y} = 0, \quad (6.154)$$

which corresponds to the above-mentioned conservation of the total momentum $P_l = \sqrt{2M}\dot{X}_l$ and the total electric charge $Q = \sqrt{2M}\dot{Y}$.

As for the relative motion in $\mathcal{M}_n^0$, we have

$$\frac{dI_1}{dt} = \left(\frac{1}{b^2} - \frac{1}{c^2}\right) I_2 I_3,$$

$$\frac{dI_2}{dt} = \left(\frac{1}{c^2} - \frac{1}{a^2}\right) I_3 I_1,$$

$$\frac{dI_3}{dt} = \left(\frac{1}{a^2} - \frac{1}{b^2}\right) I_1 I_2,$$

$$f\frac{d}{dt}\left(f\frac{dr}{dt}\right) = \frac{1}{a^3}\frac{da}{dr}I_1^2 + \frac{1}{b^3}\frac{db}{dr}I_2^2 + \frac{1}{c^3}\frac{dc}{dr}I_3^2, \quad (6.155)$$

which are the standard Euler–Poinsot equations (6.115). One can see that the last equation of the set (6.155) corresponds to the conservation of the energy of the relative motion. Indeed, multiplication of it by $v = \dot{r}$ yields

$$\frac{dE}{dt} = \frac{d}{dt}\left[\frac{M}{2}\left(f^2\dot{r}^2 + \frac{I_1^2}{a^2} + \frac{I_2^2}{b^2} + \frac{I_3^2}{c^2}\right)\right] = 0. \quad (6.156)$$

In the same manner, one can easily prove that the vector of angular momentum of the relative motion $J_k^2 = M^2(I_1^2 + I_2^2 + I_3^2)$ is another integral of motion.

We saw above from the asymptotics of the solutions (6.147) that there is an additional $SO(2)$ symmetry in the limit $r \to \infty$. An analogy with the classical top in this case corresponds to the rotation about the symmetry axis. Then the projection of angular momentum onto this axis is also an integral of motion. In our case, this additional symmetry reflects the conservation of the component of angular momentum $J_3 = MI_3$ [238, 369]. The corresponding integral of motion can be identified with the relative electric charge of the monopoles. Since in this limit, both relative and total charges of the monopoles are conserved, the individual electric charge of each monopole is also a constant. Furthermore, if the component J_3 is conserved, the square of the orbital angular momentum $L_k^2 = M^2(I_1^2 + I_2^2)$ is also an integral of motion. Obviously, this is the case of the classical monopole scattering that we briefly discussed before.

Some other interesting cases of monopole scattering, which are analogous to rigid body rotations about a principal axis, were analyzed in [39, 238]. In such a motion, only one of the three components of the vector of angular momentum J_k does not vanish. For example, if we set $J_2 = J_3 = 0$, then the monopoles have no electric charge and $I_1 = $ const. Then, the conservation of energy (6.156) implies that the radial motion is described by an equation

$$\dot{r}^2 = \frac{1}{f^2}\left(\frac{2E}{M} - \frac{I_1^2}{a^2}\right).$$

Recall that at $r = \pi$, the function $a(r)$ has a zero, i.e., the radial component of the relative velocity vanishes at some distance $r_0 > \pi$. The physical interpretation is that the monopoles approach each other down to some minimal distance r_0 and then scatter to spatial infinity. The scattering angle can be found from the metric around the "Bolt" (6.150). Since we set $J_2 = J_3 = 0$, the angular variables $\tilde{\theta}$ and $\tilde{\phi}$ are constants and $J_1 = M^2 a^2 d\tilde{\psi}/dt$. Thus, the scattering angle is

$$\Theta = \Delta\tilde{\psi}\bigg|_{t=-\infty}^{t=\infty} = I_1 \int_{-\infty}^{\infty} \frac{dt}{a(r)^2}.$$

Numerical calculations show that as the impact parameter decreases to zero, the scattering angle increases monotonously up to the limiting value $\Theta_0 = \pi/2$ [39, 238].

In the case of head-on collision, we have $J_1 = J_2 = J_3 = 0$. The equation of motion (6.155) shows that in this case all the angular variables remain constant until the radial variable reaches the value $r = \pi$. Projected onto the plane $\tilde{r}, \tilde{\psi}$, this motion corresponds to the passing of the monopole through the origin, where the derivative $d\tilde{r}/dt$ changes its sign. Since the integrals of motion are unchanged, that means the angular variable ψ on the "Bolt" must jump by $\pi/2$. Thus, in head-on collisions the monopoles are scattered through a right-angle in the plane perpendicular to the axis 1 [39, 81, 82].

6.5 Moduli Space and Low-Energy Multimonopoles Dynamics 239

Note that this is a typical feature of soliton scattering, which also appears in the scattering of skyrmions or of vortices in $2+1$-dimensional models.

Recently, more complicated processes of multimonopole scattering have been discussed [58, 294]. For example, a nice process of three-monopole scattering into a one-monopole and a two-monopole final state was investigated numerically in [294]. Another kind of monopole scattering was considered in [282]. During these processes, axial symmetry of the configuration is instantaneously attained and, in some, monopoles with discrete symmetries are formed. An interesting phenomenon is observed: the structure of nodes of the Higgs field varies during the scattering. Evidently, this corresponds to the monopole-antimonopole pair contributions.

Let us mention in conclusion that there is another approach to the computation of the monopole moduli space metric, which is formulated in terms of Nahm data [54, 394]. The idea is to calculate a metric on the space of Nahm data, i.e., to define a tangent vector to the point $T_\mu = (T^k, T^4)$. It was proved [394] that the metric on the moduli space of $SU(2)$ monopoles is equivalent to the metric on the space of Nahm data, since the transformation that relates these metrics is an isometry. For a two-monopole system, the solutions of the Nahm equation (6.114) are the elliptic functions (6.119). Thus, one can find the tangent vectors explicitly and then the Atiyah–Hitchin metric (6.143) is recovered in terms of elliptic integrals. However, a generalization of this approach to an arbitrary case of charge n multimonopoles is possible only if Nahm's equations can be solved.

7 $SU(2)$ Monopole in Quantum Theory

7.1 Field Fluctuations on Monopole Background

So far we have discussed the classical monopole solutions of the non-Abelian Yang–Mills–Higgs system that correspond to the nontrivial vacuum states of the model. It is well-known, however, that if we move one step further and consider such a solution in quantum field theory, neither mass nor any other parameter of classical configuration is an observable quantity, since at the quantum level only the sum of classical and quantum components is experimentally observable. Moreover, the separation into the quantum and classical parts is rather subtle, because the relation between them usually depends on the scale of energy. Only in some limiting cases are the quantum corrections negligible. Thus, the consistent consideration of the 't Hooft–Polyakov monopole solutions (5.41) is related to analysis of the corresponding quantum effects.

When we promote a non-Abelian monopole to the quantum field theory, the following set of problems needs to be investigated

- How can the quantum correction affect the mass of a non-Abelian monopole. If a monopole becomes heavier or it becomes lighter?
- What is the spectrum of quantum fluctuations on the monopole background? If there are bound states?
- What are the quantum numbers of the monopole sector? If a monopole is a scalar or if it is a fermion?
- Is the conjecture by Montonen and Olive [384] about the duality between topological and perturbative sectors of the gauge theory well-founded on a quantum level?

In this chapter, we will concern ourselves with only first two of these questions. Anticipating the following discussion, we shall mention here that the complete answer to the third question comes if we consider also the spectrum of fermionic fluctuations on the monopole background (see Chap. 10). The problem of Montonen–Olive duality can be solved in the supersymmetric theory, where all quantum corrections are under the control of supersymmetry. We shall discuss this situation in the last chapter.

Since in a weak coupling regime, a monopole is a very heavy object, it seems quite natural to apply the method of quasiclassical quantization (see, e.g., [68, 299] and references therein). Let us consider the quantum corrections to the $SU(2)$ Georgi–Glashow model with the Lagrangian (5.7) given in Chap. 5

$$L = -\frac{1}{4}F^a_{\mu\nu}F^{a\mu\nu} + \frac{1}{2}(D^\mu\phi^a)(D_\mu\phi^a) - V(\phi). \quad (7.1)$$

Recall that the covariant derivative and the gauge field strength tensor are defined as

$$D_\mu\phi^a = \partial_\mu\phi^a - e\varepsilon_{abc}A^b_\mu\phi^c, \qquad F^a_{\mu\nu} = \partial_\mu A^a_\nu - \partial_\nu A^a_\mu - e\varepsilon_{abc}A^b_\mu A^c_\nu, \quad (7.2)$$

respectively, and the Higgs field potential is

$$V(\phi) = \frac{\lambda}{4}(\phi^a\phi^a - v^2)^2. \quad (7.3)$$

The standard procedure of the quasiclassical quantization can be implemented, if we define the explicit form of the propagators. It can be derived if we consider one-particle equations, which describe the perturbative fluctuations of different fields in the background of a classical non-Abelian monopole. However, because of the non-linearity of the classical non-Abelian Yang–Mill–Higgs system, this problem is rather complicated and the consistent analysis is forced to be numerical almost everywhere. Nevertheless, some important features of the quantum processes in a monopole field can be studied analytically.

As a first step, we consider small fluctuations of the vector and scalar fields around the classical spherically symmetrical 't Hooft–Polyakov solution (5.41)

$$\phi^a = \frac{r^a}{er^2}H(ver) + \chi^a(x_\mu), \qquad A^a_n = \varepsilon_{amn}\frac{r^m}{er^2}[1 - K(ver)] + a^a_n(x_\mu),$$
$$A^a_0 = a^a_0(x_\mu). \quad (7.4)$$

The problem becomes even simpler, if we consider only the fluctuations outside the monopole core, where the fields take asymptotic values

$$\mathbf{A} = -\frac{[\hat{\mathbf{r}} \times \mathbf{T}]}{er}, \qquad \boldsymbol{\phi} = v\hat{\mathbf{r}}. \quad (7.5)$$

Here we make use of the vector-isovector notations, $\mathbf{A}, \boldsymbol{\phi}$, to write the expressions in a compact form.

The assumption of smallness of the quantum corrections is justified in the weak coupling regime, where the gauge coupling $e \ll 1$ and the monopole is definitely much heavier than all the other excitations in the spectrum. However, the coupling constant dynamically depends on the scale of energy and, in an asymptotically free non-Abelian model of type (5.7), the infrared regime would correspond to a massless monopole. Very interesting information about the effects, which take place in the strong coupling regime, has

7.1 Field Fluctuations on Monopole Background

been obtained recently in supersymmetric theories [469], where these light monopoles play an important role.

Now let us derive the equations for small fluctuations of the scalar and vector fields. The first variation of the action (7.1) vanishes and then the second variation yields the fluctuation Lagrangian

$$\delta^2 S = \int dt \int d^3x \Big[-\frac{1}{2}(D_\mu a_\nu^a)(D^\mu a^{\nu a}) + \frac{1}{2}(D_\mu \chi^a)(D^\mu \chi^a)$$

$$+ e\varepsilon_{abc} F^{\mu\nu a} a_\mu^b a_\nu^c + \frac{3e^2}{2}(\phi^2 \delta_{ab} - \phi^a \phi^b) \chi^a \chi^b + \frac{e^2}{2}(\phi^2 \delta_{ab} - \phi^a \phi^b) a_\mu^a a^{\mu b}$$

$$- \frac{\lambda}{2}\left[(\phi^2 - v^2)\delta_{ab} + 2\phi^a \phi^b\right] \chi^a \chi^b - 2e\varepsilon_{abc}(D^\mu \phi^a) a_\mu^b \chi^c \Big]. \quad (7.6)$$

Here we make use of integration by parts. In particular, this yields

$$(D_\mu a_\nu^a)(D^\nu a^{\mu a}) \to e^2(\phi^2 \delta_{ab} - \phi^a \phi^b) \chi^a \chi^b + e\varepsilon_{abc} F^{\mu\nu a} a_\mu^b a_\nu^c,$$

if we assume the standard background gauge condition

$$D^\mu a_\mu^a - e\varepsilon_{abc} \phi^b \chi^c = 0. \quad (7.7)$$

We shall explain the physical meaning of this condition in Sect. 7.2 below, when we analyse the separation of physical degrees of freedom in such a system of interacting fields. This is a crucial step in the construction of a consistent scheme of the Hamiltonian quantization of a non-Abelian monopole.

Clearly, there is a vacuum outside the monopole core where $D_\mu \phi^a = 0$. Then, the variables describing the fluctuations of scalar and vector fields in the background gauge are decoupled and the second variation of action becomes diagonal (recall that the metric is given by $g_{\mu\nu} = \text{diag}(-1,1,1,1)$):

$$\delta^2 S = \frac{1}{2} \int dt \int d^3x \left[a_\mu^a \left(D_v^2\right)_{ab} a_\mu^b + \chi^a \left(D_s^2\right)_{ab} \chi^b \right]$$

$$= \frac{1}{2} \int dt \int d^3x \Big[-a_0^a \left\{ (D_\mu D^\mu)_{ab} + v^2 e^2(\delta_{ab} - \hat{r}^a \hat{r}^b) \right\} a_0^b$$

$$+ 2e\varepsilon_{abc} a_m^a F_{mn}^c a_n^b + a_n^a \left\{ (D_\mu D^\mu)_{ab} + v^2 e^2(\delta_{ab} - \hat{r}^a \hat{r}^b) \right\} a_n^b$$

$$- \chi^a \left\{ (D_\mu D^\mu)_{ab} - 3v^2 e^2(\delta_{ab} - \hat{r}^a \hat{r}^b) + 2\lambda v^2 \hat{r}^a \hat{r}^b \right\} \chi^b \Big]. \quad (7.8)$$

Here we introduce the covariant operator of second derivatives

$$(D_\mu D^\mu)_{ab} = \delta_{ab}\left(-\partial_0^2 + \frac{1}{r^2}\frac{\partial}{\partial r}\left(r^2 \frac{\partial}{\partial r}\right) - \frac{\widetilde{\mathbf{L}}^2 + 1}{r^2}\right) - \frac{2i}{r^2}\varepsilon_{abc}\widetilde{L}^c - \frac{r^a r^b}{r^4}, \quad (7.9)$$

and make use of the standard angular momentum operator (cf. definition (2.11)):
$$\widetilde{L}_k = -i\varepsilon_{kmn} r_m \partial_n .$$

Thus, the problem of the calculation of the spectrum of the quantum fluctuations of the 't Hooft–Polyakov monopole on the spacial asymptotic is now reduced to the standard eigenvalue problem of radial and angular operators.

Let us identify the physical degrees of freedom. The gauge conditions (7.7) fix only three out of the 15 degrees of freedom that are presented in the second variation of action above. Six more degrees of freedom are eliminated if, in accordance with the familiar Faddeev–Popov procedure, we add to the action of the model the ghost term[1]

$$S_{ghost} = -\frac{1}{2}\int dt \int d^3x \left[(D_\mu \eta^a)(D^\mu \eta^a) + e^2(\phi^2 \delta_{ab} - \phi^a \phi^b)\eta^a \eta^b\right]. \quad (7.10)$$

The nine physical degrees of freedom left correspond to the scalar and vector fields interacting with a monopole: six degrees of freedom for two massive spin-1 bosons in combinations $A_n^\pm = (1/\sqrt{2})(a_n^1 \pm i a_n^2)$ and two degrees of freedom that correspond to a massless photon in two polarization states $(A_\pm = a_1^3 \pm i a_2^3)$, respectively. The last remaining degree of freedom corresponds to the fluctuation of the scalar field χ^3 (the Higgs boson).

Now we are ready to consider the equations on eigenvalues of the operators $\mathcal{D}_v^2$ and $\mathcal{D}_s^2$. Actually, they are one-particle relativistic wave equations for a vector and for a scalar particle that are interacting with a monopole outside of its core. The structure of the operator of second derivatives (7.9) clearly suggests that the spatial and temporal dependence of all such modes can be factorized and, therefore,

$$a_m = a_m(\mathbf{r})\, e^{i\omega t}, \qquad \chi = \chi(\mathbf{r})\, e^{i\omega t}.$$

Then the equations for the quantum fluctuations on a monopole background may be written, for a change, in vector-isovector notations as

$$D_m D_m \mathbf{a}_n - e^2[\boldsymbol{\phi},[\boldsymbol{\phi},\mathbf{a}_n]] - 2e[\mathbf{F}_{nk},\mathbf{a}_k] = -\omega^2 \mathbf{a}_n,$$
$$D_m D_m \boldsymbol{\chi} + 3e^2[\boldsymbol{\phi},[\boldsymbol{\phi},\boldsymbol{\chi}]] + 2\lambda\boldsymbol{\phi}(\boldsymbol{\phi}\cdot\boldsymbol{\chi}) = -\omega^2 \boldsymbol{\chi}. \quad (7.11)$$

These equations, which describe the vector and the scalar particles in a 't Hooft–Polyakov monopole background field, were considered from different points of view in many papers (see, for example, [121, 131, 182, 442, 444, 481]). The simplest way to analyze these equations is to transfrom the variables to the Abelian gauge $\hat{\phi} = (0,0,v)$. Rewriting the second variation of the action

---

[1] Recall that the contribution of the ghost fields to the vacuum-to-vacuum transition amplitude contains a factor -2, compared to the contribution from the fluctuations a_μ^a, χ^a. Thus, three ghost fields cancel six degrees of freedom.

(7.8) in physical variables, we obtain the terms that decribe the fluctuations of the Higgs particle and the vector bosons $A_n^\pm$, respectively:

$$A_n^- \left[(D_\mu D^\mu + e^2 v^2)\delta_{nm} - 2ieF_{nm}^3\right] A_m^+$$
$$+ A_n^+ \left[(D_\mu D^\mu + e^2 v^2)\delta_{nm} + 2ieF_{nm}^3\right] A_m^-$$
$$+ \chi^3 \left[D_\mu D^\mu + e^2 v^2 + 2\lambda v^2\right] \chi^3. \quad (7.12)$$

We thus find the equations of motion for the Higgs scalar and for the vector bosons $A_n^\pm$, respectively:

$$(D_m D_m + k_s^2)\chi^3 = 0,$$
$$\left[(D_m D_m + k_v^2)\delta_{kn} \mp 2ieF_{kn}^3\right] A_n^\pm = 0. \quad (7.13)$$

Clearly, these equations are equivalent to the one-particle equations (7.11). Here $k_s^2 = \omega^2 + m_s^2$, $k_v^2 = \omega^2 + m_v^2$ and $m_s^2 = 2\lambda v^2$, $m_v^2 = e^2 v^2$ are masses of the scalar and the vector particles, respectively.

As one might expect, the first of the equations (7.13) is the Klein–Gordon equation in a monopole external field. The meaning of the other equation becomes more transparent, if we note that the operator of unit spin is represented by a 3×3 matrix

$$(S^n)_{ij} = i\varepsilon_{nij}, \quad (7.14)$$

and the second of the equations (7.13) may be written in a form that is standard for a massive spin-1 particle in an external magnetic field **B** [176]:

$$(D_m D_m + k_v^2 \mp 2e(\mathbf{S} \cdot \mathbf{B}))A_n^\pm = 0. \quad (7.15)$$

Moreover, since these fields are characterized both by the vectors of spin **S** and isospin **T**, their motion in a non-Abelian monopole external field has a very special character [71, 213, 527]. Neither isospin nor angular momentum are integrals of motion of that system. It is a total generalized angular momentum $\mathbf{J} = \mathbf{L} + \mathbf{T} + \mathbf{S}$ that conserved for the fluctuations around the classical monopole.

7.1.1 Generalized Angular Momentum and the Spectrum of Fluctuations

We start with a simple consideration of the scalar field, that is we set $\mathbf{S} = 0$. Let us recall that the 't Hooft–Polyakov ansatz (5.41) is invariant with respect to the transformations of diagonal $SO(3)$-subgroup, which mix together the spatial and group rotations. In other words, a non-Abelian monopole is spherically symmetric up to gauge transformations, which whirl up the configuration at the usual spatial $SO(3)$-rotations. The generator of former transformation is the isospin operator **T**. Thus, the generalized angular momentum is made up of the conventional angular momentum $\widetilde{\mathbf{L}} = -i\mathbf{r} \times \nabla$ and isospin:

$$\mathbf{J} = \tilde{\mathbf{L}} + \mathbf{T}. \tag{7.16}$$

Let us consider the situation in more detail. The covariant derivative operator in monopole presence is given by (7.2) and, therefore, the operator of rotation $L_k = -i\varepsilon_{kmn} r_m D_n$ no longer satisfies the standard algebra of the operator of angular momentum. Indeed, outside the monopole core we can make use of the asymptotic (7.5), which yields

$$\mathbf{L} = -i[\mathbf{r} \times \mathbf{D}] = -i[\mathbf{r} \times (\nabla - i\frac{1}{r^2}[\mathbf{r} \times \mathbf{T}]]$$
$$= \tilde{\mathbf{L}} - [\hat{\mathbf{r}} \times [\hat{\mathbf{r}} \times \mathbf{T}]] = \tilde{\mathbf{L}} + \mathbf{T} - \hat{\mathbf{r}}(\hat{\mathbf{r}} \cdot \mathbf{T}). \tag{7.17}$$

Straightforward calculation shows that

$$[L_m, L_n] = i\varepsilon_{mnk} L_k - i\varepsilon_{mki}\varepsilon_{nlj} r_k r_l [D_i, D_j]$$
$$= i\varepsilon_{mnk} L_k + i\varepsilon_{mnk} \hat{r}_k (\hat{r}_i \cdot T_i), \tag{7.18}$$

and we can see that not $\mathbf{L}$ but rather the operator

$$\mathbf{J} = \tilde{\mathbf{L}} + \mathbf{T} = -i[\mathbf{r} \times \mathbf{D}] + \hat{\mathbf{r}}(\hat{\mathbf{r}} \cdot \mathbf{T}), \tag{7.19}$$

obeys the standard commutation relations of the angular momentum

$$[J_m, J_n] = i\varepsilon_{mnk} J_k. \tag{7.20}$$

Nontriviality of this definition is that such an angular momentum includes the isospin operator, which makes it possible to generate a "spin from isospin" [261, 298]. Indeed, if we do not take into account a possible Grassmanian deformation of a static monopole configuration (we do it later when we consider the fermionic modes in a monopole external field), the 't Hooft–Polyakov monopole itself has no intrinsic moment of rotation. Although it is a localized field configuration, its spatial rotations can be exactly compensated by the corresponding gauge transformations. However, the relation (7.20) implies that a (half)-integer angular momentum can be generated in a bound system monopole-isoscalar particle.

Recall that the operator of an electric charge (5.23) far away from a monopole core takes the asymptotic form

$$Q = \mathbf{T} \cdot \hat{\mathbf{r}} = \mathbf{J} \cdot \hat{\mathbf{r}}. \tag{7.21}$$

In general, unlike $\mathbf{J}$, this operator does not commute with the Hamiltonian of a particle in a monopole external field, thus, a quantum mechanical scattering with non-conservation of electric charge is possible. However, on the spatial boundary, the eigenstates of the asymptotic Hamiltonian may be classified according to the eigenvalues of Q.

The generalized operator of angular momentum obviously appears in the angular part of the operator of second derivatives (7.9) above. Indeed, in the adjoint representation of $SU(2)$, we have $(T^a)_{bc} = i\varepsilon_{abc}$ and, therefore,

7.1 Field Fluctuations on Monopole Background

$$(D_m D_m)_{ab} = \delta_{ab}\left[\frac{1}{r^2}\frac{\partial}{\partial r}\left(r^2\frac{\partial}{\partial r}\right) - \frac{\widetilde{\mathbf{L}}^2+1}{r^2}\right] - \frac{2}{r^2}(\mathbf{T}\cdot\widetilde{\mathbf{L}})_{ab} - \frac{r^a r^b}{r^4}$$

$$= \delta_{ab}\left[\frac{1}{r^2}\frac{\partial}{\partial r}\left(r^2\frac{\partial}{\partial r}\right) - \frac{\mathbf{J}^2-(\mathbf{T}\cdot\hat{\mathbf{r}})^2}{r^2}\right]. \qquad (7.22)$$

Then the equation for a scalar particle in a monopole field can be written as

$$(D_m D_m + k_s^2)\chi^a \equiv \left[\frac{1}{r^2}\frac{\partial}{\partial r}\left(r^2\frac{\partial}{\partial r}\right) - \frac{\mathbf{J}^2-(\mathbf{T}\cdot\hat{\mathbf{r}})^2}{r^2} + k_s^2\right]\chi^a = 0, \quad (7.23)$$

in accordance[2] with the corresponding Hamiltonian of the Abelian problem (2.21). Clearly, the separation of the angular and radial variables in the Klein–Gordon equation (7.23) occurs.

The correspondence between (7.23) and the equation of motion of a scalar particle in an Abelian monopole external field is obvious, since asymptotically the 't Hooft–Polyakov configuration is identical, up to a local gauge transformation, to the Dirac monopole. Indeed, $SU(2)$ transformation (3.88) rotates the configuration from the hedgehog gauge to the Abelian gauge. Generalized angular momentum operator $\mathbf{J}$ then transforms as

$$U^{-1}(\theta,\varphi)(\widetilde{\mathbf{L}}+\mathbf{T})U(\theta,\varphi) = [\mathbf{r}\times(\mathbf{p}-e\mathbf{A}T_3)] - \hat{\mathbf{r}}T_3$$
$$= \widetilde{\mathbf{L}} - e\,[\mathbf{r}\times\mathbf{A}T_3] - \hat{\mathbf{r}}T_3\,, \qquad (7.24)$$

where $\mathbf{A}$ is an Abelian potential (cf. the definition of the angular momentum operator (2.11) in Chap. 2). Furthermore, the electric charge operator in the Abelian gauge is

$$U^{-1}(\theta,\varphi)\,Q\,U(\theta,\varphi) = T_3\,,$$

that is, up to a gauge transformation, the operator of second derivatives (7.9) in a non-relativistic limit coincides with the Hamiltonian of the Abelian problem (2.21).

Now, an analysis of the spectrum of scalar excitations can be performed analogously to the corresponding Abelian problem considered above [131, 244]. We are looking for the solutions $\chi^a(\mathbf{r})$, which are eigenfunctions of the commuting operators $\mathbf{J}^2, J_3, \mathbf{T}^2$, as well as the charge operator $Q = \mathbf{T}\cdot\hat{\mathbf{r}}$:

$$\mathbf{J}^2\chi^a(\mathbf{r}) = j(j+1)\chi^a(\mathbf{r}), \qquad J_3\chi^a(\mathbf{r}) = m\chi^a(\mathbf{r}),$$
$$\mathbf{T}^2\chi^a(\mathbf{r}) = t(t+1)\chi^a(\mathbf{r}), \qquad Q\chi^a(\mathbf{r}) = q\chi^a(\mathbf{r})\,. \qquad (7.25)$$

Let us factorize the eigenfunction $\chi^a(\mathbf{r})$ into radial, angular and isospin components, respectively:

---

[2] Similarly, the Hamiltonian of the Abelian theory was considered by A. Goldhaber as early as in 1965 [244]. The only difference from (7.23) there is that, instead of the operator of isospin $\mathbf{T}$, an additional term of interaction between the spin $\mathbf{S}$ and an extra-momentum $eg\hat{\mathbf{r}}$ is introduced.

$$\chi^a(\mathbf{r}) = \nu^a \chi(r) Y_{jmq}(\theta, \varphi),$$

where the isovector ν^a is an eigenfunction of the operators $\mathbf{T}^2$ and Q, and the angular dependence of isoscalar excitations is given by the generalized angular harmonics (2.31) that we defined above [131]. Thus, the radial equation again has a structure of a standard Coulomb problem with a "corrupted" centrifugal potential $(j(j+1) - q^2)/r^2$ and there is no bound state of a scalar particle and a monopole.

The situation is different for a vector particle in a monopole background field. The presence of the intrinsic spin $\mathbf{S} = 1$ now modifies the operator of generalized angular momentum, which is made up of three components

$$\mathcal{J} = \widetilde{\mathbf{L}} + \mathbf{T} + \mathbf{S} = \mathbf{L} + \mathbf{S} + \hat{\mathbf{r}}(\hat{\mathbf{r}} \cdot \mathbf{T}). \qquad (7.26)$$

Thus, the problem is to find corresponding eigenfunctions of the operators $\mathcal{J}^2$ and $\mathcal{J}_3$. In general, these functions, so-called *monopole vector spherical harmonics*, can be constructed by making use of the standard Clebsch–Gordan technique of addition of momenta [407]. However, there is an ambiguity, since the composition of three vectors can be constructed on two different ways, depending on the choice of the two vectors to be composed first. If the orthonormal monopole vector spherical harmonics are supposed to be eigenfunctions of the operator $\mathbf{J} = \widetilde{\mathbf{L}} + \mathbf{T}$ and the magnitude of the orbital momentum, the procedure is similar to the one we used above for the construction of the generalized spinor harmonics in Chap. 2. Moreover, since the operator $\mathbf{J}$ is the generalized angular momentum of a scalar particle in a monopole external field, and both scalar and vector fluctuations contribute to the quantum corrections to the 't Hooft–Polyakov monopole, this approach naturally simplifies the calculations. Another way [247] is to construct the monopole vector harmonics that are eigenfunctions of the operator $\mathbf{T} + \mathbf{S}$, but there is very little advantage of such a choice.

Thus, separating the spin part of the wave function χ_s, which is supposed to be an eigenfunction of the operators S^2, S_3, defined by the matrix (7.14):

$$S^2 \chi_s = 2\chi_s, \qquad S_3 \chi_s = s\chi_s, \quad \text{where} \quad s = (-1, 0, 1),$$

we can define the generalized monopole vector harmonics by [407]

$$\mathbf{Y}^{(q)}_{jjm}(\theta, \varphi) = \sum_{m,s} \langle j1ms \mid jm \rangle Y_{jmq}(\theta, \varphi) \chi_s, \qquad (7.27)$$

where $\langle j1ms \mid jm \rangle$ are the standard Clebsch–Gordan coefficients and $Y_{jmq}(\theta, \varphi)$ are the monopole harmonics (2.31). In the limiting case $q = 0$, the functions (7.27) reduce to the ordinary vector spherical harmonics.

An alternative approach is to construct the vector monopole harmonics by applying vector differential operators to the scalar harmonics [516]. By definition, these vector harmonics can be introduced as eigenfunctions of the

7.1 Field Fluctuations on Monopole Background 249

operator of the radial component of the spin $\mathbf{S} \cdot \hat{\mathbf{r}}$, which directly appears in (7.15).

We shall not go into the details of the calculations, which are rather involved and, on the other hand, have little difference from the construction of the Kazama–Yang–Goldhaber generalized spinor harmonics described in Sect. 2.6. A similar argumentation can be applied to the case of the spin-1 particle in a monopole field.

Let us consider the structure of the spectrum of a vector particle in a monopole field. As before, we are considering the states that are eigenfunctions of the commuting operators $\mathcal{J}^2, \mathcal{J}_3, \mathbf{J}^2$ and the operator of charge $Q = \mathbf{T} \cdot \hat{\mathbf{r}}$, which appears in the definition (7.19) of the generalized angular momentum $\mathbf{J}$:

$$\mathcal{J}^2\,\mathbf{Y}^{(q)}_{\mathrm{jjm}} = \mathrm{j}(\mathrm{j}+1)\mathbf{Y}^{(q)}_{\mathrm{jjm}}, \qquad \mathcal{J}_3\,\mathbf{Y}^{(q)}_{\mathrm{jjm}} = \mathrm{m}\mathbf{Y}^{(q)}_{\mathrm{jjm}},$$
$$\mathbf{J}^2\,\mathbf{Y}^{(q)}_{\mathrm{jjm}} = j(j+1)\mathbf{Y}^{(q)}_{\mathrm{jjm}}, \qquad Q\,\mathbf{Y}^{(q)}_{\mathrm{jjm}} = q\mathbf{Y}^{(q)}_{\mathrm{jjm}}. \qquad (7.28)$$

Then the total angular momentum of the vector particle takes the values $\mathrm{j} = j \pm 1$ or $\mathrm{j} = j$. However, if the eigenvalues of the charge operator Q are $q = 0$ and $g = 1/2$, only states with $\mathrm{j} = j + 1$ and $\mathrm{j} = j$ are acceptable. If $g \geq 1$, a minimal possible value of the total angular momentum is $\mathrm{j} = q - 1$. Thus, the structure of states according to the algebra of angular momenta is [407, 516]:

- $\mathrm{j} = q - 1 \geq 0$: a multiplet with $\mathrm{j} = j + 1$;
- $\mathrm{j} = q = 0$: a spherically symmetrical trivial multiplet with $j = 1$;
- $\mathrm{j} = q > 0$: two multiplets with $\mathrm{j} = j + 1$ and $\mathrm{j} = j$, respectively;
- $\mathrm{j} > q$: three multiplets with $\mathrm{j} = j + 1$, $\mathrm{j} = j$ and $\mathrm{j} = j - 1$, respectively.

Note that there is a spherically symmetric state $\mathrm{j} = 0$ besides the trivial one $(q = 0)$, the excitation of the first type with $q = 1$ [138, 251]. Such a mode may be interpreted as a charged vector boson, thus it is a fluctuation that does not break down the initial symmetry of the monopole background configuration. Moreover, the charge quantization condition then makes it possible to preserve the spherical symmetry only if the topological charge of the monopole is unity. This is exactly the argumentation that was applied to prove the instability of the spherically symmetrical monopole with a topological charge $n > 1$ [138]. A very detailed investigation of the problem of stability of the 't Hooft–Polyakov monopole was given in [84].

As was pointed out by different authors (see, e.g., [138, 184, 251, 481]), for the states with $\mathrm{j} = q - 1$, the centrifugal potential becomes attractive and there is a bound state of a monopole and a charged vector boson in the spectrum. the simplest way to see it is to consider (7.15), which by analogy with (7.23) can be written as

$$\left[\frac{1}{r^2}\frac{\partial}{\partial r}\left(r^2\frac{\partial}{\partial r}\right) - \frac{\mathbf{J}^2 - (\mathbf{T} \cdot \hat{\mathbf{r}})^2 \pm 2(\mathbf{S} \cdot \hat{\mathbf{r}})}{r^2} + k_v^2\right] A_n^{\pm} = 0. \qquad (7.29)$$

Here we substitute the explicit form of the Coulomb magnetic field $\mathbf{B} = g\hat{\mathbf{r}}/r^2$ and use the charge quantization condition for a 't Hooft–Polyakov monopole with unit topological charge. Separating the variables, we obtain the centrifugal potential of the radial equation

$$V^{\pm}_{centr} = -\frac{q^2 \mp 2s - j(j+1)}{r^2}.$$

For vector bosons $A_n^{\pm}$, we have $q^2 = 1$, and if we consider a spherically symmetric state $j = 0$, with third components of spin $s = \mp 1$, respectively, the angular momentum is $j = 1$. Certainly, for such a state the potential is attractive: $V_{centr} = -1/r^2$. As we shall see, this effect leads to serious consequences, because, from the point of view of quantum field theory, these fluctuations of the charged vector field on a monopole background actually correspond to the generation of the quantized electric charge of the monopole. In other words, they transform a monopole into a dyon with an electric charge q. Then there is a gap in the spectrum and the mass of such a quantum dyon differs from the mass of a monopole on the mass of a vector boson m_v.

7.1.2 Quantum Correction to the Mass of a Monopole

Returning to the system of (7.11), which describes the system of an interacting scalar (or vector) particle and a monopole, we sketched the method how it can be solved above, at least outside of the monopole core. In order to do this, we must separate the angular, isotopic and radial parts of the corresponding wavefuctions and then solve the remaining radial equation. Then, making use of the solutions we can apply the standard technology to construct the explicit form of the propagators of scalar and vector fields on a monopole background [443] and estimate the quantum corrections to monopole by quasiclassical methods [312, 313, 543]. A disadvantage of this procedure is the very cumbersome structure of the solutions even far away from the monopole. However, in some limiting cases, the results become rather transparent. As an example of such calculation, we estimate here a quantum correction to the mass of a non-Abelian monopole.

Let us recall that a standard approach to the calculation of a quantum correction to the mass of a topologically non-trivial field configuration is to apply the quasiclassical expansion about the solution of the classical field equation (see, for example, [25] and references therein). Since we already have written the second variation of the action of the 't Hooft–Polyakov monopole (7.8), and have proved that outside of the monopole core it is decomposed into two diagonal matrix operators, which describe independent fluctuations of scalar and vector fields, respectively, we can try to evaluate the functional determinant by "brute force". However, this is a rather non-trivial problem, even in such a simplified case, where we neglect the internal structure of the non-Abelian monopole. Further simplification comes if we move toward the

Bogomol'nyi limit, that is if we suppose that the vector particle m_v is much heavier than the scalar particle m_s. In this approximation, the fluctuations of the vector field are frozen out and the quantum correction to the monopole mass can be estimated by calculation of the scalar functional determinant only [312].

Indeed, there are rather compact expressions for propagators of the fields on the monopole background in the Bogomol'nyi limit [66,67,391,443]. Again, the underlying reason for this simplification is the relation between the Bogomol'nyi equations and the reduced Yang–Mills self-duality equations; for the self-dual theory the contribution, which comes to the partition function of a Yang–Mills field from a vector field, as well as from a spinor field, can be expressed via a functional determinant of the scalar field alone, and therefore, this determinant is a primary quantity that completely defines the quantum corrections [146].

A very detailed calculation of the scalar determinant in the BPS monopole background field, taking into account the finite temperature effects, was given in the paper [543]. We shall not discuss these calculations here, because in order to illustrate the effect caused by quantum fluctuations of the scalar field, it is enough to restrict our consideration to the limiting case $m_s \ll m_v$. Then, the scalar field $\phi^a = \dfrac{r^a}{er^2} H(m_v r)$ rather slowly approaches its asymptotic value (5.65)[3]

$$\phi^a \to \frac{m_v}{e}\hat{r}^a - \frac{r^a}{er^2} + O(e^{-m_v r}), \qquad (7.30)$$

which is very close to the Bogomol'nyi limit, but still differs from it [313]. In other words, we are considering a very special configuration of the 't Hooft–Polyakov monopole type, where the vector field almost takes its asymptotic value on a very short distance from the origin, while the shape function of the scalar field of the monopole $H(m_v r)$ is very slowly approaching its asymptotic value, see Fig. 7.1.

The reason for making use of such a very rough approximation is that in this case, the one-loop correction to the monopole functional integral reduces to the well-known effective potential of the scalar field [170]

$$\begin{aligned}V_{eff}(\phi) &= V(\phi) + \frac{1}{2}\ln\det(-\partial_\mu^2 + m_s^2) + \frac{3}{2}\ln\det(-\partial_\mu^2 + m_v^2) \\ &= \frac{3}{32\pi^2}\left[\left(\frac{4\pi^2}{3}\frac{m_s^2}{e^2 m_v^2} - \frac{3}{2}\right)(e^2\phi^2 - m_v^2)^2 + e^4\phi^4 \ln\frac{e^2\phi^2}{m_v^2}\right. \\ &\quad \left. - m_v^2(e^2\phi^2 - m_v^2)\right], \end{aligned} \qquad (7.31)$$

---

[3] Because here and hereinafter we discuss the quantum effects, it would be convenient to write all the relations by making use of the characteristic scales, which are the masses m_s and m_v. Then, the dimensionless variable becomes $\xi = ver = m_v r$, where e is the gauge coupling.

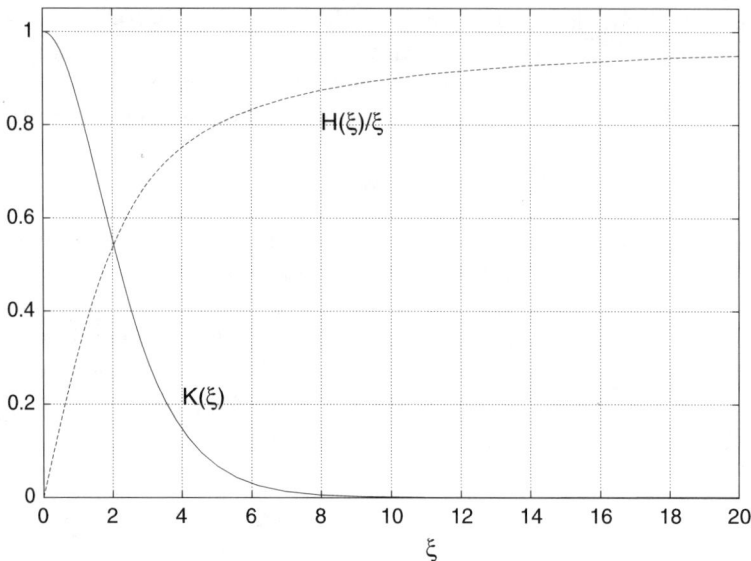

Fig. 7.1. The monopole profile functions near the Bogomol'nyi limit, $m_s \ll m_v$

where $|\phi^a| = \phi$ and the classical potential of the scalar field $V(\phi)$ is given by (5.11). Here, we take the renormalization point in such a way that the following conditions are satisfied:

$$V'_{eff}(m_v/e) = 0, \qquad V''_{eff}(m_v/e) = m_s^2. \qquad (7.32)$$

The vacuum value of the scalar field as before corresponds to $\phi_{vac} = m_v/e$.

Now we can write the energy functional in the form similar to its classical counterpart (5.57), but trading the potential of the scalar field for the effective potential (7.31):

$$\begin{aligned} E[H(m_v r)] &= \int d^3x\, B_n^a D_n \phi^a + \frac{1}{2}\int d^3x\, (B_n^a - D_n\phi^a)^2 + V_{eff}(\phi) \\ &\approx \frac{4\pi m_v}{e^2} + \frac{1}{2}\int d^3x \left(\frac{1}{er^2} - \frac{\partial}{\partial r}\left(\frac{H(m_v r)}{er}\right)\right)^2 \\ &\quad + V_{eff}[H(m_v r)]. \end{aligned} \qquad (7.33)$$

The expansion of the effective potential into functional series in powers of deviation of the scalar field from its asymptotic vacuum value $\phi_{vac} = m_v/e$, that is, in powers of difference $\frac{1}{e}(H/r - m_v)$ yields:

7.1 Field Fluctuations on Monopole Background

$$E[H(m_v r)] \approx \frac{4\pi m_v}{e^2} + \frac{4\pi}{e^2} \int dr \, r^2 \left[\frac{1}{2} \left(\frac{1}{r^2} - \frac{\partial}{\partial r} \left(\frac{H(m_v r)}{r} \right) \right)^2 \right.$$
$$\left. + \frac{m_s^2}{2} \left(\frac{H}{r} - m_v \right)^2 + \frac{1}{6e} V'''_{eff} \left(\frac{m_v}{e} \right) \left(\frac{H}{r} - m_v \right)^3 \right] \quad (7.34)$$
$$+ O(m_v),$$

where we take into account the renormalization conditions (7.32) and separate the third functional derivative of the effective potential (7.31), which is

$$V'''_{eff}(m_v/e) = \frac{3em_s^2}{m_v} + \frac{3e^3 m_v}{8\pi^2}. \quad (7.35)$$

Clearly, the first term in the expression (7.34) corresponds to the classical mass of a monopole in the Bogomol'nyi limit, while the quantum corrections are given by all the other terms. In order to evaluate these corrections, let us substitute into (7.34) the asymptotic form of the shape function of the scalar field $H(m_v r) \simeq m_v r - 1$ in agreement with expression (7.30) above. Then, the first term in the integrand in (7.34) vanishes, as we could expect, as we approach the Bogomol'nyi limit.

Let us recall that all the estimations above can be justified only if we restrict our consideration to the very tail of the monopole field, that is we are at large distances from the monopole core and $r \gg m_v^{-1}$. Therefore, the lower integration limit in (7.34) must be restricted by the value m_v^{-1}. Then, the second term in the integrand, which is proportional to m_s^2, yields

$$\frac{2\pi}{e^2} \int_{1/m_v}^{1/m_s} dr \, m_s^2 \approx \frac{2\pi}{e} m_s. \quad (7.36)$$

This is obviously the contribution to the monopole mass that originates from the scalar field with a small, but still non-zero mass m_s. In our consideration, this corresponds to a tiny difference between the masses of the 't Hooft–Polyakov monopole and the BPS monopole, respectively. Thus, it is not the quantum correction we are looking for.

Now we can consider the contribution of the third derivative of the effective potential (7.35), which also appears in the integrand in (7.34). The first term in V'''_{eff} is clearly proportional to m_s^2 as well, and, therefore, it will be absorbed into the physical mass of the scalar particle after the renormalization procedure. Thus, only the last term in (7.35) yields the quantum correction to the monopole mass [313]:

$$\Delta M = -\frac{m_v}{4\pi} \int_{1/m_v}^{1/m_s} \frac{dr}{r} = -\frac{m_v}{2\pi} \ln \frac{m_v^2}{m_s^2}. \quad (7.37)$$

We already mentioned above that we could come to the same conclusion also by straightforward evaluation of the functional determinant of the scalar field (for further details of such rather involved calculations see [312]).

To conclude this section, let us briefly comment on two features of the quantum correction (7.37) to the 't Hooft–Polyakov monopole mass above. First, it is obvious that it results in a decrease of the monopole classical mass, that is the quantum corrections inflate the monopole a bit. Note that, although our approximation was very rough, it still provides some information about the effect caused by the quantum corrections in a more consistent consideration. For example, here we do not even take into account the renormalization of the gauge coupling constant e. However, it is known that the Georgi–Glashow model is an asymptotically free theory. Therefore, the corresponding β-function describes the increase of the gauge coupling in the low-energy limit and the classical expression for the monopole mass (5.42) is justified for a weak coupling regime only. As we start to approach the non-perturbative low-energy scale of energy, the monopole mass begins to decrease. Thus, we can expect massless monopoles at the scale of the Landau pole, where the coupling constant blows up. We shall see how this mechanism works in the supersymmetric theory.

Second, note that the expression (7.37) is logarithmically divergent in the limit $m_s = 0$, which is due to the long-range character of the scalar field there. Certainly, this singularity is of the same nature as the well-known infrared singularity of the Abelian electrodynamics. The analysis of such divergences and their treatment is far beyond the scope of our review. Let us note only that the one-loop correction to the BPS monopole at the final temperature is finite [543].

7.2 Non-Abelian Monopole: Quasiclassical Quantization

7.2.1 Collective Coordinates and Constraints

The standard scheme of quasiclassical quantization around topologically non-trivial field configurations is related to the Hamiltonian formulation of the theory that was modified to this aim in the papers [149, 166, 212, 236, 281, 350, 410, 499, 525]. Let us consider how this formalism works in the case of the non-Abelian monopole. A starting point here is the expression for the monopole energy (5.16), which can be written as

$$E = \frac{1}{2} \int d^3x \left\{ E_n^a E_n^a + B_n^a B_n^a + (D_0\phi^a)(D_0\phi^a) + (D_n\phi^a)(D_n\phi^a) \right\} \quad (7.38)$$
$$+ V(\phi).$$

In order to construct the canonical Hamiltonian of this configuration, we have to rewrite this expression via canonical momenta conjugated to the dynamical variables A_μ^a and ϕ^a:

7.2 Non-Abelian Monopole: Quasiclassical Quantization

$$\pi^a = \frac{\partial L}{\partial \dot\phi^a} = D_0\phi^a, \qquad \pi_n^a = \frac{\partial L}{\partial \dot A_n^a} = E_n^a. \qquad (7.39)$$

Recall that the gauge model under consideration is singular and its treatment is a classical example of the quantization of a system with constraints [202]. The singularity of the model means that the Lagrangian (7.1) does not depend on the time derivative $\dot A_0^a$ and the primary constraint must be imposed on the corresponding momenta: $G_1 = \pi_0^a = 0$. The variable that is canonically conjugated to A_0^a is $\lambda = \partial_0 A_0^a$. Thus, the density of the Hamilton function takes the following form

$$\begin{aligned}H &= \pi_n^a \dot A_n^a + \pi^a \dot\phi^a - L \qquad (7.40)\\ &= \frac{1}{2}\left(\pi_n^a\pi_n^a + \pi^a\pi^a + B_n^a B_n^a + (D_n\phi^a)(D_n\phi^a)\right)\\ &\quad + \pi_n^a(D_n A_0^a) - e\varepsilon_{abc}\pi^a A_0^b \phi^c + \lambda\pi_0^a + V(\phi).\end{aligned}$$

Furthermore, the variation of the action with respect to the variable A_0^a yields not an equation of motion, but a secondary constraint, which then must be imposed on the physical states of quantized theory (the *Gauss law*):

$$G_2 = D_n E_n^a + e\varepsilon_{abc}\phi^b D_0\phi^c = 0. \qquad (7.41)$$

Hence, the Hamiltonian of the system may be written as

$$\mathcal{H} = \frac{1}{2}\int d^3x \left\{\pi_n^a\pi_n^a + \pi^a\pi^a + B_n^a B_n^a + (D_n\phi^a)(D_n\phi^a)\right\} + V(\phi). \qquad (7.42)$$

In the following, it will be convenient to fix the Hamiltonian (or temporal) gauge $A_0^a = 0$, since then we may make use of the canonical equal time commutators.

Recall that even if we fix such a gauge, some degrees of freedom still remain, since the fields are defined up to the gauge transformations

$$A_n \to U(\mathbf{r})A_n U^{-1}(\mathbf{r}) + iU(\mathbf{r})\partial_n U^{-1}(\mathbf{r}), \qquad \phi \to U(\mathbf{r})\phi U^{-1}(\mathbf{r}), \qquad (7.43)$$

with the time-independent matrix of the transformation $U(\mathbf{r}) = e^{iw(\mathbf{r})} \approx 1 + iw(\mathbf{r})$. We will use both the local and the infinitesimal form of it.

Now we can apply the procedure of canonical quantization. A standard formalism of the quasiclassical quantization is connected with the expansion (7.4) of the quantum fields of a monopole $(A_n^a + a_n^a, \phi^a + \chi^a)$ about the classical configuration (A_n^a, ϕ^a). The latter is given by the 't Hooft–Polyakov ansatz (5.41). However, as we already mentioned in Sect. 6.5.1, a peculiarity of the situation is that there are four normalizable zero modes in the spectrum of quantum fluctuations around a monopole of unit topological charge. We shall denote these modes as $\zeta_\mu^{a(\alpha)} \equiv (\zeta_\mu^{a(0)}, \zeta_\mu^{a(l)})$, where the index α labels the collective coordinates $X^\alpha \equiv (\mathbf{X}(t), \Upsilon(t))$ that parameterize the four-dimensional

256   7 $SU(2)$ Monopole in Quantum Theory

moduli space $\mathcal{M}_1 = \mathbb{R}^3 \times S^1$ of a single monopole. The compact abbreviation ζ_μ^a composes zero modes of scalar and vector fields, where μ is the Lorentz index of each mode.

The origination of three translation zero modes $\zeta_\mu^{a(l)} = (\chi^{a(l)}, a_n^{a(l)})$ is quite clear. They are defined according to (6.133) as special solutions of the equations (7.11) with zero eigenvalues $\omega^{(l)} = 0$. Thus, they appear because a localized monopole configuration breaks down the initial translation invariance of the Georgi–Glashow model. Translations of a monopole in any three directions of Euclidean space $\mathbb{R}^3$ restore this symmetry on a quantum level. Indeed, the configuartion will be displaced on a distance $\mathbf{X}$, if we act on a monopole by the operator of translation $U_{tr}(\mathbf{X}) = \exp\{-\mathbf{X} \cdot \boldsymbol{\nabla}\}$:

$$U_{tr}(\mathbf{X}) A_n{}^a(\mathbf{r}) = A_n{}^a(\mathbf{r} - \mathbf{X}(t))\,, \qquad U_{tr}(\mathbf{X}) \phi^a(\mathbf{r}) = \phi^a(\mathbf{r} - \mathbf{X}(t))\,. \quad (7.44)$$

Clearly this operator commutes with the Hamiltonian of the system.

In Sect. 5.2.1 we also considered the fourth zero mode $\zeta_\mu^{a(0)}$, which is related to time-dependent gauge transformations of the 't Hooft–Polyakov monopole configuration. Recall that this gauge mode is a cyclic variable. Excitation of such a mode leads to the generation of an electric charge of a monopole (cf. discussion on p. 163). Indeed, the classical solution (5.41) is invariant with respect to the time-dependent transformations $U_g(\Upsilon) = \exp\{\Upsilon(t)\,\phi(\mathbf{r})\}$. The particular choice $\Upsilon = 0$, which clearly corresponds to the purely monopole configuration with zero electric charge, breaks this invariance in the same manner as the translation symmetry became broken by a monopole localized in $\mathbb{R}^3$.

As is well-known, the presence of zero modes in the spectrum leads to some trouble with the calculation of the functional determinant. If we to naively consider these modes on an equal footing with all other fluctuations, the functional integration over them would result in a meaningless infinity[4]. These modes need a special treatment, that is, we must separate the corresponding collective coordinates from all normal oscillations of the fields around some classical background. Then, in order to discard these zero modes completely, we must impose the orthogonality condition between them and all other fluctuations.

Thus, we should restrict the integration domain in the functional space by some subspace orthogonal to ζ_μ^a. The situation looks similar to that of the case of the gauge fields quantization, where infinity appears from the functional integration over gauge equivalent configurations. There the Faddeev–Popov procedure is used to separate the integration over the group volume and

---

[4] However, the trouble caused by zero modes does not, in fact, affect the approximate evaluation of the quantum correction to the monopole mass above, if we suppose that zero modes are completely decoupled from other parts of the spectrum, the normal vector and scalar modes. In other words, we have to neglect the effects of bremsstrahlung of scalar and vector fields, and radiative friction of the monopole.

7.2 Non-Abelian Monopole: Quasiclassical Quantization

pull it out from the functional integral. Therefore, taking this analogue into account, we can say that to exclude the zero modes from the path integral, it is necessary to extract the integration over the collective coordinates in $\mathbb{R}^3 \times S^1$ explicitly.

Specific to the quasiclassical quantuzation of 't Hooft–Polyakov monopole is that we are dealing with gauge theory, that is the constraints of two different types must be imposed together and both the Faddeev–Popov procedure and the separation of the zero modes must be implemented. Doing that, we must fix some special gauge keeping zero mode to be orthogonal to the gauge transformation. The background gauge (7.7) exactly satisfies this condition.

Let us make use of the freedom that we still have in our choice of the gauge. The quantum dynamical variables now are the fluctuations of the monopole field $(a_n^a(\mathbf{r}, t), \chi^a(\mathbf{r}, t))$ and the conjugated canonical momenta (Π^a, Π_n^a). As before, the Lagrangian describing these quantum fluctuations is singular and we impose the first-order constraint

$$G_1 = \Pi_0^a = 0. \tag{7.45}$$

Again, we can fix the Hamiltonian gauge for the fluctuations: $a_0^a = 0$. The Gauss law, therefore, is the secondary constraint, which we shall impose on the quantum canonical momenta

$$\tilde{E}_n^a \approx E_n^a + D_0 a_n^a, \qquad D_0 \tilde{\phi}^a \approx D_0 \phi^a + D_0 \chi^a. \tag{7.46}$$

If, in addition, we still want the same Gauss law (7.41) to hold for the classical variables $E_n{}^a, D_0 \phi^a$, the straightforward substitution of the expansion (7.46) into (7.41) yields the condition of the background gauge (7.7) already familiar to us:

$$G_2 = D_n a_n^a - e\varepsilon_{abc}\phi^b \chi^c = 0.$$

Clearly, this secondary constraint can be also derived from the canonical equations of motion.

Note that the compact four-dimensional notations that we used to label zero modes of scalar and gauge fields are also useful to compose the corresponding canonical momenta in an abbreviated multiplet

$$\Pi_\mu^a = (\Pi^a, \Pi_n^a). \tag{7.47}$$

Then, the Gauss law and the background gauge conditions (7.7) take very simple forms

$$D_\mu \Pi_\mu^a = 0, \qquad D_\mu a_\mu^a = 0. \tag{7.48}$$

Here we actually make use of the Julia–Zee correspondence $\phi^a \rightleftharpoons A_0^a$. Indeed, since the Hamiltonian gauge is imposed, the fluctuations of the scalar and vector fields can be composed into the four-component field $a_\mu^a = (\chi^a, a_n^a)$, as well as the fields: $A_\mu^a = (\phi^a, A_n^a)$. This notation will be used in the following.

7.2.2 Quantum Mechanics on the Moduli Space

We have mentioned already that a standard treatment of the zero modes is to extract the integration over corresponding collective coordinates from the functional integral and then to impose the condition of orthogonality between these modes and all other fluctuations as a constraint. For the moment, let us ignore the presence of the gauge zero mode, that is we first consider how to treat the translational zero modes.

The collective coordinates describing the displacements of a monopole are $\mathbf{X}(t)$ and the expansion (7.4) can be written in the form

$$\tilde{A}^a_\mu = A^a_\mu(\mathbf{r} - \mathbf{X}(t)) + a^a_\mu(t, \mathbf{r}). \tag{7.49}$$

Since the set of orthonormal eigenfunctions of the operator of second derivatives (7.9) forms a complete basis, an arbitrary quantum fluctuations $a^a_\mu(t, \mathbf{r})$ can be expanded in a series

$$a^a_\mu(t, \mathbf{r}) = \sum_{i=1}^\infty C_i(t) a^a_{\mu i}(\mathbf{r}). \tag{7.50}$$

Note that the shift of radial variable $\mathbf{r} \to \mathbf{r} - \mathbf{X}(t)$ excludes the contribution of the translational zero mode

$$\zeta^{a(l)}_\mu = C_0(t) a^{a(l)}_\mu(\mathbf{r}) = C_0(t) \begin{pmatrix} \chi^{a(l)} \\ a^{a(l)}_n \end{pmatrix}, \tag{7.51}$$

from the sum (7.50). Separating also the integration over the remaining gauge zero mode, and substituting the decomposition (7.50) back into the functional of the second variation of action (7.8), yields a structure that is identical to the action of an infinite set of weakly coupled harmonic oscillators. Hence, the problem is reduced to the calculation of the functional integral over these normal modes, which can easily be evaluated according to the usual formalism.

Thus the problem is to evaluate the integral over the collective coordinates $\mathbf{X}(t)$, which are defined implicitly. The standard trick is to change the variable of integration to be the coefficient of the expansion C_0, rather than $\mathbf{X}(t)$. However, the corresponding Jacobian is unknown and it is generally a very complicated mathematical problem to calculate it. The procedure of separation of the translation zero modes becomes much easier, if we consider the functional integral over the entire phase space, that is, we have to integrate over both dynamical variables and canonical momenta. In this case, the Hamiltonian should be expressed in terms of collective coordinates $\mathbf{X}(t)$, canonically conjugated vector of momentum $\mathbf{P}(t)$ and all the residual phase space variables.

According to definition of the canonical momentum, the vector $\mathbf{P}(t)$ must satisfy the commutation relation

7.2 Non-Abelian Monopole: Quasiclassical Quantization

$$[U_{tr}, \mathbf{P}] = -iU_{\text{tr}} \cdot \boldsymbol{\nabla} . \tag{7.52}$$

The physical meaning of this variable is clear: the canonical momentum $\mathbf{P}(t)$ corresponds to the motion of the monopole center of mass. In other words, it describes the motion of a monopole as a point-like particle of mass M. We can see this immediately, if we recall that the translational zero modes arise as a result of the infinitesimal displacement of a monopole:

$$A_n^a \to A_n^a + (\partial_l A_n^a) X^l , \qquad \phi^a \to \phi^a + (\partial_l \phi^a) X^l .$$

However, it would be a misleading conclusion, if we were to identify the monopole translational zero modes with these simple displacements, since they do not satisfy the gauge condition (7.7). The proper thing to do is to complement such a pure translation in $\mathbb{R}^3$ with a gauge transformation with a special choice of the parameter, which is the classical gauge potential A_n^a itself:

$$\begin{aligned} a_n^{a\,(l)} &= \partial_l A_n^a - D_n A_l^a = F_{ln}^a , \\ \chi^{a\,(l)} &= \partial_l \phi^a - e\varepsilon_{abc} A_l^b \phi^c = D_l \phi^a . \end{aligned} \tag{7.53}$$

We can check that, defined in such a way, translational zero modes $\zeta_{(l)}^a = (a_n^{a\,(l)}, \chi^{a\,(l)})$ satisfy the gauge condition (7.7).

Note that these modes are orthogonal and normalizable. Moreover, the corresponding normalization factor is directly related to the monopole mass m. Indeed, a straightforward calculation gives the result[5] [314, 410]

$$\begin{aligned} N_{(k)(l)} &= \int d^3x\, \zeta_\mu^{a\,(k)} \zeta_\mu^{a\,(l)} = \int d^3x \left(a_n^{a\,(k)} a_n^{a\,(l)} + \chi^{a\,(k)} \chi^{a\,(l)} \right) \\ &= \int d^3x \left(F_{kn}^a F_{ln}^a + D_k\phi^a D_l\phi^a \right) = M\delta_{kl} . \end{aligned} \tag{7.54}$$

We can now separate the contribution of the translational zero modes and write the time derivative of the monopole field as[6]

$$\partial_0 A_\mu^a \approx \dot{X}^l(t)\, \zeta_{(l)}^a . \tag{7.55}$$

Therefore the Lagrangian of the system includes the kinetic term

---

[5] Hereafter, we consider the BPS monopole. Recall also that there is no summation over the indices (l).
[6] This approximation can be justified, if the monopole of classical mass M is much heavier than the excitations around it. As we mentioned above, in this case, the normal oscillation modes are decoupled, that is, we neglect the effects of radiative friction and bremsstrahlung.

7 $SU(2)$ Monopole in Quantum Theory

$$L_X = \frac{1}{2}\int d^3x \left(E_n^a E_n^a + D_0\phi^a D_0\phi^a\right)$$
$$= \frac{\dot{X}_k \dot{X}_l}{2}\int d^3x \left(a_n^{a\,(k)} a_n^{a\,(l)} + \chi^{a(k)}\chi^{a(l)}\right) = \frac{M\dot{X}_l^2}{2}. \tag{7.56}$$

The canonical momentum, which is conjugated to the collective coordinates $\mathbf{X}(t)$, is, therefore

$$\mathbf{P} = \frac{\delta L_X}{\delta \dot{\mathbf{X}}} = M\dot{\mathbf{X}}, \tag{7.57}$$

and we obtain the Hamiltonian of the translational zero modes

$$H_X = \frac{\mathbf{P}^2}{2M}, \tag{7.58}$$

which obviously describes a free moving classical particle of mass M.

Since the complete set of the eigenfunctions of the operator of second derivatives includes one more gauge zero mode $\zeta_\mu^{a\,(0)}$, we cannot just forget about it. By analogy with translational zero modes, we must separate the integration over the corresponding cyclic collective coordinate Υ and its conjugated momentum.

Recall[7] that this normalizable zero mode is defined as

$$\zeta_\mu^{a\,(0)} = \begin{pmatrix} \partial_0 A_n^a \\ \partial_0 \phi^a \end{pmatrix} = \begin{pmatrix} -D_n\omega \\ e\varepsilon_{abc}\phi^b\omega^c \end{pmatrix}, \tag{7.59}$$

where $\omega^a(\mathbf{r})$ is a parameter of the time-dependent gauge transformation $U_g(\Upsilon)$ (5.84). As we have seen in Chap. 5, such a transformation induces an electric charge of the field configuration, which is proportional to the velocity of collective motion along cyclic collective coordinate Υ, that is, $Q = g\dot{\Upsilon}$. The boundary conditions on the spatial asymptotic then allows us to define

$$\omega^a(\mathbf{r}) = \frac{Q}{g}\phi^a.$$

Straightforward calculation of the normalization factor by analogy with (7.54) now yields

$$N_{(0)(0)} = \int d^3x\, \zeta_\mu^{a\,(0)}\zeta_\mu^{a\,(0)} = \int d^3x \left\{(\partial_0 A_n^a)^2 + (\partial_0\phi^a)^2\right\}$$
$$= \int d^3x (D_n\omega^a)^2 = \frac{Q^2}{g^2}\int d^3x(D_n\phi^a)^2 = M\frac{Q^2}{g^2}. \tag{7.60}$$

Once again, separating the contribution of this mode, we arrive at the already known kinetic term (5.86) of the Lagrangian of collective motion along the cyclic collective coordinate, which supplements the term (7.56):

---
[7] cf. the discussion in Sect. 5.2.1.

7.2 Non-Abelian Monopole: Quasiclassical Quantization

$$L_\Upsilon = \frac{1}{2}\int d^3x\, \Pi_n^a \Pi_n^a = \frac{\dot\Upsilon^2}{2}\int d^3x\, (D_n\phi)^2 = \frac{M\dot\Upsilon^2}{2} = \frac{MQ^2}{2g^2}. \quad (7.61)$$

Then the canonical momentum P_0, which is conjugated to the collective coordinate $\Upsilon(t)$ is

$$P_0 = \frac{\delta L_\Upsilon}{\delta \dot\Upsilon} = M\dot\Upsilon = \frac{M}{g}Q = vQ, \quad (7.62)$$

and the corresponding Hamiltonian describes the free motion of a classical particle of mass M along the cyclic coordinate Υ:

$$H_\Upsilon = \frac{P_0^2}{2M}. \quad (7.63)$$

The physical meaning of such a motion can be clarified, if we note that the contribution of the gauge zero mode to the kinetic energy can be represented as $L_\Upsilon = I\dot\Upsilon^2/2$, where the monopole mass M set into correspondence with the moment of inertia I of a "quasi rigid body".

Both the zero modes, which we described above, and all normal oscillations are supposed to be perturbative fluctuations around a classical field configuration. The separation of the contribution of zero modes into the total energy functional yields

$$E = M + \frac{P_0^2}{2M} + \frac{\mathbf{P}^2}{2M} = M\left(1 + \frac{M^2\dot{\mathbf{X}}^2 + v^2 Q^2}{2M}\right)$$
$$\approx \sqrt{M^2 + \mathbf{P}^2 + v^2 Q^2}. \quad (7.64)$$

Here, the electric charge of a dyon is considered as a fourth component of the generalized four-momentum of collective coordinates P_α.

Although we discuss the spectrum of quantum fluctuations of the fields around a monopole, the Hamiltonian of collective coordinates (7.64) has a classical form. Thus, we can establish a correspondence between the formalism of quantization in terms of one-particle excitations and a classical motion over the moduli space $\mathcal{M}_1$. Clearly, the next step is to quantize the Hamiltonian of collective coordinates, which would correspond to the second quantization of the original theory [238, 370]. In this picture, the quantum mechanical wave function $\Psi(\mathbf{X}, \Upsilon)$ corresponds to the wave functional of the quantum field theory.

Quantum mechanics on the moduli space can be constructed by a standard procedure of replacing the collective coordinates $\mathbf{X}, \Upsilon$ and momenta $\mathbf{P}, P_0$ with corresponding quantum mechanical operators. Here it would be convenient to define these operators as

$$\mathbf{P} \to -i\frac{\partial}{\partial \mathbf{X}}, \qquad P_0 \to -i\frac{m_v}{2\pi}\frac{\partial}{\partial \Upsilon},$$

where $m_v = ve$ is a mass of the vector particle. Then the quantum mechanical Hamiltonian operator on the moduli space is[8]:

$$H = -\frac{1}{2M}\left(\frac{\partial^2}{\partial \mathbf{X}^2} + \frac{m_v^2}{4\pi^2}\frac{\partial^2}{\partial \Upsilon^2}\right). \tag{7.65}$$

Such a structure of the Hamiltonian operator implies that its eigenfunctions can be factorized as $\Psi(\mathbf{X}, \Upsilon) = \Psi(\mathbf{X})\Psi(\Upsilon)$. We can easily see that the eigenvalues of the operator $\mathbf{P}$ are components of momentum in $\mathbb{R}^3$. Also, the corresponding component of the factorized wave function is just a plane wave $\Psi(\mathbf{X}) = e^{i\mathbf{P}\cdot\mathbf{X}}$. However, the situation with eigenfunctions of the operator P_0, which describes a periodic motion along a circle S^1 parameterized by a coordinate Υ, is a little bit more subtle. Of course, we may write it as $\Psi(\Upsilon) = e^{2\pi in\Upsilon}$, where n is an integer. Then, making use of the charge quantization condition which, since a minimal electric charge in the $SU(2)$ theory is $q_0 = e/2$, now takes the form $4\pi n = eg$, we can write

$$P_0\Psi(\Upsilon) = -i\frac{m_v}{2\pi}\frac{\partial}{\partial \Upsilon}e^{2\pi in\Upsilon} = m_v n\Psi(\Upsilon) = \frac{e^2}{4\pi}M\Psi(\Upsilon). \tag{7.66}$$

Now observe that this operator of momentum P_0 is proportional to the operator of the electric charge of a dyon that has the eigenvalues $q = ne$:

$$Q = \frac{1}{v}P_0, \qquad Q\Psi(\Upsilon) = -i\frac{e}{2\pi}\frac{\partial}{\partial \Upsilon}e^{2\pi in\Upsilon} = ne\Psi(\Upsilon), \tag{7.67}$$

which is, of course, not a coincidence. Thus, the eigenvalues of the operators P_0 and Q are as before related by the relation (7.62), but the electric charge of a dyon is no longer an arbitrary parameter. According to (7.67), by second quantization it is promoted to be quantizible in the units of the vector boson charge.

Let us note that the eigenvalues of the operator of the kinetic energy of internal rotation, associated with gauge zero mode, are given by

$$E_{rot} = \frac{P_0^2}{2M} = \frac{v^2 e^2 n^2}{2M} = \frac{ve^3}{8\pi}n^2. \tag{7.68}$$

Hence, the energy of the dyon excitation is quantized and the non-relativistic correction to the energy of the classical ground state is $\Delta M = m_v e^2/(8\pi)$. Taking into account that the relativistic correction is given by the relation (7.64), we can conclude that the energy of the first excited state is different from the classical monopole mass on $vQ = ve = m_v$.

Thus, the excitation of the cyclic gauge collective coordinate associated with the quantizable electric charge of a dyon can be interpreted as the formation of a bound system monopole-charged vector boson. Indeed, in the

---

[8] Note that an obvious reparameterization makes it identical to the Laplacian on the space $\mathcal{M}_1$ [238, 370, 373]

previous section, we have seen that there is such a spherically symmetric bound state in the spectrum of charged vector bosons coupled with a monopole of unit topological charge (cf. the discussion on page 250).

In the secondary quantized theory, we have to consider not only one-particle excitations of that type, but all possible processes of creation and annihilation of different particles and interaction between them. Usually, the computation of the amplitudes of these processes is rather involved. However, if we restrict our consideration to the transitions between dyonic excitations within the same topological sector of the model, the quantum field theory may be truncated to the quantum mechanics on the moduli space, and then the corresponding amplitudes may be evaluated by making use of simple quantum mechanical description. The moduli space approach becomes tailor-made to describe these processes. For example, the effect of radiation at the scattering of two monopoles was considered in the papers [296, 372].

To complete our discussion, let us recall that the Lagrangian of the Georgi–Glashow model can be modified by including the θ-term, which also changes the electric charge of a dyon in accordance with Witten formula (5.110), which in our case becomes

$$Q = en + \frac{e\theta}{2\pi} m \,, \qquad (7.69)$$

where n, m are integers. Clearly, the first term here still arises as the contribution of a one-particle boson excitation coupled with a monopole. As for the second term, it also admits a similar interpretation in quantum field theory, if we extend the Georgi–Glashow model by incorporation of the spinor fields coupled with a non-Abelian monopole. Then the shift of the electric charge of a dyon given by (7.69) is caused by the contribution of the fermionic zero modes on the monopole background, which appears in accordance with the index theorem. We shall discuss this very interesting effect in Chap. 10.

7.2.3 Evaluation of the Generating Functional

Let us return to the evaluation of the generating functional for the Green functions on a monopole background. Note that when we extract from the functional integral the explicit integration over four collective coordinates and corresponding canonically conjugated momenta, the total number of dynamical variables does increase. If we wish our formulation to be equal to the initial model, we have to impose some additional constraints to compliment the conditions G_1 (7.45) and G_2 (7.41). Therefore, extracting the integration over the collective coordinates from the functional integral, we must impose the orthogonality condition between the zero modes and all other normal oscillatatory modes:

$$F_1^{(\alpha)} = \int d^3x \, a_\mu^a \zeta_\mu^{a(\alpha)} \equiv \begin{cases} \int d^3x \, [\chi^a(\partial_0 \phi^a) + a_n^a(\partial_0 A_n^a)] \\ \int d^3x \, [\chi^a \chi^{a(l)} + a_n^a a_n^{a(l)}] \end{cases} = 0 \,.$$

Moreover, since we are integrating over the entire configuration space of variables (p, q), we must also redefine the canonical momenta in such a way that they will be orthogonal to the momenta of zero modes. For this purpose, we write [410]

$$\Pi_\mu^a = \Pi_{\mu\perp}^a + P_\alpha N_{\alpha\beta}^{-1} \zeta_\mu^{a(\beta)}, \tag{7.70}$$

where we compose the normalization factors N_{kl} (7.54) and $N_{(0)(0)}$ (7.60) into the normalization matrix

$$N_{\alpha,\beta} = \int d^3x \, \zeta_\mu^{a(\alpha)} \zeta_\mu^{a(\beta)} = \begin{pmatrix} N_{(0)(0)} & 0 \\ 0 & N_{(k)(l)} \end{pmatrix}. \tag{7.71}$$

Here the orthogonality condition holds:

$$F_2^{(\alpha)} = \int d^3x \, \Pi_{\mu\perp}^a \zeta_\mu^{a(\alpha)} \equiv \begin{cases} \int d^3x \, [\Pi_\perp^a(\partial_0 \phi^a) + \Pi_{n\perp}^a(\partial_0 A_n^a)] \\ \int d^3x \, [\Pi_\perp^a \chi^{a(l)} + \Pi_{n\perp}^a a_n^{a(l)}] \end{cases} = 0. \tag{7.72}$$

According to the classification by Dirac [202], the constraints F_i, $i = 1, 2$ are *primary constraints*, since in quantum theory all dynamical variables, including collective coordinates X_α and the momenta conjugated to them, obey the canonical equal time commutation relations

$$[\Pi_\mu^a(\mathbf{r}, t), a_\nu^b(\mathbf{r}', t)] = -i\delta_{ab}\delta_{\mu\nu}\delta(\mathbf{r} - \mathbf{r}'), \quad [P_\alpha, X_\beta] = -i\delta_{\alpha\beta},$$
$$[\Pi_\mu^a(\mathbf{r}, t), \Pi_\nu^b(\mathbf{r}', t)] = [a_\mu^a(\mathbf{r}, t), a_\nu^b(\mathbf{r}', t)] = 0. \tag{7.73}$$

Then it is easy to check that the relations holds

$$[F_i, F_j] = 0; \quad [H_{eff}, F_i] = 0, \tag{7.74}$$

where $H_{eff}(\Pi, a, P, X)$ is the Hamiltonian operator that we redefined by separating the collective coordinates. The canonical commutation relations for the orthogonal momenta $\Pi_{\mu\perp}^a$ that we defined above take the form [410]

$$[\Pi_{\mu\perp}^a(\mathbf{r}, t), a_\nu^b(\mathbf{r}', t)] = -i\delta_{ab}\delta_{\mu\nu}\delta(\mathbf{r} - \mathbf{r}') + \zeta_\mu^{a(\alpha)} N_{\alpha\beta}^{-1} \zeta_\nu^{b(\beta)}, \tag{7.75}$$

where the operator $\zeta_\mu^{a(\alpha)} N_{\alpha\beta}^{-1} \zeta_\nu^{b(\beta)}$ is a projector onto the space of collective coordinates.

Hence, in order to write the generating functional of our system, by analogy with the Faddeev–Popov procedure we apply a similar trick [166, 236, 410, 499]:

$$Z = \int \mathcal{D}a\mathcal{D}\Pi\mathcal{D}X\mathcal{D}P \prod_{i=1}^{2} \delta[F_i] \prod_{j=1}^{2} \delta[G_j] \det\{F_i, G_j\}$$
$$\times \exp\left\{ i \int dt(\mathbf{P}\dot{\mathbf{X}} + P_0 \dot{\Upsilon}) + \int d^3x dt \, [\Pi_\mu^a \dot{a}_\mu^a - H(a, \Pi, X)] \right\},$$

7.2 Non-Abelian Monopole: Quasiclassical Quantization

where we suppose that the constraints are taken in such a way that the determinant of the extended matrix of Poisson brackets $\det\{F_i, G_j\}$ is not singular.

We may simplify the form of the effective Hamiltonian, if we take into account some physical arguments. Indeed, separation of the zero modes in (7.70) actually means that the term of interaction between these modes and all other normal oscillations is suppressed by the factor $N^{-1} \sim 1/M$. Therefore, the canonical momenta $P_{(\alpha)}$ that is conjugated to the collective coordinates of a monopole $X_{(\alpha)}$, can be represented as an expansion about the space of collective coordinates with a metric $N_{\alpha\beta}$, the *moduli space*:

$$P_{(\alpha)} \to \tilde{P}_\alpha = P_\beta (1 - N^{-1}O)_{\beta\alpha} \,. \tag{7.76}$$

Here we make use of the auxiliary matrix

$$O_{\alpha\beta} = \int d^3x \, \frac{\partial \zeta_\mu^{a(\beta)}}{\partial X_{(\alpha)}} \, a_\mu^a \,, \tag{7.77}$$

which describes the interaction between the zero modes and all other excitations.

As the next step, let us go to the reference frame moving with a monopole. Then the argument of the quantum fluctuations is shifted as $a_\mu^a(t, \mathbf{r}) \to a_\mu^a(t, \mathbf{r} - \mathbf{X})$. Substitution of the expansion (7.49) into expression (7.42) yields the following form of the effective Hamiltonian [410, 499]:

$$H = H_{cl} + H_X + H_q \,. \tag{7.78}$$

Here H_{cl} corresponds to the classical component (7.42) of the Hamiltonian operator, thus its eigenvalue is simply the monopole mass M. The second term in (7.78) is the Hamiltonian of the collective coordinates

$$H_X = \frac{1}{2} \tilde{P}_\alpha (1 - N^{-1}O)^{-1}_{\alpha\beta} N^{-1}_{\beta\gamma} (1 - N^{-1}O)^{-1}_{\gamma\delta} \tilde{P}_\delta \,. \tag{7.79}$$

Clearly, if the zero modes are completely separated from all other massive fluctuations, the low-energy dynamics of a monopole lies entirely within the moduli space. In such a case, the monopole motion is described by the Hamiltonian

$$H_X \approx \frac{1}{2} \tilde{P}_\alpha N^{-1}_{\alpha\beta} \tilde{P}_\beta \,, \tag{7.80}$$

in agreement with (7.64). This delightfully simple form of the Hamiltonian actually tells us that the low-energy monopole dynamics is completely defined by some metric on the moduli space (cf. the discussion in Sect. 6.3.3).

The last term in (7.78) is due to the contribution of the quantum fluctuations

$$H_q = \frac{1}{2} \int d^3x \left(\Pi_{\mu\perp}^a \Pi_{\mu\perp}^a + a_\mu^a \mathcal{D}_{\mu\nu}^{ab} a_\nu^b \right) \,, \tag{7.81}$$

where the explicit form of the matrix of the second variation of the action $\mathcal{D}_{\mu\nu}^{ab}$ is given by (7.8).

Note that the effective Hamiltonian (7.78) is not manifestly Lorentz-invariant, although that was an initial property of the Lagrangian (5.7). However, relation (7.64) shows that in each order of expansion in parameter P/M, the relativistic invariance must actually be restored.

To complete the quasiclassical quantization of a monopole, we must consider the elimination of the non-physical degrees of freedom of the quantum fluctuations, which is a common problem of any gauge field theory. Actually we already did this when we incorporated the constraints G_1 and G_2 into the generating functional (7.76); they may be sent back to the exponent to reproduce the ghost Lagrangian (7.10) that we considered above.

The scheme of quasiclassical quantization, which we briefly sketched above, can be applied to calculate different quantum effects in a monopole external field. The renormalization of the divergences there is quite standard and we shall not discuss the details. For example, the renormalization of quantum correction to the mass of a monopole can be done if we introduce the counterterm [312]

$$\delta H_{c.t.} = -\frac{1}{2}\delta m^2 \int d^3x \delta L_{c.t.} = -\frac{1}{2}\delta m^2 \int d^3x (e^2\phi^2 - m_v^2), \qquad (7.82)$$

where the quantity δm^2 is defined by the condition

$$\left.\frac{d}{d\phi}(\delta L_{c.t.} + \delta V_{eff})\right|_{\phi=v} = 0,$$

with δV_{eff} being a one-loop correction to the effective potential of the scalar field (7.31).

Recall that the physical meaning of such counterterms[9] consists in the subtraction of the divergent energy of the trivial vacuum without a monopole. Indeed, we have seen that under the assumption that we neglect the structure of the monopole core, the second variation of action becomes diagonal and the contribution of the quantum fluctuations (7.81) can be written as

$$H_q = \frac{1}{2}\int d^3x \sum_i \left(\Pi_\perp^{a,i\,2} + \omega_i^2 \chi_i^{a\,2}\right) + \frac{1}{2}\int d^3x \sum_i \left(\Pi_{n\perp}^{a,i\,2} + \omega_i^2 a_{n,i}^{a\,2}\right). \qquad (7.83)$$

In this case, taking into account the contribution of the collective coordinates (7.64), the ground state energy is

$$E \approx M + \frac{P_0^2}{2M} + \frac{\mathbf{P}^2}{2M} + \frac{3}{2}\sum_i \omega_i, \qquad (7.84)$$

---

[9] The structure of the Lagrangian (5.7) suggests that, generally speaking, in this model we have to consider four types of counterterms, namely $(F_{\mu\nu}^a)^2$, $(D_\mu\phi^a)^2$, ϕ^2 and ϕ^4. To simplify our brief overview we do not go into details here.

where the coefficient 3 corresponds to the number of physical degrees of freedom.

To complete the calculation of the finite quantum correction to the monopole mass ΔM, we need to subtract the energy of the vacuum fluctuations of the scalar and vector fields with eigenfrequencies $\omega_i^{(0)}$ from the divergent expression (7.84). This is the effect of the counterterm (7.82). Then we can write the one-loop correction to the monopole energy as

$$E \approx \sqrt{(M + \Delta M)^2 + \mathbf{P}^2 + v^2 Q^2}, \tag{7.85}$$

where

$$\Delta M = \frac{3}{2} \sum_i \left(\omega_i - \omega_i^{(0)} \right). \tag{7.86}$$

Here we run into a problem of purely technical nature. Clearly, further calculations would be possible only if we know the spectrum of the quantum fluctuations on a monopole background, that is, if we can perform an explicit summation over all quantum numbers in (7.86) by analogy with the calculation of the vacuum polarization by an Abelian dyon in Sect. 4.5. However, there is no general analytical solution of this problem, and therefore, such a straightforward calculation of the quantum correction to the monopole mass can be done only numerically. This is why we choose a bypass to evaluate the quantity ΔM (7.37) above. Moreover, although the scheme of the quasiclassical quantization about the topologically non-trivial field configurations has been known since the 1970s, it was actually applied in very few models, basically in $(1+1)$ dimensions, like, e.g., quantization of the kink solutions of the sin-Gordon model [212]. Even the question about the renormalization of the magnetic charge of the non-Abelian monopole, which seems to be principal from many points of view, was analyzed only in a qualitative way [43]. Actually, with respect to a monopole, besides the estimation of the quantum correction to the monopole mass above, there is maybe one more interesting application of the quasiclassical quantization. This is the process of monopole pair creation in an external magnetic field, which was analyzed by N. Manton and J. Affleck [68].

7.3 Quasiclassical Quantization and Evaluation of the Monopole Pair Creation Amplitude in an External Magnetic Field

7.3.1 Dynamics of Non-Abelian Monopole in Weak External Field

Before considering of the creation of monopoles in an external magnetic field, we consider the mechanism of excitation of the translational zero modes $\zeta_\mu^{a(l)}$

by an external homogeneous magnetic field B_k^{ext} [314]. We can use the explicit form (6.40) of the Lagrangian of interaction:

$$L_{int} = \frac{1}{2v}\varepsilon_{abc}F_{mn}^a\phi^a B_k^{ext}.$$

Then the field equations of the Georgi–Glashow model (5.14) are modified as

$$D_\nu F^{a\mu\nu} = -e\varepsilon_{abc}\phi^b D^\mu\phi^c + \mathcal{F}_\mu^a,$$
$$D_\mu D^\mu \phi^a = -\lambda\phi^a(\phi^b\phi^b - v^2) + \mathcal{F}^a, \qquad (7.87)$$

where the perturbation is related to a weak external force acting on the configuration:

$$\mathcal{F}_0^a = 0, \qquad \mathcal{F}_n^a = \frac{1}{v}\varepsilon_{nmk}D_m\phi^a B_k^{ext}, \qquad \mathcal{F}^a = \frac{1}{2v}\varepsilon_{abc}F_{mn}^a B_k^{ext}.$$

Thus, the system of equations (7.8) on the eigenfunctions of the operator of second variations of the action outside of the monopole core becomes

$$\begin{pmatrix} (\mathcal{D}_v^2)_{ab} & 0 \\ 0 & (\mathcal{D}_s^2)_{ab} \end{pmatrix} \begin{pmatrix} \chi^b \\ a_n^b \end{pmatrix} = \begin{pmatrix} \mathcal{F}^a \\ \mathcal{F}_n^a \end{pmatrix}. \qquad (7.88)$$

We may make use of the expansion (7.50) in eigenfunctions of the operator of second functional derivatives (7.9) again, but now we will not separate the contribution of zero modes:

$$a_\mu^a(t,\mathbf{r}) = \sum_{i=0}^\infty C_i(t) a_{\mu i}^a(\mathbf{r}).$$

Substitution of this expansion into (7.88) gives the system

$$\sum_{i=0}^\infty \left(\ddot{C}_i + \Omega_i^2 C_i\right) a_n^a(\mathbf{r})_i - 2e\varepsilon_{abc}\chi^b D_n\phi^c = \mathcal{F}_n^a(\mathbf{r})^{(l)},$$
$$\sum_{i=0}^\infty \left(\ddot{C}_i + \omega_i^2 C_i\right)\chi^a(\mathbf{r})_i + 2e\varepsilon_{abc}a_n^b D_n\phi^c = \mathcal{F}^a(\mathbf{r})^{(l)}, \qquad (7.89)$$

where the index (l) labels direction of the external magnetic field. Thus, the monopole dynamics in external magnetic field B^{ext} is completely defined by the time evolution of the expansion coefficients $C_i(t)$.

Now we can employ the orthogonality of the eigenfunctions $a_{\mu i}^a$. Projection of the (7.89) onto the translational zero modes $\zeta_{(k)}^a = (a_n^{a\,(k)}, \chi^{a(k)})$, which describe a displacement of the configuration, gives

$$\ddot{C}_0 N_{(l)(k)} = \int d^3x \left\{ \mathcal{F}^a(\mathbf{r})^{(l)} \chi^{a(k)} + \mathcal{F}_n^a(\mathbf{r})^{(l)} a_n^{a\,(k)} \right\}. \qquad (7.90)$$

Evaluation of the integrals on the right-hand side of this expression then yields

$$\int d^3x\, \mathcal{F}^a(\mathbf{r})^{(l)} \chi^{a(k)} = \frac{1}{v}\int d^3x\, B_l^{ext} B_l^a D_k \phi^a = \frac{1}{3} g B_l^{ext} \delta_{kl},$$
$$\int d^3x\, \mathcal{F}_n^a(\mathbf{r})^{(l)} a_n^{a(k)} = \frac{2}{3v}\int d^3x\, B_k^{ext} B_m^a D_m \phi^a = \frac{2}{3} g B_l^{ext} \delta_{kl}.$$
(7.91)

Thus, making use of the relation (7.54), we finally arrive at the quite expectable conclusion: $\ddot{C}_0 = gB^{ext}/M = w$, where w is the acceleration of a monopole. In other words, the 't Hooft–Polyakov monopole interacts with an external field, as it should in the case of a point-like magnetic charge of mass M and the magnetic counterpart of the Lorentz force is just $F = gB^{ext}$. Corrections to this simple expressions arise as a result of the radiative friction effects. They can be evaluated in the next order of perturbation expansion in external field B^{ext}.

Generalization of this classical formula to the relativistic case immediately yields the relation (4.1), which we already introduced *ad hoc* on a level of quantum electrodynamics in Chap. 4:

$$M \frac{du_\mu}{d\tau} = g \widetilde{F}_{\mu\nu} u^\nu.$$
(7.92)

Recall that $u_\mu = dx_\mu/ds$ is the relativistic 4-velocity of a monopole whose trajectory $x_\mu(\tau)$ is parameterized by a proper time τ. We take the homogeneous magnetic field to be directed along the z-axis, that is we have non-vanishing components of the dual field strength tensor $\widetilde{F}_{30} = B^{ext}$.

7.3.2 Metastable Vacuum Decay and Monopole Pair Creation in an External Field

Let us discuss further the quantum effects related to monopoles. As is known, evaluation of the functional integral can be simplified in Euclidean formulation of the theory. Thus, we shall consider the Euclidean vacuum-to-vacuum transition amplitude for a monopole in a homogeneous magnetic field. This problem was analyzed by I. Affleck and N. Manton in [68].

The starting point of the discussion is the observation about the net result of the consideration above. This is actually the replacing of the electric notations with magnetic ones in the standard problem of motion of an electric charge in a homogeneous electric field. The solution of this problem is well-known: a world line of such a particle moving in Minkowski space is a hyperbola (see, e.g., [19]). This is also a trajectory of a monopole in a homogeneous magnetic field.

For our purposes, it is sufficient to regard the Euclidean time x_4 replacing Minkowski time as $t \to x_4 = it$. Then the monopole trajectory is a closed circular world line, which lies entirely in a plane formed by the direction of the external field and the Euclidean time:

$$x_1 = x_2 = 0, \qquad x_3 = \frac{1}{w}\cos ws, \qquad x_4 = \frac{1}{w}\sin ws. \qquad (7.93)$$

Let us recall now that there is a one-to-one correspondence between the classical solution of the Euclidean equations of motion and quantum mechanical tunnelling through a barrier in Minkowski space. In other words, interpretation of the solution (7.93) in Minkowski space is as follows: the circular world line corresponds to the creation of a monopole-antimonopole pair at the moment of Euclidean time $x_4 = -w^{-1} = -M/(gB^{ext})$. Then monopole and anti-monopole are moving away from each other along the z-axis, up to the maximal separation $D = 2R = 2M/(gB^{ext})$, which corresponds to the moment of time $x_4 = 0$. The radius of this circle is just the inverse monopole acceleration:

$$R = \frac{M}{gB^{ext}} = w^{-1}. \qquad (7.94)$$

Since we suppose that the magnetic field is weak, this distance is very big, at least much bigger than the classical radius of a monopole g^2/M.

There is a difference from the axially symmetric magnetic dipole configuration that we considered above (cf. the discussion in Chap. 6, p. 185). In the latter case, the separation parameter D remains relative small; it corresponds to the scale of a few inverse vacuum expectation values of the Higgs field, thus the magnitude of the related external field B^{ext} then has to be of the same order to create such an equilibrium state The analysis by Affleck and Manton is given for the different case $|B^{ext}| \ll v^2$ [68].

We can easily see that the modified field equations (7.87) really describe the motion of a monopole on the circular trajectory (7.93). To simplify our consideration, let us consider the Bogomol'nyi limit when these equations, being written in Euclidean space, have an especially simple form [314, 366]:

$$D_4 F_{4n}^a + D_m F_{mn}^a - \frac{1}{v} B_m^{ext} F_{mn} = -e\varepsilon_{abc}\phi^b D_n \phi^c,$$

$$D_m F_{m4}^a = -e\varepsilon_{abc}\phi^b D_4 \phi^c,$$

$$D_4 D_4 \phi^a + D_m D_m \phi^a - \frac{1}{v} B_m^{ext} D_m \phi^a = 0. \qquad (7.95)$$

We supposed that the external magnetic field is directed along the negative direction of the z-axis. Since in the Bogomol'nyi limit a monopole mass is simply $M = gv$ and for a particle moving along the classical trajectory (7.93) the relation (7.94) is fulfilled, the solution for an accelerated monopole must satisfy the relations[10]

---

[10] Recall that the Gauss law (7.41) and the background gauge conditions (7.7) are here imposed as constraints.

7.3 $g\bar{g}$ Pair Creation in an External Magnetic Field

$$D_m F^a_{mn} + \frac{1}{R} F_{3n} = -e\varepsilon_{abc}\phi^b D_n \phi^c, \qquad D_m D_m \phi^a + \frac{1}{R} D_3 \phi^a = 0. \quad (7.96)$$

In the first-order of the expansion in $1/R$, these equations coincide with the system obtained in [68][11].

Clearly, the solution of this system can be obtained by making use of the perturbation expansion in $1/R$. Here the zero-order equation corresponds to the static monopole with vanishing non-Abelian electric field $E^a_n = 0$ and, in principle, we can evaluate the first-order corrections as above. However, if we neglect the internal structure of a monopole and consider only the asymptotic form of the monopole fields, the solution that describes a monopole moving along circular world line (7.93) can be constructed via simple rotation of the static solution on the angle

$$\alpha = \arctan \frac{x_4}{x_3} = \arctan\left(\frac{gB^{ext}}{M}\tau\right),$$

in the two-dimensional x_3, x_4-plane fixed by the directions of the external magnetic field and Euclidean time, respectively.

It is convenient to use the auxiliary three-dimensional coordinates, introduced in [68]: $\rho = (x, y, z) \equiv (x_1, x_2, r - R)$, where $r = \sqrt{x_3^2 + x_4^2}$. Then the Euclidean components of the Abelian electric and magnetic field of a moving monopole are

$$E_1 = \frac{gy}{\rho^3} \sin\alpha, \qquad B_1 = \frac{gx}{\rho^3} \cos\alpha,$$

$$E_2 = -\frac{gx}{\rho^3} \sin\alpha, \qquad B_2 = \frac{gy}{\rho^3} \cos\alpha,$$

$$E_3 = 0, \qquad B_3 = \frac{gz}{\rho^3}. \quad (7.97)$$

Let us return now to the analysis of the mechanism of interaction between a monopole and an external Abelian homogeneous magnetic field (see expression (6.40) above). We have already pointed out that there is a clear analogy between the monopole dynamics and evolution of a metastable system after the false vacuum decay: the external magnetic field lifts the degeneration of the Higgs vacuum and the accelerated motion of a monopole is related with its asymptotic approach to the unique true minimum of the potential:

$$\phi^a_{min} = v\hat{r}^a \left(1 + e\frac{\hat{r}_n B^{(ext)}_n}{r^2 m_s^2 m_v^2}\right),$$

---

[11] Recall that the monopole loop radius is related to the magnitude of the external magnetic field according to (7.94). This relation can be obtained, if we promote R to be a parameter of the system that is included in the Lagrangian. The first variation of action with respect to R vanishes if the relation (7.94) is fulfilled (cf. the discussion in [68] and relation (7.99) below).

which lies in the direction of the external perturbation. In other words, a ground state of the monopole field configuration that interacts with a weak external magnetic field turns out to be a metastable vacuum, whose energy can be decreased via the simultaneous creation of a monopole-antimonopole pair, provided that they are strongly separated to avoid possible annihilation. Since we suppose that the monopole cores do not overlap, we can neglect the short-range non-Abelian character of the interaction, thus the height of the barrier in that approximation is just a double of the monopole mass M [68, 483].

The problem of metastable vacuum decay was originally investigated in the pioneering paper by Langer [342] and then considered in the quantum field theory context in [173, 505]. The observation of the paper [68] is that a monopole loop (7.93) can also be considered as a bubble of true vacuum $\mid \phi_{min}^a \mid$ in a false vacuum $\mid \phi \mid = v$. Static energy of such a bubble is composed of the energy of the transitional domain (so-called *wall*), where the scalar field interpolates between false and true vacua, the energy of the domain inside the bubble (that is, the energy of interaction with external magnetic field) and the energy of Coulomb magnetic interaction between the monopoles. In the case under consideration, we can clearly write the energy of the wall as $2\pi R M$ and the energy of interaction with external field as $-g B^{ext} \pi R^2$.

The Coulomb energy piece is a bit subtle, since it diverges logarithmically for point-like magnetic charges with a naive potential of interaction $V_{coulomb} = g/R$. However, the finiteness of the core improves the situation, since then the contribution of the Coulomb interaction to the static energy of the bubble becomes $-g^2/4$ [68]. Nontrivial independence of this result on the loop radius R appears, because the integral of the Coulomb energy of the monopole interaction along the loop is proportional to g^2/R, while the consequent integration over the Euclidean time yields a period proportional to R.

Thus, the classical action of the monopole loop on the circular trajectory (7.93) is

$$S = 2\pi R M - g B^{ext} \pi R^2 - \frac{g^2}{4}. \qquad (7.98)$$

We can now consider the loop radius R as a variational parameter. Variation over it yields the saddle point of the action and we can see that it corresponds to the values

$$R_{cr} = \frac{M}{g B^{ext}} = w^{-1} \quad \text{and} \quad S_0 = S(R_{cr}) = \frac{\pi M^2}{g B^{ext}} - \frac{g^2}{4}. \qquad (7.99)$$

Let us make some remarks here. First, note that the second derivative of the action S (7.98) with respect to the loop radius R is negative. Therefore, the solution is unstable, the sphaleron-like trajectory (7.93) in the functional space corresponds to the top of the barrier between the vacua. Evidently, the monopole-antimonopole pair solution [327, 448], that we constructed by making use of the ansatz (6.15) and (6.16) in the previous chapter, also corresponds to such a configuration. The appearance of the negative mode in the spectrum of fluctuation reflects in the imaginary phase of the functional integral at the saddle point. Therefore, depending on the value of R, the configuration either returns to the original false vacuum, or a monopole pair is created (true vacuum). Then the amplitude of the metastable vacuum decay, which actually defines the probability of the monopole pair creation, can in the leading order of WKB approximation be evaluated as an exponent of Euclidean action on the classical trajectory:

$$\Gamma = K e^{-S_0} = K \exp\left\{-\frac{\pi M^2}{g B^{ext}} + \frac{g^2}{4}\right\},$$

where the pre-exponential factor K is connected with the one-loop quantum correction.

Second, the problem of metastable vacuum decay has an analytical solution in the thin wall limit, when the difference between values of true and false vacua is small in comparison with other parameters of the model [173]. Clearly, in our case, this approximation is justified in the limit of the weak external magnetic field, namely when $\mid B^{ext} \mid \ll v^2$. Then the pre-exponential factor can be evaluated by making use of standard methods (see, e.g., [24, 68, 430]):

$$K = \frac{1}{2} J \frac{\det'(\delta^2 S)}{\det(\delta^2 S_0)}, \qquad (7.100)$$

where $\delta^2 S$ and $\delta^2 S_0$ are the second variations of the action in the monopole and vacuum sectors, respectively. The notation "'" here means that the contribution of zero modes is excluded from the functional determinant of a monopole. This is reflected in the coefficient J, which stands for the corresponding normalization factor of zero modes and arises when we integrate them out.

There is another difference from standard calculation, since there is a negative mode in the spectrum of fluctuations. Therefore, the relation of two functional determinants in (7.100) must be modified in order to take into account the effect of the corresponding fluctuations of the loop radius R. From a technical point of view, the appearance of an imaginary phase of the functional integral leads to a shift of the integration contour in the complex plane. This yields an additional coefficient $1/2$ in expression (7.100).

Without going into detail, let us give the result obtained in [68]:

$$K = \frac{M^2}{16\pi^3 R^2} e^{-2\pi R \Delta M}, \qquad (7.101)$$

where ΔM is the quantum correction to the monopole mass (7.37) that we evaluated above. Thus, the amplitude of the monopole pair creation in the external homogeneous magnetic field is

$$\Gamma = \frac{(gB^{ext})^2}{8\pi^3} e^{-\frac{\pi M^2}{gB^{ext}} + \frac{g^2}{4}}, \qquad (7.102)$$

where M is now a monopole physical mass including the one-loop correction ΔM.

Clearly, this formula can immediately be introduced by analogy with its electrodynamical counterpart, which corresponds to the process of creation of two charged scalar particles e^+e^- in an external homogeneous electric field:

$$\Gamma_{e^+e^-} = \frac{(eE)^2}{8\pi^3} e^{-\frac{\pi m^2}{eE}}.$$

Thus, the good news is that one more time we have proved the naive argumentation of duality. The bad news is that the probability of the monopole pair creation is totally suppressed for any physical values of the magnetic field, since in a weak coupling regime a monopole mass must be huge.

8 Monopoles Beyond $SU(2)$ Group

So far, we have discussed the monopole field configurations that arise as solutions of the classical field equations of the simple non-Abelian Yang–Mills–Higgs $SU(2)$ theory. However, it was realized, almost immediately after discovering the 't Hooft–Polyakov solution that there are other possibilities beyond the simplest non-Abelian model. Indeed, the topological analysis shows [383,502] that the existence of monopole solutions is a general property of a gauge theory with a semi-simple gauge group G, which becomes spontaneously broken down by the Higgs mechanism to a residual vacuum subgroup H containing an explicit $U(1)$ factor. Thus, in general case, the unbroken subgroup H is non-Abelian, that is, in addition to the standard electric charge, which is associated with the generator of $U(1)$ subgroup, such a monopole should also possess some non-Abelian charges.

In the series of papers by E. Weinberg [513–515] (see also reviews [518,519] and recent publications [344–346,517]), many of the aspects of the monopole solutions in the gauge theory with a gauge group of higher rank were discussed. Unexpectedly, it turns out that some of these solutions correspond to massless monopoles. On the other hand, the description of monopoles with non-Abelian charges provides a new understanding of the duality between electric charges and monopoles; it becomes transformed into the idea of Montonen–Olive duality, which would establish a correspondence between two different gauge theories, with conventional electrically charged particles being treated as physical degrees of freedom within one of these models, and monopoles, being considered as fundamental objects within its dual.

The obvious possibility of moving one step beyond the $SU(2)$ gauge theory is to consider an extended $SU(3)$ model [161,177,279,374]. This is a very interesting subject, because the corresponding non-Abelian pure gauge theory, the QCD, is generally accepted as a theory of strong interactions. The topologically nontrivial solutions here could in some way be related to non-perturbative effects, which may drive the model to the confinement phase.

Another very interesting possibility is connected with the group of unification $SU(5)$ [203]. This is the simplest possible way to unite the electroweak and strong interactions[1]. A special interest in the monopole solutions in this

---

[1] Unfortunately, for some still unknown reasons, Nature did not choose such a primitive way of unification.

model caused the process of interaction of the monopole with fermions. The point is that, because of the non-trivial topology of the fields, there is a fermionic condensate on the monopole background [151, 445]. Since the fermionic multiplet of $SU(5)$ theory includes both quarks and leptons, the interaction of such a condensate with quarks, confined into the nucleon, effectively may lead to the proton decay into mesons. This is the so-called *Rubakov–Callan effect* or monopole catalysis of the proton decay. We shall discuss this effect in Chap. 10.

There are many other generalizations of the monopoles connected with the further extension of the gauge group. It is an interesting fact that there are monopole solutions of supersymmetrical models, like, for example, the N = 2 SUSY Yang–Mills theory [65]. In the following we will discuss this remarkable model in more detail.

8.1 $SU(N)$ Monopoles

8.1.1 Generalization of the Charge Quantization Condition

Generalization of the theory of monopoles to the case of a higher rank gauge group, say $SU(N)$, immediately poses the following question. What does the corresponding charge quantization condition, which will generalize the Dirac condition (2.6), look like? Indeed, let us consider the $SU(3)$ model. For the states of the fundamental triplet, the minimal electric charge is $e/3$. Does this mean that the minimal charge of the magnetic monopole must be triple the minimal magnetic charge of Abelian electrodynamics?

A naive answer to this question is that the phase factor of the quark wave function, which may be defined by analogy with (2.8), is

$$\exp\left\{i\frac{e}{3}\oint d\mathbf{x}\,\mathbf{A}\right\} = \exp\left\{\frac{4i\pi}{3}eg\right\} = \exp\left\{\frac{2i\pi n}{3}\right\} \neq 1, \qquad (8.1)$$

if the standard Dirac quantization condition is satisfied. Thus, we would have to modify it as $eg = 3n/2$.

However, the situation is not so trivial, because there are no quarks around – they are confined inside the hadrons. As we move toward a corresponding scale, we cannot neglect the color degrees of freedom that are related with a non-Abelian (color) charge of quarks e_{color}. This is a coupling constant, which describes an interaction between a quark and the short-range color fields $\mathbf{A}_{color}$ inside of monopole core. Thus, in addition to (8.1), we have to take into account an additional phase factor

$$\exp\left\{ie_{color}\oint d\mathbf{x}\,\mathbf{A}_{color}\right\} = \exp\left\{\frac{2i\pi n'}{3}\right\}. \qquad (8.2)$$

If the integers n, n' are connected as $n + n' = 0 \mod 3$, the phase factors (8.1) and (8.2) equilibrate each other.

Thus, in the $SU(3)$ gauge theory, which describes a quark interacting with a non-Abelian monopole, the charge quantization condition becomes self-consistent by taking into consideration an additional integer chromo-magnetic charge of a monopole, rather than by increasing of Abelian magnetic charge [177, 273]. Such a color field must be screened by some non-perturbative effects on the distance larger than the hadronic mass scale [43, 274]. We shall consider this conclusion in this chapter.

8.1.2 Towards Higher Rank Gauge Groups

Before we start to consider any particular model, let us make some general remarks about a generalization of the Georgi–Glashow model. We are working with a Yang–Mills–Higgs theory with an arbitrary simple gauge group G and the scalar field in the adjoint representation. The corresponding Lagrangian (5.7) has been written before in Chap. 5

$$L = -\frac{1}{2}\text{Tr}\, F_{\mu\nu}F^{\mu\nu} + \text{Tr}\, D_\mu \phi D^\mu \phi - V(\phi) \qquad (8.3)$$

$$= -\frac{1}{4}F^a_{\mu\nu}F^{a\mu\nu} + \frac{1}{2}(D^\mu \phi^a)(D_\mu \phi^a) - V(\phi),$$

where $F_{\mu\nu} = F^a_{\mu\nu}T^a$, $\phi = \phi^a T^a$, and we use the standard normalization of the Hermitian generators of the gauge group: $\text{Tr}(T^a T^b) = \frac{1}{2}\delta_{ab}$. The Lie algebra of the generators reads

$$[T^a, T^b] = i f_{abc} T^c. \qquad (8.4)$$

The non-zero vacuum expectation value of the scalar field corresponds to the symmetry breaking Higgs potential $V(\phi)$, which generalizes the structure of (5.11)

$$V(\phi) = \lambda(|\phi|^2 - v^2)^2, \qquad (8.5)$$

where the group norm of the scalar field is defined as $|\phi|^2 = 2\,\text{Tr}\,\phi^2 = \phi^a \phi^a$.

Thus, the vacuum manifold $\mathcal{M}$ is defined by the relation $|\phi_0|^2 = v^2$. The stationary subgroup of invariance of the vacuum is H and the topological classification of the solutions is connected with a map of the space boundary S^2 onto the coset space $\mathcal{M} = G/H$. In other words, the topological charge of a magnetic monopole is given by the elements of the homotopy group $\pi_2(G/H)$. The problem is to define the stationary subgroup of the vacuum.

Let us recall that in the simplest case $G = SU(2)$, the residual $U(1)$ symmetry was fixed by the asymptotic of the scalar field; in the unitary gauge we have $\phi \to \phi_0 = vT^3 = vQ$. This subgroup is identified as an electromagnetic one, i.e., the generator of the diagonal Cartan subgroup T^3 is set to be identical to the operator of the electric charge Q. Indeed, recall that the potential of the $SU(2)$ monopole could be constructed by simple embedding of the Dirac potential into a non-Abelian gauge group (cf. (5.51))

$$\mathbf{A} = Q\frac{1-\cos\theta}{\sin\theta}\hat{\mathbf{e}}_\varphi = Q\mathbf{A}^{Dirac},$$

and the color magnetic field of a non-Abelian monopole on the spatial asymptotic is given by

$$B_n = Q\frac{r_n}{r^3}.$$

Therefore, the problem of the construction of monopole solutions for a higher rank gauge group could be re-formulated as the problem of the definition of the matrix Q for a given group [177].

For $G = SU(N)$, the solution of this problem seems to be rather obvious, because one can choose the $U(1)$ charge operator from the elements of the corresponding Cartan subalgebra generated by the operators $\vec{H} = (H_1, H_2 \ldots H_{N-1})$ [241]. In other words, the vacuum value of the scalar field in some fixed direction, for example, in the direction of the z-axis, can be taken to lie in the diagonal Cartan subalgebra of $SU(N)$ [241,514]. In this case,

$$\phi_0 = v\vec{h}\cdot\vec{H}, \qquad (8.6)$$

where $\vec{h}$ is some $N-1$ component vector in the space of Cartan subalgebra. Thus, the boundary condition on the Higgs field on infinity is that up to a gauge transformation it is equal to a diagonal matrix of the form

$$\phi_0 = \text{diag}(v_1, v_2, \ldots v_N). \qquad (8.7)$$

Sometimes, this matrix is called a *mass matrix*, since the vacuum expectation value for the scalar field defines the monopole mass. Note that because of the definition of the trace of generators of the $SU(N)$ group, the sum of all elements v_i is zero.

Let us consider the pattern of spontaneous symmetry breaking, which is determined by the entries of the mass matrix v_i. Indeed, the invariant subgroup H consists of the transformations that do not change the vacuum ϕ_0. If all the values v_i are different, the gauge symmetry is *maximally broken* and the residual symmetry group is a maximal torus $U(1)^{N-1}$. In this case, it can be thought that in the vacuum we have $N-1$ "electrodynamics", not just a single one. We can see that (see the corresponding description in Sect. 5.3.2)

$$\pi_2\left(\frac{SU(N)}{U(1)^{N-1}}\right) = \pi_1\left(U(1)^{N-1}\right) = \mathbb{Z}^{N-1}, \qquad (8.8)$$

thus, these monopoles are classified by the topological charge[2] $n = 0, 1 \ldots N-1$.

Another limiting case is the so-called *minimal symmetry breaking*. This corresponds to the situation when all but one element of the mass matrix coincide. Then the group of invariance of the Higgs vacuum is the unitary

---

[2] Compare this with the result for the case $n = 2$ given on page 165.

group $U(N-1)$ and one can see that there is a single topological charge given by

$$\pi_2\left(\frac{SU(N)}{U(N-1)}\right) = \mathbb{Z}. \tag{8.9}$$

Nevertheless, in the Bogomol'nyi limit, such a configuration also possesses a set of $N-1$ integers, whose origination could be clarified by a more detailed analysis of the asymptotic behavior of the scalar field [241, 514]. Indeed, the long-range character of this field means that its asymptotic behavior, which generalizes (8.7) in some gauge, is of the form

$$\phi_0 = \text{diag}\,(v_1, v_2, \ldots v_N) - \frac{1}{r}\,\text{diag}\,(k_1, k_2, \ldots k_N) + O(r^{-2}). \tag{8.10}$$

Here the matrix diag $(k_1, k_2, \ldots k_N)$ is called the *charge matrix*.

In the case of maximal symmetry breaking, all the numbers k_i can be connected with the topological charges as $g_m = \sum_{i=1}^{m} k_i$. However, in the case of minimal symmetry breaking, only the first such number corresponds to the topological charge $g = k_1$. All other integers are called *magnetic weights* or *holomorphic charges*. We shall explain the meaning of these additional charges below, in a particular example of $SU(3)$ monopoles

An intermediate case of symmetry breaking is that some of the entries of the mass matrix are identical. Then the gauge group G is spontaneously broken to $K \times U(1)^r$, where K is a rank $N-1-r$ semi-simple Lie group. Such a monopole has r topological charges associated with each $U(1)$ subgroup, respectively.

8.1.3 Montonen–Olive Conjecture

Shortly after discovering the monopole solutions of non-Abelian gauge theory, Montonen and Olive proposed a new and highly non-trivial idea [384]. They pointed out that in the Bogomol'nyi limit, all the particles from the spectrum of states of the Georgi–Glashow model can be composed into the table

Particle	Mass	Charges (q,g)	Spin
Higgs	0	(0,0)	0
γ	0	(0,0)	1
$A_\pm$	ve	$(e,0)$	1
g	vg	$(0,g)$	0

Moreover, all these states are saturated the Bogomol'nyi bound (5.60)

$$M = v\sqrt{q^2 + g^2} = v|\,q + ig\,|.$$

The vacuum expectation value of the Higgs field v provides a natural scale of the model. Therefore, because the magnetic charge of a 't Hooft–Polyakov monopole is $g = 1/e$, in a weak coupling regime (i.e., such as $e \ll 1$), it is a very heavy object: $M = vg \gg v$. At the same time, the gauge bosons $A_\pm$ are very light: $m_v = ve \ll v$. Thus, there are two different sectors in the spectrum of states: heavy monopoles, which are non-perturbative large-scale topological configurations of the fields, and "conventional" light particles, which are perturbative excitations over the vacuum, respectively.

Montonen and Olive noted that under the transformation of duality, the coupling constant transforms as $e \to 4\pi/e$. Thus, the dual symmetry could be considered as a way to connect the regimes of weak and strong coupling by permutation of the topological and perturbative sectors. Indeed, in the strong coupling non-perturbative regime, where $e \gg 1$, the gauge bosons become very heavy objects, while the monopole states from the topological sector are light. Thus, the idea is that the latter probably may be considered as perturbative fluctuations of some dual fields, whereas the "conventional" particles are related with the non-perturbative sector of such a dual theory. Sometimes, the strong coupling regime is called a "magnetic" formulation of the theory and the weak coupling regime is called an "electric" one.

Thus, we may believe that there are two, maybe even completely different Lagrangians, which describe strong and weak coupling regimes in terms of different sets of fundamental fields. The theory is called *self-dual*, if the Lagrangian remains invariant by permutation of the solitons and elementary excitations. As we shall see, this is the case of the Georgi–Glashow model we are considering here.

Another argument to back up the idea of duality is that the energy of interaction of all the states that saturate the Bogomol'nyi bound is zero. We already mentioned that a two-monopole configuration is static, because there are two long-range forces mediated by electromagnetic and scalar fields, respectively. In the dual sector, an analog of this effect would be a static state of two $A_\pm$ bosons.

Unlike the monopole pair, in the weak coupling regime we cannot apply the semi-classical expansion, but the standard perturbative calculation of the scattering amplitude can be performed [519]. There are two graphs that contribute at the Born approximation, one with a massless scalar exchange (recall that in the Bogomol'nyi limit, the scalar field is massless), and one with a photon exchange. The result of the calculation shows that there is an exact cancellation of both forces in the static limit.

One could imagine the duality transformation being the rotation on a complex plane of charges $q + ig$ (see Fig. 8.1)

Note that all the states, both from the soliton and the perturbative spectrum, have found a place on this plane; particles that correspond to the fundamental quanta in the electric formulation are placed on the real axis,

8.1 $SU(N)$ Monopoles

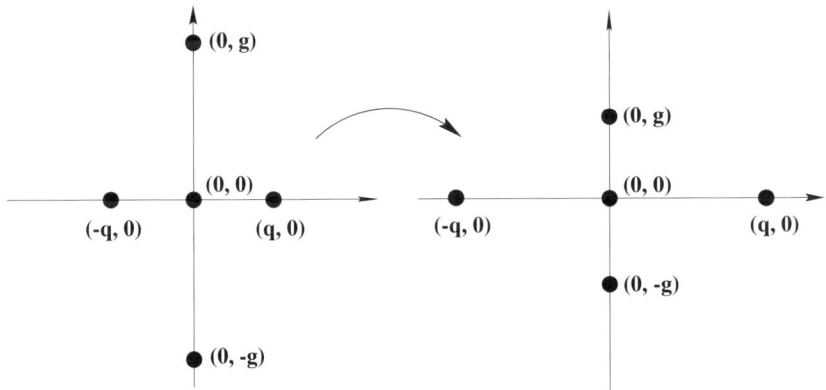

Fig. 8.1. Transformations of duality as rotations in a complex charge plane

and monopoles are placed on the imaginary axes[3]. Because all these states saturate the Bogomol'nyi bound (5.60), the mass of any state in units of the vacuum value of the scalar field in Fig. 8.1 can be defined as a length of the radius-vector. Thus, Montonen–Olive duality corresponds to the simple $SO(2)$ rotation on such a plane by the angle $\theta = \pi/2$. We have already discussed this discrete transformation in Sect. 1.6.

There is no doubt that the Montonen–Olive conjecture anticipated the modern development decades ago. However, at the end of 70s it caused a lot of objections, well-known to the authors of the paper [384] themselves. The most important in this list of questions are the following

- The gauge bosons have a spin 1. However, the rotational invariance of the 't Hooft–Polyakov solution means that the monopoles are scalars. If the duality conjecture is correct, what about the monopole spin?
- Which role do the dyons play in this picture?
- What do we have to do about the Coleman–Weinberg effect? Even if the potential of the model is vanishing on a classical level, quantum corrections could generate a non-zero effective potential of the scalar field. Therefore, there is no obvious reason to believe that the dual invariance manages to survive in a quantum theory.
- What are the consequences of the dual invariance? Can the Noether theorem be applied in this case?

We will see that the self-consistent solution of all these problems about Montonen–Olive duality can be obtained within the framework of teh supersymmetrical Yang–Mills theory. As for the last question, the answer is rather

---

[3] We do not consider dyons here because, on a classical level, their electric charge is not quantized. However, in quantum theory we shall consider Fig. 2.1 instead of Fig. 8.1.

non-trivial. It looks like it is impossible to consider the duality as a standard symmetry at all, because it is not a symmetry of a single theory, but rather a way to connect two different theories that describe weak and strong coupling constant regimes, respectively.

Thus, if we suppose that the group of gauge symmetry of the original theory is G and that a dual theory will be invariant with respect to a dual group G^*, there has to be a way to establish a correspondence between these two groups by making use of some transformation of duality. This approach was formulated in the paper [241] and related to transformations of the Cartan–Weyl basis of G in the space of simple roots.

8.1.4 Cartan–Weyl Basis and the Simple Roots

The discussion of the properties of monopoles in a gauge theory of higher rank is closely connected with notion of the Cartan–Weyl basis [11, 241]. Let us introduce some notations for the Lie algebra of an arbitrary[4], simple Lie group of the rank $r > 1$ and the dimension d. The Cartan–Weyl basis is constructed by addition $d - r$ raising and lowering generators $E_{\vec{\beta}}$ to the r commuting generators $\vec{H}$ of the diagonal Cartan subalgebra, each for one of the *roots* $\vec{\beta}_i = (\vec{\beta}_1, \vec{\beta}_2, \ldots \vec{\beta}_{d-r})$:

$$[H_i, E_{\vec{\beta}}] = \beta_i E_{\vec{\beta}}; \qquad [E_{\vec{\beta}}, E_{-\vec{\beta}}] = 2\vec{\beta} \cdot \vec{H} . \qquad (8.11)$$

Here we make use of the internal (vector) product operation in r dimensional Euclidean *root space* $\mathbb{R}^r$.

The advantage of this approach is to put a simple Euclidean geometry into correspondence to the algebra of Lie group generators. The roots $\vec{\beta}_i$ correspond to the structure constants of a Lie group. These roots, being considered as vectors in $\mathbb{R}^r$, form a lattice with the following properties [11, 515].

- A semisimple Lie algebra corresponds to every root system.
- The set of roots $\vec{\beta}_i$ is finite, it spans the entire space $\mathbb{R}^r$ and does not contain zero elements.
- If $\vec{\beta}$ and $\vec{\alpha}$ are the roots, the quantity $2\vec{\beta} \cdot \vec{\alpha}/\vec{\beta}^2$ is an integer number.
- If $\vec{\alpha}$ is a root, the only multiplies of $\vec{\alpha}$ that are roots are $\pm \vec{\alpha}$
- For a root $\vec{\beta}$ from the set $\vec{\beta}_i$ and an arbitrary positive root $\vec{\alpha} \neq \vec{\beta}$, the Weyl transformation is defined as

$$\vec{\beta} \cdot \vec{\sigma}(\vec{\alpha}) = -\vec{\beta} \cdot \vec{\alpha}, \quad \text{where} \quad \vec{\sigma}(\vec{\alpha}) = \vec{\alpha} - 2\vec{\beta}\frac{\vec{\beta} \cdot \vec{\alpha}}{\vec{\beta} \cdot \vec{\beta}} . \qquad (8.12)$$

---

[4] Recall, that in the case under consideration $G = SU(N)$, thus the rank of G is $r = N - 1$ and $d = N(N^2 - 1)$, that is $d - r = N(N - 1)$.

The set of roots is invariant with respect to this transformation. Geometrically, the Weyl transformations is a reflection in the hyperplane orthogonal to $\vec{\beta}$.

For non-simple Lie algebra, the roots are split into sets that are orthogonal to each other. Thus, it is sufficient to restrict the consideration to the case of simple Lie groups.

The third property essentially restricts the ambiguities with the choice of the root vectors. Indeed, if $\vec{\alpha}$ and $\vec{\beta}$ are any two roots with $\vec{\alpha}^2 \leq \vec{\beta}^2$, then the angle γ between these vectors is no longer arbitrary, because we have

$$\cos\gamma = \pm\frac{n}{2}\frac{|\vec{\beta}|}{|\vec{\alpha}|}, \qquad n \in \mathbb{Z}. \tag{8.13}$$

This is possible only if (i) $\vec{\beta}^2 = \vec{\alpha}^2$, (ii) $\vec{\beta}^2 = 2\vec{\alpha}^2$, and (iii) $\vec{\beta}^2 = 3\vec{\alpha}^2$. Therefore, the root diagram for a simple Lie group consists of the vectors of different lengths with possible values of the angle between these vectors $\pi/6$, $\pi/4$, $2\pi/3$, π, $3\pi/4$, and $5\pi/6$.

Note that all these roots can be separated into positive and negative ones, according to the sign in (8.13). One can choose a suitable basis that spans the root system in such a way that any root $\vec{\beta}_i$ can be represented as a linear combination of *simple roots* with integer coefficients of the same sign, positive or negative. Thus, the commutative relations of the algebra are determined by the system of the corresponding simple roots.

The properties of simple roots can be depicted graphically in the form of a flat graph[5] as Fig. 8.2. The circles here depict the simple roots. For any pair of simple roots, we have $\vec{\beta}_i \cdot \vec{\beta}_j \leq 0$. Therefore, there are four possibilities for the angle between the simple roots: $\gamma = \pi/2$ (no lines on the graph), $2\pi/3$ (one line), $3\pi/4$ (two lines) and $5\pi/6$ (three lines). The sign ">" indicates the length of the simple roots, on one side of it they are $\sqrt{2}$ times longer than on the other.

Note that the properties of the simple root basis is related with the symmetry of the model. If the group of symmetry G is broken down to the maximal Cartan subalgebra, the choice of the simple root basis is unique. However, in the case of non-maximal symmetry breaking, an alternative basis may be obtained by action of the Weyl reflection. This reflection is actually a global gauge transformation from an unbroken non-Abelian subgroup.

A particular choice of the simple roots basis can be specified by means of a vector $\vec{h}$ that lies on the root lattice. If this vector is not orthogonal to any of the simple roots $\vec{\beta}_i$, the basis is fixed by the condition $\vec{h} \cdot \vec{\beta}_i > 0$.

---

[5] It is hard to resist the temptation to quote V.I. Arnold who coined a very nice comment concerning the origination of the related terminology: *"Diagrams of this kind were certainly used by Coxeter and Witt, that is why they are usually called Dynkin diagrams"* [77].

284   8 Monopoles Beyond $SU(2)$ Group

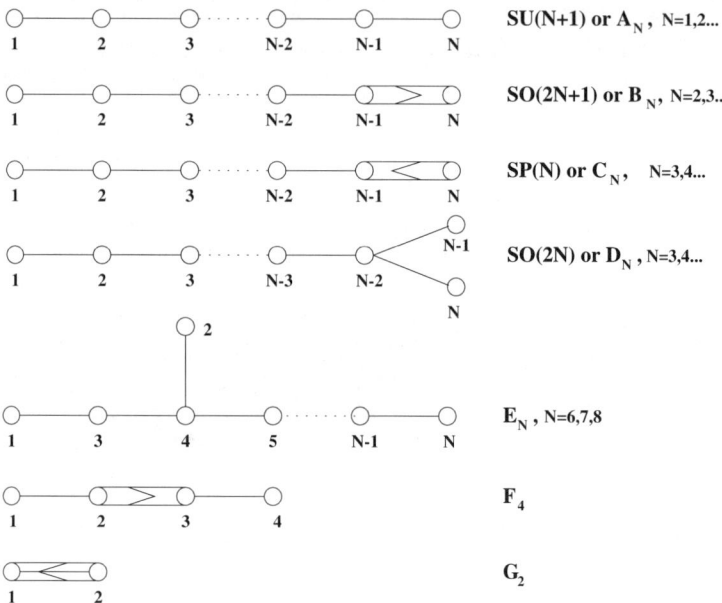

Fig. 8.2. Root diagrams

To sum up, the problem of the classification of complex simple (and hence, semi-simple) algebras is reduced to the problem of the classification of all non-splittable linearly independent r-dimensional systems of root vectors. This allows us to define a *dual* Lie algebra by means of *dual* transformation of the root lattice.

The dual of a root $\vec{\beta}$ is defined as $\vec{\beta}^* = \vec{\beta}/\vec{\beta}^2$ and the duals of the entire set of simple roots form a dual root lattice of a dual Lie group G^*. The dual lattice is isomorphic to the initial lattice. It is easy to see from the root diagram (Fig. 8.2) that, up to rescaling of the root length, the groups $SU(N)$, $SO(2N)$ and all the exceptional groups are self dual. The only non-trivial exceptions are the groups $SO(2N+1) \rightleftharpoons Sp(N)$, which are dual to each other.

We illustrate this general description on a particular example of the $SU(3)$ group below.

8.1.5 $SU(3)$ Cartan Algebra

Let us briefly review the basic elements of the $su(3)$ Lie algebra. It is given by a set of traceless Hermitian 3×3 matrices

$$T^a = \lambda^a/2, \qquad a = 1, 2 \ldots 8,$$

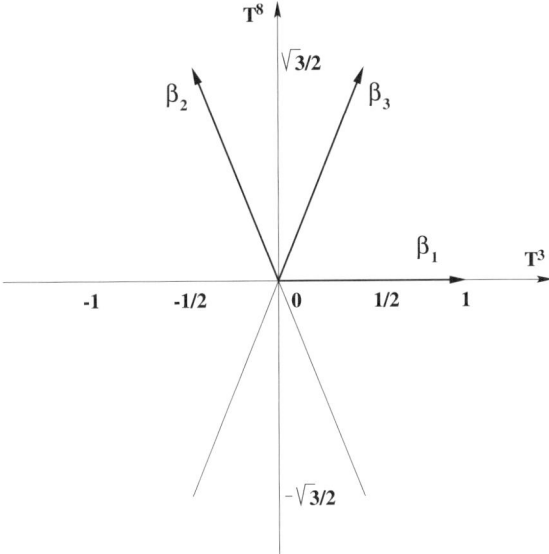

Fig. 8.3. $SU(3)$ simple root self-dual basis

where λ^a are the standard Gell-Mann matrices. Recall that they are normalized as $2 \operatorname{Tr} T^a T^b = \delta^{ab}$. The structure constants of the Lie algebra are $f^{abc} = \frac{1}{4}\operatorname{Tr}[\lambda^a, \lambda^b]\lambda^c$ and in the adjoint representation $(T^a)_{bc} = f^{abc}$.

In the following, we will be especially interested in the diagonal, or Cartan subalgebra of $SU(3)$. It is given by two generators

$$H_1 \equiv T^3 = \frac{1}{2}\begin{pmatrix} 1 & 0 & 0 \\ 0 & -1 & 0 \\ 0 & 0 & 0 \end{pmatrix}, \quad H_2 \equiv T^8 = \frac{1}{2\sqrt{3}}\begin{pmatrix} 1 & 0 & 0 \\ 0 & 1 & 0 \\ 0 & 0 & -2 \end{pmatrix}, \quad (8.14)$$

which are composed into the vector $\vec{H} = (H_1, H_2)$. Because the dimension of the group is $d = 8$, the number of positive roots is 3. Taking into account the restrictions on the angle between the vectors $\vec{\beta}_i$ and their length, we can take the basis of simple roots as (see Fig. 8.3)

$$\vec{\beta}_1 = (1, 0), \quad \vec{\beta}_2 = (-1/2, \sqrt{3}/2). \quad (8.15)$$

The third positive root is given by the composition of the first two roots $\vec{\beta}_3 = \vec{\beta}_1 + \vec{\beta}_2 = (1/2, \sqrt{3}/2)$. Since all these roots have a unit length, our choice corresponds to the self-dual basis: $\vec{\beta}_i^* = \vec{\beta}_i$. This allows us to simplify the following consideration.

Note that for any given root $\vec{\beta}_i$ the generators $\vec{\beta} \cdot \vec{H}$, $E_{\pm \beta_i}$ form an $su(2)$ algebra. The generators $E_{\pm \beta_i}$ are mentioned above the raising and lowering operators. Let us write these generators explicitly in the above-defined basis of the simple roots (8.15). For $\vec{\beta}_1$, we have

8 Monopoles Beyond $SU(2)$ Group

$$T^3_{(1)} = \vec{\beta}_1 \vec{H} = \frac{1}{2}\begin{pmatrix} 1 & 0 & 0 \\ 0 & -1 & 0 \\ 0 & 0 & 0 \end{pmatrix}, \quad (8.16)$$

$$E_{\vec{\beta}_1} = \begin{pmatrix} 0 & 1 & 0 \\ 0 & 0 & 0 \\ 0 & 0 & 0 \end{pmatrix}, \quad E_{-\vec{\beta}_1} \equiv \begin{pmatrix} 0 & 0 & 0 \\ 1 & 0 & 0 \\ 0 & 0 & 0 \end{pmatrix}.$$

For the second simple root $\vec{\beta}_2$, we have

$$T^3_{(2)} = \vec{\beta}_2 \vec{H} = \frac{1}{2}\begin{pmatrix} 0 & 0 & 0 \\ 0 & 1 & 0 \\ 0 & 0 & -1 \end{pmatrix}, \quad (8.17)$$

$$E_{\vec{\beta}_2} = \begin{pmatrix} 0 & 0 & 0 \\ 0 & 0 & 1 \\ 0 & 0 & 0 \end{pmatrix}, \quad E_{-\vec{\beta}_2} = \begin{pmatrix} 0 & 0 & 0 \\ 0 & 0 & 0 \\ 0 & 1 & 0 \end{pmatrix}.$$

The generators of the $su(2)$ subalgebra that correspond to the third composite root are given by the set of matrices

$$T^3_{(3)} = \vec{\beta}_3 \vec{H} = \frac{1}{2}\begin{pmatrix} 1 & 0 & 0 \\ 0 & 0 & 0 \\ 0 & 0 & -1 \end{pmatrix}, \quad (8.18)$$

$$E_{\vec{\beta}_3} = \begin{pmatrix} 0 & 0 & 1 \\ 0 & 0 & 0 \\ 0 & 0 & 0 \end{pmatrix}, \quad E_{-\vec{\beta}_3} = \begin{pmatrix} 0 & 0 & 0 \\ 0 & 0 & 0 \\ 1 & 0 & 0 \end{pmatrix}.$$

Clearly, the set of matrices $T^a_{(k)}$, $k = 1, 2, 3$, which includes $T^3_{(k)}$ of (8.16), (8.17) and (8.18), and

$$T^1_{(k)} = \frac{1}{2}\left(E_{\vec{\beta}_k} + E_{-\vec{\beta}_k}\right), \quad T^2_{(k)} = \frac{1}{2i}\left(E_{\vec{\beta}_k} - E_{-\vec{\beta}_k}\right) \quad (8.19)$$

satisfy the commutation relations of the $su(2)$ algebras associated with the simple roots $\vec{\beta}_1$, $\vec{\beta}_2$ and $\vec{\beta}_3$, respectively.

Let us consider the $su(2)$ subalgebra associated with the first simple root. If we supplement it by the $U(1)$ hypercharge operator, which is connected with the element of the Cartan subalgebra as

$$Y = \frac{2}{\sqrt{3}}T^8 = \frac{1}{3}\begin{pmatrix} 1 & 0 & 0 \\ 0 & 1 & 0 \\ 0 & 0 & -2 \end{pmatrix},$$

we arrive to the $u(2)$ algebra generated by operators $T^a_{(1)}, Y$. By analogy with the Euler parameterization of the $SU(2)$ group, an element of corresponding $U(2)$ transformation can be written as

$$R_{\beta_1}(\gamma,\varphi,\theta,\psi) = R_Y(\gamma)R_3(\varphi)R_2(\theta)R_3(\psi) = e^{i\gamma Y}e^{i\varphi T^3_{(1)}}e^{i\theta T^2_{(1)}}e^{i\psi T^3_{(1)}}$$

$$= \begin{pmatrix} e^{i\frac{\gamma}{3}} & 0 & 0 \\ 0 & e^{i\frac{\gamma}{3}} & 0 \\ 0 & 0 & e^{-\frac{2i\gamma}{3}} \end{pmatrix} \begin{pmatrix} \cos\frac{\theta}{2}e^{\frac{i}{2}(\varphi+\psi)} & \sin\frac{\theta}{2}e^{\frac{i}{2}(\varphi-\psi)} & 0 \\ -\sin\frac{\theta}{2}e^{\frac{i}{2}(\psi-\varphi)} & \cos\frac{\theta}{2}e^{-\frac{i}{2}(\varphi+\psi)} & 0 \\ 0 & 0 & 1 \end{pmatrix},$$

(8.20)

where the angular variables are changing within the intervals $0 \le \gamma < 2\pi$, $0 \le \varphi < 2\pi$, $0 \le \theta < \pi$, and $0 \le \psi < 4\pi$. Here the points corresponding to the values γ and $\gamma + \pi$; ψ and $\psi + 2\pi$ are pairwise identified, which corresponds to the $\mathbb{Z}_2$ subgroup.

An alternative choice is

$$R_{\beta_2}(\gamma,\varphi,\theta,\psi) = e^{i\gamma Y}e^{i\varphi T^3_{(2)}}e^{i\theta T^2_{(2)}}e^{i\psi T^3_{(2)}} \quad (8.21)$$

$$= \begin{pmatrix} e^{i\frac{\gamma}{3}} & 0 & 0 \\ 0 & e^{i\frac{\gamma}{3}} & 0 \\ 0 & 0 & e^{-\frac{2i\gamma}{3}} \end{pmatrix} \begin{pmatrix} 1 & 0 & 0 \\ 0 & \cos\frac{\theta}{2}e^{\frac{i}{2}(\varphi+\psi)} & \sin\frac{\theta}{2}e^{\frac{i}{2}(\varphi-\psi)} \\ 0 & -\sin\frac{\theta}{2}e^{\frac{i}{2}(\psi-\varphi)} & \cos\frac{\theta}{2}e^{-\frac{i}{2}(\varphi+\psi)} \end{pmatrix}.$$

In other words, the basis of the simple roots $\vec{\beta}_1$, $\vec{\beta}_2$ corresponds to two different ways to embed the $SU(2)$ subgroup into $SU(3)$. The upper left and lower right 2×2 blocks correspond to the subgroups generated by the simple roots β_1 and β_2, respectively. The third composite root $\vec{\beta}_3$ generates the $SU(2)$ subgroup, which lies in the corner elements of the 3×3 matrices of $SU(3)$.

Note that there is also so-called maximal embedding, which is given by the set of matrices

$$\tilde{T}_1 = \frac{1}{\sqrt{2}}\begin{pmatrix} 0 & 1 & 0 \\ 1 & 0 & 1 \\ 0 & 1 & 0 \end{pmatrix}, \quad \tilde{T}_2 = \frac{1}{\sqrt{2}}\begin{pmatrix} 0 & -i & 0 \\ i & 0 & -i \\ 0 & i & 0 \end{pmatrix},$$

$$\tilde{T}_3 = \begin{pmatrix} 1 & 0 & 0 \\ 0 & 0 & 0 \\ 0 & 0 & -1 \end{pmatrix},$$

which satisfy the usual $su(2)$ algebra, as well as the relations $(\tilde{T}_i)^3 = \tilde{T}_i$. Up to unitary transformation these matrices are equivalent to the vector representation of $SU(2)$. A very detailed analysis of the corresponding monopole solutions is presented in [341]. We shall not consider the maximal embedding here.

8.1.6 $SU(3)$ Monopoles

We consider the Yang–Mills–Higgs system, that is governed by the Lagrangian (8.3) with a gauge group $SU(3)$, as an explicit example of construction

8 Monopoles Beyond $SU(2)$ Group

of the monopole solutions in a model with a large symmetry group. Thus, the Higgs field $\phi = \phi^a T^a$ is taken in the adjoint representation of $SU(3)$, which is given by the set of Hermitian matrices T^a.

Unlike the original $SU(2)$ 't Hooft–Polyakov monopole solution, the vacuum manifold $\mathcal{M}$ of the $SU(3)$ Yang–Mills–Higgs theory is a sphere S^7_{vac} in eight-dimensional space. Thus, the topological classification of the solutions is related with the mapping of the spatial asymptotic S^2 onto coset space $\mathcal{M} = SU(3)/H$, where H is a residual symmetry of the vacuum. Another, not so obvious, difference is that now all the points of the vacuum manifold $\mathcal{M}$ are not identical up to a gauge transformation, because the action of the gauge group $SU(3)$ is not transitive.

Thus, in order to classify the solutions, we have to define the unbroken subgroup H. According the general relation (8.6), the asymptotic value of the scalar field in some fixed direction can be chosen to lie in the Cartan subalgebra, i.e.,

$$\phi_0 = v\vec{h} \cdot \vec{H}. \tag{8.22}$$

Clearly, this is a generalization of the $SU(2)$ boundary condition $\phi_0 = v\sigma_3/2$. To fix the basis of simple roots, we suppose that all these roots have a positive inner product with $\vec{h}$.

Furthermore, if the monopole solution obeys the Bogomol'nyi equations, in the direction chosen to define ϕ_0, the asymptotic magnetic field of a BPS monopole is also of the form

$$B_n = \vec{g} \cdot \vec{H} \frac{r_n}{r^3}. \tag{8.23}$$

Here the magnetic charge $g = \vec{g} \cdot \vec{H}$ is defined as a vector in the root space [210, 241].

The principal difference from the $SU(2)$ model is that the magnetic charge and the topological charge are now no longer identical. Indeed, it is seen that the $SU(3)$ magnetic charge is labeled by two integers. The charge quantization condition may be obtained from the requirement of topological stability in the way already discussed in Chap. 5 [210, 241]. In other words, the phase factor must be restricted by

$$\exp\{ie\vec{g} \cdot \vec{H}\} = 1. \tag{8.24}$$

General solution of this equation is given by a condition that the vector charge $\vec{g}$ lies on the dual root lattice [178, 210, 241]:

$$\vec{g} = \frac{4\pi}{e} \sum_{i=1}^{r} n_i \vec{\beta}_i^* = \frac{4\pi}{e} \left(n_1 \vec{\beta}_1^* + n_2 \vec{\beta}_2^* \right) = g_1 \vec{\beta}_1^* + g_2 \vec{\beta}_2^*, \tag{8.25}$$

where n_1 and n_2 are non-negative integers, and $\vec{g}_1, \vec{g}_2$ are the magnetic charges associated with the corresponding simple roots.

Recall that a special feature of the basis of simple roots (8.15) is that it is self-dual: $\vec{\beta}_1^* = \vec{\beta}_1$; $\vec{\beta}_2^* = \vec{\beta}_2$. Thus, in terms of the explicitly defined roots (8.15), we have

$$g = \vec{g} \cdot \vec{H} = \frac{4\pi}{e} \left[\left(n_1 - \frac{n_2}{2}\right) H_1 + \frac{\sqrt{3}}{2} n_2 H_2 \right]. \tag{8.26}$$

These relations show that a magnetic charge is not so trivially quantized as in the $SU(2)$ model; the latter has a little bit too much symmetry. Thus, the question is, if both of the numbers n_1, n_2 can be set into correspondence with some topological charges. Evidently, the answer depends on the pattern of the symmetry breaking.

$SU(3) \to U(1) \times U(1)$: Maximal Symmetry Breaking

Let us consider two situations that are possible for the $G = SU(3)$ [514]. If the Higgs vector $\vec{h}$ is not orthogonal to any of the simple roots $\vec{\beta}_i$ (8.15), there is a unique set of simple roots with positive inner product with $\vec{h}$. Thus, the symmetry is maximally broken to the maximal Abelian torus $U(1) \times U(1)$ (see the root diagram of Fig. 8.4, left).

If the inner product of $\vec{h}$ and either of the simple roots is vanishing (see Fig. 8.4, right, where $\vec{h} \cdot \vec{\beta}_1 = 0$), there are two choices of the basis of simple roots with positive inner product with $\vec{h}$, which are related by Weyl reflections. We shall discuss this type of minimal symmetry breaking below.

In the case of maximal symmetry breaking, the topological consideration shows that

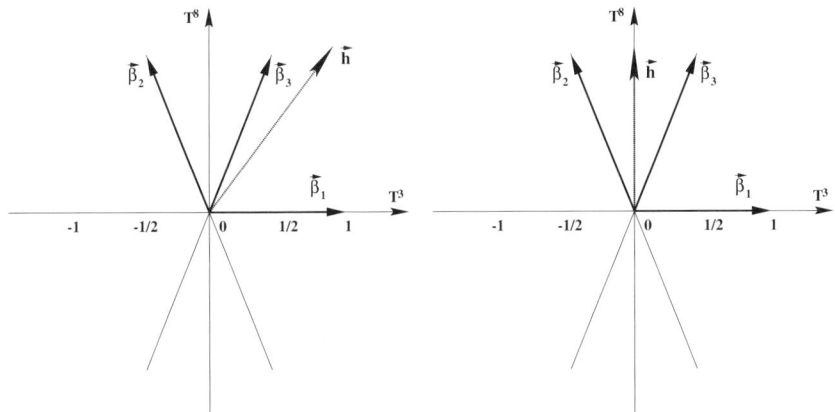

Fig. 8.4. Orientation of the Higgs field in the $SU(3)$ root space, which corresponds to the pattern of the maximal symmetry breaking $SU(3) \to U(1) \times U(1)$ (*left*) and minimal symmetry breaking $SU(3) \to U(2)$ (*right*)

$$\pi_2\left(\frac{SU(3)}{U(1)\times U(1)}\right) = \mathbb{Z}_2. \tag{8.27}$$

Thus, both of the numbers n_1, n_2 have the meaning of topological charges. Indeed, we define a magnetic charge as a winding number given by the mapping from a loop in an arbitrary Lie group into the circle on the spatial asymptotic [355]. Then the topological charge of a non-Abelian monopole is given by the integral over the surface of sphere S^2 (cf. the definition (5.46))

$$G = \frac{1}{v}\int dS_n \mathrm{Tr}(B_n\phi_0) = \vec{g}\cdot\vec{h}. \tag{8.28}$$

Therefore, if $\vec{h}$ is orthogonal to the root $\vec{\beta}_1$, only one component of $\vec{g}$ may be associated with the topological charge. Otherwise, there are two topological integers that are associated with a monopole.

The definition of topological charge (8.28) can be used to generalize the Bogomol'nyi bound (5.60) for the $SU(N)$ monopoles. If we do not consider degrees of freedom that are related with electric charges of the configuration, it becomes

$$M = v|G| = \frac{4\pi v}{e}\sum_{i=1}^{r} n_i\left(\vec{h}\cdot\vec{\beta}_i^*\right) = \sum_{i=1}^{r} n_i M_i, \tag{8.29}$$

where $M_i = \frac{4\pi v}{e}\vec{h}\cdot\vec{\beta}_i^*$ and we suppose that the orientation of the Higgs field uniquely determines a set of simple roots that satisfies the condition $\vec{h}\cdot\vec{\beta}_i^* \geq 0$ for all i. Thus, it looks like there are r individual monopoles of masses M_i.

Moreover, there is an obvious analogy between the relation (8.25) and the definition of a magnetic charge of a multimonopole configuration of the $SU(2)$ model, which is given by the sum over separate monopoles with a minimal charge. Thus, the question arises, if the monopole solutions of a higher rank gauge theory may also be understood as a composite system of a few single monopoles with a minimal charge, masses M_i and characteristic sizes of cores

$$R_c^i \sim (ve\vec{h}\cdot\vec{\beta}_i)^{-1} \tag{8.30}$$

correspondingly. A very strong argument in support of this conclusion is given by a direct calculation of the number of zero modes on the monopole background [513–515].

To analyse the situation better, let us return to a system of spherically symmetric $SU(3)$ monopoles in the basis of simple roots (8.15). Such a configuration can be constructed by a simple embedding [87, 177, 341]. The recipe is obvious: we have to choose one of the simple roots having a positive inner product with the scalar field, e.g., $\vec{\beta}_1$, and embed the 't Hooft–Polyakov solution into the corresponding $SU(2)$ subgroup. For example, embedding into the left upper corner $SU(2)$ subgroup defines the β_1-monopole that is characterised by the vector charge $\vec{g} = (1,0)$ and the mass M_1, while the embedding into the lower right corner $SU(2)$ subgroup defines the β_2-monopole with the vector charge $\vec{g} = (0,1)$ and the mass M_2. Similarly, one can embed the $SU(2)$ axially symmetric monopole-antimonopole saddle point configuration of [327, 448], which yields the state $\vec{g} = (0,0)$, or the two $SU(2)$ monopoles configuration of [507] which, depending on the root we choose, yields the states $\vec{g} = (2,0)$ or $\vec{g} = (0,2)$, respectively.

Embedding of the spherically symmetric $SU(2)$ monopole along composite root β_3 gives a $\vec{g} = (1,1)$ monopole with the magnetic charge

$$g = \vec{g} \cdot \vec{H} = \frac{1}{2} H_1 + \frac{\sqrt{3}}{2} H_2.$$

Moreover, its mass is equal to the sum of masses of the β_1-monopole and β_2-monopole: $M_1 + M_2$.

The analysis based on the index theorem shows [514] that this configuration is a simple superposition of two other fundamental solutions and can be continuously deformed into a solution that describes two well-separated single β_1 and β_2 monopoles. We shall check this conclusion by making use of another arguments below. Note that if the Higgs field is oriented along the composite root, i.e., if $\vec{h} = \vec{\beta}_3$, two fundamental BPS monopoles have the same mass:

$$M_1 = M_2 = \frac{2\pi}{e},$$

which is half of the mass of the β_3 monopole. In all other cases, this degeneration is lifted and one of the monopoles is heavier than the other one.

Spherically Symmetric $SU(3)$ Non-BPS Monopoles

To construct the embedded monopoles, we must take into account that the generators $T^a_{(i)}$ of an $SU(2)$ subgroup commute with the invariant component of the Higgs field

$$\phi^{(h)} = \left(\vec{h} - \frac{\vec{h} \cdot \vec{\beta}_i}{\beta^2} \vec{\beta}_i\right) \vec{H}, \qquad [T^a_{(i)}, \phi^{(h)}] = 0.$$

Thus, an embedded $SU(2)$ monopole is defined as [87]:

$$A_n = A_n^a T^a_{(i)}, \qquad \phi = \phi^a T^a_{(i)} + v\phi^{(h)}. \tag{8.31}$$

The additional invariant term $\phi^{(h)}$ is added to the Higgs field to satisfy the boundary conditions on the spatial asymptotic. In our basis of the simple roots, we can write

$$\vec{\beta}_1 : \quad \phi^{(h)} = \frac{h_2}{2\sqrt{3}} \begin{pmatrix} 1 & 0 & 0 \\ 0 & 1 & 0 \\ 0 & 0 & -2 \end{pmatrix},$$

$$\vec{\beta}_2 : \quad \phi^{(h)} = \frac{1}{4}\left(h_1 + \frac{h_2}{\sqrt{3}}\right) \begin{pmatrix} 2 & 0 & 0 \\ 0 & -1 & 0 \\ 0 & 0 & -1 \end{pmatrix}, \quad (8.32)$$

$$\vec{\beta}_3 : \quad \phi^{(h)} = \frac{1}{4}\left(h_1 - \frac{h_2}{\sqrt{3}}\right) \begin{pmatrix} 1 & 0 & 0 \\ 0 & -2 & 0 \\ 0 & 0 & 1 \end{pmatrix}.$$

Clearly, the embedding (8.31) is very convenient for obtaining spherically symmetric monopoles [515]. It is also helpful for examing the fields and low-energy dynamics of the charge two BPS monopoles [295]. Depending on the boundary conditions and pattern of the symmetry, some other ansätze can be implemented to investigate static monopole solutions, such as, for example, the harmonic map ansatz [294] that was used to construct non-Bogomol'nyi $SU(N)$ BPS monopoles.

In our consideration [467], which is not restricted to the case of BPS limit, we shall consider ansätze for the Higgs field of a spherically symmetric β_i monopole configuration. Depending on the way of the $SU(2)$-embedding, it can be taken[6] as a generalization of the the embedding (8.31)

$$\vec{\beta}_i : \quad \phi(r) = \Phi_1(r)\tau_r^{(i)} + \frac{\sqrt{3}}{2}\Phi_2(r)D^{(i)},$$
$$A_r = 0; \quad A_\theta = [1 - K(r)]\tau_\varphi^{(i)}, \quad A_\varphi = -\sin\theta[1 - K(r)]\tau_\theta^{(i)}, \quad (8.33)$$

where $i = 1, 2, 3$, and we make use of the $su(2)$ matrices $\tau_r^{(i)} = \left(T^a_{(i)}\hat{r}^a\right)$, $\tau_\theta^{(i)} = \left(T^a_{(i)}\hat{\theta}^a\right)$ and $\tau_\varphi^{(i)} = \left(T^a_{(i)}\hat{\varphi}^a\right)$. The diagonal matrices $D^{(i)}$, which define the embedding along the corresponding simple root, are just the $SU(3)$ hypercharge

$$D^{(1)} \equiv Y = \frac{2}{\sqrt{3}}H_2 = \frac{1}{3}\operatorname{diag}(1, 1, -2), \quad (8.34)$$

the $SU(3)$ electric charge operator

$$D^{(2)} \equiv Q = T^3 + \frac{Y}{2} = H_1 + \frac{H_2}{\sqrt{3}} = \frac{1}{3}\operatorname{diag}(2, -1, -1), \quad (8.35)$$

---

[6] The first of these ansätze (in a different basis of the simple roots) was already used in [142, 341].

and its conjugated operator

$$D^{(3)} \equiv \tilde{Q} = T^3 - \frac{Y}{2} = \frac{1}{3}\text{diag}(1, -2, 1).$$

The normalization of the ansätze (8.33) corresponds to the $su(3)$-norm of the Higgs field $|\phi|^2 = \Phi_1^2 + \Phi_2^2$ for any embedding.

Inserting the ansatz (8.33) into the rescaled Lagrangian (8.3), we can obtain the variational equations in terms of the profile functions:

$$0 = \partial_r^2 K - \frac{K(K^2 - 1)}{r^2} - \Phi_1^2 K = 0,$$

$$0 = 2\Phi_1 K^2 + 4\lambda r^2 \Phi_1(\Phi_1^2 + \Phi_2^2 - 1) - r^2\partial_r^2\Phi_1 - 2r\partial_r\Phi_1, \quad (8.36)$$

$$0 = 4\lambda r^2 \Phi_2(\Phi_1^2 + \Phi_2^2 - 1) - r^2\partial_r^2\Phi_2 - 2r\partial_r\Phi_2.$$

Clearly, these equations are identical for any $SU(2)$ embedding. However, the boundary conditions that we have to impose on the Higgs field, depend on the type of the embedding.

Let us consider the behavior of the scalar field of the configurations (8.33) along the positive direction of the z-axis. We obtain

$$\vec{\beta}_1: \quad \phi(r,\theta)\Big|_{\theta=0} = \Phi_1 H_1 + \Phi_2 H_2 = (\vec{h}\cdot\vec{H}),$$

$$\vec{\beta}_2: \quad \phi(r,\theta)\Big|_{\theta=0} = \frac{1}{2}\left[(\sqrt{3}\Phi_2 - \Phi_1)H_1 + (\sqrt{3}\Phi_1 + \Phi_2)H_2\right] = (\vec{h}\cdot\vec{H}),$$

$$\vec{\beta}_3: \quad \phi(r,\theta)\Big|_{\theta=0} = \frac{1}{2}\left[(\sqrt{3}\Phi_2 + \Phi_1)H_1 + (\sqrt{3}\Phi_1 - \Phi_2)H_2\right] = (\vec{h}\cdot\vec{H}).$$

This yields the components of the vector $\vec{h}$, which determines the nature of the symmetry breaking.

The boundary conditions that we can impose on configurations, which minimize the action (8.3) are of different types. First, the rescaled Higgs potential vanishes on the spatial asymptotic, that is, as $r \to \infty$

$$|\phi|^2 = \Phi_1^2 + \Phi_2^2 = 1.$$

Second, the inner product of the vector $\vec{h}$ with all roots has to be non-negative for any embedding. This yields

$\vec{\beta}_1:\quad (\vec{\beta}_1 \cdot \vec{h}) = \Phi_1 \geq 0,$

$$(\vec{\beta}_2 \cdot \vec{h}) = -\frac{\Phi_1}{2} + \frac{\sqrt{3}}{2}\Phi_2 \geq 0; \quad (\vec{\beta}_3 \cdot \vec{h}) = \frac{\Phi_1}{2} + \frac{\sqrt{3}}{2}\Phi_2 \geq 0,$$

$\vec{\beta}_2:\quad (\vec{\beta}_1 \cdot \vec{h}) = -\frac{\Phi_1}{2} + \frac{\sqrt{3}}{2}\Phi_2 \geq 0,$

$$(\vec{\beta}_2 \cdot \vec{h}) = \Phi_1 \geq 0, \quad (\vec{\beta}_3 \cdot \vec{h}) = \frac{\Phi_1}{2} + \frac{\sqrt{3}}{2}\Phi_2 \geq 0, \tag{8.37}$$

$\vec{\beta}_3:\quad (\vec{\beta}_1 \cdot \vec{h}) = \frac{\Phi_1}{2} + \frac{\sqrt{3}}{2}\Phi_2 \geq 0,$

$$(\vec{\beta}_2 \cdot \vec{h}) = \frac{\Phi_1}{2} - \frac{\sqrt{3}}{2}\Phi_2 \geq 0, \quad (\vec{\beta}_3 \cdot \vec{h}) = \Phi_1 \geq 0.$$

Third, the covariant derivatives of the Higgs field have to vanish at spatial infinity, that is

$$D_r \phi = r\partial_r \Phi_1 \tau_r^{(i)} + \frac{\sqrt{3}}{2}\partial_r \Phi_2 D^{(i)} = 0,$$
$$D_\theta \phi = (K-1)\Phi_1 \tau_\theta^{(i)} = 0, \tag{8.38}$$
$$D_\varphi \phi = \sin\theta (K-1) K \Phi_1 \tau_\varphi^{(i)} = 0.$$

Finally, the solution has to be regular at the origin. The condition on the short distance behavior implies

$$K(r) \to 1, \quad \Phi_1(r) \to 0, \quad \partial_r \Phi_2(r) \to 0,$$

as $r \to 0$. The energy density also goes to 0 in this limit.

The solution of (8.36) becomes very simple in the BPS limit. Then the third equation is decoupled and its solution, which is regular at the origin, is just a constant $\Phi_2 = C$ where $C \in [0; 1]$. The shape functions of the scalar and gauge field are well-known rescaled Bogomol'nyi solutions with a long-range field Φ_1

$$K(r') = \frac{r'}{\sinh r'}, \quad \Phi_1(r') = \sqrt{1-C^2}\coth r' - \frac{1}{r'}, \tag{8.39}$$

where $r' = r\sqrt{1-C^2}$.

For a non-zero scalar coupling λ, the system of equations (8.36) may be solved numerically for the range of values of vacuum expectation values $(\Phi_2)_{vac}$. According to the boundary conditions (8.37), the increasing constant $(\Phi_2)_{vac}$ results in rotation of the vector $\vec{h}$ in the root space, as shown in Fig. 8.5. However, for a single fundamental $\vec{\beta}_i$ monopole, $(\vec{h} \cdot \vec{\beta}_i) \geq 0$ if $(\Phi_2)_{vac} \geq 1/2$, and in this case $(\Phi_2)_{vac}$ have to be restricted as $(\Phi_2)_{vac} \in [1/2; 1]$, whereas for a configuration embedded along the composite root $\vec{\beta}_3$, we have $(\Phi_2)_{vac} \leq 1/2$ [467].

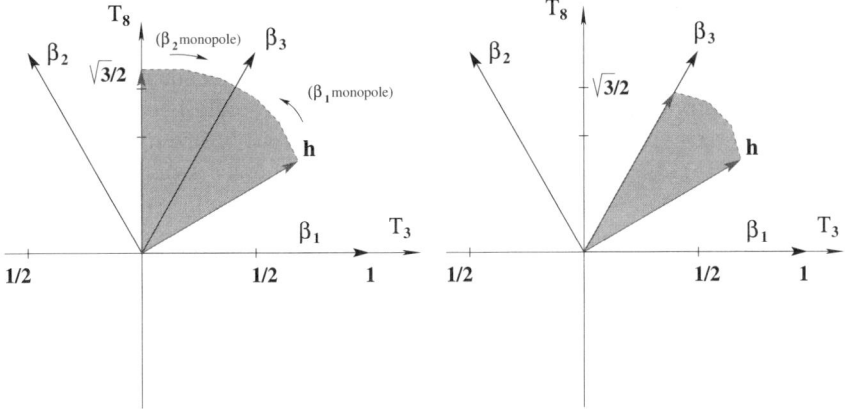

Fig. 8.5. Domains of rotations of the vector $\vec{h}$ for fundamental and composite $SU(3)$ monopoles

The physical meaning of the third of the ansätze for the scalar field (8.33) becomes clearer, if we note that on the spatial asymptotic this configuration really corresponds to the Higgs field of two distinct fundamental monopoles, $(1,0)$ and $(0,1)$, respectively. Indeed, outside of cores of these monopoles in the Abelian gauge, the scalar field can be written as the superposition:

$$\phi(r \to \infty) = v_1 T^3_{(1)} + v_2 T^3_{(2)} = \frac{1}{2}\begin{pmatrix} v_1 & 0 & 0 \\ 0 & v_2 - v_1 & 0 \\ 0 & 0 & -v_2 \end{pmatrix},$$

where the Higgs field of the β_1 and β_2 monopoles takes the vacuum expectation values v_1, v_2 respectively.

Rotation of this configuration by the matrices of the $SU(2)$ subgroup, which is defined by the third composite root $\vec{\beta}_3$

$$U = \begin{pmatrix} \cos\frac{\theta}{2} & 0 & \sin\frac{\theta}{2}e^{-i\phi} \\ 0 & 1 & 0 \\ -\sin\frac{\theta}{2}e^{i\phi} & 0 & \cos\frac{\theta}{2} \end{pmatrix},$$

yields

$$U^{-1}\phi U = \frac{1}{2}[v_1 + v_2]\tau_r^{(3)} + \frac{3}{4}[v_1 - v_2]\tilde{Q}. \tag{8.40}$$

Up to the obvious reparameterization of the shape functions of the scalar field

$$\Phi_1 \to \frac{1}{2}[F_1(r) + F_2(r)], \qquad \Phi_2 \to \frac{\sqrt{3}}{2}[F_1(r) - F_2(r)], \tag{8.41}$$

where the functions F_1, F_2 have the vacuum expectation values v_1, v_2, respectively, the configuration (8.41) precisely corresponds to the third of the ansätze (8.33). Because the $su(3)$-norm of the scalar field is set to be unity, the vacuum values must satisfy the condition $v_1^2 + v_2^2 - v_1 v_2 = v$.

Moreover, the reparameterization (8.41) allows us to write the scalar field of the β_3 monopole along positive direction of the z-axis as

$$\vec{\beta}_3: \quad \phi(r \to \infty, \theta)\Big|_{\theta=0} = \left(v_1 - \frac{v_2}{2}\right) H_1 + \frac{\sqrt{3}}{2} v_2 H_2$$
$$= (v_1 \vec{\beta}_1 + v_2 \vec{\beta}_2) \cdot \vec{H} = (\vec{h} \cdot \vec{H}),$$

and we conclude that the asymptotic values v_1 and v_2 are the coefficients of the expansion of the vector $\vec{h}$ in the basis of the simple roots and on the spatial asymptotic, the fields $F_1(\vec{\beta}_1 \cdot \vec{H})$ and $F_2(\vec{\beta}_2 \cdot \vec{H})$ can be identified with the Higgs fields of the first and second fundamental monopoles, respectively.

Thus, the embedding along the composite simple root $\vec{\beta}_3$ gives two fundamental monopoles, which in the case of maximal symmetry breaking, are charged with respect to different $U(1)$ subgroups and are on top of each other. The configuration with minimal energy corresponds to the boundary condition $(\Phi_1)_{vac} = 1, (\Phi_2)_{vac} = 0$. We can interpret it by making use of (8.41), as two identical monopoles of the same mass. This degeneration is lifted as the value of the constant solution $\Phi_2 = C$ increases, the vector of the Higgs field $\vec{h}$ smoothly rotates in the root space and the boundary conditions begin to vary.

According to the parameterization (8.41), increasing of $(\Phi_2)_{vac}$ results in the splitting of the vacuum values of the scalar fields of the first and second fundamental monopoles; the β_1-monopole is becoming heavier than the β_2-monopole. The maximal vacuum expectation value of the second component of the Higgs field of the β_3-monopole is $(\Phi_2)_{vac} = 1/2$ or $v_2 = v_1/2$. This is a border value which, according to (8.37), separates the composite β_3-monopole from a single fundamental β_i-monopole, for which $(\vec{h} \cdot \vec{\beta}_i) \geq 0$ if $(\Phi_2)_{vac} \geq 1/2$.

As $(\Phi_2)_{vac}$ varies from $(\Phi_2)_{vac} = 1/2$ to $(\Phi_2)_{vac} = 1$, the vector $\vec{h}$ rotates clockwise for the β_2-monopole and anti-clockwise for the β_1-monopole within the same domain of the root space (Fig. 8.5, left). The configuration smoothly moves to the limit $(\Phi_2)_{vac} = 1$ when the vector $\vec{h}$ becomes orthogonal to one of the simple roots. The numerical solution of the system of equations (8.36) is displayed in Figs. 8.6, and 8.7.

Let us consider the behavior of a single fundamental monopole solution as the vacuum expectation value $(\Phi_2)_{vac}$ approaches this limit [341, 467]. Then, the "hedgehog" component $(\Phi_2)_{vac}$ tends to vanish and the monopole core spreads out as $(\Phi_2)_{vac}$ is approaches the limit $C = 1$. This is the case of minimal symmetry breaking.

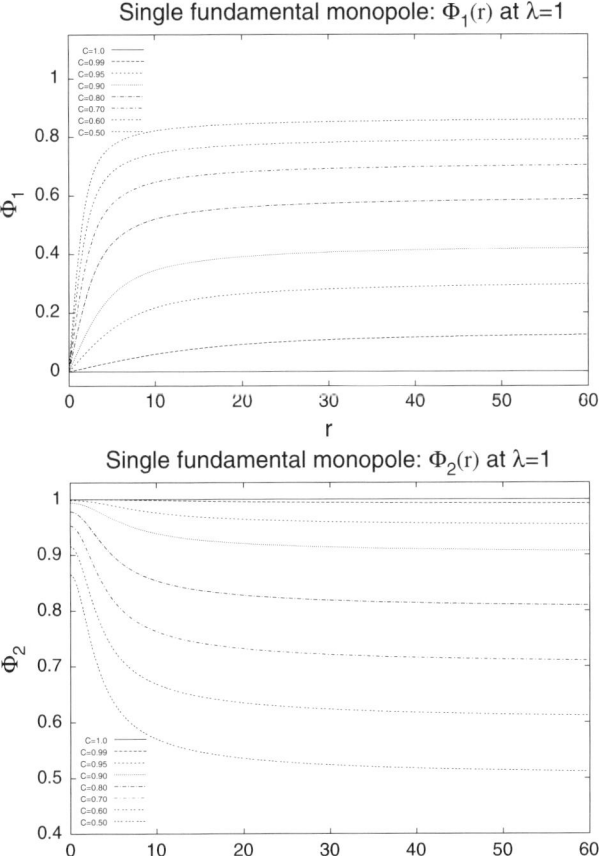

Fig. 8.6. Structure functions of the Higgs field components $\Phi_1(r)$ and $\Phi_2(r)$ of the single fundamental monopole with different vacuum expectation values $(\Phi_2)_{vac} = C$ at $\lambda = 1$

$SU(3) \to U(2)$: Minimal Symmetry Breaking

Let us analyze what happens if the scalar field becomes orthogonal to one of the simple roots. Suppose, for example, that $\vec{h} \cdot \vec{\beta}_1 = 0$, that is, $\vec{h} = (0, 1)$. Then, as $r \to \infty$,

$$\phi \to \phi_0 = vH_2 = \frac{v}{2\sqrt{3}} \begin{pmatrix} 1 & 0 & 0 \\ 0 & 1 & 0 \\ 0 & 0 & -2 \end{pmatrix}. \tag{8.42}$$

As we have already mentioned, in this case, two eigenvalues of the mass matrix coincide. Clearly, the mass matrix then commutes with the generators of $SU(2)$ subalgebra $T^a_{(1)}$, which correspond to the $\vec{\beta}_1$ simple root and mixes the degenerated eigenvalues of ϕ_0:

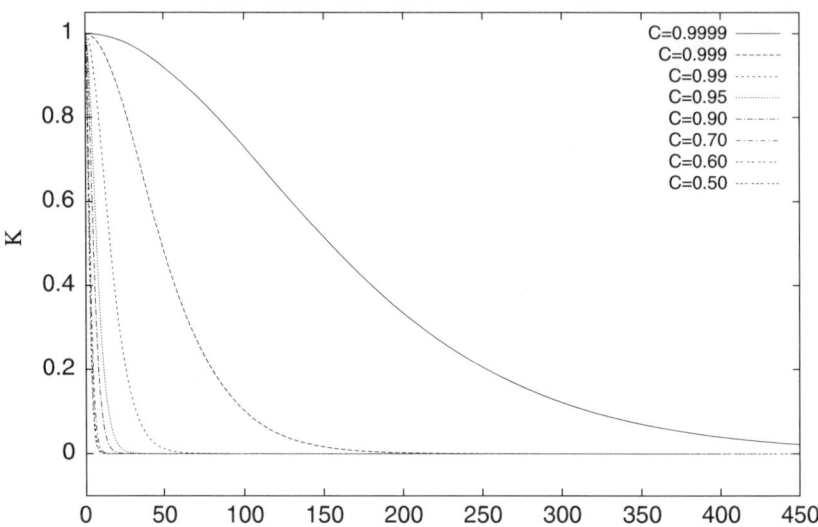

Fig. 8.7. Structure function of the gauge field $K(r)$ of the single fundamental monopole for different vacuum expectation values $(\Phi_2)_{vac} = C$ at $\lambda = 1$

$$[\phi_0, T^a_{(1)}] = 0. \tag{8.43}$$

Furthermore, there is the $U(1)$ invariant subgroup. Indeed, the diagonal matrix ϕ_0 (8.42), as before, commutes with the electric charge operator Q.

Let us comment on the last statement. Recall that the electromagnetic subgroup of $SU(3)$ is not just one of the Abelian subgroups generated by the elements of Cartan subalgebra $\vec{H}$. The electric charge operator Q is defined by (8.35) as $Q = H_1 + H_2/\sqrt{3} = \mathrm{diag}(2/3, -1/3, -1/3)$ and the eigenvalues of the matrix Q correspond to the electric charges of the fundamental $SU(3)$ triplet (quarks). Thus, the electromagnetic subgroup of spontaneously broken $SU(N)$ theory is compact. Indeed, the elements of this subgroup are given by $U = e^{i\alpha Q}$, and there are two points of the group manifold parameterized by the angles α and $\alpha + 2\pi N$, where $N = 3$, which are identical[7].

However it would be not correct to conclude that the invariant subgroup H of the minimally broken $SU(3)$ model is a direct product $SU(2) \times U(1)$. This is correct only with respect to the local structure of H, because the

---

[7] There is a principal difference between the $SU(3)$ gauge theory and the $SU(2) \times U(1)$ unified model of electroweak interaction. In the latter, the electric charge operator is defined as a linear combination

$$Q_{ew} = \sin\theta_W T^3 + \cos\theta_W Y, \qquad \sin^2\theta_W = 0.230,$$

i.e., the electromagnetic subgroup of the Standard Model is non-compact. Therefore, there is no topologically stable monopole solution within electroweak theory.

transformation of the electromagnetic $U(1)$ subgroup generated by the electric charge operator contains the elements of the center $\mathbb{Z}_2 = [-1, 1]$ of the $SU(2)$ subgroup:

$$e^{3\pi i Q} = \begin{pmatrix} 1 & 0 & 0 \\ 0 & -1 & 0 \\ 0 & 0 & -1 \end{pmatrix}.$$

Hence, the group of residual symmetry of the vacuum is $H = SU(2) \times U(1)/\mathbb{Z}_2 \approx U(2)$, and there are two different classes of the topologically non-trivial paths in H: the closed contours that encircle the $U(1)$ subgroup of H and the loops, which are traveling from the identity to the element of center $\mathbb{Z}_2$ through the $U(1)$ subgroup, and back to the identity through the $SU(2)$ subgroup [55]. The monopoles with a minimal $U(1)$ magnetic charge correspond to the contour travel only half-way around the $U(1)$ subgroup, from the identity to the unit element of the center of $SU(2)$. Such a monopole has a non-Abelian $\mathbb{Z}_2$ charge, as well as a non-Abelian $SU(2)$ charge [55].

Thus, unlike (8.27), the second homotopy group for minimal symmetry breaking is

$$\pi_2 \left(\frac{SU(3)}{U(2)} \right) = \mathbb{Z},$$

and there is only one topological charge. Indeed, for given orientation of the Higgs field in the simple root basis, the topological charge, which is defined by formula (8.28), becomes

$$G = \vec{g} \cdot \vec{h} = \frac{4\pi}{e} \left(n_1 \vec{\beta}_1^* + n_2 \vec{\beta}_2^* \right) \cdot \vec{h} = n_2 \frac{4\pi}{e} \frac{\sqrt{3}}{2}.$$

Thus, only the integer n_2, which corresponds to the non-orthogonal to the vector $\vec{h}$ simple root $\vec{\beta}_2$, is associated with the topological charge G [488].

As was pointed out by E. Weinberg, one can understand the origin of this reduction by taking into account the residual gauge freedom, which still exists within the chosen Cartan subalgebra $\vec{H}$. The point is that the vector magnetic charge $\vec{g}$ is defined up to a transformation from the Weyl subgroup, which does not take the vacuum ϕ_0 out of the Cartan subgroup [514]. In the case of maximal $SU(3)$ symmetry breaking to $H = U(1) \times U(1)$, there is just one fixed basis of simple roots given by the vector $\vec{h}$ and both integers n_1, n_2 are topological charges (8.25). If the symmetry is broken minimally to $H = SU(2) \times U(1)$, the condition that requires the inner product of the simple roots with $\vec{h}$ to be positive, does not uniquely determine the basis of the roots. There are two possible sets related by the Weyl reflection of the root diagram that result from the global gauge transformations of the unbroken $SU(2)$ subgroup. In the case under consideration, $\vec{h}$ is orthogonal to $\vec{\beta}_1$ and we can choose between two possibilities: $(\vec{\beta}_1, \vec{\beta}_2)$ and $(\vec{\beta}_1', \vec{\beta}_2') = (-\vec{\beta}_1, \vec{\beta}_1 + \vec{\beta}_2)$.

In the alternative self-dual basis, the vector magnetic charge reads

$$\vec{g} = \frac{4\pi}{e} \left(n_1' \vec{\beta}_1' + n_2' \vec{\beta}_2' \right), \tag{8.44}$$

where $n'_1 = n_2 - n_1$; $n'_2 = n_2$. Therefore, only the invariant component of the vector magnetic charge $\text{Tr}(\vec{g} \cdot \vec{H}\phi_0) = 4\pi v n_2/e$, labeled by the integer n_2, has a topological interpretation. Values of another integer $n_1 = 0$ or $n_1 = 1$ are related to two possible orientations in the root space, which correspond to the two choices of the basis, or the two possible ways to embed the $SU(2)$ subgroup that we described above [514].

Indeed, in the case of minimal symmetry breaking, the magnetic charge $g = \vec{g} \cdot \vec{H}$ is no longer invariant under the transformation generated by the elements of the unbroken $U(2)$ subgroup. The explicit form of this transformation is given by the matrix R (see (8.20)). Such a transformation defines the so-called *magnetic orbit* of the charge G [89, 387]. An arbitrary point of the orbit could be obtained by the transformation

$$g \to R\vec{g} \cdot \vec{H} R^{-1} = \frac{4\pi}{e} \left[n_2 \frac{\sqrt{3}}{2} H_2 + \left(n_1 - \frac{n_2}{2} \right) \hat{r}_k T_{(1)}{}^k \right], \qquad (8.45)$$

where the components of the non-Abelian magnetic charge are connected with the unit vector

$$\hat{r}_k = (\sin\theta \cos\varphi, \sin\theta \sin\varphi, \cos\varphi),$$

which defines a sphere S^2 in the group space of $SU(3)$.

The magnetic orbit on the group space travels through the Cartan subalgebra in two different points, which are connected by the Weyl reflection. For the given topological charge n_2, each orbit is characterized by the radius $|n_1 - n_2/2|$ [89]. According to the general terminology we described above, a pair of integers $n_1, n_2 - n_1$ have to be connected with *magnetic weights*[8]. Configurations with a holomorphic charge $[n_1] = 0$ correspond to a simple embedding of a $SU(2)$ monopole configuration with an arbitrary topological charge n_2 into the $SU(3)$ group [510]. Nontrivial, spherically symmetrical solutions with $n_1 \neq 0$ were considered in [88, 229, 348, 349, 509, 526].

Thus, in the case under consideration, the fundamental monopoles are the states labeled by topological charge $n_2 = 0, 1$ and a holomorphic charge $[n_1] \in \mathbb{Z}$.

Finally, let us briefly comment on the statement we made at the beginning of this chapter (cf. (8.1) and (8.2)), concerning the possible contribution of a non-Abelian magnetic charge of a monopole to the charge quantization condition. Indeed, the first term in (8.45) corresponds to the topological charge, which is quantizable according to the charge quantization condition, whereas the second term may be associated with a non-Abelian charge of the monopole. However, this interpretation is not quite clear due to some ambiguities, which we discuss below.

---

[8] The integer n_1 is also called a *holomorphic charge* [387]. To distinguish it from a topological charge, the notation in square brackets is used. For example, a state with a unit topological charge and a unit holomorphic charge is labeled as $(1, [1])$.

8.2 Massive and Massless Monopoles

The question as to what could be the physical meaning of the components of the non-Abelian magnetic charge, which are abundant at non-maximal symmetry breaking, has been discussed for years. We refer the reader to the papers [89, 138, 387]. Rather convincing seems to be the argument that there is no topological restriction that forbids decay of an arbitrary configuration into the state with minimal energy, that is, with a minimal possible value of the non-Abelian magnetic charge [43, 138].

However, in the BPS limit this argument is no longer valid, because all the states, including the states with non-Abelian magnetic charge, correspond to the same absolute minimum of energy. Thus, we have to understand what happens with a massive state in the limiting case of minimal symmetry breaking.

Let us return to the $SU(3)$ model. We argued above that there are two different monopoles corresponding to the two simple roots $\vec{\beta}_1$ and $\vec{\beta}_2$, respectively. However, it follows from the Bogomol'nyi bound (8.29) that only one configuration with a non-zero topological charge n_2 remains massive in the minimal symmetry breaking case ($\vec{h} \cdot \vec{\beta}_1 = 0$):

$$M = \frac{4\pi v}{e}\left(n_1 \vec{\beta}_1 \cdot \vec{h} + n_2 \vec{\beta}_2 \cdot \vec{h}\right) = M_2 = n_2 \frac{\sqrt{3}}{2}\frac{4\pi v}{e}. \qquad (8.46)$$

Another monopole turns out to be massless:

$$M_1 = n_1 \frac{4\pi v}{e} \vec{\beta}_1 \cdot \vec{h} = 0.$$

This agrees with the results of our numerical calculations above; a single isolated fundamental monopole spreads out in space, its core radius increases as vector $\vec{h}$ approaches the limit where it becomes orthogonal to one of the simple roots (see Fig 8.7). The mass of such an "inflated" monopole decreases and tends to zero.

However, the pattern of symmetry breaking becomes more complicated for a β_3-monopole, which is a composite state of two fundamental monopoles on top of each other [467]. Recall that the idea is to treat minimal $SU(3)$ symmetry breaking as a special case of maximal symmetry breaking, i.e., to analyse the rotation of the vector $\vec{h}$ in the root space. Indeed, it yields the splitting of the fundamental monopole masses as $(\Phi_2)_{vac}$ increases. One would expect that in the limiting case of minimal symmetry breaking, the β_2 monopole is becoming massless, that is, in that limit the vacuum value of the field F_2 should vanish, $v_2 \to 0$. However, two monopoles are overlapped and the presence of the massive monopole changes the situation. Indeed, the symmetry outside the core of the β_1-monopole is broken down to $U(1)$, which also changes the pattern of the symmetry breaking by the scalar field of the second monopole. One can see that the vector $\vec{h}$ becomes orthogonal to the simple root $\vec{\beta}_2$, when $(\Phi_1)_{vac} = \sqrt{3}/2, (\Phi_2)_{vac} = 1/2$, or $v_2 = v_1/2$. Going back to (8.42), we can see that, in this case on the spatial asymptotic the scalar field along the z-axis is

$$\phi(r \to \infty, \theta)\Big|_{\theta=0} = \frac{3}{4} v_1 Q,$$

where Q is the electric charge matrix (8.35). Thus, the symmetry is still maximally broken and both monopoles are massive.

Equation (8.42) indicates that the symmetry is minimally broken if the vector $\vec{h}$ becomes orthogonal to the simple root $\vec{\beta}_1$ and $v_1 = v_2/2$. Then, the eigenvalues of the scalar field are the same as H_2, that is, the unbroken symmetry group is really $U(2)$. However, for the third composite root, such a situation corresponds to the negative value of the inner product $(\vec{h} \cdot \vec{\beta}_2)$ and it has to be excluded. Thus, the maximal value of the second component of the Higgs field of the β_3-monopole is $(\Phi_2)_{vac} = 1/2$.

This conclusion allows us to understand what happens if we consider two distinct fundamental monopoles well-separated by a distance $R_0 \gg R_c$. As the vector $\vec{h}$ approaches to the direction orthogonal to either of the simple roots, the core of the corresponding monopole tends to expand until its characteristic size approaches the scale of R_0 [344, 352]. At this stage, this monopole loses its identity as a localized field configuration. We have seen that, if this monopole were isolated, it would spread out and disappear, dissolving into the topologically trivial sector. However, as its core overlaps with the second massive monopole, it ceases to expand [185, 186, 295, 354].

Because at this stage the topological charge is resolved, this state is no longer a topological soliton. E. Weinberg [344, 517] suggested that such a configuration be interpreted as a "non-Abelian cloud" of characteristic size R_0, surrounding the massive monopole. The Coulomb magnetic field inside this cloud includes components that correspond to both Abelian and non-Abelian charges. However, on distances larger than R_0, only the Abelian component is presented. The zero modes, which correspond to the massless monopole, are transformed into the parameters of the non-Abelian global orientation and the parameter characterizing the radii of these clouds [295].

This situation leads to some modification of the Montonen–Olive conjecture. If the symmetry is maximally broken the formal difference from the situation that we discussed in Sect. 8.1.3 is that the spectrum of the states is redoubled: there are two self-dual massless photons corresponding to the two simple roots and two monopoles forming the dual pairs of massive monopoles and vector bosons, one for each simple root: $(M_{\beta_i} \rightleftharpoons A_{\beta_i})$. In the case of minimal symmetry breaking, the state dual to a massless monopole has to be a gauge boson with mass zero. If the symmetry is broken down to the subgroup $SU(3)$, we could say that such a massless monopole are states dual to gluons.

Because these massless monopoles are not connected with any topological charges, there is the problem of interpretation of such a state, or, generally speaking, the problem of interpretation of the magnetic weights. The answer can be obtained similarly to the case of maximal symmetry breaking, by analysis of the spectrum of the fluctuations on this background [513–515]. However, in this way, we have to confront some trouble connected with the problem of the definition of the global non-Abelian gauge transformations [62, 106, 396–398].

8.2.1 Pathologies of Non-Abelian Gauge Transformations

Let us recall that in the $SU(2)$ model with the unbroken Abelian subgroup, the gauge zero mode appears as a result of a time-dependent gauge transformation, which generates an electric charge of the configuration. In other words, an excitation of the Abelian gauge zero mode transforms a monopole into a dyon (see the related discussion in Sect. 5.2.1 above).

A naive generalization of the result for the case of the unbroken non-Abelian subgroup is that there are corresponding time-dependent gauge transformations that generate some set of non-Abelian zero modes. The excitation of these modes would transform a monopole into a non-Abelian dyon and it appears plausible that, in the case of minimal symmetry breaking, a configuration can possess both a non-Abelian magnetic and a non-Abelian electric charge.

However, this conclusion is not quite correct, because some of these non-Abelian gauge zero modes turn out to be non-normalizable excitations [62, 396]. The reason is that a global non-Abelian gauge transformation in principle cannot be well-defined in the presence of a monopole [106, 397, 398].

To see this, let us consider the above example minimal symmetry breaking of the $SU(3)$ model. For the sake of simplicity, we take the vector $\vec{h}$ to be orthogonal to the simple root $\vec{\beta}_1$ once again. The vacuum expectation value of the Higgs field ϕ_0 (8.42) is proportional to the element of the Cartan subalgebra H_2 and commutes with the generators of the $SU(2)$ subalgebra $T^a_{(1)}$, as well as with the second generator of the Cartan subalgebra H_1.

Note that there is a transformation given by the matrices R (8.20) of the unbroken $U(2)$ subgroup, which rotates the vacuum ϕ_0 from a singular to the "hedgehog" gauge [397]:

$$\phi_0 \xrightarrow[SU(3)]{} R\phi_0 R^{-1} = \frac{v}{\sqrt{3}} \begin{pmatrix} 2 & 0 \\ 0 & 3\sigma_k \hat{r}_k - 1 \end{pmatrix} . \tag{8.47}$$

This is a generalization of the vacuum asymptotic of $SU(2)$ model that we discussed in Chap. 5:

$$\phi_0 \xrightarrow[SU(2)]{} \frac{v}{2} \sigma_k \hat{r}_k .$$

The diagonal operators $T_{(1)}^3$ and H_1, which commute with ϕ_0, transform in the same way:

$$T_{(1)}^3 \to RT_{(1)}^3 R^{-1} = \frac{1}{4}\begin{pmatrix} 2 & 0 \\ 0 & -\sigma_k \hat{r}_k - 1 \end{pmatrix}, \quad (8.48)$$

$$H_1 \to RH_1 R^{-1} = \frac{1}{4}\begin{pmatrix} 2 & 0 \\ 0 & 3\sigma_k \hat{r}_k - 1 \end{pmatrix}.$$

Clearly, these matrices are regular on the sphere S^2 given by the unit vector $\hat{r}_k$. The problem is that there are two other $SU(2)$ generators, which commute with ϕ_0, but do not have such a property. Indeed, for example, rotation of the generator $T_{(1)}^1$ to the "hedgehog" gauge yields

$$T_{(1)}^1 \to RT_{(1)}^1 R^{-1} = \frac{1}{2}\begin{pmatrix} 0 & \cos(\theta/2) & \sin(\theta/2)e^{-i\psi} \\ \cos(\theta/2) & 0 & 0 \\ \sin(\theta/2)e^{i\psi} & 0 & 0 \end{pmatrix}. \quad (8.49)$$

Evidently, this matrix possesses a singularity at the south pole:

$$\sin(\theta/2)e^{i\psi} = \frac{x+iy}{\sqrt{2(1+z)}} \xrightarrow[\cos\theta \to \pi]{} \infty.$$

We may check in a similar way that in the spherical gauge, the generator $T_{(1)}^2$ is not regular at the north pole. More generally, it is impossible to define a regular global gauge transformation that is generated by the non-diagonal operators $T_{(1)}^1$ and $T_{(1)}^2$.

An outline of the proof of this statement is rather simple [55, 63, 397]. Note that on the spatial asymptotic S^2, we can neglect all fields but those that are related to the unbroken symmetry subgroup. Then, we may make use of some generalization of the Wu–Yang formalism once again, i.e., we can cover the sphere S^2 by two hemispheres $S^2 = R^N \cup R^S$. Both on the north and on the south hemispheres smooth functions ψ_N and ψ_S (the section of the bundle) are defined, which are connected on the equator $S^1 = R^N \cap R^S$ (the overlap region) by the transition functions $U(\varphi)$:

$$\psi_N(\theta = \pi/2, \varphi) = U(\varphi)\psi_S(\theta = \pi/2, \varphi)U^{-1}(\varphi). \quad (8.50)$$

Thus, the gauge transformation $U(\varphi)$ relates the sections of the bundle and defines a closed contour on the group space. Indeed, we may write $U(\varphi) = \exp\{iQ\varphi\}$, where Q is the electric charge generator (8.35) and the identification of the points φ and $\varphi + 2\pi N$ is supposed. The map of this loop on the equator defines the first homotopy group, which yields the magnetic charge of the non-Abelian monopole.

The local gauge transformations defined within each of the hemispheres have to be consistent with the compatibility condition (8.50), that is,

$$\psi_N(\theta, \varphi) \to U_N(\varphi)\psi_N(\theta, \varphi)U_N^{-1}(\varphi), \quad \text{as} \quad 0 \le \theta \le \pi/2,$$

$$\psi_S(\theta,\varphi) \to U_S(\varphi)\psi_S(\theta,\varphi)U_S^{-1}(\varphi), \quad \text{as } \pi/2 \le \theta \le \pi,$$
$$U_N(\theta = \pi/2,\varphi) = U(\varphi)U_S(\theta = \pi/2,\varphi)U^{-1}(\varphi). \tag{8.51}$$

To define a gauge transformation on the sphere S^2 globally, the algebra of generators T^a of unbroken symmetry subgroup H must be independent of the angular variables θ, φ up to an inner authomorphism, which preserves the Lie algebra of H:

$$T_N^a(\theta,\varphi) = g_N(\theta,\varphi)T^a g_N^{-1}(\theta,\varphi), \quad \text{as } 0 \le \theta \le \pi/2,$$
$$T_S^a(\theta,\varphi) = g_S(\theta,\varphi)T^a g_S^{-1}(\theta,\varphi), \quad \text{as } \pi/2 \le \theta \le \pi,$$

where $g \in H$. The compatibility condition (8.50) means that

$$T_N^a(\theta = \pi/2,\varphi) = U(\varphi)T_S^a(\theta = \pi/2,\varphi)U^{-1}(\varphi),$$

or

$$T^a = g_N^{-1}(\pi/2,\varphi)U(\varphi)g_S(\pi/2,\varphi)T^a g_S^{-1}(\pi/2,\varphi)U^{-1}(\varphi)g_N(\pi/2,\varphi)$$
$$= \Omega_0 T^a \Omega_0^{-1}. \tag{8.52}$$

Therefore, the transformation $\Omega_0 \equiv g_N^{-1}(\pi/2,\varphi)U(\varphi)g_S(\pi/2,\varphi)$ defines a trivial authomprphism that is an element of the center of H. Recall that in the case under consideration, $H = U(2) \approx SU(2) \times U(1)/\mathbb{Z}_2$ and the center consists of two elements[9] $\Omega_0 = [-1, 1]$. Thus, the gauge function that defines the first fundamental homotopy group is

$$U(\varphi) = g_N(\theta = \pi/2,\varphi)\Omega_0 g_S^{-1}(\theta = \pi/2,\varphi).$$

Now we can see that, if the angle θ varies from $\theta = \pi/2$ (equator) to $\theta = 0$ or $\theta = \pi$ (north and south poles of the sphere, respectively), the closed contour on the group space can be continuously deformed to a point, that is, the corresponding winding number is zero. This means that in the theory with non-Abelian monopoles, the global gauge transformations are ill-defined in general. The only exception is the trivial case when the function $U(\varphi) = \exp\{iQ\varphi\}$ is homotopically equivalent to the zero element [397]. This condition can be satisfied for an arbitrary charge matrix Q, if the charge quantization condition is fulfilled and the transition function $U(\varphi)$ maps an even number of loops on the group space, while the azimuthal variable φ on the equator S^1 varies from 0 to 2π. In other words, the generators of the $U(2)$ invariant subgroup, which commute with the vacuum of the Higgs field, can be defined globally only if the topological charge of the configuration takes even values.

On the other hand, it is impossible to define normalizable zero modes which would correspond to the generators of the non-Abelian subgroup that

---

[9] Of course, this argumentation could be easily applied to an arbitrary semi-simple group.

do not commute with the vector magnetic charge $g = \vec{g}\cdot\vec{H}$. The reason is that such a fluctuation of the fields is not compatible with the restrictions imposed by the Gauss law [62,398]. This contradiction can be removed by introducing an additional constraint on the vector magnetic charge to be parallel to the scalar field:

$$\vec{g}\cdot\vec{\beta}_i = 0,\qquad(8.53)$$

for all simple roots orthogonal to $\vec{h}$. The physical meaning of this constraint is that all long-range forces are restricted to be Abelian.

In our basis of simple $SU(3)$ roots, the constraint (8.53) yields

$$n_1 - n_2/2 = 0.$$

Thus, the simplest non-pathological configuration with a non-zero holomorphic charge $[n_1] = 1$ consists of two massive monopoles.

Thus, we conclude that we have the famous "no go theorem", which states that there are no colored dyons [106, 397, 398]. However, this theorem does not exclude another possibility, namely that we can consider the dual non-Abelian Lie group G^* as a gauge group of the configuration. Then there could be non-Abelian monopoles that transform according to some representation of the dual group [89, 90, 334].

8.3 $SU(3)$ Monopole Moduli Space

The conclusion that the multimonopole configurations appear in a rather natural way in a model with the gauge group rank greater than one caused a special interest in the investigation of the moduli space of these monopoles.

Let us recall that the idea of the moduli space approximation is to truncate the infinite-dimensional configuration space of a system to the subspace of n collective coordinates involving the final number of degrees of freedom. The low-energy dynamics of the configuration is then described as a geodesic motion of a point particle on this moduli space.

A straight way to construct a moduli space metric is to analyze the zero modes around of the full family of solutions for a given magnetic charge. This metric describes the low-energy interaction between the monopoles. However, in some cases, an explicit solution is not available. Then we can invert this logic, namely we can try to make use of the knowledge of low-energy dynamics to restore, at least asymptotically, the moduli space metric.

To contract the metric on $SU(3)$ moduli space, we make use of an analogy with the discussion of Sect. 6.3.3, where we considered the classical Lagrangian of point-like well-separated $SU(2)$ dyons. Again, we imply a naive picture of the classical interaction of two fundamental monopoles, which are obtained by embedding of the single $SU(2)$ monopole configuration along the composite root $\vec{\beta}_1 + \vec{\beta}_2$. However, since these monopoles are charged with respect to different Abelian subgroups, the character of interaction between the $SU(3)$ BPS monopoles depends on the type of embedding [346]. This is also correct for the non-BPS extension [467].

Indeed, then there is only a long-range electromagnetic field that mediates the interaction between two widely separated non-BPS monopoles, that is, they are considered as classical point-like particles with magnetic charges $g_i = \vec{g}_i \cdot \vec{H} = \vec{\beta}_i \cdot \vec{H}$. For a non-zero scalar coupling λ, the contribution of the scalar field is exponentially suppressed. The energy of the electromagnetic interaction then originates from the kinetic term of the gauge field $\sim \text{Tr}\, F_{\mu\nu}F^{\mu\nu}$ in the Lagrangian (8.3). Therefore, an additional factor $\text{Tr}[(\vec{\beta}_i \cdot \vec{H})(\vec{\beta}_j \cdot \vec{H})] = (\vec{\beta}_i \cdot \vec{\beta}_j)$ appears in the formula for the energy of electromagnetic interaction. In the case under consideration, $(\vec{\beta}_1 \cdot \vec{\beta}_2) = -1/2$, while $(\vec{\beta}_i \cdot \vec{\beta}_i) = 1$. This corresponds to the attraction of two different fundamental $SU(3)$ monopoles and repulsion of two monopoles of the same $SU(2)$ subalgebra due to the non-trivial group structure. The energy of interaction between the $\vec{\beta}_1$ and $\vec{\beta}_2$ monopoles is then:

$$V_{int} = -\frac{(\mathbf{r}_1 \mathbf{r}_2)}{r_1^3 r_2^3}.$$

We can check this conclusion by making use of an analogy with the classical electrodynamics of point-like charges. Let us suppose that both monopoles are located on the z-axis at the points $(0, 0, \pm R)$.

The electromagnetic field of this configuration can be calculated in the Abelian gauge, where the gauge field becomes additive [76]. If the monopoles are embedded along the same simple root, say $\vec{\beta}_1$, we can write the components of the gauge field as

$$A_r = A_\theta = 0, \qquad A_\varphi = (1 + \cos\theta_1)\frac{\sigma_3^{(1)}}{2} + (1 + \cos\theta_2)\frac{\sigma_3^{(1)}}{2}. \qquad (8.54)$$

Simple calculation yields the components of the electromagnetic field strength tensor

$$F_{r\theta} = 0; \qquad F_{r\varphi} = rR\sin^2\theta \left(\frac{1}{r_1^3}\frac{\sigma_3^{(1)}}{2} - \frac{1}{r_2^3}\frac{\sigma_3^{(1)}}{2}\right),$$

$$F_{\theta\varphi} = -r^2\sin\theta \left(\frac{r - R\cos\theta}{r_1^3}\frac{\sigma_3^{(1)}}{2} + \frac{r + R\cos\theta}{r_2^3}\frac{\sigma_3^{(1)}}{2}\right), \qquad (8.55)$$

where r_1, r_2 are the distances of the point r to the points at which monopoles are placed. The field energy becomes

$$\begin{aligned} E &= \text{Tr}\left(\frac{1}{r^2\sin^2\theta}F_{r\varphi}^2 + \frac{1}{r^4\sin^2\theta}F_{\theta\varphi}^2\right) \\ &= \frac{1}{2}\left[\left(\frac{\mathbf{r}_1}{r_1^3}\right)^2 + \left(\frac{\mathbf{r}_2}{r_2^3}\right)^2 + \frac{2(\mathbf{r}_1\mathbf{r}_2)}{r_1^3 r_2^3}\right], \end{aligned} \qquad (8.56)$$

that is, the potential energy of the electromagnetic interaction of two $\vec\beta_1$ monopoles is repulsive. However, for a $\vec\beta_3$ configuration with vector charge $\vec{g} = (1, 1)$, the components of the gauge fields are

$$A_r = A_\theta = 0; \qquad A_\varphi = (1 + \cos\theta_1)\frac{\sigma_3^{(1)}}{2} + (1 + \cos\theta_2)\frac{\sigma_3^{(2)}}{2}, \qquad (8.57)$$

and, because $\mathrm{Tr}\,(\sigma_3^{(1)}\sigma_3^{(2)}) = -1$, the field energy is

$$E = \frac{1}{2}\left[\left(\frac{\mathbf{r}_1}{r_1^3}\right)^2 + \left(\frac{\mathbf{r}_2}{r_2^3}\right)^2 - \frac{(\mathbf{r}_1\mathbf{r}_2)}{r_1^3 r_2^3}\right], \qquad (8.58)$$

that is the $\vec\beta_1$ and $\vec\beta_2$ monopoles attract each other with a half-force compared to the case of the repulsion of two $\vec\beta_1$ monopoles.

Let us now consider the BPS monopoles. To derive the metric on the $SU(3)$ moduli space, we shall analyze the situation in more detail [239]. Again, suppose that there are two fundamental monopoles at the points $\mathbf{r}_i$, $i = 1, 2$, separated by a large distance r. The idea is to exploit an analogy with a classical picture of the non-relativistic interaction between two point-like charges, moving with small velocities $\mathbf{v}_i$. Thus, we consider a classic, long-range interaction between two particles with magnetic charges $\vec{g}_1, \vec{g}_2$ associated with the corresponding simple roots,

$$g_i = (\vec{g}_i \cdot \vec{H}) = \frac{4\pi}{e}(\vec\beta_i \cdot \vec{H}). \qquad (8.59)$$

Because we considering the fundamental monopoles, the charge quantization condition yields $g_1 = g_2 = g = 4\pi/e$.

Recall that some of the collective coordinates of the multi-monopole system correspond to the dyonic degrees of freedom. These excitations shall transform a monopole into a dyon (see Sect. 6.5.1). Other collective coordinates correspond to the spatial translations of the configuration. We make an assumption that the electric charges of the dyons are also vectors in the root space, that is [346],

$$Q_i = q_i(\vec\beta_i \cdot \vec{H}). \qquad (8.60)$$

Thus, the potential of interaction of two identical dyons remains proportional to the inner product $(\beta_i \cdot \beta_j)$, as in the case of a purely magnetically charged configuration.

The canonical momentum of one of the dyons, which is moving in the external field of another static dyon, is given by

$$\mathbf{P} = M\mathbf{v}_1 + q_1\mathbf{A} + g\widetilde{\mathbf{A}}, \qquad (8.61)$$

where the vector potentials of the electromagnetic field generated by the static dyon are

8.3 SU(3) Monopole Moduli Space

$$\mathbf{A} = g_2 \mathbf{a}, \qquad \tilde{\mathbf{A}} = -q_2 \mathbf{a},$$

and we again make use of the definition $\mathbf{a}$ of the rescaled Dirac potential (see (6.53)):

$$[\nabla \times \mathbf{a}] = \mathbf{r}/r^3 = -\nabla(1/r).$$

Thus, far away from each of the dyons their electric and magnetic fields are

$$\mathbf{B}^{(i)} = g_i \frac{\mathbf{r} - \mathbf{r}_i}{|\mathbf{r} - \mathbf{r}_i|^3} = \frac{4\pi}{e}(\vec{\beta}_i \cdot \vec{H}) \frac{\mathbf{r} - \mathbf{r}_i}{|\mathbf{r} - \mathbf{r}_i|^3},$$

$$\mathbf{E}^{(i)} = Q_i \frac{\mathbf{r} - \mathbf{r}_i}{|\mathbf{r} - \mathbf{r}_i|^3} = q_i(\vec{\beta}_i \cdot \vec{H}) \frac{\mathbf{r} - \mathbf{r}_i}{|\mathbf{r} - \mathbf{r}_i|^3}, \tag{8.62}$$

which correspond to the scalar potentials

$$A_0^{(i)} = q_i \frac{(\vec{\beta}_i \cdot \vec{H})}{|\mathbf{r} - \mathbf{r}_i|}, \qquad \tilde{A}_0^{(i)} = g \frac{(\vec{\beta}_i \cdot \vec{H})}{|\mathbf{r} - \mathbf{r}_i|}.$$

In the evaluation of the energy of interaction between the two fundamental monopoles considered above, we take into account only the electromagnetic part and neglect the contribution of scalar fields. However, in the BPS limit, the Higgs field also becomes long-ranged and the mass of the dyon is defined as

$$M_i = \left(\vec{h} \cdot \vec{\beta}_i^*\right) \sqrt{q_i^2 + g^2}, \tag{8.63}$$

which is a generalization of (8.29). However, this formula is only correct for a single isolated dyon.

An external field of another dyon modifies the Coulomb-like tail of the scalar field (cf. (8.10)) as

$$\phi_i = v\vec{h} \cdot \vec{H} - \sqrt{q_i^2 + g^2}\sqrt{1 - \mathbf{v}_i^2} \frac{(\vec{\beta}_i \cdot \vec{H})}{|\mathbf{r} - \mathbf{r}_i|}. \tag{8.64}$$

As in the previous consideration of Sect. 6.3.3, we neglect here the difference between the masses of the monopole and the dyon, i.e., we assume for simplicity that both the velocities and the electric charges of the dyons are relatively small.

We also have to take into account the Coulomb-like potential associated with the dilatonic charge of the Higgs field, which is similar to the minimal $SU(2)$ model:

$$Q_D^{(i)} = (\vec{\beta}_i \cdot \vec{H})\sqrt{q_i^2 + g^2}. \tag{8.65}$$

The long-range tail of the Higgs field yields some distortion of the vacuum expectation value of the scalar field in the neighborhood of another monopole (see (8.64)). As a result, the size of its core is increasing a bit, whereas the mass is decreasing:

$$M \to M - \frac{(\vec{\beta}_1^* \cdot \vec{\beta}_2^*)}{r}\sqrt{1-\mathbf{v}_2^2}\sqrt{(q_1^2+g^2)(q_2^2+g^2)}\,. \tag{8.66}$$

To sum up, the formal difference from the calculations we presented in Sect. 6.3.3 consists in an additional factor $\lambda = -2(\vec{\beta}_1 \cdot \vec{\beta}_2)$, which appears as a coefficient at all terms of interactions. Thus, by making use of simple analogy with (6.59), we can immediately write the Lagrangian of relative motion of two widely separated $SU(3)$ BPS dyons of the same mass[10] M as [346]:

$$L = \frac{1}{4}\left(M + \frac{\lambda g^2}{r}\right)\left(\dot{\mathbf{r}} \cdot \dot{\mathbf{r}} - \frac{Q^2}{g^2}\right) - \frac{\lambda}{2}gQ\,\dot{\mathbf{r}} \cdot \mathbf{a}\,, \tag{8.67}$$

where a relative charge of the pair of dyons is $Q = |q_1 - q_2|$ and the relative position is defined by the vector $\mathbf{r} = \mathbf{r}_1 - \mathbf{r}_2$.

Recall that in the $SU(2)$ theory, the relative electric charge is connected with the gauge cyclic collective coordinate $\Upsilon(t)$, which parameterizes the $U(1)$ subgroup. In the case of the $SU(3)$ model, the electric charges q_1 and q_2 are related to two different $U(1)$ subgroups, which can be parameterized by two cyclic variables, the angles α_1 and α_2, respectively. Then we may interpret the charges q_1, q_2 as conserved momenta conjugated to these variables. In the basis of a self-dual simple root that we are using, the period of the variables α_i is $T = 2\pi e$ for both subgroups. The relative phase $\Upsilon = \alpha_1 - \alpha_2$ is a variable conjugated to the relative charge Q.

The moduli space Lagrangian of relative motion must be written in terms of the generalized velocities. Thus, by analogy with our discussion in Sect. 6.3.3, we perform the Legendre transformation

$$L(\mathbf{r}, \Upsilon) = L(\mathbf{r}, Q) + gQ\dot{\Upsilon}\,,$$

where

$$Q \equiv \frac{2g^3}{M + \frac{\lambda g^2}{r}}\left(\dot{\Upsilon} + (\vec{\beta}_1 \cdot \vec{\beta}_2)(\mathbf{a} \cdot \dot{\mathbf{r}})\right)\,. \tag{8.68}$$

Thus, we finally obtain the transformed Lagrangian of the relative motion of two widely separated $SU(3)$ dyons [239, 346],

$$L = \frac{1}{4}\left(M + \frac{\lambda g^2}{r}\right)\dot{\mathbf{r}} \cdot \dot{\mathbf{r}} + \frac{g^4}{M + \frac{\lambda g^2}{r}}\left(\dot{\Upsilon} - \frac{\lambda}{2}(\mathbf{a} \cdot \dot{\mathbf{r}})\right)^2\,. \tag{8.69}$$

As before, this expression does not depend explicitly on the collective coordinate Υ. This means that the corresponding equation of motion is just the condition of conservation of the relative electric charge (8.68), rather than a dynamical equation.

From this form of the Lagrangian (8.69), we can read the asymptotic metric for the moduli space of two widely separated $SU(3)$ BPS monopoles

---

[10] Of course, one may make use of the reduced mass $M = (M_1+M_2)/M_1M_2$, where M_1, M_2 are the masses of the dyons.

$$ds^2 = \left(1 + \frac{\lambda g^2}{Mr}\right) d\mathbf{r}^2 + \frac{\left(\frac{g^2}{2M}\right)^2}{1 + \frac{\lambda g^2}{Mr}} \left(d\Upsilon + (\vec{\beta}_1 \cdot \vec{\beta}_2)(\mathbf{a} \cdot d\mathbf{r})\right)^2, \qquad (8.70)$$

that is, the metric of Taub-NUT space with the length parameter $\lambda g^2/M$.

The principal difference from the asymptotic metric, which describes two $SU(2)$ dyons, is that the parameter λ remains negative only if both of these fundamental monopoles correspond to the same simple root. We can expect this, because in this case, there are two widely separated $SU(2)$ monopoles, and the corresponding moduli space is described by the singular Taub-NUT metric (6.64). However, we have seen that, if these two monopoles correspond to the different simple roots, $\vec{\beta}_1$ and $\vec{\beta}_2$, we have [346]:

$$\lambda = -2(\vec{\beta}_1 \cdot \vec{\beta}_2) = -2[(1,0) \cdot (-1/2, \sqrt{3}/2)] = 1.$$

Note that the conservation of both the total and relative electric charges results from the conservation of the individual electric charges q_1, q_2 of each monopole. This yields a $U(1)$ symmetry of the metric of relative motion, which is not a symmetry of the Atiyah–Hitchin metric (6.143), where relative charge, in general, is not an integral of motion. This symmetry simplifies the low-energy dynamics of two distinct $SU(3)$ monopoles; there is no right-angle scattering, but bounce trajectory in a head-on collision.

A generalization to the case of arbitrary numbers of interacting dyons is rather obvious [346]. Each BPS dyon is associated with a simple root $\vec{\beta}_n$ and possess electric and magnetic charges g_i, Q_i defined as by (8.59) and (8.60), respectively. Moreover, they also have a dilatonic charge $Q_D^{(i)}$ given by (8.65) and all the long-range forces are proportional to the inner product of the corresponding simple roots. Thus, we may write the Lagrangian of relative motion of a system of N well-separated BPS dyons as

$$L = \frac{1}{2} M_{nm} \dot{\mathbf{r}}_n \cdot \dot{\mathbf{r}}_m + \frac{g^4}{2} (M^{-1})_{nm} \left(\dot{\Upsilon}_n + (\mathbf{W}_{nk} \cdot \dot{\mathbf{r}}_k)\right) \left(\dot{\Upsilon}_m + (\mathbf{W}_{ml} \cdot \dot{\mathbf{r}}_l)\right), \qquad (8.71)$$

where $n, m \in [1, N]$, the mass[11] of an n-th dyon M_n is defined according to (8.63) and the matrix M_{nm}, which has the physical meaning of a reduced mass matrix, is defined as

$$M_{nn} = M_n - \sum_{k \neq n} \frac{g^2(\vec{\beta}_n^* \cdot \vec{\beta}_k^*)}{r_{nk}}, \qquad M_{nm} = \frac{g^2(\vec{\beta}_n^* \cdot \vec{\beta}_m^*)}{r_{nm}}. \qquad (8.72)$$

Here the electromagnetic piece of interaction between these widely separated monopoles is given by the superposition of the Dirac potentials $\mathbf{a}_{nm}$

---

[11] As before we neglect the difference between the mass of a monopole and that of a dyon.

$$\mathbf{W}_{nn} = \sum_{k \neq n} (\vec{\beta}_n^* \cdot \vec{\beta}_k^*) \, \mathbf{a}_{nk}, \qquad \mathbf{W}_{nm} = -(\vec{\beta}_n^* \cdot \vec{\beta}_m^*) \, \mathbf{a}_{nm}.$$

For each individual potential we have $[\nabla \times \mathbf{a}_{nk}] = \nabla \dfrac{1}{|\mathbf{r}_n - \mathbf{r}_k|}$.

Let us consider the example of a $SU(3)$ configuration consisting of three monopoles with different masses and topological charges $(2,1)$ in the case of maximal symmetry breaking. With our choice of the simple root basis $\vec{\beta}_1, \vec{\beta}_2$ given by (8.15), the corresponding mass matrix can then be written as [285]

$$M_{mn} = \begin{pmatrix} M_1 - \frac{g^2}{r_{12}} + \frac{g^2}{2r_{13}} & \frac{g^2}{r_{12}} & -\frac{g^2}{2r_{13}} \\ \frac{g^2}{r_{12}} & M_2 - \frac{g^2}{r_{12}} + \frac{g^2}{2r_{23}} & -\frac{g^2}{2r_{23}} \\ -\frac{g^2}{2r_{13}} & -\frac{g^2}{2r_{13}} & M_3 + \frac{g^2}{2r_{13}} + \frac{g^2}{2r_{23}} \end{pmatrix}.$$

Some approximations can be imposed to this system to investigate limiting cases. For example, if one of the monopoles is very heavy, say $M_3 \gg M_1 \sim M_2$, its position in the frame of the center of the mass of two other monopoles is fixed. Then, the Taub-NUT metric, which is singular in the limit $r_{12} = 0$, could be recovered from (8.71) [285].

There is very convincing argumentation in support of the conclusion that the asymptotic metric is exact, if all the interacting monopoles are associated with different roots [122, 175, 231, 345, 518, 519]. In the above example, this condition if not satisfied and, when two monopoles that are associated with the same simple root, approach each other, the moduli space metric (8.70) develops a singularity. Generally, at small monopole separation we have to take into account the interaction connected with the short-range non-Abelian fields inside of monopoles. This interaction could modify the Lagrangian (8.71).

Obviously, the complete information about the geometry of the moduli space is much more important than the asymptotic metric of the system of widely separated monopoles. There is some difference from the case of the moduli space of $SU(2)$ monopoles, which can be seen, if we consider the particular case of two distinct interacting $SU(3)$ monopoles.

Clearly, such a moduli space is eight-dimensional and it contains a three-dimensional center-of-mass subspace $\mathbb{R}^3$ spanned by the translation collective coordinates

$$\mathbf{R} = \frac{M_1 \mathbf{r}_1 + M_2 \mathbf{r}_2}{M_1 + M_2}.$$

However, in general, the moduli space of a higher rank gauge group does not factorize onto a circle S^1, since for the distinct fundamental $SU(3)$ monopoles, the total electric charge of the configuration is not associated with a periodic collective coordinate.

Indeed, in the case under consideration, the total charge is defined as a linear combination

$$\tilde{q} = \frac{M_1 q_1 + M_2 q_2}{M_1 + M_2}, \tag{8.73}$$

where q_1, q_2 denote the quantized electric charges associated with two different $U(1)$ subgroups. Thus, the relative charge $Q = |q_1 - q_2|$ is quantized, but the total charge (8.73) is not quantized, unless the ratio M_1/M_2 is rational [231, 345].

Thus, the collective variable v, which is conjugated to the total charge $\tilde{q}$, is defined on the $\mathbb{R}^1$ subspace. Then the moduli space of two distinct fundamental monopoles is of the form [231, 345]

$$\mathcal{M} = \mathbb{R}^3 \times \frac{\mathbb{R}^1 \times \mathcal{M}_0}{\mathbb{Z}},$$

where $\mathcal{M}_0$ is the relative moduli space parameterized by the relative positions and phases. The $\mathbb{Z}$ factoring over the infinite discrete group appears due to an identification on gauge variables v and Υ, which correspond to the total and relative charges $\tilde{q}$ and Q, respectively [345]. In general, it is a discrete subgroup of the isometry group of $\mathbb{R}^1 \times \mathcal{M}_0$.

It has been shown that for two distinct fundamental monopoles, the four-dimensional subspace of relative motion $\mathcal{M}_0$ is a Taub-NUT manifold with a positive length parameter. In general, the properties and dimension of the manifold $\mathcal{M}_0$ essentially depend on the character of the long-range fields. The simplest case is the maximal symmetry breaking [162]. Then the monopole vector charge is $g = (1, 1)$ and we have the system of two separated $SU(2)$ monopoles, which we discussed above in Chap. 6. Its relative moduli space $\mathcal{M}_0^{(1,1)}$ is four-dimensional and the moduli space metric must be hyperkähler, thus its geometry is shown to be smooth Taub-NUT space [175, 231, 345].

The situation is more complicated in the case of minimal symmetry breaking, due to two different possibilities that correspond to the holomorphic charges [0] and [1], respectively.

A two-monopole configuration with the magnetic charge $g = (2, [0])$ can be constructed by simple embedding of the charge two axially symmetric $SU(2)$ monopole into the $SU(3)$ theory. The corresponding moduli space $\mathcal{M}_0^{(2,[0])}$ is the charge 2 $SU(2)$ monopole moduli space fibred over S^2. The factor S^2 here corresponds to the unbroken $SU(2)$ subgroup, thus, the $\mathcal{M}_0^{(2,[0])}$ is a six-dimensional manifold. However in this case, there are long-range non-Abelian magnetic fields associated with the unbroken $SU(2)$ subgroup and the global gauge transformations are ill-defined

If the holomorphic charge is equal to [1], the condition (8.53) is satisfied and all the long-range fields are Abelian. Then the pathology of the global gauge transformation disappears and the moduli space obtains additional degrees of freedom. Indeed, a direct calculation of the number of zero modes on the $g = (2, [1])$ monopole background shows that $\mathcal{M}_0^{(2,[1])}$ is of dimension eight [132, 514].

This example shows that although, the asymptotic form of the moduli space metric is given by (8.70), its general construction is a very complicated problem. As mentioned earlier, in Sect. 6.4.5, the metric on the monopole moduli space and the metric on the space of Nahm data (6.106) are related by an isometry. Thus, solution of Nahm equations for the $SU(3)$ monopoles provides a simple way of finding the metric on the $SU(3)$ monopole moduli space.

8.3.1 $SU(3)$ Monopoles: Nahm Equations

The Nahm approach to the $SU(2)$ multimonopoles can be generalized for a general $SU(N)$ theory [392, 393]. Moreover, the matrix-valued functions $T^k(s)$, which satisfy the system of non-linear ordinary differential equations (6.106), must be modified to describe both the cases of minimal and maximal symmetry breaking.

Let us briefly describe the case of the $SU(3)$ theory. In the simplest possible case, the symmetry is broken to the maximal torus $U(1) \times U(1)$. For a single fundamental monopole, $\vec{g} = (1,0)$ or $\vec{g} = (0,1)$, embedded along corresponding simple root, the Nahm data $T^k(s)$ (6.104) are defined over the interval $s \in [0,1]$. Thus, it is a triplet of real numbers associated with the position of the monopole, as we discussed in Sect. 6.4.5 above.

For a composite $\vec{g} = (1,1)$ monopole, the solution of the Nahm equations was constructed in [175]. Then the one-dimensional functions $T^k(s)$ in the interval $s \in [-1,1]$ are given by two triplets $T^k(s_1)$ and $T^k(s_2)$, which are defined over the sub-intervals $s_1 \in [-1,0]$ and $s_2 \in [0,1]$, respectively. The Nahm data must be regular inside each sub-interval and some boundary conditions between these sub-intervals are imposed at $s = 0$ to provide analyticity over the whole interval $[-1,1]$. The corresponding Nahm equations are satisfied by a trivial solution, thus, these two triplets of real numbers correspond to the positions of two monopoles and the boundary condition at $s = 0$ fixes the relative phase.

Knowledge of the Nahm data for the charge $\vec{g} = (1,1)$ $SU(3)$ monopoles makes it possible to obtain the metric on the relative moduli space [175, 231, 345], which is the Taub-NUT metric with a positive length parameter (8.70).

The example of the $\vec{g} = (2,1)$ monopoles, which we discussed above, is more complicated. A very detailed study of the $SU(3)$ monopole moduli space have been done by Dancer [185–188], who also analyzed this particular case. Further consideration of the $SU(3)$ monopoles was presented in [295].

The Nahm matrices of these $SU(3)$ monopoles are defined over the interval[12] $s \in [-2,1]$ with two subintervals, one for each of the monopoles and some complicated boundary condition at $s = 0$.

---

[12] In the works by A. Dancer [185, 186] they were defined over the "shifted" interval $s \in [0,3]$.

8.3 $SU(3)$ Monopole Moduli Space

The experience that we have already gained from the discussion of the Nahm data of the $SU(2)$ monopoles, suggest that the Nahm data in the first subinterval, $s_1 \in [-2,0]$, are Hermitian matrices of dimension 2×2. The rotation invariance means that the proper ansatz is $T^k(s_1) = f^k(s_1)\sigma_k/2$. On the second subinterval $s_2 \in [0,1]$, the Nahm data $T^k(s_2)$ have the dimension 1×1, that is, they are of the form $T^k(s_2) = r^k/2$, where the vector r_k defines the position of the $(0,1)$ monopole.

For a minimal symmetry breaking $SU(3)$ to $U(2)$ the situation is different. Let us briefly describe the properties of the twelve-dimensional moduli space $\mathcal{M}$ of the $\vec{g} = (2,[1])$ monopole. Dancer showed [185, 186] that $\mathcal{M}$ is a hyper-Kähler manifold with commuting actions of $Spin$ (3) (which is the double cover of the $SO(3)$ group of spatial rotations), $U(2)$ (the unbroken gauge group that we discussed above), and the group of translation of the center of mass $\mathbb{R}^3$.

Let us separate the collective coordinates of the center of mass and the coordinates of the relative motion. Then the relative moduli space $\mathcal{M}_0$ is an eight-dimensional irreducible hyper-Kähler manifold obtained by quotation of $\mathcal{M}$ by $\mathbb{R}^3 \times U(1)$, where $U(1)$ is the center of $U(2)$. Therefore, the manifold $\mathcal{M}_0$ has the free commuting action of $SO(3)$ and $SU(2)/Z_2$.

By quotienting $\mathcal{M}_0$ further by $SU(2)/Z_2$, one obtains a five-dimensional space N^5, which has a non-free action of rotations $SO(3)$ [186]. This space describes monopoles with fixed center of mass, quotiented by the action of the unbroken gauge group. By analogy with the case of the $SU(2)$ monopoles, the information about the isometries of the moduli spaces allows us to write an explicit expression for the metric on $\mathcal{M}_0$ in terms of invariant one-forms [295].

Further simplification is possible, if we take the quotient of the space N^5 by the $SO(3)$ action. The rotational $SO(3)$ symmetry leads to the corresponding ansatz for the Nahm data: $T^k(s) = f^k(s)\sigma_k/2$. The Nahm equations (6.106) have a general solution (6.117), which, taking into account the boundary conditions on the boundaries of the interval $s \in [-2,1]$, are [186]

$$f_1 = -\frac{D \operatorname{cn}_k(Ds)}{\operatorname{sn}_k(Ds)}, \quad f_2 = -\frac{D \operatorname{dn}_k(Ds)}{\operatorname{sn}_k(Ds)}, \quad f_3 = -\frac{D}{\operatorname{sn}_k(Ds)}. \quad (8.74)$$

Thus, after removal of all the possible symmetries of the monopole moduli space we are left with a two-dimensional submanifold, a quotient space $N^5/SO(3)$. However, the isotropy group is not the same at all points of the space N^5; its quotient by the $SO(3)$ group is not a manifold [187]. There are six copies of the space $N^5/SO(3)$, which form the totally geodesic Dancer space Y depicted in the Fig 8.8. This space is symmetric under reflection in each of the three Cartesian axes. The metric and geodesic flow on Y, corresponding to the monopole scattering, was studied in [187, 188].

As follows from (8.74), a proper parameterization of a two-dimensional subspace $N^5/SO(3)$ is given by the parameters k, D, where $0 \leq k \leq 1$, $0 \leq D < \frac{2}{3}K(k)$, and $K(k)$ is the complete elliptic integral (6.97). The parameter D has the interpretation of the separation between the monopoles.

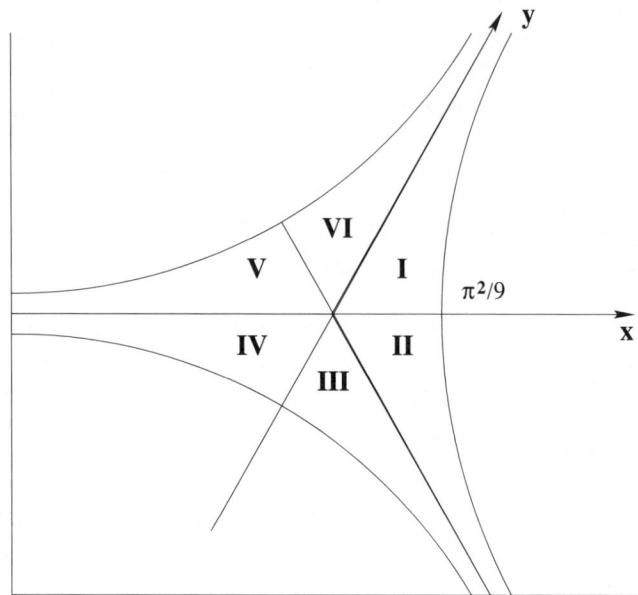

Fig. 8.8. Geodesic space Y of $SU(3)$ monopoles (Dancer space)

In each of the regions of Y, local (non-orthogonal) coordinates x, y can be introduced. For example, in the region I, the local coordinates are [187, 295]

$$x = (1-k^2)D^2, \qquad y = k^2 D^2.$$

Points of this sector of Y correspond to the different monopole configurations. Some of them we have already discussed. For example, the particular value of the separation parameter $D = 0$ corresponds to the origin of the coordinates $x = 0$, $y = 0$. This yields the spherically symmetric monopole studied in [88, 526]. If the parameter D approaches another limit $D = 2K(k)/3$, the configuration consists of two embedded $SU(2)$ monopoles.

In the limiting case $k = 0$, the elliptic integral takes value $K = \pi/2$ and the separation parameter is defined over the interval $0 \le D < \pi/3$. This corresponds to the border on the sector I defined by $0 \le x < \pi^2/9$, $y = 0$. Because in this limit the elliptic Jacobi integrals become simple trigonometric functions, the corresponding Nahm data define the *trigonometric monopoles* [185]. They are actually the axially symmetric monopoles found by Ward [509].

In the opposite limit $k = 1$, the elliptic integral K diverges logarithmically and the elliptic integrals approach the hyperbolic functions. These are the so-called *hyperbolic monopoles*. They correspond to another border of the segment I defined by the line segment $x = 0, 0 \le y < \infty$. The analysis of this configuration shows that these monopoles are also axially symmetric.

This family interpolates between the case of two widely separated embedded $SU(2)$ monopoles of charge one ($y \to \infty$) and a spherically symmetric charge two monopole at the origin.

The geodesic on Y corresponds to the monopole scattering. For example, the two line segments of the border of the sector I together correspond to the geodesic on the space Y. As we move along it, two well-separated axially symmetric $SU(2)$ monopoles approach each other, collide to form the spherically symmetrical configuration, which then deforms to the axially symmetric Ward monopoles [185]. This a generalization of the monopole scattering that we considered in Sect. 6.5.1.

There are some interesting features of the $SU(3)$ monopole scattering [187, 188]. For example, there can be double scattering of two monopoles at right-angles in two orthogonal planes. In turns out that among the geodesics are unusual ones that do not describe a trivial scattering of two monopoles, but a process where two colliding monopoles form a bound state asymptotically moving together to the border of Y characterized by $D = 2K(k)/3$. This is the case when two charge one monopoles in initial state ($\vec{g} = (1,1)$) scatter into the final state $g = (2,[1])$. Thus, there are two $SU(2)$ embedded monopoles and a massless monopole state, excitation of which carries off the kinetic energy when the monopoles stick together [58]. The parameter D in this picture corresponds to the radius of such a non-Abelian cloud [295].

From the point of view of the Montonen–Olive duality, such a process is a very intriguing dual analog of the confinement in the dual $SU(3)$ theory, that is, the Quantum Chromodynamics. Indeed, there is some obvious evidence that quark binding into colorless states, baryons and mesons, is associated with a number of non-perturbative effects. As we shall see in the next chapter, here the monopoles can really play an outstanding role.

9 Monopoles and the Problem of Confinement

9.1 Quark Confinement in QCD

One of the reasons why nowadays there is such a strong interest in the monopole problem is a subject that we have not discussed yet, namely the problem of *confinement*. It has been quite popular since the beginning of the 1980s to believe that monopoles in the QCD vacuum could be related with this remarkable phenomenon, which still remains one of the very few Big Unsolved Problems of the theoretical physics of the XXI-st century. Moreover, this is the "classic question that has resisted solution over the years", which was included by the Clay Mathematics Institute in the list of seven Millennium Prize Problems [14]. The award will amount to $1,000,000, thus, it is worth-while to account for more information about the matter.

We consider the well-known Lagrangian of the $SU(3)$ gauge theory involving three flavors of fermions in some representation of the gauge group; the quarks:

$$L = -\frac{1}{4}F^a_{\mu\nu}F^{\mu\nu a} + \overline{\psi}_i(iD_\mu\gamma^\mu - m_i)\psi_i, \tag{9.1}$$

where ψ_i is the spinor field and the index $i = 1, 2, 3$ labels the flavors[1]. This is Quantum Chromodynamics, the renormalizable asymptotically free gauge theory, which is generally accepted as a correct theory of strong interactions. This is a very remarkable theory, since the corresponding QCD Landau pole, i.e., the dynamically generated mass scale Λ_{QCD} at which the running coupling constant blows up and the model becomes completely useless, in contrast to QED lies in a low-energy domain. In other words, Nature gives us the rare possibility of seeing what happens with a theory as we approach a non-perturbative regime.

The answer is known from the experiment: on large distances (or small scales of momenta) there are neither quarks nor gluons, but colorless bound states, baryons and mesons, as physical variables. That is the so-called *weak* definition of confinement: there are no color states in the physical spectrum. Thus, the drastic difference from Abelian theory is that the quarks and gluons are no longer proper degrees of freedom of QCD over the entire range of scales.

---

[1] In the next chapter we shall discuss the structure of the fermionic action in more detail.

Weak definition of confinement of colour, which is common in popular literature, is not quite satisfactory because the mechanism of quarks binding into the hadrons remains a black box. Again, experiment tells us that the character of the gluon-mediated quark-antiquark interaction becomes different as we approach the strong coupling regime; the confining potential is linear, which can be explained by formation of a flux tube connecting quark and antiquark.

There are many phenomenological models that have successfully used this observation to describe the spectrum of hadrons. Thus, the *strong* definition of confinement is to explain, starting from conventional perturbative QCD, the origination of the linear potential between the quarks, and that is a real challenge. Actually, there are two different regimes of QCD, the perturbative region and the confinement phase, where the strong interaction forces the quarks and gluons to condense. The problem of strong confinement is the problem of phase transition between the confinement phase and deconfinement phase of QCD.

We can try to understand what happens, if we notice that a hierarchy of scales of QCD is provided by two quantities: the characteristic scale of chiral symmetry breaking $\Lambda_\chi \sim 1\ GeV$ and $\Lambda_{QCD} \sim 180\ MeV$. The perturbative QCD governs the scales of momenta above the Λ_χ. In this domain, the quarks and transverse gluons are physical asymptotic states in an exact analogy with QED, where asymptotic states are electrons and transverse photons. On the scale of the mass of the W-boson, the QCD coupling constant is ~ 0.11 and the theory irreproachably agrees with experimental data.

However, the situation is more subtle because the coupling quarks with gluons becomes non-perturbative at the scale of Λ_χ. This is the scale of another remarkable phenomenon: *spontaneous chiral symmetry breaking*. Note that, if we neglect the quark mass term, the Lagrangian (9.1) turns out to be invariant not only with respect to the $SU(3)$ gauge transformations. Since the Lagrangian (9.1) in the massless limit decomposes into two disconnected parts, which correspond to the left and right fermions coupled to the gauge field, it is also invariant under the global unitary transformations $U(3) \times U(3)$, which independently rotate these left- and right-hand components:

$$\bar{\psi}_R(x) \to \bar{\psi}_R(x) U_R^{-1}, \qquad \psi_R(x) \to U_R \psi_R(x),$$
$$\bar{\psi}_L(x) \to \bar{\psi}_L(x) U_L^{-1}, \qquad \psi_L(x) \to U_L \psi_L(x). \qquad (9.2)$$

This is so-called *chiral symmetry*, which is specific for QCD. Note that this symmetry does not survive completely on a quantum level, since classical invariance with respect to the Abelian transformations of the chiral components $\psi_L \to \psi_L e^{i\alpha}, \psi_R \to \psi_R e^{-i\alpha}$ is destroyed by an anomaly. Thus, the global symmetry of QCD in the perturbative domain, above the scale of $\Lambda_\chi \sim 1\ GeV$, remains $SU(3) \times SU(3) \times U(1)$.

The problem is that the chiral transformations mix the states with opposite P-parities. Therefore in the chiral-invariant world, the masses of two states having the same set of quantum numbers, but P-parity, are identical. Naively, we could expect that a small masses of the bare quarks ($m_u \simeq 4\ MeV, m_d \simeq 7\ MeV$) are responsible for some tiny splitting in the spectrum of mesons. However, the experimental data tells us that the real mass splitting is huge. Indeed, let us consider two states with the same quantum numbers but opposite parity, the vector ρ-meson, which has the same quark content as pions, and the light axial vector a_1-mesons. They have the masses $m_\rho = 770\ MeV$ and $m_a = 1260\ MeV$, respectively. Therefore, the chiral symmetry must be broken spontaneously by some mechanism.

Experimental values of the phenomenological parameters of the QCD sum rules, which characterize the QCD vacuum, provide some clue to the mystery of non-perturbative QCD [196]. There is a quantity that can be associated with an order parameter describing the spontaneous chiral symmetry breaking, the chiral condensate $<\bar\psi\psi> \simeq -(240\ MeV)^3$. By definition, it is a closed quark loop and its non-zero value means that a nearly massless quark propagating in QCD vacuum for some reason obtains a dynamical mass.

Another important characteristic of the QCD vacuum is the so-called gluon condensate

$$\frac{1}{32\pi^2}<F_{\mu\nu}F^{\mu\nu}> \simeq (200\ MeV)^4\,.$$

One may consider the non-zero value of this condensate as a signature of the non-trivial structure of the QCD vacuum [196]. Indeed, if we consider zero-point oscillations over a trivial vacuum, such a quantity vanishes, since the average potential energy of zero-point oscillations $<(B_n^a)^2>$ is equal to the average kinetic energy $<(E_n^a)^2>$ and, therefore, $<F_{\mu\nu}F^{\mu\nu}> = 2<(E_n^a)^2 - (B_n^a)^2> = 0$. Thus, perturbative fluctuations cannot be responsible for a non-zero value of the gluon condensate. This parameter indicates that there are some large-scale non-perturbative fluctuations of the gauge field in the QCD vacuum.

On the other hand, a non-zero value of the chiral condensate indicates that there are fermionic zero modes in the QCD vacuum, because [112]

$$<\bar\psi\psi> = -\pi\bar\nu(0)\,, \tag{9.3}$$

where $\bar\nu(\lambda)$ is an averaged spectral density of the Dirac operator: $i\gamma^\mu D_\mu \psi = \lambda\psi$. However, according to the index theorem (cf. the following discussion in Chap. 10) this means that the background gauge field must be characterized by some non-trivial topological number.

A very elegant description of the mechanism of chiral symmetry breaking by instantons was developed in 80's (the so-called *instanton liquid model*, see [195, 196, 452] and references therein). The remarkable conclusion is that although both spontaneous chiral symmetry breaking and phase transition of QCD into the confinement phase take place approximately at the same non-perturbative scale, the mechanisms of those phenomena may be completely different. Moreover, the weak confinement, that is the binding of quarks into hadrons, can be explained entirely as an effect of the instanton-induced effective interaction between light quarks [196, 452]. At the same time, it looks as thought the instanton liquid model cannot generate a linear potential of interaction between quarks that can fit the experimental dates. Thus, the problem of strong confinement remains unsolved. Actually, one must find a consistent way to combine two completely different pictures, the perturbative QCD and a low energy effective theory, e.g., the Skyrme model, which deal with quarks and gluons and hadronic degrees of freedoms, respectively. The underlying description must interpolate dynamically between these limiting cases and all non-perturbative effects must vanish into thin air as we move above Λ_χ in the energy scale.

There is a quantity that describes the interaction between the quarks and can be associated with a non-local order parameter that characterizes the phases of QCD. This is the Wilson loop operator, the trace of the parallel transport along a finite closed path in space-time:

$$W(C) = \text{Tr}\left\{P\exp\left[i\oint_C A_\mu(\xi)d\xi^\mu\right]\right\} \tag{9.4}$$

(P marks the path ordering along the contour C). If the contour is closed, the Wilson loop is a gauge invariant quantity. The Wilson loop can be interpreted as an amplitude of propagation of a quark-antiquark static pair along time direction. The quarks are separated by the distance R and, for a large value of time T, we can write

$$W(C) \simeq \exp\left[-V(R)T\right],$$

where $V(R)$ is the energy of static interaction between the quarks. If this energy for a large quark separation grows linearly, that is, $V(R) = \sigma R$, then the Wilson rectangle loop behaves as follows

$$W(C) \simeq e^{-\sigma RT}. \tag{9.5}$$

This relation is known as the *area law* and it is a signature of confinement. The parameter σ is called the *string tension*.

Note that in *quenched QCD*, that is, in the limit of infinitely heavy quarks, the string tension is a good order parameter that characterizes the confinement-deconfinement phase transition. On the other hand, the chiral condensate is a proper order parameter in the chiral limit (zero quark mass). However, current lattice calculations suggest that the chiral condensate and the string tension are non-vanishing in both phases (although both these quantities are suppressed in the deconfinement phase). Thus, the realistic QCD experiences not a phase transition, but the *crossover*.

Obviously, in the deconfinement phase, the exponent in the Wilson loop depends on the distance R in a different way. It tends to the limit that is just double the quark mass (no interaction at all). However, even if we have introduced a proper order parameter to describe the QCD phase transition, the question still remains: what is the mechanism of the confinement?

Nowadays, a common point of view is that there are some non-perturbative configurations of the gauge field that play a crucial role in the confinement of color degrees of freedom, the quarks do not have much to do with it[2]. Thus, confinement is considered to be a property of pure gluodynamics.

There are several phenomenological approaches to model non-perturbative properties of the Yang–Mills vacuum, for example, the instanton liquid model, or the stochastic QCD vacuum model, etc. (for a review see [452, 475]). Probably one of the most popular versions of the confinement model on the market today is a direction that was initiated in seventies by Mandelstam [362, 364], 't Hooft [271, 276] and Nambu [395]. This approach is based on a direct analogy with the theory of superconductivity. In this simple qualitative picture, the confinement is explained as a dual Meissner effect that prevents the electric color flux from spreading out.

The basic idea is that in the infrared limit QCD takes the form of the effective dual Abelian Higgs model. Then the confinement of color can be understood as the effect of monopole condensation in the QCD vacuum. By analogy with the Cooper condensation in a superconductor sample, the corresponding order parameter is a monopole condensate $<\varphi>\neq 0$, whose non-zero value yields squeezing of the chromoelectric field into a flux tube, just as the Abrikosov string is formed in a superconductor. That would imply a linear confining potential between quarks.

There are two subtle points in this very popular picture. First, the idea about equivalence between low-energy QCD and a dual superconductor model still remains highly speculative[3]. Second, there really are electric charges in a normal superconductor that may condense into the Cooper pairs. In the monopole condensation model, the monopoles have to be created first. This needs an additional energy. Therefore, such a model can work if the monopoles are almost massless. This is in agreement with our discussion of the properties of the non-Abelian monopole solution of the Georgi–Glashow model above: the monopole mass M is proportional to the inverse gauge coupling constant e and, therefore, M tends to zero as we approach the scale of Λ_{QCD}.

However, the worrisome question is that now we have to deal with gluodynamics and it is not clear how monopoles could appear in a pure gauge theory. Indeed, we have seen that without a contribution of the scalar field, the monopole-like solutions are singular. The approach designed to discover

---

[2] Note that Gribov argued against this paradigm [248].
[3] This type of duality can be proved in some supersymmetric generalization of QCD which, alas, is not realized in our real world. We shall consider this model in the last part of this book.

monopoles in a theory, where they cannot exist, is called *Abelian projection*. We shall discuss this technique in this chapter.

9.1.1 Dual Superconductor

Before proceeding with the description of Abelian projection, let us consider the physical content of the idea of dual superconductivity. First, we explain qualitatively what the dual superconductor itself is and what happens if we put a test charge (quark) into such a medium.

In the simplest Abelian case, we can summarize our description above in the following low-energy phenomenological form of the model

$$L = -\frac{1}{4}\widetilde{F}_{\mu\nu}\widetilde{F}^{\mu\nu} + \frac{1}{2}|D_\mu\varphi|^2 - \frac{\lambda}{4}\left(|\varphi|^2 - v^2\right)^2, \qquad (9.6)$$

where the covariant derivative of the complex scalar field φ is defined as

$$D_\mu\varphi = \partial_\mu\varphi + ig\widetilde{A}_\mu\varphi, \qquad (9.7)$$

and the dual field strength tensor is written via dual potential $\widetilde{A}_\mu$ as follows:

$$\widetilde{F}_{\mu\nu} = \partial_\mu\widetilde{A}_\nu - \partial_\nu\widetilde{A}_\mu = \frac{1}{2}\varepsilon_{\mu\nu\rho\sigma}F^{\rho\sigma}.$$

Note that the Lagrangian (9.6) is invariant with respect to $U(1)$ gauge transformation $\varphi \to e^{ie\alpha}\varphi$.

Clearly, this theory differs little from standard Landau–Ginzburg model. The only point is the permutation of the magnetic and electric fields, which can be considered as a result of the discrete transformation (1.81). The scalar complex order parameter field $<\varphi>$ in this dual picture must be treated as a many-particle macroscopic wave function that describes a hypothetic monopole condensate. Even although the properties of such a system are well-known, we will briefly recapitulate them.

Let us consider the static configuration $\partial_0\varphi = 0$, $\partial_0\widetilde{A}_\mu = 0$ in the Hamiltonian gauge $\widetilde{A}_0 = 0$. Then the energy of the system (9.6) reads

$$E = \frac{1}{2}[\varepsilon_{ijk}\partial_j\widetilde{A}_k]^2 + \frac{1}{2}\left(\partial_k\varphi - ig\widetilde{A}_k\varphi\right)\left(\partial_k\varphi^* + ig\widetilde{A}_k\varphi^*\right) + \frac{m^2}{2}|\varphi|^2 + \frac{\lambda}{4}|\varphi|^4. \qquad (9.8)$$

It is well-known that if the parameter $m^2 = \lambda v^2$ is positive, the minimum of the energy (9.8) is characterized by a vanishing condensate $<\varphi> = 0$. In this phase, the vacuum expectation value of the monopole current $j_k = -\frac{i}{2}(\varphi^*\partial_k\varphi - \varphi\partial_k\varphi^*)$ vanishes, if the condensate field slowly varies in space. Meanwhile, the dual photon associated with the fluctuations of the field $\widetilde{A}_k$ remains massless.

However, if $m^2 < 0$, the vacuum state is completely different, because the minimum of the functional (9.8) corresponds to $<\varphi> = v$. Here, we encounter the spontaneous symmetry breaking mechanism once again. Indeed, if the vacuum would be $U(1)$ invariant, a vacuum expectation value of any magnetically charged field, such as φ, would vanish. Otherwise, the symmetry of the action is no longer a symmetry of the vacuum, that is, the symmetry is spontaneously broken.

It is convenient to represent φ as $\varphi = \rho\exp[i\alpha]$. The phase, a massless field α, is actually an unphysical Goldstone mode. It may be eliminated by the gauge transformation

$$\varphi \to e^{-i\alpha}\varphi, \qquad \widetilde{A}_\mu \to \widetilde{A}_\mu(x) - \frac{1}{g}\partial_\mu\alpha.$$

The energy then becomes

$$E = \frac{1}{2}[\varepsilon_{ijk}\partial_j\widetilde{A}_k]^2 + (\partial_k\rho)^2 + g^2\rho^2\widetilde{A}_k^2 + \frac{\lambda}{4}\left(\rho^2 - v^2\right)^2,$$

and for the small excitations about the non-trivial vacuum, $\chi = \rho - v$ and $\widetilde{A}_k$, we have the equations

$$(\partial_k^2 + 2g^2v^2)\widetilde{A}_k = 0, \tag{9.9}$$

$$(\partial_k^2 + \lambda v^2)\chi = 0, \tag{9.10}$$

where, as usual, the Lorenz condition for a dual vector potential has been used. Thus, we are dealing with one Higgs field with the mass $m = \sqrt{\lambda}v$ and one massive vector field with the mass $m_v = \sqrt{2}gv$. The Goldstone mode disappears, having been transformed into the longitudinal component of the gauge field and the $U(1)$ symmetry is broken.

We can treat the (9.9) as a dual analog of the London equation. Phenomenologically, it implies the existence of a steady magnetic current at zero magnetic field. Clearly, (9.9) means that the electric field in a dual superconductor should vanish exponentially within the sample as $\exp\{-m_v r\}$. Thus, an external electric field does not penetrate into a dual superconducting sample beyond a thin boundary layer, whose thickness is about the inverse mass of the vector field: $\lambda_1 = m_v^{-1}$. This is the so-called *dual Meissner effect*. It the context of field theory, we can say that the broken gauge symmetry of the QCD vacuum in a dual superconducting phase leads to massive dual photons.

Note that we actually have two length scales in (9.9): the second scale is given by the inverse mass of the scalar field as $\lambda_2 = m_s^{-1}$. A superconductor is called type I if $\lambda_1 < \lambda_2$, otherwise we have a superconductor of type II.

Let us recall that the dual Higgs model (9.6) contains the Abrikosov–Nielsen–Olesen string solution [64, 399], which, in the context of our consideration, describes a finite electric flux tube. The topological analyses again helps us. Indeed, for the energy to be finite, it is necessary that the field φ minimizes $V[\varphi]$ at the two-dimensional spatial infinity. This means that $\varphi(|r| = \infty)$ may take vacuum values on the S^1 circle $|\varphi| = v$. However, the spatial infinity of two-dimensional space is also the circle S^1. As a point moves around this circle, the value of φ can wind an integer number of times $n \in \mathbb{Z}$ around the circle $|\varphi| = v$, that is, we have a homotopic mapping $\phi : S^1 \to S^1$, which was considered in Chap. 3 (cf. discussion on page 79). A solution that is characterized by a winding number n is possible for a configuration with the boundary conditions at $r \to \infty$

$$\varphi(r,\theta) \to v\exp(in\theta), \qquad \widetilde{A}_k \to \frac{n}{gr}\hat{e}_k, \qquad (9.11)$$

where r and α are the polar radius and the angle in the cylindrical coordinates in the two-dimensional plane, and $\hat{e}_k$ is the unitary vector tangential to the circle of radius r. Thus, we relate the orientation of $\varphi(\mathbf{x})$ on the complex plane with spatial coordinates, just as in the case of the 't Hooft–Polyakov monopole, and the covariant derivative of the scalar field vanishes asymptotically again:

$$D_r\varphi \to 0, \quad D_\theta\varphi \to \frac{1}{r}\partial_\theta\varphi + ig\widetilde{A}_\theta\varphi$$
$$= \frac{inv}{r}\exp(in\theta) - \frac{inv}{r}\exp(in\theta) = 0. \qquad (9.12)$$

As we have already mentioned, any deviation from an integer value of n results in a discontinuity of φ, which leads to an infinite energy of the configuration. The winding number is simply the topological charge.

The boundary condition (9.11) results in remarkable properties of the solution. Let us calculate the flux Φ of the electric field in the z-direction. According to the Stokes theorem, it reads

$$\Phi = \oint_{S^1} \widetilde{A}_k dl_k = \frac{2\pi n}{g}. \qquad (9.13)$$

This is the dual version of the famous condition for *flux quantization*. It states that there is a nontrivial vortex solution that describes the electric field penetrating the superconductor along a line, the so-called *flux tube*.

In order to describe this object better, let us find the localization of the electric field. It cannot be distributed over the whole area $\mathbb{R}^2/\{0\}$ of the two-dimensional plane, because the potential (9.11) does not produce any electric field for any point but the origin $\{0\}$, where it is singular. Thus, the electric field should be present near the core of the solution, where the dual potential deviates from the boundary conditions (9.12). At some point of this region, the field φ must vanish, since the homotopic classification is equivalent to the description in terms of the two-dimensional Poincaré–Hopf index. The latter is defined as a mapping of a circle surrounding an isolated point r_0, where the scalar field is vanishing, $\phi(r_0) = 0$, onto a sphere of unit radius S^1. Thus, the winding number is equal to the sum of non-degenerated zeros of the scalar field. To prove this, we have to shrink the contour S^1 on the spatial boundary down to a very small one, quite near the core. The winding number is $n = 1$ for the former and $n = 0$ for the latter. Because the winding number is an integer, the abrupt change of n can occur only at the moment when the contour crosses the point where $\varphi = 0$, at which the phase of φ is undefined. The nulling of φ means that the superconducting state is destroyed. In physical terms, we say that the electric field penetrates into the dual superconductor along the flux tube where the dual superconductivity is destroyed. The tube thickness is a result of the balance between the surface tension of the interface between the normal and the superconducting phases (which tries to compress the tube) and the pressure of fields compressed in the tube.

To complete our discussion, we recall that the relation between the characteristic scales of the vector and the scalar fields, λ_1 and λ_2, respectively, defines the dynamical properties of the system [127]. By analogy with expression (5.57), the linear energy density of the vortex, the so-called *string tension*, can be expressed as a sum of a surface term equal to $\pi^2 m_v^2/g^2$ and an integral term, the minimization of which allows us to find the solution. For $|n| \geq 2$, this still allows a choice between a "fat" single vortex with flux $n\Phi_1$ and a multi-vortex configuration composed of n widely separated vortices with unitary flux Φ_1. As to which possibility is realized depends on the ratio λ_1/λ_2. In type I superconductors, a vortex with $|n| \geq 2$ has a lower energy than n vortices with unitary winding number. These vortices tend to merge together. The situation is reversed in the type II dual superconductor, where the vortices with large n decay into many vortices with $Q = 1$.

Coming back to the properties of QCD vacuum, we can refer to the numerical calculations that were performed in the case of pure $SU(2)$ gluodynamics on a lattice [456]. The result shows that there probably the second possibility is realized, but there is also a possibility that $\lambda_1 = \lambda_2$ is not excluded (see the discussion in the review [165]).

So far, we have discussed the properties of a hypothetic medium: a dual superconductor. Now let us suppose that, in addition to the dual photon and a collective macroscopic field φ, which describe a monopole condensate, our model also includes massive electrically charged spinors q. Since the corresponding particles, the quarks, are very heavy, they do not destroy the collective dynamics of the monopole condensate. However, the properties of the vacuum state would strongly affect the character of interaction between the quarks. In a normal phase, the field of two electrically charged test particles with opposite charges is the standard electric field of a dipole. However, in the superconducting phase, monopoles condense and there are monopole currents, which form the flux tube between the quarks (see Fig. 9.1). The energy of the flux tube is proportional to its length, that is, we have a linear potential of quark-antiquark interaction.

Further development of this formalism was presented in [98, 99, 101, 102, 359]. This is a more realistic model of the dual description of long-distance QCD which, in the absence of quarks, is given by the Lagrangian of the

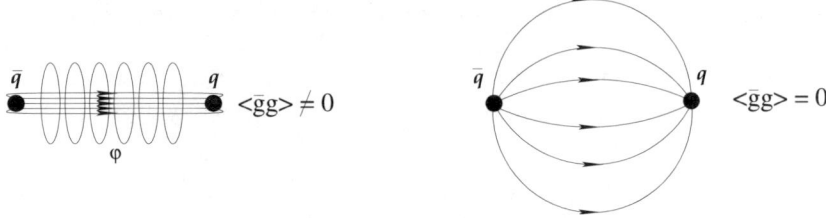

Fig. 9.1. Formation of the chromoelectric flux tube as a dual Meissner effect

non-Abelian vector $SU(3)$ field (dual gluons) coupled with the isoscalar Higgs field. In this model the latter must carry color magnetic charge. These fields are weakly coupled at long distances, in contrast to original quark-gluon interaction. The spontaneous symmetry breaking mechanism produces the mass of the dual gluons.

The model can be extended to incorporate classical heavy quarks at finite separation. Since dual coupling is weak and quarks supposed to be very heavy, the dual Lagrangian can be expanded in inverse powers of the quark mass and then various calculations can be performed to estimate the form of the potential of interquark interaction. This matter is, however, out of the scope of the present book and we refer the reader to the original papers for a detailed description of this approach. Here, we only note that it would be very interesting to see whether such a dual QCD has something in common with the low-energy effective theory of massive and massless $SU(3)$ monopoles, which we discussed in Chap. 8. The dual superconductivity will remain only a phenomenological model until it is explicitly derived from QCD. Moreover, so far we have no analytic proof of the appearance of the monopole condensate, neither in realistic QCD nor in pure gluodynamics.

Probably the reason why the model of dual superconductivity has become so popular, is mainly due to an interesting observation: at our present state of knowledge in all theories that allow us to prove the strong confinement analytically, this phenomen appears as an effect caused by the monopole condensation! Unfortunately, all these models, compact QED [429], the $SO(3)$ Georgi–Glashow model [430] and the $N = 2$ SUSY Yang–Mills theory [469] differ a lot from the realistic QCD. However, there are some evidences that may suggest that the QCD vacuum does behave as a dual superconductor. The most convincing arguments came from the lattice formulation of QCD.

9.2 Monopoles in the Lattice QCD

The most successful non-perturbative approach to QCD is the lattice formulation due to Wilson [521]. The original motivation was to construct a gauge-invariant formulation of a field theory, where the fundamental notion is the partition function[4]

$$Z = \int \mathcal{D}A_\mu \mathcal{D}\psi \mathcal{D}\bar\psi e^{-S[A_\mu,\bar\psi,\psi]}, \qquad (9.14)$$

rather than an action functional $S[A,\bar\psi,\psi]$. The functional integration here involves both the gauge A_μ and the Grassmann variables $\psi,\bar\psi$.

A consistent operational definition of the partition function can be given, if we discretize the Euclidean space in a gauge invariant way. This allows us to replace the integration over the gauge connection A_μ, which takes values in the Lie algebra, by functional integration over the group space.

Wilson used a discrete version of the path ordered phase factor that a fermion wave function ψ picks up when a fermion is moving from a point x to y in the presence of a gauge field A_μ:

$$\psi(y) = P \exp\left\{\int_x^y ieA_\mu dx^\mu\right\} \psi(x).$$

Clearly, for a closed path this phase factor is the Wilson loop (9.4).

In the lattice formulation, the fermions, which are labelled by an integer n, are placed at the nodes of a four-dimensional hypercubical lattice. The gauge fields are associated with the links joining the nearest neighbor sites. Then a parallel transport along a link is given by a unitary matrix

$$U_\mu(n) \equiv U(n, n+e_\mu) = \exp\{iaeA_\mu(n)\} \in SU(N), \qquad (9.15)$$

where a is a lattice spacing, $e_1 = (1,0,0,0)$, etc., and the vector field A_μ lies in some representation of the gauge group. In the following, we shall use the shorthand $aeA_\mu(n) \equiv \theta_\mu(n)$.

The local gauge $SU(3)$ transformation of the field variables on a lattice are defined on each link as

$$\psi(n) \to V(n)\psi(n), \quad \bar\psi(n) \to \bar\psi(n)V^\dagger(n),$$
$$U_\mu(n) \to V(n)U_\mu(n)V^\dagger(n+e_\mu),$$

where a unitary matrix $V(n)$ belongs to the same representation as $U_\mu(n)$. Note that the path ordering of the phase factor after discretization transforms into the relation

$$U_{-\mu}(n) \equiv U(n, n-e_\mu) = \exp\{-iaeA_\mu(n-e_\mu)\} = U_\mu^\dagger(n-e_\mu, n).$$

There are two types of gauge invariant quantities on a lattice:

- A path-ordered string of products such as

---
[4] Here we use the Euclidean form of the action.

$$\text{Tr } \bar\psi(n) U_\mu(n) U_\nu(n+e_\mu)\ldots U_\lambda(m-e_\lambda)\psi(m),$$

where the trace is the usual matrix trace over the group indices. The fermionic part of the lattice action

$$S_{ferm} = \frac{a^3}{2}\sum_{n,\mu}[\bar\psi(n+e_\mu)\gamma_\mu U_\mu(n)\psi(n) - \bar\psi(n)U_\mu^\dagger(n)\gamma_\mu\psi(n+e_\mu)] \quad (9.16)$$

has such a form. Note that if we set up a periodic boundary condition on a lattice, such a string will be closed by the periodicity and we do not need to cap the string by the fermions. These strings are called *Polyakov lines*.

- Let us consider a plaquette $\Pi_{\mu\nu}(n)$ an elementary square of the lattice labelled by the corner site n and two unit vectors e_μ, e_ν. Then the lattice counterpart of the Wilson loop operator (9.4) is the operator of parallel transport around the elementary plaquette

$$\Pi_{\mu\nu}(n) = \text{Tr}\left\{U_\mu(n)\, U_\nu(n+e_\mu)\, U_\mu^\dagger(n+e_\nu)\, U_\nu^\dagger(n)\right\}$$

$$= \text{Tr}\left\{U(n, n+e_\mu)U(n+e_\mu, n+e_\nu+e_\mu)U(n+e_\nu+e_\mu, n+e_\nu)U(n+e_\nu, n)\right\}.$$

The action of the gauge field has to be built up of the set of these loops. Indeed, let us consider the integral over the group variables

$$\mathcal{Z} = \int\prod_{n,\mu} dU_\mu(n)\exp\left[-\frac{2N}{e^2}\text{Re}\{1 - \Pi_{\mu\nu}(n)\}\right]. \quad (9.17)$$

Here dU is the Haar measure of the integration, which is invariant with respect to the left action of the group, namely $d(VU) = dU$ and is supposed to be normalized to unity, that is, $\int dU = 1$. The summation over the links $\mu < \nu$ must be taken under the exponent to avoid double counting.

If the lattice spacing a is small compared with the characteristic scales of the theory, the continuum limit can be recovered from the expansion of the link variable (9.15), $U_\mu(n) \approx 1 + iaeA_\mu + O(a^2)$. Indeed, in the lowest order in lattice spacing we have

$$\sum_n\sum_{\mu<\nu}\text{Re Tr}\{1-\Pi_{\mu\nu}(n)\} \approx a^4\sum_n\frac{1}{2}\sum_{\mu,\nu}\frac{1}{2N}\text{Tr}F_{\mu\nu}F^{\mu\nu} + O(a^6),$$

where, for a given plaquette, we trade the sum over directions $\mu < \nu$ from a point n for half of the sum over all directions: $\sum_{\mu<\nu} \to \frac{1}{2}\sum_{\mu,\nu}$. Thus, in the continuum limit a standard gauge action is recovered:

$$S_{gauge} \xrightarrow[a\to 0]{} \frac{1}{e^2}\int d^4x\, \text{Tr}\, F_{\mu\nu}F^{\mu\nu}.$$

9.2.1 Compact QED and Lattice Monopoles

Thus, the functional integral of the lattice Yang–Mills theory can be written as

$$\mathcal{Z} = \int \mathcal{D}U e^{-NS_{gauge}},$$

where the Euclidean lattice action of the gauge field is

$$S_{gauge} = -\frac{1}{2e^2} \sum_n \sum_{\mu,\nu} \text{Tr}\left[\Pi_\mu(n) + \Pi_\mu^\dagger(n)\right]. \tag{9.18}$$

For the sake of simplicity, let us consider the Abelian electrodynamics on a lattice. Polyakov pointed out [429] that there is a qualitative difference between such a theory formulated in terms of periodic variables $U_\mu(n)$, the elements of the Lie group, and the conventional QED, where the dynamical variable is a connection A_μ, an element of the corresponding Lie algebra. Consequently, unlike the continuum QED, its lattice counterpart has monopoles built into it. Moreover, there is a confinement phase of lattice QED associated with the monopole condensation.

The $U(1)$ theory on the lattice is called the *compact QED*, because the corresponding action can be written in terms of angular variables. Indeed, we now have $U_\mu(n) = e^{i\theta_\mu(n)}$, where θ is a phase that can be integrated over the compact domain $[-\pi;\pi]$. Moreover, each plaquette can also be characterized by an angular variable (cf. Fig. 9.2)

$$\Pi_{\mu\nu}(n) = e^{i(\theta_\mu(n) + \theta_\nu(n+e_\mu) - \theta_\mu(n+e_\nu) - \theta_\nu(n))} \equiv e^{i\theta_{\mu\nu}(n)},$$

and the theory is compact, since the range of functional integration is finite.

Let us prove that the lattice compact QED contains Abelian monopoles. Note that the plaquette variable $\theta_{\mu\nu}(n)$ describes a field flux through the plaquette. Indeed, in the limit of small lattice spacing, we have

$$\theta_{\mu\nu} = ae(\Delta_\nu \theta_\mu - \Delta_\mu \theta_\nu) \approx -a^2 e F_{\mu\nu}, \tag{9.19}$$

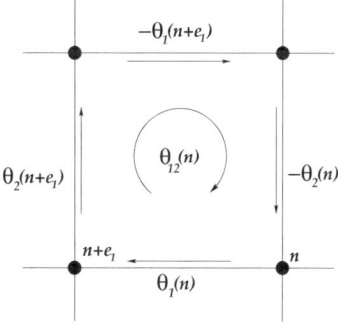

Fig. 9.2. Plaquette variables of the lattice QED

where the operator of the nearest-neighbour finite differences Δ replaces the usual derivative:

$$\Delta_\nu \theta_\mu(n) \equiv \theta_\mu(n+e_\nu) - \theta_\mu(n) \approx a\partial_\nu \theta_\mu(n). \tag{9.20}$$

We can also define the operator of backward differences

$$\Delta_\nu^d \theta_\mu(n) \equiv \theta_\mu(n) - \theta_\mu(n-e_\nu), \tag{9.21}$$

with the property $\Delta^d = -\Delta^\dagger$

Thus, the Wilson–Polyakov action of compact electrodynamics can be represented in so-called "cosine" form:

$$S_{gauge} = -\frac{1}{e^2} \sum_n \sum_{\mu,\nu} \cos\theta_{\mu\nu}(n). \tag{9.22}$$

The very simple structure of this expression was used by Villain to apply a remarkable transformation of the partition function (9.17) [504]. Indeed, in the Abelian theory, the integrand is a periodic function of variable $\theta_{\mu\nu}$, so we may expand it in Fourier series. In the weak coupling limit, this procedure yields the Villian approximation [24, 111, 423]

$$\exp\left\{\frac{1}{e^2}\cos\theta_{\mu\nu}\right\} \xrightarrow[e\to 0]{} \sum_{m_{\mu\nu}=-\infty}^{m_{\mu\nu}=\infty} \exp\left\{-\frac{1}{2e^2}|\theta_{\mu\nu} - 2\pi m_{\mu\nu}|^2\right\}. \tag{9.23}$$

The partition function with the action in the Villain form has the same symmetry properties as for the Wilson action. Moreover, it was proved that the phase structure of both models is similar.

The antisymmetric tensor $m_{\mu\nu}$ appears in the Villain action due to periodicity of the cosine in the initial formulation. It is a set of six independent integers at each lattice site, which does not contribute to the action since they are integer multiplies of 2π. However, it is possible to redefine a compact plaquette field $\theta_{\mu\nu} \to \bar\theta_{\mu\nu} = \theta_{\mu\nu} - 2\pi n_{\mu\nu}$, $\bar\theta_{\mu\nu} \in [-\pi,\pi[$, $\mu < \nu$, where $n_{\mu\nu}$ is an independent plaquette variable. Thus, the physical field $\bar\theta_{\mu\nu}$ is composed of two pieces: the contribution of the electromagnetic gauge field living on the links and the field of the string penetrating the plaquette.

Indeed, the parallel transport over the plaquette $\Pi_{\mu\nu}$, which encloses a string, contributes a phase $2\pi n_{\mu\nu}$ which remains invisible due to the compactness of the model. Note that for a static monopole field, this is a lattice analog of the continuum formula (1.49), the sum of the Coulomb field of a monopole and the quantized singular flux. In the continuum limit, $m_{\mu\nu}$ becomes a singular two-dimensional structure that represents the Dirac string world sheet.

Let us consider the Maxwell equations of the compact QED in the Villain form corresponding to the action in (9.23):

9.2 Monopoles in the Lattice QCD

$$\Delta_\mu \bar\theta_{\mu\nu} \equiv \Delta_\mu(\theta_{\mu\nu} - 2\pi m_{\mu\nu}) = 0, \quad (9.24)$$

and

$$\frac{1}{2}\varepsilon_{\mu\nu\rho\sigma}\Delta_\nu\bar\theta_{\rho\sigma} \equiv \frac{1}{2}\varepsilon_{\mu\nu\rho\sigma}\Delta_\nu(\theta_{\rho\sigma} - 2\pi n_{\rho\sigma}) = 2\pi j_\mu, \quad (9.25)$$

where we take into account that $\varepsilon_{\mu\nu\rho\sigma}\Delta_\nu\theta_{\rho\sigma} \propto \varepsilon_{\mu\nu\rho\sigma}\Delta_\nu\Delta_\rho\theta_\sigma \equiv 0$ and use the definition of the magnetic current

$$j_\mu = \frac{1}{2}\varepsilon_{\mu\nu\rho\sigma}\Delta_\nu n_{\rho\sigma}. \quad (9.26)$$

Thus, the Bianchi identity is violated on the lattice. Note that the magnetic current is conserved in the sense that $\Delta_\mu j_\mu = 0$. Then, the monopole current forms a closed loop on the periodic lattice.

The magnetic charge can be calculated if we integrate the magnetic current j_μ over three-dimensional volume V. Let us choose the time direction to be $\mu = 4$ and consider an elementary three-dimensional cube $V(n)$ at a fixed moment of time. Then the Abelian magnetic charge is defined by the flux through the surface δV, which consists of the six plaquettes Π_{ij} (see Fig. 9.3):

$$g = -\sum_{\delta V(n)} n_{ij}(n), \quad (9.27)$$

where integers n_{ij} correspond to the number of strings passing through the plaquette Π_{ij}. In Fig. 9.3 only $n_{12} = 1$. Clearly, the unit magnetic charge is placed at the centre of the cube $V(n)$.

Finally, let us note that, unlike the conventional QED, there is no problem with singularities or the physical interpretation of the solution, since both the monopoles and the string fields are defined not on the original lattice, where the fermions and the gauge fields are living, but on the *dual lattice*, where the variable m_μ corresponds to the monopoles on the dual links.

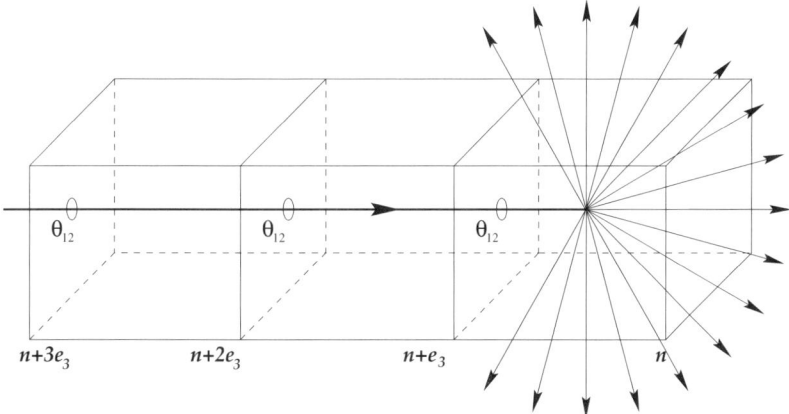

Fig. 9.3. Lattice monopoles and strings in the compact QED

9.2.2 Lattice Duality

It turns out that the idea of duality, which arose in the context of Maxwell electrodynamics, can be generalized to other models. From a modern point of view, the notion of duality means that there are two (or even more) completely equivalent formulations of the same theory, which use different sets of fundamental variables. Moreover, these dual formulations are usually related by the interchange of some parameters, such as, for example, the coupling constant e^2, and its inverse $1/e^2$. Historically this interpretation of duality goes back to the work by Kramers and Wannier [338], who introduced the transformation of the lattice duality to evaluate the temperature of the phase transition of the two-dimensional Ising model.

Dual formulation of the compact QED was first investigated by Banks, Myerson and Kogut [111]. They found a transformation that brings the lattice theory to the form describing a monopole gas with magnetic Coulomb interaction[5]. In fact we can perform a duality transformation that transforms the compact $U(1)$ gauge theory into a non-compact Abelian Higgs model [427]. The scalar field in this model describes monopoles that are condensed in the superconducting phase, where a dual photon becomes massive. This picture obviously is in a perfect agreement with the phenomenological discussion of the previous section.

Recall that at the end of Chap. 1 we briefly mentioned that a usual electromagnetic duality can be considered as a transformation that changes the variables of the functional integration. In particular, we may make use of a Gaussian integration over an auxiliary field that can be promoted to the dual variable (see the discussion on page 24). We can apply this idea to the partition function of the compact QED in the Villain form with the action (9.23) [319]

$$\mathcal{Z} = \int \prod_{n,\mu,\nu} \int_{-\pi}^{\pi} \frac{d\theta_\mu(n)}{2\pi} \sum_{n_{\mu\nu}=-\infty}^{n_{\mu\nu}=\infty} \exp\left[-\frac{1}{2e^2}|\theta_{\mu\nu}(n) - 2\pi n_{\mu\nu}(n)|^2\right]. \quad (9.28)$$

The starting point here is the Gaussian integration over an auxiliary tensor field $\Theta_{\mu\nu}$, which, up to a normalization factor, allows us to represent the Villain partition function as

$$\mathcal{Z} \simeq \int \prod_{n,\mu<\nu} \int d\Theta_{\mu\nu}(n) \int_{-\pi}^{\pi} d\theta_\mu(n) \sum_{n_{\mu\nu}=-\infty}^{\infty} \quad (9.29)$$
$$\exp\left\{-\frac{e^2}{2}\left[\Theta_{\mu\nu}^2(n) + i\Theta_{\mu\nu}(n)[\theta_{\mu\nu}(n) - 2\pi n_{\mu\nu}(n)]\right]\right\}.$$

---

[5] A very nice and compact geometrical description in terms of the differential forms on the lattice was constructed by Fröhlich and Marchetti [227] (see also [165, 427]).

Note that, unlike $\theta_{\mu\nu}$, the field $\Theta_{\mu\nu}$ is defined on the plaquette, not on the links. Now the summation over integer numbers in (9.29) can be performed using the Poisson sum formulae:

$$\sum_{l=-\infty}^{\infty} \delta(x-l) = \sum_{m=-\infty}^{\infty} e^{2\pi i m x}, \qquad \sum_{l=-\infty}^{\infty} f(l) = \sum_{m=-\infty}^{\infty} \int d\alpha f(\alpha) e^{2\pi i m \alpha}. \tag{9.30}$$

Thus, the field $\Theta_{\mu\nu}$ is forced to take integer values. Furthermore, integration over θ_μ yields the equation $\Delta_\mu^d \Theta_{\mu\nu} = 0$, which is satisfied automatically, if we define a dual vector potential (cf. definition (9.19)) as

$$\widetilde{\Theta}_{\mu\nu}(*n) \equiv \frac{1}{2}\varepsilon_{\mu\nu\rho\sigma}\Theta_{\rho\sigma}(n) = \left(\Delta_\mu^d \widetilde{\theta}_\nu(*n) - \Delta_\nu^d \widetilde{\theta}_\mu(*n)\right), \tag{9.31}$$

where $\widetilde{\theta}_\mu(*n) \in \mathbb{Z}$. . Thus, the dual transformation of the partition function yields [111, 319]

$$\mathcal{Z} \simeq \sum_{\widetilde{\theta}_\mu \in \mathbb{Z}} \prod_{*n,\mu,\nu} \exp\left[-\frac{e^2}{2}\widetilde{\Theta}_{\mu\nu}^2(*n)\right]. \tag{9.32}$$

Clearly, the weak coupling limit of the model (9.29) corresponds to the strong coupling limit of the model (9.32), that is, the duality maps these limits again. Thus, the four-dimensional compact gauge theory is dual to a $\mathbb{Z}$ gauge theory.

However, the dual variables $\widetilde{\theta}_\mu$ are defined not on the links of original lattice Λ^4, but on the cubes of the *dual lattice* $*\Lambda^4$, whose sites are labeled by an integer $*n$ (see Fig. 9.5).

The dual lattice $*\Lambda^4$ is obtained by a shift of the original lattice by half of the lattice spacing a in all four dimensions (cf. the shift of the three-dimensional analog as in Fig. 9.4). Sites of the dual lattice are set into correspondence to the original hypercubes, their links are dual to the original three-cubes, the plaquettes are dual to the original plaquettes and the sites correspond to the dual hypercubes. Clearly, if we introduce differential forms both on the original and dual lattices, this definition precisely corresponds to the Hodge star duality (3.31) of these forms.

Thus, the lattice duality supposes not only transformations of the plaquette field variables $\theta_{\mu\nu}(n) \to \widetilde{\Theta}_{\mu\nu}(*n)$, but also transition from Λ^4 to $*\Lambda^4$. The operator Δ_μ is defined on the links of the original lattice, while the operator Δ_μ^d acts on the links of dual lattice. Furthermore, the lattice monopoles are described by the magnetic current $m_\mu(*n)$ (9.26), which is dual to three-tensor $\Delta_\nu n_{\mu\nu}(n)$. While the former current is defined on the links of the dual lattice, the integers $n_{\mu\nu}(n)$ are defined on the original lattice Λ^4.

By analogy with (9.15), we can define the dual Abelian group matrices

$$\widetilde{U}_\mu(*n) = e^{iga\widetilde{\theta}_\mu(*n)}.$$

336 9 Monopoles and the Problem of Confinement

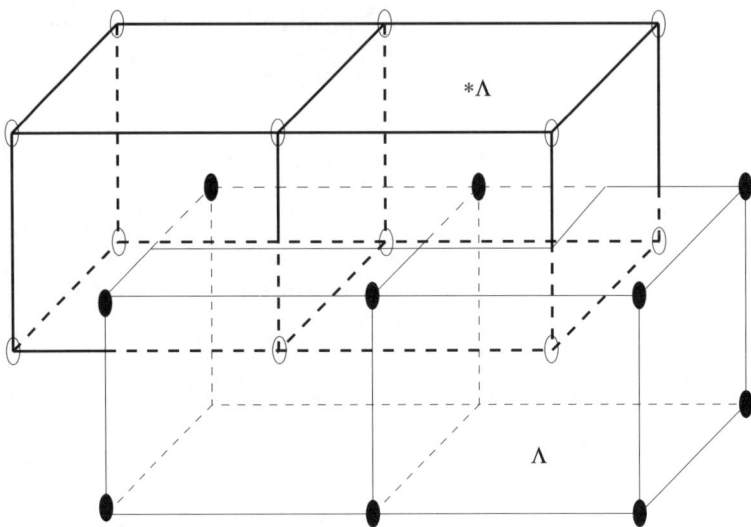

Fig. 9.4. Original and dual lattices in $d = 3$

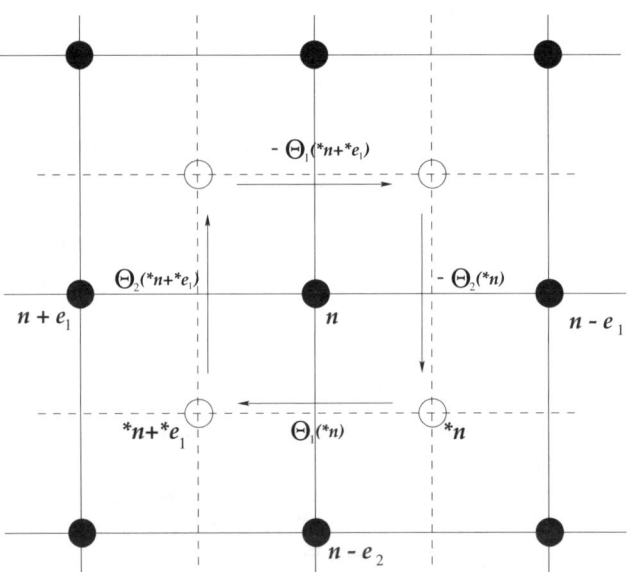

Fig. 9.5. The original lattice and the definition of the lattice dual plaquette variables $\widetilde{\Theta}_{\mu\nu}(*n)$ of the dual lattice

Then the dual action in (9.32) can be represented as a sum over dual plaquettes:
$$S^d = -\frac{e^2}{2} \sum_{*n} \sum_{\mu,\nu} \text{Tr}\left[\widetilde{\Pi}_\mu(*n) + \widetilde{\Pi}_\mu^\dagger(*n)\right], \tag{9.33}$$

where the dual plaquette field is
$$\widetilde{\Pi}_\mu(*n) = \text{Tr}\left\{\widetilde{U}_\mu(*n)\widetilde{U}_\nu(*n+e_\mu)\widetilde{U}_\mu^\dagger(*n+e_\nu)\widetilde{U}_\nu^\dagger(*n)\right\}.$$

Let us return to the evaluation of the partition function (9.32). We can transform the summation over the integers $\widetilde{\theta}_\mu$ into an integral by making use of the second of the Poisson identities (9.30) above. This yields

$$\mathcal{Z} \simeq \int \mathcal{D}\widetilde{\theta}_\mu \exp\left[-\frac{e^2}{2}\left(\sum_{*n,\mu,\nu}\widetilde{\Theta}^2_{\mu\nu}(*n)\right) + 2\pi i \sum_{*n,\mu}\widetilde{\theta}_\mu m_\mu\right], \tag{9.34}$$

where $m_\mu(*n)$ is the integer-valued magnetic current (9.26), which is coupled with a dual photon $\widetilde{\theta}_\mu(*n)$.

Since the integral over the dual gauge potential $\widetilde{\theta}_\mu(*n)$ is Gaussian, taking into account (9.31), we finally obtain [111, 319]

$$\mathcal{Z} \simeq$$
$$\sum_{n_\mu(*n)\in\mathbb{Z}}\left(\prod_{*n}\delta(\Delta_\mu m_\mu)\right)\exp\left[-\frac{2\pi^2}{e^2}\sum_{*n,*n'}m_\mu(*n)D^{-1}(*n-*n')m_\mu(*n')\right], \tag{9.35}$$

where $D^{-1}(*n - *n')$ is the lattice version of four-dimensional massless propagator that satisfies $\Box D^{-1}(n-n') = \delta_{nn'}$, and $\Box \equiv \Delta_\mu \Delta_\mu^d$ is the lattice Laplace operator.

The partition function (9.35) describes integer-valued ring currents of massless monopoles that propagate along the links of dual lattice. Actually, this is a three-dimensional lattice Coulomb gas in a plasma phase. The problem is now to sum over such configurations to take into account their contribution to the Wilson loop operator.

Thus, there is a disordering mechanism connected with monopole loops because the monopole density changes as the coupling constant varies. In the perturbative domain, the monopole ring currents are strongly suppressed and we can drop it. However, when e^2 becomes large, there are macroscopic monopole loops that disorder the system. In his pioneering work, Polyakov showed [429] that, in this case, the Wilson loop operator satisfies the area law and there is confinement in the compact QED due to monopole condensation. To check this conclusion, we must define on the links of the original lattice an external integer-valued electric current with a charge $q \in \mathbb{Z}$ [111]

$$J_\mu = \begin{cases} q & \text{if link} \quad (n, n+e_\mu) \text{ lies in the contour } C, \\ -q & \text{if link} \quad (n+e_\mu, n) \text{ lies in the contour } C, \\ 0 & \text{otherwise}. \end{cases} \quad (9.36)$$

This current represents a Wilson loop. In other words, we have a monopole gas interacting with an external magnetic field generated by the stationary ring current J_μ and the corresponding terms must be included into the action (9.35). This is a picture of dual electrodynamics in the superconducting phase, where a linear force between two well-separated charges arises due to the flux formation.

However, we have to remember that the confinement here is no more than an artifact of the model, the compact lattice QED. Indeed, in the continuum limit we are forced to take $e^2 \to 0$. Then the Coulomb self-energy of monopoles blows up, their density goes to zero exponentially and no trace of the monopoles remains [111].

Another way to establish a relation between the Polyakov result and our phenomenological understanding of confinement as a dual Meissner effect, was discussed by Fröhlich and Marchetti [227], who pointed out that the partition function (9.32) can be represented as a limit of the non-compact dual Abelian Higgs model

$$\mathcal{Z} \simeq \int_{-\infty}^{\infty} \mathcal{D}\tilde{\theta} \int_{-\pi}^{\pi} \mathcal{D}\varphi \exp\left\{ \sum_{*n,\mu} S^d + \frac{1}{2} \sum_{*n,\mu} \left| \varphi(*n) - \tilde{U}(*n)\varphi(*n+e_\mu) \right|^2 \right.$$
$$\left. + \frac{\lambda}{4} \sum_{*n} (|\varphi|^2 - |\varphi_0|^2)^2 \right\}, \quad (9.37)$$

where S^d is the dual gauge action (9.33), which is defined via dual plaquette matrices $\tilde{U}(*n)$. The dual scalar field is parametrized as usual: $\varphi(*n) = \rho(*n) e^{\tilde{\alpha}(*n)}$. It carries a magnetic charge g and this is the magnetic order parameter that we discussed in the previous section. Now we note that in the London limit $\lambda \to \infty$, the scalar field becomes infinitely heavy and its radial dependence is completely frozen out: $\rho = |\varphi_0|$ for all sites of the dual lattice $*n \in \mathbb{Z}$. Now only the phase $\tilde{\alpha}(*n) \in [-\pi, \pi]$ of the Higgs field remains a physical degree of freedom of the scalar field. In other words, in the London limit, the action of the model (9.37) is defined in terms of the compact angular variable $\tilde{\alpha}(*n)$. In the unitary gauge $\tilde{\alpha}(*n) = 0$, the angular dependence of the dual Higgs field can also be gauged out and the partition function reduces to the form

$$\lim_{\lambda \to \infty} \mathcal{Z} \simeq \int_{-\infty}^{\infty} \mathcal{D}\tilde{\theta} \exp\left\{ \sum_{*n,\mu} S^d + \frac{|\varphi_0|^2}{2} \sum_{*n,\mu} \left| 1 - \tilde{U}(*n) \right|^2 \right\}. \quad (9.38)$$

It is clear now that, in the limit of infinite dual photon mass $|\varphi_0| \to \infty$, the dominating contribution to the partition function comes from the vicinity of the configurations $\widetilde{U}(*n) = 1$, that is, $\widetilde{\theta}_\mu(*n) = 2\pi m_\mu(*n)$, where $m_\mu(*n)$ are integers that we identify with the magnetic current (9.26). Performing Gaussian integration over the dual gauge field $\widetilde{\theta}(*n)$, we finally recover the dual $\mathbb{Z}$ field theory in the form (9.32) [227, 427]. Thus, we may conclude that the compact QED is equivalent to the dual Abelian Higgs model in the double limit of infinite dual photon mass and infinite dual Higgs mass.

9.3 Abelian Projection

9.3.1 "Monopoles" from Abelian Projection

So far we have discussed a model that is very different from conventional QCD. Although a lattice model of such a type can be generalized for the case of a compact non-Abelian gauge theory, it still remains a rather phenomenological model, which, however is in fairly good agreement with phenomenological QCD (see, e.g., [98, 99]). The crucial point is that the idea of duality between the Abelian Higgs model and low-energy QCD remains a speculation that must be proved.

Indeed, a sensitive spot of the dual superconductor picture is that there are no monopoles in the gluodynamics. Moreover, there is no Higgs field that would break the symmetry spontaneously to the Abelian subgroup. Therefore a monopole "á la 't Hooft–Polyakov", that is, a localized solution of the classical field equations, does not exist in a pure gauge theory. However, there might exist some large-scale field configurations that, in some gauge, could look "like a monopole".

The idea to consider such configurations as a driving force of the confinement, was pushed forward by 't Hooft [275], who in 1981 suggested a strategy of separation of these "monopoles" in gluodynamics. The guiding idea comes from the observation that a proper treatment of any gauge theory is connected with a procedure of separation of the physical degrees of freedom. In other words, we have to fix some particular gauge that can be considered as a constraint imposed on the model.

It is quite possible, however, that, in some isolated points, the corresponding gauge condition turns out to be singular. For example, a non-physical degree of freedom of the Abelian Higgs model (9.8), the phase of the scalar field, can be eliminated by imposing the unitary gauge, as above:

$$\varphi \to U\varphi = |\varphi|, \quad U = \exp(-in\alpha), \quad \widetilde{A}_\mu \to \widetilde{A}_\mu - \frac{1}{g}\partial_\mu\alpha. \tag{9.39}$$

So far, we have paid little attention to the singularity at the origin of the corresponding parameter, an azimuthal angle of polar coordinates. At this point, the gauge fixing condition becomes ill-defined. However, this is precisely the point where such a transformation creates a singular magnetic flux that penetrates the plane.

This singularity, of course, is not physical and such a vortex is no more than an artifact of the gauge condition that we fixed. However, we have to remember that the Abelian Higgs model contains stable topologically non-trivial solutions of the field equations, the Abrikosov–Nielsen–Olesen vortices. The scalar field determines the thickness of the string, which becomes infinitely thin in the London limit. These configurations are regular everywhere on the plane and they are located precisely at the points where a singularity of the gauge fixing condition (9.39) occurs. Thus, the guess by 't Hooft was that these singularities might be a signature of the topologically non-trivial field configuration.

Let us consider how this idea can be implemented in an non-Abelian gauge theory [275, 276]. The dynamical variables are now the Yang–Mills vector potentials $A_\mu(x_\mu)$, the elements of the Lie algebra. A gauge fixing condition must be imposed as a constraint on the physical states in the Hilbert space. This is an important step since the physical situation becomes much clearer in a proper gauge. One example is the background gauge condition that we imposed in Chap. 6 to quantize the monopoles. However, we are free to choose instead the Lorentz gauge $\partial_\mu A_\mu = 0$, or the unitary gauge as above, or any other gauge fixing condition.

At the same time, the partition function must be independent of an explicit form of the gauge condition. This is the step where Faddeev–Popov ghosts appear (see, e.g., Chap. 12 of [15]). Despite their bizarre properties (they are scalars but do anticommute), the ghost fields are harmless, since they are not propagating and the only effect of the ghosts is to provide the invariance of the partition function under variation of the gauge condition. However, there is no guarantee that the separation between the physical states and ghosts, which exists in the perturbative domain, remains unchanged in the strong coupling regime. On the other hand, some gauge fixing conditions may have degenerated solutions, the Gribov copies. The latter ambiguity arises in the Yang–Mills theory. Thus, we have to analyze the gauge fixing procedure in more detail.

The conjecture by 't Hooft is to make use of the gauge fixing procedure as an instrument to trace up the monopole-like configurations in gluodynamics. Let us consider the $SU(N)$ Yang–Mills theory. Unlike the above considered Georgi–Glashow model, there is no scalar field that provides the spontaneous symmetry breaking to $U(1)$. However, there are Abelian subgroups given by the elements of the diagonal Cartan subalgebra. We may try to separate related degrees of freedom by imposing some condition that does not affect the diagonal subgroup of $SU(N)$, thus the gauge shall be not fixed completely.

Let us consider a fairly general gauge non-invariant field $X(x_\mu)$ in the adjoint representation of $SU(N)$. This is an $N \times N$ traceless matrix, which under local gauge transformations $U(x_\mu)$ transforms as

$$X(x_\mu) \to U(x_\mu) X(x_\mu) U^{-1}(x_\mu) \,. \tag{9.40}$$

The explicit meaning of the field X can be different. For example, it might be the Polyakov loop operator or the Wilson loop (9.4). The composite field operators are also possible. For example, in a model with the spinor matter field, we may consider $(X)_{ij} = \bar{\psi}^i \psi^j$, where i, j are the group indices. In the pure gluodynamics with the gauge group $SU(N)$ with $N > 2$, we may also consider $(X)_{ij} = F_i^{\mu\nu} F_j^{\mu\nu}$.

Moreover, in general, such a field X does not need to be a Lorentz scalar, thus the choice $(X)_{ij} = (F^{12})_{ij}$ is also acceptable. Only the transformation law (9.40) is essential for our choice, and therefore, the gauge potential $(A^\mu)_{ij}$ must be ruled out due to the affine character of the corresponding transformation. The trick of the Abelian projection is to treat X as an ersatz that would substitute the Higgs field.

Clearly, the proper choice of a matrix $U(x_\mu) \in SU(N)$ makes it possible to diagonalize X:

$$X \to UXU^{-1} = \begin{pmatrix} \lambda_1 & \ldots & 0 \\ \vdots & \ddots & \vdots \\ 0 & 0 & \lambda_N \end{pmatrix}, \tag{9.41}$$

where we suppose some ordering for the eigenvalues λ_i. If, for example, the field X lies in the Lie algebra of $SU(N)$, that is, $(X)_{ij} = (X^a)(T^a)_{ij}$, where $(T^a)_{ij}$ are the $SU(N)$ generators in the adjoint representation, the eigenvalues may be ordered as $\lambda_1 > \lambda_2 > \cdots > \lambda_N$, with $\prod_{i=1}^N \lambda_i = 1$. If the field X lies in the group $SU(N)$, we have

$$\lambda_i = e^{i\phi_i}, \quad \sum_{i=1}^N \phi_i = 0, \quad |\phi_i - \phi_j| \leq 2\pi, \quad \forall\ i, j,$$

and then the ordering $\phi_1 > \phi_2 > \cdots > \phi_N$ is natural [339]. Evidently, such a field X resembles the definition (8.7).

However, if we make use of the transformations (9.40) as a special gauge condition, we see that this does not fix the gauge completely. Indeed, the matrix U is defined up to the left multiplication by a diagonal $SU(N)$ matrix

$$V = \begin{pmatrix} e^{i\omega_1} & \ldots & 0 \\ \vdots & \ddots & \vdots \\ 0 & 0 & e^{i\omega_N} \end{pmatrix},$$

because $X = VXV^{-1}$ and $[X, V] = 0$. The matrix V, therefore, lies in the diagonal Cartan subgroup of $SU(N)$. Recall (see the discussion on page 278) that the corresponding subalgebra of dimension $N - 1$ is generated by the operators $\vec{H} = (H_1, H_2 \ldots H_{N-1})$. These transformations are just rotations about the axis $T^3, T^8, T^{15} \ldots$.

Thus, the residual local gauge group is $U(1)^{N-1}$, which justifies the notion of Abelian projection. Indeed, in the gauge, where the field X is diagonal, the vector-potential becomes

$$A_\mu \to A'_\mu = UA_\mu U^{-1} + \frac{i}{e} U\partial_\mu U^{-1}, \qquad (9.42)$$

and its diagonal components $a^i_\mu \equiv (A'_\mu)_{ii}$ transform as $N-1$ "photons":

$$a^i_\mu \to (a^i_\mu)' = a^i_\mu + \frac{i}{e}\partial_\mu \omega_i. \qquad (9.43)$$

The condition of tracelessness of the $SU(N)$ matrix $\operatorname{Tr} A_\mu = 0$ yields the constraint on these fields

$$\sum_{i=1}^N a^i_\mu = 0.$$

The remaining $N(N-1)$ off-diagonal components of the matrix $(A_\mu)_{ij}$ transform as the charged matter fields, that is,

$$(A_\mu)_{ij} \to e^{i(\omega_i - \omega_j)} (A_\mu)_{ij}.$$

We shall refer to them as "gluons" [339].

However, these "gluons" in such an Abelian projected gluodynamics are not massless. Indeed, recall that the matrix X, which we are trying to diagonalize, is treated as is it were an ersatz of the Higgs field. Then the mass term of the "gluons" as usual arises from the square of the covariant derivative $\operatorname{Tr}(D_\mu X)^2$, which contains the mass term $e^2(\omega_i - \omega_j)(A_\mu)^2_{ij}$. These massive states are charged with respect to the Cartan subgroup of $SU(N)$ and their electric charges are $q_i = \pm 1$. The separation between the massless "photons" and the massive "gluons" is, of course, completely artificial since it depends on the way we fixed the gauge.

Let us return to the gauge fixing condition, which can be included into the Lagrangian of the theory. In order to do this, we shall introduce the Lagrange multipliers b_{ij} for the off-diagonal components of the field X and fix the remaining $N-1$ Abelian degrees of freedom via some standard condition, e.g., the Lorentz gauge [275, 277]:

$$L_{gauge} = \sum_{i<j}^N b_{ij} X_{ij} + \sum_{i=1}^{N-1} c_i \partial_\mu a^i_\mu. \qquad (9.44)$$

Recall that the gauge fixing procedure must not destroy the gauge invariance of the partition function, which has to include the Faddeev–Popov determinant. Thus, the ghosts term also appears in the Lagrangian. This conventional procedure is self-consistent also in the Abelian projected gluodynamics, where massless ghost fields are not propagating as they should [275].

9.3 Abelian Projection

The procedure of Abelian projection makes it possible to separate so-called "QED^{N-1}" from the original non-Abelian gluodynamics. However, by analogy with the singularity of the gauge fixing condition of the Abelian Higgs model (9.39) at the origin, the gauge in which the "Higgs" field $X(x_\mu)$ is diagonalized may also be singular at some isolated points $x_\mu^{(i)}$, where the matrix of the local gauge transformation $U(x_\mu)$ rotates $X(x_\mu^{(i)})$ to the form (9.41) with two degenerated eigenvalues:

$$\lambda_i = \lambda_{i+1} \equiv \lambda. \tag{9.45}$$

If X is an element of the group $SU(N)$, that is, $\lambda_i = e^{i\phi_i}$, this relation leads to identification of the phases $\phi_i = \phi_{i+1} + 2\pi n$.

This is the precisely the point where "would-be monopoles" arise. Indeed, let us consider the neighborhood around $x_\mu^{(i)}$, where the matrix of the Abelian projection has a singularity. In order to do this we shall not diagonalize the matrix X completely, taking the 2×2 Hermitian submatrix into consideration, where two adjacent eigenvalues come close to being degenerated. In this intermediate gauge we have

$$X = \begin{pmatrix} D_1 & \vdots & 0 & \vdots & 0 \\ \cdots & & \cdots & & \cdots \\ 0 & \vdots & \begin{matrix} \lambda + \varepsilon_3 & \varepsilon_1 - i\varepsilon_2 \\ \varepsilon_1 + i\varepsilon_2 & \lambda - \varepsilon_3 \end{matrix} & \vdots & 0 \\ \cdots & & \cdots & & \cdots \\ 0 & \vdots & 0 & \vdots & D_2 \end{pmatrix} \propto \lambda + \varepsilon_a(x_\mu)\sigma_a. \tag{9.46}$$

Here D_k are diagonal blocks with non-coinciding eigenvalues other than λ and we use a standard parameterization of an $SU(2)$ matrix by the set of Pauli matrices.

Clearly, the singularity arises if all the components $\varepsilon_a(x_\mu) \to 0$ as we approach the point $x_\mu^{(i)}$. These three constraints define a world line or a point in the three-dimensional space where a "would-be monopole" is located. On the other hand, the degeneration of the eigenvalues means that at $x_\mu^{(i)}$ the residual group of symmetry is no longer the Cartan subgroup $U(1)^{N-1}$, but the non-Abelian group $U(1)^{N-3} \times U(2)$.

Note that in the neighborhood of the singularity $x_\mu^{(i)}$, the non-Abelian components $\varepsilon^a(x)$ of the "Higgs" field X may be written as an expansion into a series

$$\varepsilon^a(x) = \left(x - x^{(i)}\right)^b \partial_b \varepsilon^a(x^{(i)}) \sim \left(x - x^{(i)}\right)^a.$$

Clearly, this resembles the "hedgehog" asymptotic of the scalar field in the Georgi–Glashow model.

The last step in the diagonalization of the matrix X is to rotate this "Higgs" field within the $SU(2)$ subspace to the unitary gauge (5.50). This can be done if we apply the already familiar singular gauge transformation (5.3)

$$U(\theta,\varphi) = e^{i\sigma_3 \frac{\varphi}{2}} e^{i\sigma_2 \frac{\theta}{2}} e^{-i\sigma_3 \frac{\varphi}{2}} = \begin{pmatrix} \cos\frac{\theta}{2} & -\sin\frac{\theta}{2} e^{-i\varphi} \\ \sin\frac{\theta}{2} e^{i\varphi} & \cos\frac{\theta}{2} \end{pmatrix}, \quad (9.47)$$

which is ill-defined at the south pole of the S^2 sphere in the $SU(2)$ subspace.

We can decompose the matrix of this transformation as $U = U_{reg} U(\theta,\varphi)$, where a regular matrix $U_{reg} \in SU(N)$ is constant in the neighborhood around $x_\mu^{(i)}$. This is a singularity of $U(\theta,\varphi)$ that is responsible for the non-vanishing of the magnetic current (cf. definition (5.33) in Chap. 5):

$$m_\mu^i = \frac{1}{2}\varepsilon_{\mu\nu\rho\sigma}\partial_\nu f_{\rho\sigma}^i, \quad (9.48)$$

where the "electromagnetic" field strength $f_{\mu\nu}^i = \partial_\mu a_\nu^i - \partial_\nu a_\mu^i$. Indeed, in terms of the original fields this tensor can be written as

$$f_{\mu\nu}^i = \left(UF_{\mu\nu}U^{-1} + \left[U\left(A_\mu + i\partial_\mu\right)U^{-1}, U\left(A_\nu + i\partial_\nu\right)U^{-1}\right]\right)_{ii}. \quad (9.49)$$

Since the tensor $F_{\mu\nu}$ is supposed to be regular everywhere, the singularities in (9.49) could only originate from terms $\sim U\partial_\mu U^{-1}$.

Using the definition of the current m_μ^i, we may evaluate the magnetic charge given by the volume integration over the region around the point $x_\mu^{(i)}$:

$$\begin{aligned} g^i &= \frac{1}{4\pi}\int d^3x\, m_0^i = \frac{1}{8\pi}\int d^2S_n\, \varepsilon_{nmk} f_{mk}^i \\ &= -\frac{i}{4\pi}\int d^2S_n\, \varepsilon_{nmk}\left[U\partial_m U^{-1}, U\partial_k U^{-1}\right]_{ii} \\ &= \frac{i}{4\pi}\int d^2S_n\, \varepsilon_{nmk}\partial_m\left[U\partial_k U^{-1}\right]_{ii}. \end{aligned} \quad (9.50)$$

The integrand in the last line is written as a total derivative. However, the singularities of the matrices $U(x_\mu)$ do not allow us to apply the Gauss theorem naively.

Let us note that the magnetic currents vanish everywhere except at the point $x_\mu^{(i)}$ where two eigenvalues coincide. Hence we may restrict the area of integration to be the infinitesimal sphere $S_\varepsilon^2(x_\mu^{(i)})$ surrounding this point. Then only the singularity of the $SU(2)$ matrix $U(\theta,\varphi)$ matters and we have [339]

$$g^i = \frac{1}{8\pi}\int_{S_\varepsilon^2} dS_{\mu\nu}\, \varepsilon_{\mu\nu\rho\sigma}\partial_\rho(1-\cos\theta)\partial_\sigma\varphi\,[\sigma_3]_{ii}. \quad (9.51)$$

This expression can be considered by analogy with the definition of the topological charge (5.37) as a counterpart of the Brouwer index of the "genuine" monopole configuration in the Georgi–Glashow model. In the latter case, the zeros of the scalar field are associated with the positions of the monopoles. Indeed, the integrand in (9.51) is the Jacobian of the transformation from local coordinates on sphere $S^2_\varepsilon(x^{(i)}_\mu)$ to group coordinates θ, φ. In other words, it defines the map $S^2_\varepsilon(x^{(i)}) \to S^2(\theta, \varphi)$, which is characterized by the second group of homotopy $\pi_2(S^2) = \mathbb{Z}$. Thus, our "would-be monopoles" are carrying the charges $g^i = \pm 1/2$ with respect to the Abelian subgroup.

Generalization of this scheme to a "multimonopole" configuration is obvious: now two eigenvalues λ_i and λ_{i+1} become degenerated in a number of points $\{x^{(i)}_\mu\}$, where label i takes any values from 0 to N. Then we have to sum over all monopole locations in a volume V. The constraint, which follows from our definition of the magnetic charge, is that $\sum_{i=1}^{N} g^i = 0$. Furthermore, because there are $U(N)^{N-1}$ electric charges of massive "gluons" $q_i = \pm 1$, a generalization of the charge quantization condition with respect to the Abelian subgroup is

$$\sum_{i=1}^{N} e_i g_i = \frac{n}{2}, \qquad n \in \mathbb{Z}.$$

Let us recall once again that the "monopoles" that we are discussing are not real field configurations, but rather fictive objects that appear as an artifact of the formalism of the gauge fixing and Abelian projection. The conjecture by 't Hooft and others is that the proper choice of the gauge fixing condition may separate such field configurations that dominate in the partition function of gluodynamics in the strong coupling regime. Then these configurations look like $U(1)$ "monopoles" in the Abelian gauge, but actually they are rather the large scale fluctuations of the non-Abelian gauge field. In the Landau gauge, for example, the dual Meissner effect is observed without monopole condensation, but an operator of mass dimension 2 appears instead, providing the gluon condensate and the mass generation of the Abelian electric fields [485].

Indeed, we know already that an infinite chain of instantons along the Euclidean time direction is identical to a single BPS monopole. Other field configurations can also produce a similar effect. This conjecture can be proved numerically, if we apply the scheme of Abelian projection to the gluodynamics on a lattice [339]. These calculations showed (see, e.g., [265, 464, 484] or [165] and references therein) that the monopole currents appear in the Abelian projected gluodynamics. However, we are not so optimistic to claim that monopoles were really copiously observed. More likely we have to say that these monopoles are artifacts of the gauge fixing condition that was imposed to reduce gluodynamics to the Abelian gauge theory. Moreover, the monopole dominance in the confinement phase of gluodynamics was observed up to now only in the special case of the so-called *maximal Abelian gauge*.

9.3.2 Maximal Abelian Gauge

Clearly, the formalism of the Abelian projection has many ambiguities. There is no natural candidate that could play the role of the "Higgs" field X. Moreover, it is not clear what is supposed to be the right choice of the gauge fixing condition. An argument of renormalizability of the gauge fixing condition was used by 't Hooft [275], when he introduced the maximal Abelian gauge, which is both Lorentz covariant and gauge covariant, with respect to the Cartan subgroup. Let us introduce such a condition as follows: we decompose the gauge field into purely diagonal and purely off-diagonal parts:

$$A_\mu = a_\mu + A_\mu^{off}.$$

The maximal Abelian gauge is defined as [275, 339]

$$D_\mu(a) A_\mu^{off} \equiv \partial_\mu A_\mu^{off} + ie\left[A_\mu^{off}, a_\mu\right] = 0. \quad (9.52)$$

Thus, this is a differential equation rather than a straightforward condition of diagonalization. However, it also separates a residual $U(1)^{N-1}$ symmetry with respect to the transformations from the Cartan subgroup, since the diagonal elements $(a_\mu)_{ii}$ transform as a massless "photon" field, whereas A_μ^{off} corresponds to the massive charged "gluons". For example, in the simplest case of $SU(2)$ gauge theory the gauge condition (9.52) is

$$\left(\partial_\mu \pm ieA_\mu^3\right) A_\mu^\pm = 0, \quad \text{where} \quad A_\mu^\pm = A_\mu^1 \pm iA_\mu^2. \quad (9.53)$$

This equation can be treated as a condition of minimization of the functional $F[A^\pm] = \int d^4x [(A_\mu^1)^2 + (A_\mu^2)^2]$, that is, it makes the non-Abelian field A_μ as diagonal as possible.

The disadvantage of the gauge fixing condition (9.52) is that in such a gauge, the massless ghosts shall propagate [275] and this problem is not resolved yet. Another problem is that for a fixed configuration A_μ, there could be Gribov's copies among solutions of the equations (9.53), which must be excluded [109, 255]. As yet there is no clear understanding of how to deal with this problem in a continuum theory, although in the lattice model Gribov's copies can be eliminated [255].

Moreover, we can see that the regular 't Hooft–Polyakov monopole solution in the hedgehog gauge itself does not satisfy the maximal Abelian gauge condition. Indeed, let us consider its spatial asymptotic $A_k^a = \varepsilon_{amk} x_k/(er^2)$. Then we have

$$\left(\partial_\mu \pm ieA_\mu^3\right) A_\mu^\pm = \mp \frac{\cos\theta}{er^2} e^{\pm i\varphi}.$$

Thus, the 't Hooft–Polyakov monopole has too many non-Abelian degrees of freedom to fit the condition (9.53). However, in the unitary gauge (5.50) it has only the Abelian component A_μ^3, that is, the unitary gauge for a monopole is the maximal Abelian gauge as well.

The maximal Abelian gauge is becoming very popular, since it makes possible to discover monopoles in a number of unusual places. Indeed, we can expect in advance that the field of an infinite chain of instantons satisfies the maximal Abelian projection, since this configuration, up to the gauge transformation, is identical to the BPS monopole. However, the miracle of Abelian projection allows us to produce a monopole current even from a single instanton!

Following [163], let us consider the field of an $SU(2)$ instanton, a topologically nontrivial solution of the pure Yang–Mills self-duality equations:

$$A_\mu^a = \frac{2}{e} \frac{\eta_{a\mu\nu} x_\nu}{r^2 + \tau^2 + \rho^2}, \qquad (9.54)$$

where τ is Euclidean time and an arbitrary parameter ρ is called the size of an instanton. The self-dual 't Hooft tensor $\eta_{a\mu\nu}$ is defined as

$$\frac{1}{2}\varepsilon_{\mu\nu\rho\sigma}\eta_{a\rho\sigma} = \eta_{a\mu\nu} = \begin{cases} \varepsilon_{a\mu\nu}, & \mu,\nu = 1,2,3 \\ \delta_{a\nu}, & \mu = 4 \\ -\delta_{a\mu}, & \nu = 4 \end{cases}.$$

The field of an instanton can be written in components as

$$A_k = -\frac{1}{e} \frac{(\tau\sigma_k - \varepsilon_{kma} x_m \sigma_a)}{r^2 + \tau^2 + \rho^2},$$

$$A_4 = \frac{1}{e} \frac{x_a \sigma_a}{r^2 + \tau^2 + \rho^2}. \qquad (9.55)$$

Now recall that this solution is defined up to the gauge transformations (5.2)

$$A_\mu \to U^{-1} A_\mu U - \frac{i}{e} U^{-1} \partial_\mu U, \qquad (9.56)$$

which particularly can be used to set the gauge $A_4 = 0$ [27]. We may consider the general form of $SU(2)$ transformation

$$U = n_0 + i n_i \sigma_i, \quad \text{where} \quad n_\mu = \frac{x_\mu}{\sqrt{r^2 + \tau^2}},$$

which rotates the instanton field (9.54) to the form

$$A_\mu^a = \frac{2\rho^2}{e} \frac{\bar\eta_{a\mu\nu} x_\nu}{(r^2 + \tau^2)(r^2 + \tau^2 + \rho^2)}, \qquad (9.57)$$

where $\bar\eta_{a\mu\nu}$ is the anti-selfdual 't Hooft tensor with the properties $\frac{1}{2}\varepsilon_{\mu\nu\rho\sigma}\bar\eta_{a\rho\sigma} = -\bar\eta_{a\mu\nu}$.

Recall that the instanton is defined as a mapping $S^3 \to S^3$. Thus, such a transformation unwinds the singularity at infinity, but creates a singularity at the origin [163]. Clearly, this field obeys the condition of the maximal Abelian projection (9.53), but the gauge transformation produces a singularity whose origination is discussed in Appendix C in detail.

We already know that a bit more careful treatment of such a singularity allows us to see that the field strength tensor $F_{\mu\nu}$ does not change under these transformations. However, in the formalism of the Abelian projection, we restrict our consideration to the diagonal components of the potential A_μ^3 only. Then the "Abelian" field strength $f_{\mu\nu}^i = \partial_\mu A_\nu^3 - \partial_\nu A_\mu^3$ contains the singularity of string type and monopole current j_μ^i (9.48) does not vanish. In the case of a single instanton, placed at the origin of a coordinate system, this current is a straight line that passes through the center of the instanton.

Thus, the claim is that, in some gauge, the field of a single instanton after the "Abelization" procedure produces singularities of monopole type [163]. It was shown that, in the case of the dilute gas of instantons, the mechanism of Abelian projection produces ring monopole currents whose characteristic scale is of order of instanton radius ρ [256]. However, the numerical calculations on the lattice showed [257] that the string tension is reproduced by a single monopole current of another type that permeates the whole lattice volume. Other monopole currents are localized and they do not contribute to the Wilson loop operator. Since they are scale-invariant at small distances, their connection to the instantons seems to be doubtful.

This looks a bit suspicious and suggests that these "monopoles" again do not have much in common with the original field configuration, but rather appear as an artifact of the gauge condition that transforms a non-Abelian gauge theory into the model with Abelian electric and magnetic charges. These configurations are not solutions of the field equations and their similarity with a monopole is based only on the local behaviour of the fields in the neighborhoods of the singularities.

Such a formalism can in principle produce a monopole current from a very arbitrary configuration. However, we cannot exclude the possibility that Abelian projection can be used as a practical tool to identify large scale fluctuations in a non-perturbative regime. This point of view is supported by numerical calculations, which show that the value of the string tension in the confinement phase, the most important phenomenological parameter of low-energy QCD, can be reproduced in the method of Abelian projection, with an error smaller than 8 percent, if only the contribution of the monopole currents into the Wilson loop operator is taken into account [465, 482]. This is so-called *monopole dominance*, which is probably the strongest argument in favour of the formalism of Abelian projection. However, the monopole dominance disappears if we change the gauge fixing condition [164, 192].

To complete this section, we note that some modification of the gauge fixing condition (9.52) can cure the problem of Gribov copies [463]. This idea of dynamical Abelian projection is to make use of an analogy with the Faddeev–Popov trick. We may insert an identity

$$\sqrt{\det(-D_\mu^2)} \int \mathcal{D}\phi e^{i\int d^4 x \operatorname{Tr}(D_\mu\phi)^2} = 1,$$

into the functional integral over the gauge fields, where an auxiliary real scalar field ϕ lies in the adjoint representation of the gauge group. The interaction between this field and the gauge potential A_μ arises from the standard covariant derivative $D_\mu\phi$ and now we can select the diagonal components of the gauge field if off-diagonal components of ϕ are set to zero.

9.4 Polyakov Solution of Confinement in the $d = 3$ Georgi–Glashow Model

We have already seen in Sect. 9.2 how the mechanism of the strong confinement in the compact QED is connected with the condensation of monopoles. This conclusion supports the model of dual superconductivity even if the realistic QCD is a much more complicated theory. On the other hand, the compact QED is formulated on the lattice where a strong coupling regime is natural and the existence of the phase transition into the confinement phase can be proved numerically. However, Polyakov noted that in $d = 3$, similar results can be obtained also in the continuum Georgi–Glashow model, where condensation of monopoles also leads to the area-low behaviour of the Wilson loop operator [430].

9.4.1 Dilute Gas of Monopoles in the $d = 3$ Georgi–Glashow Model

The phenomenon of confinement means that a parameter with the dimension of mass somehow has to be generated in the model. This parameter actually defines the string tension and, in the model of dual superconductivity, it is precisely the vacuum expectation value of the monopole condensate. However, a qualitative description of the confinement is also possible in terms of the Debye plasma screening of the Coulomb potential. Indeed, the Poisson equation for an electrostatic potential A_0 is

$$\nabla^2 A_0 = -4\pi \sum_{i=1}^N e_i n_i, \qquad (9.58)$$

where the sum is taken over all particles with a charge e_i, which are distributed with a density n_i. If the dilute gas of these charges is in equilibrium state, the density is given by the well-known Boltzmann formula

$$n_i = n_{i0}\, e^{-E_i/kT} = n_{i0}\, e^{-e_i A_0/kT},$$

where n_{i0} is the density of charges with no potential. Because plasma as a whole is neutral, we have $\sum_{i=1}^{N} e_i n_{i0} = 0$. If we suppose now that $e_i A_0/kT \ll 1$, the Poisson equation becomes

$$\nabla^2 A_0 - \mu^2 A_0 = 0, \tag{9.59}$$

where $\mu^2 = 4\pi \sum_{i=1}^{N} n_{i0} \, e_i^2/kT$ defines the so-called *Debye mass*. Clearly, the solution for the potential then becomes $A_0 \sim r^{-1} e^{-\mu r}$, that is a photon, propagating in plasma, obtains the magnetic, or so-called Debye mass, and a standard Coulomb potential becomes screened by a short-range exponent.

As was shown by Polyakov in the 70s, a similar effect of screening by dilute gas of monopoles yields a magnetic mass of photon in the 3d Georgi–Glashow model [429, 430].

Let us begin a brief review of Polyakov's work by considering the truncated Lagrangian (5.7) of the Georgi–Glashow model in $d = 3$ Euclidean space:

$$L = \frac{1}{4} F_{mn}^a F_{mn}^a + \frac{1}{2} (D_m \phi^a)^2 + \frac{\lambda}{4} (\phi^a \phi^a - v^2)^2. \tag{9.60}$$

Here both Lorentz and group indices take the values $m, n, a = 1, 2, 3$.

Note that the expression (9.60) is identical to the potential energy of the original $3 + 1$ Georgi–Glashow model. In other words, a static monopole configuration, which is a solution of the system of the field equations (5.14), corresponds to the motion of a classical particle in $\mathbb{R}^3$. The similarity of the respective models suggests that the solitons of the theory (9.60) are instantons in $d = 3$, rather than dynamical monopoles. The gauge field of the former configuration is asymptotically decaying[6] as $\sim r^{-2}$.

Other essential difference between the $d = 3$ Georgi–Glashow theory (9.60) and its 4-dimensional version is that the gauge coupling constant e of the former model already has dimension of mass. Thus, there is a characteristic scale already on the classical level and the dynamically generated mass is proportional to it. Furthermore, the classical Coulomb potential of three-dimensional model is $V_{Coulomb} \sim e^2 \ln r$, that is the energy of interaction between the charges increases with increasing r, although not linearly. We shall see that quantum corrections could improve the situation and, for a large separation, a linear potential $V \sim \alpha r$ is generated.

Let us consider the 't Hooft–Polyakov solution (5.41) outside the monopole core, that is at the distances $r \gg m_v^{-1} \sim (ve)^{-1}$ where

$$\phi^a \approx v\hat{r}^a - \frac{r^a}{er^2} e^{-m_s r}; \quad A_n^a \approx \varepsilon_{amn} \frac{r_m}{er^2}, \tag{9.61}$$

---

[6] Recall that in $d = 4$ there is also a difference between the field of a monopole, which has the Coulomb asymptotic $\sim r^{-2}$, and the field of an instanton, which decays as $\sim r^{-4}$. Such a long-range asymptotic is used to identify the corresponding field configurations in numerical lattice simulations.

9.4 Polyakov Solution of Confinement in the $d = 3$ Georgi–Glashow Model

where we do not neglect the term which is exponentially suppressed by the mass of the scalar field m_s. As we have seen, this asymptotic behaviour corresponds to the Abelian field of a monopole of unit topological charge

$$B_k = \frac{1}{ev}\varepsilon_{kmn}\phi^a F^a_{mn} \approx \frac{r_k}{er^2}\,.$$

Suppose that there is a number of well separated monopoles and anti-monopoles in the vacuum state[7]. Then the magnetic field is given by a simple superposition of the Abelian fields of several monopoles with charges g_i, which are placed at the points $r_k^{(i)}$:

$$B_k = \frac{1}{e}\sum_{i=1}^{N} g_i \frac{r_k - r_k^{(i)}}{|r - r^{(i)}|^3}\,, \tag{9.62}$$

where we suppose that for any pair of monopoles $m_v^{-1} \ll |r - r^{(i)}|$. The action of the system then can be written as

$$S = NS_1 + S_{int}\,, \tag{9.63}$$

where the first term is the multiple of the classical static one-monopole action (5.48)

$$S_1 = \frac{4\pi m_v}{e^2}f\left(\frac{m_s}{m_v}\right)\,,$$

which we expressed via the masses of the vector and scalar fields. Recall that the smooth function $f(m_s/m_v)$ weakly depends on the scalar coupling and in the Bogomol'nyi limit we have $f(0) = 1$.

The second term in (9.63) is the energy of interaction between the monopoles

---

[7] In general, this assumption of dilute Abelian monopole gas may cause some doubts and its correctness can be questioned. The point is that the vacuum state by definition must yield the minimum of free energy of the system. Thus, to prove if the picture of dilute monopole gas is valid, we have to account for the effect of entropy and check if the interaction between the monopoles remains weak. The most simple solution is, of course to consider the Bogomol'nyi limit where monopoles do not interact at all due to the long-range tail of the scalar force and there are the multi-monopole solutions that we discussed in Chap. 6 above. It turns out, however, that in the limit of the vanishing mass of the scalar field, the monopole gas undergoes the phase transition, which destroys the initial approximation and kills the mechanism of the generation of magnetic mass [198]. On the other hand, the energy of interaction between non-BPS monopoles can be large enough to produce a strong overlap of the cores of monopoles and destroy the picture of dilute monopole gas. The same effect causes excitation of dyonic degrees of freedom and, therefore, the Debye mass in a monopole plasma can be generated only if there is some mechanism that keeps monopoles well-separated.

$$S_{int} = \frac{2\pi}{e^2} \sum_{i\neq j}^{N} \frac{g_i g_j}{|r^{(i)} - r^{(j)}|} - \frac{2\pi}{e^2} \sum_{i\neq j}^{N} \frac{e^{-m_s|r^{(i)}-r^{(j)}|}}{|r^{(i)} - r^{(j)}|}. \qquad (9.64)$$

Here, unlike Polyakov's original work [430], we still keep the exponentially suppressed terms that describe the energy of the additional Yukawa interaction mediated by the massive Higgs field quanta. In the Bogomol'nyi limit, the action (9.63) is reduced to

$$S = NS_1 + \frac{2\pi}{e^2} \sum_{i\neq j}^{N} \frac{g_i g_j - 1}{|r^{(i)} - r^{(j)}|}. \qquad (9.65)$$

As we have already seen (cf. discussion in Sect. 6.3), this corresponds to the familiar effect of non-interaction between two BPS (anti-)monopoles ($g_i = g_j = \pm 1$) and doubling of the energy of attraction between a monopole and an anti-monopole ($g_i = -g_j = \pm 1$). Note that in the opposite limit of a very heavy scalar field, the second term in (9.64) becomes negligible.

The problem is to define the photon propagator on the monopole background taking into account the quantum fluctuations of the fields. To evaluate the corresponding functional integral we must separate the monopole zero modes and diagonalize the second variation of the action. Actually, we have to apply the technique discussed in Chap. 7. Moreover, it is not necessary to evaluate the one-loop quantum correction explicitly.

Let us consider the one-monopole partition function $\mathcal{Z}_1$ defined by (7.76). Separating the integration over the coordinates of the position of a monopole $r_k^{(1)}$, which are translational zero modes, we can write [430]

$$\mathcal{Z}_1 = \int \mathcal{D}r_k^{(1)} \, \Delta \, e^{-S_1} \equiv \int dr_k^{(1)} \, \zeta, \qquad (9.66)$$

where the quantity ζ absorbs both the measure of integration over the translational zero modes, the one-loop determinant Δ and the exponent of the classical action of a monopole. This quantity can be calculated in principle, at least numerically.

Using this definition, we can rewrite the partition function of the dilute monopole gas in $d = 3$ as [24, 430]

$$\mathcal{Z} = \sum_{N, g_i} \frac{\zeta^N}{N!} \int \prod_{k=1}^{N} d^3 r^{(k)} \, e^{-S_{int}} \qquad (9.67)$$

$$= \sum_{N, g_i} \frac{\zeta^N}{N!} \int \prod_{k=1}^{N} d^3 r^{(k)} \exp\left\{ -\frac{2\pi}{e^2} \sum_{i\neq j} \frac{g_i g_j}{|r^{(i)} - r^{(j)}|} + \frac{2\pi}{e^2} \sum_{i\neq j}^{N} \frac{e^{-m_s|r^{(i)}-r^{(j)}|}}{|r^{(i)} - r^{(j)}|} \right\}.$$

Interpretation of this expression in terms of statistical physics is obvious: this is the grand canonical partition sum of the Coulomb gas of charged particles that is modified by an additional attractive Yukawa interaction.

9.4 Polyakov Solution of Confinement in the $d = 3$ Georgi–Glashow Model

We can now see a direct correspondence to the effect of Debye screening: the quantity $\zeta^{-1/3}$, which corresponds to the density of the monopole gas, is actually the finite correlation length that yields the scale of the magnetic mass of the photon.

We can see that this effect really occurs by straightforward calculation of the correlation functions of dilute monopole gas[8]. For the sake of simplicity, we consider the limit of a very heavy Higgs field, where the second term in (9.67) vanishes. By making use of a well-known trick, the partition function (9.67) then may be written as a functional integral over an auxiliary field $\xi(r)$:

$$Z = \int \mathcal{D}\xi \, \exp\left\{\frac{e^2}{32\pi^2} \int d^3r \, (\nabla \xi)^2 \right\} \sum_N \sum_{g_i} \frac{\zeta^N}{N!} \prod_{i=1}^{N} d^3 r^{(i)} e^{i\sum_i g_i \xi(r^{(i)})} \tag{9.68}$$

However, recall that $SU(2)$ monopoles with only the unit topological charge $g_i = \pm 1$ are stable. Then the sum over magnetic charges takes the simple form

$$e^{i\sum_i g_i \xi(r^{(i)})} = e^{i\xi} + e^{-i\xi} = 2\cos\xi \,.$$

Now performing summation over all monopoles in (9.68), we have

$$Z = \int \mathcal{D}\xi \, \exp\left\{-\frac{e^2}{32\pi^2} \int d^3r \left[(\nabla \xi)^2 - 2M^2 \cos\xi\right]\right\}, \tag{9.69}$$

where the Debye mass is defined as

$$M^2 = \frac{32\pi^2}{e^2}\zeta \,. \tag{9.70}$$

Indeed, expansion of $\cos\xi$ in the functional integral (9.69) yields the action

$$S_\xi = \int d^3r \left[(\nabla \xi)^2 - M^2 \xi^2 + \alpha \xi^4 + \ldots\right]$$

and the equation of motion obtained by varying the auxiliary field $\xi(r)$ is the sin-Gordon equation

$$\nabla^2 \xi - 2M^2 \sin\xi = 0 \,. \tag{9.71}$$

This is a non-linear generalization of the Debye equation (9.59) for the dilute gas of monopoles. Its solution defines the parameters of the magnetic screening by the monopole plasma.

The selfconsistency of the model is preserved by weakness of the non-linear interaction in (9.71), since the contribution of the terms of all higher powers

---

[8] Note that we are following here the original paper by Polyakov [430]. An alternative and more simple description, based on the variational approach, was presented in [331].

in the field ξ are negligible compared with the leading quadratic term [430]. Indeed, the coupling at ξ^4 is weak because

$$g = \frac{4\pi^2}{3e^2}\zeta \sim e^{-S_1} \ll 1\,.$$

The correlation functions of the fields can be obtained from the functional integral (9.69), if we introduce an external source $\eta(x)$ that couples to the monopoles as

$$S_\eta = \int d^3r \sum_i g_i \eta(r)\delta(r-r^{(i)}) \equiv \int d^3r\, \rho(r)\eta(r)\,. \tag{9.72}$$

Here the density of monopole plasma is defined as $\rho(r) = \sum_i g_i \delta(r - r^{(i)})$. Then evaluation of the partition function with external current $\eta(x)$ yields

$$Z = \int \mathcal{D}\xi\, \exp\left\{-\frac{e^2}{32\pi^2}\int d^3r\left[(\nabla\xi - \nabla\eta)^2 - 2M^2\cos\xi\right]\right\}. \tag{9.73}$$

We can now evaluate average values of any function of the fields according to the standard procedure. We are interested in the correlation function of the field operator (9.62), which can be written as

$$B_k(r) = \frac{1}{e}\int d^3r'\, \frac{(r-r')_k}{|r-r'|^3}\, \rho(r')\,, \tag{9.74}$$

or, performing the Fourier transformation,

$$B_m(k) = \frac{4i\pi k_m}{ek^2}\rho(k)\,.$$

Evidently, the expectation value of the operator of the electromagnetic field strength is related to the vacuum average of ξ as

$$<B_m(r)> = \frac{1}{e}\int d^3r'\, \frac{(r-r')_m}{|r-r'|^3}<\rho(r')> \tag{9.75}$$

$$= -\frac{i}{e}\int d^3r'\, \frac{\partial}{\partial r'}\left(\frac{1}{|r-r'|}\right)\frac{\delta Z}{\delta \eta}\bigg|_{\eta=0} \tag{9.76}$$

$$= i\frac{e}{4\pi}\partial_m <\xi(r)>\,.$$

The procedure of the evaluation of other correlators appears very similar. For example, the operator of monopole density reads [280]

$$<\rho(r)> = \frac{ie^2}{16\pi^2}<\nabla^2\xi(r)> = iM^2\frac{e^2}{8\pi^2}<\sin\xi(r)>\,. \tag{9.77}$$

Then the Fourier transform yields

9.4 Polyakov Solution of Confinement in the $d = 3$ Georgi–Glashow Model

$$< \rho(k) > = \frac{e^2}{16\pi^2} k^2 < \xi(k) >,$$

and the two-point correlation function can be calculated by analogy with (9.77):

$$< \rho(k)\rho(-k) > = \left(\frac{e}{4\pi}\right)^2 k^2 \cdot k^4 \left(\frac{e}{4\pi}\right)^4 < \xi(k)\xi(-k) >.$$

Since non-linearity of (9.71) is almost negligible, the propagator of the auxiliary field is simply $< \xi(k)\xi(-k) > \sim (k^2 + M^2)^{-1}$ and we can write

$$< \rho(k)\rho(-k) > \left(\frac{e}{4\pi}\right)^2 \left(k^2 - \frac{k^4}{k^2 + M^2}\right) = \left(\frac{e}{4\pi}\right)^2 \frac{M^2 k^2}{k^2 + M^2}.$$

Thus, the two-point correlation function of the operator of the electromagnetic field is

$$< B_m(k)B_m(-k) > \; = \; < B_m(k)B_m(-k) >_0 + 16\pi^2 \frac{k_m k_n}{e^2 k^4} < \rho(k)\rho(-k) >$$

$$= \left(\delta_{mn} - \frac{k_m k_n}{k^2}\right) + \frac{k_m k_n}{k^2} \frac{M^2}{k^2 + M^2} \quad (9.78)$$

$$= \delta_{mn} - \frac{k_m k_n}{k^2 + M^2}.$$

Here $< B_m(k)B_m(-k) >_0 = (\delta_{mn} - k_m k_n / k^2)$ is the standard propagator of a massless photon. The correction to it, which comes from the second term, means that a photon, propagating in a dilute gas of monopoles, indeed acquires the magnetic mass M of (9.70).

9.4.2 Wilson Loop Operator in $d = 3$ Georgi–Glashow Model

We have already mentioned that an order parameter for the strong confinement phase transition may be associated with the expectation value of the Wilson loop operator (9.4)

$$< W(C) > \; = \; < \text{Tr}\left\{\exp\left[i \oint_C A_\mu(\xi) d\xi^\mu\right]\right\} >. \quad (9.79)$$

In the confinement phase this operator satisfies the area-law behavior $< W(C) > \simeq e^{-\sigma RT}$, while the deconfinement phase is characterized by perimeter law.

The Wilson loop can be written as an integral over the area S bounded by the contour C:

$$W(S) = <\exp\left\{i\int_S B_m(r)dS_m\right\}> = <\exp\{i\eta(r)*\rho(r)\}>, \quad (9.80)$$

where we make use of the definition (9.74) and introduce the external field of a pair of external sources of unit electric charge as

$$\eta(r) = \int dS(r')\frac{r-r'}{|r-r'|^3} = -\int dS(r')\frac{\partial}{\partial r'}\left(\frac{1}{|r-r'|}\right).$$

The explicit form of the Wilson loop operator up to a normalization factor is

$$W(S) = \sum_{N,g_i}\frac{\zeta^N}{N!}\int\prod_{k=1}^N d^3r^{(k)}\exp\left\{-\frac{2\pi}{e^2}\sum_{i\neq j}\frac{g_ig_j}{|r^{(i)}-r^{(j)}|} + \int_S dS\frac{r-r^{(i)}}{|r-r^{(i)}|^3}\right\}.$$
(9.81)

The similarity of the corresponding functional integration in (9.81) and (9.67) is obvious. Thus, by complete analogy with the calculation of the effective action (9.69) above, we arrive at [430]

$$W(S) = \int \mathcal{D}\xi \exp\left\{-\frac{e^2}{32\pi^2}\int d^3r\left[(\nabla(\xi-\eta))^2 - 2M^2\cos\xi\right]\right\}, \quad (9.82)$$

which corresponds to (9.73).

In spite of the similarity of (9.82) to (9.73), in the former case we do not suppose that the non-linearity of the field ξ is negligible, because an external field $\eta(r)$ can be strong. To calculate the Wilson loop operator (9.82) in such a system, we note that the field of the test electric charges must satisfy the Poisson equation

$$\nabla^2\beta(x) = 4\pi\delta(z)\theta_S,$$

where $\theta_S = 1$ on the Wilson loop and $\theta_S = 0$ otherwise. Since the value of $W(S)$ does not depend on the form and orientation of the contour, it can be chosen to be in the xy-plane. Then the non-linear Debye equation (9.71), which is modified by the external sources

$$\nabla^2(\xi-\eta) - 2M^2\sin\xi = 0, \quad (9.83)$$

far away from the loop C turns out to be the one-dimensional sin-Gordon equation with the kink solution [430]

$$\xi = 4\arctan\left(e^{-Mz}\right), \quad \text{as} \quad z > 0,$$
$$\xi = -4\arctan\left(e^{-Mz}\right), \quad \text{as} \quad z < 0. \quad (9.84)$$

From the exponent of the Wilson loop operator (9.82) we can now separate the integral over the xy-plane, which gives the area of the loop. This leaves

9.4 Polyakov Solution of Confinement in the $d = 3$ Georgi–Glashow Model

$$W(S) = \exp\left\{-\frac{e^2 M^2}{32\pi^2}\int dz\,(\sin\xi - 2\cos\xi)\int dx dy\right\} = e^{-\sigma RT},$$

where the string tension is defined as

$$\sigma = \frac{e^2 M^2}{32\pi^2}\int dz\,(\sin\xi - 2\cos\xi) \sim \frac{4e\sqrt{2\zeta}}{\pi}.$$

Here we used the definition of the Debye mass (9.70).

Thus, the electric field of two test charges forms a string with tension given by the one-loop monopole action ζ (9.66). This conclusion can be proved by a direct calculation of the expectation value of the electric field operator [280], which shows that the electric string with unit flux is formed in the monopole plasma. Evidently, we may conclude that there is strong confinement in the $d = 3$ Georgi–Glashow model under the assumption that the monopole plasma vacuum state exists.

Finally, let us make some remarks. First, note that it is very instructive to see the similarity of the discussion above to the dual description of the familiar Josephson effect [280]. Recall that the Josephson effect appears in system of two superconductors separated by a barrier. A quantity that characterizes the effect is the change of the phase of the order parameter as we cross the barrier. Its counterpart in the Polyakov solution is the field ξ, which defines the character of interaction between monopoles. Further, the electric charge of the Cooper pair is replaced by the topological charge of the monopole and supercurrents of correlated electrons through the barrier correspond to the density of monopoles in the dilute gas vacuum.

Second, embedding of the model into four-dimensional Euclidean space changes the situation drastically [24]. The solution, which was localized in $d = 3$, corresponds to the world line in $d = 4$. Since the magnetic flux is conserved, these world lines can be either infinite or form the closed loops. We may evaluate the contribution of the monopole-antimonopole loops to the functional integral. The conclusion is that the loops of small size do not introduce a disorder in the system and the electric field of two test charges remains Coulomb-like.

The confinement phase is associated with the large-scale monopole loops, their search is actually the main subject of works directed to evaluating the string tension on the lattice. However, the correspondence between realistic QCD in $d = 4$ and the Georgi–Glashow model in $d = 3$ remains questionable, although the effect of finite temperature leads to the dimensional reduction effectively truncating gluodynamics to be $d = 3$ theory [240].

We have already noted at the evaluation of the correlators above that we make use the of form of the generating functional (9.67) with the scalar field splitted out completely. This approximation fails as we approach the Bogomol'nyi limit $m_s = 0$, where the effect of additional long-range interaction leads to the first-order phase transition in the monopole plasma. Then the picture of the magnetic mass generation that we discussed above is not quite correct. Unfortunately, the same effect gives the excitation of dyonic degrees of freedom of monopoles, because the strong interaction in a model of $d = 3$ plasma of dyons [476, 477] inevitably destroys the initial approximation of dilute gas. Thus, the model of confinement "á la Polyakov" remains self-consistent only in the limit of a very massive Higgs field and zero electric charge of monopoles.

Our last remark concerns the differences between the mechanism of confinement in the $d = 3$ Georgi–Glashow model and in the Seiberg–Witten theory [469]. The most obvious difference is that the BPS limit is natural for the latter supersymmetric theory, while the former model is not applicable there. Although in both theories confinement is closely connected with contributions of monopoles, in the Georgi–Glashow model, the monopoles are not propagating and cannot condense. In contrast, the condensation of BPS states is a key to the description of confinement within the $N = 2$ SUSY model. We shall discuss this model at the end of Chap. 13.

10 Fermions in the Field of Non-Abelian Monopole and Rubakov–Callan Effect

So far, we have restricted ourselves to a minimal set of bosonic fields entering the initial Lagrangian of the Georgi–Glashow model. Including interaction with fermionic fields modifies the theory in an essential way. Such an extended system, can be, for example, a model of the unification of electroweak and strong interactions, which includes both leptons and quarks. The monopole solutions naturally arise in the bosonic sector of this theory as well.

As we shall see in this chapter, the coupling of fermions with a monopole background field yields a number of highly non-trivial effects. On the other hand, including the fermions in a model can be considered as a first step toward a consistent analysis of the properties of monopoles in some supersymmetrical models, for example, the $N = 2$ SUSY Yang–Mills theory that we shall consider in the last part.

10.1 Dirac Hamiltonian on the Non-Abelian Monopole Background

The difference between the problem under consideration and the simplest model of the Abelian interaction between the fermions and the Dirac monopole, which was considered in Chap. 2, is that from now on both the fermionic and the bosonic fields lie in some Lie algebra of the gauge group.

Note that even in the simplest non-Abelian example of the $SU(2)$ theory coupled with fermions, the physical contents of the model depends on the representation of the group. One may consider, for example, either the adjoint (isovector or triplet), or the fundamental (isospinor or complex-doublet) representations of the $SU(2)$ Lie-algebra-valued fermions.

Let us consider the isospinor fermions, that is, we define an eight-component wavefunction ψ^a_α labeled by two indices. The isotopic, or "colour" index a, takes the values $1, 2$ and the Lorentz index α takes the values $1, \ldots, 4$. Here the four-dimensional Dirac matrices γ^μ act on the Lorentz components of the wave function, while the isospinor Pauli matrices τ^a are coupled via the group indices:

10 Rubakov–Callan Effect

$$(\gamma^\mu \psi)^a_\alpha = (\gamma^\mu)_{\alpha\beta}\, \psi^a_\beta\,, \qquad (\tau^a \psi)^b_\alpha = (\tau^a)_{bc}\, \psi^c_\alpha \tag{10.1}$$

that is, the function ψ^a_α is considered as a matrix of dimension 4×2.

The Lagrangian of interaction between a fermion and a monopole can be written as[1] [27, 298, 373, 376]

$$L_\psi = \bar\psi^i_\alpha \left\{ i(\gamma^\mu)_{\alpha\beta} \left[\delta_{ij}\partial_\mu + ie(T^a)_{ij} A^a_\mu \right] - ih(\gamma^5)_{\alpha\beta}(T^a)_{ij}\phi^a \right\} \psi^j_\beta\,. \tag{10.2}$$

The above expression describes both the Yukawa-type pseudoscalar interaction with a coupling h between the fermions and the scalar fields, and the interaction between the gauge and the isospinor fields, which are coupled in the covariant derivative as:

$$D_\mu \psi^i_\alpha = \left[\delta_{ij}\partial_\mu + ie(T^a)_{ij} A^a_\mu \right] \psi^j_\alpha\,. \tag{10.3}$$

Hence the mass of the fermions is completely defined by the vacuum expectation value of the Higgs field v. Evidently, the restoration of the original symmetry means nullification of the fermionic mass term in (10.2). For the sake of completeness, we do not drop out the indices in (10.2) and (10.3), both the Lorentz and the isospinor ones. In the following, however, we shall use compact matrix notations, although in some cases, the indices will be written explicitly as above.

First, let us consider the solutions of the Dirac equation within the topologically trivial sector, where the symmetry is spontaneously broken down to the electromagnetic subgroup and the Higgs field is constant everywhere in space and it is directed along the third axis in the isospace: $\phi = (0, 0, v)$.

The two-dimensional fundamental representation of $SU(2)$ is provided by the set of familiar traceless Hermitian generators $T^a = \tau^a/2$ and only the third component of the gauge potential A^3_μ is involved:

$$\left[\gamma^\mu \left(\partial_\mu + \frac{ie}{2}\tau^3 A^3_\mu \right) - \gamma^5 \frac{hv}{2}\tau^3 \right] \psi = 0\,. \tag{10.4}$$

Thus, the equations for the top and the bottom components of the isotopic doublet in trivial vacuum are decoupled, respectively, as

$$\left[\gamma^5\gamma^\mu \left(\partial_\mu + \frac{ie}{2} A^3_\mu \right) - \frac{hv}{2} \right] \psi^1 = 0\,,$$

$$\left[\gamma^5\gamma^\mu \left(\partial_\mu - \frac{ie}{2} A^3_\mu \right) + \frac{hv}{2} \right] \psi^2 = 0\,. \tag{10.5}$$

---

[1] Here we consider the pseudoscalar coupling between a Higgs field and fermions. It is instructive for the reader to carry out the calculations with the conventional choice of the scalar coupling (cf., for example, the related discussion in [27]). Modifications of this Lagrangian also are possible. We can, for example, include an extra term $L_m = m\bar\psi\psi$, which describes the bare fermion mass m. Henceforth we suppose that $m = 0$.

10.1 Dirac Hamiltonian on the Non-Abelian Monopole Background

Thus it is seen that the electric charges of the components ψ^1 and ψ^2 are $q = \pm e/2$, respectively. At the same time, they are of the same mass[2] $m = hv/2$.

To make our discussion correlated with the problem of the scattering of a Dirac fermion on an Abelian monopole considered in Chap. 2, we shall use the Dirac representation of the γ matrices once again:

$$\gamma_0 = \begin{pmatrix} 1 & 0 \\ 0 & -1 \end{pmatrix}, \quad \gamma_k = \begin{pmatrix} 0 & \sigma_k \\ -\sigma_k & 0 \end{pmatrix}, \quad \gamma_5 = \begin{pmatrix} 0 & 1 \\ 1 & 0 \end{pmatrix}. \tag{10.6}$$

It is well known that the theory is invariant with respect to the fermion number conjugation symmetry, or C-conjugation. The usual matrix of the $U(1)$ fermion charge conjugation in the Dirac representation (10.6) is

$$C = i\gamma^2\gamma^0 = \begin{pmatrix} 0 & -\varepsilon \\ -\varepsilon & 0 \end{pmatrix} = -C^{-1},$$

where $\varepsilon = i\sigma_2$ is a 2×2 matrix with entries ε_{ij}, $\varepsilon_{01} = 1$. Thus, for a standard Dirac fermion we would have $\psi^C = C\bar\psi^T$ up to a phase factor. However, the fermionic field now carries a non-Abelian $SU(2)$ charge and such an operation must be supplemented by the multiplication of the wavefunction ψ_α^a by the matrix $(\tau^2)_{ab}$, which acts on the isotopic degrees of freedom. Indeed, $\tau^2\tau^a\tau^2 = -(\tau^a)^T$ and the conjugated isospinor is defined as

$$(\psi^C)_\alpha^a = (\tau^2)_{ab} C_{\alpha\beta}(\bar\psi^T)_\beta^b. \tag{10.7}$$

Let us consider now the Dirac equation for a fermion in the external field of a 't Hooft–Polyakov monopole. This system is described by the Lagrangian (10.2). Substituting (5.41) into (10.2) leaves the fermionic equation of motion in the form of the Schrödinger equation (cf. its Abelian analog (2.105)):

$$i\frac{\partial\psi}{\partial x_0} = \mathcal{H}\psi \equiv \begin{pmatrix} 0 & \mathcal{D} \\ \mathcal{D}^\dagger & 0 \end{pmatrix} \begin{pmatrix} \zeta \\ \eta \end{pmatrix}, \tag{10.8}$$

where we introduce the operators

$$\mathcal{D} = i\sigma^k D_k - ih\phi^a T^a = \sigma^k \left(i\frac{\partial}{\partial x_k} + A(r)\varepsilon_{akn}T^a\hat{r}^n \right) - i\tau^a\hat{r}^a F(r),$$

$$\mathcal{D}^\dagger = i\sigma^k D_k + ih\phi^a T^a = \sigma^k \left(i\frac{\partial}{\partial x_k} + A(r)\varepsilon_{akn}T^a\hat{r}^n \right) + i\tau^a\hat{r}^a F(r),$$

$$\tag{10.9}$$

and decompose the Dirac wavefunction into up and down spin-isospinor components $\psi = \begin{pmatrix} \zeta \\ \eta \end{pmatrix}$, which are complex 2×2 matrices. For the sake of compactness, we used here the shorthand

---

[2] The unusual sign of the mass term in the second of the equations (10.5) can be inverted by a unitary transformation [27].

$$A(r) = \frac{1 - K(r)}{2r}, \qquad F(r) = \frac{h}{2e}\frac{H(r)}{r}.$$

In studying the behaviour of these fermions in the presence of the monopole, we have to investigate the chiral properties of the fermions. For a given representation of the Dirac matrices the left-hand and the right-hand fermions are defined, respectively, as

$$\psi_L = \frac{1+\gamma_5}{2}\psi = \frac{\zeta+\eta}{2}\begin{pmatrix}1\\1\end{pmatrix}, \qquad \psi_R = \frac{1-\gamma_5}{2}\psi = \frac{\zeta-\eta}{2}\begin{pmatrix}1\\-1\end{pmatrix}, \qquad (10.10)$$

for those $\gamma_5\psi_L = \psi_L$, $\gamma_5\psi_R = -\psi_R$.

The peculiarity of the dynamics of such a fermion is that its mass entirely arises due to coupling with the Higgs field. Therefore, it varies with the distance from the monopole core. Clearly, on the spatial asymptotic, where the symmetry is spontaneously broken down to the $U(1)$ subgroup, as a limiting case we have the system (10.5) once again. However, if a fermion somehow manages to filter down to the monopole centre, where the initial gauge symmetry is completely restored, its mass there vanishes. The question is, if there are such states in the spectrum of fluctuations for which the centrifugal barrier is lifted. Thus, to prove if a fermion can fall down onto the monopole centre, we must analyze the properties of the operator of an angular momentum.

We have seen already in Sect. 7.1.1 that the corresponding integral of motion is not a standard operator of the angular momentum of a fermion, but the combination $\mathbf{J} = \mathbf{L} + \mathbf{T} + \mathbf{S}$. Now the operator of spin is

$$S_k = \frac{1}{2}\begin{pmatrix}\sigma_k & 0\\0 & \sigma_k\end{pmatrix},$$

where standard Pauli matrices $(\sigma^k)_{\alpha\beta}$ are coupled with the Lorentz indices of the left and right components of the field ψ_α^a, and then the operator $\mathbf{J}$ acts as

$$\mathbf{J} = \mathbf{L} + \boldsymbol{\sigma} \otimes I + I \otimes \boldsymbol{\tau}. \qquad (10.11)$$

Indeed, straightforward calculation of the commutator of this operator and the radially symmetric Hamiltonian operator $\mathcal{H}$, which we defined above in (10.8), shows that they do commute[3].

Clearly, there is a certain correspondence between this system and the solutions of the Dirac equation, which describes an electron in the external field of an Abelian monopole (cf. related discussion in Chap. 2). This analogy is emphasized by the similarity between the definitions of the operator $\mathbf{J}$ and its quantum-mechanical counterpart (2.61). Indeed, as before, the addition of three components of the angular momentum $\mathbf{J}$ may yield a spherically symmetric s-wave state in two possible cases:

---

[3] For the axially symmetric multimonopole ansatz (6.15) and (6.16) the Dirac Hamiltonian commutes with the operator $\mathbf{J} = \mathbf{L} + \boldsymbol{\sigma} \otimes I + n I \otimes \boldsymbol{\tau}$, where n is the topological charge of the background configuration.

10.1 Dirac Hamiltonian on the Non-Abelian Monopole Background

- a state with zero orbital momentum $l = 0$ and a vanishing sum of spin and isospin momenta: $\mathbf{S} + \mathbf{T} = 0$;
- a state with unit orbital momentum $l = 1$, which is compensated by the total contribution of spin and isospin momenta.

However, a principal difference is that in the Abelian case, in order to avoid the problem of self-adjointness of the spin 1/2 Dirac Hamiltonian over the complete space of its eigenfunctions, we are forced to impose a very special boundary condition (2.115) at the origin, which actually connects the states with opposite chirality. In contrast to the Abelian theory, the 't Hooft–Polyakov monopole is regular everywhere and there is no reason to believe that the boundary condition of such type is unique.

Indeed, for an s-wave fermion, we have the condition $Q = \hat{r}^a T^a = -\hat{r}^a S^a$, which relates the operator of the electric charge Q and the operator of spirality. In other words, the interaction between the fermion and the monopole may inverse the electric charge of the fermion. This inversion must of course, be compensated by the excitation of the corresponding dyonic degrees of freedom of a monopole:

$$\psi_+ \; + \; \text{Monopole} \; \to \; \psi_- \; + \; \text{Dyon},$$

and there is no violation of the total electric charge conservation.

However, the situation is not so obvious because this naive picture is definitely beyond the initial relativistic quantum mechanical approximation, which supposes that the monopole field does not change in the course of the fermion scattering. As we have seen above (cf. the discussion in Sect. 7.2.2) in a quantum theory the electric charge of a monopole is quantized. This quantization can be interpreted as the coupling of a monopole with a charged vector boson. In other words, the excitation of the dyonic degrees of freedom should be supplied with the increase of the monopole mass by $\sim m_v = ev$. Therefore, for a fermion falling down onto the monopole with an energy lower than such a scale, this process is forbidden. Thus, the one-particle approximation is no longer justified and we have to investigate the process of monopole-fermion interaction within the complete quantum field theory framework. The first step toward this description is to analyze zero modes of the Dirac operator on the non-Abelian monopole background.

10.1.1 Fermionic Zero Modes

Since the operator of generalized angular momentum $\mathbf{J}$ does not mix the left and the right chiral components of the spin-isospin wave function ψ neither in the trivial nor in the monopole sectors, in order to solve the Dirac equation (10.8), as before, we shall decompose it into two independent equations for these components. Now it is convenient to make use of parameterization [27]

$$\psi^i_\alpha = \widetilde{\psi}^j_\alpha \varepsilon_{ji}, \tag{10.12}$$

which is especially useful for the consideration of the interaction between the fermions and an external non-Abelian bosonic field. An obvious advantage of parameterization (10.12) is that we can get rid of the transposed Pauli matrices via

$$(\tau^a)_{ij}\psi_\alpha^j = \psi(\tau^a)^T = \widetilde{\psi}_\alpha^k \varepsilon_{kj}(\tau^a)_{ij} = -\widetilde{\psi}_\alpha^k(\tau^a)_{ij}\varepsilon_{jk} \equiv -\widetilde{\psi}\tau^a\varepsilon. \quad (10.13)$$

The structure of the Dirac Hamiltonian in (10.8) allows us to factorize the time-dependent part of the wave function as $\psi(x_\mu) = \psi(\mathbf{r})e^{-iEt}$ again, so this leads the system (10.8) to the pair of coupled equations

$$i\sigma^k\partial_k\widetilde{\eta} - A(r)\varepsilon_{akn}\hat{r}^n\left(\sigma^k\widetilde{\eta}\sigma^a\right) + iF(r)\hat{r}^a\left(\widetilde{\eta}\sigma^a\right) = -E\widetilde{\zeta},$$
$$i\sigma^k\partial_k\widetilde{\zeta} - A(r)\varepsilon_{akn}\hat{r}^n\left(\sigma^k\widetilde{\zeta}\sigma^a\right) - iF(r)\hat{r}^a\left(\widetilde{\zeta}\sigma^a\right) = -E\widetilde{\eta}. \quad (10.14)$$

Here we used the compact matrix notations and make no difference between spin and isospin Pauli matrices: $\tau^a \equiv \sigma^a$.

The system (10.14) was first considered by Jackiw and Rebbi [298] and afterwards it was analyzed in [155]. Some modification of this system to the case of the fermion of constant bare mass, which is not affected by the interaction with the Higgs field, was investigated in [376]. Here we shall not consider the latter situation.

The situation here is very similar to what happens with Dirac's fermion in the external field of an Abelian monopole. The analysis of the problem is related with the usual rules for adding angular momenta according to the standard Clebsch–Gordan technique.

Recall that there are different options to compose three terms $\mathbf{L}$, $\mathbf{S}$ and $\mathbf{T}$ into the vector of the generalized angular momentum $\mathbf{J}$ (cf. the discussion in Chap. 7). In the case under consideration, it is more convenient to compose first the spin and the isospin momenta of a fermion. Since for an $SU(2)$ fermion, the sum $\mathbf{S} + \mathbf{T}$ can be equal either to 1, or 0, the corresponding eigenfunctions of the Dirac Hamiltonian, the matrix-valued spin-isospinors η and ζ, can be decomposed into scalar and vector components as:

$$(\widetilde{\eta})_{im} = u_1(\mathbf{r})\delta_{im} + v_1^a(\mathbf{r})(\sigma^a)_{im}, \quad \left(\widetilde{\zeta}\right)_{im} = u_2(\mathbf{r})\delta_{im} + v_2^a(\mathbf{r})(\sigma^a)_{im},$$

where $u_{1,2}(\mathbf{r})$ and $v_{1,2}^a(\mathbf{r})$ are scalar and vector functions of spatial coordinates, respectively. Substituting this decomposition into the system (10.14) and comparing the expressions at δ_{im} and $(\sigma^a)_{im}$, we obtain the following set of two pairs of equations

$$\{\partial_a + [2A + F]\hat{r}_a\}u_1 + \varepsilon_{abc}\{i\partial_b - F\hat{r}_b\}v_1^c = iEv_2^a,$$
$$\{\partial_a - [2A - F]\hat{r}_a\}v_1^a = iEu_2,$$
$$\{\partial_a + [2A - F]\hat{r}_a\}u_2 + \varepsilon_{abc}\{i\partial_b + F\hat{r}_b\}v_2^c = iEv_1^a,$$
$$\{\partial_a - [2A + F]\hat{r}_a\}v_2^a = iEu_1.$$

10.1 Dirac Hamiltonian on the Non-Abelian Monopole Background

Subsequent analysis is related to the expansion of the functions $u_{1,2}$ and $v_{1,2}^a$ in the scalar and vector spherical harmonics. We shall not go into details of these calculations and refer the reader to the original publications [155,298]. Here, we only comment on the remarkable properties of the spherically symmetrical s-wave solutions with vanishing angular momentum. First, we shall show that, if the coupling constant h is not vanishing, there is a normalizable eigenfunction of the Dirac Hamiltonian (10.8) with zero eigenvalue $E = 0$, the *fermionic zero mode*.

Evidently, the use of properties of symmetry considerably simplifies our consideration. In principle, the above consideration suggests two possible s-wave wavefunctions, $(\widetilde{\psi})_{00}$ and $(\widetilde{\psi})_{01}$ with orbital momentum $l = 0$ and $l = 1$, respectively. Note that the former function is obviously independent on angular coordinates and corresponds to the state with zero sum of spin and isospin momenta.

The operator $\mathbf{J}$, according to the definition (10.11), acts on the spin-isospinors $(\widetilde{\eta})$ and $(\widetilde{\zeta})$ as

$$J_i\widetilde{\psi} = J_i\begin{pmatrix}\widetilde{\zeta}\\ \widetilde{\eta}\end{pmatrix} = \begin{pmatrix}-i\varepsilon_{ijk}r_j\partial_k\widetilde{\zeta} + \frac{1}{2}\left(\sigma_i\widetilde{\zeta} - \widetilde{\zeta}\sigma_i\right)\\ -i\varepsilon_{ijk}r_j\partial_k\widetilde{\eta} + \frac{1}{2}\left(\sigma_i\widetilde{\eta} - \widetilde{\eta}\sigma_i\right)\end{pmatrix}. \quad (10.15)$$

This implies that an s-wave state with zero orbital momentum satisfies:

$$[\sigma^i, \widetilde{\zeta}] = 0, \qquad [\sigma^i, \widetilde{\eta}] = 0\,.$$

Hence we conclude that $(\widetilde{\eta})_{00} = g_1(r) \cdot \mathbb{I}$ and $(\widetilde{\zeta})_{00} = g_2(r) \cdot \mathbb{I}$, where $g_{1,2}(r)$ are some (complex) functions just of the radial variable.

Another s-wave state with unit orbital momentum must obviously be proportional to the vector $\hat{r}^a$. Since such a wavefunction has to be invariant with respect to the combination of spatial and isotopical rotations, for both spin-isospinor components we are left with $(\widetilde{\eta})_{01} = (\sigma^a\hat{r}^a)h_1(r)$ and $(\widetilde{\zeta})_{01} = (\sigma^a\hat{r}^a)h_2(r)$, where $h_{1,2}(r)$ are some functions of the radial variable.

Thus, it is seen that in the sector with zero angular momentum, we obtain

$$\widetilde{\eta} = g_1(r) \cdot \mathbb{I} + (\sigma^a\hat{r}^a)h_1(r)\,, \quad \widetilde{\zeta} = g_2(r) \cdot \mathbb{I} + (\sigma^a\hat{r}^a)h_2(r)\,. \quad (10.16)$$

We are interested in a regular, spherically symmetric solution of the Dirac equation that describes a spin-isospin fermion falling down onto the center of the monopole. For such a state, the ansatz (10.16) becomes particularly simple because the symmetry is completely restored at the origin where the Higgs field is vanishing and the Dirac Hamiltonian coincides with the Hamiltonian of free motion. Clearly, there is a centrifugal barrier for the states with $h_{1,2}(r) \neq 0$, since they depend on the angular coordinates. However, this barrier is lifted for the spherically symmetric states with zero orbital momentum of the reduced form

$$\tilde{\eta} = g_1(r) \cdot \mathbb{I}, \qquad \tilde{\zeta} = g_2(r) \cdot \mathbb{I}. \tag{10.17}$$

which are the s-wave functions that do not vanish at the origin.

Note that the system of equations (10.15) still contains an angular dependence via vector $\hat{r}^a$. Substitution of the ansatz (10.17) immediately yields two equations

$$\left(\frac{\partial}{\partial r} + \frac{1-K}{r} + \frac{h}{2e}\frac{H}{r} \right) g_1 = 0,$$

$$\left(\frac{\partial}{\partial r} + \frac{1-K}{r} - \frac{h}{2e}\frac{H}{r} \right) g_2 = 0 \tag{10.18}$$

(recall that we are considering the states with zero energy).

Evidently, there are two solutions of this system (10.18):

$$g_1 = C \exp\left\{ -\int_0^r \left[\frac{1-K(r')}{r'} + \frac{h}{2e}\frac{H(r')}{r'} \right] dr' \right\},$$

$$g_2 = C \exp\left\{ -\int_0^r \left[\frac{1-K(r')}{r'} - \frac{h}{2e}\frac{H(r')}{r'} \right] dr' \right\}, \tag{10.19}$$

where C is a normalization constant.

Thus it is seen from the asymptotic behaviour of the monopole profile functions (5.44) that both these solutions are regular at the origin as $r \to 0$. However, as $r \to \infty$, the former solution exponentially decreases as $g_1 \sim e^{-m_f r}/r$, whereas the latter wave function grows exponentially as $g_2 \sim e^{m_f r}/r$. Here $m_f = hv/2$ is the mass of the fermion on the spatial asymptotic. Hence there is a unique normalizable zero mode g_1 of the Dirac Hamiltonian on the monopole background.

Such a mode must exist due to the Atiyah–Singer index theorem, which we shall discuss in the next section. Clearly, if the coupling between Higgs and spinor fields is switched off, that is, we set $h = 0$ everywhere, the system (10.18) has no normalizable solutions.

Experience with the BPS monopoles would suggest that our description becomes especially simple in the Bogomol'nyi limit. Indeed, in this case we can find an analytical solution for the fermionic zero mode of the BPS monopole. Substituting the solutions (5.63)

$$K = \frac{\xi}{\sinh \xi}, \qquad H = \xi \coth \xi - 1, \tag{10.20}$$

where the dimensionless variable $\xi = ver$ is used again, into the system (10.18), we obtain the ordinary differential equation for the normalizable wave functions with correct asymptotic behaviour

10.1 Dirac Hamiltonian on the Non-Abelian Monopole Background

$$\left(\frac{\partial}{\partial \xi} + \frac{1}{\xi} - \frac{1}{\sinh \xi} + \frac{h}{2e}\left[\frac{\cosh \xi}{\sinh \xi} - \frac{1}{\xi}\right]\right) g_1 = 0. \quad (10.21)$$

The solution depends on the relation between the Yukawa and gauge coupling constants. For the sake of simplicity, let us set, for example, $h = 2e$, that is $m_f = m_v$. Then the (10.21) is solved by

$$g_1 = \frac{C}{\cosh^2(\xi/2)}, \quad (10.22)$$

with asymptotic behaviour $g_1 \sim e^{-\xi}$ at a large distance from the monopole core. Here C is a normalization constant. It can be seen that the wavefunction of the zero mode is strongly localized around the origin, thus, we may interpret it as a spherically symmetric bound state of a monopole and a fermion. This conclusion leads to a number of non-trivial consequences.

Using a different parameterization $A(r) \to -A(r)$ and fixing the relation between the couplings as $h = e$, that is, $m_f = 2m_v$, Manton and Schroers [373] obtained another analytical solution of the (10.21):

$$g_1 = C' \xi^{3/2} \sqrt{\frac{\cosh(\xi/2)}{\sinh^3(\xi/2)}},$$

which decays asymptotically as $g_1 \sim \xi^{3/2} e^{-\xi/2}$. Clearly, this zero mode is also localized at the origin.

In summary, we get the following expression for the explicit form of the spherically symmetric fermionic zero mode (up to a normalization factor):

$$\psi^i_{\alpha(0)} = \begin{pmatrix} 0 \\ \eta^i_m \end{pmatrix}, \quad \eta^i_m = \begin{pmatrix} 0 & g_1 \\ -g_1 & 0 \end{pmatrix}. \quad (10.23)$$

Finally, let us note that the operation of C-conjugation (10.7) yields

$$\psi^i_{\alpha(0)} \to C\psi^i_{\alpha(0)} = e^{-i\pi/2} \psi^i_{\alpha(0)}.$$

Thus, the zero mode is invariant with respect to C-conjugation.

10.1.2 Zero Modes and the Index Theorem

The existence of zero modes of the Dirac operator is closely related to the topological properties of the background gauge field. This is a subject of the *Atiyah–Singer index theorem*.

Let us consider the Dirac equation (10.8) again. We can write the system of equations (10.14) as

$$\begin{aligned} \not{D}\eta &\equiv (\sigma_k D_k - \phi)\eta = iE\zeta, \\ \not{D}^\dagger \zeta &\equiv (\sigma_k D_k + \phi)\zeta = iE\eta, \end{aligned} \quad (10.24)$$

where we used the definition of the covariant derivative (10.3) and the rescaled Higgs field is written as $\phi = h\phi^a T^a$.

We shall call the matrix

$$\mathcal{D} = \begin{pmatrix} 0 & \slashed{D} \\ \slashed{D}^\dagger & 0 \end{pmatrix} \tag{10.25}$$

the Dirac operator. The first-order differential operators $\slashed{D}$ and $\slashed{D}^\dagger$ that appear in these equations are acting on open space $\mathbb{R}^4$. Therefore, the Atiyah–Singer index theorem [78], which was proved on the compact spaces, cannot be used straightforwardly. Its modification for the system under consideration was given by Callias [155]. He noted that the differential operator $\slashed{D}$ is elliptic and bounded with respect to the Sobolev norm [491]. Extended to the Fredholm operator, it is adjoint to $\slashed{D}^\dagger$ with respect to the inner product in the Hilbert space.

For zero energy states, the equations (10.24) are decoupled and, as mathematicians say, the eigenfunction η lies in the kernel of the operator $\slashed{D}$ and ζ lies in the kernel of the operator $\slashed{D}^\dagger$ [373]. Thus, ker $\slashed{D}^\dagger = \{0\}$, but ker $\slashed{D}$ is non-zero on the monopole background.

Indeed, we can prove the first statement even without any knowledge about the explicit zero-mode solution of the Dirac equation above. The analysis is simplified by consideration of two Hermitian (self-adjoint) operators $\slashed{D}\slashed{D}^\dagger$ and $\slashed{D}^\dagger\slashed{D}$, which are elliptic too. Note that the zero modes of the operator $\slashed{D}$ are also zero modes of the operator $\slashed{D}^\dagger\slashed{D}$ and vice versa. Indeed, on the space of square integrable spinors ψ, which are the eigenfunctions of the operator $\slashed{D}$ with the product

$$(\bar{\psi}, \psi) \equiv \int dx\, \bar{\psi}^\dagger(x)\psi(x)\,,$$

we have

$$\left(\eta,\, \slashed{D}^\dagger \slashed{D}\eta\right) = (\slashed{D}\eta,\, \slashed{D}\eta) = 0\,,$$

that is, ker $\slashed{D} = $ ker $\slashed{D}^\dagger\slashed{D}$.

For a BPS monopole, we have $D_k\phi = B_k$ and then this quadratic elliptic operator can be written as

$$\slashed{D}^\dagger\slashed{D} = -D_k^2 - \sigma_k B_k - \phi^2\,.$$

On the other hand,

$$\ker \slashed{D}^\dagger \subset \ker \slashed{D}\slashed{D}^\dagger = \{0\}\,,$$

because $\slashed{D}\slashed{D}^\dagger$ is a positively defined operator with no normalizable zero modes. For a BPS monopole it takes the form

$$\slashed{D}\slashed{D}^\dagger = -D_k^2 - \phi^2\,.$$

10.1 Dirac Hamiltonian on the Non-Abelian Monopole Background

Now note that the two sets of non-zero eigenvalues of the operators $\slashed{D}\slashed{D}^\dagger$ and $\slashed{D}^\dagger \slashed{D}$ are identical. Indeed, let ψ_λ be an eigenfunction of the operator $\slashed{D}^\dagger \slashed{D}$ with eigenvalues $\lambda \neq 0$:

$$(\slashed{D}^\dagger \slashed{D})\psi_\lambda = \lambda \psi_\lambda \, .$$

The left action of the operator $\slashed{D}$ on this relation yields

$$(\slashed{D}\slashed{D}^\dagger)\psi'_\lambda = \lambda \psi'_\lambda \, ,$$

where the eigenfunction $\psi'_\lambda = \slashed{D}\psi_\lambda$ corresponds to the same eigenvalue λ. For non-zero values of λ this relation is invertible. However, in the case of zero modes the situation is different.

The kernel of the elliptic operator $\slashed{D}^\dagger \slashed{D}$ is a real vector space with inner product. The dimension of this space is given by the number of zero modes k. The following difference

$$\begin{aligned} k &= \dim \ker (\slashed{D}) - \dim \ker (\slashed{D}^\dagger) \\ &= \dim \ker (\slashed{D}^\dagger \slashed{D}) - \dim \ker (\slashed{D}\slashed{D}^\dagger) = \mathrm{Ind}\,(\slashed{D}) \, , \end{aligned} \quad (10.26)$$

is a characteristic integer attached to the differential operator. It is called the *index* of the operator $\slashed{D}$.

Let us now outline the proof of the index theorem following the argumentation by E. Weinberg [513]. To calculate the index of a differential operator we need an explicit formula. The general idea is to set the index k into correspondence to the functional trace in the Hilbert space [155, 513] (here M^2 is a non-negative real parameter)

$$\mathcal{I}(M^2) = \mathrm{Tr}\left[\frac{M^2}{\slashed{D}^\dagger \slashed{D}} - \frac{M^2}{\slashed{D}\slashed{D}^\dagger}\right]. \quad (10.27)$$

Indeed, we can see that $\lim_{M^2 \to 0} \mathcal{I}(M^2) = \mathrm{Ind}\,(\slashed{D})$. Now, taking into account the form of the Dirac operator $\mathcal{D}$ (10.25), we can write

$$\mathcal{I}(M^2) = -\mathrm{Tr}\,\gamma_0 \frac{M^2}{\mathcal{D}^2 + M^2} = -\int d^3x \,\mathrm{tr}\,\langle x | \gamma_0 \frac{M^2}{\mathcal{D}^2 + M^2} | x \rangle, \quad (10.28)$$

where "tr" is now a standard matrix trace over both isotopic and Dirac indices.

There is a certain similarity of this expression and the familiar procedure of the Pauli–Villars regularization of propagators. E. Weinberg noted [513] that the integrand in (10.28) can be written as divergence of the current

$$J_k(x) = \frac{1}{2}\,\mathrm{tr}\langle x | \gamma_0 \gamma_k \frac{1}{\mathcal{D} + M} | x \rangle = \frac{1}{2}\,\mathrm{tr}\langle x | \gamma_0 \gamma_k \mathcal{D} \frac{1}{\mathcal{D}^2 + M^2} | x \rangle, \quad (10.29)$$

that is,

$$\mathcal{I}(M^2) = -\int d^3x \, \partial_k J_k(x) = \int_{S^2} dS_k \, J_k \, .$$

Here the last integration is over the sphere S^2 on the spatial infinity. An analogy with the standard technique of evaluation of propagators of massive scalar particle suggests to make use of the expansion in powers of external field strength[4] $G_{\mu\nu}$:

$$\frac{1}{D^2 + M^2} = \frac{1}{D_k^2 + \phi^2 + M^2}$$
$$+ \frac{1}{D_k^2 + \phi^2 + M^2} \left(\frac{1}{2} \gamma^\mu \gamma^\nu G_{\mu\nu} \right) \frac{1}{D_k^2 + \phi^2 + M^2} + \ldots$$

Substituting this expansion into the definition of the current (10.29), we can see that the contribution from the first, free term vanishes when we take the trace over the spinor indices. Furthermore, because the BPS monopole fields decay asymptotically as $G_{\mu\nu} \sim r^{-2}$, the only term that contributes to the integral over the sphere S^2 on the spatial infinity is

$$J_k = \varepsilon_{k\mu\nu\rho} \, \text{tr} \, \langle x | D_\mu \frac{1}{D_k^2 + \phi^2 + M^2} G_{\nu\rho} \frac{1}{D_k^2 + \phi^2 + M^2} | x \rangle, \quad (10.30)$$

where the trace is now taken over the group indices. The remaining step is to substitute into this expression the asymptotic form of the monopole field

$$F_{km} = \varepsilon_{kmn} \frac{r^n r^a}{r^4} T^a, \qquad D_k \phi^a = \frac{r^k r^a}{r^4} T^a,$$

and evaluate the "propagator" by using the Fourier transformation [513]. The result of the calculation for the fields in two-dimensional fundamental representation of $SU(2)$ is

$$\hat{r}^k J_k = \frac{1}{4\pi r^2} \frac{v}{\sqrt{v^2 + M^2}},$$

where v is the vacuum expectation value of the Higgs field. It follows that $\mathcal{I}(M^2) = v/\sqrt{v^2 + M^2}$ and, therefore, Ind $(\not{D}) = 1$, that is, there is only one fundamental fermionic zero mode in the background bosonic field of a single $SU(2)$ monopole.

The Callias index theorem [155] for the n-monopole configuration generally relates the index of the operator $\not{D}$ to the topological charge of the background field as

---

[4] There is a certain inconsistency in notation here. Imposing the Hamiltonian gauge, we adopt pseudo four-dimensional notations for the "field strength tensor" $G_{\mu\nu}$, which actually composes the covariant derivative of the Higgs field ($D_k \phi = G_{k0}$) and spatial components of the electromagnetic field strength ($F_{km} = G_{km}$). In other words, we make use of the Julia–Zee correspondence $A_0 \rightleftharpoons \phi$ once again.

10.1 Dirac Hamiltonian on the Non-Abelian Monopole Background

$$\text{Ind }(\slashed{D}) = \dim \ker (\slashed{D}^\dagger \slashed{D}) - \dim \ker (\slashed{D}\slashed{D}^\dagger) = Cn, \tag{10.31}$$

where the constant C depends on the representation of the fermion field. For the fermions in the fundamental representation of $SU(2)$, we have $C = 1$ and for the adjoint fermions $C = 2$.

To complete our brief discussion of the properties of fermionic zero modes, let us note that the excitation of the fermionic zero mode does not change the energy of the vacuum state of the $SU(2)$ monopole. Therefore, these modes can be interpreted as some kind of Grassmannian deformations of the bosonic monopole configuration. In other words, such a mode may be considered as a collective coordinate[5] alongside with four bosonic collective coordinates that parameterize the one-monopole moduli space $\mathcal{M}_1$. An arbitrary fluctuation of the isospinor field on the monopole background can be expanded in the complete set of modes

$$\psi = a_0 \psi_0 + \text{ contribution of non}-\text{zero modes}, \tag{10.32}$$

where we separated the contribution of this Grassmannian collective coordinate ψ_0.

The anticommutation relations for the field ψ mean that the coefficient of the zero mode expansion satisfies the algebra of fermionic harmonic oscillator:

$$\{a_0^\dagger, a_0\} = 1; \quad \{a_0^\dagger, a_0^\dagger\} = \{a_0, a_0\} = 0.$$

To construct a vacuum state of a monopole we shall take into account that the action of the operator of the creation of zero mode $a_0^\dagger$ on the vacuum without a fermion $|\Omega\rangle$ does not change the vacuum energy while it is annihilated by $a_0|\Omega\rangle = 0$. Thus, there are two vacuum states: $|\Omega\rangle$ and $a_0^\dagger|\Omega\rangle$ that correspond to the two-fold degenerated monopole ground state.

However, the latter is obtained from the former by action of the operator $a_0^\dagger$, which changes the fermionic charge of the vacuum as $\Delta N_f = 1$. On the other hand, the invariance of the vacuum with respect to the C-conjugation means that these two degenerated vacuum states must have opposite

---

[5] Recall that zero modes of the n-monopole configuration lie in the kernel of the Dirac operator, which is an n-dimensional real vector space with inner product [373]. It consists of n single zero modes localized at one of the monopoles with unit topological charge. This allows us to define an $O(n)$ vector bundle over the moduli space $\mathcal{M}_n$. For a single monopole the bundle $O(1)$ is flat, but it has a non-trivial holonomy around the gauge collective coordinate on S^1. Thus, the fermionic zero mode corresponds to the collective coordinate along the fibre.

fermionic numbers. Therefore, we have to admit[6] that these vacuum states have fermionic charges $\pm 1/2$, respectively [298].

An even more interesting situation occurs if we consider a set of N_f fermions of different flavors coupled with a monopole. Then the initial $U(1) \sim O(2)$ invariance of the Lagrangian (10.2) related to the conservation of the fermionic charge is extended up to the $SO(2N_f)$ symmetry. To see this we must represent the corresponding generalization of the Lagrangian (10.2) via $2N_f$ Weyl fermions $\chi^i, i = 1, 2 \ldots 2N_f$ transforming according to the vector representation of $SO(2N_f)$. Hence the fluctuations of these fermionic fields on the monopole background can be expanded as

$$\chi^i = a_0^i \chi_0 + \text{contribution of non-zero modes}. \qquad (10.33)$$

To understand the effect of this zero mode, let us note that, in the case of just two Weyl fermions ($N_f = 1$), the operators of annihilation and creation of the zero mode, a_0 and $a_0^\dagger$, respectively, can be represented via a pair of self-conjugated operators b_0^1, b_0^2, as

$$a_0 = \frac{1}{\sqrt{2}}(b_0^1 + i b_0^2), \qquad a_0^\dagger = \frac{1}{\sqrt{2}}(b_0^1 - i b_0^2), \qquad (10.34)$$

where the operators $b_0^a, a = 1, 2$ satisfy the Clifford algebra $\left\{b_0^i, b_0^j\right\} = \delta^{ij}$. Thus, the monopole vacuum state is isomorphic to the spinor that lies in the two-dimensional representation of this algebra. Hence such a state is two-fold degenerated and the background monopole configuration transforms as a spinor.

Evidently, for an arbitrary number of flavors N_f, we have $2N_f$ operators b_0^a, which correspond to the 2^{N_f}-dimensional representation of the Clifford algebra. Thus, the monopole ground state is promoted to be a spinor of $SO(2N_f)$.

Let us make one more comment. Recall that the starting point of our analysis was to consider spin-isospinor fermions coupled to a monopole. If we take the fermions in isovector representation of the gauge group [298], the Callias index theorem tells us that there are two fermionic zero modes on the monopole background. Another difference from the fundamental fermions is that the angular momentum of the isovector fermions $\mathbf{J} = \mathbf{L} + \mathbf{S} + \mathbf{T}$ in the vacuum state is not vanishing; it takes the values $\pm 1/2$. Thus we have two isovector fermionic zero modes that carry spin $1/2$ and instead of the expansion (10.32) we obtain

---

[6] This argumentation remains correct if we restrict our consideration to the classical theory. However, the fermion number conjugation symmetry is anomalous and becomes destroyed by the quantum anomaly. Then CP invariance of the full quantum theory, which exists if the vacuum instanton angle θ is set equal to zero, can be used to assign the fermionic numbers $\pm 1/2$ to the monopole vacuum states.

10.1 Dirac Hamiltonian on the Non-Abelian Monopole Background

$$\psi = a_0^{1/2}\psi_0^{1/2} + a_0^{-1/2}\psi_0^{-1/2} + \text{contribution of non-zero modes}. \quad (10.35)$$

Therefore the monopole vacuum state becomes four-fold degenerated:

Table 10.1. Quantum numbers of the monopole vacuum state

State	Spin	Fermionic charge
$\|\Omega\rangle$	0	-1
$a_{0,1/2}^\dagger \|\Omega\rangle$	$1/2$	0
$a_{0,-1/2}^\dagger \|\Omega\rangle$	$-1/2$	0
$a_{0,1/2}^\dagger a_{0,-1/2}^\dagger \|\Omega\rangle$	0	1

Thus the Grassmannian deformations of the monopole configuration may provide the monopole with a spin, in this particular case $s = \pm 1/2$.

Let us mention in conclusion that the three bosonic zero modes arise due to violation of the translational symmetry of the model by the localized monopole configuration. A similar interpretation also has the gauge zero mode of a monopole. It turns out that we may also interpret the fermionic zero modes of a monopole in the same way. This is possible if we consider a supersymmetric generalization of the Georgi–Glashow model. Then these fermionic zero modes arise as a result of partial supersymmetry breaking by the monopole configuration. We shall discuss this mechanism in Chap. 12. Note only that this observation suggests that a self-consistent duality "á la Montonen–Olive", which is related with a consideration of the fermionic degrees of freedom of a monopole, may be an intrinsic property of a supersymmetric model.

10.1.3 S-Wave Fermion Scattering on a Monopole

We can proceed further by discussing the process of scattering of the fermions on a non-Abelian monopole. For the sake of simplicity, let us consider first the asymptotic states. In this case, we can neglect the processes that are mediated by the non-Abelian short-range interactions inside of monopole core. Our second approximation is to consider the massless fermions.

Let us recall that in the Dirac representation of the γ-matrices that we are using, the left-handed Dirac fermion is defined as $\psi_L = \chi_L \begin{pmatrix} 1 \\ 1 \end{pmatrix}$, while the right-handed Dirac fermion is $\psi_R = \chi_R \begin{pmatrix} 1 \\ -1 \end{pmatrix}$, where $\chi_L = (\zeta + \eta)/2$ and $\chi_R = (\zeta - \eta)/2$, respectively. Thus, $\psi = \psi_L + \psi_R = \begin{pmatrix} \zeta \\ \eta \end{pmatrix}$.

Let us turn off the interaction between the scalar and the spinor fields, that is we set the coupling $h = 0$ in the Lagrangian (10.2). Then the chiral components of the spin-isospin fermion wavefunction are decoupled, two of the equations (10.14) become identical and we can consider the scattering of just one, say left Weyl two-spinor χ_L. For the sake of brevity, we shall drop the chiral index below.

As in the previous section, we shall consider the spherically symmetric fermionic states with zero angular momentum at the large distance from the monopole core. In spite of the similarity of these scattering states to the fermionic zero mode (10.22) and (10.23), the latter, unlike the former, is localized at the monopole. On the other hand, we have seen that in the case of the massless fermions, the Dirac Hamiltonian has no normalizable zero mode.

Intuitively it is clear that for the fermions with zero angular momentum, the Dirac operator reduces to two dimensions. Indeed, let us look at the Dirac equation for a left-handed fermion in the background monopole field:

$$i\sigma^\mu (\partial_\mu + ieA_\mu)\psi_L = 0, \quad \text{where} \quad \sigma_\mu = (\mathbb{I}, \sigma_k). \tag{10.36}$$

Here the s-wave states of isospinor chiral fermions, which are coupled with the gauge field on the non-Abelian monopole, are given by the Jackiw–Rebbi ansatz (10.16) [298]. For the sake of completeness, let us finally write the explicit normalized form of this spin-isospinor matrix:

$$\chi = \widetilde{\chi}\varepsilon = \frac{1}{\sqrt{8\pi}\,r}\left[v_1(r,t) + (\sigma^a \cdot \hat{r}^a)v_2(r,t)\right]\varepsilon$$
$$= \frac{1}{\sqrt{8\pi}\,r}\begin{pmatrix} -v_2 \sin\theta\, e^{-i\phi} & v_1 + v_2 \cos\theta \\ -v_1 + v_2 \cos\theta & v_2 \sin\theta\, e^{i\phi} \end{pmatrix}, \tag{10.37}$$

where the rows correspond to the isotopic indices and the lines correspond to the Lorentz indices.

Now recall that on a large distance from a monopole its field is purely electromagnetic. The operator of electric charge in a regular gauge is defined as $Q = (\hat{\phi}^a \cdot T^a)$ and on the spatial asymptotic it takes the form $Q = (\hat{r}^a \cdot \sigma^a)/2$. It acts on the wavefunction $\widetilde{\chi}$ as $Q\widetilde{\chi} = -(\widetilde{\chi}\sigma^a)\hat{r}^a/2$ and we shall see that its eigenfunctions, just as in the trivial vacuum, correspond to the states with electric charges $q = \pm 1/2$, respectively.

However, the interaction with the charged non-Abelian vector fields inside the monopole core mixes the states with different electric charges. Indeed, substitution of the matrix (10.37) into the Dirac equation for the s-wave left fermions (10.36) yields the system of two coupled equations, which obviously corresponds to the reduced system (10.15):

$$\partial_0 v_2 + \partial_r v_1 - \frac{K}{r}v_1 = 0,$$
$$\partial_0 v_1 + \partial_r v_2 + \frac{K}{r}v_2 = 0. \tag{10.38}$$

10.1 Dirac Hamiltonian on the Non-Abelian Monopole Background

These equations may be written in a compact two-dimensional matrix form

$$\left(\tilde{\gamma}_0\partial_0 + \tilde{\gamma}_1\partial_r - \frac{K}{r}\tilde{\gamma}_5\right)\upsilon = 0, \tag{10.39}$$

where the scalar and the vector components of the spherically symmetric wave function, υ_1 and υ_2, respectively, are composed into the doublet

$$\upsilon = \begin{pmatrix} -i\upsilon_1 \\ \upsilon_2 \end{pmatrix}, \tag{10.40}$$

and we introduced the notation for two-dimensional Dirac matrices

$$\tilde{\gamma}_0 = \sigma_3, \qquad \tilde{\gamma}_1 = -i\sigma_1, \qquad \tilde{\gamma}_5 = \tilde{\gamma}_0\tilde{\gamma}_1 = -i\sigma_3\sigma_1 = \sigma_2. \tag{10.41}$$

Clearly, with this identification (10.39) has the structure of the two-dimensional Dirac equation for a fermion coupled with a gauge field and having a space-dependent dynamical mass term [151]

$$M(r) = \frac{K}{r}\tilde{\gamma}_5.$$

Such an effective mass $M(r)$ has nothing to do with the physical mass of the fermion, which arises due to coupling with the scalar field.

Let us consider the two-dimensional Dirac equation (10.39) on the spatial asymptotic. For large r, the monopole profile function $K \approx 0$ and we have

$$(\tilde{\gamma}_0\partial_0 + \tilde{\gamma}_1\partial_r)\upsilon = 0. \tag{10.42}$$

Note that the corresponding action of the massless Weyl fermion is invariant with respect to the chiral rotations, which in this case coincide with the $U(1)$ gauge transformations

$$\upsilon \to e^{-i\alpha(r,t)\tilde{\gamma}_5}\upsilon. \tag{10.43}$$

However, this symmetry is broken inside the monopole core.

The (10.42) has two linearly independent solutions

$$\upsilon_+ = \frac{1}{\sqrt{2}}e^{-i\omega(t-r)}\begin{pmatrix} 1 \\ i \end{pmatrix}, \qquad \upsilon_- = \frac{1}{\sqrt{2}}e^{-i\omega(t+r)}\begin{pmatrix} 1 \\ -i \end{pmatrix}, \tag{10.44}$$

whose interpretation is quite obvious: the state υ_+ corresponds to the spherical wave that falls down to the monopole, while υ_- corresponds to the outgoing s-wave. Transitions between an incoming s-wave and an outgoing s-wave are possible only inside the monopole core.

In the original notations of the ansatz (10.37), the asymptotic solutions (10.44) may be written as

$$\tilde{\chi}_+ = \frac{ie^{-i\omega(t-r)}}{4\sqrt{\pi}\,r}[1+(\sigma^a\cdot\hat{r}^a)], \qquad \tilde{\chi}_- = \frac{ie^{-i\omega(t+r)}}{4\sqrt{\pi}\,r}[1-(\sigma^a\cdot\hat{r}^a)]. \tag{10.45}$$

It is straightforward to see that these wavefunctions are the eigenstates of the electric charge operator Q with the eigenvalues

$$q = \frac{1}{2} \quad \rightarrow \quad \tilde{\chi}_- \,, \qquad q = -\frac{1}{2} \quad \rightarrow \quad \tilde{\chi}_+ \,. \tag{10.46}$$

Furthermore, we have to consider also the states of opposite, right chirality, which are also electrically charged doublets.

Thus, we have to conclude that the s-wave left-hand fermions with a negative charge can only fall down on the monopole, while the left-hand spherically symmetric fermions with a positive charge can only be emitted. For the right-hand fermions the situation is reversed and generally we have the following possibilities:

Table 10.2. Quantum numbers of the s-wave fermions asymptotic states

Charge	Chirality	Direction	Fermion Number
$-1/2$	Left	in	$-$
$+1/2$	Left	out	$-$
$-1/2$	Right	out	$-$
$+1/2$	Right	in	$-$
$-1/2$	Left	out	$+$
$+1/2$	Left	in	$+$
$-1/2$	Right	in	$+$
$+1/2$	Right	out	$+$

Not everything that we have discussed so far is related with the non-Abelian nature of the monopole. Indeed, we can come to the same conclusions by analyzing the spectrum of the Dirac Hamiltonian in the field of an Abelian monopole [193, 306] (cf. the discussion in Sect. 2.6). Indeed, the case of the spherically symmetric wave functions of the third type (2.110) that we considered there, exactly corresponds to our solutions of two-dimensional Dirac equation (10.42). Recall that, as we have seen in Chap. 2, the Hamiltonian operator of an s-wave fermion coupled with an Abelian monopole can be defined on a semi-infinite line $0 \le r < \infty$ only if we impose the special boundary condition (2.115) $\chi_L(0) = e^{i\theta}\chi_R(0)$ at the origin. This boundary condition breaks chirality of the massless fermion that passes through the monopole[7]. An alternative can be a boundary condition

$$\chi_L^+(0) = e^{i\theta}\chi_L^-(0) \,, \tag{10.47}$$

---

[7] Since we are considering the massless left-handed fermions, the state χ_R is also the wavefunction of an anti-fermion. Therefore, this condition also violates the fermion number conservation.

which conserves the chirality but does violate the electric charge [155, 246, 250, 533].

In the case of the 't Hooft–Polyakov monopole there is no singularity at the origin and the Dirac operator is well-defined everywhere on the interval $0 \le r < \infty$. Therefore, the quantum-mechanical transition amplitude between the *in* and *out* states can be straightforwardly evaluated. From this point of view, the boundary condition at the origin can be considered as a result of some processes taking place inside the monopole core.

In the limiting case of the very heavy monopole $M \gg \lambda^{-1}$, where M is the monopole mass and λ is the Compton wavelength of a fermion, these processes effectively give rise to the boundary conditions at the origin that summarize the core effect. For example, the process mediated by the charged vector boson obviously may yield the charge exchange inside the core. Moreover, if we are discussing the processes at the quantum mechanical level, both the chirality and the fermion number cannot be violated and only the charge exchange process is acceptable.

Indeed, the boundary condition at the origin that we have to impose on the two-dimensional action functional, which corresponds to the Dirac equation (10.39), means that Kv/r is regular as $r \to 0$. Because $K(0) = 1$, this implies that

$$v(0) = 0. \qquad (10.48)$$

Evidently, this boundary condition mixes the left chiral states (10.44) with positive and negative electric charges and it agrees with (10.47) for $\theta = 0$.

To understand heuristically what is a physical interpretation of such a boundary condition, recall that there is the monopole-charged boson bound state in the spectrum of one-particle fluctuations around the monopole. This state corresponds to the excitation of dyonic degrees of freedom of the monopole. Thus, a charge exchange between the fermions and the monopole core could be possible. However, we already have noted that such a process is forbidden unless the energy of the initial fermion, which scatters on the monopole, is larger than the scale given by the mass of the vector boson m_v.

Therefore, the one-particle quantum mechanical description of the low-energy scattering of the s-wave fermion on the monopole is not self-consistent. Nevertheless, it gives some clue to the character of the interaction. Indeed, the Table 10.2 above implies that there could also be some processes in the monopole core that may preserve the electric charge, but violate the chirality or the fermion number.

Indeed, the remarkable observation by V. Rubakov and C. Callan [151, 152, 445, 446] is that in the quantum field theory, the non-perturbative processes effectively give rise to the unsuppressed amplitude of the fermion scattering on the monopole with non-conservation of the fermion number or the chirality. Also, we shall see that a consistent quantum field theoretic consideration of such a process shall restore the conservation of the electric charge.

10.2 Anomalous Non-Conservation of the Fermion Number

10.2.1 Axial Anomaly and the Vacuum Structure

Let us proceed by recalling some facts about the vacuum of the classical gauge theory. Recall that in Sect. 5.3.4 we already considered the topological properties of the vacuum state in monopole (dyon) sectors of the Yang–Mills–Higgs $SU(2)$ theory. It has a non-trivial structure, which is rather typical for any non-Abelian gauge theory (see, e.g., [25, 299]). A key point is that the gauge field A_μ^a on the spatial asymptotic is defined up to a transformation (5.95)

$$U(\mathbf{r}) = e^{2i\pi\hat{r}^a T^a} = e^{i\pi\hat{r}^a \sigma^a}, \tag{10.49}$$

where we restrict our consideration to the case of a fundamental representation of the $SU(2)$ group once again. Given the fact that our model is gauge invariant, it is clear that the unitary operator of the corresponding time-independent transformation

$$G[U]\, A_\mu\, G^{-1}[U] \equiv U A_\mu U^{-1}(\mathbf{r}) - \frac{i}{e} U \partial_\mu U^{-1}$$

commutes with the Hamiltonian operator. The action of the operator $G[U]$ on the trivial classical vacuum $A_k = 0$ generates an infinite tower of the gauge equivalent degenerated vacuum states

$$|n\rangle = G[U]\,|0\rangle,$$

which can be labeled by the Pontryagin index (5.96)

$$n = \frac{e^2}{8\pi^2} \int d^4x \operatorname{Tr} F_{\mu\nu} \tilde{F}^{\mu\nu}. \tag{10.50}$$

However, the degeneration is lifted by the tunnelling transitions between these states and the eigenfunctions of the Hamiltonian operator are not the states $|n\rangle$ but the superposition

$$|\theta\rangle = \sum_{n=-\infty}^{\infty} e^{-in\theta} |n\rangle,$$

which is called the θ-*vacuum*. The vacuum angle $\theta \in [0, 2\pi]$ becomes one of the parameters of the quantum theory and the different θ-sectors of the theory are orthogonal to each other.

As we briefly mentioned in Sect. 5.4, this topologically non-trivial structure of the vacuum is reflected by adding the corresponding θ-term (5.104) to the Lagrangian of the model:

10.2 Anomalous Non-Conservation of the Fermion Number

$$L_\theta = -\frac{\theta e^2}{32\pi^2} F^a_{\mu\nu}\widetilde{F}^{a\mu\nu}. \tag{10.51}$$

The value of the tunnelling amplitude between the different topological sectors defines how the physical observables depend on the vacuum angle θ.

On the other hand, in the theory with massless fermions, the Pontryagin index is related to the effects of non-conservation of chirality or, depending on the model, the anomalous non-conservation of the fermion number. The well-known relation for the Adler–Bell–Jackiw triangle anomaly

$$\partial^\mu j^5_\mu = \frac{e^2}{32\pi^2} F^a_{\mu\nu}\widetilde{F}^{a\mu\nu}, \tag{10.52}$$

shows, for example, that the axial current $j^5_\mu = \bar\psi\gamma_\mu\gamma_5\psi$ is not conserved in the quantum field theory. Integration of this formula over the spatial volume relates the Pontryagin index (5.96) to the difference between the numbers of the left-handed fermionic modes and right-handed modes[8] $n = N_L - N_R$.

The analysis by Rubakov and Callan [151, 152, 445, 446] is closely related to the fact that, by analogy with the anomalous non-conservation of the axial current, in the $SU(2)$ model with left-handed fermion doublet ψ_L the chiral rotations coincide with the gauge transformations. Hence the gauge invariant current $j_\mu = \bar\psi_L\gamma_\mu\psi_L$ is not conserved because of anomaly and the Pontryagin index is related to the non-conservation of the fermion number $N_f = \int d^3x\, j_0$:

$$\Delta N_f = \frac{e^2}{32\pi^2}\int d^3x dt\, F^a_{\mu\nu}\widetilde{F}^{a\mu\nu}. \tag{10.53}$$

In other words, the eigenvalues of the Dirac operator in the monopole background field are not constant while this field varies between the different topological sectors [27, 150, 167]. The final and the initial spectra are the same. However, the spectral flow is non-trivial, that is, there are some levels that cross zero. The number of the levels that crosses zero from above, N_+, is not equal to the number of levels N_- that crosses zero from below. Thus, the difference $\Delta N_f = N_+ - N_-$ is not equal to zero. If there are fermions of different flavors in the model, this picture is correct for each given flavor.

10.2.2 Effective Action of Massless Fermions

Coming back to the investigation of the fermionic states in the monopole external field we shall concentrate our discussion on the dynamics of the s-wave fermions that are responsible for the most interesting effects in the interaction between the bosonic and the fermionic fields. Indeed, due to the absence of the centrifugal barrier, the s-wave fermions easily feel the inner

---

[8] Since we have seen that for a monopole configuration the Pontryagin index coincides with the topological charge of a monopole, this relation is just another form of the Callias index theorem that we discussed above.

structure of the monopole, while the higher partial waves are reflected long before they approach the core.

At the level of quantum field theory we have to integrate out the fermions and derive an effective action of the bosonic fields. A natural zero-order approximation to the monopole-fermion dynamics is to restrict our consideration to the s-wave fermions and to consider the spherically symmetric quantum fluctuations about the 't Hooft–Polyakov static configuration in the gauge sector, which are given by (7.4):

$$A_0^a = \frac{r^a}{er} a_0(r,t), \qquad A_n^a = \varepsilon_{amn} \frac{r^m}{er^2}[1-K(\xi)] + \frac{r_n r^a}{er^2} a_1(r,t), \qquad (10.54)$$

with the boundary condition $a_0(r, \pm\infty) = a_1(r, \pm\infty) = 0$. Here we use the dimensionless rescaled radial variable ξ once again. We do not consider the fluctuation of the Higgs field here, it remains decoupled from the fermionic sector and, as before, the fermions are massless.

In fact, this configuration describes the quantum fluctuations of the electric field

$$E_n^a = \frac{r_n r^a}{er^2}(\partial_0 a_1 - \partial_r a_0) - \left(\frac{\delta_{na}}{r} - \frac{r_n r^a}{er^3}\right) a_0 K. \qquad (10.55)$$

Thus, the radial component of the electric field arises entirely due to the quantum fluctuations, $eE_r = \partial_0 a_1 - \partial_r a_0$, while the transversal component $eE_\perp = a_0 K/r$ depends on the monopole profile function $K(\xi)$.

The magnetic field also is modified because of the quantum fluctuations in the gauge sector. It becomes (cf. (5.45))

$$B_n^a = \frac{r_n r^a}{er^4}\left(1 - K^2 + \xi \frac{dK}{d\xi}\right) - \xi \frac{dK}{d\xi} \frac{\delta_{an}}{er^2} - a_1 K\, \varepsilon_{ank} \frac{r^k}{er^2}. \qquad (10.56)$$

Then the kinetic term of the action becomes two-dimensional

$$S = S_{(0)} + \frac{1}{2}\int d^3x dt \, \text{Tr}\, F_{\mu\nu} F^{\mu\nu} \qquad (10.57)$$

$$= MT + \frac{2\pi}{e^2}\int dt dr \left[r^2 (\partial_0 a_1 - \partial_r a_0)^2 + 2(a_0^2 + a_1^2)K^2\right].$$

We can also see that the Pontryagin index is entirely defined by the quantum fluctuations in the gauge sector of the model:

$$n = \frac{e^2}{32\pi^2}\int d^3x dt \, F_{\mu\nu}^a \widetilde{F}^{a\mu\nu}$$
$$= \frac{1}{\pi}\int dr dt \left(\partial_r[a_0(1-K^2)] - \partial_0[a_1(1-K^2)]\right). \qquad (10.58)$$

Further simplification occurs, if we neglect the structure of the monopole core. This corresponds to the situation when the vacuum expectation value

10.2 Anomalous Non-Conservation of the Fermion Number

of the Higgs field is very large and the vector bosons are very heavy. Outside the core the gauge profile function $K(\xi)$ is vanishing, $K = 0$, but the action of the quantum fluctuations of the gauge fields (10.57) remains finite. Moreover, the normalized transition amplitude between the different topological sectors

$$\frac{\langle n \mid n \mid 0 \rangle}{\langle 0 \mid 0 \rangle} = \int \mathcal{D}\bar{\psi}\mathcal{D}\psi\mathcal{D}A \; n e^{-(S-S_{(0)})}$$

is not exponentially suppressed. This anomalous behavior does not occur, of course, in the vacuum sector, where $K(\xi) = 1$ everywhere in space and the action of the quantum fluctuations is divergent.

Let us now turn to the interaction of s-wave massless fermions (10.37) with the background field of a monopole taking into account the quantum fluctuations in the gauge sector. We shall consider, as before, the left-handed Weyl fermions χ parameterized by the Jackiw–Rebbi ansatz (10.37), but disregard the effects of the monopole core. The boundary condition (10.48) has to be imposed instead.

By analogy with the expression (10.39) above we can write the corresponding two-dimensional fermionic action as

$$S_f = i \int dr dt \; \bar{v}\tilde{\gamma}_i \left(\partial_i + i\tilde{\gamma}^5 a_i \right) v , \qquad (10.59)$$

where $i = 0, 1$ and $\tilde{\gamma}_i, \tilde{\gamma}^5$ are the two-dimensional γ-matrices (10.41) introduced above. Recall that the gauge transformation (10.43) acts on the spinor v like a chiral rotation.

Evidently, in this simplified consideration only the radial component of the electric field contributes to the action of the fluctuations in the gauge sector (10.57):

$$E_n^a = \frac{r_n r^a}{er^2}(\partial_0 a_1 - \partial_1 a_0) = \frac{r_n r^a}{er^2}\varepsilon_{ij}\partial_i a_j . \qquad (10.60)$$

A key point is that such a reduced two-dimensional model with the action (10.59), the massless axial electrodynamics, is analogous to the Schwinger model and, like the latter, it admits an exact solution [151, 152, 445, 446]. The only difference is that now the system is defined on the semi-infinite line $r \in \,]0, \infty]$, that is, the field equations must be supplemented by the boundary condition (10.48). Thus, the Rubakov–Callan model can be solved by analogy with two-dimensional electrodynamics.

To see this, let us decompose the fluctuations of the gauge field as

$$a_i = \varepsilon_{ij}\partial_j \rho + \partial_i \lambda , \qquad (10.61)$$

where the scalar function $\rho(r,t)$ is related to the magnitude of the radial electric field

$$E = \mid E_n \hat{r}_n \mid = \mid \hat{r}_n E_n^a T^a \mid = \frac{\sigma^a \hat{r}^a}{2e}\varepsilon_{ij}\partial_i a_j ,$$

as
$$E = \frac{1}{2e}\Box\rho, \quad \text{where} \quad \Box \equiv \partial_0^2 + \partial_1^2.$$

Furthermore, $\lambda(r,t)$ is obviously a pure gauge function because under the gauge transformation (10.43) the fields transform as
$$v \to e^{-i\lambda\tilde{\gamma}^5} v, \quad a_i \to a_i + \partial_i \lambda. \tag{10.62}$$

Thus, this function does not enter any observable quantity and should be expressed in terms of ρ after gauge fixing.

Note that neither the function λ nor the function ρ are defined uniquely. Some freedom still remains in addition to the gauge rotation, because we can add to λ an arbitrary harmonic function $\alpha(r,t)$ and simultaneously add to ρ a function $\beta(r,t)$ such that $\partial_0 \beta = \partial_1 \alpha$ [56, 445, 446]:
$$\lambda \to \lambda + \alpha, \quad \rho \to \rho + \beta, \quad \text{where} \quad \Box\alpha = 0.$$

One can easily show that this transformation does not change the radial electric field (10.60). This allows us to impose an extra condition on the gauge function $\lambda(r,t)$:
$$\lambda(0,t) = 0,$$
which eliminates the problem of the singularity of the gauge transformation (10.62) at the origin and provides a smooth definition of the topological charge (10.58). Note that this relation can also be obtained as a condition that we have to impose to establish a correspondence between the asymptotic states of s-wave fermions on the one hand, for which neuther of the components of the spin-isospinor (10.40) are equal to zero, and, on the other hand, the spherically symmetric fermionic zero mode (10.23), which is localized at the monopole [45].

The well-known Schwinger trick, which allows us to evaluate the functional integral over the fermions, is to make a change of the fermionic variables:
$$v \to e^{-\rho - i\lambda\tilde{\gamma}^5} v_0, \quad \bar{v} \to \bar{v}_0 e^{\rho - i\lambda\tilde{\gamma}^5}. \tag{10.63}$$

This transformation effectively absorbs the term of interaction and brings the fermionic action (10.59) to the form of free theory:
$$S_f \to i\int drdt \bar{v}_0 \left\{ \tilde{\gamma}^i \partial_i - i\tilde{\gamma}^i \tilde{\gamma}^5 \left(a_i - \partial_i \lambda - i\tilde{\gamma}^5 \partial_i \rho \right) \right\} v_0$$
$$= i\int drdt \bar{v}_0 \tilde{\gamma}_i \partial_i v_0 = i\int drdt \bar{v}_0 G_f^{-1} v_0, \tag{10.64}$$

where we take into account that $i\tilde{\gamma}^5 \partial_i \rho = \varepsilon_{ij}\partial_j \rho$ and make use of the decomposition (10.61). Thus, the spinor v_0 satisfies the free Weyl equation and the functional integration over the fermions becomes trivial. The only problem is to check whether the proper boundary condition, which is independent

10.2 Anomalous Non-Conservation of the Fermion Number

of λ and ρ, is imposed on the fields v_0. Actually, we already consider this when we set the condition of nullification of the gauge function $\lambda(0,t)$ at the origin. Then the boundary condition (10.48) holds also for the free fermions: $i\tilde{\gamma}^5 v_0(0) = 0$.

Let us now evaluate the functional integral over the fluctuations of the gauge fields a_i. The regularity at the origin requires special care once again; to satisfy this condition we can set $a_0(0) = 0$. Then the decomposition (10.61) means that the action functional is regular as $r \to 0$ if $\partial_r \rho(0,t) = 0$. The combined gauge and chiral transformation (10.63) significantly simplifies the integration over the gauge fields, since in terms of the remaining physical variable $\rho(r,t)$, the action S_g reads

$$S_g = \frac{1}{2} \int d^3x dt \, \mathrm{tr} F_{\mu\nu} F^{\mu\nu} \tag{10.65}$$

$$= \frac{2\pi}{e^2} \int dt dr r^2 \left(\partial_0 a_1 - \partial_r a_0\right)^2 = \frac{4\pi}{e^2} \int dt dr \rho \Box r^2 \Box \rho, \tag{10.66}$$

where we integrated by parts taking into account the boundary conditions on the variable ρ.

To complete our evaluation of the effective action of the model, we must also consider the contribution that comes from the measure of the functional integration over the fermions, which is not invariant with respect to the transformation (10.63). The Jacobian of this transformation,

$$J = \det\left[\tilde{\gamma}^i \left(i\partial_i - \tilde{\gamma}^5 a_i\right)\right] = \exp\left\{\frac{1}{2\pi} \int dr dt \rho \Box \rho\right\},$$

can be evaluated in the standard way, for example, by the Fujikawa method [45,56] or by the point-splitting method by analogy with the Schwinger model [445,446]. Note that this Jacobian is related to the topological charge of the background field. Indeed, $\Box \rho = 2eE$, where $E = E_n \hat{r}_n$ and, therefore,

$$\ln J = \frac{e}{\pi} \int dr dt \, \rho E = \frac{e^2}{8\pi^2} \int d^3x dt \, E_n B_n = n \, .$$

Thus the bosonic part of the action is diagonal in the fields ρ:

$$S_{eff} = \frac{1}{2} \int dr dt \, \rho \left(-\frac{1}{\pi}\Box + \frac{8\pi}{e^2}\Box r^2 \Box\right)\rho = \frac{1}{2}\int dr dt \, \rho G_g^{-1} \rho, \tag{10.67}$$

as well as the fermionic part, which is diagonal in fermions v, and the model is solved. Furthermore, a simple transformation yields

$$G_g = \left(\Box \frac{8\pi r^2}{e^2}\Box - \frac{1}{\pi}\Box\right)^{-1} = \frac{1}{\pi}\left\{\left(\Box + \frac{k}{r^2}\right)^{-1} + \Box^{-1}\right\}, \tag{10.68}$$

where $k = e^2/(8\pi)$. This decomposition tells as that the corresponding propagator can be evaluated as the sum of two components: the two-dimensional propagator of the free massless scalar field

$$\Box \mathcal{D}(r,t) = \delta(r)\delta(t), \qquad \mathcal{D}(r,t) = \frac{1}{4\pi}\ln[\mu^2(r^2+t^2)],$$

which we make dimensionless by introducing the parameter μ having a dimension of mass[9], and the Green function $\mathcal{R}_k(r,t,r',t')$ of the equation

$$\left(\Box + \frac{k}{r^2}\right)\mathcal{R}_k(r-r',t-t') = \delta(r-r')\delta(t-t'). \tag{10.69}$$

Its solution is the Legendre function $Q_{d(k)}$

$$\mathcal{R}_k(r-r',t-t') = \frac{1}{2\pi}Q_{d(k)}\left(1 + \frac{(r-r')^2 + (t-t')^2}{2rr'}\right), \tag{10.70}$$

where $d(k) = \frac{1}{2}\left(\sqrt{1+4k^2}-1\right)$. The asymptotic behavior of this function as $z \to 1$ and as $z \to \infty$ is, respectively,

$$Q_d(z) \stackrel{z\to\infty}{\sim} \sqrt{\pi}\frac{\Gamma(d+1)}{2^{d+1}\Gamma(d+3/2)} z^{-(1+d)} + \ldots,$$

$$Q_d(z) \stackrel{z\to 1}{\sim} \frac{1}{2}\ln\frac{z+1}{z-1} + \ldots \tag{10.71}$$

The boundary conditions on the function $\mathcal{R}_k(r-r', t-t')$ are

$$\begin{aligned}\partial_r \mathcal{R}_k(r-r',t-t') &= 0, &\text{as}\quad r\to 0,\\ \mathcal{R}_k(r-r',t-t') &\to 0, &\text{as}\quad r\to\infty.\end{aligned} \tag{10.72}$$

The explicit expression for the Legendre function for $d(k) = 0$ also tells us that

$$\mathcal{R}(r,t,r',t') \xrightarrow[d(k)\to 0]{} \frac{1}{4\pi}\ln\frac{(r+r')^2 + (t-t')^2}{(r-r')^2 + (t-t')^2}.$$

Thus, we finally can write the propagators of the spherically symmetrical left-handed massless fermions and gauge field, respectively [445, 446],

$$G_f(r,t;r',t') = (-i\tilde{\gamma}^0\partial_0 + i\tilde{\gamma}^1\partial_1)\left(\mathcal{D}(r-r',t-t') - \mathcal{D}(r+r',t-t')\tilde{\gamma}^1\right),$$

$$G_g(r,t;r',t') = -\pi\Big(\mathcal{D}(r-r',t-t') + \mathcal{D}(r+r',t-t') - \mathcal{R}(r,t,r',t')\Big).$$

Clearly, these functions satisfy the boundary conditions that we imposed above.

---
[9] Actually, this parameter plays the role of an infrared regulator, which does not enter the gauge invariant Green function.

10.2 Anomalous Non-Conservation of the Fermion Number

Recall that, for the sake of simplicity, we restrict our consideration to the $SU(2)$ model with just one left-handed doublet of the Weyl fermions. However, it turns out that on a quantum level such a model is not self-consistent because of the global anomaly. To get rid of this anomaly, we must consider an extended model with two or more flavors of fermions. In this case, the model describes an anomalous non-conservation of the fermion number again, but all the results are modified by picking up a dependence on the number of the fermion flavors [151, 152, 445, 446].

10.2.3 Properties of the Anomalous Fermion Condensate

The explicit form of the bosonic and the fermionic Green functions of the spherically symmetric massless excitations in the system of fermions coupled to a monopole allows us to evaluate the anomalous fermion number violating matrix element. As usual, the problem is to calculate the vacuum-to-vacuum transition amplitude in the monopole background field. In the model with two left-handed $SU(2)$ doublets of fermions

$$v^{(i)} = \begin{pmatrix} v_1^{(i)} \\ v_2^{(i)} \end{pmatrix}, \quad i = 1, 2,$$

we can consider, for example, the vacuum average of the gauge-invariant bilocal operator

$$F(r,t) = v_1^{(1)}(r,t)v_1^{(2)}(r,t) + v_2^{(1)}(r,t)v_2^{(2)}(r,t) \tag{10.73}$$

between the gauge invariant monopole states $|M, \theta\rangle$ (here the index M composes the monopole quantum numbers). Coming back to the parameterization (10.37) we can represent this operator in terms of the physical left-handed spinors $\chi_\alpha^{i\,(i)}$ as

$$\frac{1}{4\pi r^2} F(r,t) = \varepsilon_{ij}\varepsilon_{\alpha\beta}\chi_{\alpha i}^{(1)}(r,t)\chi_{\beta j}^{(2)}(r,t). \tag{10.74}$$

Clearly, this correlation function describes the process that violates the fermion number. In other words, the anomalous condensate

$$\mathcal{F} = \langle M, \theta \mid F(r,t) \mid M, \theta \rangle \tag{10.75}$$

is related to the transitions between the various fermionic doublets.

In order to understand the situation better, let us consider the s-wave fermion scattering on the monopole that we already discussed above on page 376. We now have two $SU(2)$ doublets $\psi^{(i)} = \begin{pmatrix} \psi_\uparrow^{(i)} \\ \psi_\downarrow^{(i)} \end{pmatrix}_L$, $i = 1, 2$. The eigenvalues of the operator of the electric charge are positive for the states $\chi_\uparrow^{(i)}, \bar\chi_\downarrow^{(i)}$ and negative for the states $\bar\chi_\uparrow^{(i)}, \chi_\downarrow^{(i)}$.

The anomaly in the sector with unit topological charge means that the fermion number of both doublets changes in the same way: $\Delta N_f^{(i)} = -1$. Thus, the conservation of the electric charge means that the allowed process is [153, 470]

$$\psi_{\uparrow,L}^{(1)} + \text{Monopole} \to \bar{\psi}_{\downarrow,R}^{(1)} + \text{Monopole},$$

although the scattering into the final state $\bar{\psi}_{\downarrow,R}^{(2)}$ + pairs of $\bar{\psi}_{\downarrow}^{(i)}\psi_{\downarrow}^{(i)}$ is also possible. The fermion in the initial state $\psi^{(2)}$ scatters in a similar way and all these processes are described by the matrix element (10.74).

Consistent evaluation of the amplitude (10.74) is related with the functional integration over the s-wave fermion fields in an arbitrary topologically non-trivial background field [183]. The two-dimensional model under consideration is solvable and $2n$-point fermionic correlation functions $\mathcal{F}(r_1, t_1, \ldots r'_n, t'_n)$ can be directly calculated as follows

$$\mathcal{F}(r_1, t_1, r'_1, t'_1 \ldots r_n, t_n, r'_n, t'_n)$$

$$= \int \mathcal{D}\rho \prod_{i=1}^{2} \mathcal{D}\bar{v}_0^{(i)} \mathcal{D}v_0^{(i)} v_0^{(i)}(r_1, t_1) \bar{v}_0^{(i)}(r'_1, t'_1) \ldots v_0^{(i)}(r_n, t_n) \bar{v}_0^{(i)}(r'_n, t'_n)$$

$$\times \exp\left\{-\rho(r_1, t_1) - i\lambda(r_1, t_1)\tilde{\gamma}^5 + \rho(r'_1, t'_1) - i\lambda(r'_1, t'_1)\tilde{\gamma}^5 \ldots\right.$$

$$\left. - \rho(r_n, t_n) - i\lambda(r_n, t_n)\tilde{\gamma}^5 + \rho(r'_n, t'_n) - i\lambda(r'_n, t'_n)\tilde{\gamma}^5 \right\} \exp\{-(S_{eff} + S_f)\}. \tag{10.76}$$

However, the calculation of correlation function $\mathcal{F}$ (10.75) in such a way is technically rather complicated because we must sum over all sectors with different values of the θ-angle. To avoid this problem, Rubakov suggested to apply the cluster decomposition [446]. The idea is to consider not the vacuum average (10.75), but a composite two-point correlator

$$\mathcal{F}(r_1, t_1; r_2, t_2) \equiv \langle M, \theta \mid F(r_1, t_1) F^\dagger(r_2, t_2) \mid M, \theta \rangle, \tag{10.77}$$

where the operator $F(r, t)$ is given by the relation (10.74). The advantage of this construction is that the gauge invariant operator $F(r_1, t_1) F^\dagger(r_2, t_2)$ carries zero fermion number and its vacuum expectation value in the sector with an arbitrary vacuum angle θ coincides with the vacuum average in the sector $\theta = 0$.

In this case, the general formula (10.76) yields

$$\mathcal{F}(r_1, t_1; r_2, t_2) = \int \mathcal{D}\rho \, e^{-S_{eff}[\rho] - 2\rho(r_1, t_1) + 2\rho(r_2, t_2)}$$

$$\times \int \prod_{i,a=1}^{2} \mathcal{D}\bar{v}_0^{(i)} \mathcal{D}v_0^{(i)} v_0^{(i)}(r_a, t_a) \ldots \bar{v}_0^{(i)}(r'_a, t'_a) e^{-S_f}, \tag{10.78}$$

where the bosonic (S_{eff}) and the fermionic (S_f) parts of the effective action are given by the expressions (10.64) and (10.67), respectively. Since the fermionic action is free, that is, there is no interaction between bosonic and fermionic sectors, such a two-point correlation function can be written as

$$\mathcal{F}(r_1,t_1;r_2,t_2)$$
$$= \mathrm{Tr}\Big[G_f(r_1,t_1;r_2,t_2)G_f(r_2,t_2;r_1,t_1)\Big] \int \mathcal{D}\rho e^{-S_{eff}[\rho]-2\rho(r_1,t_1)+2\rho(r_2,t_2)}.$$

The remaining integration over the field ρ is Gaussian and we can see that

$$\mathcal{F}(r_1,t_1;r_2,t_2) = \mathrm{Tr}\Big[G_f(r_1,t_1;r_2,t_2)G_f(r_2,t_2;r_1,t_1)\Big]$$
$$\times \exp\Big\{-8G_g(r_1,t_1;r_2,t_2)+4G_g(r_1,t_1;r_1,t_1)+4G_g(r_2,t_2;r_2,t_2)\Big\}. \tag{10.79}$$

The leading order contribution to the functional integral comes from the neighborhood of the saddle point configuration, which can be splitted out as

$$\rho_{saddle}(r,t) = \rho_+ + \rho_- = 2G_g(r,t;r_1,t_1) - 2G_g(r,t;r_2,t_2).$$

Because the explicit forms of the Green functions of the gauge and fermion fields are already known, we can just substitute these propagators in (10.79). Making use of the asymptotic behaviour of the functions G_f and G_g in the limit $\mid t_1 - t_2 \mid \to \infty$, we can see that the contribution of the fermionic factor decreases as $\sim (t_1-t_2)^{-2}$, while the contribution of the bosonic fields increases as $(t_1-t_2)^2$. Thus, in this approximation, these two factors cancel each other and the final result is [445, 446]:

$$\lim_{|t_1-t_2|\to\infty} \mathcal{F}(r_1,t_1;r_2,t_2) = \frac{1}{16\pi^2 r_1 r_2}. \tag{10.80}$$

We can now return to the evaluation of the amplitude of the process (10.75) with non-conservation of the fermion number. The sectors with different values of the vacuum angle θ are splitted out, so we have

$$\lim_{|t_1-t_2|\to\infty} \mathcal{F}(r_1,t_1;r_2,t_2) = \langle M,\theta \mid F(r_1,t_1) \mid M,\theta\rangle\langle M,\theta \mid F^\dagger(r_2,t_2) \mid M,\theta\rangle. \tag{10.81}$$

Here the leading contribution to the matrix elements $\mathcal{F}$ and $\mathcal{F}^\dagger$ comes from the neighborhood of the saddle points values ρ_+ and ρ_- respectively.

Since the field ρ is related with the radial electric field of a monopole E, we can estimate its value on the saddle point of the functional integration. The calculation shows that these values correspond to the Pontryagin index $n[\rho_\pm] = \pm 1$, that is, such configurations describe the instanton transitions between the distinct monopole vacuum states, $\mid M,0\rangle$ and $\mid M,2\pi\rangle$, which are localized around the core [445, 446].

It is interesting to compare the two-point correlation function (10.80) with its analog in the original Schwinger model [400]. In the latter case, the saddle point configuration is saturated by a superposition of a vortex and an antivortex located at r_1, t_1 and r_2, t_2, respectively, so the non-zero fermionic condensate appears non-perturbatively as well.

Thus, the anomalous fermion-number violating condensate in the monopole external field on the large distances is

$$\mathcal{F} = \langle M, \theta \mid F(r,t) \mid M, \theta \rangle = \frac{e^{i\theta}}{4\pi r}. \tag{10.82}$$

This matrix element is not suppressed by a standard exponential factor, such a behavior is quite typical for a non-perturbative process. Therefore, we conclude that the fermion number is not conserved in the monopole background.

10.2.4 Properties of Other Condensates

Let us return to the analysis of the boundary conditions that we imposed on the states of the s-wave fermions coupled to the monopole. Not only the fermion number, but, in principle, all the quantum numbers, which are different for the up and down components of the massless Dirac doublets of fermions $\psi^{(i)} = \begin{pmatrix} \psi^{(i)}_\uparrow \\ \psi^{(i)}_\downarrow \end{pmatrix}$, may change due to interaction with the monopole. For these fermions, the boundary conditions above can be reformulated as the conditions that fix the values of the $4i$ chiral and vector charges of the final fermion state [471]

$$C^{(i)}_\uparrow = \int d^3x\, \bar\psi^{(i)}_\uparrow \gamma^0 \psi^{(i)}_\uparrow, \qquad C^{(i)}_\downarrow = \int d^3x\, \bar\psi^{(i)}_\downarrow \gamma^0 \psi^{(i)}_\downarrow, \tag{10.83}$$

$$C^{(i),5}_\uparrow = \int d^3x\, \bar\psi^{(i)}_\uparrow \gamma^0 \gamma_5 \psi^{(i)}_\downarrow, \qquad C^{(i),5}_\downarrow = \int d^3x\, \bar\psi^{(i)}_\downarrow \gamma^0 \gamma_5 \psi^{(i)}_\uparrow.$$

Each of these charges may change in the course of fermion scattering on the monopole.

However, the qualitative discussion above suggests that some of the corresponding matrix elements may vanish, in particular the electric-charge violating fermion condensate must be equal to zero. Let us discuss the situation in more detail [211, 309, 538]. In the Abelian gauge we can write the operator of the electric charge of the s-wave fermions as

$$Q_f = \frac{1}{2} \sum_i \left(\bar\psi^{(i)} \sigma_3 \gamma^0 \psi^{(i)} \right) = \int d^3x \sum_i \left(\bar\psi^{(i)}_\uparrow \gamma^0 \psi^{(i)}_\uparrow - \bar\psi^{(i)}_\downarrow \gamma^0 \psi^{(i)}_\downarrow \right)$$

$$= \frac{1}{2} \sum_i \left(C^{(i)}_\uparrow - C^{(i)}_\downarrow \right) = -\frac{i}{2} \sum_i \int dr\, \bar v^{(i)} \tilde\gamma^0 \tilde\gamma^5 v^{(i)}, \tag{10.84}$$

10.2 Anomalous Non-Conservation of the Fermion Number

where in the last step we change the notations from two-dimensional γ matrices (10.41).

Obviously, if we have two left-handed fermionic doublets, the problem is to evaluate the vacuum expectation value of the two-point operator

$$Q(r,t) = v_1^{(1)}(r,t)v_2^{(2)}(r,t) + v_2^{(1)}(r,t)v_1^{(2)}(r,t), \quad (10.85)$$

which could break the conservation of the electric charge.

However, if we want to interpret the corresponding condensate as an order parameter, which is related with non-conservation of the electric charge, we have to consider the gauge invariance of such a quantity. Unfortunately, the operator Q (10.85) is not gauge invariant.

To see how this happens, let us recall that a correct definition of the operator product of the bilocal functions of type (10.83) needs some regularization scheme, for example, the well-known point-splitting method. From this point of view, the gauge invariance of the operator (10.85) can be effectively restored if we trade the "bare" function $\psi(x)$ for the so-called "dressed" fermion wave functions in the external gauge field [211]

$$\psi(x) \to \psi(x) \exp\left\{\frac{ie}{2} \int_0^x dy A(y)\right\} \equiv \psi(x) V(x).$$

The price we have to pay is that now the operator $V(x)$ will also contribute to the matrix element of the charge operator (10.85):

$$\langle M,0 | Q_{phys} | M,0\rangle \propto \langle M,0 | \bar{v}(i)\tilde{\gamma}^0\tilde{\gamma}^5 v^{(i)} | M,0\rangle\langle M,0 | V^2 | M,0\rangle. \quad (10.86)$$

We can interpret this modification as a contribution of the Coulomb energy of interaction. Since in the two-dimensional model the potential is decaying as $U(r) \sim \ln r$, this energy is infinite and the contribution of the phase factor is supressed as $\sim \lim_{r\to\infty} \exp\{-\ln r\}$. Therefore, the gauge-invariant condensate with non-conservation of the electric charge is vanishing [211, 309, 538, 540]. This conclusion was confirmed by a more detailed investigation, which takes into account the effects of excitation of charged fields inside the monopole core [107, 108, 310, 426].

However, there are many other non-trivial processes with fermions coupled to the monopole apart from the appearance of the fermion number violating condensate (10.82). The triangle anomaly in the conservation of the fermionic current is significant only for that particular condensate, while all other amplitudes appear entirely due to interaction of the s-wave fermions with the non-Abelian fields in the monopole core. Actually, the monopole presence catalysts the processes, which, in principle, are not forbidden, but are related to an exchange of a heavy vector boson and, therefore, suppressed by the inverse power of its mass m_v.

As an example, let us consider the scattering of a zero charge "meson" state $\bar\chi^{(2)}_\uparrow \chi^{(1)}_\uparrow$ on the monopole in the $SU(2)$ model with two chiral fermions. The boundary conditions of the fermion wave functions, which we imposed at the origin, mean that the process

$$\psi^{(1)}_{\uparrow,L} + \bar\psi^{(2)}_{\uparrow,R} + \text{Monopole} \;\to\; \psi^{(1)}_{\downarrow,L} + \bar\psi^{(2)}_{\downarrow,R} + \text{Monopole} \qquad (10.87)$$

without violation of the fermion number is allowed. Clearly, such a process can also be mediated by the exchange of a "normal" charged vector boson. The difference is that the presence of a monopole removes the corresponding suppressing factor, that is, we can talk about the *monopole catalysis* of such a process. Indeed, let us consider the gauge-invariant operator that corresponds to the non-anomalous flavour-mixed process (10.87) [309]

$$\Gamma(r,t) = \bar v^{(1)}(r,t)(1+\tilde\gamma^5) v^{(1)}(r,t) \bar v^{(2)}(r,t)(1-\tilde\gamma^5) v^{(2)}(r,t). \qquad (10.88)$$

In this case, both the initial and the final states are electrically neutral and there is no transition between the vacua with different values of the θ-angle. Thus, this matrix element can be evaluated in the pure monopole sector

$$\langle M, 0 \mid \Gamma(r,t) \mid M, 0 \rangle.$$

The calculations of this amplitude can be performed by making use of the definition (10.76) in complete analogy with the evaluation of the matrix element (10.82) above. The cluster decomposition then yields [56, 309]

$$\lim_{t\to\infty} \langle M, 0 \mid \Gamma(r,t) \mid M, 0 \rangle \sim \frac{1}{r^6},$$

that is, there is indeed no suppression by the mass of the vector boson.

10.3 Monopole-Fermion Scattering in the Bosonisation Technique

A number of simplifications that we assumed in the calculations of the previous section may cause some doubts as to whether the Rubakov–Callan effect could survive in a more realistic picture of the monopole-fermion interaction. Recall that the fermion-number violating correlator (10.82) was evaluated under the following assumptions:

- the effect of the monopole core is reduced to some boundary conditions on the fermion wave functions at the origin;
- only s-wave excitations are taking into consideration;
- the fermions are supposed to be massless.

10.3 Monopole-Fermion Scattering in the Bosonisation Technique

Thus, a consistent investigation of the problem is related with the analysis of the contribution of higher partial waves, the contribution of the processes of interaction with non-Abelian fields of the monopole core and the effect of the fermion mass. Note that the bare masses of the fermions of different flavors are different.

Let us briefly describe some of the results. As we mentioned above, the effects of the monopole core excitations and their contribution to the vacuum condensates were investigated in a few papers, see, e.g., [107,108,183,297,310, 426]. The conclusion is that the contribution of the non-Abelian interaction to the Green function of the S-wave massless fermion does not eliminate the condensate, which violates the conservation of the fermion-number, yet some novel features appears.

Already in his pioneering work [446] (see also the discussion in the review [56]) Rubakov considered the effect of the corrections to the s-wave approximation. He demonstrated that the contribution to the effective action of the higher partial waves, which was estimated as a correction to the saddle point value of the action, is proportional to e^4, that is, it is almost negligible in a weak coupling regime. Also negligible is the one-loop correction to the s-wave fermion propagator, which can be related to the inversion of the chirality of the fermion: evidently such a process leads to reflection of the incoming wave from the monopole at the distances far away from the core [418,446]. Actually, this is the mechanism that generates the anomalous magnetic moment and that precisely corresponds to our discussion in Chap. 2, where such an effective term in the Dirac Abelian Hamiltonian was used as a regulator of the theory [306].

The group of problems which is related to the effect of the fermion mass seems to be most complicated in the Green function formalism, especially if we would like to take into consideration the processes inside the monopole core. However, if we restrict our consideration to the pure electromagnetic interaction, the massive fermion condensates can be easily evaluated by making use of the well-known trick of bosonization [363], which was modified by Callan to the case of the monopole-fermion system [151,152].

10.3.1 Vertex Operator and Bosonization of the Free Model

Let us recall that the dynamics of the spherically symmetric fermions coupled to the monopole can be reduced to some form of the two-dimensional Schwinger model. Callan used the fact that this model can be represented as a bosonic theory where the relation between the chiral fermion doublet $\chi(r,t)$ and the scalar field $\phi(r,t)$ is given by the *vertex operator*[10]

---

[10] In all relations of this type we suppose the normal ordering with respect to the normalization parameter μ.

$$v = \begin{pmatrix} v_1 \\ v_2 \end{pmatrix} = \sqrt{\frac{a\mu}{2\pi}} \begin{pmatrix} \exp\left\{i\sqrt{\pi}\left(\phi(r,t) - \int_0^r dr' \partial_0 \phi(r',t)\right)\right\} \\ i\exp\left\{i\sqrt{\pi}\left(\phi(r,t) + \int_0^r dr' \partial_0 \phi(r',t)\right)\right\} \end{pmatrix},$$

where a is some constant and μ is an arbitrary parameter having the dimension of mass, which is used as a normalization point at the normal ordering of operators, and for the sake of compactness, we have omitted the anti-commuting Klein factors, which must be introduced to provide the correct commutation relations for the left- and right-handed fermions [253].

The difference from the original model considered by Coleman [172] and Mandelstam [363] is that we have to impose the boundary conditions[11] on the scalar field $\phi(r,t)$, which correspond to the boundary conditions on the fermion field $v(r,t)$ above:

$$\partial_r \phi(0,t) = 0.$$

The vertex operator (10.89) provides a one-to-one correspondence between the set of correlation functions of the scalar field $\phi(r,t)$ and the Green functions of the free fermions $v(r,t)$. One can verify directly that the action of the massless fermions (10.64) then transforms to the two-dimensional action of free bosons

$$S_b = \frac{1}{2} \int dr dt \left[(\partial_0 \phi)^2 - (\partial_r \phi)^2\right]. \tag{10.89}$$

This action obviously corresponds to the free field equation

$$\Box \phi(r,t) = 0. \tag{10.90}$$

Other bilocal fermionic operators, in particular the charges (10.83) and corresponding currents, can also be expressed in terms of the boson field. For example, we can consider the fermionic current $j_i^5 = \bar{v}\tilde{\gamma}_i\tilde{\gamma}^5 v$, which corresponds to the charge $C^{(i)}$. This current is coupled with the vector field of a monopole $a_i(r,t)$ in the two-dimensional action (10.59). After some calculations, in which take into account the normal ordering of the operators[12], this current can be rewritten locally as

$$j_i = \frac{1}{\sqrt{\pi}} \varepsilon_{ij} \partial_j \phi. \tag{10.91}$$

In the language of the scalar field $\phi(r,t)$, the operators of the densities of the states with positive ($C^{(i)}_\uparrow$) and negative ($C^{(i)}_\downarrow$) electric charge are, respectively,

---

[11] In general, the consistent scheme of bosonization of the model with massless fermions is related to the consideration of the model on the line of finite length $r \in [r_0, R]$. Then we can take the limit $r_0 \to 0$, $R \to \infty$. In this case, the boundary conditions must be implemented both at the point $r = r_0$ and at the point $r = R$. To simplify our discussion we shall not discuss this procedure.

[12] A possible alternative to this approach is to apply the well-known point-splitting method.

10.3 Monopole-Fermion Scattering in the Bosonisation Technique

$$C_\uparrow^{(i)} = \frac{1}{2\sqrt{\pi}}(\partial_r\phi + \partial_0\phi), \qquad C_\downarrow^{(i)} = \frac{1}{2\sqrt{\pi}}(\partial_r\phi - \partial_0\phi). \qquad (10.92)$$

Since the canonical momentum of the bosonized model is $\pi = \partial_0\phi$, these densities may also be written as

$$C_\uparrow^{(i)} = \frac{1}{2\sqrt{\pi}}(\partial_r\phi + \pi), \qquad C_\downarrow^{(i)} = \frac{1}{2\sqrt{\pi}}(\partial_r\phi - \pi), \qquad (10.93)$$

and the free scalar Hamiltonian is $H_b = \frac{1}{2}\left(\pi^2 + (\partial_r\phi)^2\right)$.

The fermion number density excitations can also be separated, if we note that a general solution of the free field (10.90) can be decomposed as (cf. (10.44))

$$\phi(r,t) = \phi_\uparrow(r+t) + \phi_\downarrow(r-t).$$

The chiral components $\phi_\uparrow(r+t)$ and $\phi_\downarrow(r-t)$ correspond to the operators of the charge densities

$$C_\uparrow^{(i)} = \frac{1}{\sqrt{\pi}}\partial_r\phi_\uparrow, \qquad C_\downarrow^{(i)} = \frac{1}{\sqrt{\pi}}\partial_r\phi_\downarrow \qquad (10.94)$$

respectively. Thus, the total electric and fermion charges in the boson language are

$$Q_f = \int_0^\infty dr(C_\uparrow^{(i)} - C_\downarrow^{(i)}) = \frac{1}{\sqrt{\pi}}[\phi(r=\infty) - \phi(r=0)],$$

$$N_f = \int_0^\infty dr(C_\uparrow^{(i)} + C_\downarrow^{(i)}) = \frac{1}{\sqrt{\pi}}[\phi(r=\infty) + \phi(r=0)]. \qquad (10.95)$$

Clearly, we can interpret these charges as a kink-like configuration of the bosonized model on a semi-infinite line, which interpolates between the vacuum values $\sqrt{\pi}$ and 0, as shown in Fig. 10.1. Indeed, the fermion number is conserved on the spatial asymptotic, and we have to fix $\phi(r=\infty) = $ const. However, the value $\phi(r=0)$ can change in the course of the scattering and the corresponding fermion number may change when the fermion penetrates the monopole. For example, the process

$$\psi_{\uparrow,L}^{(1)} + \text{Monopole} \to \bar\psi_{\downarrow,R}^{(1)} + \text{Monopole},$$

which we discussed above, is shown in Fig. 10.1.

The form of the operator of the electric charge (10.95) may suggest that the value $\phi(r=0)/\sqrt{\pi}$ could have a meaning of the electric charge at the monopole core, that is, a charge exchange between the core and fermions is possible. This is correct until the interaction with the gauge field is included; as we have seen above, it has a very large Coulomb energy, which is infinite in the limit of the point-like monopole.

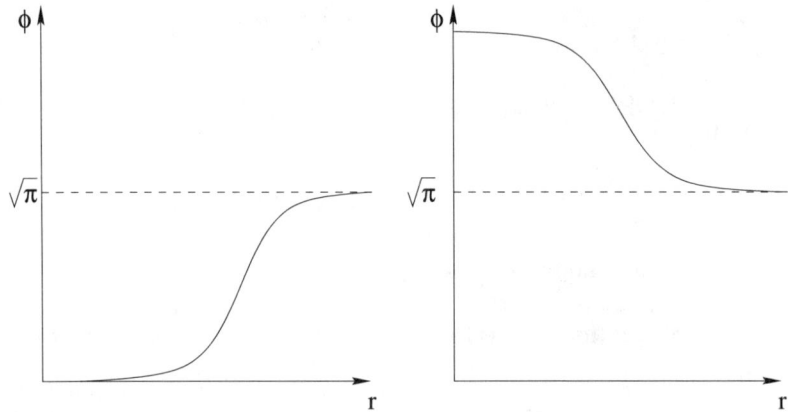

Fig. 10.1. Fermion-monopole scattering in the bosonized theory

The peculiarity of the bosonized model is that, according to the definition (10.95), it contains the asymptotic states with fractional fermion number N_f which do not appear in the spectrum of the original fermion-monopole system. However, these states disappear from the model, if we introduce the mass term of the fermions.

Indeed, let us now discuss the effect of the bare fermion mass. Here we have to recover the index of helicity. In the model with Dirac fermions, we have to deal both with the left-handed ($\psi_L(r,t)$) and right-handed ($\psi_R(r,t)$) components[13] of the Dirac field ψ, which can be bosonized by introducing two boson fields

$$\bar{v}_L \tilde{\gamma}_i v_L = \frac{1}{\sqrt{\pi}} \partial_i \phi_L, \qquad \bar{v}_R \tilde{\gamma}_i v_R = \frac{1}{\sqrt{\pi}} \partial_i \phi_R.$$

The contribution to the mass term of the fermionic action comes from the operators

$$\bar{v}_R v_L = \mu \, \exp\left\{ i\sqrt{\pi}(\phi_L - \phi_R) \cos\sqrt{\pi} \int_0^r dr' [\partial_0 \phi_R(r',t) + \partial_0 \phi_L(r',t)] \right\},$$

$$\bar{v}_R \tilde{\gamma}^5 v_L = i\mu \, \exp\left\{ i\sqrt{\pi}(\phi_L - \phi_R) \sin\sqrt{\pi} \int_0^r dr' [\partial_0 \phi_R(r',t) + \partial_0 \phi_L(r',t)] \right\}.$$

The somewhat awkward form of this expression becomes more transparent if we recall that the canonical momenta that are conjugated to ϕ_L and ϕ_R are $\pi_L = \partial_0 \phi_L$; $\pi_R = \partial_0 \phi_R$, respectively. Then a canonical transformation of the dynamical variables yields [152]

---

[13] Recall that we are working in non-diagonal representation of $\tilde{\gamma}_5$.

10.3 Monopole-Fermion Scattering in the Bosonisation Technique

$$\phi_\uparrow = \frac{1}{2}(\phi_L - \phi_R) + \frac{1}{2}\int_0^r dr'(\pi_L + \pi_R),$$

$$\phi_\downarrow = -\frac{1}{2}(\phi_L - \phi_R) + \frac{1}{2}\int_0^r dr'(\pi_L + \pi_R),$$

$$\Pi_\uparrow = \frac{1}{2}(\pi_L - \pi_R) + \frac{1}{2}(\partial_r\phi_L + \partial_r\phi_R),$$

$$\Pi_\downarrow = -\frac{1}{2}(\pi_L - \pi_R) + \frac{1}{2}(\partial_r\phi_L + \partial_r\phi_R). \tag{10.96}$$

This transformation does not change the free part of the boson action (10.89), which in the case under consideration can be written as

$$S_b = \frac{1}{2}\int drdt \left[(\Pi_\uparrow)^2 + (\Pi_\downarrow)^2 - (\partial_r\phi_\uparrow)^2 - (\partial_r\phi_\downarrow)^2\right]. \tag{10.97}$$

Then the mass term takes an especially simple form

$$S_m = \frac{m\mu}{2}\int drdt \left[(1 - \cos 2\sqrt{\pi}\phi_\uparrow) + (1 - \cos 2\sqrt{\pi}\phi_\downarrow)\right]. \tag{10.98}$$

Thus, the bosonized version of the original model takes the form of a direct sum of two sin-Gordon models, which are related by the boundary conditions on the fields ϕ

$$\partial_r\phi_\uparrow(0,t) = \partial_r\phi_\downarrow(0,t).$$

The fermionic states are now represented by the kink solitons of the sin-Gordon model, which are the only asymptotic states in the spectrum of the massive theory. The pathological states with fractional fermion numbers are eliminated because now they have infinite energy.

Let us now return to the interaction of the fermions with the gauge field in the bosonized theory. In the model with one left-handed doublet and one right-handed doublet the fermionic current, which generalizes (10.91), is

$$j_i^5 = \frac{1}{\sqrt{\pi}}\varepsilon_{ij}\partial_j(\phi_L - \phi_R),$$

and, because the potential of the gauge field a_i (10.61), which is coupled with this current, in the gauge $\partial_n\lambda = 0$ can be expressed as $a_i = \varepsilon_{ij}\partial_j\rho$, the corresponding interaction term becomes

$$S_{int} = \frac{1}{\sqrt{\pi}}\int drdt\, \partial_i\rho\, \partial_i(\phi_L - \phi_R) = \frac{1}{\sqrt{\pi}}\int drdt\, \Box\rho\, (\phi_L - \phi_R).$$

We can now add this term to the kinetic part of the effective action of the gauge field (10.67), which is quadratic in the fields ρ:

$$S_g + S_{int} = \frac{1}{2}\int drdt\left[\rho\left(-\frac{1}{\pi}\Box + \frac{8\pi}{e^2}\Box r^2\Box\right)\rho + \Box\rho(\phi_L - \phi_R)\right]. \quad (10.99)$$

Performing Gaussian integration over the field ρ and making use of the canonical transformations (10.96), we obtain the bosonized form of the effective action of the massive theory

$$S_{eff} = \frac{1}{2}\int drdt\Big[(\partial_n\phi_\uparrow)^2 + (\partial_n\phi_R)^2 + m\mu(1 - \cos 2\sqrt{\pi}\phi_\uparrow)$$

$$+ m\mu(1 - \cos 2\sqrt{\pi}\phi_\downarrow) - \frac{e^2}{8\pi r^2}(\phi_\uparrow - \phi_\downarrow)^2\Big]. \quad (10.100)$$

The last term here has an interpretation of the Coulomb energy of the interaction of the fermions with the fluctuation of the gauge field. According to (10.95), the electric charge density in the units of the charge of the gauge boson $e/2$ is $Q_f(r) = (\phi_\uparrow - \phi_\downarrow)/\sqrt{\pi}$. However, in a self-consistent theory, the energy functional must be finite everywhere in space. This means that the Coulomb energy of interaction must vanish at the origin, that is,

$$\phi_\uparrow(0,t) = \phi_\downarrow(0,t).$$

Thus, there is no charge exchange inside the monopole core and the electric charge of fermions is conserved.

However, the effect of the vacuum angle changes the situation. Indeed, we can add to the bosonic part of the action (10.67) the θ-term (5.104)

$$S_\theta = \frac{\theta e^2}{32\pi^2}\int d^3x dt\, F^a_{\mu\nu}\tilde{F}^{a\mu\nu} = \frac{\theta e^2}{2\pi}\int dr\, r^2 E_n B_n.$$

Since the magnetic field of the 't Hooft–Polyakov monopole of unit topological charge is $B_n = \frac{r_n}{er^3}$, and the radial electric field can be expressed as $E = \Box\rho/(2e)$, the inclusion of the θ-term leads to the shift of the linear in $\Box\rho$ term in (10.99):

$$S_g + S_{int} + S_\theta = \frac{1}{2}\int drdt\left[\rho\left(-\frac{1}{\pi}\Box + \frac{8\pi}{e^2}\Box r^2\Box\right)\rho + \Box\rho\left([\phi_\uparrow - \phi_\downarrow] + \frac{\theta}{2\pi}\right)\right]. \quad (10.101)$$

Therefore, the Coulomb energy becomes

$$E_{Coul} = \frac{e^2}{8\pi r^2}\left(\phi_\uparrow - \phi_\downarrow + \frac{\theta}{2\sqrt{\pi}}\right)^2.$$

The physical interpretation of this result is quite clear: it takes into account the energy of interaction with the electric charge of the monopole generated

10.3 Monopole-Fermion Scattering in the Bosonisation Technique

by the instanton transitions. Thus, the boundary condition at $r = 0$ also changes,

$$\phi_\uparrow(0,t) = \phi_\downarrow(0,t) + \frac{\theta}{2\sqrt{\pi}},$$

which corresponds to the definition of the electric charge of the system (5.110) above with contribution of the Witten angle : $q = en + \frac{e\theta}{2\pi}$.

10.3.2 Monopole Catalysis of the Proton Decay

Without any doubt, the catalysis of proton decay by a monopole, which a is known as the *Rubakov–Callan effect* [151, 152, 445, 446], is one of the most interesting phenomena, that may be possible in the presence of a monopole. If such a process is detected experimentally (and there are many different groups of scientists working in this direction), it will be one of the most important discoveries that can lift the veil from the very fundamental laws of our universe.

The possibility that the monopole catalysis mechanism of the proton decay may really work, was initially proved within the framework of the minimal $SU(5)$ model of unification in the processes of the type

$$\text{p} + \text{Monopole} \rightarrow e^+ + \text{Monopole} + \text{mesons}.$$

Although we know that, for some unclear reasons, this type of unification was not the first choice of Nature, the Rubakov–Callan effect is also inherent in the $SO(10)$ GUT [435] and some other, more realistic models. In our discussion, we shall consider the $SU(5)$ model, which is a simpler example.

Let us briefly recapitulate some basic information about the structure of the $SU(5)$ model of unification. First generation of fermions is composed of 15 particles, leptons and quarks, which are placed in two multiplets, the quintet state

$$(\psi_L^C)^a = (d_1^C, d_2^C, d_3^C, e^-, \nu)_L, \qquad (\psi_R)_a = (d_1, d_2, d_3, e^+, \nu^C)_R, \quad (10.102)$$

where upper and lower indices $a = 1, \ldots 5$ refer to the representation 5 and its conjugated $\bar{5}$, respectively, and index C labels the charge conjugation. The decouplet is formed as the antisymmetrized product of two quintets

$$\psi_L^{ab} = \frac{1}{\sqrt{2}}(\psi_L^a \psi_L^b - \psi_L^b \psi_L^a) = \begin{pmatrix} 0 & u_3^C & -u_2^C & -u_1 & -d^1 \\ -u_3^C & 0 & u_1^C & -u_2 & -d^2 \\ u_2^C & -u_1^C & 0 & -u_3 & -d^3 \\ u_1 & u_2 & u_3 & 0 & -e^+ \\ d_1 & d_2 & d_3 & e^+ & 0 \end{pmatrix}.$$

There are 24 generators of the gauge group $SU(5)$, which include four diagonal generators of the Cartan subalgebra. These correspond to the three operators of hypercharges and one operator of the electric charge. Thus, the adjoint representation of the gauge sector of the $SU(5)$ model includes 24 vector bosons:

$$A_a^b = \begin{pmatrix} A_{11} - \frac{B_0}{\sqrt{30}} & A_{12} & A_{13} & X_1^C & Y_1^C \\ A_{21} & A_{22} - \frac{2B_0}{\sqrt{30}} & A_{23} & X_2^C & Y_2^C \\ A_{31} & A_{32} & A_{33} - \frac{3B_0}{\sqrt{30}} & X_3^C & Y_3^C \\ X_1 & X_2 & X_3 & \frac{W^3}{\sqrt{2}} - \frac{3B_0}{\sqrt{30}} & W^+ \\ Y_1 & Y_2 & Y_3 & W^- & -\frac{W^3}{\sqrt{2}} + \frac{3B_0}{\sqrt{30}} \end{pmatrix}.$$

Here 8 gauge bosons A_{ij}, $i,j = 1,2,3$ mediate the interaction between the quarks within the $SU(3)$ subgroup, that is, they are the usual gluons. Three vector bosons $W^\pm, W^3$ are responsible for the interactions in the lepton sector and the B_0-boson corresponds to the $U(1)$ subgroup of the electroweak part of the model.

The most interesting components of the model are the two color triplets of vector bosons X_i, Y_i with electric charges $\pm 4/3$ and $\pm 1/3$, respectively. The transitions between the quarks and the lepton states becomes possible due to the exchange of these bosons, which makes the proton unstable with respect to decay into mesons and leptons. Thus, in the $SU(5)$ model neither baryonic nor leptonic numbers are conserved separately, but their difference is conserving. However, these processes are strongly suppressed by the masses of X, Y-bosons, which are of the order of $M_X \sim M_Y \geq 10^{14}$ GeV.

Note that the spectrum of the model is not completed by these perturbative states. The non-perturbative sector of the model with the $SU(5)$ group of symmetry contains the monopole solutions and, therefore, the presence of a monopole can catalyse the proton decay, effectively removing the X, Y-boson mass-depending factor in the corresponding amplitudes. We have to note that, to our knowledge, there is no complete analysis of the interaction between the fermions and the different types of monopole configurations, even within the minimal $SU(5)$ gauge theory. Recall that, depending on the pattern of the symmetry breaking, these monopoles can be both massive and massless. The simplest possibility, which was considered in [203], is to analyze the situation when the $SU(5)$ is spontaneously broken to $SU(3) \times U(1)$ and the monopole solution is actually the 't Hooft–Polyakov monopole, which is embedded in the leptoquark $SU(2)_M$ subgroup of $SU(5)$ as

$$T^i_{SU(2)_M} = \left(0, 0, \frac{\tau^i}{2}, 0\right).$$

In this case, the only generator of the unbroken subgroup of $SU(5)$ is given by the composition of the elements of the Cartan subalgebra (cf. expression (8.35)):

$$T^3_{SU(2)_M} = \left(0, 0, \frac{\tau^3}{2}, 0\right) = -\frac{1}{2}Q + \frac{1}{2}Y,$$

10.3 Monopole-Fermion Scattering in the Bosonisation Technique

where $Q = \text{diag}\,(1/3, 1/3, 1/3, -1, 0)$ is the electric charge generator and $Y = \text{diag}\,(1/3, 1/3, -2/3, 0, 0)$ is the matrix of the color hypercharge. In other words, the monopole configuration has both the usual magnetic charge and the chromo-magnetic charge. Thus, there are both the color and the Abelian magnetic fields outside the core and the interaction of the monopole with the quarks and the leptons, which also have both the usual electric charge and the color charges, is characterized by an effective charge, which is actually the isospin of the $SU(2)_M$ leptoquark subgroup. Thus, all fermionic states from the 5-plet and 10-plet are labeled by the eigenvalue $\bar q$ of the operator $T^3_{SU(2)_M}$.

We can see that the states u_1, u_2, d_3^C, e^+ have the same isospin $\bar q = 1$, while the states u_1^C, u_2^C, d_3, e^- correspond to $\bar q = -1$. Other fermions presented in the model have zero isospin with respect to the $SU(2)_M$ leptoquark subgroup, that is, they do not interact with the monopole.

In this case, the model is reduced to the familiar $SU(2)$ theory with four left-handed doublets of the fermions of the first generation

$$\psi_\alpha^{i\,(a)} = \left\{ \begin{pmatrix} d^3 \\ e^+ \end{pmatrix}_L,\; \begin{pmatrix} e^- \\ -d_3^C \end{pmatrix}_L,\; \begin{pmatrix} u_1^C \\ u^2 \end{pmatrix}_L,\; \begin{pmatrix} -u_2^C \\ u^1 \end{pmatrix}_L \right\},$$

where $a = 1, 2, 3, 4$. Then we can make use of the analysis of the s-wave interaction between the fermions and the 't Hooft–Polyakov monopole, as described in the previous section. By analogy with (10.74), the problem is to evaluate the vacuum average of the operator[14]

$$\frac{1}{2^8 \pi^4} F(r) = \varepsilon_{ij} \varepsilon_{\alpha\beta} \psi_{\alpha i}^{(a)} \psi_{\beta j}^{(b)} \varepsilon_{mn} \varepsilon_{\rho\sigma} \psi_{\rho m}^{(c)} \psi_{\sigma n}^{(d)}, \qquad (10.103)$$

which corresponds to the anomalous condensates

$$\langle (e^+ e^- - d^3 d_3^C)(u_1^C u^1 - u_2^C u^2) \rangle \sim \frac{1}{r^6},$$

$$\langle (u^2 e^- - u_1^C d_3^C)(u^1 d^3 - u_2^C e^+) \rangle \sim \frac{1}{r^6},$$

$$\langle (u_2^C d_3^C + u^1 e^-)(u_1^C e^+ + u^2 d^3) \rangle \sim \frac{1}{r^6}. \qquad (10.104)$$

These matrix elements do not violate either the electric charge or the difference between the baryon and the lepton numbers. However, the baryon number itself is no longer conserved. Indeed, there is the baryon number violating condensate

$$\langle u_1^C u_2^C d_3^C e^+ \rangle \sim \frac{1}{r^6}, \qquad (10.105)$$

whose quantum numbers correspond to the Rubakov–Callan process of the proton decay

---

[14] Note that the vacuum average (10.103) is not invariant with respect to the transformations of the color subgroup $SU(3)$. Here, we again encounter the pathology of the global color transformation that we briefly discussed in Chap. 8.

$$p + \text{Monopole} \rightarrow e^+ + \text{Monopole} + (\text{pions}).$$

Other condensates, which are unsuppressed in the presence of the monopole, can be evaluated on the same way. Note that the non-anomalous condensates also contribute to the amplitude of the monopole-induced proton decay. An analog of the process (10.87), for example is

$$d_L^3 + u_R^1 + \text{Monopole} \rightarrow u_{2R}^C + e_L^+ + \text{Monopole},$$

which obviously violates the baryon number.

10.3.3 Monopole Catalysis of the Proton Decay: Semiclassical Model

So far we discuss the microscopic picture of the monopole catalysis and evaluate the related cross-sections for monopole induced processes. We mentioned that there are a lot of assumptions in these calculations, in particular, we neglect the structure of the monopole core which in the Grand Unified Theory, has a very tiny size of the order of 10^{-16} GeV. However, a realistic calculation of the cross-section of the monopole catalysis should involve several widely separated scales, the smallest of which is the size of the monopole core. The largest scale is given by the hadronic scale associated with the pion decay constant $f_\pi = 93$ MeV, also the electroweak anomaly may be related to the monopole catalysis, the corresponding length scale is given by the vacuum expectation value of the Standard Model ~ 100 GeV. The consideration above concentrated only around the smallest size of the core. Moreover, the core itself was replaced by the boundary conditions (10.48) at the origin.

Actually, it was proposed that the monopole catalyzed decay of a nucleon may occur already at a long distance associated with hadronic scale [154]. In other words, one can consider not a quantum field theory of quarks coupled with a monopole, but a classical effective model of a nucleon interacting with the 't Hooft–Polyakov monopole [143, 144, 154, 326].

Callan and Witten suggested to invoke the Skyrme-type model, which describes a baryon as a soliton configuration of the low-energy effective theory [154]. In the simplest situation of the $SU(2)$ flavor group, the effective action of the model is given by the Lagrangian [478]:

$$L_{Sk} = \frac{f_\pi^2}{4} \text{Tr}\left(R_\mu R^\mu\right) + \frac{\kappa^2}{8} \text{Tr}\left([R_\mu, R_\nu]\right)^2, \quad (10.106)$$

where κ is a dimensionless constant and the current $R_\mu = iU^{-1}\partial_\mu U$ is related to the Cartan $SU(2)$ one-forms (A.3). The first term here corresponds to the non-linear sigma model and the second term stabilizes the solution. The matrix U stands for the pion chiral field:

$$U = \exp\left\{\frac{i}{f_\pi}\sigma^a \phi^a\right\},$$

10.3 Monopole-Fermion Scattering in the Bosonisation Technique

which represents a topological mapping $S^3 \to S^3$ from coordinate space to the $SU(2)$ group space[15]. The scalar field $\phi = \phi^a \sigma^a /2$ represents the triplet of pions. This mapping can be characterized by the third homotopy group $\pi_3(S^3) = \mathbb{Z}$, and the stability of the configuration is guaranteed by the winding number, which within this framework is associated with the baryon number:

$$B = \frac{\varepsilon_{ijk}}{24\pi^2} \int d^3x \, \text{Tr} \, (R_i R_j R_k) \,. \tag{10.107}$$

The nucleon is defined by yet another static hedgehog ansatz $U = \exp\{if(r)(\sigma^a r^a)\}$, which defines a chiral angle $f(r)$. It runs from $f(0) = 0$ to $f(\infty) = \pi$, which corresponds to the simplest spherically symmetric soliton, a *skyrmion* with baryon number $B = 1$. We do not discuss the properties of the Skyrme model here. The curious reader may find a full account of the theory and applications of the model in the low-energy hadron physics, e.g., in [20].

Because $R_\mu \sim U^{-1}(\partial_\mu \phi)U$, in terms of the $su(2)$ algebra-valued rescaled scalar field ϕ, the Skyrme Lagrangian (10.106) takes the form of the non-linear sigma-model

$$L_{NLSM} = \frac{1}{2}\text{Tr}\left[(\partial_\mu \phi)(\partial^\mu \phi) + \frac{1}{2}[\partial_\mu \phi, \, \partial_\nu \phi]^2\right]. \tag{10.108}$$

A monopole can be incorporated in this model, if we consider a *gauged Skyrme model* [154, 249, 326], which describes a gauge invariant system of coupled scalar and vector fields with a fourth-order Skyrme term with derivatives replaced by covariant derivatives D_μ:

$$L = -\frac{1}{2}\text{Tr}\left(F_{\mu\nu}F^{\mu\nu} + D_\mu \phi D^\mu \phi + \frac{1}{2}[D_\mu \phi, D_\nu \phi]^2\right) - V(\phi). \tag{10.109}$$

Evidently, this is a generalization of (5.7). The solutions of this model are referred to as *Skyrmed monopoles* [249]. Briefly speaking, the presence of the Skyrme term leads to additional attraction between the skyrmed monopoles, thus, in contrast to skyrmions, the lowest energy bound states possess the axial symmetry. A similar model was proposed in [154], where the gauge field was taken to be the Abelian potential of the static Dirac monopole and the electric charge matrix $Q = \text{diag}(2/3, -1/3)$ corresponds to the doublet of the light quarks.

Recall that for a scalar particle in the field of a point-like monopole the centrifugal potential is $\sim (j(j+1)-e^2)/r^2$ (see page 248). Thus, the boundary conditions we have to impose at the origin on the isotopic components of the pion field ϕ in the monopole are different; while the wavefunction of

---

[15] Since the finite energy solution satisfies $U(r) \to 1$ as $r \to \infty$, all the points of the spatial infinity are identified, thus, the space $\mathbb{R}^3$ is compactified to a sphere S^3.

the charged pions has to vanish there, the neutral component $\pi_0 \sim \phi^3 \sigma^3/2$ may penetrate the monopole. In other words, at the monopole location the boundary condition on the neutral chiral field becomes [154]

$$U(0) = \exp\{if(0)\sigma^3/2\}, \qquad (10.110)$$

where the function $f(0)$ is not necessarily constrained.

On the other hand, the presence of the electromagnetic potential A_μ modifies the baryonic current, which may be explicitly written as

$$\begin{aligned} B_\mu &= \frac{\varepsilon_{\mu\nu\rho\sigma}}{24\pi^2} \operatorname{Tr} \{U^{-1}D^\nu U\, U^{-1}D^\rho U\, U^{-1}D^\sigma U\} \\ &+ \frac{\varepsilon_{\mu\nu\rho\sigma}}{16\pi^2} \operatorname{Tr} \{F^{\rho\sigma}(D^\nu U U^{-1} + U^{-1}D^\nu U)\}, \end{aligned} \qquad (10.111)$$

where the non-zero component of the electromagnetic field strength tensor is $F_{\theta\varphi} = 1/er^2$. This gauge invariant conserved current includes a second term that is a total divergence and represents the anomaly (10.52).

It was suggested in [154] that the neutral pion field may vary with time. Evidently, this corresponds to the physical picture of wrapping a skyrmion around a monopole, which is a model of monopole-proton scattering. Then the radial component of the baryonic current (10.111) on the monopole background becomes

$$B_r = \frac{\dot{f}(t)}{8\pi^2 r^2}, \qquad (10.112)$$

that is, there is a flux of the baryon number of magnitude $\dot{f}/2\pi$ into the monopole. Since in the $SU(5)$ Grand Unified Theory, the non-conservation of the baryon number comes with the non-conservation of the lepton number together, the leptons must be radiated out while the nucleon is squeezed down into a monopole. Indeed, the boundary condition at the origin (10.110) cannot prevent the skyrmion from decaying.

A disadvantage of such a model is that the monopole is treated as pointlike and the non-conservation of the baryon number is entirely related to the boundary condition on the pion field at the origin. It was suggested [143, 144, 326] to consider an extendend model that has $SU(2)_L \times SU(2)_R$ global symmetry and contains the gauged Skyrme field coupled both with the left and right sectors. Thus, monopole and skyrmion are regular classical solutions of such a model. Numerical analysis shows that there is a radial baryonic current that coincides with (10.112) at large distances and there are no static solutions with baryon number one on top of the monopole [143,144]. A more refined analysis, which takes into acocunt the properties of the system at short distances, reveals a sequence of skyrmion-monopole bound states separated by finite energy barriers. Thus, this picture resembles the periodic structure of the topological vacua [144]. However, the barrier height is very small compared to the nucleon mass and the monopole catalysis takes place semiclassically.

Let us mention in conclusion that the experimental observation of the monopole catalysis of the proton decay would be a very remarkable event that could touch the very basis of QFT. If this process is discovered, the consequences may be unpredictable. For example, some optimists mentioned that the energy of the monopole-induced baryon decay could be used to solve the future energy crisis [209] and, maybe, the power stations of the XXII century would produce not dirty nuclear energy, but would use just a "pinch" of monopoles as a power element. However, the monopole catalysis is not a general property of any theory of Grand Unification, but it is a strongly model-dependent phenomenon [56, 189, 190, 454, 455]. Since it is not clear today as to which of the models of unification could become a "Standard GUT", the beautiful monopole catalysis remains endangered.

Part III

Supersymmetric Monopoles

11 Construction of Supersymmetric Yang–Mills Theories

The discussion of the different aspects of the non-Abelian magnetic monopoles in the previous chapters becomes very transparent in the wider class of four-dimensional supersymmetric (SUSY) gauge theories. As we shall see below, these models yield a very nice frame to reveal most of the remarkable features of the monopoles from a single point point of view.

Indeed, a supersymmetrical theory by very definition includes both bosonic and fermionic variables. Thus, there is a nice natural interplay between the localized monopole configuration and the fermions in such a model. Also the Montonen–Olive conjecture of duality turns out to be proved in supersymmetric models. A clue to the latter solution is the observation that an exact duality on a quantum level could only survive if the Bogomol'nyi bound is not violated by the quantum corrections. This becomes possible if the bosonic and fermionic loop diagrams, which could affect the effective potential, are mutually cancelling. This mechanism of cancellation is a familiar property of supersymmetric models.

Although the monopole solutions of the supersymmetric extension of the simple $SU(2)$ Yang–Mills–Higgs model were discussed as early as in 1978 [65], it was the celebrated work by N. Seiberg and E. Witten [469] that revitalized the real interest in the properties of SUSY monopoles and provide a significant progress in the understanding of non-perturbative dynamics in these theories.

This fascinating development initiated by N. Seiberg and E. Witten caused a real burst of interest. Although the original paper [469] is written very explicitly and clearly, one could trace out a special genre in the vast flow of publications related to this subject, the "Pedagogical introduction to the work of Seiberg and Witten" [123], see, e.g., [73, 199, 311, 347, 424, 473]. Our presentation follows the line of these excellent reviews.

11.1 What is Supersymmetry?

In this chapter, we would like to discuss some technical aspects of supersymmetry that are relevant to our discussion. Actually, our goal is to provide an introduction, which is necessary to understand recent exciting developments in supersymmetric gauge theories. There are many good reviews on supersymmetry that give a thorough introduction into the basis of supersymmetry.

11.1.1 Poincaré Group and Algebra of Generators

The direct way to introduce supersymmetry is to consider an extension of the usual space-time symmetries. Let us consider a conventional change of coordinates under action of the Poincaré group:

$$x^\mu \to x^{\mu'} = \Lambda^\mu_\nu x^\nu + a^\mu,$$

where Λ^μ_ν is an ordinary Lorentz transformation (or its Euclidean counterpart, a matrix of $SO(4)$ rotations) and a^μ is a vector of space-time translation. Thus, the infinitesimal generators of the Poincaré group, which act on the space of fields, are six components of the antisymmetric momentum tensor

$$M^{\mu\nu} \equiv M^{\mu\nu 0} = x^\mu P^\nu - x^\nu P^\mu + S^{\mu\nu} = L^{\mu\nu} + S^{\mu\nu}$$

(cf. (4.26) in Chap. 4) and four components of the vector of momentum $P^\mu = -i\partial^\mu$. Here, we decomposed the tensor of momentum into the orbital and the spin parts, $L^{\mu\nu}$ and $S^{\mu\nu}$, respectively. These generators form a Lie algebra, which is defined by the commutation relations

$$[M^{\mu\nu}, M^{\rho\sigma}] = i(g^{\nu\rho}M^{\mu\sigma} + g^{\mu\sigma}M^{\nu\rho} - g^{\mu\rho}M^{\nu\sigma} - g^{\nu\sigma}M^{\mu\rho}),$$
$$[P^\mu, P^\nu] = 0, \quad [M^{\mu\nu}, P^\rho] = i(g^{\nu\rho}P^\mu - g^{\mu\rho}P^\nu), \tag{11.1}$$

where we use, as before, the convention $g_{\mu\nu} = g^{\mu\nu} = (-1, 1, 1, 1)$ for the Minkowski metric in the flat space.

Recall some properties of the Lorentz subgroup $SO(3,1)$ of the Poincaré group. First, let us show that its covering group is $SL(2,\mathbb{C})$. Indeed, because the spin generators $S^{\mu\nu}$ commute both with P^μ and $L^{\mu\nu}$, they generate the finite dimensional representations of the Lorentz group. Then the spin operators $S_i = \frac{1}{2}\varepsilon_{ijk}S^{jk}$ and the operators of the Lorentz boost $K^i = S^{0i}$ form the algebra

$$[S^i, S^j] = i\varepsilon^{ijk}S^k, \quad [K^i, K^j] = -i\varepsilon^{ijk}S^k, \quad [K^i, S^j] = i\varepsilon^{ijk}K^k, \tag{11.2}$$

which reproduces the third set of the commutation relations (11.1) in a non-covariant form.

Let us consider the complex combinations

$$J_+ = \frac{1}{2}(S_i + iK_i), \qquad J_- = \frac{1}{2}(S_i - iK_i).$$

Then the relations (11.2) can be rewritten in the form that coincides with two commuting algebras of $SU(2)$ group[1]:

---

[1] Actually we are now not discussing a full Lorentz group, but its so-called *orthochronous part* defined by $\Lambda^0{}_0 > 0$, $\det \Lambda = 1$. Of course, the Lorentz group is not compact and its local structure is not $SU(2) \times SU(2)$, because the operators $J_\pm$ are not Hermitian: $(J^k_\pm)^\dagger = J^k_\mp$.

11.1 What is Supersymmetry?

$$[J_\pm^i, J_\pm^j] = i\varepsilon^{ijk} J_\pm^k, \quad [J_\pm^i, J_\mp^j] = 0.$$

Clearly, J_+^2 and the J_-^2 are two Casimir operators of the Lorentz group with half-integer eigenvalues $j_+(j_+ + 1)$ and $j_-(j_- + 1)$, respectively. Thus, a finite dimension irreducible representation of dimension $(2j_+ + 1)(2j_- + 1)$ can be labeled by (j_+, j_-). In other words, unlike the generators $M^{\mu\nu}$, the operators $J_\pm$ can be represented by finite dimensional Hermitian matrices. For the fundamental representations we can write

$$\begin{aligned}(\tfrac{1}{2},0) \;&:\; J_+^i \;\to\; \tfrac{i}{2}\sigma^i, \\ (0,\tfrac{1}{2}) \;&:\; J_-^i \;\to\; \tfrac{i}{2}\bar\sigma^i = -\tfrac{i}{2}\sigma^i,\end{aligned} \qquad (11.3)$$

where $\bar\sigma^i$ is dual to σ^i in the sense that

$$(\sigma^i)_{\alpha\dot\beta}(\bar\sigma^j)^{\dot\beta\alpha} = \mathrm{Tr}(\sigma^i\bar\sigma^j) = 2\delta_{ij}.$$

Thus, we have two basis vectors in the space of complex 2×2 matrices:

$$\begin{aligned}\sigma^\mu &= (\sigma^\mu)_{\alpha\dot\beta} = (\mathbb{I}, \sigma^i), \\ \bar\sigma^\mu &= (\bar\sigma^\mu)_{\dot\alpha\beta} = (\mathbb{I}, -\sigma^i),\end{aligned} \qquad (11.4)$$

and we can define $(\tfrac{1}{2},0)$ generators of the covering $SL(2,\mathbb{C})$ group as

$$(\sigma^{\mu\nu})_\alpha{}^\beta = \frac{i}{2}\left[(\sigma^\mu)_{\alpha\dot\gamma}(\bar\sigma^\nu)^{\dot\gamma\beta} - (\sigma^\nu)_{\alpha\dot\gamma}(\bar\sigma^\mu)^{\dot\gamma\beta}\right], \qquad (11.5)$$

while $(0,\tfrac{1}{2})$ generators of the covering $SL(2,\mathbb{C})$ are

$$(\sigma^{\mu\nu})^{\dot\alpha}{}_{\dot\beta} = \frac{i}{2}\left[(\bar\sigma^\mu)^{\dot\alpha\gamma}(\sigma^\nu)_{\gamma\dot\beta} - (\bar\sigma^\nu)^{\dot\alpha\gamma}(\sigma^\mu)_{\gamma\dot\beta}\right]. \qquad (11.6)$$

A spinor may be introduced as a two-component field that lies in the fundamental representation of $SL(2,\mathbb{C})$. However, as we have seen, unlike for the $SU(2)$ group, for $SL(2,\mathbb{C})$ a matrix of representation and its conjugated are not equivalent. This implies that there are two types of Weyl spinors: the left-handed ψ_α that transforms according to the fundamental representation $(\tfrac{1}{2},0)$, while the right-handed spinors $\bar\psi^{\dot\alpha}$ transform according to the fundamental representation $(0,\tfrac{1}{2})$ (here $\alpha,\dot\alpha = 1,2$). Clearly, $\bar\psi^{\dot\alpha} = (\psi_\alpha)^*$.

The four-component Dirac bispinor in four-dimensional space consists of both chiral components:

$$\psi = \begin{pmatrix} \zeta_\alpha \\ \bar\eta^{\dot\alpha} \end{pmatrix}.$$

Actually, we already made use of these notations in the previous chapter. The Dirac bispinor evidently lies in the reducible $(\tfrac{1}{2},0)\oplus(0,\tfrac{1}{2})$ representation.

There is another possibility of composing two chiral spinors into the four-component *Majorana bispinor*:

$$\psi = \begin{pmatrix} \zeta_\alpha \\ \bar\zeta^{\dot\alpha} \end{pmatrix},$$

that is, the Majorana spinor can be thought of as a Dirac spinor with $\zeta \equiv \eta$.

The indices of the right-handed spinors are traditionally dotted to emphasize the difference between those transformation laws. The raising and lowering of the indices is given by the action of the Levi-Civita tensor $\varepsilon = (\varepsilon)^{\alpha\beta} = i\sigma_2$, $\varepsilon^{-1} = (\varepsilon)_{\alpha\beta}$, which in the spinor space plays an analogous role to the metric tensor in Minkowski space:

$$\psi^\alpha = \varepsilon^{\alpha\beta}\psi_\beta \;, \qquad \bar\psi^{\dot\alpha} = \varepsilon^{\dot\alpha\dot\beta}\bar\psi_{\dot\beta} \;. \tag{11.7}$$

The usual conventions are

$$\varepsilon^{\dot\alpha\dot\beta} = \varepsilon^{\alpha\beta} \;, \qquad \varepsilon^{\alpha\gamma}\varepsilon_{\gamma\beta} = \delta^\alpha_\beta \;,$$
$$\varepsilon^{12} = \varepsilon^{\dot 1\dot 2} = 1 \;, \qquad \varepsilon_{12} = \varepsilon_{\dot 1\dot 2} = -1 \;, \tag{11.8}$$

so, $\bar\psi_{\dot\alpha} = \psi^\dagger_\alpha$ and

$$(\psi\chi)^\dagger = \bar\chi_{\dot\alpha}\bar\psi^{\dot\alpha} = \bar\chi\bar\psi = \bar\psi\bar\chi \;.$$

By analogy with the common matrix indices, the spinor indices can be dropped out in the widely used shorthand notations. Then we can write, for instance for the Lorentz scalars

$$\psi\chi = \psi^\alpha\chi_\alpha = \varepsilon^{\alpha\beta}\psi_\beta\chi_\alpha = -\varepsilon^{\alpha\beta}\psi_\alpha\chi_\beta = -\psi_\alpha\chi^\alpha = \chi\psi \;,$$

and

$$\bar\psi\,\bar\chi = \bar\psi_{\dot\alpha}\bar\chi^{\dot\alpha} = -\bar\psi^{\dot\alpha}\bar\chi_{\dot\alpha} = \bar\chi\,\bar\psi \;.$$

The ordering of anticommuting Grassmanian variables is important. In convolving the undotted indices we write first the spinor with the upper index, while for the dotted indices we write first the spinor with the lower index.

The vector and tensor quantities can also be constructed from these Grassmanian variables. For example, $\psi_\alpha\chi_{\dot\alpha}$ in the $(\frac{1}{2},\frac{1}{2})$ representation transforms as a Lorentz vector.

Let us say a few worlds about two Casimir operators of the Poincaré group, which commute with all ten generators of the latter. They are the operator of square mass $m^2 = P^2 = P_\mu P^\mu$ and the square of the Pauli–Lubanski spin vector

$$W^\mu = -\frac{1}{2}\varepsilon^{\mu\nu\rho\sigma}P_\nu M_{\rho\sigma} \equiv -\frac{1}{2}\varepsilon^{\mu\nu\rho\sigma}P_\nu S_{\rho\sigma} \;.$$

Clearly, $W^2 = W_\mu W^\mu = -m^2 S_i S^i$. Hence this operator has eigenvalues $-m^2 s(s+1)$ and $W_\mu = \lambda P_\mu$ for the massive and massless states, respectively, where the corresponding half-integer quantum number λ is the *helicity*. In the rest frame of a massive state we have $P_\mu = (m, 0, 0, 0)$. For the massless states we obviously have $P^2 = W^2 = 0$.

So far we have discussed the Lorentz transformations. However, we are not finished yet with the description of the symmetries of a consistent physical theory. Besides the space-time symmetry, there also are some internal symmetries, like, for example, gauge or flavor symmetry. Moreover, we suppose that the Lorentz invariant theory must be invariant with respect to the discrete CPT transformations, otherwise we shall get a lot of problems related with non-locality and violation of the spin-statistics theorem[2].

In 1967 Coleman and Mandula showed [169] that in a local relativistic quantum field theory the most general symmetry group of the S-matrix must be a direct product of the Poincaré group and a compact internal Lie group, provided that: (i) there is only one zero-mass state and a finite number of excitations associated with a massive one-particle states; and (ii) the vacuum state is unique and there is a finite energy gap between it and lowest one-particle state.

The latter assumption is introduced to avoid infrared problem, while the former allows us to exclude the models with spontaneous symmetry breaking from consideration. Actually, the first restriction can be relaxed, that is, we can consider a tower of massless one-particle excitations over the ground state [252]. The price paid for such an extension is that now the extended group of conformal symmetry must be admitted instead of the Poincaré group.

However, the Coleman–Mandula theorem involves only bosonic symmetries, that is, the algebra of generators of symmetry, namely four momenta P^μ, six Lorenz generators $M^{\mu\nu}$ and generators of Hermitian internal symmetry T^a, involves only commutators. Moreover, the direct product structure means that

$$[T^a, P^\mu] = [T^a, M^{\mu\nu}] = 0,$$

and, therefore, the Lie algebra generators T^a are scalars with respect to the Lorentz group.

Furthermore, the Casimir operators commute not only with all generators of the Poincaré group, but also with the generators of internal symmetry:

$$[T^a, P^2] = [T^a, W^2] = 0.$$

The consequences of these relations are:
(i) *O'Raifeartaigh theorem*: all particles belonging to an irreducible multiplet of the internal symmetry group must have the same mass.
(ii) These particles must have the same spin.

Clearly, for the massless excitations we have $[T^a, P_\mu] = [T^a, W_\mu] = 0$, that is, the transformations of the internal symmetry cannot change the helicity.

---

[2] Actually we are restricting our consideration to the conventional local QFT. We are not discussing here the recent idea that there might be minuscule violations of Lorentz and CPT invariance, which would arise as suppressed effects from a more fundamental theory.

The idea of supersymmetry is to extend the Lie algebra, which is formed by the conventional generators P^μ, $M^{\mu\nu}$, in order to include anticommuting generators on an equal footing with these commuting generators.

11.1.2 Algebra of Generators of Supersymmetry

Let us see how this approach works. Speaking more formally, we must supplement the set of the bosonic generators by a set of N anticommuting generators Q_α^I, $\bar{Q}_{\dot\alpha}^I$, which transform as spin-half operators in the $(\frac{1}{2},0)$ and $(0,\frac{1}{2})$ representations of the Lorentz group, respectively, that is,

$$\bar{Q}_{\dot\alpha}^I = Q_\alpha^{I\dagger} = \varepsilon_{\alpha\beta}(Q^{\beta I})^\dagger \ .$$

Here the index $I = 1, 2, \ldots, N$ labels all the different spinors Q_α^I. It is the total number of supersymmetries of the model. The case $N=1$ corresponds to *simple supersymmetry*, whereas $N > 1$ is referred to as *extended supersymmetry*.

Since these generators are not scalars with respect to the Lorentz group, they cannot be introduced as some operators of an internal symmetry. They rather represent an extension of the Poincaré space-time symmetry algebra by anticommuting generators[3].

The uniqueness of such an extension is related to the so-called *Haag–Lopuszański–Sohnius* theorem [252], which states that under the assumption of positivity of the metric of the underlying Hilbert space[4], no representation with the spin higher than $1/2$ is possible for fermionic supersymmetry generators.

To define the supersymmetry algebra we have to consider the anticommutator $\{Q, \bar{Q}\}$, which transforms as $(\frac{1}{2},0) \otimes (0,\frac{1}{2}) = (\frac{1}{2},\frac{1}{2})$ and, therefore, must be proportional to the only spin-1 operator, P^μ:

$$\{Q_\alpha^I, \bar{Q}_{\dot\beta}^J\} = -2\delta^{IJ} \sigma^\mu_{\alpha\dot\beta} P_\mu \ . \qquad (11.9)$$

The factor of two is introduced here as a convention and the minus sign is due to the metric signature.

Note that the SUSY generators Q_α^I satisfy the commutation relations

$$[M^{\mu\nu}, Q_\alpha^I] = i(\sigma^{\mu\nu})_\alpha{}^\beta Q_\beta^I, \quad [M^{\mu\nu}, \bar{Q}_I^{\dot\alpha}] = i(\bar\sigma^{\mu\nu})^{\dot\alpha}{}_{\dot\beta} \bar{Q}_I^{\dot\beta} \ . \qquad (11.10)$$

These relations clarify the physical meaning of the generators of supersymmetry. Indeed, let us consider the component of the momentum tensor $M_{12} = S_3$. Then, the commutation relations (11.10) yield

---

[3] From a mathematical point of view, the key idea of supersymmetry is to promote the Lie algebra of the Poincaré generators to a *graded Lie algebra*.

[4] That is, the anticommutator of the operators $\{Q_\alpha^I, \bar{Q}_{\dot\alpha J}\}$ is supposed to be a positively defined operator. This condition excludes the states of negative mass (tachyons) from all possible irreducible representations of the SUSY algebra.

11.1 What is Supersymmetry?

$$[S_3, Q_1^I] = \frac{1}{2}(\sigma^3)_1{}^\beta Q_\beta^I = \frac{1}{2}Q_1^I,$$
$$[S_3, Q_2^I] = \frac{1}{2}(\sigma^3)_2{}^\beta Q_\beta^I = -\frac{1}{2}Q_2^I,$$
(11.11)

and we also have similar relations for the operators $\bar{Q}_{\dot\alpha}^I = Q_\alpha^{I\dagger}$. In other words, the operators Q_1^I and $Q_2^{I\dagger}$ raise the z-component of the spin by a half, while the operators Q_2^I and $Q_1^{I\dagger}$ lower it by a half.

On the other hand, the generators of supersymmetry commute with the operators of translations [480]:

$$[Q_\alpha^I, P^\mu] = [\bar{Q}_{\dot\alpha I}, P^\mu] = 0,$$
(11.12)

which can be proved by making use of the Jacobi identities for the operators of the graded Lie algebra

$$[[Q_\alpha^I, P^\mu], P^\nu] + [[P^\nu, Q_\alpha^I], P^\mu] + [[P^\mu, P^\nu], Q_\alpha^I] = 0,$$
$$\{[Q_\alpha^I, Q_\beta^J\}, P^\mu] + \{[P^\mu, Q_\alpha^I], Q_\beta^J\} - \{[Q_\beta^J, P^\mu], Q_\alpha^I\} = 0.$$
(11.13)

which supplement the usual bosonic and fermionic Jacobi identities

$$[[P^\mu, P^\nu], P^\rho] + [[P^\nu, P^\rho], P^\mu] + [[P^\rho, P^\mu], P^\nu] = 0,$$
$$\{\{Q_\alpha^I, Q_\beta^J\}, Q_\gamma^K\} + \{\{Q_\beta^J, Q_\gamma^K\}, Q_\alpha^I\} + \{\{Q_\gamma^K, Q_\alpha^I\}, Q_\beta^J\} = 0.$$
(11.14)

Here, the operator of translation P^μ and the generator of supersymmetry Q_α^I can be replaced by an arbitrary bosonic and a fermionic generator, respectively.

To prove the relations (11.12), let us note that the commutator $[Q_\alpha^I, P^\mu]$ must transform as $(\frac{1}{2},0) \otimes (\frac{1}{2},\frac{1}{2}) = (0,\frac{1}{2}) \oplus (1,\frac{1}{2})$ under the Lorentz group. Since spin-3/2 generators are excluded due to the Haag–Lopuszański–Sohnius theorem, the only possibility is

$$[Q_\alpha^I, P^\mu] = C^{IJ}(\sigma^\mu)_{\alpha\dot\beta}\bar{Q}^{\dot\beta J}, \quad [\bar{Q}_{\dot\alpha}^I, P^\mu] = (C^{IJ})^* Q_\beta^J (\bar\sigma^\mu)_{\dot\alpha}^{\beta},$$

where C^{IJ} are complex Lorentz scalar coefficients. Substitution of this relations into the first of the Jacobi identities (11.13) yields

$$(CC^*)^{IJ}(\sigma^{\mu\nu})_\alpha^\beta Q_\beta^J = 0,$$

where C is a matrix with matrix elements C^{IJ}. Thus, the matrix (CC^*) vanishes. However, it is not enough to conclude that the entries C^{IJ} themselves must be equal to zero. To see this we have to consider the anti-commutator $\{Q_\alpha^I, Q_\beta^J\}$, which enters second of the Jacobi identities (11.13). It lies in the representation $(\frac{1}{2},0) \otimes (0,\frac{1}{2}) = (0,0) \oplus (1,0)$ and generally can be written as

$$\{Q_\alpha^I, Q_\beta^J\} = 2\varepsilon_{\alpha\beta} Z^{IJ} + \varepsilon_{\beta\gamma} M_{\mu\nu}(\sigma^{\mu\nu})_\alpha^\gamma X^{IJ},$$
(11.15)

where Z^{IJ} and X^{IJ} are anti-symmetric and symmetric Lorentz scalars, respectively. Therefore, they may be represented via some linear combinations of the generators of internal symmetries.

Now, contracting the second of the Jacobi identities (11.13) with $\varepsilon_{\alpha\beta}$, we can see that, because Z^{IJ} commutes with P^μ, the matrix C is symmetric:

$$(C^{IJ} - C^{JI})P^\mu = 0.$$

Since $CC^* = CC^\dagger = 0$, this condition implies that all numbers C^{IJ} are vanishing and we can now see that the relation (11.12) is fulfilled. Its physical meaning is quite clear: the space-time translations do not affect the spin degrees of freedom. Furthermore, the nullification of the commutator $[Q_\alpha^I, P^\mu]$ excludes the symmetric term in the right-hand side of the anti-commutator (11.15), because $M_{\mu\nu}$ do not commute with P^μ, whereas the anti-commutator $\{Q,Q\}$ does. Thus, $X^{IJ} = 0$ and the algebra of the supersymmetry operators becomes simply

$$\{Q_\alpha^I, Q_\beta^J\} = 2\varepsilon_{\alpha\beta}Z^{IJ}, \qquad \{\bar{Q}_{\dot\alpha}^I, \bar{Q}_{\dot\beta}^J\} = 2\varepsilon_{\dot\alpha\dot\beta}Z^{*IJ}. \tag{11.16}$$

The complex coefficients $Z^{IJ} = -Z^{JI}$ are called *central charges* because they actually commute with all the generators, both bosonic and fermionic. The usual convention is $Z^{IJ} = -Z_{IJ}$.

The situation becomes even simpler for $N = 1$ SUSY algebra, because the symmetry of the anti-commutator $\{Q_\alpha^I, Q_\beta^J\}$ with respect to permutation of pairs of indices $(\alpha, I) \rightleftharpoons (\beta, J)$, together with the antisymmetry of the Levi-Civita tensor $\varepsilon_{\alpha\beta}$, mean that the central charges are excluded for $N = 1$ supersymmetry.

Let us note that the relations (11.9) and (11.16) are invariant under $U(N)$ transformations

$$Q_\alpha^I \to U_{IJ}Q_\alpha^J, \qquad \bar{Q}_{\dot\alpha}^I \to \bar{Q}_{\dot\alpha}^J U_{IJ}^\dagger.$$

We can see that, if the central charges are vanishing, this internal symmetry of the supersymmetry generators, which does not change the algebra of SUSY, is higher degree – the presence of the central charges reduces this unitary symmetry to a smaller subgroup.

Clearly, for $N = 1$ SUSY, there is only one supercharge Q_α. Then the internal global symmetry group is $U(1)$, which is known as *R-symmetry*:

$$[Q_\alpha, \mathcal{R}] = Q_\alpha, \qquad [\bar{Q}_{\dot\alpha}, \mathcal{R}] = -\bar{Q}_{\dot\alpha}. \tag{11.17}$$

Actually, this is the familiar chiral symmetry of the spinor field, since under the transformations of parity $Q \to \bar{Q}$, $\bar{Q} \to Q$ and the Abelian charge $\mathcal{R}$ transforms as $\mathcal{R} \to -\mathcal{R}$. Thus, the $N = 1$ SUSY generators have a chiral charge $+1$ and -1, respectively.

We shall see how the central charges are related with the topology of the supersymmetric Yang–Mills theory, in particular with the magnetic charge of the field configurations.

11.2 Representations of SUSY Algebra

Let us consider the unitary irreducible representation of SUSY algebra, which is constructed by analogy with the representation theory of the Poincaré group. One can think that, since any representation of the SUSY algebra contains a representation of the Poincaré algebra, its representations (that is, the multiplets of asymptotic on-shell physical states) as before may be labeled by eigenvalues of the mass-square operator $P^\mu P_\mu = m^2$ and the square of the Pauli–Lubanski vector $W^\mu W_\mu$. However, the SUSY generators commute only with former operator, whereas W^2 is no longer a Casimir operator ($Q, \bar{Q}$ do not commute with $M^{\mu\nu}$). Therefore, representations of SUSY algebra contain states of the same mass but different spins (or helicities in the massless case).

11.2.1 $N = 1$ Massive Multiplets

Let us now examine the simple $N = 1$ SUSY with one supercharge Q_α. From the consideration above, it is clear that we have to consider two different cases of massive and massless states.

Unlike the classification of irreducible multiplets of the Poincaré group, the square of the Pauli–Lubanski vector W_μ has to be replaced by the second Casimir operator of superalgebra [357]

$$G^2 = G^{\mu\nu}G_{\mu\nu}, \qquad G_{\mu\nu} \equiv V_\mu P_\nu - V_\nu P_\mu, \tag{11.18}$$

where

$$V_\mu = W_\mu - \frac{1}{4}\bar{Q}_{\dot\alpha}(\bar\sigma_\mu)^{\dot\alpha\alpha}Q_\alpha.$$

Let us consider the massive $N = 1$ supermultiplets first. To construct the representations of SUSY algebra, we can use the Wigner trick, namely go to the fixed Lorentz frame, which is characterized by a certain value of momenta P^μ and study the "little group" of this vector. For the massive states, we can always choose the rest frame, where $P^\mu = (m, 0, 0, 0)$, $m \neq 0$. In this frame, the $N = 1$ supersymmetry algebra (11.9) is reduced to

$$\{Q_\alpha, \bar{Q}_{\dot\beta}\} = 2m\sigma^0_{\alpha\dot\beta} = 2m\begin{pmatrix} 1 & 0 \\ 0 & 1 \end{pmatrix}, \tag{11.19}$$

and the Casimir operators can be written as

$$P^2 = m^2, \qquad G^2 = 2m^4 J^2, \tag{11.20}$$

where

$$J_k = S_k - \frac{1}{4m}\bar{Q}_{\dot\alpha}\bar\sigma_k^{\dot\alpha\alpha}Q_\alpha.$$

Since both the spin vector S_k and the Pauli matrices $\bar\sigma_k^{\dot\alpha\alpha}$ satisfy the $SU(2)$ algebra, the operator J^2 has eigenvalues $j(j+1)$, integer or half-integer. Thus, a state of an $N = 1$ SUSY multiplet can be labeled by the quantum numbers, m, j, j_3.

The operators of $N = 1$ supersymmetry $Q_\alpha, \bar Q_{\dot\beta}$ are actually two pairs of creation and annihilation operators that act on the vacuum state $|\Omega\rangle$. Indeed, a simple rescaling

$$Q_\alpha \to a_\alpha = \frac{Q_\alpha}{\sqrt{2m}}, \qquad \bar Q_{\dot\alpha} \to a_\alpha^\dagger = \frac{\bar Q_{\dot\alpha}}{\sqrt{2m}}, \qquad (11.21)$$

reduces the algebra of supercharges (11.19) to the conventional Clifford algebra of the ordinary fermionic creation/annihilation operators

$$\{a_\alpha, a_\beta^\dagger\} = \delta_{\alpha\beta}. \qquad (11.22)$$

Recall that there are no central charges in $N = 1$ SUSY, thus, all other anti-commutators vanish, which completely defines the $N = 1$ superalgebra.

The Clifford vacuum $|\Omega\rangle$ can be defined with respect to consequent action of the fermionic annihilation operators Q_α on any state of definite m and j:

$$|\Omega\rangle = Q_1 Q_2 |m, j\rangle, \qquad Q_1 |\Omega\rangle = Q_2 |\Omega\rangle = 0.$$

Evidently, the state $|\Omega\rangle$ is $(2j+1)$-fold degenerated. Note that the Clifford vacuum is not a vacuum of field theory, since $|\Omega\rangle$ is a one-particle state with a mass m and a given spin (helicity), but not a ground state.

This definition allows us to label the states of the SUSY multiplets by spin s rather than the eigenvalues j. Indeed, making use of the algebra of the supersymmetry generators $Q, \bar Q$, it is straightforward to see that the action of the corresponding operator J_k on the state $|\Omega\rangle$ reduces to the spin operator S_k, that is, the Clifford vacuum is an eigenstate of spin rather than J_k: $|\Omega\rangle = |m, s, s_3\rangle$.

Action of two creation operators on such a vacuum state yields four independent states of the same mass:

$$|\Omega\rangle, \quad a_1^\dagger |\Omega\rangle, \quad a_2^\dagger |\Omega\rangle, \quad a_1^\dagger a_2^\dagger |\Omega\rangle.$$

Taking into account the $(2j+1)$-fold degeneration of each state, we conclude that there are a total of $4 \cdot (2j+1)$ states in the massive $N = 1$ SUSY multiplet.

Because the SUSY generators have spin $1/2$, these four states are different in spin. For $|\Omega\rangle = |m, s, s_3\rangle$, we have the states of spin $s_3 = j_3, j_3 - \frac{1}{2}, j_3 + \frac{1}{2}, j_3$.

Let us consider the $j = 0$ vacuum state first. The corresponding fundamental massive multiplet includes two bosonic and two fermionic degrees of freedom, namely one real scalar, one massive Weyl fermion and one real pseudoscalar. Indeed, the parity operator interchanges $a_1^\dagger$ with $a_2^\dagger$ and one of the spin-0 states has to be a pseudoscalar. This irreducible representation of $N = 1$ SUSY is known as *chiral massive multiplet*. In the simplest supersymmetrical generalization of QCD this multiplet includes the left-handed quark and its superpartner, the squark.

11.2 Representations of SUSY Algebra

In this simple example we can see a very universal result: the number of fermionic states in each SUSY multiplet must be equal to the number of bosonic states.

Indeed, let us consider $j = 1/2$, $N = 1$ massive *vector multiplet*. arguments similar to those considered above allow us to conclude that this multiplet contains four bosonic and four fermionic degrees of freedom. The corresponding particle spectrum can be defined by using standard momentum addition rules. It corresponds to a gauge field, a Dirac fermion and a scalar field, all having the same mass.

Finally, let us note that formally there are no algebraic restrictions on the number of generators of SUSY. However, the consideration above tells us that increasing N leads to the appearance of higher spin states in the corresponding SUSY multiplets. Thus, there is a physical condition of renormalizability of any consistent quantum field theory that does not admit any spin states greater than one. This restricts the number of supersymmetries $N \leq 4$.

11.2.2 $N = 1$ Massless Multiplets

Let us proceed further by considering the massless irreducible SUSY multiplets. For this special case, we fix a time-like reference frame where $P^\mu = (E, 0, 0, E)$. Then the second Casimir operator of the SUSY algebra (11.18) vanishes, because

$$G^2 = -2E^2(V_0 - V_3)^2 = \frac{E^2}{2} \bar{Q}_{\dot{2}} Q_2 \bar{Q}_{\dot{2}} Q_2 = 0. \qquad (11.23)$$

On the other hand, we obtain

$$\{Q_\alpha, \bar{Q}_{\dot\alpha}\} = 2P_\mu (\sigma^\mu)_{\alpha\dot\alpha} = \begin{pmatrix} 0 & 0 \\ 0 & 4E \end{pmatrix}, \qquad (11.24)$$

that is, one of the anticommutators in the algebra of the operators $\{Q_1, \bar{Q}_{\dot 1}\}$ is vanishing in the massless case.

Let us recall that a norm of a physical state in a unitary theory has to be positive. Requiring that this condition be satisfied, we must set the corresponding creation operator $a_1^\dagger = \bar{Q}_{\dot 1}/\sqrt{2E}$ equal to zero. Therefore, only one generator of supersymmetry survives and, rescaling it as $a_2 = Q_2/\sqrt{2E}$, $a_2^\dagger = \bar{Q}_{\dot 2}/\sqrt{2E}$, we are left with only two independent physical states in the Clifford vacuum rather than four, as in the massive case:

$$|\Omega\rangle, \quad a_2^\dagger |\Omega\rangle. \qquad (11.25)$$

Thus, there is a tower of states of the massless multiplets, each of which is labeled by the helicity λ. One may suppose that, by analogy with the fundamental massive SUSY multiplet, the fundamental massless vacuum state $|\Omega\rangle$

may have zero helicity. However, the operator $a_2^\dagger$ transforms as $(0, \frac{1}{2})$, that is, it increases the helicity λ of the vacuum state $|\Omega\rangle$ by $1/2$. In this case, a doublet (11.25) is not an eigenstate of a CPT symmetry operator. Thus, we conclude that in any Lorentz invariant local quantum field theory we have to consider the massless multiplets formed by the doublet of states with opposite helicities, for example, pairing two doublets with helicities $\lambda, \lambda+1/2$ and $-\lambda, -\lambda - 1/2$, respectively. The $N = 1$ massless $\lambda = 1/2$ doublet contains a vector particle and a Majorana spinor. In the SUSY QCD this is the so-called *gauge multiplet* and these one-particle states are set into correspondence with a gluon and its superpartner, the gluino.

11.2.3 $N = 2$ Extended SUSY

We can proceed further by generalization of the scheme considered above to the case of $N = 2$ SUSY irreducible multiplets. For the sake of simplicity, let us consider first the algebra (11.16) without central charge, $Z = 0$.

In the massive case, we now have four annihilation operators a_α^I, $I = 1, 2$, which are set into correspondence with supercharges. Hence there are $2^{2N} \cdot (2j + 1)$ states in a massive irreducible multiplet. The fundamental $(j = 0)$ multiplet therefore contains 16 one-particle states. If the vacuum state $|\Omega\rangle$ has spin zero, these states transform as five scalars, eight spinors and three spin-1 states.

As before, the number of states of the massless multiplet is reduced, since we have N creation operators $a^{I\dagger}$, which generate a total of 2^N states. For the $N = 2$ SUSY we have four massless states

$$|\Omega\rangle, \quad (a_2^1)^\dagger |\Omega\rangle, \quad (a_2^2)^\dagger |\Omega\rangle, \quad (a_2^1)^\dagger(a_2^2)^\dagger |\Omega\rangle.$$

If the vacuum state $|\Omega_0\rangle$ has the helicity 0, the states of $\lambda = 0$ massless supermultiplet have helicities $0, 1/2$ and 1, respectively. To form a CPT eigenstate, it has to be supplemented by its partner, the conjugated multiplet, which corresponds to the helicity -1 vacuum state $|\Omega_{-1}\rangle$. The latter multiplet includes the states of helicity $-1, -1/2, 0$.

These two multiplets Ω_0 and Ω_{-1} form the so-called $N = 2$ *on-shell vector multiplet*. In the QCD context, it is also referred to as the $N = 2$ *gauge multiplet*, which includes a massless gluon (2 vector states) and its superpartners, two Weyl spinors (four spin-1/2 states) and a complex scalar (two spin-0 states). Note that $N = 1$ on-shell scalar multiplet and $N = 1$ vector multiplet together contain the same set of fields as an $N = 2$ on-shell vector multiplet.

The $\lambda = 1/2$ $N = 2$ multiplet, which corresponds to the vacuum state $\Omega_{-\frac{1}{2}}$ with helicity $\lambda = 1/2$, is CPT self-conjugated. It includes two scalars and two fermionic states. This is the so-called massless $N = 2$ on-shell *hypermultiplet*, which consists of two copies of the $N = 1$ massless scalar multiplet.

In the presence of central charges, the situation changes. In particular, the algebra of $N = 2$ SUSY (11.16) with with central charge Z possesses an internal global $SU(2) \times U(1)$ symmetry. This is $\mathcal{R}$-symmetry, where the $U(1)$ subgroup is related with the unitary rotations of the operators (11.17), while the matrix of the transformation of the $SU(2)$ subgroup acts on the spinor indices of the operators Q^I_α.

We can make use of this symmetry to diagonalize the annihilation/creation operators. Indeed, in the presence of the central charges, the algebra of supersymmetry (11.16) in the rest frame is

$$\{Q^I_\alpha, \bar{Q}^J_{\dot{\beta}}\} = 2m\delta^{IJ}\sigma^0_{\alpha\dot{\beta}},$$
$$\{Q^I_\alpha, Q^J_\beta\} = 2\varepsilon_{\alpha\beta}Z^{IJ}; \quad \{\bar{Q}_{\dot\alpha I}, \bar{Q}_{\dot\beta J}\} = -2\varepsilon_{\dot\alpha\dot\beta}Z^*_{IJ},$$
(11.26)

and the generators of supersymmetry Q^I_α, $\bar{Q}^I_{\dot\alpha}$ cannot be treated as the creation and annihilation operators, respectively. One has to diagonalize the matrix Z^{IJ} to the form where its eigenvalues are real and non-negative.

Indeed, in the $N=2$ theory, the antisymmetric matrix of central charges Z^{IJ} has the off-diagonal element $Z = Z^{12}$. Then the diagonalization of the basis is given by an appropriate unitary transformation of the generators of supersymmetry to the linear combinations

$$a_\alpha = \frac{1}{\sqrt{2}}\left(Q^1_\alpha + \varepsilon_{\alpha\beta}(Q^{2\beta})^\dagger\right), \quad b_\alpha = \frac{1}{\sqrt{2}}\left(Q^1_\alpha - \varepsilon_{\alpha\beta}(Q^{2\beta})^\dagger\right), \quad (11.27)$$

and their conjugated $(a_\alpha)^\dagger, (b_\alpha)^\dagger$. These operators create a state of definite spin and, making use of (11.16), in the rest frame we have the anti-commutation relations

$$\{a_\alpha, a^\dagger_\beta\} = 2\delta_{\alpha\beta}(m+Z), \quad \{b_\alpha, b^\dagger_\beta\} = 2\delta_{\alpha\beta}(m-Z), \quad (11.28)$$

whereas all other anti-commutators vanish.

Let us recall now that all physical states must have a positive norm. Since the anticommutators of (11.28) are positively defined operators, for the massless states the central charge of any SUSY irreducible representation must be equal to zero, $Z = 0$. However, for the massive case, this relation means that the mass of the states m of a given multiplet must be bounded below by the central charge as

$$\mid Z \mid \leq m. \quad (11.29)$$

The corresponding massive multiplets contain the same set of states as the $N=2$ multiplets without central charges.

If the boundary is saturated, $m = \mid Z \mid$, one set of the creation operators becomes projections onto states of zero norm, that is, we actually have lost half of the generators of SUSY. Then the situation becomes completely identical to the case of massless multiplets that we discussed above. In other words, there are $4 \cdot (2j+1)$ reduced massive $N=2$ SUSY multiplets, which are referred as *short multiplets*.

Let us mention in conclusion that the relation (11.29) is nothing but a supersymmetric generalization of the Bogomol'nyi equation (5.60), which we have already encountered in the various chapters of this book. In fact, the central charges of the supersymmetric extension of the Georgi–Glashow model are related with magnetic and electric charges of the configuration [523]. However, before considering such an extension, we would like to discuss how a field theory can be constructed from the representation of supersymmetry.

11.3 Local Representations of SUSY

Armed with the results of the previous section, we are now prepared to formulate the supersymmetric field theory, that is, to construct a supersymmetric Lagrangian in terms of the multiplets of fields, which form the off-shell representations of the SUSY algebra discussed above.

There is a very elegant and straightforward way to obtain such a theory. This is related with a generalization of the Poincaré group acting on the space-time coordinates. The key idea of the related conception of *superspace* is to define an extension of the conventional space-time by consideration of a manifold on which the transformations of supersymmetry act. Then we may introduce the *superfields*, which are defined as functions of the coordinates of such a superspace.

11.3.1 $N = 1$ Superspace

Let us now consider the simple $N = 1$ supersymmetry. Since $N = 1$ superfields must lie in some representation of a graded Lie group, the first step is to turn from the SUSY algebra to a Lie group by exponentiating the generators of supersymmetry. However, the action of elements of the graded Lie group by definition must not leave the group manifold, that is a SUSY Lie algebra must be written entirely in terms of commutators. This can be done if we introduce two space-time independent Grassmann spinors θ_α, $\bar\theta_{\dot\alpha} = \theta_\alpha^*$, which satisfy the algebra

$$\{\theta_\alpha, \theta_\beta\} = \{\bar\theta_{\dot\alpha}, \bar\theta_{\dot\beta}\} = \{\theta_\alpha, \bar\theta_{\dot\beta}\} = 0 \tag{11.30}$$

and commute with all bosonic SUSY generators and anticommute with all fermionic generators. Clearly, θ_α is a left-handed spinor, while $\bar\theta_{\dot\alpha}$ is a right-handed spinor.

The advantage of the introduction of these Grassmann variables is that contracting them with the SUSY generators Q_α, $\bar Q_{\dot\alpha}$, we may replace the anticommutators of the $N = 1$ SUSY algebra (11.9) and (11.16) with the commutators:

$$[\theta^\alpha Q_\alpha, \theta^\beta Q_\beta] = [\bar\theta_{\dot\alpha} \bar Q^{\dot\alpha}, \bar\theta_{\dot\beta} \bar Q^{\dot\beta}] = 0\,,$$
$$[\theta^\alpha Q_\alpha, \bar\theta_{\dot\alpha} \bar Q^{\dot\beta}] = 2\theta^\alpha (\sigma^\mu)_{\alpha\dot\beta} \bar\theta^{\dot\beta} P_\mu\,. \tag{11.31}$$

11.3 Local Representations of SUSY

As mentioned above, it is convenient to use shorthand notations for the spinor multiplication. We can write, for example, $\theta^2 = \theta\theta = \theta^\alpha\theta_\alpha = -2\theta^1\theta^2 = 2\theta_2\theta_1$, and similarly, $\bar\theta^2 = \bar\theta\bar\theta = \bar\theta_{\dot\alpha}\bar\theta^{\dot\alpha} = 2\bar\theta_{\dot 1}\bar\theta_{\dot 2}$. For the sake of brevity, we shall employ this spinor summation convention.

In addition, there are some useful spinor identities:

$$(\theta\psi)(\theta\chi) = -\frac{1}{2}(\theta\theta)(\psi\chi), \quad (\theta\sigma^\mu\bar\theta)(\theta\sigma^\nu\bar\theta) = \frac{1}{2}g^{\mu\nu}(\theta\theta)(\bar\theta\bar\theta),$$

$$(\bar\theta\bar\psi)(\bar\theta\bar\chi) = -\frac{1}{2}(\bar\psi\bar\chi)(\bar\theta\bar\theta), \quad (\theta\psi)(\bar\theta\bar\chi) = \frac{1}{2}(\theta\sigma^\mu\bar\theta)(\psi\sigma_\mu\bar\chi),$$

$$(\sigma^\mu\bar\theta)_\alpha(\theta\sigma^\nu\bar\theta) = \frac{1}{2}g^{\mu\nu}\theta_\alpha(\bar\theta\bar\theta) - i(\sigma^{\mu\nu}\theta)_\alpha(\bar\theta\bar\theta), \qquad (11.32)$$

$$(\bar\theta\bar\psi)(\theta\chi) = \frac{1}{2}(\theta\sigma^\mu\bar\theta)(\chi\sigma_\mu\bar\psi) = (\theta\chi)(\bar\theta\bar\psi).$$

These are the basic relations for deriving well-known Fierz identities.

The advantage of such a reformulation of the SUSY algebra in terms of only commutators (11.31) is that now the Baker–Campbell–Hausdorff formula

$$e^A e^B = e^{A+B+\frac{1}{2}[A,B]+\dots}$$

can be used to define the action of SUSY Lie algebra group element:

$$G(x,\theta,\bar\theta,\omega) = \exp(i\theta Q + i\bar\theta\bar Q + ix^\mu P_\mu)\exp(-\frac{1}{2}\omega^{\mu\nu}M_{\mu\nu}). \qquad (11.33)$$

Clearly, this is a unitary operator, since $(\theta Q)^\dagger = \bar\theta\,\bar Q$.

The set of coordinates $(x_\mu, \theta, \bar\theta)$ parameterize the coset space, which is defined as $N = 1$ super Poincaré group mod Lorentz group. This space is actually a direct sum of four-dimensional Minkowski space and four-dimensional space spanned by the Grassmann coordinates $\theta, \bar\theta$. It is referred to as $N = 1$ *rigid superspace*, since the supersymmetry is global.

Then the Baker–Campbell–Hausdorff formula allows us to define the transformation of these coordinates under action of the left multiplication with a group element $G(a, \xi, \bar\xi')$:

$$G(x', \theta', \bar\theta') = G(a, \xi, \bar\xi)G(x, \theta, \bar\theta), \qquad (11.34)$$

which determines the translations in $N = 1$ rigid superspace:

$$\begin{aligned}
x^{\mu'} &= x^\mu + a^\mu + i\theta\sigma^\mu\bar\xi - i\xi\sigma^\mu\bar\theta, \\
\theta' &= \theta + \xi, \\
\bar\theta' &= \bar\theta + \bar\xi,
\end{aligned} \qquad (11.35)$$

where $a, \xi, \bar\xi$ are the parameters of the transformation.

We can now define a superfield as a function on the $N = 1$ rigid superspace $\Sigma(x, \theta, \bar\theta)$. According to (11.35), a SUSY transformation of a superfield is given by the left multiplication of a generator of supertranslations:

$$G(a,\xi,\bar\xi)\,\Sigma(x,\theta,\bar\theta) = \Sigma(x',\theta',\bar\theta')\,.$$

For the infinitesimal SUSY transformation, an expansion in Taylor series yields an explicit linear representation of the generators of $N=1$ SUSY as differential operators acting on the superspace:

$$\begin{aligned}
P_\mu &= -i\partial_\mu\,,\\
Q_\alpha &= -i\frac{\partial}{\partial\theta^\alpha} - (\sigma^\mu)_{\alpha\dot\beta}\bar\theta^{\dot\beta}\partial_\mu\,,\\
\bar Q_{\dot\alpha} &= i\frac{\partial}{\partial\bar\theta^{\dot\alpha}} + \theta^\beta(\sigma^\mu)_{\beta\dot\alpha}\partial_\mu\,.
\end{aligned} \qquad (11.36)$$

The generator Q and its Hermitian conjugated $\bar Q$, therefore, generate both translations in θ and translations in x as well. We can easily see that these operators enjoy the SUSY algebra. Indeed, we can check that, for example, $\{Q_\alpha,\bar Q_{\dot\alpha}\} = 2i(\sigma^\mu)_{\alpha\dot\alpha}\partial_\mu = -2\sigma^\mu P_\mu$ as in (11.9).

However, using the chain rule, we can see that the usual operator of translation $i\partial_\mu$ is not invariant[5] with respect to the transformation (11.35). The super-covariant fermionic derivatives are

$$D_\alpha = \frac{\partial}{\partial\theta^\alpha} + i(\sigma^\mu)_{\alpha\dot\beta}\bar\theta^{\dot\beta}\partial_\mu\,,\quad \bar D_{\dot\alpha} = \frac{\partial}{\partial\bar\theta^{\dot\alpha}} + i\theta^\beta(\sigma^\mu)_{\beta\dot\alpha}\partial_\mu\,, \qquad (11.37)$$

where $\bar D_{\dot\alpha} = D_\alpha^\dagger$. These derivatives satisfy the algebra

$$\begin{aligned}
\{D_\alpha, D_\beta\} &= \{\bar D_{\dot\alpha}, \bar D_{\dot\beta}\} = 0\,,\\
\{D_\alpha, \bar D_{\dot\beta}\} &= 2i(\sigma^\mu)_{\alpha\dot\beta}\partial_\mu = -2(\sigma^\mu)_{\alpha\dot\beta}P_\mu\,,
\end{aligned} \qquad (11.38)$$

and anticommute with all generators of supersymmetry $Q,\bar Q$.

For the sake of completeness, let us briefly recapitulate the basic relations of the analysis on the $N=1$ superspace. The derivatives in θ and $\bar\theta$ are defined as[6]

$$\begin{aligned}
\partial_\alpha\theta^\beta &\equiv \frac{\partial\theta^\beta}{\partial\theta^\alpha} = \delta_\alpha^\beta\,,\quad & \partial^\alpha\theta^\beta &\equiv -\varepsilon^{\alpha\gamma}\partial_\gamma\theta^\beta = -\varepsilon^{\alpha\beta}\,,\\
\partial^{\dot\alpha}\bar\theta_{\dot\beta} &\equiv \frac{\partial\bar\theta_{\dot\alpha}}{\partial\bar\theta_{\dot\beta}} = \delta_{\dot\beta}^{\dot\alpha}\,,\quad & \partial_{\dot\alpha}\bar\theta_{\dot\beta} &\equiv -\varepsilon_{\dot\alpha\dot\gamma}\partial^{\dot\gamma}\bar\theta_{\dot\beta} = -\varepsilon_{\dot\alpha\dot\beta}\,,
\end{aligned} \qquad (11.39)$$

respectively. Other useful relations, for example, are

$$\begin{aligned}
\partial_\alpha\theta^\beta\theta^\gamma &= \delta_\alpha^\beta\theta^\gamma - \delta_\alpha^\gamma\theta^\beta\,,\quad \partial_\alpha(\theta^\beta\theta_\beta) = 2\theta_\alpha\,,\quad \partial_{\dot\alpha}(\bar\theta_{\dot\beta}\bar\theta^{\dot\beta}) = -2\bar\theta_{\dot\alpha}\,,\\
\partial^2\theta^2 &\equiv \partial^\alpha\partial_\alpha(\theta^\beta\theta_\beta) = 4\,,\quad \partial^2\bar\theta^2 \equiv \partial_{\dot\alpha}\partial^{\dot\alpha}(\bar\theta_{\dot\beta}\bar\theta^{\dot\beta}) = -4\,.
\end{aligned} \qquad (11.40)$$

---

[5] The underlying reason is that the flat $N=1$ rigid superspace has non-zero torsion.

[6] To simplify our discussion, we do not make a difference between the definitions of the left and the right derivatives.

11.3 Local Representations of SUSY

These derivatives anticommute and, therefore, the operation of differentiation over the coordinates of the superspace is nilpotent:

$$\left(\frac{\partial}{\partial \theta^\alpha}\right)^2 = \left(\frac{\partial}{\partial \bar\theta_{\dot\alpha}}\right)^2 = 0. \qquad (11.41)$$

Thus, there is no inverse operation and an integration over the anticommuting variables cannot be defined by analogy with conventional integration. Furthermore, there is no notion of distance between two elements of the Grassmann algebra. In other words, we cannot define an appropriate topology and so it is not reasonable to speak about definite integrals, because it is impossible to construct the corresponding integral sum.

Nevertheless, one can introduce integration axiomatically as an algebraic operation. This is the so-called *Berezin integration*, which is defined from the requirement that the Grassmannian integration must be close to the usual one in some sense. For instance, since in quantum theory we usually deal with integrals with infinite limits, it is natural to require that the integral we are constructing is invariant with respect to shifts of the single integration variable θ:

$$\int d\theta\, f(\theta + \xi) = \int d\theta\, f(\theta).$$

Since a function of one variable is linear, $f(\theta) = a + b\theta$, we immediately obtain a formal integration rule for Grassmannian variables

$$\int d\theta = 0, \quad \int d\theta\, \theta = 1, \qquad (11.42)$$

and

$$\int d\theta\, \frac{\partial f(\theta)}{\partial \theta} = 0.$$

In other words, Berezin integration is identical to differentiation:

$$\frac{\partial f(\theta)}{\partial \theta} = \int d\theta\, f(\theta).$$

These definitions can be easily generalized for the case of $N = 1$ superspace. Clearly,

$$\int d\theta^\alpha\, \theta_\beta = \delta^\alpha{}_\beta \quad \text{and} \quad \int d\theta^1 d\theta^2 \theta^2 \theta^1 = 1.$$

Since $\theta\theta = -2\theta^1\theta^2$, we can define

$$d^2\theta = -\frac{1}{4} d\theta^\alpha d\theta^\beta\, \varepsilon_{\alpha\beta}, \quad d^2\bar\theta = -\frac{1}{4} d\bar\theta_{\dot\alpha} d\bar\theta_{\dot\beta}\, \varepsilon^{\dot\alpha\dot\beta},$$

so that

$$\int d^2\theta\, \theta\theta = \int d^2\bar\theta\, \bar\theta\bar\theta = 1.$$

Evidently, we also have $d^4\theta = d^2\theta d^2\bar\theta$ and $\int d^4\theta\, \theta\theta\bar\theta\bar\theta = 1$.

11.3.2 $N = 1$ Superfields

An ordinary field is defined as a function of the space-time coordinates x_μ. Similarly, an $N = 1$ superfield is a function of the superspace coordinates $x_\mu, \theta, \bar\theta$. Hence, it is a function of anticommuting Grassmann coordinates and, therefore, its expansion into a power series in $\theta, \bar\theta$ is restricted to be finite, because $\theta_1^2 = \theta_2^2 = \bar\theta_1^2 = \bar\theta_2^2 = 0$. Thus, a most general scalar $N = 1$ superfield can always be written as an expansion in components

$$\Sigma(x, \theta, \bar\theta) = f(x) + \theta\phi(x) + \bar\theta\bar\chi + \theta\theta m(x) + \bar\theta\bar\theta n(x) \qquad (11.43)$$
$$+ \theta\sigma^\mu\bar\theta A_\mu(x) + \theta\theta\bar\theta\bar\lambda(x) + \bar\theta\bar\theta\theta\psi(x) + (\bar\theta\bar\theta)(\theta\theta)D(x),$$

where some redundant terms, for example, $\theta\bar\sigma^\mu\theta A_\mu$, are removed by making use of the Fierz identities (11.32).

Thus, the superfield Σ contains the complex scalar fields $f(x)$, $m(x)$, $n(x)$ and $D(x)$, two $(\frac{1}{2}, 0)$-spinors $\phi(x)$ and $\psi(x)$, two $(0, \frac{1}{2})$-spinors $\bar\chi(x)$ and $\bar\lambda(x)$ and the complex vector field $A_\mu(x)$, altogether 16 degrees of freedom, 8 bosonic and 8 fermionic. The leading component of the expansion is the scalar filed $f(x)$. However, to describe a physical system we do not need all these components, because some constraint on the superfield can be imposed. Indeed, a general superfield lies in a reducible representation of the SUSY algebra. An irreducible off-shell representation can be selected, if we impose a certain set of constraints that eliminates some of the components.

First consider the so-called $N = 1$ *chiral superfield* Φ, which is defined by the covariant superspace constraint

$$\bar D_{\dot\alpha}\Phi(x, \theta, \bar\theta) = 0. \qquad (11.44)$$

This equation can be solved in shifted bosonic coordinates $y^\mu = x^\mu + i\theta\sigma^\mu\bar\theta$. Indeed, then $\bar D_{\dot\alpha} y^\mu = D_\alpha \bar\theta = 0$ and the constraint (11.44) takes the form

$$\frac{\partial}{\partial\bar\theta^{\dot\alpha}}\Phi(y, \theta, \bar\theta) = 0.$$

A general solution of (11.44), then, is an arbitrary function $\Phi(y, \theta)$ of the variables y^μ and θ_α, but not of $\bar\theta_{\dot\alpha}$. Thus, the expansion in θ yields the $N = 1$ chiral scalar superfield

$$\Phi(y, \theta) = \phi(y) + \sqrt{2}\theta\psi(y) + \theta\theta F(y), \qquad (11.45)$$

where the factor $\sqrt{2}$ is a convention. Its components, therefore, are two complex scalar fields $\phi(y)$ and $F(y)$ and a complex left-handed Weyl spinor, altogether 4+4 real off-shell components.

Recovering now the original coordinates of the superspace $(x, \theta, \bar\theta)$, we see that the full expansion of the chiral superfield is given by

11.3 Local Representations of SUSY

$$\Phi(y,\theta,\bar\theta) = \phi(x) + \sqrt{2}\theta\psi(x) + \theta\theta F(x) + i\theta\sigma^\mu\bar\theta\partial_\mu\phi(x)$$
$$- \frac{i}{\sqrt{2}}\theta\theta\partial_\mu\psi(x)\sigma^\mu\bar\theta - \frac{1}{4}(\bar\theta\bar\theta)(\theta\theta)\partial_\mu\partial^\mu\phi(x)\,, \qquad (11.46)$$

where the leading component of the expansion is the usual scalar field $\phi(x)$.

Let us consider the supersymmetry variations (11.36), which are generated by the operator

$$\delta_\xi = i\xi^\alpha Q_\alpha + i\bar\xi_{\dot\alpha}\bar Q^{\dot\alpha}\,.$$

Then the chiral superfield Φ transforms as $\Phi \to \Phi + \delta_\xi\Phi$. Using the definition of the SUSY generators (11.36) and the component expansion (11.46), we obtain the supersymmetry variations of the components

$$\begin{aligned}
\delta\phi &= \sqrt{2}\xi^\alpha\psi_\alpha\,,\\
\delta\psi_\alpha &= \sqrt{2}\xi_\alpha F + i\sqrt{2}(\sigma^\mu)_{\alpha\dot\alpha}\bar\xi^{\dot\alpha}\partial_\mu\phi\,,\\
\delta F &= -i\sqrt{2}\bar\xi^{\dot\alpha}(\bar\sigma^\mu)_{\dot\alpha\alpha}\partial_\mu\psi^\alpha\,.
\end{aligned} \qquad (11.47)$$

The common factor of $\sqrt{2}$ can be eliminated by rescaling the constant spinors, the parameters of the transformation ξ and $\bar\xi$.

Thus, the transformation of supersymmetry maps a scalar field into a spinor field and vice versa.

Note that $F(x)$ is an auxiliary field and the corresponding term in the expansion of the chiral superfield transforms into a space-time derivative under the transformations of supersymmetry. This field is required to close off-shell SUSY algebra, i.e., only the whole set of the fields ϕ,ψ,F forms a representation of the superalgebra.

The *anti-chiral* scalar superfield is defined by analogy with (11.44) by the constraint

$$D_\alpha\Phi^\dagger(x,\theta,\bar\theta) = 0\,, \qquad (11.48)$$

which is the Hermitian conjugate of Φ. Generally speaking, chiral superfields and anti-chiral superfields are annihilated by $\bar D_{\dot\alpha}$ and D_α, respectively. We can consider, for example, a spinor chiral superfield Φ_α which is defined similarly to (11.44).

Clearly, any product of (anti)-chiral superfields is also an (anti)-chiral superfield and any arbitrary function of an (anti)-chiral superfield is an (anti)-chiral superfield. However, the Hermitian operators given by the product $\Phi^\dagger\Phi$ and the sum $\Phi + \Phi^\dagger$ are not (anti)-chiral superfields.

Let us consider now the $N=1$ *vector superfield*, which is defined by the covariant reality constraint:

$$V(x,\theta,\bar\theta) = V^\dagger(x,\theta,\bar\theta)\,. \qquad (11.49)$$

Taking into account the general form of the expansion (11.43), we can easily see that this constraint restricts the vector field A_μ, as well as the scalar fields

$f(x)$ and $D(x)$, to being real. Other consequences of (11.49) are $\bar{\chi} = \phi^*, m = n^*, \bar{\lambda} = \psi^*$.

Thus, $N = 1$ vector superfield has 16 real components, which corresponds to four real scalars, two complex Weyl spinors and one real vector field. There are too many of them to describe a single supermultiplet. However, there is a possibility of eliminating some degrees of freedom by making use of the supersymmetric generalization of a gauge transformation.

To see how this works, let us note that the vector superfield may be viewed as the reducible representation of supersymmetry, since it contains the chiral and anti-chiral superfields. Indeed, the particular case of the vector field is just a combination $i(\Phi - \Phi^\dagger)$ of the given chiral (Φ) and anti-chiral ($\Phi^\dagger$) superfields. Moreover, any vector field can be decomposed as

$$V(x, \theta, \bar{\theta}) = V_{WZ} + i(\Phi^\dagger - \Phi).$$

Indeed, we can define an infinitesimal $U(1)$ gauge transformation of a vector superfield V to be

$$V(x, \theta, \bar{\theta}) \to V(x, \theta, \bar{\theta}) + i(\Phi^\dagger - \Phi), \qquad (11.50)$$

since it gives an expected form of the infinitesimal gauge transformation of the vector field components in (11.43): $A_\mu \to A_\mu + i\partial_\mu \Lambda$, where the gauge function is defined by the scalar components of the expansion (11.46): $\Lambda = \phi(x) - \phi^*(x) = 2i \operatorname{Im} \phi$. Thus, we can gauge out some of the components.

In the *Wess–Zumino gauge*, the $N = 1$ vector superfield reduces to the simple form[7]

$$V_{WZ}(x, \theta, \bar{\theta}) = \theta \sigma^\mu \bar{\theta} A_\mu(x) + i\theta\theta\bar{\theta}\bar{\lambda}(x) - i\bar{\theta}\bar{\theta}\theta\lambda(x) + \frac{1}{2}(\bar{\theta}\bar{\theta})(\theta\theta)D(x). \quad (11.51)$$

Thus, the Wess–Zumino superfield V_{WZ} contains an Abelian massless gauge field A_μ, that is the leading component of the expansion, its superpartner gaugino λ and a real auxiliary field $D(x)$.

It is clear that the vector superfield V may be considered as a supersymmetrical generalization of the conventional vector-potential. Note that all powers of the $N = 1$ Wess–Zumino superfield other than squared, are vanishing because each term in (11.51) contains at least one θ:

$$\begin{aligned} V_{WZ}^2 &= \theta\sigma^\mu\bar{\theta}\theta\sigma^\nu\bar{\theta} A_\mu A_\nu = \frac{1}{2}(\bar{\theta}\bar{\theta})(\theta\theta)A_\mu(x)A^\mu(x), \\ V_{WZ}^n &= 0, \qquad n \geq 3. \end{aligned} \qquad (11.52)$$

Let us consider now the supersymmetry transformation of the vector superfield $V \to V + \delta_\varepsilon V$. We read the transformations of the component fields:

---

[7] Note that fixing the Wess–Zumino gauge breaks supersymmetry, but still allows the conventional gauge transformations.

11.3 Local Representations of SUSY

$$\delta A_\mu = i\bar{\lambda}_{\dot\alpha}(\bar\sigma_\mu)^{\dot\alpha\alpha}\xi_\alpha,$$
$$\delta\lambda_\alpha = \frac{1}{2}(\sigma^\mu\bar\sigma^\nu)_\alpha{}^\beta \xi_\beta F_{\mu\nu} + i\xi_\alpha D, \qquad (11.53)$$
$$\delta D = \bar\xi_{\dot\alpha}(\bar\sigma^\mu)^{\dot\alpha\alpha}\partial_\mu\lambda_\alpha + \partial_\mu\bar\lambda_{\dot\alpha}(\bar\sigma^\mu)^{\dot\alpha\alpha}\xi_\alpha.$$

where $F_{\mu\nu} = \partial_\mu A_\nu - \partial_\nu A_\mu$ is the familiar Abelian field strength.

These transformations supplement the SUSY variations of the component of the scalar superfield given by (11.47). Note that the variation of the auxiliary D-field, which has the highest mass dimension among the components of the vector superfield, turns out to be a total derivative by analogy with the variation of the auxiliary field F in (11.47). Note that the latter field is also a component of highest dimension two in the scalar multiplet. This is quite a general property of a supersymmetric theory, for any given multiplet the component of highest dimension is an auxiliary field that transforms into a space derivative.

However, the transformation of supersymmetry (11.53) does not respect the Wess–Zumino gauge fixing decomposition. Therefore, the vector superfield V, a generalization of the conventional Yang–Mills vector potential, is not a proper object in superfield formalism.

In accordance with the ordinary gauge field theory, we can introduce a counterpart of the field strength tensor, which allows us to construct kinetic terms for the vector field. The *supersymmetric field strength* is defined as a superfield that is invariant with respect to the gauge transformations (11.50) and contains only the components of the Wess–Zumino superfield $A_\mu(x), \lambda(x), D(x)$.

Let us consider the left-handed and the right-handed spinor superfields

$$W_\alpha = -\frac{1}{4}\bar D_{\dot\beta}\,\bar D^{\dot\beta} D_\alpha, \qquad \bar W_{\dot\alpha} = -\frac{1}{4}D^\alpha D_\alpha\,\bar D_{\dot\alpha} V. \qquad (11.54)$$

Clearly, W_α is a chiral superfield

$$\bar D_{\dot\alpha} W_\alpha = -\frac{1}{4}\bar D_{\dot\alpha}(\bar D_{\dot\beta}\,\bar D^{\dot\beta})D_\alpha V \equiv 0, \qquad (11.55)$$

since the product of three operators $\bar D$ vanishes, whereas $\bar W_{\dot\alpha}$ is an anti-chiral superfield.

Note that the definition (11.54) means that the fields W_α and $\bar W_{\dot\alpha}$ are related by the additional covariant constraint

$$\bar D_{\dot\alpha}\,\bar W^{\dot\alpha} = D^\alpha W_\alpha, \qquad (11.56)$$

which in particular means that

$$\mathrm{Im}(D^\alpha W_\alpha) = 0. \qquad (11.57)$$

Using the algebra of the super-covariant derivatives (11.38) and the constraint on the chiral superfields Φ, we can easily see that the spinor superfield

W_α satisfies the condition of invariance with respect to the Abelian gauge transformations (11.50). Thus, changing the variables to the shifted coordinates y_μ as above, we can write the component representation of this chiral field in the Wess–Zumino gauge as

$$W_\alpha = -i\lambda_\alpha(y) + \theta_\alpha D(y) + i(\sigma^{\mu\nu}\theta)_\alpha F_{\mu\nu}(y) + \theta\theta(\sigma^\mu \partial_\mu \bar{\lambda}(y))_\alpha. \qquad (11.58)$$

Thus, the irreducible off-shell spinor multiplet, or field strength multiplet contains eight real components. We can see that the superspace constraint (11.57) for a component $F_{\mu\nu}$ is just the Bianci identity for the field strength tensor.

11.3.3 Non-Abelian Multiplets

First, we shall define the supersymmetric generalization of the gauge transformations. Indeed, in the non-Abelian case, the vector superfield necessarily belongs to the adjoint representation of the gauge group: $V = V^a T^a$, $[T^a, T^b] = if_{abc}T^c$, where $(T^a)_{bc} = -if_{abc}$ are Hermitian generators of the Lie algebra and f^{abc} are real structure constants. It is now convenient to normalize these generators by $\text{tr}(T^a T^b) = \delta^{ab}$.

Since the fields take values in the Lie algebra, the basic object is an exponential e^{-V} rather than V. It transforms as

$$e^{-V} \to e^{-i\Lambda^\dagger} e^{-V} e^{i\Lambda}, \qquad (11.59)$$

where Λ is a chiral superfield.

Note that in the Wess–Zumino gauge the expansion of the exponent yields

$$e^V = 1 + V + \frac{V^2}{2}, \qquad (11.60)$$

since all terms of powers higher than two are vanishing. Evidently, to first order in Λ the (11.59) reproduces the infinitesimal Abelian gauge transformation (11.50) with identification $\Phi = \Lambda$.

A natural requirement for the non-Abelian field strength is that it must be covariant with respect to this transformation. To define such an object, we consider the transformation properties of the expression

$$e^V D_\alpha e^{-V} \to e^{-i\Lambda}(e^V D_\alpha e^{-V})e^{i\Lambda} + e^{-i\Lambda}(D_\alpha e^{i\Lambda}),$$

where we used the constraint $D_\alpha \Lambda^\dagger = 0$ (see, for example, [480]). The first term in the r.h.s. transforms covariantly, whereas the second term can be annihilated by action of the operator $\bar{D}_{\dot\alpha}$, which commutes with Λ. Thus, the non-Abelian generalization of the supersymmetric field strength (11.54) is

$$W_\alpha = -\frac{1}{4}\bar{D}_{\dot\beta}\,\bar{D}^{\dot\beta}e^{-V}D_\alpha e^V; \qquad \bar{W}_{\dot\alpha} = \frac{1}{4}D^\alpha D_\alpha\,e^V \bar{D}_{\dot\alpha}e^{-V}, \qquad (11.61)$$

which transforms as

$$W_\alpha \to e^{-i\Lambda}W_\alpha e^{i\Lambda}, \qquad \bar{W}_{\dot\alpha} \to e^{-i\Lambda^\dagger}\bar{W}_{\dot\alpha}e^{i\Lambda^\dagger}. \qquad (11.62)$$

To obtain the component expansion of the non-Abelian superfield strength, we substitute (11.60) into the definition (11.61):

$$W_\alpha = -\frac{1}{4}\bar{D}_{\dot\beta}\,\bar{D}^{\dot\beta}D_\alpha V + \frac{1}{8}\bar{D}_{\dot\beta}\,\bar{D}^{\dot\beta}[V, D_\alpha V].$$

Clearly, the first term here is identical to the Abelian case, while the second term yields

$$[V, D_\alpha V] = \theta\bar\theta(\sigma^{\nu\mu}\theta)_\alpha[A_\mu, A_\nu] + i\theta\theta\bar\theta\bar\theta(\sigma^\mu)_{\alpha\dot\beta}[A_\mu, \bar\lambda^{\dot\beta}].$$

Then the non-Abelian superfield strength takes a form similar to (11.58):

$$W_\alpha = -i\lambda_\alpha(y) + \theta_\alpha D(y) + i(\sigma^{\mu\nu}\theta)_\alpha F_{\mu\nu}(y) + \theta\theta(\sigma^\mu\nabla_\mu\bar\lambda(y))_\alpha, \qquad (11.63)$$

where

$$F_{\mu\nu} = \partial_\mu A_\nu - \partial_\nu A_\mu - \frac{i}{2}[A_\mu, A_\nu]$$

can be recognized as the familiar non-Abelian field strength tensor, whereas

$$\nabla_\mu\bar\lambda_{\dot\alpha} = \partial_\mu\bar\lambda_{\dot\alpha} - \frac{i}{2}[A_\mu, \bar\lambda_{\dot\alpha}]$$

is the Yang–Mills covariant derivative[8]. Since the fields take values in the Lie algebra with the structure constants f_{abc}, we can write $F^a_{\mu\nu} = \partial_\mu A^a_\nu - \partial_\nu A^a_\mu + \frac{1}{2}f_{abc}A^b_\mu A^c_\nu$. To restore the standard form of the non-Abelian field strength (5.12) and covariant derivative (5.10) we can introduce the gauge coupling constant e and rescale the vector superfield V as $V \to 2eV$. Then all the components A_μ, λ^α and D must be rescaled in the same way.

11.4 $N = 1$ SUSY Lagrangians

So far we have discussed the superfields that are defined on the superspace; the component content of these corresponds to some representations of the

---

[8] Recall that all the fields of the vector multiplet, including fermions, belong to the same adjoint representation of the gauge group, that is, $\lambda = \lambda^a T^a$. To avoid possible misunderstanding let us also note that in this chapter we denote the usual gauge covariant derivative as ∇_μ, whereas the symbol D_μ is reserved for the super-derivative. The standard notation for the auxiliary field $D(x)$ can only complete the mess.

11 Supersymmetric Yang-Mills Theories

superalgebra. Since we would like to construct a supersymmetric quantum field theory, the next step, of course, is to find the corresponding $N=1$ SUSY Lagrangians for the scalar and vector superfields, respectively.

Some clue to the explicit structure of the proper Lagrange densities may be given by the observation that the F-term in the component expansion of an arbitrary chiral scalar superfield, and the D-term in the component expansion of an arbitrary vector superfield, are invariant with respect to the transformations of supersymmetry (up to a total derivative). Therefore, these terms can be considered as prototypes of the Lagrangians of the corresponding superfields.

The related components can be projected out by the integration over the Grassmann variables as:

$$\int d^2\theta \, \Phi(x,\theta,\bar\theta) = F(x), \qquad \int d^2\theta d^2\bar\theta \, V(x,\theta,\bar\theta) = D(x).$$

We also have to take into account the conjugated component $F^\dagger$. Thus, we are interested in the combination $FF^\dagger$, which appears as the $\theta\theta\bar\theta\bar\theta$-component in the expansion of an appropriate superfield.

Let us consider the chiral scalar supermultiplets. The product of chiral and anti-chiral scalar superfields $\Phi\Phi^\dagger$ is a Hermitian real superfield that contains such a component. Indeed, substitution of the expansion (11.46) and subsequent integration over the Grassmann variables yields

$$\int d^4x d^2\theta d^2\bar\theta \, \Phi^\dagger\Phi = \int d^4x \left(\partial_\mu \phi^\dagger \partial^\mu \phi + i\bar\psi \bar\sigma_\mu \partial^\mu \psi + F^\dagger F \right), \quad (11.64)$$

where the total derivatives are dropped out. This is the free Lagrangian for a massless scalar field $\phi(x)$, a massless fermion and an auxiliary non-propagating field $F(x)$. The latter components can be eliminated from the Lagrangian. Indeed, such an action contains no derivatives of F and the equation of motion sets it identically to zero. As mentioned above, auxiliary fields are needed in supersymmetric theories to provide a manifestly covariant formulation.

Thus, we have obtained the kinetic term of a single chiral superfield. More generally, we may consider a model with several different superfields Φ_i. Evidently, the corresponding generalization of (11.64) is

$$\int d^4x d^2\theta d^2\bar\theta \, \Phi_i^\dagger \Phi_i = \int d^4x \left(\partial_\mu \phi_i^\dagger \partial^\mu \phi_i + i\bar\psi_i \bar\sigma_\mu \partial^\mu \psi_i + F_i^\dagger F_i \right). \quad (11.65)$$

Apart from the kinetic term (11.65), a general SUSY invariant Lagrangian of the chiral superfield also includes supersymmetric terms of non-derivative self-interaction. Most generally they can be constructed in terms of the *superpotential* $\mathcal{W}(\Phi_i)$, which is defined as an arbitrary *holomorphic* function of chiral superfields Φ_i.

11.4 $N = 1$ SUSY Lagrangians

Clearly, the superpotential is also a chiral superfield. It can be written, for example, as[9]

$$\mathcal{W}(\Phi_i) = b_i \Phi_i + \frac{1}{2} m_{ij} \Phi_i \Phi_j + \frac{1}{3} g_{ijk} \Phi_i \Phi_j \Phi_k , \qquad (11.66)$$

where the coefficients b, m, g are invariant symmetric tensors of the given, say, fundamental representation, that is, the superpotential is gauge invariant quantity. Higher powers of Φ are not taken into account, since they would have a mass dimension greater than four, which would lead us to a non-renormalizable theory.

Because

$$\int d^2\theta \, \Phi_i \Phi_j = \phi_i F_j + \phi_j F_i - \psi_i \psi_j ,$$

$$\int d^2\theta \, \Phi_i \Phi_j \Phi_k = \phi_i \phi_j F_k + \phi_i F_j \phi_k + F_i \phi_j \phi_k \qquad (11.67)$$

$$\qquad\qquad - \psi_i \psi_j \phi_k - \psi_i \psi_k \phi_j - \psi_j \psi_k \phi_i ,$$

the potential part of the Lagrangian is of the form [29]

$$\int d^2\theta \, \mathcal{W}(\Phi_i) + \int d^2\bar\theta \, \bar{\mathcal{W}}(\Phi_i^\dagger) = b_i F_i + m_{ij} \phi_i F_j - \frac{1}{2} m_{ij} \psi_i \psi_j$$

$$+ g_{ijk} \phi_i \phi_j F_k - g_{ijk} \psi_i \psi_j \phi_k \quad + \quad \text{h.c.} \qquad (11.68)$$

Therefore, the equation of motion for the auxiliary field F, which has no kinetic term, is simple

$$F_i^\dagger = \frac{\partial \mathcal{W}}{\partial \phi_i} = b_i + m_{ij} \phi_j + g_{ijk} \phi_j \phi_k . \qquad (11.69)$$

We can substitute this algebraic equation back into the action in order to eliminate F. Then we obtain the scalar potential, which is determined in terms of the superpotential $\mathcal{W}$ as

$$U_{boson} = \sum_i \left| \frac{\partial \mathcal{W}}{\partial \phi_i} \right|^2 = \text{Tr} \left| \frac{\partial \mathcal{W}}{\partial \phi} \right|^2 . \qquad (11.70)$$

If there is a single superfield Φ, we have the simplest case of a cubic superpotential $\mathcal{W}(\Phi) = \frac{m}{2} \Phi^2 + \frac{g}{3} \Phi^3$. Then the scalar potential becomes $U_{boson} = m\Phi + g\Phi^3$ and the action of the Wess–Zumino model can be written as

$$\int d^4x \Big\{ \partial_\mu \phi^\dagger \partial^\mu \phi + i\bar\psi \bar\sigma_\mu \partial^\mu \psi - m^2 \phi \phi^\dagger - \frac{m}{2} (\bar\psi\,\bar\psi + \psi\psi)$$

$$- mg[\phi^\dagger \phi^2 + (\phi^\dagger)^2 \phi] + g(\phi^\dagger \bar\psi\,\bar\psi + \phi\psi\psi) - g^2(\phi\phi^\dagger)^2 \Big\} . \qquad (11.71)$$

---

[9] This corresponds to the simple *Wess–Zumino model*.

Thus, we can see that this model describes the Yukawa interaction between the scalar and spinor fields.

In general, an $N=1$ SUSY Lagrangian of chiral superfield can be written as

$$L = \int d^2\theta d^2\bar{\theta}\, K(\Phi_i, \Phi_j^\dagger) + \int d^2\theta\, \mathcal{W}(\Phi_i) + \int d^2\bar{\theta}\, \bar{\mathcal{W}}(\Phi_i^\dagger), \qquad (11.72)$$

where a general real function $K(\Phi_i, \Phi_j^\dagger)$ of an arbitrary number of chiral superfields is called the *Kähler potential*. Clearly, this is a vector superfield.

The structure of the kinetic term in (11.72) is typical for the supersymmetrical version of the four-dimensional $N=1$ *non-linear sigma-model* (see, e.g., [16]). Indeed, a bosonic sigma-model is defined by the action

$$S = \int d^4x\, g_{ij}(\phi)\partial_\mu \phi_i \partial^\mu \phi_j,$$

where ϕ_i are real scalar fields and the functional g_{ij} is defined as a metric of a Riemannian manifold that is parameterized by these fields.

In the case of the model (11.72), the component expansion gives the kinetic term of the leading bosonic component precisely of this form with the metric on the field space given by the Kähler potential in a similar way:

$$g_{ij} = \frac{\delta^2 K}{\delta\phi_i\, \delta\phi_j^*}. \qquad (11.73)$$

Thus, the chiral superfields can be thought of as coordinates on a complex Kähler manifold (see (3.39) in Chap. 3 and the related discussion). Recall that the condition of kählerity is very restrictive, in particular the Kähler structure of the metric (11.73) means that it is invariant with respect to transformations

$$K(\Phi_i, \Phi_j^\dagger) \to K(\Phi_i, \Phi_j^\dagger) + \Lambda(\Phi_i) + \Lambda^\dagger(\Phi_i^\dagger),$$

where Λ is an arbitrary holomorphic function of Φ_i.

The advantage of this property of the supersymmetric action is that its monstrous structure with a frightening amount of component fields can now be described in terms of algebraic geometry. Actually, this conception is already familiar to us, because exactly the same formalism was already used in Chap. 6, where the dynamics of the multimonopoles was described in terms of the metric on the moduli space $\mathcal{M}_n$ (see the discussion on page 89). Recall that the condition of kählerity defines the group of holonomy of the manifold with real dimension $2n$ to be $U(n)$. This restriction, together with the condition of holomorphicity of supersymmetric structures, like for example the coupling constant and superpotential, allows us to obtain a variety of very strong and non-trivial results.

11.4 $N = 1$ SUSY Lagrangians

To complete our construction of the $N = 1$ SUSY Lagrangian, we have to write the Lagrangian of the vector multiplet. Making use of the non-Abelian superfield strength of (11.63), we can choose a gauge invariant supersymmetric Lagrangian

$$L = \int d^2\theta \, \text{Tr} \, W^\alpha W_\alpha \qquad (11.74)$$

$$= \text{Tr}\left(-2i\lambda^\alpha (\sigma^\mu)_{\alpha\dot\alpha} \nabla_\mu \bar\lambda^{\dot\alpha} + D^\alpha D_\alpha - \frac{1}{2} F^{\mu\nu} F_{\mu\nu} + \frac{i}{2} F^{\mu\nu} \widetilde{F}_{\mu\nu}\right).$$

Let us now rescale the vector superfield, as mentioned at the end of the previous section to recover the usual form of the gauge covariant derivative and the gauge field strength: $V \to 2eV$. Then the factor $4e^2$ appears in the Lagrangian $W^\alpha W_\alpha \to 4e^2 W^\alpha W_\alpha$. It can be eliminated by the redefinition $L \to L/4e^2$. From now on we adopt the rescaled notations.

Evidently, we can get rid of the $F\widetilde{F}$-term if we consider another, slightly more symmetric definition[10]

$$L = \frac{1}{4e^2} \text{Tr} \left(\int d^2\theta \, W^\alpha W_\alpha + \int d^2\bar\theta \, \bar{W}_{\dot\alpha} \bar{W}^{\dot\alpha} \right), \qquad (11.75)$$

which in the Abelian case corresponds to the well-known Maxwell electrodynamics:

$$L^{N=1}_{Maxwell} = -\frac{1}{4e^2} F^{\mu\nu} F_{\mu\nu} + \frac{i}{2e^2} \bar\lambda \sigma^\mu \overleftrightarrow{\partial}_\mu \lambda + \frac{1}{2e^2} D^2.$$

However, in the non-Abelian theory the subject of interest is rather the Yang–Mills Lagrangian of the form (11.74), because it can be related with the θ-term (5.104).

Indeed, so far we have paid little attention on the gauge coupling. In supersymmetric theory we shall use the same complex coupling τ (3.17) that was introduced in Chap. 2 and was used to represent the whole Lagrangian of the Georgi–Glashow model in the form (5.111):

$$\tau = \frac{\theta}{2\pi} + \frac{4\pi i}{e^2}. \qquad (11.76)$$

Thus, by complete analogy with (5.111), the $N = 1$ SUSY Yang–Mills Lagrangian can be written as

$$L^{N=1}_{YM} = \frac{1}{32\pi} \text{Im} \left(\tau \, \text{Tr} \int d^2\theta \, W^\alpha W_\alpha \right) \qquad (11.77)$$

---

[10] Another possible Hermitian combination is of the form

$$\frac{i}{4e^2} \text{Tr} \left(\int d^2\theta W^\alpha W_\alpha - \int d^2\bar\theta \bar{W}_{\dot\alpha} \bar{W}^{\dot\alpha} \right).$$

It contains the $F\widetilde{F}$-term, but breaks the parity explicitly and has to be excluded.

$$= \text{Tr}\left(-\frac{1}{4}F^{\mu\nu}F_{\mu\nu} + \frac{\theta e^2}{32\pi^2}F^{\mu\nu}\tilde{F}_{\mu\nu} + \frac{1}{2}D^2 - i\lambda\sigma^\mu \nabla_\mu \bar{\lambda}\right).$$

This is one of the miracles of the supersymmetry: a single term $W^\alpha W_\alpha$ in the supersymmetric Lagrangian generates both the gauge kinetic term, which is normalized on the gauge coupling constant, and the topological term, which multiplies the θ-angle.

The last step in the procedure of the construction of $N=1$ SUSY is to define the term of interaction between the chiral scalar superfield $\Phi = \Phi^a(t^a)_{ij}$ and the vector field V. Clearly, this has to be done in a gauge-invariant way. However, the kinetic term of the Wess–Zumino Lagrangian of the free scalar superfield (11.64) is not invariant with respect to the local gauge transformations

$$\Phi \to e^{-i\Lambda}\Phi, \qquad \Phi^\dagger \to \Phi^\dagger e^{i\Lambda^\dagger}, \tag{11.78}$$

where $\Lambda = \Lambda^a(T^a)_{ij}$. It is easy to see that this transforms as

$$\Phi^\dagger \Phi \to \Phi^\dagger \Phi \, e^{i(\Lambda^\dagger - \Lambda)}.$$

However, taking into account the transformation properties of e^{-2V} (11.59), we can check that the combination $\Phi^\dagger e^{-V}\Phi$ is gauge invariant. Thus, we can write the kinetic term of the Lagrangian of scalar superfield as

$$\int d^2\theta d^2\bar\theta\, \Phi^\dagger e^{-V}\Phi = \int d^2\theta d^2\bar\theta(\Phi^\dagger \Phi - \Phi^\dagger V \Phi + \frac{1}{2}\Phi^\dagger V^2 \Phi)$$
$$= (\nabla_\mu \phi)^\dagger(\nabla^\mu \phi) - i\bar\psi \bar\sigma^\mu \nabla_\mu \psi + F^\dagger F \tag{11.79}$$
$$+ \frac{1}{2}\phi^\dagger D^a t^a \phi - \frac{i}{\sqrt{2}}\phi^\dagger t^a \lambda^a \psi + \frac{i}{\sqrt{2}}\bar\psi t^a \phi \bar\lambda^a,$$

where the covariant derivative is $\nabla_\mu \phi = \partial_\mu \phi + \frac{i}{2}A^a_\mu t^a \phi$, the total space derivative terms are dropped out and we make use of the shorthand notations for the coupling terms, e.g., $\phi^\dagger t^a \lambda^a \psi = \phi^{b\dagger}(t^a)_{bc}\lambda^a \psi^c$, etc.

Note that we have not yet rescaled the vector superfield as $V \to 2eV$. It is necessary to do such a scaling if we would like to treat the Lagrangians of the scalar chiral superfield Φ and the vector superfield V on the same footing. Then the Lagrangian (11.79) takes the form

$$\int d^2\theta d^2\bar\theta\, \Phi^\dagger e^{-V}\Phi \to \int d^2\theta d^2\bar\theta\, \Phi^\dagger e^{-2eV}\Phi = (\nabla_\mu \phi)^\dagger(\nabla^\mu \phi) - i\bar\psi\bar\sigma^\mu \nabla_\mu \psi$$
$$+ F^\dagger F + e\phi^\dagger D^a t^a \phi - ie\sqrt{2}\phi^\dagger t^a \lambda^a \psi + ie\sqrt{2}\bar\psi t^a \phi \bar\lambda^a, \tag{11.80}$$

where the covariant derivative is now redefined as $\nabla^\mu \phi = \partial \phi + ieA^a_\mu t^a \phi$.

Collecting together all the terms (11.68) and (11.74) and (11.79), we finally arrive at the whole $N = 1$ SUSY Lagrangian that describes the super-Yang–Mills field minimally coupled to a chiral field Φ:

$$L^{N=1} = \frac{1}{32\pi}\text{Im}\left(\tau \,\text{tr}\int d^2\theta \, W^\alpha W_\alpha + \int d^2\theta d^2\bar{\theta}\, \Phi^\dagger e^{-V}\Phi \right.$$
$$\left. + \int d^2\theta \, \mathcal{W}(\Phi) + \int d^2\bar{\theta}\, \bar{\mathcal{W}}(\Phi^\dagger)\right).$$

It should be noted that the $N = 1$ superfields V and Φ in this expression belong to different supermultiplets and transform separately. Moreover, they are taken in different representations of the gauge group whose generators are the matrices T^a and t^a, respectively. We shall refer to the scalar supermultiplet as the matter multiplet of the $N = 1$ gauge theory. Usually such a field transforms in the fundamental representation of the gauge group.

In component notation, the rescaled supersymmetrical and gauge invariant $N = 1$ Lagrangian can be written as

$$L^{N=1} = \text{tr}\left(-\frac{1}{4}F^{\mu\nu}F_{\mu\nu} + \frac{\theta e^2}{32\pi^2}F^{\mu\nu}\tilde{F}_{\mu\nu} + \frac{1}{2}D^2 - i\lambda\sigma^\mu\nabla_\mu\bar{\lambda}\right)$$
$$+ (\nabla_\mu\phi)^\dagger(\nabla^\mu\phi) - i\bar{\psi}\bar{\sigma}^\mu\nabla_\mu\psi + F_k^\dagger F_k + e\phi^\dagger D^a t^a \phi$$
$$- ie\sqrt{2}\phi^\dagger t^a \lambda^a \psi + ie\sqrt{2}\bar{\psi}t^a\phi\bar{\lambda}^a + \frac{\partial \mathcal{W}}{\partial \phi_k}F_k + \frac{\partial \bar{\mathcal{W}}}{\partial \phi_k^\dagger}F_k^\dagger$$
$$- \frac{1}{2}\frac{\partial^2 \mathcal{W}}{\partial \phi_i \partial \phi_j}\psi_i\psi_j - \frac{1}{2}\frac{\partial^2 \bar{\mathcal{W}}}{\partial \phi_i^\dagger \partial \phi_j^\dagger}\bar{\psi}_i\bar{\psi}_j. \tag{11.81}$$

Note that the gauge kinetic part of this Lagrangian contains the term $\phi^\dagger D\phi$, which describes coupling of the D-field with the scalar field. Thus, we can eliminate the auxiliary field D by solving the corresponding algebraical equation. This gives

$$D^a = -e\phi t^a \phi^\dagger, \tag{11.82}$$

and there is a contribution to the effective potential of the scalar field other than (11.70). If the superpotential is set to be vanishing, the scalar potential, which appears in the action instead of the auxiliary non-propagating fields, becomes

$$U[\phi] = F_k^\dagger F_k + \frac{1}{2}D^2 = \sum_k \left|\frac{\partial \mathcal{W}}{\partial \phi^k}\right|^2 + \frac{e^2}{2}\left|\phi^\dagger t^a \phi\right|^2. \tag{11.83}$$

With all this information about the structure of the supersymmetric theories, we can examine the classical spectrum of the topologically non-trivial configurations that are present there.

12 Magnetic Monopoles in the $N=2$ Supersymmetric Yang–Mills Theory

So far, we have described the $N=1$ scalar and vector multiplets that contain particles of spin $(0, 1/2)$ and spin $(1/2, 1)$, respectively. This particle content may be employed to extend our consideration to the supersymmetrical generalization of the $SU(2)$ Yang–Mills–Higgs theory coupled with spinor fields in $d=4$.

Actually, this model can be considered as a supersymmetric cousin of the Georgi–Glashow theory. The elegant and beautiful structure of the $N=2$ SUSY Yang–Mills theory has very little difference from the conventional model, but there is the miracle of supersymmetry, which makes the model exactly solvable. The hope of theoreticians is that there be some properties of the SUSY Yang–Mills theory that in the non-perturbative regime can shed a new light on the problem of QCD confinement. Therefore, it looks very interesting that the monopole-like configuration plays a crucial role in the low-energy supersymmetric dynamics. Moreover, the Montonen–Olive conjecture of duality becomes an exact property of such a theory, where the gauge and the monopole sectors are related to each other by some transformation of duality. However, before we briefly describe the low-energy non-perturbative dynamics of the $N=2$ SUSY gauge theory, we have to consider the Lagrangian structure of this theory on the classical level.

12.1 $N=2$ Supersymmetric Lagrangian

First, let us recall that there are two $N=1$ on-shell multiplets, the scalar multiplet (spins 0 and $1/2$) and the vector multiplet (spins $1/2$ and 1). The corresponding fields, which appear in the $N=1$ SUSY Yang–Mills Lagrangian (11.81), are the chiral scalar superfield Φ and the real vector superfield V, respectively.

Let us note that these two $N=1$ multiplets together contain the same set of fields as $N=2$ massless vector multiplet. However, the Lagrangian (11.81) is not yet $N=2$ supersymmetric, although the extended symmetry can be restored.

A first step toward this higher supersymmetry is to take all the component fields that appear in the Lagrangian (11.81), both the $N=1$ vector and the scalar multiplets, in the same adjoint representation of the gauge group. Then, these fields can be considered as the components of the single $N = 2$ vector supermultiplet, which necessarily lies in the adjoint representation. Thus, we set

$$(T^a)_{bc} = (t^a)_{bc} = -if_{abc},$$

where f_{abc} are the group structure constants. Then the Yukawa coupling terms in the Lagrangian (11.79) can be represented as

$$\begin{aligned}\phi^\dagger t^a \lambda^a \psi &= -i\phi^{b\dagger} f_{abc} \lambda^a \psi^c \\ &= \phi^{b\dagger} \lambda^a \psi^c \text{ tr } T^b[T^a, T^c] = \text{tr } (\phi^\dagger \{\lambda, \psi\}),\end{aligned} \quad (12.1)$$

etc.

Second, the invariance of the theory with respect to the transformations of $N = 2$ supersymmetry means that both generators of the SUSY algebra enter the model equally. In other words, there is a rotational $SU(2)_R$ symmetry of the Lagrangian that relates the supercharges Q_α^1 and Q_α^2. However, the same symmetry transformation relates the spinor components of the $N = 1$ scalar and the $N = 1$ vector multiplets, λ and ψ, respectively. On the other hand, the superpotential $\mathcal{W}$ that appears in (11.81) is only coupled with ψ. Hence extended $N = 2$ supersymmetry of the model is possible, if the superpotential is vanishing, thus we have to set $\mathcal{W} = 0$.

Finally, the kinetic energy terms of both $N = 1$ components must have the same normalization. We encountered this condition already, when we rescaled the vector superfield[1] as $V \to 2eV$. Then there is no difference between the normalization of the Yang–Mills component and the scalar part of the Lagrangian (11.81), and the complete $N = 2$ SUSY Yang–Mills Lagrangian can be written as

$$\begin{aligned}L^{N=2} &= \frac{1}{32\pi}\text{Im}\left[\tau \text{ tr}\left(\int d^2\theta\, W^\alpha W_\alpha + 2\int d^2\theta d^2\bar{\theta}\, \Phi^\dagger e^{-2eV}\Phi\right)\right] \\ &= \text{tr}\Bigg(-\frac{1}{4}F^{\mu\nu}F_{\mu\nu} + \frac{\theta e^2}{32\pi^2}F^{\mu\nu}\widetilde{F}_{\mu\nu} + (\nabla^\mu\phi)^\dagger(\nabla_\mu\phi) - \frac{e^2}{2}[\phi^\dagger, \phi]^2 \\ &\quad - i\lambda\sigma^\mu\nabla_\mu\bar{\lambda} - i\bar{\psi}\bar{\sigma}^\mu\nabla_\mu\psi - ie\sqrt{2}\phi^\dagger\{\lambda, \psi\} + ie\sqrt{2}\{\bar{\psi}, \bar{\lambda}\}\phi\Bigg),\end{aligned}$$

(12.2)

where the trace is taken in the adjoint representation of the gauge group $SU(2)$. In this expression, the auxiliary fields have been eliminated by solving the corresponding algebraic equations of motion. Recall that this procedure yields the scalar self-coupling term $U = \dfrac{e^2}{2}[\phi^\dagger, \phi]^2$. The structure of the

---

[1] An alternative would be an overall factor $1/e^2$ multiplying the Lagrangian of the gauge fields and the consequent rescaling of the scalar superfield as $\Phi \to \Phi/e$.

Lagrangian (12.2) is highly balanced by the condition of $SU(2)_R$ symmetry, which guarantees it to be $N=2$ supersymmetric.

It looks natural now to extend the definition of $N=1$ rigid superspace above to $N=2$ superspace by adding four Grassmannian degrees of freedom $\tilde{\theta}_\alpha, \bar{\tilde{\theta}}_{\dot\alpha}$ to the $N=1$ anticommuting spinor coordinates θ_α, $\bar{\theta}_{\dot\alpha}$. The global $SU(2)_R$ symmetry of the model is related with chiral rotations of these doublets by opposite phases.

A generic $N=2$ superfield is defined as a function on the $N=2$ rigid superspace $F(x, \theta, \bar\theta, \tilde\theta, \bar{\tilde\theta})$. We are interested in a particular $N=2$ superfield for which the component expansion in Grassmannian coordinates reproduces the content of the $N=2$ vector multiplet. By analogy with the definition of the $N=1$ chiral superfield Φ (11.44), there is a set of constraints that allows us to project these components out. Thus, the $N=2$ *chiral superfield* Ψ is defined as a singlet with respect to the $SU(2)_R$ rotations that satisfy the covariant constraints

$$\bar{D}_{\dot\alpha}\Psi(x,\theta,\bar\theta,\tilde\theta,\bar{\tilde\theta}) = 0, \qquad \bar{\tilde{D}}_{\dot\alpha}\Psi(x,\theta,\bar\theta,\tilde\theta,\bar{\tilde\theta}) = 0, \qquad (12.3)$$

where the superderivative $\tilde{D}_\alpha$ is defined by analogy with D_α up to a replacement $\theta \to \tilde\theta$. Then, introducing the shifted bosonic coordinates by

$$\tilde{y}^\mu = y^\mu + i\tilde\theta\sigma^\mu\bar{\tilde\theta} = x^\mu + i\theta\sigma^\mu\bar\theta + i\tilde\theta\sigma^\mu\bar{\tilde\theta},$$

we can easily see that a general solution of the constraints (12.3) is an arbitrary function of variable $\tilde{y}^\mu$ and Grassmann coordinates $\theta, \tilde\theta$. Then the expansion of Ψ in powers of $\tilde\theta$ yields (cf. (11.45)):

$$\Psi = \Phi(\tilde{y},\theta) + \sqrt{2}\,\tilde\theta\, W(\tilde{y},\theta) + \tilde\theta\tilde\theta\, G(\tilde{y},\theta)\,.$$

This expansion relates the components of the $N=2$ chiral superfield to the $N=1$ chiral superfields.

Clearly, the component $\Phi(\tilde{y},\theta)$ corresponds to the scalar superfield (11.45), while dimensional arguments allows us to identify the component $W(\tilde{y},\theta)$ with the $N=1$ chiral spinor superfield (11.54). When this expansion is substituted into the constraints (12.3), it leads to the identification of the third component as a gauge-covariant chiral function of the coupled $N=1$ scalar and vector superfields (see, e.g., [357]):

$$G(\tilde{y},\theta) = \int d^2\bar\theta\, \Phi^\dagger(\tilde{y} - i\theta\sigma^\mu\bar\theta, \theta, \bar\theta) e^{-2eV(\tilde{y}-i\theta\sigma^\mu\bar\theta,\theta,\bar\theta)}\,.$$

We have seen already that the $N=2$ chiral superfield Ψ has the same field content as the off-shell vector multiplet: ϕ, A_μ and global $SU(2)$ spinor doublet ψ, λ. All these fields are presented in the $N=2$ SUSY Yang–Mills Lagrangian (12.2), which becomes extremely compact when written in terms of the superfield Ψ:

$$L^{N=2} = \frac{1}{16\pi}\text{Im tr}\int d^2\theta d^2\tilde{\theta}\,\frac{1}{2}\tau\Psi^2\,. \qquad (12.4)$$

More generally, we can write the Lagrangian of the $N=2$ superfield, which satisfies the covariant constraints (12.3), as

$$L^{N=2} = \frac{1}{16\pi}\text{Im}\int d^2\theta d^2\tilde{\theta}\,\text{tr}\mathcal{F}(\Psi)\,, \qquad (12.5)$$

where the holomorphic function $\mathcal{F}(\Psi)$ is called the *$N=2$ prepotential*. Actually, (12.4) is the only possible form of the classical prepotential that is fixed by the condition of the renormalizability.

In terms of the component fields in $N=1$ superspace, the Lagrangian (12.5) can be represented as follows

$$L^{N=2} = \frac{1}{32\pi}\text{Im}\left(\int d^2\theta \mathcal{F}_{ab}(\Phi)W^{a\alpha}W^b_\alpha + 2\int d^2\theta d^2\bar\theta\,(\Phi^\dagger e^{2eV})^a\mathcal{F}_a(\Phi)\right), \qquad (12.6)$$

where $\mathcal{F}_a(\Phi) \equiv \partial\mathcal{F}/\partial\Phi^a$, $\mathcal{F}_{ab}(\Phi) \equiv \partial^2\mathcal{F}/\partial\Phi^a\partial\Phi^b$ and a,b are global $SU(2)$ gauge indices. Comparing this Lagrangian with its general $N=1$ counterpart (11.72), we conclude that the non-holomorphic Kähler potential of the $N=2$ SUSY Yang–Mills model becomes

$$K = \text{Im}\,[(\Phi^\dagger e^{2eV})^a \mathcal{F}_a(\Phi)]\,, \qquad (12.7)$$

and the Kähler metric on the space of fields is given by $g_{ab} = \text{Im}\,\partial_a\partial_b\mathcal{F}(\Phi)$.

Let us note that the general structure of the action (12.5) can be considered as an effective *macroscopic theory* that is valid in the low-energy regime. Then we may forget about the restriction of renormalizability and analyze this effective model more closely. Evidently, the structure of the action (12.5) precisely corresponds to the supersymmetric non-linear *sigma model* whose Kähler potential is written in terms of a derivative of a holomorphic function. The complex scalar fields ϕ^a are treated as the local complex coordinates on a target space, a special Kähler manifold.

Having found the complete $N=2$ Lagrangian (12.2), we can proceed further by finding an explicit form of the supersymmetrical variations of the component fields of the $N=2$ chiral superfield $\Psi \to \Psi + \delta_\xi\Psi$. These variations are similar to the infinitesimal transformations of the component $N=1$ fields given by (11.47) and (11.53), up to a replacement of the usual derivative by the covariant derivative. A rather involved calculation yields the related $N=2$ *supercurrent* which corresponds to these transformations (for more details see, e.g., [73]):

$$\begin{aligned}S^\mu_{(1)} = &-\frac{i}{2}\left(\bar\lambda^a\bar\sigma^\mu\sigma^{\rho\nu}\xi + \bar\xi\bar\sigma^{\rho\sigma}\bar\sigma^\mu\lambda^a\right)F^a_{\rho\sigma} - \left(\bar\xi\bar\sigma^\mu\lambda^a + \bar\lambda^a\bar\sigma^\mu\xi\right)\phi^\dagger T^a\phi \\ &+ \sqrt{2}\xi\sigma^\nu\bar\sigma^\mu\psi^a\nabla_\nu\phi^{a\dagger} + \sqrt{2}\bar\psi^a\bar\sigma^\mu\sigma^\nu\bar\xi\nabla_\nu\phi^a\,.\end{aligned} \qquad (12.8)$$

There is another $N = 2$ supercurrent that can be easily written, if we take into account the invariance of the model with respect to the Weyl reflection considered above. This discrete transformation is an element of the $SU(2)_R$ symmetry that acts on the spinor components as $\lambda \to \psi$ and $\psi \to -\lambda$. It is now straightforward to obtain the second supercurrent by a simple replacement:

$$S^\mu_{(2)} = -\frac{i}{2}\left(\bar\psi^a \bar\sigma^\mu \sigma^{\rho\nu}\zeta + \bar\zeta \bar\sigma^{\rho\sigma}\bar\sigma^\mu \psi^a\right)F^a_{\rho\sigma} - \left(\bar\zeta\bar\sigma^\mu\psi^a + \bar\psi^a\bar\sigma^\mu\zeta\right)\phi^\dagger T^a \phi \qquad (12.9)$$
$$-\sqrt{2}\zeta\sigma^\nu\bar\sigma^\mu\lambda^a \nabla_\nu \phi^{a\dagger} - \sqrt{2}\bar\lambda^a \bar\sigma^\mu\sigma^\nu\bar\zeta \nabla_\nu\phi^a,$$

where ξ, ζ are the parameters of two sets of the $N = 2$ SUSY transformations.

12.1.1 Praise of Beauty of $N = 2$ SUSY Yang–Mills

Actually, there is nothing strange in the model with the action (12.2), whose field content is already familiar to us. Indeed, let us compose the two-component Weyl spinors λ and ψ that appear in the Lagrangian (12.2), into the Dirac bispinor

$$\chi = \begin{pmatrix} \psi_\alpha \\ -i\bar\lambda^{\dot\alpha} \end{pmatrix}, \qquad \bar\chi = (i\lambda^\alpha, \bar\psi_{\dot\alpha}),$$

and recall that the four-dimensional γ-matrices are defined as $\gamma^\mu \equiv \begin{pmatrix} 0 & \sigma^\mu \\ \bar\sigma^\mu & 0 \end{pmatrix}$. In this representation, $\gamma^5 = i\gamma^0\gamma^1\gamma^2\gamma^3 = \begin{pmatrix} -1 & 0 \\ 0 & 1 \end{pmatrix}$. Then the $N = 2$ SUSY Yang–Mills Lagrangian (12.2) can be represented as

$$L^{N=2} = -\frac{1}{4}F_a^{\mu\nu}F^a_{\mu\nu} + \frac{\theta e^2}{32\pi^2}F_a^{\mu\nu}\tilde F^a_{\mu\nu} + (\nabla_\mu \phi^a)^\dagger(\nabla^\mu \phi^a) - \frac{e^2}{2}(\phi^\dagger T^a \phi)^2$$
$$- i\bar\chi^a \gamma^\mu(\nabla_\mu\chi)^a - \sqrt{2}e\left[\phi(\bar\chi^a T^a)\frac{1+\gamma_5}{2}\chi + \phi^\dagger(\bar\chi^a T^a)\frac{1-\gamma_5}{2}\chi\right].$$

Furthermore, the complex scalar field can be replaced by two real fields

$$\phi^a = \frac{1}{\sqrt{2}}(\phi_1^a + i\phi_2^a). \qquad (12.10)$$

Hence, the Lagrangian (12.11) is takes the form

$$L^{N=2} = -\frac{1}{4}F_a^{\mu\nu}F^a_{\mu\nu} + \frac{\theta e^2}{32\pi^2}F_a^{\mu\nu}\tilde F^a_{\mu\nu} + \frac{1}{2}(\nabla_\mu \phi_1^a)^2 \qquad (12.11)$$
$$+ \frac{1}{2}(\nabla_\mu \phi_2^a)^2 - \frac{e^2}{2}[\phi_1^a, \phi_2^b]^2 - i\bar\chi\gamma^\mu\nabla_\mu\chi - e\phi_1 \bar\chi\chi - ie\gamma_5 \phi_2 \bar\chi\chi.$$

Thus, this is a Lagrangian of the system of interacting bosonic and fermionic fields that is very similar to the Georgi–Glashow model coupled with fermions. The condition of supersymmetry only imposes the restriction on the masses of the particles of the same multiplet and requires the coupling constant of the vector and scalar multiplets to be unique. So, there is nothing strange in the model (12.11) and, as noted in short review [215], what really looks unusual there, is the perfection and the universality of such a model, which actually incorporates almost all non-trivial elements of the modern quantum field theory and, therefore, deserves to be used as a *"Model to Teach Quantum Gauge Theory"*. The properties of this very remarkable theory include the following.

- It describes non-Abelian gauge fields coupled with the matter fields.
- It includes scalar and pseudoscalar fields with minimal Yukawa couplings to the fermions.
- It is renormalizable.
- It is asymptotically free.
- The Higgs mechanism is used to generate the masses of the particles softly.
- The non-perturbative sector of the model contains both monopoles and instantons.
- There is a chiral symmetry between scalar and pseudoscalar fields of the classical Lagrangian. The global symmetries of the classical $N = 2$ SUSY Yang–Mills Lagrangian include the $SU(2)_R$-symmetry, which arises from the automorphism of the algebra of supersymmetry and an additional $U(1)_R$ symmetry

$$\theta \to e^{i\alpha}\theta; \quad \tilde{\theta} \to e^{i\alpha}\tilde{\theta}; \quad \Psi \to e^{2i\alpha}\Psi. \tag{12.12}$$

Quantum anomaly breaks the latter symmetry to the discrete subgroup $\mathbb{Z}_4$. The discussion in the next chapter will clarify this.
- The corresponding algebra of SUSY contains the central charges.
- The action of the model is scale-invariant but there is a quantum anomaly in the trace of the energy-momentum tensor.
- The model provides a realization of the Montonen–Olive conjecture of duality.
- The model arises from the string theory in the point-particle limit.

Let us choose $SU(2)$ as the gauge group of the model. It makes it even closer to the Georgi–Glashow model. The differences now are in the structure of the scalar potential $U[\phi]$, which takes the form (11.83), and that all the fields, including the fermions, take values in the adjoint representation of the gauge group. However, unlike the genuine Georgi–Glashow model (5.7), we are now dealing with the complex scalar field ϕ^a (12.10) and the main difference is that the potential of the scalar field

$$U = \frac{e^2}{2}(f_{abc}\phi^{b\dagger}\phi^c)^2$$

does not fix the Higgs vacuum uniquely. Indeed, this potential vanishes if ϕ and $\phi^\dagger$ commute with each other.

For a given choice of the $SU(2)$ gauge group, this condition is satisfied if these fields take values in the Abelian subgroup of $SU(2)$:

$$\phi_0 = vT^3 = v\frac{\sigma^3}{2}, \qquad (12.13)$$

where v is now an arbitrary complex number. More generally, for a model with the gauge group G, the Higgs vacuum is defined by the relation

$$[\phi_0, \phi_0^\dagger] = \varepsilon_{abc}\phi^{b\dagger}\phi^c = 0,$$

which means that ϕ_0 lies in the diagonal Cartan subalgebra $\vec{H} = (H_1, H_2 \ldots H_{N-1})$ of G.

Let us note that we already considered a similar definition of the Higgs vacuum when we discussed the properties of the $SU(N)$ monopoles (cf. (8.6) and the following discussion on page 278). Thus, the Higgs vacuum breaks the gauge group to H and there is a set of gauge inequivalent classical vacua of the $N = 2$ SUSY Yang–Mills theory. However, the $N = 2$ SUSY is still in force and it remains a symmetry of these vacua.

The very important point is that each of these vacua corresponds to a different physical situation. Indeed, we know, for example, that the vacuum value of the scalar field generates the masses of particles via the Higgs mechanism and the physical observables are, therefore, directly related with ϕ_0. Thus, unlike the simple Georgi–Glashow model, there is a space of different vacuum values of the scalar field of dimension equal to the rank r of the gauge group G. We shall discuss this subject below, within the framework of the full quantum theory, where the moduli space is parameterized by the vacuum expectation values of the Higgs field $<\phi>$. In the context of this chapter, our primary subject is the spectrum of the monopole solutions of the classical $N = 2$ SUSY Yang–Mills theory.

12.2 $N = 2$ Supersymmetric $SU(2)$ Magnetic Monopoles

12.2.1 Construction of $N = 2$ Supersymmetric SU(2) Monopoles

Let us show now that there are monopole solutions of the model with the action (12.2). As before, in the case of the 't Hooft–Polyakov monopole solution of the Georgi–Glashow model, we can consider the static, time-independent configurations first, and ignore the topological θ-term for a while.

Recall that the difference of the bosonic sector of the supersymmetric action (12.2) from the Georgi–Glashow model is that there are two real scalar fields, ϕ_1 and ϕ_2, which correspond to the scalar and pseudoscalar coupling to the Dirac fermions. For the sake of simplicity, let us suppose that asymptotically the complex scalar field lies on the unit sphere, that is, $|\phi| = 1$ as $r \to \infty$.

Then, by analogy with expression (5.57), the bosonic Hamiltonian of the static $SU(2)$ super Yang–Mills theory can be written in the form

$$E = \frac{1}{2}\int d^3x \left\{ (E_n^a)^2 + (B_n^a)^2 + (\nabla_n \phi_1^a)^2 + (\nabla_n \phi_2^a)^2 + e^2(\varepsilon_{abc}\phi_1^b \phi_2^c)^2 \right\}$$

$$= \frac{1}{2}\int d^3x \left\{ [E_n^a - \nabla_n \phi_1^a \sin\delta - \nabla_n \phi_2^a \cos\delta]^2 + [B_n^a - \nabla_n \phi_1^a \cos\delta + \nabla_n \phi_2^a \sin\delta]^2 \right\}$$

$$+ 2\int d^3x \left\{ E_n^a(\nabla_n \phi_1^a \sin\delta + \nabla_n \phi_2^a \cos\delta) + B_n^a(\nabla_n \phi_1^a \cos\delta - \nabla_n \phi_2^a \sin\delta) \right\}$$

$$+ \frac{e^2}{2}\int d^3x \, (\varepsilon_{abc}\phi_1^b \phi_2^c)^2 \,,$$

where δ is an arbitrary parameter and, according to our convention from the previous chapter, the symbol ∇_n denotes a covariant derivative.

Thus, if the scalar potential is vanishing, the lower energy bound is given by the system of equations

$$\begin{aligned} E_n^a &= \nabla_n \phi_1^a \sin\delta + \nabla_n \phi_2^a \cos\delta \equiv \nabla_n \tilde{\phi}_1^a \,, \\ B_n^a &= \nabla_n \phi_1^a \cos\delta - \nabla_n \phi_2^a \sin\delta \equiv \nabla_n \tilde{\phi}_2^a \,, \end{aligned} \quad (12.14)$$

where we define the linear combinations of the scalar fields

$$\tilde{\phi}_1^a = \phi_1^a \sin\delta + \phi_2^a \cos\delta, \qquad \tilde{\phi}_2^a = \phi_1^a \cos\delta - \phi_2^a \sin\delta \,, \quad (12.15)$$

which appear in this supersymmetric counterpart of the BPS equations (5.58). On the other hand, a proper parameterization of the vacuum manifold of the model is still given in terms of the asymptotic values of the fields ϕ_i, rather than the rotated fields $\tilde{\phi}_i$, because they are invariant with respect to the transformations of the modular group $SL(2,\mathbb{Z})$.

A particular representation of these two scalar fields as

$$\phi_1^a = \phi^a \cos\delta, \qquad \phi_2^a = -\phi^a \sin\delta \,,$$

immediately yields the BPS equations (5.58) for the field $\tilde{\phi}_1 = 0$, $\tilde{\phi}_2 = \phi$, up to an obvious identification of the free angular parameters. Thus, the general static spherically symmetric monopole solution in the bosonic sector may be obtained as a generalization of the BPS solution:

$$\phi_1^a = \frac{r^a}{er^2}H(r)\cos\delta, \qquad \phi_2^a = -\frac{r^a}{er^2}H(r)\sin\delta \,,$$

$$A_n^a = \varepsilon_{amn}\frac{r^m}{er^2}(1 - K(r)) \,, \quad (12.16)$$

where the structure functions are the well-known analytical solutions (5.63), which are expressed via the dimensionless parameter $\xi = er$:

12.2 $N = 2$ Supersymmetric $SU(2)$ Magnetic Monopoles

$$K = \frac{\xi}{\sinh \xi}, \qquad H = \xi \coth \xi - 1. \tag{12.17}$$

Asymptotic behavior of the complex Higgs field is now

$$\phi^a \to e^{-i\delta} \hat{r}^a \left(1 - \frac{1}{er}\right) \quad \text{as} \quad \xi \to \infty. \tag{12.18}$$

Let us note that the real components, ϕ_1 and ϕ_2, may have different behavior while they are approaching this regime.

It is natural to define two-component electric and magnetic charges as [224, 232, 233]:

$$q^i = \int d^3x \, \partial_n (E_n^a \phi_i^a), \qquad g^i = \int d^3x \, \partial_n (B_n^a \phi_i^a). \tag{12.19}$$

Then the bound on the mass of $N = 2$ supersymmetric monopoles becomes

$$M \geq [(g^1 + q^2) \cos \delta + (q^1 - g^2) \sin \delta], \tag{12.20}$$

which implies that the Bogomol'nyi bound is saturated if

$$\tan \delta = \frac{q^1 - g^2}{g^1 + q^2}, \tag{12.21}$$

and then $E_n^a = \nabla_n \tilde{\phi}_1^a$, $B_n^a = \nabla_n \tilde{\phi}_2^a$.

So far, we brutally set $\tilde{\phi}_1$ to be identically zero and make use of the straightforward analogy with construction of the BPS monopoles in the previous chapter. However, one can construct a wider class of monopoles by relaxing this restriction. Indeed, the Gauss law (5.75) for such a static field configuration with a two-component scalar field takes the form

$$\nabla_n E_n - ie[\tilde{\phi}_1, \nabla_0 \tilde{\phi}_1] - ie[\tilde{\phi}_2, \nabla_0 \tilde{\phi}_2] = 0.$$

Let us now impose a gauge $A_0 = \tilde{\phi}_1$. Then the Gauss law becomes a covariant Laplace equation for the field $\tilde{\phi}_1$:

$$\nabla^2 \tilde{\phi}_1 - e^2 [\tilde{\phi}_2, [\tilde{\phi}_1, \tilde{\phi}_2]] = 0, \tag{12.22}$$

which is referred to as the *"secondary BPS equation"* [232, 233, 353, 501]. Certainly, this equation becomes even more restrictive if we only consider the configurations that satisfy $[\tilde{\phi}_2, \tilde{\phi}_1] = 0$.

The *"primary BPS equation"* is obviously $B_n = \nabla_n \tilde{\phi}_2$. For a given solution of this equation, the secondary BPS equation describes a large gauge transformation of the fields of the BPS monopole $A_k, \tilde{\phi}_2$ [353]. Thus, the solution of the secondary BPS equation yields the gauge zero modes about the original monopole configuration. This zero mode exists for each solution of the primary BPS equation. We already know that these gauge transformations generate an electric charge of the configuration that transforms a monopole into a dyon.

Taking into account the Bianchi identity for the magnetic field and the equation of motion for the electric field, we can now recover the lower bound on the mass of dyon by analogy with our evaluation of the BPS mass bound (5.60) in Chap. 5 [224, 353]:

$$M \geq | (q^1 - g^2) + i(g^1 + q^2) | . \quad (12.23)$$

So far, we are discussing the model with the gauge group $SU(2)$. In this case, the vanishing of the scalar potential, $[\phi_1, \phi_2]^2 = 0$, means that both ϕ_1 and ϕ_2 belong to the unique Cartan subgroup of the gauge group. In other words, ϕ_1 must be proportional to ϕ_2 and, therefore, the components of the electric and magnetic charge vectors are also proportional to each other. Thus, $g^1 q^2 = g^2 q^1$ and the bound (5.60) is recovered from (12.23):

$$M \geq \sum_{i=1,2} \sqrt{q_i^2 + g_i^2} .$$

However, the principal difference from the Georgi–Glashow model is that now the BPS bound is directly related with the algebra of supersymmetry. Moreover, the electric and the magnetic charges (12.19) are the real and the imaginary components of the $N = 2$ central charge Z, respectively [523].

12.3 Central Charges in the $N = 2$ SUSY Yang–Mills

We have mentioned already that the $N = 2$ SUSY algebra (11.16) includes a complex central charge Z. Let us now construct this charge explicitly.

Note that the generators of $N = 2$ supersymmetry Q_α^1, Q_α^2 by definition are the charges of the spinor components of the supercurrents (12.8) and (12.9), respectively. Indeed, expanding these currents in the parameter of the infinitesimal SUSY transformation ξ, we can write

$$S_\mu = S_\mu^\alpha \xi_\alpha + \bar{S}_{\mu\dot{\alpha}} \bar{\xi}^{\dot{\alpha}} ,$$

and then

$$S^{(1)}_{\mu\ \alpha} = -(\sigma^\nu)_{\alpha\dot{\alpha}} \bar{\lambda}^{a\dot{\alpha}} (iF^a_{\mu\nu} + \widetilde{F}^a_{\mu\nu}) + \sqrt{2}(\sigma_\nu \bar{\sigma}_\mu \psi^a)_\alpha \nabla^\nu \phi^{a\dagger} + (\sigma_\mu)_{\alpha\dot{\alpha}} \bar{\lambda}^{a\dot{\alpha}} \phi^\dagger T^a \phi ,$$

$$S^{(2)}_{\mu\ \alpha} = -(\sigma^\nu)_{\alpha\dot{\alpha}} \bar{\psi}^{a\dot{\alpha}} (iF^a_{\mu\nu} + \widetilde{F}^a_{\mu\nu}) - \sqrt{2}(\sigma_\nu \bar{\sigma}_\mu \lambda^a)_\alpha \nabla^\nu \phi^{a\dagger} + (\sigma_\mu)_{\alpha\dot{\alpha}} \bar{\psi}^{a\dot{\alpha}} \phi^\dagger T^a \phi .$$
(12.24)

Here we make use of the property $\varepsilon \sigma^\mu \bar{\psi} = -\bar{\psi} \bar{\sigma}^\mu \varepsilon$ and the following identities involving the Pauli matrices

12.3 Central Charges in the $N = 2$ SUSY Yang–Mills

$$\sigma^\mu \bar\sigma^\nu \sigma^\rho = g^{\mu\nu}\sigma^\rho - g^{\mu\rho}\sigma^\nu + g^{\nu\rho}\sigma^\mu + i\varepsilon_{\mu\nu\rho\omega}\sigma^\omega,$$
$$\bar\sigma^\mu \sigma^\nu \bar\sigma^\rho = g^{\mu\nu}\bar\sigma^\rho - g^{\mu\rho}\bar\sigma^\nu + g^{\nu\rho}\bar\sigma^\mu - i\varepsilon_{\mu\nu\rho\omega}\bar\sigma^\omega. \quad (12.25)$$

Recall that we use the metric $g_{\mu\nu} = (-1, 1, 1, 1)$.

Therefore, the anticommutator (11.16), which yields the central charge, can be written as the equal-time anticommutator of the volume integrals over the temporal components of these currents:

$$\{Q^1_\alpha, Q^2_\beta\} = \left\{\int d^3x\, S^{(1)}_{0\alpha}(\mathbf{x}), \int d^3x'\, S^{(2)}_{0\beta}(\mathbf{x}')\right\} = 2\varepsilon_{\alpha\beta} Z. \quad (12.26)$$

Olive and Witten noticed [523] that this commutator non-vanishes due to the boundary terms, which are just the electric and the magnetic charges of the configuration. Indeed, let us note that $\bar\lambda_{\dot\alpha} = \lambda^\dagger_\alpha$ and

$$\bar\lambda^{\dot\alpha} = \varepsilon^{\dot\alpha\dot\beta}\bar\lambda_{\dot\beta} = i(\sigma_2\lambda^\dagger)^{\dot\alpha}.$$

It is clear that the temporal components of the supercurrents (12.24) are

$$S^{(1)}_{0\alpha} = -i(\sigma_k\sigma_2\lambda^{a\dagger})_\alpha(iF^a_{0k} + \widetilde{F}^a_{0k}) + \sqrt{2}(\sigma_k\psi^a)_\alpha \nabla_k\phi^{a\dagger} + i(\sigma_2\lambda^{a\dagger})_\alpha \phi^\dagger T^a \phi,$$

$$S^{(2)}_{0\alpha} = -i(\sigma_k\sigma_2\psi^{a\dagger})_\alpha(iF^a_{0k} + \widetilde{F}^a_{0k}) - \sqrt{2}(\sigma_k\lambda^a)_\alpha \nabla_k\phi^{a\dagger} + i(\sigma_2\psi^{a\dagger})_\alpha \phi^\dagger T^a \phi.$$

To evaluate the central charge Z, we substitute these expressions into anticommutator (12.26) and make use of the anticommutation relations for the fields ψ and λ. The relevant terms in (12.26) are

$$\{Q^1_\alpha, Q^2_\beta\} = i\sqrt{2}\int d^3x \left[(\sigma_i\sigma_2\sigma_j^T)_{\alpha\beta} - (\sigma_i\sigma_2\sigma_j^T)_{\beta\alpha}\right](iF^a_{0k} + \widetilde{F}^a_{0k})\nabla_k\phi^{a\dagger}. \quad (12.27)$$

Next, making use of the algebra of the Pauli matrices

$$(\sigma_i\sigma_2\sigma_j^T)_{\alpha\beta} = \left(\sigma_2(-\delta_{ij} + i\varepsilon^{ijk}\sigma_k^T)\right)_{\alpha\beta},$$

we obtain

$$\{Q^1_\alpha, Q^2_\beta\} = -2i\sqrt{2}\int d^3x\, (\sigma_2)_{\alpha\beta}\delta_{ij}(iF^a_{0k} + \widetilde{F}^a_{0k})\nabla_k\phi^{a\dagger}$$
$$= -2\sqrt{2}\varepsilon_{\alpha\beta}\int d^3x\, (iF^a_{0k} + \widetilde{F}^a_{0k})\nabla_k\phi^{a\dagger}. \quad (12.28)$$

By complete analogy with our previous discussion of the definition of the electric and magnetic charges (5.53) and (5.46) in Chap. 5, the volume integrals here can be written as the integrals over the surface of the sphere S^2 on the spatial infinity:

$$\int d^3x\, F^a_{0k}\nabla_k\phi^{a\dagger} = \int d^2S_k\, E^a_k \phi^{a\dagger} = \int d^3x\, \partial_k(E^a_k \phi^{a\dagger}),$$
$$\int d^3x\, \widetilde{F}^a_{0k}\nabla_k\phi^{a\dagger} = \int d^2S_k\, B^a_k \phi^{a\dagger} = \int d^3x\, \partial_k(B^a_k \phi^{a\dagger}). \quad (12.29)$$

Hence, the anticommutator (12.26) is of the form

$$\{Q_\alpha^1, Q_\beta^2\} = -2\sqrt{2}\varepsilon_{\alpha\beta}\int d^3x\, \partial_k(iE_k^a + B_k^a)\phi^{a\dagger} = 2\varepsilon_{\alpha\beta}Z\,. \qquad (12.30)$$

By the same token, we find that

$$\{\bar{Q}_{\dot\alpha}^1, \bar{Q}_{\dot\beta}^2\} = -2\sqrt{2}\varepsilon_{\dot\alpha\dot\beta}\int d^3x\, \partial_k(-iE_k^a + B_k^a)\phi^{a} = 2\varepsilon_{\dot\alpha\dot\beta}Z^*\,. \qquad (12.31)$$

Here we make use of the definition of the complex central charge Z given by (11.16). Thus the central charge satisfies

$$Z = \sqrt{2}\int d^3x\, \partial_k(iE_k^a + B_k^a)\phi^{a\dagger}\,. \qquad (12.32)$$

The final step is to recall that the electric and the magnetic charges are given by (12.19), where the complex scalar field is decomposed into two real components as $\phi^a = \frac{1}{\sqrt{2}}(\phi_1^a + i\phi_2^a)$. Therefore, the central charge of the $N=2$ SUSY Yang-Mills theory is simply [523]

$$\begin{aligned} Z &= \int d^3x\, \partial_k(E_k^a\phi_2^a + B_k^a\phi_1^a) + i\int d^3x\, \partial_k(E_k^a\phi_1^a - B_k^a\phi_2^a) \\ &= [(q^1 - g^2) + i(g^1 + q^2)]\,. \end{aligned} \qquad (12.33)$$

Thus, the algebra of supersymmetry (11.16), which includes the central charge, according to (11.29) yields a mass bound

$$|Z| = |(q^1 - g^2) + i(g^1 + q^2)| \le M\,, \qquad (12.34)$$

which is precisely the BPS bound (12.23). Thus, in the $N=2$ SUSY Yang–Mills theory the magnetic and the electric charges of the bosonic monopole configuration appear in the explicit form of the central charge of the $N=2$ supersymmetry algebra and the Bogomol'nyi bound is a direct consequence of the extended supersymmetry. There is a difference from its classical counterpart (5.60), because if the $N=2$ supersymmetry is not broken by the one-loop quantum corrections, the Bogomol'nyi bound (12.34) is not modified. Note that in that case, the BPS states with magnetic and electric charges will be presented in the spectrum of physical states of the quantum supersymmetric theory.

It must be kept in mind, however, that this statement is correct if the vacuum expectation value of the scalar field is large. Seiberg and Witten pointed out [469] that in the strong coupling limit, the expression (12.23) must be modified. We shall discuss this issue in the last chapter.

12.4 Fermionic Zero Modes in Supersymmetric Theory

Let us consider the $N = 2$ Lagrangian (12.11), where for the sake of simplicity, we choose $\theta = 0$ first and make use of the relations of the type of (12.1):

$$L^{N=2} = \text{tr}\left\{-\frac{1}{4}F^{\mu\nu}F_{\mu\nu} + \frac{1}{2}(\nabla_\mu \phi_1)^2 + \frac{1}{2}(\nabla_\mu \phi_2)^2 \right.$$
$$\left. - i\bar{\chi}\gamma^\mu \nabla_\mu \chi + e\bar{\chi}[\phi_1, \chi] + ie\gamma_5\bar{\chi}[\phi_2, \chi]\right\} - \frac{e^2}{2}[\phi_1^a, \phi_2^b]^2 . \quad (12.35)$$

As we have seen, the BPS bound is saturated if the scalar field belongs to the Cartan subalgebra of the gauge group. Then the potential of the scalar field is vanishing for any particular value of the vacuum expectation value v. In other words, there are *flat directions* in the configuration space of the scalar field, a line of degenerated local minima for which the BPS bound is saturated by definition. This is one more difference from the non-supersymmetric non-Abelian monopole solutions for which the BPS limit corresponds to the nullification of the scalar coupling λ.

For the fields saturating the BPS bound (12.23), we can write the infinitesimal transformations of the $N = 2$ supersymmetry with the parameter of supertranslation ξ:

$$\delta\chi^a = \left(\sigma^{\mu\nu}F_{\mu\nu}^a + \gamma^\mu \nabla_\mu(\phi_1^a + \gamma_5\phi_2^a)\right)\xi ,$$
$$\delta A_\mu^a = i\bar{\xi}\gamma_\mu \chi^a - i\bar{\chi}^a \gamma_\mu \xi , \quad (12.36)$$
$$\delta\phi_1^a = i\bar{\xi}\chi^a - i\bar{\chi}^a \xi ,$$
$$\delta\phi_2^a = i\bar{\xi}\gamma_5 \chi^a - i\bar{\chi}^a \gamma_5 \xi .$$

It is instructive to apply these transformations to see that they change the $N = 2$ Lagrangian (12.11) by a total derivative. Indeed, the transformations (12.36) are actually the variations of the properly rescaled $N = 1$ Lagrangians (11.47) and (11.53), where the auxiliary fields are eliminated.

Let us assume that the fields entering the expressions (12.36) satisfy the BPS equations (12.14). Thus, we shall consider the supersymmetry variations on the $SU(2)$ monopole background. This is the classical solution of (12.16) with no fermions, which we choose as an initial configuration.

Let us consider, how the trnsformations of supersymmetry act on a simple static BPS monopole. For such an intial configuration, we set $\phi_2^a = 0$ and there are no ferminos around, i.e., $\chi^a = 0$. Then the supersymmetry variations of the bosonic fields in (12.36) are vanishing and the supersymmetry variation of the spinor field becomes

$$\delta\chi^a = \left(\sigma^{\mu\nu}F_{\mu\nu}^a + \gamma^\mu (\nabla_\mu \phi_1^a)\right)\xi .$$

For a static field configuration ($E_n^a = 0$), the BPS equations become simply $B_n^a = \nabla_n \phi_1^a$. Then the supersymmetry variation takes the form

$$\delta\chi^a = \left(\varepsilon_{mnk}\sigma^{mn}B_k^a + \gamma^n B_n^a\right)\xi = B_n^a\left(\frac{i}{2}\varepsilon_{nmk}\gamma^m\gamma^k + \gamma^n\right)\xi,$$

where we make use of the definition (11.5) and related properties of the γ-matrices. Thus, we obtain

$$\delta\chi^a = \gamma^n B_n^a \left(1 + \frac{i}{3!}\varepsilon_{nmk}\gamma^n\gamma^m\gamma^k\right)\xi = \gamma^n B_n^a (1 + i\gamma_0\gamma_5)\xi. \quad (12.37)$$

This means that if the parameter of the supersymmetry transformation ξ satisfies the equation

$$(1 + i\gamma_0\gamma_5)\xi = (1 - \Gamma_5)\xi = 0,$$

where

$$\Gamma_5 = -i\gamma_0\gamma_5 = \gamma_1\gamma_2\gamma_3 = \begin{pmatrix} 0 & -i \\ i & 0 \end{pmatrix},$$

and $\Gamma_5^2 = 1$, $\Gamma_5^\dagger = \Gamma_5$, the variation of the spinor field vanishes identically. This corresponds to unbroken supersymmetry, since we suppose that there are no fermions in the initial configuration. Evidently, if we decompose $\xi = \xi_+ + \xi_-$, where

$$\xi_\pm = \frac{1}{2}(1 \pm \Gamma_5)\xi,$$

the transformation of the supersymmetry generated by the parameter ξ_+ acts on the bosonic monopole background trivially, i.e., $\delta_{\xi_+}\chi^a = 0$. On the other hand, the supersymmetry variation generated by ξ_- breaks down half of the supersymmetry[2] and drives the configuration from $\chi^a = 0$ to

$$\delta_{\xi_-}\chi^a \equiv \chi_{(0)}^a = -2\gamma^n B_n^a \xi_-. \quad (12.38)$$

These zero energy Grassmannian variations of the bosonic monopole solution (12.16) are two fermionic zero modes whose properties we already discussed in Chap. 10. One can prove that these modes are time-independent solutions of the Dirac equation for a fermion coupled with a supersymmetric monopole. Indeed, variation of the Lagrangian (12.35) with respect to the field $\bar{\chi}$ yields the Dirac equation

$$i\gamma^\mu \nabla_\mu \chi - e[\phi_1, \chi] = 0.$$

It is straightforward now to substitute the explicit form of the fermionic zero modes (12.38) into this equation to ensure that they are the solutions with zero eigenvalues.

---

[2] This is why this solution sometimes is referred to as the *1/2-BPS monopole*.

Let us note that we can expect this effect in advance, since the Callias index theorem [156] predicts exactly two fermion zero modes for the fermions in the adjoint representation of the gauge group. Actually, from the $N = 2$ supersymmetry algebra with central charge, we already encounter the partial breaking of supersymmetry for the states that belong to the short multiplets and saturate the BPS bound (11.29) (see the discussion on page 419).

Recall that the presence of two fermionic zero modes on the monopole bosonic background implies that there is *$N = 2$ BPS monopole multiplet*, which can be constructed starting from the vacuum spin-0 state $|\Omega\rangle$ by consequent action of the operators of creation of these zero modes $a^\dagger_{\pm 1/2}$ (cf. our remarks on page 371). This monopole multiplet contains four states: two scalars and two fermions [409]. Note that these states are dual to the states of the massive short $N = 2$ chiral multiplet of four helicity states.

Furthermore, the remaining half of the supersymmetry of the $N = 2$ BPS monopoles allows us to set a correspondence between two fermionic zero modes and four bosonic zero modes [230], which form a supermultiplet with respect to the unbroken supersymmetry. Indeed, according to (12.36), for each fermionic zero mode $\chi_{(0)}$ of (12.38) the remaining half of the supersymmetry transformation generated by the supertranslations ξ_+ yields

$$\delta A_n^a = i\xi^\dagger_+ \gamma_n \chi^a_{(0)} - i\chi^{a\dagger}_{(0)} \gamma_n \xi_+ , \qquad (12.39)$$
$$\delta \phi_1^a = i\xi^\dagger_+ \gamma_0 \chi^a_{(0)} - i\chi^{a\dagger}_{(0)} \xi_+ .$$

12.5 Low Energy Dynamics of Supersymmetric Monopoles

To establish a correspondence with our discussion of the collective coordinates of a non-supersymmetric monopole, which we encountered before in Chap. 6, let us recall that the Bogomol'nyi equations are equivalent to the self-duality equations of the pure Yang–Mills theory with identification of the connection $A_\mu = (\phi_1, A_k)$. The moduli space approach[3] is to consider the bosonic collective coordinates X_α that parameterize the n-monopole moduli space $\mathcal{M}_n$ (cf. Chap. 6).

The normalizable bosonic zero modes define the vector $(\delta_\alpha A_\mu) = \partial_\alpha A_\mu - \nabla_\mu \omega_\alpha$, which is tangent to the monopole moduli space $\mathcal{M}_n$. Here, ordinary translations are complemented by a gauge transformation with a parameter ω_α, which is chosen to satisfy the relations (6.139), and the contribution of the gauge zero mode is taken into account. Then the Riemannian metric on $\mathcal{M}_n$ takes the form (6.137):

---

[3] To avoid possible misunderstanding, let us note that, in the context of this section, the notion of the moduli space is not directly related to the properties of the physical vacuum of the model. Here, we make use of it to describe the low-energy dynamics of the monopoles as we did before.

$$g_{\alpha\beta} = \int d^3x \, \text{Tr} \left(\delta_\alpha A_\gamma \delta_\beta A_\gamma \right) .$$

It was mentioned previously that the moduli space $\mathcal{M}_n$ has a quaternionic structure in the tangent space, that is, this manifold is hyper-Kähler. Geometrically, the gauge parameter ω_α defines a natural connection on $\mathcal{M}_n$ with the covariant derivative $\nabla_\alpha^{(\omega)}$. The Christoffel connection associated with the Riemannian metric on the moduli space is

$$\Gamma_{\alpha\beta\gamma} = \int d^3x \text{Tr} \left(\delta_\alpha A_\rho \nabla_\beta^{(\omega)} \delta_\gamma A_\rho \right) . \tag{12.40}$$

As was already briefly mentioned, the presence of the fermions modifies this picture [373]. In addition to the bosonic collective coordinates, we have to introduce some set of fermionic collective coordinates, which are the complex Grassmann numbers $\lambda_\alpha(t)$, and the time-dependent coefficients of the zero-mode components of the expansion of the spinor field $\chi = \lambda_\alpha(t)\chi_{(0)}^\alpha$. However, the relations (12.39) imply that the fermionic and the bosonic collective coordinates of the supersymmetric monopole are no longer independent.

It is convenient to introduce Euclidean Hermitian matrices Γ_μ

$$\Gamma_n = \gamma_0 \gamma_n, \quad \Gamma_4 = \gamma_0, \quad \Gamma_5 = \Gamma_1 \Gamma_2 \Gamma_3 \Gamma_4 ,$$

which satisfy the algebra $\{\Gamma_\mu, \Gamma_\nu\} = 2\delta_{\mu\nu}$. Then the fermionic and bosonic zero modes are paired as [230, 260]

$$\chi_{(0)}^a = (\delta_\alpha A_\mu^a) \, \Gamma^\mu \, \xi_- \lambda^\alpha , \tag{12.41}$$

that is, two bosonic zero modes are paired with one fermionic zero mode [260]. Note that the gauge zero mode in this case is included automatically.

Following our earlier discussion, we can derive the effective Lagrangian that governs the low-energy dynamics of the $N = 2$ supersymmetric monopoles [230]:

$$L_{X,\lambda} = \frac{M}{2} g_{\alpha\beta}(\dot{X}^\alpha \dot{X}^\beta + 4i\lambda^{\alpha\dagger} \nabla_0 \lambda^\beta) , \tag{12.42}$$

where

$$\nabla_0 \lambda^\alpha = \dot{\lambda}^\alpha + \Gamma_{\beta\gamma}^\alpha \dot{X}^\beta \lambda^\gamma$$

is the temporal component of the covariant derivative on the moduli space with the Christoffel connection (12.40) acting on the fermionic collective coordinates λ_α.

The presence of the fermionic degrees of freedom in the Lagrangian of the collective coordinates reveals some novelty. Unlike its bosonic counterpart of (6.138) of Chap. 6, the Lagrangian (12.42) now corresponds to the low-energy quantum mechanical theory, rather than to a classical effective Lagrangian. Nevertheless, the metric on the moduli space $\mathcal{M}_n$ of supersymmetric monopoles remains hyper-Kählerian and there are three almost complex structures

$(I^{(m)})_{\alpha\beta}$ on $\mathcal{M}_n$ that play the role of the imaginary units of quaternions. One of these structures can be used to introduce the complex coordinates on the moduli space. Thus, the effective Lagrangian (12.42) is invariant under transformations of unbroken supersymmetry [230]

$$\delta X^\alpha = i\xi_0 \lambda^\alpha + i\xi_m (I^{(m)})^{\alpha\beta} \lambda_\beta,$$
$$\delta \lambda^\alpha = -\dot{X}^\alpha \xi_0 - \xi_m (I^{(m)})^{\alpha\beta} \dot{X}_\beta, \qquad (12.43)$$

where ξ_0, ξ_m are four real Grassmannian parameters. Clearly, these four supersymmetries of the moduli space originate from the unbroken supersymmetries of the field theory.

Since the pseudoscalar Higgs field was eliminated by chiral rotation, further consideration is remarkably similar to the non-supersymmetric case. For a single monopole, the moduli space $\mathcal{M}_1$ is simply $\mathbb{R}^3 \times S^1$ with a flat metric, but non-trivial group of holonomy again. The n-monopole moduli space $\mathcal{M}_n$ is asymptotically isomorphic locally to the set of n copies of $\mathcal{M}_1$, which correspond to the picture of n well-separated single monopoles. Separating the collective coordinates of the centre of mass and the total electric charge of the configuration by analogy with (6.142), we obtain the hyper-Kähler manifold, $\mathcal{M}_n^0$ which determines the relative motion of the monopoles.

However, if the second component of the Higgs field has a non-vanishing vacuum expectation value, the consideration turns out to be more involved. If, in addition, the rank of the gauge group is greater than one, the low-energy monopole dynamics becomes rather different from our discussion above. This interesting situation was described in more detail recently [94, 96, 97, 233, 353, 441, 501].

12.6 $N = 2$ Supersymmetric Monopoles beyond $SU(2)$

To generalize our previous discussion, let us analyze the properties of $N = 2$ supersymmetric monopoles in the model with gauge group $G = SU(N)$.

Since we are considering the BPS monopoles, the classical potential of the scalar field must vanish. Recall that for $N = 2$ SUSY, this condition is satisfied, if the two components of the Higgs field ϕ_1, ϕ_2 are in the Cartan subalgebra of G and on the spatial asymptotic

$$\phi_i = (\vec{h}_i \cdot \vec{H}), \qquad (12.44)$$

where $\vec{h}_i$, $i = 1, 2$, are vectors in the root space of the Cartan subalgebra of dimension $r = $ rank (G). Then, for a given vacuum, we can also define the electric and magnetic charges as vectors in the root space (cf. (8.25) and (12.19)) by

$$q_i = \int d^3x \, \partial_n (E_n^a \phi_i^a) = (\vec{q} \cdot \vec{h}_i),$$

$$g_i = \int d^3x \, \partial_n (B_n^a \phi_i^a) = (\vec{g} \cdot \vec{h}_i). \tag{12.45}$$

In this notation, the BPS bound (12.23) can be written as

$$M = \left| (\vec{h}_1 + i\vec{h}_2) \cdot (\vec{q} + i\vec{g}) \right| = \sqrt{2} \, |\vec{h} \cdot (\vec{q} + i\vec{g})|, \tag{12.46}$$

where, according to the definition (12.44), the complex vector of the Higgs field is

$$\vec{h} = \frac{1}{\sqrt{2}} \left(\vec{h}_1 + i\vec{h}_2 \right).$$

So far, we have paid little attention to the fact that the electric and magnetic vectors lie on two different lattices: the former can be expanded in the basis of simple roots $\vec{\beta}_i$ of the given Lie group G, while the latter is defined in terms of the expansion in the basis of dual co-root vectors $\vec{\beta}_i^*$ of the dual lattice:

$$\vec{q} = e \sum_{i=1}^{r} m_i \vec{\beta}_i; \qquad \vec{g} = \frac{4\pi}{e} \sum_{i=1}^{r} n_i \vec{\beta}_i^*. \tag{12.47}$$

Here the integers $m_i, n_i \in \mathbb{Z}$ are the electric[4] and the magnetic quantum numbers[5].

This expansion allows us to write the BPS mass bound in the form

$$M = \left| e \sum_{i=1}^{r} m_i \left(\sqrt{2} \vec{\beta}_i \cdot \vec{h} \right) + \frac{4i\pi}{e} \sum_{i=1}^{r} n_i \left(\sqrt{2} \vec{\beta}_i^* \cdot \vec{h} \right) \right|$$

$$\equiv \left| \sum_{i=1}^{r} (m_i \Phi_i + n_i \Phi_i^*) \right|, \tag{12.48}$$

where we introduce the rescaled scalar field and its dual as

$$\Phi_i = e\sqrt{2} \left(\vec{\beta}_i \cdot \vec{h} \right), \qquad \Phi_i^* = \frac{4i\pi}{e} \sqrt{2} \left(\vec{\beta}_i^* \cdot \vec{h} \right). \tag{12.49}$$

These fields are expanded over the basis of simple roots and co-roots, respectively. This form of the BPS mass bound is evidently symmetric with respect to the dual transformations. Actually this is the Seiberg–Witten form of the BPS boundary, which appears in the quantum $N = 2$ supersymmetric theory.

---

[4] If we restrict our consideration to the classical limit, the electric charge remains non-quantizable and there is no charge lattice, which we discussed at the end of Chap. 2.

[5] For a special unitary group, we can always choose the self-dual basis of the simple roots: $\vec{\beta}_i^* = \vec{\beta}_i$.

12.6 $N = 2$ Supersymmetric Monopoles beyond $SU(2)$

We shall consider this remarkable relation in more detail when we discuss the quantum vacuum moduli space of the $N = 2$ SUSY Yang–Mills theory.

Let us recall now that the BPS equation for the $N = 2$ supersymmetric monopoles is written in terms of the $SO(2)$ rotated Higgs fields $\tilde{\phi}_i^a$ (12.15). Since the scalar fields lie in the root space of the Cartan subalgebra, we can describe this transformation as a rotation of the vectors $\vec{h}_i$, that is, $\tilde{\phi}_i \to \tilde{\phi}_i = \vec{h}_i' \cdot \vec{H}$, where

$$\vec{h}_1' = \vec{h}_1 \sin\delta + \vec{h}_2 \cos\delta, \qquad \vec{h}_2' = \vec{h}_1 \cos\delta - \vec{h}_2 \sin\delta. \qquad (12.50)$$

Asymptotically, these fields decay as

$$\tilde{\phi}_1(r) = \vec{h}_1' \cdot \vec{H} - \frac{\vec{q} \cdot \vec{H}}{r} + O(r^{-2}),$$
$$\tilde{\phi}_2(r) = \vec{h}_2' \cdot \vec{H} - \frac{\vec{g} \cdot \vec{H}}{r} + O(r^{-2}), \qquad (12.51)$$

and the angle of rotation of the Higgs fields is restricted by the constraint (12.21)

$$\tan\delta = \frac{q^1 - g^2}{g^1 + q^2}.$$

Note that for a purely magnetically charged state, the long-range scalar interaction is entirely given by the asymptotic behavior of the field $\tilde{\phi}_2$, while the second component $\tilde{\phi}_1(r)$ has no Coulomb tail at all. However, in the strong coupling limit, the roles of the components of the Higgs field are inverted.

Taking into account the definitions of the charges (12.45), we can easily see that the constraint on the angle of rotation becomes simply

$$\vec{g} \cdot \vec{h}_1' = \vec{q} \cdot \vec{h}_2',$$

and the BPS mass formula (12.20) can be written as

$$M = |\vec{q} \cdot \vec{h}_1' + \vec{g} \cdot \vec{h}_2'|.$$

These two contributions to the mass are referred to as the *magnetic mass* ($\vec{g} \cdot \vec{h}_2'$) and the *electric mass* ($\vec{q} \cdot \vec{h}_1'$), respectively. In the weak coupling regime $e \ll 1$, the electric mass is obviously much smaller than the magnetic mass. This observation justifies the use of the semiclassical low-energy approximation.

Let us consider the classical limiting case of vanishing electric mass. From our previous discussion of the non-supersymmetric $SU(N)$ monopoles described in Chap. 8, we know that the physical situation strongly depends on the character of the symmetry breaking. If the vector $\vec{h}_2'$ is not orthogonal to any of the simple co-roots $\vec{\beta}_i^*$, the vacuum expectation value of the scalar field $\tilde{\phi}_2$ breaks the $SU(N)$ symmetry down to the residual group $U(1)^{N-1}$

and each of the integers n_i appearing in (12.47) has the meaning of a topological charge. In this classical limit, there are no electric charges of the BPS states and the field $\tilde\phi_1$ does not have a long-range Coulomb tail.

In the particular case of the $N = 2$ SUSY Yang-Mills theory with the gauge group $SU(3)$, this situation corresponds to the (n_1, n_2) 1/2-BPS monopole discussed in [232, 353]. In the more general case of the $SU(N)$ gauge group, there are $r = N - 1$ types of monopoles and the magnetic mass of the corresponding configuration is of the form

$$ M = \left| \frac{4\pi}{e} \sum_{i=1}^{r} n_i \left(\vec\beta_i^* \cdot \vec h_2' \right) \right| . $$

Due to triangle inequality, this mass obeys

$$ M \leq \sum_{i=1}^{r} n_i M_i , $$

where $M_i = 4\pi(\vec\beta_i^* \cdot \vec h_2')/e$ and, as before, we suppose that the set of simple co-roots satisfies the condition $(\vec\beta_i^* \cdot \vec h_i') \geq 0$ for all i. This corresponds to the expression (8.29) in Chap. 8.

Again, we can interpret M_i as a mass of a single fundamental monopole with a minimal magnetic charge. Indeed, let us recall that the magnetic charge satisfies the condition of the topological quantization (8.24): $\exp\{2i\pi e\vec g \cdot \vec H\} = 1$ and the charge matrix is

$$ \vec g \cdot \vec H = \sum_{i=1}^{r} n_i (\vec\beta_i^* \cdot \vec H) = \operatorname{diag}(k_1, k_2, \ldots k_{N-1}) , $$

with non-negative integers k_r, which, in the case of the maximal symmetry breaking, are related to the corresponding topological charges. Hence, the configuration of the mass M is stable with respect to decay into $N - 1$ species of the fundamental monopoles, each of the mass M_i which is associated with the simple co-root $\vec\beta_i^*$. Even in the special case of so-called *marginal stability*, when $M = \sum_{i=1}^{r} n_i M_i$, there is no phase space for a physical decay. We are already familiar with a similar conclusion in the particular case of the composite $SU(3)$ monopole, which is also valid for non-BPS monopoles [467].

So far we have discussing the solutions of the primary BPS equation. However, there are solutions of the secondary BPS equation (12.22), which we can find for each solution of the primary BPS equation.

Let us recall now that the secondary BPS equation is actually the equations for the gauge-orthogonal zero modes, which corresponds to the large gauge transformations of the fields of the BPS monopole. The latter solution of the primary BPS equation corresponds to an electrically neutral configuration. If the gauge symmetry is broken maximally, there are $N - 1$ such

12.6 $N=2$ Supersymmetric Monopoles beyond $SU(2)$

gauge zero modes. Solving the secondary BPS equation (12.22), we can recover the electric charges of the monopoles from the asymptotic of the Higgs fields (12.51), in other words, we are "dressing the monopole electrically".

However, the vacuum expectation value of the "electric" component of the Higgs field $\tilde{\phi}_1$ is no longer obliged to be proportional to the "magnetic" component $\tilde{\phi}_2$, as happens in the case of the $SU(2)$ supersymmetric Yang–Mills theory. In other words, the electric charge vector $\vec{q}$ of an $SU(N)$ dyon is no longer aligned with the magnetic charge vector $\vec{g}$.

Typically, in theories with extended $N=4$ supersymmetry, the dyonic BPS states, which are solutions both of the primary and the secondary BPS equations, break $3/4$ of the supersymmetry, while in the model with $N=2$ supersymmetry, they still preserve half of the supersymmetry [94, 96, 97, 233, 353, 441, 501]. Nevertheless, somewhat inconsistently, they are referred to as the *1/4-BPS states*. The interpretation of the solutions of $SU(N)$ BPS equations as a composite system of $N-1$ fundamental monopoles, suggests that such a 1/4-BPS configuration can be thought of as a static system of a few 1/2-BPS monopoles. These solutions correspond to the composite root vectors of the Cartan–Weyl basis.

Indeed, a fundamental magnetic monopole, which corresponds to a simple co-root $\vec{\beta}_i^*$, is a solution of the primary BPS equation. Each such a monopole could have only its own type of electric charge, which corresponds to the root $\vec{\beta}_i$. The self-duality of the basis means that for a fundamental $SU(N)$ monopole, the electric and the magnetic charge vectors are aligned.

The situation is different in the case of the composite monopoles, which correspond to the composite roots: they consist of two or more fundamental monopoles on top of each other. Recall that this configuration is static, because the electric (Coulomb) part of the interaction between the monopoles is precisely compensated for by the long-range scalar force. Then the electric charges of the different monopoles are functions of their relative orientation. Hence the low-energy dynamics of these BPS states becomes more complicated than in the case of the simple non-supersymmetric $SU(2)$ gauge theory, because the low-energy Lagrangian of the composite monopoles picks up an additional term, which is associated with different orientations of two scalar fields. We shall consider this situation below.

To recover the low-energy effective Lagrangian of the supersymmetric monopoles we can make use of the same approach as in Sect. 8.3 above. It was argued [96, 353, 501] that the corresponding potential is simply half of the electric mass of the configuration, that is,

$$V_{eff} = \frac{1}{2}\left(\vec{q}\cdot\vec{h}_1'\right).$$

Recall that in the weak coupling regime this is a small correction to the magnetic mass. Taking into account the definition of the electric charge vector (12.45) and the equation of motion of the field $\tilde{\phi}_1^a$ (the secondary BPS equation (12.22)), we can write the electric mass as [501]

$$\left(\vec{q}\cdot\vec{h}_1'\right) = \int d^3x\, \partial_n\left(E_n^a \tilde{\phi}_1^a\right) = \int d^3x\, \partial_n\left(\tilde{\phi}_1^a \nabla_n \tilde{\phi}_1^a\right)$$
$$= \operatorname{tr}\int d^3x\, \left\{(\nabla_n \tilde{\phi}_1)^2 - e^2[\tilde{\phi}_1, \tilde{\phi}_2]\right\}. \tag{12.52}$$

However, $(\nabla_n \tilde{\phi}_1)$ is a large gauge transformation of the monopole field with the gauge parameter $\tilde{\phi}_1$. These transformations correspond to the set of gauge zero modes of the configuration:

$$\delta A_n = \nabla_n \tilde{\phi}_1 = \sum_\alpha \left(\vec{h}_1' \cdot \vec{K}^\alpha\right) \delta_\alpha A_n \equiv G^\alpha \delta_\alpha A_n,$$
$$\delta \tilde{\phi}_2 = ie[\tilde{\phi}_1, \tilde{\phi}_2] = \sum_\alpha \left(\vec{h}_1' \cdot \vec{K}^\alpha\right) \delta_\alpha \tilde{\phi}_2 \equiv G^\alpha \delta_\alpha \tilde{\phi}_2, \tag{12.53}$$

where $\vec{K}^\alpha$ are the components of the Killing vector field $G = (\vec{h}_1' \cdot \vec{K}^\alpha)$ on the moduli space $\mathcal{M}$, which are generated by the $U(1)^r$ gauge transformations. Thus, the electric mass of the 1/4-BPS state can be written as

$$\left(\vec{q}\cdot\vec{h}_1'\right) = g_{\alpha\beta}\left(\vec{q}\cdot\vec{K}^\alpha\right)\left(\vec{q}\cdot\vec{K}^\beta\right), \tag{12.54}$$

where $g_{\alpha\beta}$ is the hyper-Kähler metric on $\mathcal{M}$. If the form of this metric is known, as in the particular case of the $SU(3)$ 1/4-BPS monopole, the electric mass can be calculated directly from the metric. An alternative approach is to write the metric in terms of the Nahm data [287].

In the next section, we review the modification of the low-energy effective Lagrangian, which describes the motion on the moduli space of the $SU(3)$ supersymmetric monopoles.

12.6.1 $SU(3)$ $N = 2$ Supersymmetric Monopoles

We can proceed further by analogy with our previous discussion of Chap. 8, where we concentrated on the particular case of the $SU(3)$ gauge theory. Then the corresponding Cartan subalgebra is given by two generators (8.14)

$$H_1 = \frac{1}{2}\begin{pmatrix} 1 & 0 & 0 \\ 0 & -1 & 0 \\ 0 & 0 & 0 \end{pmatrix}, \quad H_2 = \frac{1}{2\sqrt{3}}\begin{pmatrix} 1 & 0 & 0 \\ 0 & 1 & 0 \\ 0 & 0 & -2 \end{pmatrix}, \tag{12.55}$$

and the self-dual basis of the simple roots can be chosen in the form (8.15) as before:

$$\vec{\beta}_1 = (1, 0), \quad \vec{\beta}_2 = (-1/2, \sqrt{3}/2). \tag{12.56}$$

In addition, there is the third, composite root, which is defined as $\vec{\beta}_3 = \vec{\beta}_1 + \vec{\beta}_2$, as shown in Fig. 12.1. Thus, $\vec{\beta}_i^* = \vec{\beta}_i$.

Let us consider the case of the maximal symmetry breaking: $SU(3) \to H = U(1) \times U(1)$. Then there are are two fundamental monopoles with

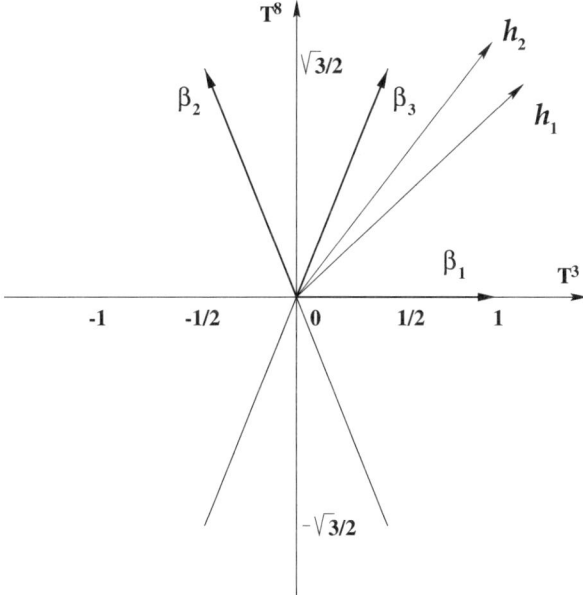

Fig. 12.1. Positive simple roots of the $SU(3)$ supersymmetric theory. The vectors $\vec{h}_i$ define the orientation of two scalar fields

vector magnetic charges $\vec{g} = (1,0)$ and $\vec{g} = (0,1)$, which are aligned along fundamental roots $\vec{\beta}_1$ and $\vec{\beta}_2$, respectively. The corresponding charge matrices $\vec{g} \cdot \vec{H}$ are

$$\frac{1}{2}\,\text{diag}\,(1,-1,0), \qquad \frac{1}{2}\,\text{diag}\,(0,1,-1),$$

while the charge matrix of the composite $(1,1)$ monopole is $\frac{1}{2}\,\text{diag}\,(1,0,-1)$. The latter state is the 1/4-BPS configuration that we would like to consider.

Let us take the electric numbers of these states to be $m_i = (q_1/e, q_2/e)$, where q_i are arbitrary numbers. The magnetic charge of the fundamental monopole is $g = 4\pi/e$ and, according to (12.48), these states have the masses

$$M_i = \sqrt{2}|(q_i + ig)(\vec{\beta}_i \cdot \vec{h})| = \sqrt{2}\,|\Phi_i|\,\sqrt{q_i^2 + g^2}\,, \qquad (12.57)$$

which evidently corresponds to (8.63) of Chap. 8. Recall that in the weak coupling limit, the masses (12.57) are slightly different from the mass of a fundamental electrically neutral BPS-state, which is now given by $\sqrt{2}g\,|\Phi_i|$.

The complex scalar fields Φ_1 and Φ_2 defined as in (12.49) are now disaligned, that is,

$$\Phi_i = e\sqrt{2}\,\left|(\vec{\beta}_i \cdot \vec{h})\right| e^{i\omega_i}\,,$$

where

$$|\Phi_i| = e\sqrt{2}\left|(\vec{\beta}_i \cdot \vec{h})\right| = e\sqrt{(\vec{\beta}_i \cdot \vec{h}_1)^2 + (\vec{\beta}_i \cdot \vec{h}_2)^2},$$

and the argument of the complex Higgs field Φ_i is

$$\tan \omega_i = \frac{(\vec{\beta}_i \cdot \vec{h}_2)}{(\vec{\beta}_i \cdot \vec{h}_1)}.$$

In addition, there is a dilatonic charge of the scalar field, which is defined according to (8.65) as

$$Q_D^{(i)} = (\vec{\beta}_i \cdot \vec{H})\sqrt{q_i^2 + g^2}. \qquad (12.58)$$

Hence, the scalar part of the long-range interaction between the monopoles depends on the relative orientation of the Higgs fields: it vanishes if the fields Φ_1 and Φ_2 are anti-parallel, while its magnitude becomes maximal if they are aligned. Thus, by analogy with (8.66), we can now write the total Coulomb potential of the composite $(1,1)$ monopole, which consists of two static components [96, 441]:

$$V_{eff} = -\frac{1}{r}\left(q_1 q_2 + g^2 - \cos(\omega_1 - \omega_2)\sqrt{(q_1^2 + g^2)(q_2^2 + g^2)}\right). \qquad (12.59)$$

Here we also take into account electrostatic and magnetostatic contributions.

Note that the mass of the composite $\vec{g} = (1,1)$ 1/4-BPS state

$$M_{(1,1)} = \sqrt{2}|(q_1 + ig)(\vec{\beta}_1 \cdot \vec{h}) + (q_2 + ig)(\vec{\beta}_2 \cdot \vec{h})| \qquad (12.60)$$

becomes additive, i.e., $M_{(1,1)} = M_1 + M_2$, only if [441]

$$\tan(\omega_1 - \omega_2) = \frac{g(q_1 - q_2)}{q_1 q_2 + g^2}. \qquad (12.61)$$

In this case, the dilatonic part of the Coulomb interaction in (12.59) precisely balances the long-range electromagnetic interaction and the potential vanishes. In other words, two constituent monopoles are static and the 1/4-BPS configuration can be regarded as a superposition of two individual 1/2-BPS states.

If the difference $\Delta\omega = \omega_1 - \omega_2$ of the arguments of the complex fields Φ_i is small, we can easily see that the expansion of the Coulomb potential (12.59) in q_i/g yields

$$V_{eff} \approx \frac{1}{2r}\left[Q^2 - g^2(\omega_1 - \omega_2)^2\right],$$

where a relative electric charge of the BPS states is $Q = q_1 - q_2$. Clearly, this potential is repulsive if $|g\Delta\omega/Q| < 1$, that is, in this case the supersymmetric 1/4-BPS monopole does not exist. The net interaction is vanishing if

12.6 $N = 2$ Supersymmetric Monopoles beyond $SU(2)$

$$\Delta\omega = \omega_1 - \omega_2 = \frac{Q}{g}.$$

This is a condition of stability of the 1/4-BPS monopole, which evidently agrees with (12.61).

Let us briefly describe the low-energy dynamics of the 1/4-BPS monopoles. We have seen that, for an arbitrary orientation of the component of the Higgs field, there is a non-vanishing potential of the interaction V_{eff}. This potential may be attractive and a 1/4-BPS monopole exists as a bounded system of two fundamental monopoles. Thus, the low-energy approximation can be applied only if the potential energy is small compared to the rest mass of the fundamental monopole.

We now consider the low-energy dynamics of the $(1,1)$ $SU(3)$ monopole [232, 501]. Recall that the electrically neutral classical solution of the primary Bogomol'nyi equation has a mass $M_{(1,1)} = M_1 + M_2$ (cf. our discussion in Chap. 8) and the corresponding eight-dimensional moduli space can be decomposed as [231, 345]

$$\mathcal{M} = \mathbb{R}^3 \times \frac{\mathbb{R}^1 \times \mathcal{M}_0}{\mathbb{Z}}.$$

The factor-space $\mathbb{R}^3 \times \mathbb{R}^1$ is parameterized by the collective coordinates of the center of mass and the global gauge transformation parameter $\tilde{\Upsilon}$, which is related to the internal $U(1)$ angles of the two fundamental monopoles α_i as:

$$\tilde{\Upsilon} = \alpha_1 + \alpha_2.$$

The conjugated momentum along the internal coordinate $\tilde{\Upsilon}$ corresponds to the overall charge (8.73):

$$\tilde{Q} = \frac{M_1 q_1 + M_2 q_2}{M_1 + M_2}$$

(in units of $e(\vec{\beta}_1 + \vec{\beta}_2)$). The metric of the subspace $R^3 \times R^1$ is flat.

The $\mathcal{M}_0$ is the relative moduli space with the positive Taub-NUT metric (8.70)

$$ds^2 = g_{\alpha\beta} dX^\alpha dX^\beta$$
$$= \left(M_r + \frac{g^2}{r}\right) dr^2 + \frac{g^4}{M_r + \frac{g^2}{r}} (d\Upsilon + (\mathbf{a} \cdot d\mathbf{r}))^2, \qquad (12.62)$$

where $M_r = M_1 M_2/(M_1 + M_2)$ is the reduced mass of the monopole and $\mathbf{a}$ is the vector potential of a static dyon (cf. our discussion on page 201). The collective coordinates $X^\alpha = (\mathbf{r}, \Upsilon)$ correspond to the relative position ($\mathbf{r} = \mathbf{r}_1 - \mathbf{r}_2$) and the relative internal phase (Υ) of two fundamental monopoles:

$$\Upsilon = 2\frac{M_2 \alpha_1 - M_1 \alpha_2}{M_1 + M_2}.$$

The momentum along the internal coordinate Υ corresponds to the relative charge $Q = (q_1 - q_2)$ in units of $e(\vec{\beta}_1 - \vec{\beta}_2)$. Recall also that factor $\mathbb{Z}$ denotes a discrete identification for the charge coordinates

$$(\Upsilon, \tilde{\Upsilon}) = \left(\Upsilon + \frac{4\pi M_2}{M_1 + M_2}, \; \tilde{\Upsilon} + 2\pi\right).$$

If the ratio M_1/M_2 is rational, the asymptotic geometry has a compact factor $S^1 \times S^1$ that corresponds to the two completely separated fundamental monopoles with the conserved electric charges q_1 and q_2.

Thus, there are two isometries of the eight-dimensional moduli space $\mathcal{M}_{(1,1)}$ that are generated by the Killing vector fields, $\partial/\partial \Upsilon$ and $\partial/\partial \tilde{\Upsilon}$, respectively. Then a vector field on the moduli space $\mathcal{M}_{(1,1)}$ can be decomposed into two orthogonal components

$$\left(\vec{h}'_1 \cdot \vec{K}\right) = \left(h_+ \frac{\partial}{\partial \tilde{\Upsilon}} + h_- \frac{\partial}{\partial \Upsilon}\right), \quad (12.63)$$

where $h_\pm = \left(\vec{h}'_1 \cdot (\vec{\beta}_1 \pm \vec{\beta}_2)\right)$.

Physically, this decomposition means that the potential associated with the Killing vector $\partial/\partial \tilde{\Upsilon}$ is a constant. It corresponds to the total electric energy of the configuration. The second Killing vector $\partial/\partial \Upsilon$ gives rise to the position dependent potential, which is responsible for the appearance of the dyonic bound states [96, 232, 501].

To sum up, the long-distance tail of the Higgs field $\tilde{\phi}_1$, which is associated with the electric charge vector $\vec{q}$, gives rise to the potential of interaction V_{eff}. Then the low-energy effective Lagrangian which describe the relative bosonic collective coordinates X^α of the 1/4-BPS state, besides the usual kinetic term includes the potential piece: [96, 501]

$$L = \frac{1}{2} g_{\alpha\beta} \left(\dot{X}^\alpha \dot{X}^\beta - G^\alpha G^\beta\right), \quad (12.64)$$

where $g_{\alpha\beta}$ is the Taub-NUT metric (12.62) and $G^\alpha = (\vec{h}'_1 \cdot \vec{K}^\alpha)$ is the vector field generated by the Killing vector $\partial/\partial \Upsilon$. Thus, the interaction between two fundamental monopoles is given in terms of the low-energy supersymmertic quantum mechanics on the Taub–NUT space twisted by the vector field G [232, 233].

Note that the non-relativistic Lagrangian (12.64) has a "secondary" Bogomol'nyi bound for the 1/4-BPS state, which is saturated when $\dot{X}^\alpha = G^\alpha$. The total energy of the corresponding states includes the magnetic mass $(\vec{g} \cdot \vec{h}'_2)$, and we can write

$$M = g_{\alpha\beta} G^\alpha G^\beta + (\vec{g} \cdot \vec{h}'_2). \quad (12.65)$$

Indeed, the electric field of the monopole is $E_n = \dot{X}^\alpha \delta_\alpha A_n$, and for the configurations that saturate the "secondary" BPS bound, we have $E_n = G^\alpha \delta_\alpha A_n = \nabla_n \tilde{\phi}_1$. Thus, the first term in (12.65) is simply the electric mass.

So far, we have been concerned about maximal symmetry breaking. The case of minimal $SU(3)$ symmetry breaking was considered recently in the paper [287]. Recall that then the Higgs field $\vec{h}'_2$ is orthogonal to one of the simple roots, say $\vec{\beta}_1$, and its asymptotic value breaks the symmetry to $U(2)$. The corresponding spectrum of states includes the massive fundamental $\vec{\beta}_2$-monopole with a topological charge n and the massless $\vec{\beta}_1$-monopole, which is labeled by the holomorphic charge $[k]$.

The configuration of the $(2, [1])$ supersymmetric monopole was considered in [287]. This is an example of two massive monopoles and one massless monopole, which can be thought of as a non-Abelian cloud surrounding the two massive monopoles. We briefly discussed this system at the end of Chap. 8. The difference from the non-supersymmetric case is that there is now a long-range tail of the second scalar field $\vec{h}'_1$, which breaks the symmetry further to the minimal subgroup $U(1) \times U(1)$. It was shown that the potential of the $(2, [1])$ supersymmetric monopole is attractive and the massless monopole is confined to one of these two massive monopoles.

Let us stop our discussion at this point. Recent developments in the understanding of the low-energy dynamics of the supersymmetric monopoles, which basically used the same simple picture of geodesic motion on the underlying moduli space suggested by N. Manton in 1982 [368], have greatly improved our understanding of the structure of the vacuum of supersymmetric theories. The restricted volume of our review does not allow us to go into detail of many remarkable works. In particular, the general description of the low-energy dynamics of the supersymmetric monopoles was given recently in [233], where the complete effective Lagrangian of bosonic and fermionic collective coordinates was derived. We also do not discuss here the powerful Nahm formalism, which allows us to obtain many results in a very simple and elegant way [287, 353]. In this rapidly developing situation, we direct the reader to the original works [95–97, 118, 119, 232, 233, 287, 353, 441, 501].

Finally, let us note that the effective Lagrangian of the collective coordinates can be treated as a first step towards a complete quantum theory. The moduli space approximation is actually a truncation of the whole functional space of the quantum monopole to the finite-dimension subspace of zero modes. However, a supersymmetric theory has a great advantage over all other models: in the former, due to the fine balance between the bosonic and fermionic degrees of freedom, the quantum corrections are controlled. Thus, one could expect that there is a possibility of seeing what happens with a quantum supersymmetric theory in a low-energy limit where the non-perturbative effects play a special role.

13 Seiberg–Witten Solution of $N = 2$ SUSY Yang–Mills Theory

In the previous two chapters we briefly discussed the so-called *microscopic* form of the $N = 2$ supersymmetric Yang–Mills theory which is formulated in terms of the component fields of the $N = 2$ chiral superfield Ψ. This theory is renormalizable and it is asymptotically free. However, it is known that at sufficiently low energy not all of the fields appear in the spectrum as physical asymptotic states. At this scale, one can try to construct a *macroscopic* effective theory dealing with physical degrees of freedom that are (almost) massless. Such a theory may have properties which are completely different from those of the microscopic theory. A famous example is the chiral effective Lagrangian of the low-energy QCD, whose spectrum includes baryons and pions rather than quarks and gluons, which are degrees of freedom of the perturbative QCD.

Technically, the procedure is to perform the integration over the massive states of the microscopic theory in the corresponding functional integral. In principle, this allows us to construct a low-energy effective action. However, the non-perturbative character of the fluctuations at the low-energy scale makes this approach rather complicated, even in the supersymmetric theories, where all quantum corrections are under control. In the real world with no supersymmetry, the situation becomes much more complicated. We mentioned already in Chap. 9 that the problem of strong confinement in the conventional QCD remains unsolved.

In 1994, Seiberg and Witten suggested a new way to avoid the technically complicated, direct procedure of the calculation of the low-energy effective action of the supersymmetric theories [469]. This seminal work deserves to be considered as one of the milestones of theoretical physics of the XX-th century. The point is that the following breakthrough is not restricted to the problem of the determination of the exact form of the four-dimensional non-perturbative low-energy effective action of the $N = 2$ supersymmetrical Yang–Mills theory, but it initiated quite a radical change of the paradigm of theoretical physics.

For longer than a century, it was believed that all varieties of physical theories can be in some way derived from the yet unknown, but absolutely perfect, Theory of Everything, the Holy Grail of generations of physicists. This mainstream was supported by the consequent successes of the unification approach, which spans from Maxwell electrodynamics to the united $SU(3) \times SU(2) \times U(1)$ theory of electroweak and strong interactions. The matter of the "revolution of 1994" is that such an ideology has drastically changed: nowadays it is common to believe that there is no unique fundamental (string) theory operating with some fixed set of fields, but that there are different formulations of it that are related by some transformations of duality.

We are not in the position to discuss here neither the aspects of the dualities of strings and branes nor even the details of the Seiberg–Witten solution of the $N = 2$ supersymmetric Yang–Mills theory. We recommend many introductions to this subject, see, e.g., [73, 123, 199, 311, 347, 378, 401, 424, 473] and the references therein. Our interest is restricted to the role of the magnetic monopoles and the idea of duality in the Seiberg-Witten theory.

In the remaining part of our review, we shall discuss these topics, considering the pure $N = 2$ supersymmetric Yang–Mills theory with the gauge group $SU(2)$.

13.1 Moduli Space

13.1.1 Moduli Space and its Parameterization

*"Every theory has a hole.
One has only to look for it carefully."*

Mark Twain[1]

We are looking for an effective action that describes the behavior of the model at energies lower than a scale of some infrared cutoff. This parameter has to be much smaller than the masses of the lightest particles in the spectrum.

Note that there are two different definitions of the effective action.

(i) The conventional generating functional of the one-particle irreducible diagrams $\Gamma[\Psi]$ that can be obtained by a straightforward summation of the contributions of the loop diagrams of all orders, where the integration over the loop momenta is performed over all ranges of values, from 0 to ∞. Clearly, the effective action $\Gamma[\Psi]$ depends on the normalization point μ. In the models with spontaneous symmetry breaking, the normalization point is usually taken to be identical with the vacuum expectation value of the Higgs field.

(ii) The Wilsonian effective action $S_W[\Psi]$ is defined in a similar way, however, the integration domain in this case is restricted from below by the scale of μ, which is the infrared cut-off parameter.

---

[1] Attribution according to I.B. Khriplovich, who used this nice quotation in his book *Parity Nonconservation in Atomic Phenomena* (Gordon and Breach, 1991). I am thankful for his kind permission to reproduce the quotation here, in a different context.

If the spectrum of the model contains no massless states and the normalization point is taken to be much smaller than the mass of the lightest particle, there is no difference between $\Gamma[\Psi,\mu]$ and $S_W[\Psi]$. However, if massless states are present, only the Wilsonian effective low-energy action remains self-consistently defined, and this is the case that we are discussing here. Note that the difference between these two definitions of the effective action exists also in the supersymmetric gauge models, because of the *Konishi anomaly*, which implies that the beta-function depends on the definition we are using. The most important difference between the two types of the effective action is that, unlike $\Gamma[\Psi,\mu]$, the Wilsonian effective action has a coupling which holomorphically depends on the normalization point μ. In the context of the discussion below, we shall use well-defined Wilsonian effective action.

The object of our study is given by the microscopic $N=2$ supersymmetric Lagrangian (12.5):

$$L^{N=2} = -\frac{1}{4}F_a^{\mu\nu}F_{\mu\nu}^a + \frac{\theta e^2}{32\pi^2}F_a^{\mu\nu}\tilde{F}_{\mu\nu}^a + \frac{1}{2}(\nabla_\mu \phi_1^a)^2 \quad (13.1)$$
$$+ \frac{1}{2}(\nabla_\mu \phi_2^a)^2 - \frac{e^2}{2}[\phi_1^a,\phi_2^b]^2 - i\bar\chi\gamma^\mu\nabla_\mu\chi - e\phi_1\bar\chi\chi - ie\gamma_5\phi_2\bar\chi\chi.$$

We have mentioned already that there is a continuous set of classical vacua that corresponds to the nullification of the potential of the complex Higgs field: tr $[\phi,\phi^\dagger]^2 = 0$. In the context of the $SU(2)$ gauge theory, this condition is fulfilled if $\phi_0 = v\sigma^3/2$ and the complex number v can be considered as a coordinate along the flat direction in the functional space that parameterizes physically inequivalent vacua.

This situation is remarkably similar in form to our consideration of the moduli space of the monopoles. The latter manifold is parameterized by the collective coordinates and the low-energy approximation allows us to truncate the infinite dimensional functional space to the moduli space $\mathcal{M}$. By analogy, the space of all physically inequivalent vacua of the $N=2$ SUSY Yang–Mills theory also corresponds to the lowest energy bound and an effective theory can be determined by consideration of the soft massless excitations, which, at least in the low-energy approximation, can be separated from the massive modes. This similarity justifies the reason why the former space is also referred to as the *moduli space*. We hope that the reader can distinguish between both notions from the context of our discussion. Generally, the notion of the *moduli* is defined as a set of parameters labeling the geometry of a manifold.

The analogy with the monopole moduli space can be broadened, because in the $N=2$ supersymmetric Yang–Mills theory, the quantum moduli space turns out also to be hyper-Kähler, i.e., it reveals a quaternionic structure in the tangent space and the corresponding metric is then of the form of (11.73). This opens the possibility of applying the very powerful methods of differential and algebraic geometry to solve the problem of the construction of the low-energy effective action of the $N = 2$ SUSY Yang–Mills theory. In the following we shall discuss this remarkable approach.

Let us note that the classical moduli space of the $N = 2$ super Yang–Mills theory $\mathcal{M}$, which is defined as the space of vacua parameterized by the complex coordinate v along a flat direction, the moduli, includes a special point $v = 0$. To understand the physical meaning of this singularity, let us recall that a non-zero vacuum expectation value of the scalar field breaks the $SU(2)$ gauge symmetry to the Abelian subgroup. The structure of the Lagrangian (13.1) means that then two of the vector bosons, $A_\pm = \frac{1}{\sqrt{2}}(A^1 \pm iA^2)$, become massive, with a mass $m_v = \sqrt{2}v$. Furthermore, the supersymmetry of the model means that the same mass will obtain the components of the spinor field χ^1 and χ^2, while the gauge boson A^3, a photon, and the spinor χ^3 remain massless. Clearly, the mode of the fluctuations of the scalar field in the direction σ^3 also remains massless.

Thus, all these fields can be considered as dynamical variables of the Wilsonian low-energy effective action, which is invariant with respect to the unbroken $U(1)$ symmetry and still possesses the original $N = 2$ supersymmetry.

The situation is completely different if $v = 0$, because then the original gauge symmetry is unbroken and all the fields remain massless. Since the vacuum expectation value of the complex Higgs field parameterizes the moduli space, this is a singular point on $\mathcal{M}$. In general, a singularity on the moduli space corresponds to additional massless excitations that were not taken into account originally[2].

Generally, the dimension of the moduli space is given by the rank r of the gauge group G. The vacuum expectation value of the scalar field breaks the symmetry to the subgroup H, which is generated by the elements of the Cartan subalgebra. The elements in the coset G/H do not leave the vacuum invariant, but act as gauge transformations that relate the physically equivalent vacua. However, the coset G/H contains some elements that do not take ϕ out of the Cartan subalgebra, since for a given basis of the Cartan roots, there are the Weyl reflections whose action does not change the set of roots $\vec{\beta}_i$ (see the discussion of the Weyl transformations (8.12) in Chap. 8). Therefore, the correct parameterization of the moduli space must be given in terms of the Weyl invariant functions.

Let us consider now the simplest case of the $SU(2)$ gauge symmetry, which breaks down to the residual subgroup $U(1)$ by the non-zero value of the complex constant v in (12.13). Evidently, the $N = 2$ supersymmetry remains unbroken. Besides these symmetries of the vacuum, there is the above-mentioned Weyl reflections that induce an inversion of the isovector ϕ^a in the group space: $R: \phi^a \to -\phi^a$. It is clear that for the configuration (12.13) these transformations are the $SU(2)$ rotations around the second axis by π

---

[2] Also the dimension of the moduli space changes there. Thus, the moduli space is not the usual differentiable manifold, but an *orbifold*.

13.1 Moduli Space

Fig. 13.1. "Yin–Yang" structure of the classical moduli space $\mathcal{M}_{cl}$ of the $N = 2$ supersymmetric Yang–Mills model

produced by the generator $i\sigma^2$: $R\,\sigma^3 R^\dagger = -\sigma^3$. Hence the vacuum values v and $-v$ are gauge equivalent and we have to identify these points on the moduli space. Thus, the Weyl invariant of the $SU(2)$ group is

$$u = \frac{1}{2}\,\mathrm{tr}\,\phi^2 = \frac{1}{2}\,v^2\,, \tag{13.2}$$

and this is the proper quantity needed for a gauge invariant parameterization of the moduli space. Nevertheless, in some cases, it is convenient to make use of the variable v, which labels the different vacua.

Let us now summarize the information about the structure of the classical moduli space $\mathcal{M}_{cl}$ of the $N = 2$ supersymmetric Yang–Mills model with the gauge group $SU(2)$. It is parameterized by a complex coordinate v, which labels gauge inequivalent vacua. The symmetry with respect to the Weyl $\mathbb{Z}_2$ reflections means that the moduli space is identical to the upper half $\mathbb{H}^+$ of a complex plane punctured at the origin, where $v = 0$. The weak coupling regime corresponds to the large values of the variable v.

Thus, the classical moduli space $\mathcal{M}_{cl}$ reveals a structure that can be nicely illustrated by the famous "Yin–Yang" symbol of duality (see Fig. 13.1). However, the effect of the quantum fluctuations can drastically change the situation and the *quantum moduli space* $\mathcal{M}_q$ can be quite different from its classical counterpart $\mathcal{M}_{cl}$. To see how this happens, let us recall that in the classical electrodynamics, for example, there is no stable ground state of the system of two particles of opposite charge, that is, $\mathcal{M}_{cl} = 0$. However, the quantum mechanical description of the same system yields a unique ground state ($\mathcal{M}_q = 1$). There are other examples of the reshaping of the vacuum.

- A particle in a double-well potential $V(x) = \lambda(x^2 - v^2)^2$ in 1+1-dimensions. The classical ground state is two-fold degenerated ($\mathcal{M}_{cl} = \mathbb{Z}_2$). However, the quantum mechanical effect of tunnelling "stirs" up the ground states and lifts the initial degeneration: $\mathcal{M}_q = 1$. This nice example allows us to see how the non-perturbative (instanton) effects on a quantum mechanical level rebuild the vacuum.
- The scalar field with periodic potential $V(\phi) = 1 - \cos(\beta\phi)$. The classical vacuum is n-fold degenerated, that is, $V(\phi_0) = 0$ if $\phi_0 = 2\pi n \beta^{-1}$, $n \in \mathbb{Z}$, and therefore, $\mathcal{M}_{cl} = \mathbb{Z}_n$. Quantum mechanical consideration shows that the non-perturbative instanton transitions between the neighboring vacua remove the degeneracy again, and the corresponding wave functions are the linear combinations of non-perturbed wave functions at every minimum. Thus, the band structure appears and there is only one quantum ground state, $\mathcal{M}_q = 1$.
- Coleman–Weinberg effect. Unlike the previous examples, this is the effect that appears in the quantum field theory Clearly, the vacuum state of the classical scalar electrodynamics with a potential $\lambda\phi^4$ is trivial and unique: $\phi_0 = 0$. The effect of the one-loop quantum corrections results in the dynamically generated additional effective potential $\sim \phi^2 \ln(\phi^2/\Lambda^2)$, which reshapes the vacuum: the quantum moduli space is $\mathcal{M}_q = 1$, as well as its classical counterpart, $\mathcal{M}_{cl} = 1$, but in the former case, the vacuum expectation value of the scalar field is not equal to zero and the photon, coupled with ϕ, becomes massive.

These simple examples demonstrate that the properties of the quantum vacuum strongly depend on the values of parameters of the model. Especially interesting is the last example, which shows that there are different domains of the moduli space that correspond to the different phases of the theory. On the other hand, the example of the Coleman–Weinberg effect suggests that a singularity of the classical moduli space may be not a singularity of $\mathcal{M}_q$.

We have already noted that the $N = 2$ supersymmetric Yang–Mills theory (13.1) closely resembles the conventional QCD, whose vacuum has different phases. We have mentioned already that the properties of the QCD vacuum can be tested, if we include a very heavy charged field in the model. The potential of the interaction between these probe charges, which are separated by a large distance R, defines the character of the corresponding phases. In the context of OCD, we are discussing the phase of confinement (low-energy regime) and the perturbative regime of the asymptotically free theory.

Let us recapitulate the possibilities that may arise in the the microscopic model of type (13.1), both on the classical and on the quantum levels.

- *Coulomb branch*: $V(R) \sim \alpha/R$, where α is the coupling constant. In this phase, the gauge field is massless, which implies the vanishing of the vacuum expectation value of the scalar field. Hence, it is a limiting case in which the long-range forces are not screened by the vacuum condensate

of the scalar field. In this phase, the coupling constant α does not change with a distance.
- *Higgs phase*: $V(R) \sim const$. This regime occurs if the symmetry is spontaneously broken and there are massive vector and scalar fields of the same mass that mediate the short-range Yukawa interaction: $V(R) \sim e^{-m_v R}/R \to 0$ at the long distances. A celebrated example of the theory, which admits that regime, is the standard electroweak $SU(2) \times U(1)$ model. Note that both the Coulomb phase and the Higgs phase are presented on the classical level, where the mass scale given by v is sufficiently large. Note that even in this phase, the remaining Abelian symmetry of the model still gives rise to the long-range Abelian interaction mediated by the massless photon.
- *Free phase*: $V(R) \sim \alpha(R)/R \sim (R \ln R)^{-1}$. This is a phase which is entirely related to the quantum effects. For this case, there is another type of screening of the electric charges of the test particles due to renormalization effects. Indeed, there are massless charged fluctuations in the vacuum and the photon propagator becomes dressed by the virtual pairs of these quanta. The behavior of the coupling constant $\alpha(R) \sim 1/\ln R$ depends on the value of the vacuum scalar condensate. If it is not vanishing, $\alpha(R)$ becomes frozen at the scale of m_v^{-1}. However, in the massless case this running is not restricted to some scale and the coupling constant at the large distances becomes asymptotically small. Thus, the electric charges of the test particles in this limit are completely screened. This is known as the *Landau zero-charge effect*.
- *Confinement phase*: $V(R) \sim \sigma R$, where the parameter σ is the string tension. This regime does not exist in Abelian theory. We can suppose that, as was discussed above in Chap. 9, the appearance of the mass gap in this phase is related to the mechanism of the monopole condensation, which may squeeze the field between two test particles into a tube.

Taking into consideration the effect of the virtual monopoles, we can consider one more phase.

- *Free magnetic phase*: $V(R) \sim \ln R/R$. In this case, the effective coupling constant can be written as $\alpha(R) \sim \ln R$. Thus, its behavior is opposite to that of the free electric phase. This running of the coupling corresponds to the effect of antiscreening of the test charges, which we briefly discussed in Chap. 4. If the monopoles, for some reason were to become massless, the effective electric charge of the test particle would be renormalized to infinity: $e^2(R) \sim \ln R$. Otherwise, the effect of the virtual pairs of monopoles becomes frozen at the scale of the monopole inverse mass. In the semiclassical approximation, we have $M^{-1} \ll m_v^{-1}$ and this type of screening does not occur.

Let us make two more remarks. Of course, we could imagine a dual situation when there are two magnetically charged test particles that are placed in the vacuum of the model (13.1). Then the behavior of the magnetic potential is described by the same pattern of phases as in the electric case, up to the obvious interchange of the free electric and magnetic phases. The Higgs and confinement phases are also dual: there are magnetic flux tubes in the Higgs phase, while the electric flux tubes appear in the regime of confinement. However, the relation between the Higgs and confinement phases can be less obvious. In principle, a smooth variation of the parameters of a model could relay both regimes and therefore, strictly speaking, there is no invariant distinction between them.

Note that there are N_f hypermultiplets of the matter fields that can be coupled to the chiral field Φ in (13.1). The $N = 2$ four-dimensional hypermultiplet on-shell field components are a complex scalar $SU(2)_R$ doublet q and a Dirac spinor. The latter component sometimes is referred to as "quark" although we already have spinor particles among field components of the vector multipet V. Generally, the properties of the vacuum of the extended model depend on the relation between the number of flavors N_f and the number of colors N, and the vacuum expectation values of the scalar fields ϕ and q. There are various types of scenarios of transitions between different phases. To understand what happens with the theory, which in the semiclassical limit is governed by the Lagrangian (13.1), we have to analyze how the quantum fluctuations affect the vacuum of the $N = 2$ supersymmetric Yang–Mills theory. We restrict our consideration to the simplest case of the model (13.1) without the matter fields.

13.1.2 Quantum Moduli Space of $N = 2$ SUSY Yang–Mills Theory

Since quantum corrections of any kind cannot destroy the $N = 2$ supersymmetry of the full quantum theory, there is a guarantee that the overall form (12.5) of the general Lagrangian cannot be spoiled by some new counterterms:

$$L^{N=2} = \frac{1}{16\pi} \text{Im} \int d^2\theta d^2\tilde{\theta} \text{ tr } \mathcal{F}(\Psi). \tag{13.3}$$

However, the $N = 2$ holomorphic prepotential $\mathcal{F}(\Psi)$ now receives both the perturbative and the non-perturbative corrections to it classical form $\mathcal{F}_{class}(\Psi) = \frac{1}{2}\tau\Psi^2$. Hence, in order to find the low-energy effective action, we have to define the form of the prepotential, which becomes the primary object of investigation.

Recall that the expansion (12.6) of (13.3) in $N = 1$ components, the gauge superfield $W^{a\alpha}$ and the chiral superfield Φ^a, yields

$$L^{N=2} = \frac{1}{32\pi} \text{Im} \left(\int d^2\theta \mathcal{F}_{ab}(\Phi) W^{a\alpha} W^b_\alpha + 2 \int d^2\theta d^2\bar{\theta} \, (\Phi^\dagger e^{2eV})^a \mathcal{F}_a(\Phi) \right), \tag{13.4}$$

where $\mathcal{F}_a(\Phi) \equiv \partial \mathcal{F}/\partial \Phi^a$, $\mathcal{F}_{ab}(\Phi) \equiv \partial^2 \mathcal{F}/\partial \Phi^a \partial \Phi^b$ and the Kähler metric on the space of fields $d^2 s = g_{ab} \Phi^a \Phi^b$ is given by

$$g_{ab} = \text{Im}\, \partial_a \partial_b \mathcal{F}(\Phi).$$

Let us now turn to the infrared regime. At the low-energy scale, only the massless fields appear as physical states and the gauge symmetry is broken to an Abelian subgroup. The corresponding Wilsonian effective action still remains invariant with respect to the $N = 2$ supersymmetry and, therefore, must have the form of (13.3), where the group indices take only the values $a, b = 3$. In addition, for the term of interaction, we have $e^{2eV} = 1 + 2eV + 2e^2 V^2$, and a simple calculation similar to that of Chap. 11 shows that in the Abelian case, only the trivial term of unity survives. Thus, the effective Largangian that governs the low-energy $U(1)$ theory is

$$L_{eff} = \frac{1}{32\pi} \text{Im} \left(\int d^2\theta \frac{\partial^2 \mathcal{F}}{\partial \Phi^2} W^\alpha W_\alpha + 2 \int d^2\theta d^2\bar{\theta}\, \Phi^\dagger \frac{\partial \mathcal{F}}{\partial \Phi} \right). \quad (13.5)$$

In the component fields, the effective Lagrangian (13.5) is given by (cf. (12.2))

$$L_{eff} = \frac{1}{4\pi}\, \text{Im}\, \tau(\Psi) \left[(\partial_\mu \phi)^\dagger (\partial^\mu \phi) - \frac{1}{4}\left(F^{\mu\nu} F_{\mu\nu} - i\, F_{\mu\nu} \widetilde{F}^{\mu\nu} \right) \right.$$
$$\left. - i(\lambda \sigma^\mu \partial_\mu \bar{\lambda} + i \bar{\psi} \bar{\sigma}^\mu \nabla_\mu \psi) \right]. \quad (13.6)$$

Thus, the theory has been effectively reduced to a four-dimensional non-linear sigma-model with the identification of the scalar fields ϕ^a as the local complex coordinates on a target space with the Kähler metric. On the other hand, the moduli space is parameterized by the complex number $v = \phi_0$, the vacuum expectation value of the Higgs field. Thus, the Lagrangian (13.5) also yields the Kähler metric in the moduli space[3]. Furthermore, the $N = 2$ supersymmetry of the vacuum requires that the coefficients at the kinetic terms of the gauge field, the fermions and the scalars, must be related to the same prepotential. This means that the metric on the moduli space of the $N = 2$ supersymmetric Yang–Mills theory is identical for all kinetic terms of the effective action:

$$d^2 s = \text{Im}\, \frac{\partial^2 \mathcal{F}}{\partial v^2} dv d\bar{v} = \text{Im}\, \tau(v)\, dv d\bar{v}, \quad (13.7)$$

where $\bar{v}$ denotes the complex conjugate of v and $\tau(v) \equiv \mathcal{F}''(v) = \dfrac{\partial^2 \mathcal{F}}{\partial v^2}$ is an effective scale-dependent coupling constant of the low-energy theory, which now replaces its classical limit (11.76). Recall that in the latter case, τ is

---

[3] In the context of the non-linear sigma-model, such a metric is known as the Zamolodchikov metric.

defined as a constant which is supposed to be completely independent of the value of v.

Thus, the quantum moduli space of the $SU(2)$ theory is parameterized by the vacuum expectation value of the corresponding classical Weyl invariant $u = \langle \text{tr}(\phi)^2 \rangle$, which in the classical limit tends to (13.2). To evaluate this quantity in the quantum theory, we have to examine the effects of quantum corrections that will affect the prepotential $\mathcal{F}$. In particular, we can start from the microscopic Lagrangian (13.1) and perform the perturbative calculations.

First, recall that on the classical level, the model has a global $\mathcal{R}$-symmetry $SU(2)_R \times U(1)_R$, where the $U(1)$ subgroup includes the usual chiral symmetry of the massless spinor theory (cf. (11.17) and (12.12)). This symmetry acts as

$$\theta \to e^{i\alpha}\theta, \quad \tilde{\theta} \to e^{i\alpha}\tilde{\theta}, \quad \Psi \to e^{2i\alpha}\Psi,$$

thus, in the classical theory, we have

$$\Phi \to e^{2i\alpha}\Phi, \quad \phi \to e^{2i\alpha}\phi, \quad W \to e^{i\alpha}W,$$

and the action remains invariant if $\mathcal{F} \to e^{4i\alpha}\mathcal{F}$.

However, in the quantum theory, the latter symmetry becomes broken to a discrete subgroup due to chiral anomaly. Namely, in the general case of the model with the gauge group $SU(N)$, a chiral rotation of the fermions (12.12) through the angle α changes the effective Lagrangian by the anomalous term

$$\delta L_{eff} = \frac{\alpha N}{8\pi^2} F_{\mu\nu}\tilde{F}^{\mu\nu}. \tag{13.8}$$

Note that this causes a shift in the value of the θ-parameter and in the classical theory the chiral rotation can be used to eliminate it. However, the invariance of the quantum action with respect to the $U(1)_R$ symmetry is broken by the instanton power-like corrections to the prepotential $\mathcal{F}$, although not completely: the Pontryagin index (5.96)

$$n = \frac{1}{8\pi^2}\text{tr}\int d^4x F_{\mu\nu}\tilde{F}^{\mu\nu}$$

is an integer, thus the remaining unbroken symmetry is $\mathbb{Z}_{4N}$ [468]. This is in agreement with the evaluation of the number of fermionic zero modes on the instanton background: there are $2N$ zero modes for each chiral fermion, both λ and ψ in the adjoint representation. Then the first non-vanishing $2 \cdot 2N$-point fermionic correlation function

$$\langle \lambda(x_1) \ldots \lambda(x_{2N})\psi(y_1) \ldots \psi(y_{2N}) \rangle$$

includes the integration over zero modes and, under the chiral rotations, it picks up the factor $e^{4iN\alpha}$. The latter breaks the $U(1)_R$ to the discrete subgroup $\mathbb{Z}_{4N}$, whose generators are $e^{2i\pi\alpha}$ with parameters $\alpha = n/4N$, $n = 1, \ldots, 4N$.

13.1 Moduli Space

Thus, the $U(1)_R$-symmetry of the quantum $SU(2)$ theory is broken by the non-perturbative instanton corrections to a discrete subgroup $\mathbb{Z}_8$, which is a symmetry of the moduli space. We shall see, however, that there is an additional $\mathbb{Z}_2$ symmetry for a given point of the vacuum.

The property of $U(1)_R$-symmetry can be used to define the prepotential of the quantum theory. Indeed, not only the leading scalar component ϕ, but the entire $N=2$ chiral superfield Ψ transforms as $\Psi \to e^{2i\alpha}\Psi$ under the $U(1)_R$ chiral rotation. Then the chiral rotation of the gauge field kinetic term[4] in the Lagrangian (13.5) yields

$$\frac{1}{16\pi} \operatorname{Im} \mathcal{F}''(e^{2i\alpha}\Psi) \left[-F^{\mu\nu}F_{\mu\nu} + i\, F_{\mu\nu}\widetilde{F}^{\mu\nu} \right]$$
$$= \frac{1}{16\pi} \operatorname{Im} \mathcal{F}''(\Psi) \left[-F^{\mu\nu}F_{\mu\nu} + i\, F_{\mu\nu}\widetilde{F}^{\mu\nu} \right] + \frac{\alpha N}{8\pi^2} F_{\mu\nu}\widetilde{F}^{\mu\nu},$$

where we take into account the explicit form of the shift (13.8) and make use of the expression (13.6). Then the structure of the prepotential of the quantum $N=2$ supersymmetric pure Yang–Mills theory may be determined from the equation

$$\mathcal{F}''(e^{2i\alpha}\Psi) \approx \mathcal{F}''(\Psi + 2i\alpha\Psi) = \mathcal{F}''(\Psi) + \frac{2\alpha N}{\pi}. \tag{13.9}$$

Thus, expansion in α yields

$$\mathcal{F}'''(\Psi) = \frac{iN_c}{\pi \Psi}. \tag{13.10}$$

Solving this equation, we find the prepotential of the $SU(2)$ theory:

$$\mathcal{F}_{per}(\Psi) = \frac{i}{2\pi} \Psi^2 \ln \frac{\Psi^2}{\Lambda^2}. \tag{13.11}$$

Here, Λ is a dynamically generated constant with the dimension of mass, a counterpart of the familiar Λ_{QCD}. This quantity defines the scale where the theory becomes strongly coupled.

As we have already mentioned, the supersymmetry guarantees that no corrections of higher order change this result. This is related to the structure of the perturbative β-function of the $N=2$ supersymmetric Yang–Mills theory, which is restricted to the one-loop correction.

To obtain the explicit form of the perturbative β-function, we need to consider a general structure of the $U(1)_R$-invariant perturbative $N=2$ prepotential that includes both (13.11) and the classical limiting case:

$$\mathcal{F}(\Psi) = \Psi^2 \left(c_1 + c_2 \ln \frac{\Psi^2}{\Lambda^2} \right), \tag{13.12}$$

---
[4] Recall that the $N=2$ supersymmetry provides that all other kinetic terms transform in the same way.

where $c_1 = \frac{1}{2}\tau_{class}$ and $c_2 = i/2\pi$. In this expression, tree-level and one-loop corrections are combined. The complex classical coupling constant τ_{class} is defined as in (11.76):
$$\tau = \frac{\theta}{2\pi} + \frac{4\pi i}{e^2}.$$

The left-hand side of (13.12) corresponds to the scale of v, while the right-hand side of this expression describes the coupling at the scale of renormalization point μ. Indeed, the double differentiation of the prepotential (13.12) yields the renormalization group equation

$$\frac{4\pi}{e^2(v)} = \frac{4\pi}{e^2(\mu)} + \frac{1}{\pi}\ln\frac{v^2}{\mu^2} \equiv \frac{1}{\pi}\ln\frac{v^2}{\Lambda^2}, \qquad (13.13)$$

where
$$\Lambda^2 = \mu^2 \exp\left(-\frac{4\pi^2}{e^2(\mu)}\right).$$

Thus, the running gauge constant at the scale μ is $e(\mu) = -\beta \ln(\mu^2/\Lambda^2)$, where
$$\beta = \mu\frac{de}{d\mu} = -\frac{e^3(\mu)}{4\pi^2}. \qquad (13.14)$$

This is quite expectable expression for the perturbative β-function of the asymptotically free $SU(2)$ theory with two Weyl fermions and a complex scalar field in the adjoint representation of the gauge group. Indeed, the effective gauge coupling constant of the low-energy theory with Wilsonian effective action corresponds to the scale $\mu \to v$. If this scale is sufficiently large, the gauge coupling becomes very small and in the limit $v \to \infty$, where the theory becomes asymptotically free, the perturbative expansion is well-justified. This is the semiclassical limit where

$$\begin{aligned}\mathcal{F}(v) &\sim \frac{i}{2\pi}v^2 \ln\frac{v^2}{\Lambda^2}, \\ \tau(v) &= \frac{\partial^2 \mathcal{F}}{\partial v^2} \sim \frac{i}{\pi}\left(\ln\frac{v^2}{\Lambda^2} + 3\right),\end{aligned} \qquad (13.15)$$

and $u \sim \frac{1}{2}v^2$. The massless fields of the model are free at the large distances and this is the free electric phase of the theory.

Let us note now that the form (13.12) of the prepotential breaks the remaining $\mathbb{Z}_8$ symmetry further. Indeed, under transformations of the original $U(1)_R$-symmetry, the prepotential (13.11) transforms as

$$\mathcal{F}_{per}(\Psi) \to e^{4i\alpha}\left[\frac{i}{2\pi}\Psi^2 \ln\frac{\Psi^2}{\Lambda^2} + \left(\frac{1}{2} - \frac{2\alpha}{\pi}\right)\Psi^2\right].$$

Hence, the corresponding shift of the action is irrelevant if $\dfrac{4\alpha}{\pi} = n$, where n is an integer. Then $\alpha = \frac{1}{4}\pi n$ and the original $U(1)_R$-symmetry breaks to $\mathbb{Z}_8$:

$\Psi \to e^{i\pi n/2}\Psi$, thus for odd n, we have the inversion $\Psi \to -\Psi$. On the other hand, the vacuum is parameterized by the variable $u = \frac{1}{2}v^2$ (13.2), not by the vacuum expectation value of the scalar field v. Evidently, a non-vanishing value of u breaks $\mathbb{Z}_8$ further to $\mathbb{Z}_4$, which is the symmetry of a given point on the moduli space.

Another observation is that the perturbative quantum corrections affect the global structure of the quantum moduli space $\mathcal{M}_q$. Indeed, the classical moduli space $\mathbb{H}^+$ has only one singularity at $u = 0$, at which the massless gauge bosons appear in the spectrum. As we may expect, the quantum moduli space has another singularity at $u = \infty$, which is related to the branch cut of the logarithm of $u = v^2/2$ in the one-loop corrected prepotential (13.15), and $\tau(v)$ is a multivalued function of u.

The latter singularity means that, if we consider a closed path around this point, the effective coupling $\tau(v)$, which is a multivalued function of u, will change as $\tau \to \tau - 2$ as we cross the cut. This means that there is a non-trivial *monodromy* on the moduli space. We shall discuss this property below. Taking into account the definition of the coupling τ of (11.76) in the semiclassical limit, we can easily see that this shift yields an irrelevant change of the θ angle: $\theta \to \theta - 4\pi$. However, when monopoles are present in the spectrum of the states, this transformation yields a shift in the electric charge of a monopole according to (5.110).

Thus, the existence of non-trivial monodromies of the $N = 2$ supersymmetric Yang–Mills theory suggests that there is no unique form of the low-energy effective action written in terms of some given degree of freedom: the loop around $u = \infty$ affects the physical spectrum of the monopole states. This was an initial observation by Seiberg and Witten [469].

We can proceed further by noting that the moduli space can be compactified by adding the point $u = \infty$. Hence, the quantum moduli space of the model can be thought of as a Riemann sphere S^2 with two singular points at $u = 0$ and $u = \infty$, respectively. However, this conclusion is not correct, because it would be misleading to extend the semiclassical relations (13.15), which are correct only in the region close to $u = \infty$, to the whole range of values of the variable u.

Indeed, the metric on the moduli space

$$\mathrm{Im}\,\tau(u) \sim \frac{1}{\pi} \ln \frac{u}{\Lambda^2},$$

is single-valued and positive in the weak-coupling region, although the effective coupling τ of (13.15) is a multi-valued function. Thus, the physical condition of unitarity is satisfied in the vicinity of $u = \infty$. On the other hand, the restriction imposed by unitarity means that the metric must be positive not only in this part of the moduli space, but everywhere:

$$\mathrm{Im}\,\tau(u) = \mathrm{Im}\,\mathcal{F}'' > 0, \quad \text{for any } u. \tag{13.16}$$

However, the prepotential $\mathcal{F}$ by its definition is a holomorphic function and, as is well-known from complex analysis, the metric $\mathrm{Im}\,\tau(u)$ is, therefore, a harmonic function: $\partial\bar\partial\,\mathrm{Im}\,\tau(u) = 0$. However, a globally defined harmonic function cannot have a minimum. The only possibility is that τ remains constant through the moduli space and that is the classical case.

Thus, we have to conclude that neither the prepotential nor the coordinates v, which parameterise the quantum moduli space, can be defined globally. Then the expressions (13.15) remain well-defined only in the semi-classical patch in the vicinity of $u = \infty$. Some other variables, say v_D, must be introduced to parameterize the region of the complex plane, where $\tau(v)$ becomes negative. This situation to a certain extent resembles the non-singular description of the Abelian monopole that we discussed in Chap. 3.

13.2 Global Parametrization of the Quantum Moduli Space

13.2.1 Transformation of Duality for $N = 2$ Low-Energy Effective Theory

The analysis above leads to the conclusion that the coordinates v and v_D are appropriate only in a certain region of $\mathcal{M}_g$, around the corresponding singularity and, in order to parameterize the entire quantum moduli space of the $N = 2$ supersymmetric theory, we have to answer the primary question: how many and what kind of singularities does the space $\mathcal{M}_g$ possess?

The partial answer to this question is, in fact, already known. As discussed above, the description in terms of the fields W and Φ, which enter the effective Lagrangian (13.5), is well-defined in the weak coupling limit, where the vacuum expectation value of the Higgs field is large. This is the patch around the singularity $u = \infty$. What about the strong coupling limit? Then the parameter v becomes small and we are definitely beyond the domain of definition of the microscopic theory. However, we know that there is a way to establish a connection between the regimes of weak and strong coupling, namely, there are transformations of duality. The guiding idea of the work by Seiberg and Witten [469] is to apply the arguments of duality to define a proper set variables, which may provide an adequate formulation of the theory in the strong coupling limit.

Since the low-energy effective action (13.5) is Abelian, the dual formulation of the theory can be constructed by analogy with the usual electrodynamics[5]. We already considered this transformation at the end of Chap. 1 (cf. (1.89) and the related discussion). Let us now define the *dual chiral superfield* as

---

[5] An attempt to introduce a dual formulation of a non-Abelian theory in the same way would be incorrect.

13.2 Global Parametrization of the Quantum Moduli Space

$$\Phi_D = \mathcal{F}'(\Phi), \tag{13.17}$$

and introduce a *dual prepotential* as a quantity that satisfies

$$\mathcal{F}'_D(\Phi_D) \equiv \frac{d\mathcal{F}_D(\Phi_D)}{d\Phi_D} = -\Phi. \tag{13.18}$$

Clearly, this is simply the Legendre transformation of the prepotential

$$\mathcal{F}_D(\Phi_D) = \mathcal{F}(\Phi) - \Phi\mathcal{F}'(\Phi) = \mathcal{F}(\Phi) - \Phi\Phi_D.$$

Then the inverse transformation is given by the (13.17).

Introduced in such a way, the dual transformations (13.17) and (13.18) are very similar in form to the usual canonical transformations. Indeed, the quantity $\mathcal{F}'(\Phi)$ can be thought of as a sort of "canonical momentum" that is conjugated to the "coordinate" Φ and the Jacobian of these transformations is equal to unity.

Let us see how the dual transformations of (13.17) and (13.18) affect the low-energy effective action (13.5). Making use of the dual variables, we can rewrite the second term there as

$$\mathrm{Im}\int d^2\theta d^2\bar\theta\ \Phi^\dagger \mathcal{F}'(\Phi) = \mathrm{Im}\int d^2\theta d^2\bar\theta\ (-\mathcal{F}'_D(\Phi_D))^\dagger \Phi_D$$
$$= \mathrm{Im}\int d^2\theta d^2\bar\theta\ \Phi_D^\dagger \mathcal{F}'_D(\Phi_D). \tag{13.19}$$

Thus, this term is invariant with respect to the dual transformations defined in (13.17) and (13.18).

Next, we have to find how the supersymmetric field strength W_α changes under the dual transformations. The difference from the duality between the chiral fields Φ and Φ_D is that the relation between the superfield strength W_α and its dual $(W_D)^\alpha$ cannot be local since it includes the conventional duality between the Abelian field strengths $F_{\mu\nu}$ and $\widetilde{F}_{\mu\nu}$. This observation suggests a way to define the duality transformation of W_α.

Recall that the supersymmetric field strength W_α is defined as a covariantly constant quantity: $\mathrm{Im}(D^\alpha W_\alpha) = 0$. This constraint contains, as a component, the usual Bianci identity. Then, by analogy with the dual transformation of the variables of the functional integration in the electrodynamical path integral (1.89), we can trade the functional integral over the vector superfield V for the integral over W_α, subject to constraint (11.57), which is incorporated into the Lagrangian (13.5). Introducing a real Lagrange multiplier vector superfield V_D, we obtain

$$Z \sim \int \mathcal{D}V \exp\left[-\frac{1}{32\pi}\mathrm{Im}\int d^4x d^2\theta \mathcal{F}''(\Phi)W^\alpha W_\alpha\right]$$

$$\simeq \mathcal{D}W\mathcal{D}V_D \exp\left[-\frac{1}{32\pi}\mathrm{Im}\int d^4x \left(\int d^2\theta \mathcal{F}''(\Phi)W^\alpha W_\alpha \right.\right.$$
$$\left.\left. + \frac{1}{2}\int d^2\theta d^2\tilde\theta\, V_D D^\alpha W_\alpha\right)\right]. \tag{13.20}$$

Integrating by parts and using the relation (11.55), $\bar{D}_{\dot\alpha}W^\alpha = 0$, we find

$$\int d^2\theta d^2\tilde\theta\, V_D D^\alpha W_\alpha = -\int d^2\theta d^2\tilde\theta\, (D^\alpha V_D)W_\alpha = \int d^2\theta\, \bar{D}_{\dot\beta}\bar{D}^{\dot\beta}(D^\alpha V_D W_\alpha)$$
$$= \int d^2\theta \left(\bar{D}_{\dot\beta}\bar{D}^{\dot\beta}D^\alpha V_D\right) W_\alpha \equiv -4\int d^2\theta (W_D)^\alpha W_\alpha, \tag{13.21}$$

where a dual vector superfield strength $(W_D)^\alpha$ is defined as

$$(W_D)^\alpha \equiv \frac{1}{4}\bar{D}_{\dot\beta}\bar{D}^{\dot\beta}D^\alpha V_D. \tag{13.22}$$

The functional integration over the superfield W_α is Gaussian, thus we have:

$$Z \sim \int \mathcal{D}V_D \exp\left[\frac{1}{32\pi}\mathrm{Im}\int d^4x d^2\theta \left(-\frac{1}{\mathcal{F}''}(W_D)^\alpha W_{D\alpha}\right)\right].$$

Hence, the dual transformation yields a form of the $N = 2$ supersymmetric low-energy effective Lagrangian, that is an alternative to (13.5):

$$\mathcal{L}_{eff} = \frac{1}{32\pi}\mathrm{Im}\left(\int d^2\theta \mathcal{F}''_D(\Phi_D)(W_D)^\alpha W_{D\alpha} + 2\int d^2\theta d^2\bar\theta\, \Phi^\dagger_D \mathcal{F}'_D(\Phi_D)\right). \tag{13.23}$$

Here the effective coupling $\tau = \mathcal{F}''$ is replaced by its dual, which is defined as

$$\tau_D \equiv \mathcal{F}''_D(\Phi_D) = -\frac{1}{\mathcal{F}''(\Phi)} = -\frac{1}{\tau}. \tag{13.24}$$

Clearly, if we identify the coupling τ with the parameter (3.17), which we considered in the context of the action of the modular group, the relation (13.24) represents the transformation of S-duality (2.144). Together with the transformation of T-duality $\tau \to \tau + 1$, which acts on the complex coupling τ producing the shift of the angular parameter $\theta \to \theta + 2\pi$, it generates the modular group $SL(2,\mathbb{Z})$. The latter is the full group of duality of the model.

Indeed, the low-energy effective Lagrangian (13.5) can be written in the "mixed" form:

$$\mathcal{L}_{eff} = \frac{1}{32\pi}\left[\mathrm{Im}\int d^2\theta \frac{d\Phi_D}{d\Phi}W^\alpha W_\alpha + i\int d^2\theta d^2\bar\theta\, \left(\Phi^\dagger_D \Phi - \Phi^\dagger \Phi_D\right)\right], \tag{13.25}$$

13.2 Global Parametrization of the Quantum Moduli Space

which is evidently invariant with respect to the dual transformations (13.17) and (13.18).

It is instructive to see how these transformation act on the two-dimensional vector $\begin{pmatrix} \Phi_D \\ \Phi \end{pmatrix}$:

$$\begin{pmatrix} \Phi_D \\ \Phi \end{pmatrix} \longrightarrow \begin{pmatrix} \Phi'_D \\ \Phi' \end{pmatrix} = \begin{pmatrix} a & b \\ c & d \end{pmatrix} \begin{pmatrix} \Phi_D \\ \Phi \end{pmatrix},$$

which are familiar transformations of the modular group (2.145) and (3.16). The S-duality corresponds to the transformation

$$\begin{pmatrix} \Phi_D \\ \Phi \end{pmatrix} \longrightarrow \begin{pmatrix} 0 & 1 \\ -1 & 0 \end{pmatrix} \begin{pmatrix} \Phi_D \\ \Phi \end{pmatrix} = S \begin{pmatrix} \Phi_D \\ \Phi \end{pmatrix}, \quad (13.26)$$

while the T-duality is generated by

$$\begin{pmatrix} \Phi_D \\ \Phi \end{pmatrix} \longrightarrow \begin{pmatrix} 1 & 1 \\ 0 & 1 \end{pmatrix} \begin{pmatrix} \Phi_D \\ \Phi \end{pmatrix} = T \begin{pmatrix} \Phi_D \\ \Phi \end{pmatrix}. \quad (13.27)$$

We can easily see that the latter transformation of (13.27) does not change the second term of the Lagrangian (13.25). Thus, the corresponding variation of the action is given by the shift of the first term:

$$\delta S = \frac{1}{32\pi} \operatorname{Im} \int d^4x d^2\theta W^\alpha W_\alpha = \frac{1}{16\pi} \int d^4x F_{\mu\nu} \tilde{F}^{\mu\nu} = 2\pi n,$$

where $n \in \mathbb{Z}$ is the Pontryagin index (5.96). This transformation is simply a translation of the vacuum angle $\theta \to \theta + 2\pi$ which does not affect the physical quantities. Therefore, the partition function remains invariant under transformations (13.26) and (13.27), which generate the modular group $SL(2, \mathbb{Z})$.

The similarity with the dual transformation of the Abelian electrodynamics suggests that the dual theory governed by the Lagrangian (13.23) describes the $N = 2$ Abelian vector multiplet containing the "magnetic" photon and magnetically charged "dual Higgs" field. The magnetic monopoles and dyons are the states of another dual hypermultiplet of the matter fields that can be locally coupled to the dual field Φ_D, just in the same way as the "normal" hypermultiplet of the matter fields can be coupled to the field Φ in the original formulation. This interpretation means that the dual theory is the $N = 2$ supersymmetric Abelian electrodynamics, which however, unlike (13.5), is not an asymptotically free theory. The corresponding "magnetic" β-function of the dual $U(1)$ theory is different from its "electric" counterpart (13.14) by a sign and a factor $1/2$, which arises because the Abelian gauge fields no longer contribute to the running coupling:

$$\beta_D = \mu \frac{de_D}{d\mu} = \frac{e_D^3(\mu)}{8\pi^2}. \quad (13.28)$$

The transformation of duality relates the regimes of weak and strong coupling: the perturbative description in terms of the dual coupling τ_D corresponds to the non-perturbative regime in terms of the original coupling τ and vice versa. Hence, the vacuum expectation value of the dual scalar component ϕ_D of the dual chiral superfield Φ_D, the complex number v_D, can be considered as a proper coordinate, which may parameterize the region of the moduli space in the patch near $u = 0$.

Following the suggestion by Seiberg and Witten [469], we suppose that the gauge invariant quantity $u = \langle \text{Tr}\, \phi^2 \rangle$ provides a global parameterization of the moduli space with regard both to the variables $v(u)$ and

$$v_D(u) \equiv \frac{\partial \mathcal{F}}{\partial v}.$$

This equation can be used to express the effective coupling in terms of the variables v, v_D. Indeed, the scale-dependent effective coupling τ is defined as $\tau(v) = \dfrac{\partial^2 \mathcal{F}}{\partial^2 v}$. Then, taking into account the definition of the dual field (13.17), we can write the formula

$$\tau(v) = \frac{dv_D}{dv}, \tag{13.29}$$

which implies the following dual-invariant form of the metric on the moduli space (cf. (13.7)):

$$\begin{aligned} d^2 s &= \text{Im}\, \tau(v)\, dv d\bar{v} = \text{Im}\, \frac{dv_D}{dv} dv d\bar{v} \\ &= \text{Im}(dv_D d\bar{v}) = \frac{i}{2}(dv d\bar{v}_D - dv_D d\bar{v}). \end{aligned} \tag{13.30}$$

To understand the geometrical meaning of this metric, let us recall our discussion of the complex spaces in Chap. 3. The moduli space $\mathcal{M}$ is parameterized by a holomorphic coordinate u. The functions $v(u)$ and $v_D(u)$ provide a map $f : \mathcal{M} \to X$, where the vector space $X \simeq \mathbb{C}^2$ is covered by two patches with local coordinates v and v_D, respectively. Thus, these functions define a section of the space X, a vector bundle $\mathcal{M} \otimes X$ over the moduli space $\mathcal{M}$ with the structure group $SL(2, \mathbb{Z})$.

Indeed, the space X can be endowed with the holomorphic form $\omega_h = dv_D \wedge dv$ and the symplectic differential form

$$\omega = \text{Im}\, dv_D \wedge d\bar{v} = \frac{i}{2}(dv \wedge d\bar{v}_D - dv_D \wedge d\bar{v}).$$

The condition of holomorphicity means that we have to consider the sections $V = \begin{pmatrix} v_D \\ v \end{pmatrix}$ of the bundle $f : \mathcal{M} \to X$ for which the pullback of ω_h vanishes: $f^*(\omega_h) = 0$.

13.2 Global Parametrization of the Quantum Moduli Space

The Kähler metric is associated with the symplectic form ω. Thus, the pullback $f^*(\omega)$ of the symplectic form ω yields the metric on the moduli space

$$d^2s = \operatorname{Im} \frac{dv_D}{du}\frac{d\bar{v}}{d\bar{u}} du d\bar{u} = \frac{i}{2}\left(\frac{d\bar{v}_D}{d\bar{u}}\frac{dv}{du} - \frac{dv_D}{du}\frac{d\bar{v}}{d\bar{u}}\right) du d\bar{u}, \quad (13.31)$$

which in the full quantum theory is invariant with respect to the transformations of the modular group $SL(2,\mathbb{Z})$. Although such a metric is not positive for an arbitrary range of values $v(u)$, $v_D(u)$, the explicit solution obtained by Seiberg and Witten provides this property. The discussion in the next section will clarify this.

13.2.2 BPS Bound Reexamined

Let us return to the analysis of the BPS mass bound (12.23). First, from (5.110) we know that the effect of the vacuum angle is to change the electric charge vector (12.47) as

$$\vec{q} = e \sum_{i=1}^{2}\left(m_i \vec{\beta}_i + n_i \frac{\theta}{2\pi}\vec{\beta}_i^*\right), \quad (13.32)$$

while the magnetic charge vector remains unchanged:

$$\vec{g} = \frac{4\pi}{e}\sum_{i=1}^{2} n_i \vec{\beta}_i^*. \quad (13.33)$$

Here, $m_i, n_i \in \mathbb{Z}$ are the electric and magnetic quantum numbers, respectively. Recall that the $SU(N)$ simple roots are self-dual, that is, $\vec{\beta}_i^* = \vec{\beta}_i$. Thus, the BPS mass bound (12.48) becomes

$$M = \left| e\sum_{i=1}^{2} m_i \left(\sqrt{2}\vec{\beta}_i \cdot \vec{h}\right) + \frac{e\theta}{2\pi}\sum_{i=1}^{2} n_i \left(\sqrt{2}\vec{\beta}_i^* \cdot \vec{h}\right) + \frac{4\pi i}{e}\sum_{i=1}^{r} n_i \left(\sqrt{2}\vec{\beta}_i^* \cdot \vec{h}\right)\right|$$

$$= \left|\sum_{i=1}^{2}\left\{m_i + n_i\left(\frac{4\pi i}{e^2} + \frac{\theta}{2\pi}\right)\right\}\left(\sqrt{2}e\vec{\beta}_i \cdot \vec{h}\right)\right|$$

$$= \left|\sum_{i=1}^{2}\{m_i + \tau n_i\}\Phi_i\right|. \quad (13.34)$$

For the sake of simplicity, we have restricted our consideration to the case of the $SU(2)$ gauge group. Then both scalar fields belong the same unique Cartan subalgebra and the components of the two-dimensional vectors of the electric and magnetic charges are proportional to each other: $q_1/q_2 = g_1/g_2$. This implies that the central charge of the microscopic $N = 2$ supersymmetric $SU(2)$ pure Yang–Mills theory is simply

$$Z = v(m + \tau n). \tag{13.35}$$

However, this relation is valid only in the weak coupling regime, where τ approaches to its classical limit (11.76). To obtain a dual-invariant modification of the expression for the central charge, we can note that in the classical limit $\tau = v_D/v$ and, therefore,

$$Z = vm + v_D n = (n, m) \begin{pmatrix} v_D \\ v \end{pmatrix}. \tag{13.36}$$

Seiberg and Witten suggested that this expression can be used as an exact BPS mass formula that is correct both in weak and strong coupling regimes.

The physical interpretation of the mass formula (13.36) is rather obvious. Recall that all states saturating the BPS bound belong to the short massive multiplet of the $N = 2$ supersymmetry. The mass of these states is simply $M = |Z|$ and two components of (13.36) are precisely the electric and magnetic masses, whose definition we discussed above on page 455.

Indeed, let us consider a massive hypermultiplet of the matter $N = 1$ fields $A, A^\dagger$ with a charge m. The $N = 2$ supersymmetry of the low-energy effective action fixes the coupling of these fields with the $N = 1$ chiral scalar superfield uniquely: $\sqrt{2}m\Phi AA^\dagger$. Then, using the definition of the central charge, we can easily find that for such a state $Z = vm$.

Similarly, in the dual formulation of the theory, the short massive dual hypermultiplet is represented by the $N = 1$ chiral superfields $B, B^\dagger$ with a magnetic charge n. These fields are coupled to the dual chiral supefield Φ_D as

$$\sqrt{2}n\Phi_D BB^\dagger. \tag{13.37}$$

We can see that the corresponding additive central charge is $Z = nv_D$, so that for a mixed dyon state, we obtain the relation (13.36).

Clearly, the mass formula (13.36) must be dual invariant, because it yields the masses of the physical states. The invariance with respect to the transformation of S-duality is obvious, since it acts as a permutation $v \rightleftharpoons v_D$, which also interchanges the electric and magnetic quantum numbers as $m \rightleftharpoons n$. The transformations of the full group of duality act on the section $V = \begin{pmatrix} v_D \\ v \end{pmatrix}$ and the charge vector as:

$$\begin{pmatrix} v_D \\ v \end{pmatrix} \to M \begin{pmatrix} v_D \\ v \end{pmatrix}, \qquad (n, m) \to (n, m) M^{-1}. \tag{13.38}$$

Let us note that, although the Zamolodchikov metric (13.31) is formally invariant under the symplectic transformations $V \to MV + C$, $M \in SP(2, \mathbb{R})$, where C is a constant vector[6], only the transformations of the discrete subgroup $SL(2, \mathbb{Z})$ survive in the quantum theory. Indeed, the charge vector

---

[6] In the absence of the matter fields, this vector is vanishing.

(n, m) has the integer entries and a transformed vector must have integer entries as well. Thus, $M = \begin{pmatrix} a & b \\ c & d \end{pmatrix}$ is an integral matrix of $SL(2, \mathbb{Z})$. If there are singularities on the moduli space $\mathcal{M}$, the section $\begin{pmatrix} v_D \\ v \end{pmatrix}$ is transformed under action of the elements of the monodromy group Γ_M, a subgroup of $SL(2, \mathbb{Z})$.

Furthermore, under action of the $SL(2, \mathbb{Z})$ group, the effective coupling τ (13.29) is transforming as

$$\tau \longrightarrow \frac{a\tau + b}{c\tau + d}$$

Recall that we already encountered this relation in Chap. 3, on that occasion in the context of the transformations of the periodic matrix of a torus, a Riemann surface of genus 1, under the action of the transformation of the modular group that changes the basis of homology (cf. (3.18) and the related discussion). The logic of Seiberg and Witten is to identify the moduli space of the $N = 2$ supersymmetric theory with a certain Riemann surface and to consider the action of the modular group in this geometrical context. The variables v, v_D in this frame turn out to be related to the periods on this surface. Then they can be recovered from the structure of the given monodromy on the moduli space.

13.3 Seiberg–Witten Explicit Solution

13.3.1 Monodromies on the Moduli Space

Let us analyze the geometry of the quantum moduli space of the $N = 2$ SUSY $SU(2)$ Yang–Mills theory. We assume that there is only a finite set of isolated singularities. First, we have to identify these and calculate the corresponding monodromies.

The practical method of calculation is to consider the behavior of the multi-valued functions v and v_D, as the variable u, which parameterizes the moduli space $\mathcal{M}$, is varying. More precisely, we have to consider a closed contour on $\mathcal{M}$. If this contour does not encircle a singularity, the functions v and v_D do not change as u goes around a loop. However, if this contour encircles some singularity, these functions do not return to their original values, but get transformed into certain linear combinations. This is what is called *monodromy*.

Recall that the singularities of the moduli space are associated with the appearance of certain additional massless physical states, which destroy the description in terms of the Wilsonian effective action. The classical singularity at $u = 0$, for example, is due to the massless vector bosons and their spinor superpartners, which are the components of the vector superfield. They become massless as the Higgs mechanism does not work at this point and the gauge symmetry is restored there from $U(1)$ to $SU(2)$. However, as we shall see below, the global structure of the quantum moduli space turns out to be different. Thus, the question is how many singularities there are and what kind of massless excitations are associated with them.

As we have already mentioned, the quantum moduli space has a singularity at $u = \infty$, which is related to the branch cut of the logarithm of the one-loop corrected effective coupling. This singularity shall be present in the semiclassical limit of the full quantum theory, because of the asymptotic freedom. Indeed, at large values of $u = v^2/2$, the model is well-defined by the microscopic Lagrangian (13.1). The prepotential and the effective coupling then are given by the one-loop corrected expressions (13.15) above, and this implies

$$v_D(u) = \frac{\partial \mathcal{F}}{\partial v} = \frac{i}{\pi}\sqrt{2u}\left(\ln\frac{2u}{\Lambda^2} + 1\right),$$

$$v(u) = \sqrt{2u},\qquad(13.39)$$

$$\tau(u) = \frac{i}{\pi}\left(\ln\frac{2u}{\Lambda^2} + 3\right).$$

Let us define the matrix of monodromy around $u = \infty$. As discussed above, we can consider a counterclockwise closed contour of very large radius in the complex u-plane, which yields $u \to e^{2\pi i}u$. Since this plane is compactified into a Riemann sphere, this contour is equivalent to a clockwise path around the singular point $u = \infty$. Then $\ln u \to \ln u + 2\pi i$ and the functions (13.39) are affected by monodromy as

$$v(u) \to -v(u);\quad v_D(u) \to \frac{i}{\pi}(-v)\left(\ln\frac{e^{2\pi i}u}{\Lambda^2} + 1\right) = -v_D + 2v,$$

$$\tau(u) \to \frac{i}{\pi}\left(\ln\frac{e^{2\pi i}u}{\Lambda^2} + 3\right) = \tau(u) - 2.\qquad(13.40)$$

We see that this monodromy is defined by the value of the coefficient, which appears in front of the logarithm in the expression for the effective coupling τ. Recall that it is directly related to the perturbative β-function of the model.

From this we find that the matrix of monodromy, acting on the section V, is of the form:

$$\begin{pmatrix} v_D \\ v \end{pmatrix} \to M_\infty \begin{pmatrix} v_D \\ v \end{pmatrix} = \begin{pmatrix} -1 & 2 \\ 0 & -1 \end{pmatrix}\begin{pmatrix} v_D \\ v \end{pmatrix} = S^2 T^{-2}\begin{pmatrix} v_D \\ v \end{pmatrix},\qquad(13.41)$$

where $T^{-1} = \begin{pmatrix} 1 & -1 \\ 0 & 1 \end{pmatrix}$ is the inverse of the matrix of T-duality (13.27) and S is a matrix of S-duality (13.26). Note that S^2 is proportional to the element of unity of $SL(2,\mathbb{Z})$, $S^2 = -\mathbb{I}$.

On the other hand, the monodromy matrix at $u = \infty$ also acts on the charge vector as:

13.3 Seiberg–Witten Explicit Solution

$$(n, m) \to (n, m)M_\infty^{-1} = (-n, -m - 2n), \quad (13.42)$$

that is, these states do not form a left eigenvector of the matrix of monodromy M_∞. This can be understood, if we notice that the phase rotation in the complex u-plane $u \to e^{2\pi i}u$ results in the shift of the vacuum angle $\theta \to \theta + \delta\theta = \theta - 4\pi$. Then, according to (13.42), the electric quantum number of a monopole under the monodromy M_∞ is shifted as

$$m \to -m + n\frac{\delta\theta}{2\pi} = -m - 2n.$$

Let us consider now the structure of the quantum moduli space at the finite values of u. As u decreases, we approach the strong coupling regime, where the expressions (13.15) for the prepotential $\mathcal{F}(v)$ and the effective coupling $\tau(v)$ are no longer valid. Thus, in this region $u \neq v^2/2$, and we have to continue our consideration of the singularities on the moduli space in order to understand the structure of the low-energy effective action beyond the semi-classical limit in the vicinity of the singular point at $u = \infty$.

Note that this singularity appears due to the branch cut of the logarithmic function of v_D in (13.39). Clearly, this cut has to originate at some point, thus, the singularity at $u = \infty$ cannot be alone.

At a first glance, there is the classical singularity at $u = 0$, which we discussed above. However, it would be premature to conclude that this singularity survives on the quantum moduli space. Indeed, if there were only two singularities on the Riemann sphere, at $u = 0$ and $u = \infty$, respectively, the monodromy M_0 around $u = 0$ would be the same as the monodromy M_∞ around $u = \infty$. This means that the monodromy group would be Abelian and then the quantity v^2 would not change under either monodromy. In that case, $u = v^2/2$ may be considered as a well-defined global coordinate on $\mathcal{M}$.

However, it is clear that then we would immediately run into the same conflict with restriction imposed by the unitarity. Indeed, recall that in that case, the coupling $\text{Im}\,\tau$ has to be a harmonic function that has to be positive definite throughout the moduli space (see, e.g., [123]). Thus, it cannot have a minimum there and the singularity of the classical moduli space at $u = 0$ is excluded in the quantum theory.

Moreover, the gauge bosons, which are the components of the spin-1 multiplet, in a full quantum theory never become massless and the spontaneously broken gauge symmetry of the quantum theory is never restored; it remains Abelian over the entire space $\mathcal{M}_q$. Seiberg and Witten pointed out that the presence of the massless gauge bosons would imply a superconformal invariance in the infrared limit, which is not present at any scale. Thus, we have to conclude that we have to look for the strong coupling singularities not at the origin, but in some other places.

Note that the global symmetry of the moduli space with respect to the $\mathbb{Z}_2$ reflections $u \to -u$ means that the singularities on $\mathcal{M}_q$ come pairwise: for a singularity at u_0 there has to be a counterpart at $-u_0$. The fixed points of this discrete transformation on the Riemann sphere are $u = \infty$ and $u = 0$, but the latter singularity has to be excluded. Thus, the minimal assumption about the global structure of the quantum moduli space $\mathcal{M}_q$ is that there are at least two additional singularities at $u = \pm u_0$ (see Fig. 13.2).

An intriguing question regards the physical origin of these singularities. What are the states that become massless at these points and what is the physical meaning of the non-zero value of u_0? Since the gauge bosons, as well as other states of the $N = 2$ multiplet, are excluded, we have to consider other possibilities. Evidently, the value of u_0 can be associated with some non-perturbative scale of theory, a counterpart of the Λ_{QCD}.

Recall that there is the massive matter multiplet with spin $1/2$. This is the short $N = 2$ multiplet, which contains non-perturbative BPS states, the monopoles. We have seen that the matter fields of the monopoles (dyons) cannot be locally coupled to the fundamental fields appearing in the formulation (13.5). However, in the dual description (13.23), the dual fields Φ_D are coupled to the magnetically charged hypermultiplet locally as in (13.37). Thus, the monopole becomes massless as $v_D = 0$. Then we may associate the strong coupling singularity at u_0 with the point on the moduli space where v_D vanishes, but $v \neq 0$.

Furthermore, we suppose that the mass formula (13.36) can be applied both to the pure electrically charged states $(0, m)$ and to the monopoles $(n, 0)$. Then the dyons become massless as $vm + v_D n = 0$ for some non-zero values of the electric and magnetic quantum numbers. We can associate these massless dyonic excitations with the third singularity of the quantum moduli space at $-u_0$ (see Fig. 13.2).

Let us investigate the consequences of these assumptions. First, we may calculate the monodromy matrix M_{u_0} at $u = u_0$. It may be defined by making use of the arguments of duality. Namely, we have to calculate a monodromy for a massless electric hypermultiplet $(0, 1)$ coupled to the field Φ. Evidently, its mass vanishes as $v = 0$. Then we may apply the transformation of duality to find the monodromy for a massless monopole $(1, 0)$.

However, we consider here another, more straightforward way of calculating the monodromy M_{u_0}, which is related with the form of the $U(1)$ magnetic β-function of (13.28). In the Abelian theory, we can set $\theta_D = 0$ and then

$$\tau_D = -\frac{\partial v}{\partial v_D} = \frac{4\pi i}{e_D^2}. \tag{13.43}$$

The structure of the renormalization group equation (13.13) suggests setting a correspondence between the quantum singularities and the non-perturbative scale Λ. Near this singularity $u_0 \sim \Lambda^2$, the renormalization point μ is proportional to $v_D \to 0$, which is the only scale in this region. Thus, the "magnetic" β-function (13.28) can be written as

$$v_D \frac{d\tau_D}{dv_D} = -\frac{i}{\pi}.$$

13.3 Seiberg–Witten Explicit Solution

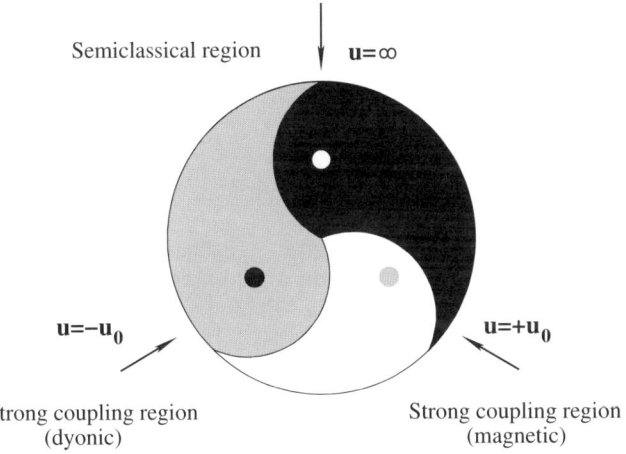

Fig. 13.2. The structure of singularities on the quantum moduli space $\mathcal{M}_q$ of the $N = 2$ supersymmetric Yang–Mills model

Integrating this equation, we obtain

$$\tau_D = -\frac{i}{\pi} \ln v_D. \tag{13.44}$$

Further integration of (13.43) allows us to see that, in the region near u_0,

$$v \approx v_0 + \frac{i}{\pi} v_D \ln v_D, \tag{13.45}$$

where a constant v_0 is not equal to zero, otherwise both v_D and v would vanish simultaneously. In the latter case, both the electrically and the magnetically charged BPS states would be massless. In the vicinity of u_0, this situation corresponds to an effective theory of the light monopoles interacting with the light electric states.

Since v_D is supposed to be a good coordinate near the point u_0, it can be expanded there as

$$v_D \sim c_0(u - u_0). \tag{13.46}$$

Then taking u around a loop encircling the singularity u_0 counterclockwise, so that $(u - u_0) \to e^{2\pi i}(u - u_0)$, we obtain:

$$\begin{aligned} v(u) &\to v(u) - 2v_D(u), & v_D(u) &\to v_D(u), \\ \tau_D(u) &\to \tau_D(u) + 2. \end{aligned} \tag{13.47}$$

The last of these equations means that the corresponding monodromy transformation of the coupling τ is of the form

$$\tau(u) = -\frac{1}{\tau_D} \to \frac{\tau(u)}{1 - 2\tau(u)}.$$

Then the monodromy matrix M_{u_0}, which by acting on the section V yields the transformations (13.47), is

$$\begin{pmatrix} v_D \\ v \end{pmatrix} \to M_{u_0} \begin{pmatrix} v_D \\ v \end{pmatrix} = \begin{pmatrix} 1 & 0 \\ -2 & 1 \end{pmatrix} \begin{pmatrix} v_D \\ v \end{pmatrix} = ST^2 S^{-1} \begin{pmatrix} v_D \\ v \end{pmatrix}. \quad (13.48)$$

We can easily see that this this matrix is related to the matrix of monodromy M_∞ of (13.41) as $M_{u_0} = S M_\infty^{-1} S$.

The last remaining monodromy at $u = -u_0 \sim -\Lambda^2$ is related to the monodromy at $u = u_0 \sim \Lambda^2$ by the above-mentioned $\mathbb{Z}_2$-symmetry that acts on the parameter u as $u \to e^{i\pi} u$. This symmetry is a property of the entire moduli space. Then we can start from the semiclassical region, where u is large and the perturbative relations (13.39) are well-defined. The $\mathbb{Z}_2$ reflection in this region yields

$$v = \sqrt{2u} \to \tilde{v} = iv,$$

$$v_D = \frac{i}{\pi}\sqrt{2u}\left(\ln\frac{2u}{\Lambda^2} + 1\right) \quad \to \quad \tilde{v}_D = -\frac{\sqrt{2u}}{\pi}\left(\ln\frac{2u}{\Lambda^2} + 1\right) - i\sqrt{2u} \quad (13.49)$$

$$= i(v_D - v).$$

We can now make use of the analogy with the previous case to determine the monodromy at the strong coupling singularity $u = -u_0$. In the vicinity of this point, $\tilde{v}_D$ should now be a good coordinate that admits a linear expansion in u, that is,

$$\tilde{v}_D \sim \tilde{c}_0(u + u_0).$$

The dual effective coupling in this patch becomes

$$\tilde{\tau}_D = -\frac{\partial \tilde{v}}{\partial \tilde{v}_D} = -\frac{i}{\pi}\ln \tilde{v}_D. \quad (13.50)$$

Thus, we have

$$\tilde{v} = \tilde{v}_0 + \frac{i}{\pi}\tilde{v}_D \ln \tilde{v}_D. \quad (13.51)$$

To obtain the monodromy matrix, we have to consider a contour encircling the singularity at $u = -u_0$, so that $(u + u_0) \to (u + u_0)e^{2\pi i}$. By analogy with (13.47), this gives

$$\tilde{v}(u) \to \tilde{v}(u) - 2\tilde{v}_D(u); \qquad \tilde{v}_D(u) \to \tilde{v}_D(u),$$
$$\tilde{\tau}_D(u) \to \tilde{\tau}_D(u) + 2. \quad (13.52)$$

This description may be written in terms of the original variables v, v_D, which are related to $\tilde{v}, \tilde{v}_D$ by $\mathbb{Z}_2$-symmetry as in (13.49). Then, as u goes around a loop encircling $u = -u_0$,

$$v(u) \to 3v(u) - 2v_D(u); \qquad v_D(u) \to 2v(u) - v_D(u),$$
$$\tau(u) \to \frac{2 - \tau(u)}{3 - 2\tau(u)}. \quad (13.53)$$

13.3 Seiberg–Witten Explicit Solution

Thus, the third monodromy matrix M_{-u_0} is acting as

$$\begin{pmatrix} v_D \\ v \end{pmatrix} \to M_{-u_0} \begin{pmatrix} v_D \\ v \end{pmatrix} = \begin{pmatrix} -1 & 2 \\ -2 & 3 \end{pmatrix} \begin{pmatrix} v_D \\ v \end{pmatrix} = ST^{-2}ST^{-2} \begin{pmatrix} v_D \\ v \end{pmatrix}. \quad (13.54)$$

The simple check of self-consistency of these calculations above is to prove that a loop, which encircles both strong coupling singularities, does reproduce the monodromy at $u = \infty$. Indeed, we can easily see that the factorization condition

$$M_{u_0} M_{-u_0} = M_\infty,$$

is fulfilled. Note that the matrices $M_{u_0} M_{-u_0}$ do not commute and it looks like the factorization formula does not respect the $\mathbb{Z}_2$ symmetry of the moduli space. However, the definition of the composition of monodromies requires a choice of a base point u on the moduli space. Then for the reflected base point $-u$, we obtain

$$M'_{-u_0} M_{u_0} = M_\infty,$$

where

$$M'_{-u_0} = \begin{pmatrix} 3 & 2 \\ -2 & -1 \end{pmatrix} = T^{-2}ST^{-2}S. \quad (13.55)$$

Thus, the $\mathbb{Z}_2$ symmetry exchanges the monodromy matrices (13.54) and (13.55).

The three monodromy matrices of (13.41), (13.48) and (13.54) (or, equivalently (13.55)) span a monodromy subgroup $\Gamma(2)$ of the modular group $SL(2, \mathbb{Z})$. The moduli space, the u-plane with three singularities, is the quotient of the upper half $\mathbb{H}^+$ of a complex plane by $\Gamma(2)$. Note that the matrix of S-duality (13.26) is not an element of the monodromy group, thus the $N = 2$ Yang–Mills theory is not self-dual[7].

Finally, let us recall that a monodromy transformation acts on the charge vector (n, m) by the right multiplication with M^{-1}. The corresponding BPS state, which becomes massless at the point of singularity, must be invariant with respect to this transformation of mondromy, that is, (n, m) has to be an eigenvector of the monodromy matrix with unit eigenvalue. Indeed, the monopole state with a unit magnetic quantum number $(1, 0)$ is a left eigenvector of the monodromy matrix M_{u_0}:

$$(1, 0) \begin{pmatrix} 1 & 0 \\ -2 & 1 \end{pmatrix} = (1, 0).$$

This state becomes massless at the point $v_D = 0$, where $u = u_0$. On the other hand, the $(1, -1)$ dyon state becomes massless at the point $u = -u_0$. Indeed,

---

[7] There is a difference from the $N = 4$ supersymmetric Yang–Mills theory where the group of monodromy includes the transformations of S-duality and, therefore, that model is a special case of a self-dual theory.

there the BPS mass formula yields $v - v_D = 0$, thus $\tilde{v}_D = 0$. We can easily see that this state is the left eigenvector of the monodromy matrix M_{-u_0}:

$$(1, -1) \begin{pmatrix} -1 & 2 \\ -2 & 3 \end{pmatrix} = (1, -1).$$

As we noted above, the $\mathbb{Z}_2$ reflections on the moduli space change the matrix of monodromy as $M_{-u_0} \to M'_{-u_0}$. We can see that the left eigenvector of the matrix M'_{-u_0} (13.55) is the $(1,1)$ dyon state:

$$(1, 1) \begin{pmatrix} 3 & 2 \\ -2 & -1 \end{pmatrix} = (1, 1).$$

Hence the reflection $u \to -u = e^{i\pi} u$ inverts the sign of the electric charge of the $(1, -1)$-dyon. We may anticipate this effect in advance, since the $u \to -u$ reflection yields the related shift of the θ-angle.

Thus, to summarize, the key message of the Seiberg–Witten work is that the exact quantum moduli space of the $N = 2$ supersymmetric Yang–Mills theory is covered by three patches; in the center of each of those, the local weakly coupled theory is described in terms of the properly chosen local variables. There is a well-defined corresponding local low-energy effective Lagrangian in each path, but none of these three Lagrangians, which are related by the transformations of duality[8], is more fundamental than the other two.

13.3.2 Solution of the Monodromy Problem

The analogy that obviously exists between the quantum moduli space of the $N = 2$ supersymmetric Yang–Mills theory and the moduli space of genus one Riemann surfaces, a torus, suggests application of the related geometrical technique to find the multivalued functions $v(u)$ and $v_D(u)$ that form a holomorphic section V of the bundle $\mathcal{M} \otimes X$, where $X \simeq \mathbb{C}^2$. Indeed, we have mentioned that the effective coupling $\operatorname{Im} \tau(u)$ can be identified with the τ-parameter of a torus whose homology basis transforms under the action of the group of monodromy $\Gamma(2)$. Thus, the problem of finding a solution of the $N = 2$ low-energy effective action can be reduced to the Riemann–Hilbert problem of finding the functions with a given monodromy around the singularities.

Recall that a torus (an elliptic curve) is defined by an algebraic equation (cf. (3.14) and (6.93), Fig. 3.5 and the related discussion in Chaps. 3 and 6) of the form

$$y^2 = x^3 + \alpha x + \beta, \tag{13.56}$$

which generates a family of the genus one Riemann surfaces S. Here $\alpha, \beta \in \mathbb{C}$ are arbitrary parameters. This equation defines a two-sheet function that has

---

[8] In this context, it would be more correct to use the notion of the electro-magnetic-dyonic *triality*, see Fig. 13.2.

four branch points: three are corresponding to the zeros of the function $y(x)$ and one branch point is located at $x = \infty$. These branch points are pairwise connected by two cuts.

There are two fundamental one-cycles of the torus. The cycle a is a contour that encircles the cut between first two branch points, while the cycle b is a contour that starts from the upper side of the second cut, goes to the first cut, continues across the cut to the second sheet and returns to the original sheet across the first cut. These cycles are normalized such that their intersection number is equal to one:

$$a \circ b = 1. \tag{13.57}$$

The fundamental circles on the complex u-plane allows us to give a new interpretation to the singularities on the moduli space. Indeed, these cycles continuously vary with u and if two of the branch points coincide, a cycle shrinks to zero and the elliptic curve becomes singular. Thus, the singularities are the points where a curve in the family $S(u)$ develops a vanishing cycle. This corresponds to the singularity of the complex u-plane with a non-trivial monodromy around it. Recall that, physically, these singularities are associated with the appearance of massless particles in the spectrum.

To parameterize the torus in terms of the variables y, x, it is convenient to consider the basis for one-forms on $S(u)$

$$\lambda_1 = \frac{dx}{y}, \qquad \lambda_2 = \frac{xdx}{y}.$$

An arbitrary element of the first homology group λ is a meromorphic one-form with vanishing residue. The elements of the first cohomology group λ can be set into correspondence to an element of the first homology group on the Riemann surface, an integral along the closed path γ

$$\gamma \to \oint_\gamma \lambda.$$

By definition, the one-form λ_1 is a unique holomorphic differential on $S(u)$ up to the action of the operation of the scalar multiplication. Then the integration along two periods of the torus, which are defined as

$$\omega_1 = \oint_a \frac{dx}{y}, \qquad \omega_2 = \oint_b \frac{dx}{y},$$

allows us to identify the parameter of the torus τ of (3.17) as the ratio

$$\tau = \frac{\omega_2}{\omega_1} = \frac{\oint_b \omega}{\oint_a \omega}.$$

Let us note that this form is very similar to the definition (13.29) of the effective coupling constant:

$$\tau \equiv \frac{dv_D}{dv} = \frac{dv_D}{du} \bigg/ \frac{dv}{du}. \qquad (13.58)$$

Then we can identify

$$\frac{dv}{du} = \oint_b \frac{d\lambda}{du}, \qquad \frac{dv_D}{du} = \oint_a \frac{d\lambda}{du},$$

where λ is an arbritrary meromorphic one-form. The identification of the functions v, v_D as the integrals of a one-form, $v = \oint_b \lambda$, and $v_D = \oint_a \lambda$ respectively, allows us to prove that this form depends on λ_1 only [469]. Thus, the period of the torus is defined by the ratio of two periodic integrals

$$\frac{dv}{du} = C \oint_a \frac{dx}{y}, \qquad \frac{dv_D}{du} = C \oint_b \frac{dx}{y}, \qquad (13.59)$$

where C is some common dimensionless constant.

Note that these periodic integrals (13.59) satisfy a second-order Picard–Fuchs differential equation in the complex plane u. Indeed, monodromies usually arise in the context of the solution of a differential equation with periodic coefficients (see, for example, [12]). Thus, one can write the corresponding differential equation for the multi-valued functions v, v_D and solve it explicitly [124, 311]. This approach is different from the original paper by Seiberg and Witten, where the geometrical language of the elliptic curves was applied. We shall briefly sketch this solution.

Note that the vanishing cycles on a torus are related to the BPS bound (13.36). Let us consider a path γ, which can be expanded in the fundamental cycles of the one-torus as $\gamma = ma + nb$, where $m, n \in \mathbb{Z}$. If the corresponding path shrinks to zero, assuming that the meromorphic one-form λ has a vanishing residue, we can write

$$\oint_\gamma \lambda = m \oint_b \lambda + n \oint_a \lambda = mv + nv_D \equiv Z \underset{\gamma \to 0}{\to} 0.$$

This is a massless BPS state with electric and magnetic quantum numbers (m, n), respectively. In other words, the charges of the dyon have a nice geometrical interpretation as the coordinates of the corresponding vanishing cycle in the holonomy basis.

We can see that the changes of the homology basis are precisely the transformations of the modular group $SL(2, \mathbb{Z})$, which acts on the electric and magnetic quantum numbers, as discussed above in Chap. 2 (cf. (2.145)). Then the dual invariant quantity, the Schwinger–Zwanziger quantization condition

(2.131), has a nice geometrical interpretation of the intersection number of the basis one-cycles

$$a_i \circ b_j = q_i g_j - q_j g_i = n, \quad n \in \mathbb{Z}, \tag{13.60}$$

which obviously generalizes the relation (13.57).

To obtain the effective action explicitly by making use of the arguments of symmetry, Seiberg and Witten suggested considering the curve [469]

$$y^2 = (x - u_0)(x + u_0)(x - u), \tag{13.61}$$

which corresponds to the strong coupling singularities at the points $\pm u_0$. One can prove that the monodromies of the periods of the torus, which are defined by the auxiliary spectral curve of that type, are identical to the monodromies of the quantum moduli space and generate the monodromy group $\Gamma(2)$ [73, 424, 469].

Substituting this formula into (13.59) and integrating over u, we obtain

$$\begin{aligned}v(u) &= -2C \oint_a dx \frac{\sqrt{x-u}}{\sqrt{x^2 - u_0^2}} = -4 \int_{-u_0}^{u_0} dx \frac{\sqrt{x-u}}{\sqrt{x^2 - u_0^2}}, \\ v_D(u) &= -2C \oint_b dx \frac{\sqrt{x-u}}{\sqrt{x^2 - u_0^2}} = -4 \int_{u_0}^{u} dx \frac{\sqrt{x-u}}{\sqrt{x^2 - u_0^2}},\end{aligned} \tag{13.62}$$

where we make use of the equivalence of the contributions from the intergation over the square-root branch cut and under the square-root branch cut. The explicit formulae for $v(u)$ and $v_D(u)$ can be given in terms of the elliptic integrals E and K, or in terms of the hypergeometric functions [469].

We can see that the functions (13.62) are indeed singular at the points $\pm u_0, \infty$. Furthermore, the known asymptotic behavior in the semi-classical regime: $v(u) \to \sqrt{2u}$, as $u \to \infty$, allows us to define the constant C. Indeed, in this limit we have

$$v \approx -4C\sqrt{u} \int_{-u_0}^{u_0} \frac{dx}{\sqrt{x^2 - u_0^2}} = -4\pi C \sqrt{u},$$

which yields $C = -\sqrt{2}/(4\pi)$. We can write the second solution for the dual variable in the weak coupling regime as an integral

$$v_D = \frac{\sqrt{2u}}{\pi} \int_{1/u}^{u_0} dz \frac{\sqrt{z-1}}{\sqrt{z^2 - (u_0/u)^2}},$$

where we changed a variable of integration as $z = x/u$. Note that in the limit $u \to \infty$, this expression has a logarithmic divergence at $z = 0$:

$$v_D \approx \frac{i}{\pi}\sqrt{2u}\,\ln\frac{u}{u_0},$$

which coincides with the one-loop formulae (13.39) above.

On the other hand, as $u \to u_0$, we obtain

$$v_D \approx \frac{1}{\pi}\int_{1/u}^{u_0} dz\,\frac{\sqrt{z-1}}{\sqrt{z-(u_0/u)}} = \frac{i}{2}(u-u_0), \tag{13.63}$$

which justifies the relation (13.46). The strong coupling limit of the variable v is then simply

$$v(u_0) = \frac{\sqrt{2}}{\pi}\int_{-u_0}^{u_0}\frac{dx}{\sqrt{x+u_0}} = \frac{4\sqrt{u_0}}{\pi}.$$

This relation yields the value of the constant v_0 in (13.45). Furthermore, according to (13.59), the derivative of v with respect to u is

$$\frac{dv}{du} = -\frac{\sqrt{2}}{2\pi}\int_{-u_0}^{u_0}\frac{dx}{\sqrt{(x-u_0)(x+u_0)(x-u)}}.$$

Thus, the expansion of the variable v near u_0 becomes

$$v \approx \frac{4\sqrt{u_0}}{\pi} - \frac{1}{2\pi}(u-u_0)\ln(u-u_0),$$

which is in agreement with (13.45) and (13.46), and produces the expectable monodromy around the strong coupling singularity.

Let us finally note that the solution at the second strong coupling singularity is determined by the $\mathbb{Z}_2$ symmetry as we described above. This yields the Seiberg–Witten solution of the $N=2$ supersymmetric Yang–Mills theory, because the relation $v_D = d\mathcal{F}/dv$ means that the prepotential $\mathcal{F}(v)$ can now be calculated by inverting the first of the equations in (13.62) to obtain u as a function of v and then, inserting the result into the second equation (13.62), to obtain v_D as a function of v. Then the integration of the result with respect to v gives $\mathcal{F}(v)$, and hence, the low-energy effective action that is valid within a certain domain.

13.3.3 Confinement and the Monopole Condensation

The Seiberg–Witten solution also allows us to prove the conjecture about the possible role of the monopole condensation in the mechanism of confinement. However, instead of the realistic QCD, we shall consider the $N=1$ gauge supersymmetric theory, which describes the chiral superfield Φ interacting

13.3 Seiberg–Witten Explicit Solution

with the vector superfield W_α. These fields can be thought of as being the components of the $N = 2$ chiral superfield Ψ. The idea is that the low-energy effective action of the $N = 2$ super Yang–Mills theory may provide some information about the non-perturbative structure of the $N = 1$ theory, the supersymmetric counterpart of the QCD.

Indeed, the $N = 2$ supersymmetry is not compatible with phenomenology, both because of the exact chiral symmetry and equal masses of the bosonic and fermionic components of the $N = 2$ supermultiplet. Thus, confinement and chiral symmetry breaking may be possible after the $N = 2$ SUSY breaking, either softly or spontaneously.

To break $N = 2$ supersymmetry down to $N = 1$ softly, we have to include the superpotential $W = m\text{Tr}\,\Phi^2$ in the action (12.6) by hand. This term gives a bare mass m to the states of the chiral multiplet and, since the scalar field ϕ is a leading component of the chiral superfield, the superpotential lifts the degeneration of the $N = 2$ vacuum. This is a way to define $N = 1$ supersymmetric theory as the low-energy effective limit of the N=2 gauge theory.

The motivation of the soft supersymmetry breaking is an assumption about the existence of the mass gap. Indeed, it was shown in the 1980s that there is a mass gap of the $N = 1$ microscopic theory, which is related with a non-vanishing vacuum expectation value of the gaugino condensate $\langle \bar\lambda \lambda \rangle$ [402]. We expect that the quantum moduli space of the $N = 1$ theory is similar to that of the $N = 2$ Yang–Mills theory, that is, there are two singularities at the strong coupling regime. Then the existence of the gaugino condensate of the original theory implies that in a dual formulation, which is valid in the vicinity u_0, a dual magnetic photon also becomes massive.

This mass gap may exist if there are either some massless gauge fields, which would give rise to a strongly coupled non-Abelian gauge theory, or if there are dual light charged fields, that are the excitations of the vacuum of the dual sector which corresponds to the strong coupling singularity. Both options mean that there must be some additional massless states somewhere on the quantum moduli space $\mathcal{M}$.

We have already mentioned that the existence of the massless gauge bosons, which would correspond to the singularity at $u = 0$, and recovering of the original non-Abelian symmetry, is not compatible with the structure of the quantum moduli space and has to be excluded. Therefore, taking into account the physical interpretation of the strong-coupling singularities at $u = \pm u_0$, we can consider the effect of the light monopoles (dyons).

In the strong coupling limit $u \to u_0$, a proper description can be given in terms of the dual superfields Φ_D and W_D^α. This is the theory of the light monopoles weakly coupled to dual photons. The corresponding dual Abelian $N = 1$ chiral superfields $B, B^\dagger$ are coupled to the dual chiral supefield Φ_D as in (13.37) and the $N = 1$ superpotential of the low-energy effective theory becomes

$$\tilde{W} = \sqrt{2}\Phi_D BB^\dagger + m\,\mathrm{Tr}\,\Phi^2. \tag{13.64}$$

Here the second term appears because of the superpotential, which beaks the $N=2$ supersymmetry of the original low-energy effective action. This is an exact form that is restricted by the non-renormalization theorem [469].

Recall that the vacuum expectation value of the scalar component of Φ_D, which appears in the low-energy effective action, is v_D and $u = \mathrm{Tr}\,\Phi^2$. Up to a gauge transformation, the vacuum state has to be a solution of the equation

$$d\tilde{W} = 0,$$

which satisfies the condition $|B| = |B^\dagger|$, so that the D-term vanishes. If $m=0$, that vacuum corresponds to $|B| = |B^\dagger| = 0$ and v_D is an arbitrary parameter. Thus, this is the $N=2$ moduli space that we considered above. However, if $m \neq 0$, we obtain

$$\sqrt{2}BB^\dagger + m\frac{du}{dv_D} = 0,$$

and $v_D B = v_D B^\dagger = 0$. If we assume that $du/dv_d \neq 0$, we get $B, B^\dagger \neq 0$. The second equation then requires $v_D = 0$ and the solution is given by

$$B = B^\dagger = \left(-mu'(0)/\sqrt{2}\right)^{1/2}. \tag{13.65}$$

Since this field is charged, its non-zero vacuum expectation value generates a mass for the dual photon via the Higgs mechanism. However, these fields that form a condensate are the monopoles, and that is precisely the picture of the dual Meissner effect, which we already discussed in Chap. 9.

Finally, let us recall that we restrict our overview to the simplest case of the softly broken $N = 2$ $SU(2)$ gauge theory with fundamental quarks of one flavor. Evidently, there are different possibilities of extending the model. One may consider, for example, the gauge groups $SU(N)$, $SO(N)$ or $USp(2N)$, as well as an arbitrary number N_f of the quark flavors compatible with the restriction of asymptotic freedom. These models exhibit a rich variety of vacua properties depending on the relation between the rank of the gauge group and N_f. However, in most cases, monopole condensation occurs. A very detailed microscopic investigation of the various regimes for the monopole condensation related to non-Abelian string junction in the Higgs phase of $N=2$ two flavor QCD was presented recently [474]. The restricted book volume does not allow us to discuss this interesting development. We refer the reader to the short overviews [333, 334], the recent paper [335] and references therein.

13.4 Concluding Remarks

A variety of results has been obtained in recent years, expanding the Seiberg–Witten approach to different models with gauge groups of higher rank and

13.4 Concluding Remarks

various multiplets representing the matter fields. However, the most remarkable achievements are probably related with the so-called *second superstring revolution*. The underlying idea here is related to the notion of the spectral curve, which was considered as an auxiliary construction for describing the quantum moduli space of the low-energy theory, with some physical object. Actually, this is the same technique that was used for more than a decade to construct the multimonopole solutions. We briefly discussed this subject in Chap. 6. It turns out that, in the context of the string theory, one can consider the spectral curves as compactification manifolds that produce a connection between a "normal" ten-dimensional string theory, and four-dimensional low-energy physical models. Furthermore, the implication of the idea of duality, which arises from the Montonen–Olive conjecture, allows us to explain the relations between all perturbative string theories, which represent different expansions around different singularities of the string moduli space. The physical meaning of the corresponding singularities is that some non-perturbative BPS states there become massless in accordance to the interpretation of the singularities of the $N = 2$ SUSY Yang–Mills moduli space. In string theory, these BPS states are the *branes*, which are classical solutions of the model. They are extended along p spatial directions and localized in remaining $d-p$ transverse directions. However, any attempt to discuss the related topics would definitely take us far beyond our main subject, the monopoles. Let us stop our brief discussion at this point.

A Representations of $SU(2)$

In this appendix we provide details of the parameterization of the group $SU(2)$ and differential forms on the group space.

An arbitrary representation of the group $SU(2)$ is given by the set of three generators T_k, which satisfy the Lie algebra

$$[T_i, T_j] = i\varepsilon_{ijk} T_k, \quad \text{with} \quad \varepsilon_{123} = 1.$$

The element of the group is given by the matrix

$$U = \exp\{i\mathbf{T} \cdot \boldsymbol{\omega}\}, \tag{A.1}$$

where in the fundamental representation $T_k = \frac{1}{2}\sigma_k$, $k = 1, 2, 3$, with

$$\sigma_1 = \begin{pmatrix} 0 & 1 \\ 1 & 0 \end{pmatrix}, \quad \sigma_2 = \begin{pmatrix} 0 & -i \\ i & 0 \end{pmatrix}, \quad \sigma_3 = \begin{pmatrix} 1 & 0 \\ 0 & -1 \end{pmatrix},$$

standard Pauli matrices, which satisfy the relation

$$\sigma_i \sigma_j = \delta_{ij} + i\varepsilon_{ijk}\sigma_k.$$

The vector $\boldsymbol{\omega}$ has components ω_k in a given coordinate frame.

Geometrically, the matrices U are generators of spinor rotations in three-dimensional space $\mathbb{R}^3$ and the parameters ω_k are the corresponding angles of rotation. The Euler parameterization of an arbitrary matrix of $SU(2)$ transformation is defined in terms of three angles θ, φ and ψ, as

$$U(\varphi, \theta, \psi) = U_z(\varphi) U_y(\theta) U_z(\psi) = e^{i\sigma_3 \frac{\varphi}{2}} e^{i\sigma_2 \frac{\theta}{2}} e^{i\sigma_3 \frac{\psi}{2}}$$

$$= \begin{pmatrix} e^{i\frac{\varphi}{2}} & 0 \\ 0 & e^{-i\frac{\varphi}{2}} \end{pmatrix} \begin{pmatrix} \cos\frac{\theta}{2} & \sin\frac{\theta}{2} \\ -\sin\frac{\theta}{2} & \cos\frac{\theta}{2} \end{pmatrix} \begin{pmatrix} e^{i\frac{\psi}{2}} & 0 \\ 0 & e^{-i\frac{\psi}{2}} \end{pmatrix}$$

$$= \begin{pmatrix} \cos\frac{\theta}{2} e^{\frac{i}{2}(\psi+\varphi)} & \sin\frac{\theta}{2} e^{-\frac{i}{2}(\psi-\varphi)} \\ -\sin\frac{\theta}{2} e^{\frac{i}{2}(\psi-\varphi)} & \cos\frac{\theta}{2} e^{-\frac{i}{2}(\psi+\varphi)} \end{pmatrix}. \tag{A.2}$$

Thus, the $SU(2)$ group manifold is isomorphic to three-sphere S^3. The Euler angles θ, φ and ψ take values within the intervals $0 \leq \theta \leq \pi$, $0 \leq \varphi \leq 2\pi$ and $0 \leq \psi \leq 4\pi$. Note that the reduction to the parametric space of the orthogonal group $SO(3)$ can be achieved, if we fix the range of values of the angle ψ to be restricted to the interval $0 \leq \psi \leq 2\pi$ and make the identification $\psi \sim \psi + 2\pi$.

Using the matrices (A.2), we can define so-called canonical left and right one-forms on the group $SU(2)$, which are also called *Maurer–Cartan one-forms* (note that $dU^{-1} U = -U^{-1} dU$)

$$R = U^{-1} dU = \frac{i}{2}\sigma_k R_k, \qquad L = dU\, U^{-1} = \frac{i}{2}\sigma_k L_k. \tag{A.3}$$

Since $\det U = 1$, we have the condition on these forms

$$d \det U = d e^{\operatorname{tr} \ln U} = \operatorname{tr} L = \operatorname{tr} R = 0.$$

The components of the Maurer–Cartan forms in the basis given by the Pauli matrices are written as

$$\begin{aligned}
R_1 &= -\sin\psi\, d\theta + \cos\psi \sin\theta\, d\varphi, & L_1 &= \sin\varphi\, d\theta - \cos\varphi \sin\theta\, d\psi, \\
R_2 &= \cos\psi\, d\theta + \sin\psi \sin\theta\, d\varphi, & L_2 &= \cos\varphi\, d\theta + \sin\varphi \sin\theta\, d\psi, \\
R_3 &= d\psi + \cos\theta\, d\varphi, & L_3 &= d\varphi + \cos\theta\, d\psi.
\end{aligned} \tag{A.4}$$

Clearly, they satisfy the *Maurer–Cartan equations*

$$dR_n = \frac{1}{2}\varepsilon_{nmk} R_m \wedge R_k, \qquad dL_n = -\frac{1}{2}\varepsilon_{nmk} L_m \wedge L_k.$$

In the same way, we can define the set of angular coordinates $\tilde\psi, \tilde\theta, \tilde\varphi$, that parameterizes the sphere $SO(3)$ and the one-forms on the space of parameters of this group.

The left and right forms on the group $SU(2)$ are dual to the vector field ξ_k, components of which form the standard basis of the Lie algebra on the group $SU(2)$:

$$\langle \xi_k^{(R)}, R_m \rangle = \delta_{km}, \qquad \langle \xi_k^{(L)}, L_m \rangle = \delta_{km}.$$

Here the right and left Killing vectors are related with generators of rotations about the corresponding axis of Cartesian coordinates. They can be written in terms of the Euler parameterization as

$$\begin{aligned}
\xi_1^{(R)} &= -\cot\theta \cos\psi \frac{\partial}{\partial \psi} - \sin\psi \frac{\partial}{\partial \theta} + \frac{\cos\psi}{\sin\theta} \frac{\partial}{\partial \varphi}, \\
\xi_2^{(R)} &= -\cot\theta \sin\psi \frac{\partial}{\partial \psi} + \cos\psi \frac{\partial}{\partial \theta} + \frac{\sin\psi}{\sin\theta} \frac{\partial}{\partial \varphi}, \\
\xi_3^{(R)} &= \frac{\partial}{\partial \psi},
\end{aligned} \tag{A.5}$$

and

$$\xi_1^{(L)} = -\frac{\cos\varphi}{\sin\theta}\frac{\partial}{\partial\psi} + \sin\varphi\frac{\partial}{\partial\theta} + \cot\theta\cos\varphi\frac{\partial}{\partial\varphi},$$

$$\xi_2^{(L)} = \frac{\sin\varphi}{\sin\theta}\frac{\partial}{\partial\psi} + \cos\varphi\frac{\partial}{\partial\theta} - \cot\theta\sin\varphi\frac{\partial}{\partial\varphi},$$

$$\xi_3^{(L)} = \frac{\partial}{\partial\varphi}, \qquad (A.6)$$

for the left and right Killing vector field, respectively. The vector fields on the parameter space of the $SO(3)$ group can be constructed in the same way.

Note that the generators of the left and right rotations commute, while left and right Killing vectors satisfy the $SU(2)$ Lie algebra

$$[\xi_m^{(R)}, \xi_n^{(R)}] = -\varepsilon_{mnk}\xi_k^{(R)}, \quad [\xi_m^{(L)}, \xi_n^{(L)}] = \varepsilon_{mnk}\xi_k^{(L)}, \quad [\xi_m^{(L)}, \xi_n^{(R)}] = 0.$$

Thus, the right one-form R_n is invariant with respect to the left action of the $SU(2)$ group while the left one-form L_n is invariant with respect to the right action of the group $SU(2)$, i.e., the corresponding Lie derivative with respect to $\xi_n^{(L)}$ and $\xi_n^{(R)}$ vanishes. The metric on the group manifold, which is constructed using the one-forms R_n, by definition is left-invariant with the Killing vectors $\xi_n^{(L)}$.

The group space of $SU(2)$ group is isomorphic to one of the "remarkable" spheres S^0, S^1, S^3 and S^7, which are characterized by the left × right parallelism.

The vector fields on the sphere S^3 are related with the angular momentum operator as

$$L_n^{(R)} = -i\xi_n^{(R)}, \qquad L_n^{(L)} = i\xi_n^{(L)}.$$

It follows from the relation (A.7) that the components of the operator of angular momentum satisfy the usual commutation relation, which does not distinguish between left and right rotations:

$$[L_n, L_m] = i\varepsilon_{nmk}L_k.$$

Eigenfunctions of the operator of angular momentum are known as *Wigner functions*

$$D_{m\mu}^l(\varphi, \theta, \psi) \equiv e^{im\varphi}d_{m\mu}^l(\theta)e^{i\mu\psi}, \qquad (A.7)$$

where $d_{m\mu}^l(\theta)$ are defined as [10]:

$$d_{m\mu}^l(\theta) = \left(\frac{(l-m)!(l+m)!}{(l-\mu)!(l+\mu)!}\right)^{\frac{1}{2}} (1-x)^{\frac{m+\mu}{2}}(1+x)^{-\frac{m-\mu}{2}}$$

$$\times P_{l+m}^{(-m-\mu, -m+\mu)}(x), \qquad (A.8)$$

$x = \cos\theta$ and $P_n^{(a,b)}(x)$ is a Jacobi polynomial

$$P_n^{(a,b)}(x) = \frac{(-1)^n}{2^n n!}(1-x)^{-a}(1+x)^{-b}\frac{d^n}{dx^n}\left[(1-x)^{a+n}(1+x)^{b+n}\right].$$

The Wigner function is related to the generalized spherical harmonics as

$$Y_{\mu l m}(\theta, \varphi) = D^l_{\mu m}(-\varphi, \theta, \varphi).$$

The matrices (A.2), which correspond to the fundamental representation of the group $SU(2)$, are particular cases of the Wigner functions: $D^{1/2}_{\mu m}(\varphi, \theta, \psi) = U(\varphi, \theta, \psi)$.

However, the difference between left and right rotations on the sphere S^3, which is hidden behind the general definition of the operator of angular momentum, reappears if we consider the ladder operators $L_\pm = L_1 \pm L_2$. Then the Wigner functions satisfy the equations

$$\begin{aligned} L_\pm^{(R)} D^l_{m\mu} &= \sqrt{l(l+1) - \mu(\mu \pm 1)} D^l_{m\mu \pm 1}, & L_3^{(R)} D^l_{m\mu} &= \mu D^l_{m\mu}; \\ L_\pm^{(L)} D^l_{m\mu} &= -\sqrt{l(l+1) - m(m \mp 1)} D^l_{m \mp 1 \mu}, & L_3^{(L)} D^l_{m\mu} &= -m D^l_{m\mu}. \end{aligned} \quad (A.9)$$

B Quaternions

Four-dimensional Euclidean space $\mathbb{R}^4$ is quite special, since it admits a natural multiplicative structure. This becomes very important in clarifying the description of the moduli spaces of the monopoles. In this Appendix, we briefly give addition material to that used in Sect. 6.5.1.

Let us consider a set of 2×2 complex matrices $\mathcal{R}^4$. It is closed under matrix addition and multiplication by real scalars and, therefore, may be considered as a real vector space. The bases of the space $\mathcal{R}^4$ are given by the set of matrices

$$e_1 = \begin{pmatrix} 1 & 0 \\ 0 & 1 \end{pmatrix} = \mathbb{I}_2, \qquad e_2 = \begin{pmatrix} 0 & -i \\ -i & 0 \end{pmatrix} = -i\sigma_1,$$

$$e_3 = \begin{pmatrix} 0 & -1 \\ 1 & 0 \end{pmatrix} = -i\sigma_2, \qquad e_4 = \begin{pmatrix} -i & 0 \\ 0 & i \end{pmatrix} = -i\sigma_3, \qquad (\text{B.1})$$

which satisfy the algebra

$$e_4 e_\mu = e_\mu e_4 = e_\mu, \qquad e_n e_m = -\delta_{nm} + \varepsilon_{nmk} e_k \qquad (n,m,k = 1,2,3). \qquad (\text{B.2})$$

The basis $\{e_\mu\}$ provides a natural isomorphism from $\mathcal{R}^4$ to $\mathbb{R}^4$ given by the mapping

$$X = e_1 x_1 + e_2 x_2 + e_3 x_3 + e_4 x_4 \to x_\mu = (x_1, x_2, x_3, x_4).$$

Since the basis $\{e_\mu\}$ is orthonormal, this mapping does not change the norm $\| X \|^2 = x_1^2 + x_2^2 + x_3^2 + x_4^2$ and such an isomorphism is an isometry. Note that a matrix X of the space $\mathcal{R}^4$ can be written as

$$X = \begin{pmatrix} x_1 - ix_4 & -ix_2 - x_3 \\ -ix_2 + x_3 & x_1 + ix_4 \end{pmatrix}, \qquad (\text{B.3})$$

and then the norm $\| X \|^2 = \det X$.

The commutation relations (B.2) is a particular case of the so-called *algebra of quaternions*. The space of quaternions $\mathcal{H}$ can be viewed as the set of complex matrices $\mathcal{R}^4$ equipped with a standard set of matrix operations, or as the vector space $\mathbb{R}^4$ with multiplicative structure.

Since e_1 is a multiplicative identity, we can drop it and write an arbitrary quaternion as $X = x_0 + x_n e_n$. The operation of the quaternionic conjugation is defined as

$$X =\longrightarrow \bar{X} = x_\mu \bar{e}_\mu = x_0 - x_n e_n .$$

Thus, if a quaternion X is considered as a matrix in $\mathcal{R}^4$, its conjugated $\bar{X}$ is the conjugated transpose matrix.

The product of two quaternions can be computed using the relations (B.2). In particular, we have $\bar{X}X = X\bar{X} = \| X \|^2$ and $\overline{XY} = \bar{Y}\bar{X}$. The real and imaginary parts of a quaternion are

$$\operatorname{Re} X = \frac{1}{2}(X + \bar{X}) = X_4, \qquad \operatorname{Im} X = \frac{1}{2}(X - \bar{X}) = X_n e_n.$$

Quaternions whose imaginary part is equal to zero are called real quaternions. A *unit quaternion* satisfies the relation $\| X \|^2 = 1$. Clearly, these quaternions correspond to the elements of $\mathcal{R}^4$ with unit determinant, that is, the group of unit quaternions is actually the group $SU(2)$. Its group space, a sphere S^3 naturally arises as a subspace of $\mathbb{R}^4$.

The quaternionic notions make many relations compact and transparent. For example, Euclidean Dirac matrices γ_μ simply become

$$\gamma_\mu = \begin{pmatrix} 0 & e_\mu \\ \bar{e}_\mu & 0 \end{pmatrix}, \qquad \{\gamma_\mu, \gamma_\nu\} = 2\delta_{\mu\nu} .$$

Since the set of unit quaternions forms the group $SU(2)$, the transformation properties of vectors and spinors can also be written in quaternionic notations. Recall that the transformations of the $SU(2)$ group can be decomposed into left and right rotations as $SU(2)_L \times SU(2)_R$. The unit quaternions X and Y can be set into correspondence with elements of these subgroups: $X \to x \in SU(2)_L$, $Y \to y \in SU(2)_R$. Then a vector quaternion transforms as $v \to Xv\bar{Y}$, while the spinor quaternions s, c, which correspond to representations of the Lorentz group $(0, \frac{1}{2})$, $(\frac{1}{2}, 0)$, respectively, transform as

$$s \to Xs; \qquad c \to Yc.$$

In this notation, the Euclidean Dirac equation for a massless spinor reads

$$\begin{pmatrix} 0 & D \\ \bar{D} & 0 \end{pmatrix} \begin{pmatrix} s \\ c \end{pmatrix} = 0,$$

where $D \equiv e_\mu D_\mu$ is the quaternionic Dirac operator and $\bar{D} = \bar{e}_\mu D_\mu$. It is decoupled into a pair of Weyl equations that descibe the massless fermion of a given chirality:

$$Dc = 0, \qquad \bar{D}s = 0.$$

Note that the operator $\bar{D}D = D_\mu D_\mu$ is the usual Laplace operator.

The (pseudo)-scalar and (pseudo)-tensor quaternions may be constructed by multiplication of the spinor and vector quaternions. Let us take, for example, two vectors $v = v_\mu e_\mu$ and $w = w_\mu e_\mu$. One easily finds that the real parts of the quaternionic products $v\bar{w}$ and $\bar{v}w$ transform like scalars while their imaginary parts transform like a self-dual and anti-self-dual antisymmetric tensor of second rank, respectively. In particular, the quaternionic equation

$$\bar{D}v = 0,$$

which defines the vector of tangent space T_v, can be written in component notation as

$$D_\mu v_\mu = 0, \qquad D_\mu v_\nu - D_\nu v_\mu = \frac{1}{2}\varepsilon_{\mu\nu\rho\sigma}D_\rho v_\sigma. \qquad (B.4)$$

The second of these equations is a self-duality equation for the tensor $F_{\mu\nu} = D_\mu v_\nu - D_\nu v_\mu$.

C SU(2) Transformations of the Monopole Potential

Let us consider the transformations that relate the monopole potential in the Abelian gauge and the hedgehog gauge, respectively. On the spatial asymptotic, the potential of the non-Abelian $SU(2)$ monopole becomes

$$A_n = A_n^a \frac{\sigma^a}{2} = \varepsilon_{amn} \frac{r_m}{er^2} \frac{\sigma^a}{2} = -\frac{1}{er^2} [\mathbf{r} \times \mathbf{T}]_n \,, \tag{C.1}$$

where the isospin operator is taken in the fundamental representation of the $SU(2)$ group: $T^a = \frac{1}{2}\sigma^a$. Cartesian components of the monopole vector potential are:

$$A_x = \frac{1}{2re} \begin{pmatrix} -\sin\theta \sin\varphi & -i\cos\theta \\ i\cos\theta & \sin\theta\sin\varphi \end{pmatrix}, \tag{C.2}$$

$$A_y = \frac{1}{2re} \begin{pmatrix} \sin\theta \cos\varphi & -\cos\theta \\ -\cos\theta & -\sin\theta\cos\varphi \end{pmatrix}, \quad A_z = \frac{\sin\theta}{2re} \begin{pmatrix} 0 & ie^{-i\varphi} \\ -ie^{i\varphi} & 0 \end{pmatrix}, \tag{C.3}$$

where we used the standard parameterization in terms of azimuthal and polar angles

$$x = r\sin\theta\cos\varphi, \quad y = r\sin\theta\sin\varphi, \quad z = r\cos\theta.$$

Clearly, the non-Abelian magnetic field, which corresponds to the potential (C.1), is regular everywhere in $\mathbb{R}^3$ but the origin $\{0\}$:

$$B_n = B_n^a \frac{\sigma^a}{2}, \quad B_n^a = \frac{1}{2}\varepsilon_{nmk} F_{mk}^a = \frac{r^a r_n}{er^4}, \tag{C.4}$$

where the field strength tensor is

$$F_{mn}^a = \partial_m A_n^a - \partial_n A_m^a - e\varepsilon_{abc} A_m^b A_n^c. \tag{C.5}$$

The matrix of $SU(2)$ transformations, which unwraps the "hedgehog" from the spherically symmetric form (C.1) to the third axis, is

$$U(\theta,\varphi) = e^{-i(\sigma\hat{\varphi})\theta/2} = e^{-i\sigma_3 \frac{\varphi}{2}} e^{-i\sigma_2 \frac{\theta}{2}} e^{i\sigma_3 \frac{\varphi}{2}} = \begin{pmatrix} \cos\frac{\theta}{2} & -\sin\frac{\theta}{2} e^{-i\varphi} \\ \sin\frac{\theta}{2} e^{i\varphi} & \cos\frac{\theta}{2} \end{pmatrix}. \tag{C.6}$$

C $SU(2)$ Transformations of the Monopole Potential

This transformation also rotate the Pauli matrices as

$$U^{-1}\sigma_k U = (\cos\varphi\hat{\theta}_k - \sin\varphi\hat{\varphi}_k)\sigma_1 + (\cos\varphi\hat{\varphi}_k + \sin\varphi\hat{\theta}_k)\sigma_2 + \hat{r}_k\sigma_3.$$

However, this transformation is singular at the south pole $\theta = \pi$. To understand the situation better, let us define a regularized polar angle

$$\Theta = \theta\frac{1+\cos\theta}{1+\cos\theta+\varepsilon^2},$$

where the parameter ε removes the singularity [49,131]. Then, the regularized matrices $\tilde{U} = U(\Theta,\varphi)$ rotate the monopole potential as

$$\tilde{U}^{-1}A_x\tilde{U} = \frac{\sin\varphi}{2er}\begin{pmatrix} \sin(\Theta-\theta) & [\cos(\Theta-\theta)-i\cos\theta\cot\varphi]e^{-i\varphi} \\ [\cos(\Theta-\theta)+i\cos\theta\cot\varphi]e^{i\varphi} & -\sin(\Theta-\theta) \end{pmatrix},$$

$$\tilde{U}^{-1}A_y\tilde{U} = -\frac{\cos\varphi}{2er}\begin{pmatrix} \sin(\Theta-\theta) & [\cos(\Theta-\theta)+i\cos\theta\tan\varphi]e^{-i\varphi} \\ [\cos(\Theta-\theta)-i\cos\theta\tan\varphi]e^{i\varphi} & -\sin(\Theta-\theta) \end{pmatrix},$$

$$\tilde{U}^{-1}A_z\tilde{U} = \frac{i\sin\theta}{2er}\begin{pmatrix} 0 & e^{-i\varphi} \\ -e^{i\varphi} & 0 \end{pmatrix}, \qquad (C.7)$$

and the affine part of the gauge transformation is

$$-\frac{i}{e}\tilde{U}^{-1}\partial_x\tilde{U} = -\frac{\sin\varphi}{er\sin\theta}\begin{pmatrix} \sin^2\frac{\Theta}{2} & \frac{1}{2}\sin\Theta e^{-i\varphi} \\ \frac{1}{2}\sin\Theta e^{i\varphi} & -\sin^2\frac{\theta}{2} \end{pmatrix}$$

$$+\frac{i\Theta'}{2er}\cos\theta\cos\varphi\begin{pmatrix} 0 & e^{-i\varphi} \\ -e^{i\varphi} & 0 \end{pmatrix},$$

$$-\frac{i}{e}\tilde{U}^{-1}\partial_y\tilde{U} = \frac{\cos\varphi}{er\sin\theta}\begin{pmatrix} \sin^2\frac{\Theta}{2} & \frac{1}{2}\sin\Theta e^{-i\varphi} \\ \frac{1}{2}\sin\Theta e^{i\varphi} & -\sin^2\frac{\theta}{2} \end{pmatrix}$$

$$+\frac{i\Theta'}{2er}\cos\theta\sin\varphi\begin{pmatrix} 0 & e^{-i\varphi} \\ -e^{i\varphi} & 0 \end{pmatrix},$$

$$-\frac{i}{e}\tilde{U}^{-1}\partial_z\tilde{U} = -\frac{i\Theta'}{2er}\sin\theta\begin{pmatrix} 0 & e^{-i\varphi} \\ -e^{i\varphi} & 0 \end{pmatrix}, \qquad (C.8)$$

where

$$\Theta' = d\Theta/d\theta = \frac{1+\cos\theta}{1+\cos\theta+\varepsilon^2}\left(1+\theta\varepsilon^2\frac{1-\sin\theta}{1+\cos\theta+\varepsilon^2}\right)$$

is singular in the limit $\varepsilon^2 \to 0$.

Thus, the smoothed gauge transformation of the $SU(2)$ monopole potential on the spatial asymptotic

$$A_n^{\text{Abelian}} = \frac{1}{2}A_n^{a\,\text{Str}}\sigma^a = \tilde{U}^{-1}A_n\tilde{U} - \frac{i}{e}\tilde{U}^{-1}\partial_n\tilde{U},$$

gives the regularized form of the potential in the Abelian gauge $A_n{}^{\text{Abelian}}$:

$$A_n{}^{\text{Abelian}} = -\frac{1}{2er}\left\{\hat{\varphi}_n\left(\frac{\cos\Theta-1}{\sin\theta}+\sin(\Theta-\theta)\right)\sigma_3\right.$$
$$\left.+\left[\left(\cos(\Theta-\theta)-\frac{\sin\Theta}{\sin\theta}\right)\hat{\varphi}_n\sigma_1+(\Theta'-1)\hat{\theta}_n\sigma_2\right]\begin{pmatrix}e^{i\varphi}&0\\0&e^{-i\varphi}\end{pmatrix}\right\} \quad (\text{C.9})$$

or

$$A_n^{a\,\text{Abelian}} = \frac{1}{er}\left\{\left[\frac{1-\cos\Theta}{\sin\theta}-\sin(\Theta-\theta)\right]\hat{\varphi}_n\delta_{a3}\right.$$
$$-\left[(1-\Theta')\hat{\theta}_n\sin\varphi+\left(\cos(\Theta-\theta)-\frac{\sin\Theta}{\sin\theta}\right)\hat{\varphi}_n\cos\varphi\right]\delta_{a1}$$
$$\left.+\left[(1-\Theta')\hat{\theta}_n\cos\varphi-\left(\cos(\Theta-\theta)-\frac{\sin\Theta}{\sin\theta}\right)\hat{\varphi}_n\sin\varphi\right]\delta_{a2}\right\}.$$

The same transformation $\tilde{U}$ of the non-Abelian magnetic field $B_n = B_n^a\sigma^a/2$, where B_n^a is given by (C.4), yields

$$B_n \to \tilde{U}^{-1}B_n\tilde{U} = \frac{r_n}{2er^3}\begin{pmatrix}\cos(\Theta-\theta)&-\sin(\Theta-\theta)e^{-i\varphi}\\-\sin(\Theta-\theta)e^{i\varphi}&-\cos(\Theta-\theta)\end{pmatrix},$$

that is

$$B_n^a \to -\frac{r_n}{er^3}\left(\delta_{a3}\cos(\Theta-\theta)-\sin(\Theta-\theta)\left(\delta_{a1}\cos\varphi+\delta_{a2}\sin\varphi\right)\right).$$

In the naive limit $\Theta \to \theta$, we would obviously recover the Coulomb field of the Abelian monopole without any singular pieces, while the potential (C.9) would take the form of a Dirac monopole potential embedded into $SU(2)$ group:

$$A_n^{\text{Abelian}} \to \frac{1}{2er}\frac{1-\cos\theta}{\sin\theta}\hat{\varphi}_n\sigma_3. \quad (\text{C.10})$$

However, the singularity at $\theta = \pi$ requires more careful treatment. Indeed, although the isotropic components A_n^1, A_n^2 of the non-Abelian vector potential vanish as we take the limit $\Theta \to \theta$, they still contribute to the field strength tensor F_{mn}^a. Indeed, the third component of the non-Abelian magnetic field (C.5) is

$$B_n^3 = \varepsilon_{nmk}\partial_m A_k^3 - e\varepsilon_{nmk}A_m^1 A_k^2. \quad (\text{C.11})$$

Clearly, the differentiation of the singular Dirac potential at the first term here produces not only Coulomb magnetic field we expected, but also a singular flux of the Dirac string (see (1.49)):

$$B_n = \frac{r_n}{er^3} - \frac{4\pi}{e}\hat{z}\,\theta(-z)\delta(x)\delta(y). \quad (\text{C.12})$$

512   C $SU(2)$ Transformations of the Monopole Potential

The non-Abelian nature of the potential we are considering modifies this result, because the contribution of the second term in (C.11) is also non-zero. Indeed, the piece

$$\Delta B_n = \lim_{\varepsilon^2 \to 0} e\varepsilon_{nmk} A_m^1 A_k^2 = \lim_{\varepsilon^2 \to 0} \frac{r_n}{er^3}(1 - \Theta')\left(\cos(\Theta - \theta) - \frac{\sin\Theta}{\sin\theta}\right)$$

does not vanish at $\theta = \pi$ due to the singularity of the derivative Θ'.

Let us consider it at the vicinity of this point as a distribution on the volume measure $r^2 \sin\theta d\theta d\varphi$. Then the non-vanishing contribution of ΔB_n takes the form

$$\lim_{\delta \to 0} \lim_{\varepsilon^2 \to 0} r^2 \int_{\pi-\delta}^{\pi} \sin\theta d\theta \int_0^{2\pi} d\varphi \Delta B_n = -\frac{2\pi \hat{z}}{e} \lim_{\delta \to 0} \lim_{\varepsilon^2 \to 0} \int_{\pi-\delta}^{\pi} d\theta \Theta' \sin\theta$$

$$= \frac{2\pi \hat{z}}{e} \lim_{\delta \to 0} \lim_{\varepsilon^2 \to 0} \cos\Theta(\theta)\Big|_{\pi-\delta}^{\pi} = \frac{4\pi}{e} \hat{z}\,. \qquad (C.13)$$

Therefore,

$$\Delta B_n = \frac{4\pi}{e} \hat{z}\, \theta(-z)\delta(x)\delta(y)\,,$$

which precisely cancels the string singularity of the field of the Dirac monopole. Thus, the field of the $SU(2)$ monopole contains no singularity in the Abelian gauge [131].

References

References of General Character

1. A.I. Akhiezer and V.B. Berestetskii: *Quantum Electrodynamics* (Interscience, New York 1965)
2. M. Abramowitz and I.A. Stegun: *Handbook of Mathematical Functions* (Government Printing Office, Washington 1972)
3. M. Audun: *Spinning Tops* (Cambridge University Press 1996)
4. A.P. Balachandran, G. Marmo, B.-S. Skagerstam and A. Stern: *Gauge Symmetries and Fibre Bundles* (Springer, Berlin Heidelberg New York 1983)
5. V.G. Bagrov and D.M. Gintman: *Exact Solutions of Relativistic Wave Equations* (Kluwer Academic Publishers, Dordrecht Boston London 1990)
6. V.B. Berestetskii, E.M. Lifshitz and L.P. Pitaevskii: *Quantum Electrodynamics* (Pergamon Press, Oxford 1982)
7. R. Bertlmann: *Anomalies in Quantum Field Theory* (Clarendon Press, Oxford 1996)
8. P.F. Byrd and M.D. Friedman: *Handbook of Elliptic Integrals for Engineers and Scientists* (Springer, Berlin Heidelberg New York 1971)
9. B.A. Dubrovin, A.T. Fomenko and S.P. Novikov: *Modern Geometry – Methods and Applications*, 3 v, (Springer, Berlin Heidelberg New York 1984, 1985, 1990)
10. A.R. Edmonds: *Angular Momentum in Quantum Mechanics* (Princeton University Press, Princeton 1960)
11. J.E. Humphreys: *Introduction to Lie Algebras and Representation Theory* (Springer, Berlin Heidelberg New York 1972)
12. H. Jeffreys and B. Jeffreys: *Methods of Mathematical Physics* (Cambridge University Press, Cambridge 1962)
13. J.D. Jackson: *Classical Electrodynamics*, 3rd edn (Wiley, New York 1998)
14. A.M. Jaffe and E. Witten: *Quantum Yang-Mills Theory*, Clay Mathematics Institute Millenium Prize Problem, 2000
15. V.G. Kiselev, Ya.M. Shnir and A.Ya. Tregubovich: *Introduction to Quantum Field Theory* (Gordon and Breach Science Pulishers, Amsterdam 2000)
16. S.V. Ketov: *Quantum Non-linear Sigma-Model* (Springer, Berlin Heidelberg New York 2000)
17. S. Kobayashi and K. Nomizu: *Foundations of Differential Geometry*, Vols. I, II (Wiley-Interscience, New York 1963, 1969)
18. L.D. Landau and E.M. Lifshitz: *Mechanics* (Pergamon Press, New York 1969)

19. L.D. Landau and E.M. Lifshitz: *The Classical Theory of Fields*, 4th edn (Pergamon Press, New York 1975)
20. V.G. Makhankov, Y.P. Rybakov and V.I. Sanyuk: *The Skyrme model: Fundamentals, Methods, Applications* (Springer, Berlin Heidelberg New York 1993)
21. G.L. Naber: *Topology, Geometry, and Gauge Fields: Foundations* (Springer, Berlin Heidelberg New York 1997)
22. M. Nakahara: *Geometry, Topology and Physics* (A. Hilger, Bristol 1990)
23. C. Nash: *Differential Topology and Quantum Field Theory* (Academic Press, London 1991)
24. A.M. Polyakov: *Gauge Fields and Strings* (Harwood Academic Publishers 1987)
25. R. Radjaraman: *Solitons and Instantons* (North-Holland Publishers 1982)
26. G. Ripka: *Dual Superconductor Models of Color Confinement*, Lecture Notes in Physics, Vol. 639 (Springer, Berlin Heidelberg New York 2004)
27. V.A. Rubakov: *Classical Theory of Gauge Fields* (Princeton University Press 2002)
28. A.S. Schwarz: *Quantum Field Theory and Topology* (Springer, Berlin Heidelberg New York 1993)
29. J. Wess and J. Bagger: *Supersymmetry and Supergravity* (Princeton University Press 1983)
30. R.S. Ward and R.O. Wells: *Twistor Geometry and Field Theory* (Cambridge University Press 1990)
31. G. Wentzel: *Quantum Theory of Fields* (Interscience, New York 1949)

Other Magnetic Monopole Bibliographies

32. R.A. Carrigan: *Magnetic Monopole Bibliography: 1973-1976*, Fermilab preprint 77/42 (1977)
33. R.E. Craven and W.P. Trower: *Magnetic Monopole Bibliography: 1981–1982*, Fermilab preprint 82/96 (1982)
34. G. Giacomelli et al.: *Magnetic Monopole Bibliography*, Bologna preprint DFUB-2000-09 (2000); hep-ex/0005041
35. A.S. Goldhaber and W.P. Trower: *Magnetic Monopoles* (American Association of Physics Teachers, College Park 1990)
36. J. Ruzicka and V.P. Zrelov: *Fifty Years of Dirac Monopole: Complete Bibliography*, Dubna preprint JINR-1-2-80-850 (1980)
37. D.M. Stevens: *Magnetic Monopoles: an Updated Bibliography*, Virginia preprint VPI-EPP-73-5 (1973)

Reviews, Books and Conference Proceedings

38. J. Arafune and H. Sugawara (eds):*Monopoles and Proton Decay. Proceedings of the Workshop on Monopoles and Proton Decay* (Tsukuba 1983)
39. M.F. Atiyah and N.J. Hitchin: *The Geometry and Dynamics of Magnetic Monopoles* (Princeton University Press 1988)
40. F.A. Bais: *To be or not to be? Magnetic Monopoles in Non-Abelian Gauge Theories*. In *Fifty Years of Yang–Mills Theory* (World Scientific, Singapore 2004).

41. M. Blagojević and P. Senjanović: Phys. Rep. **157** (1988) 233
42. R.A. Carrigan and W.P. Trower (eds):*Magnetic Monopoles. Proceedings of the NATO Advanced Study Institute on Magnetic Monopoles* (Plenum Press, New York London 1983)
43. S. Coleman: *The Magnetic Monopole Fifty Years Later*. In *Proceedings of the 19th International School of Subnuclear Physis, Erice, Italy* ed by A. Zichichi (Plenum Press, New York London 1983) pp 21–117
44. N.S. Craigie, P. Goddard and W. Nahm (eds):*Monopoles in Quantum Field Theory. Proceedings of the ICTP Monopole Meeting* (Trieste 1982)
45. N.S. Craigie, G. Giacomelli, W. Nahm and Q. Shafi: *Theory and Detection of Magnetic Monopoles in Gauge Theories* (World Scientific, Singapore 1986)
46. G. Giacomelli: *Magnetic Monopoles* Riv. Nuovo Cim. **7N12** (1984) 1
47. G. Giacomelli: *Magnetic Monopole Scarches*, Bologna preprint DFUB-20-94 (1994) In *Lake Louise 1994, Proceedings, Particle physics and cosmology* pp 150–191
48. G. Giacomelli and L. Patrizii: *Magnetic Monopole Searches*, Bologna preprint DFUB-2003-1 (2003) In *Trieste 2002, Astroparticle physics and cosmology* (ICTP, Trieste 2003) p. 121
49. P. Goddard and D. Olive: Rep. Prog. Phys. **41** (1978) 1357
50. D.E. Groom: Phys. Rep. **140** (1986) 323
51. N.J. Hitchin: *Monopoles, Minimal Surfaces and Algebraic Curves* (Les Presses de l'Université de Montréal, 1987)
52. A. Jaffe and C.H. Taubes: *Vortices and Monopoles* (Birkhäuser, Boston 1980)
53. I.G. Koh: *Magnetic Monopoles in Gauge Unified Theories*. In *Monopoles, Solitons and Nonlinear Phenomena: Proceedings of the 2nd Summer Symposium on Theoretical Physics* ed H.S. Song (Min Eum Sa, Seoul 1983) pp 39–81
54. N. Manton and P. Sutcliffe: *Topological Solitons* (Cambridge University Press 2004)
55. J. Preskill: *Magnetic Monopoles*, Ann. Rev. of Nucl. and Part. Science, **34** (1984) 461
56. V.A. Rubakov: Rep. Prog. Phys. **51** (1988) 189
57. J.L. Stone (ed): *Monopole '83. Proceedings of NATO Advanced Research Workshop on Monopole* (Plenum Press, New York London 1984)
58. P.M. Sutcliffe: Int. J. of Mod. Phys. **A12** (1997) 4663
59. V.I. Strazhev and L.M. Tomilchik: Sov. J. Part. Nucl., **4** (1973) 78
60. V.I. Strazhev and L.M. Tomilchik: *Electrodynamics with a Magnetic Charge* (Nauka i Tekhnika Publ., Minsk 1975)

Research Papers

61. A. Aharonov and D. Bohm: Phys. Rev. **115** (1959) 485
62. A. Abouelsaood: Nucl. Phys. **B226** (1983) 309
63. A. Abouelsaood: Phys. Lett. **B125** (1983) 467
64. A.A. Abrikosov: Sov. Phys. JETP **5** (1957) 1174
65. A. D'Adda, R Horslay and P. Di Vecchia: Phys. Lett. **B76** (1978) 298
66. S.L. Adler: Phys. Rev. **D18** (1978) 411

67. S.L. Adler: Phys. Rev. **D19** (1979) 2997
68. I.K. Affleck and N.S. Manton: Nucl. Phys. **B194** (1982) 38
69. I.J.R. Aitchison: Acta Phys. Pol. **B18** (1986) 207
70. R. Akhoury, J.-H. Jun and A.S. Goldhaber: Phys. Rev. **D21** (1980) 454
71. A.I. Alekseev: Theor. Math. Phys. **77** (1988) 1273
72. A.I. Alekseev: *Motion of Color Charge in the Field of Chromoelectric "Hedgehog"*, Protvino preprint IHEP/85-89 (1985) unpublished
73. L. Alvarez-Gaumé and S.F. Hassan: Fortsch. Phys. **45** (1997) 159
74. M.M. Ansourian: Phys. Rev. **D14** (1976) 2732
75. S.M. Apenko: Sov. Phys. JETP **67** (1988) 1995
76. J. Arafune, P.G.O. Freund and C.J. Goebel: Journal of Math. Phys. **16** (1975) 433
77. V.I. Arnold: Phys. Usp. **42** (1999) 1205
78. M.F. Atiyah and I. Singer: Ann. Math. **87** (1968) 484
79. M.F. Atiyah and R.S. Ward: Comm. Math. Phys. **55** (1977) 117
80. M.F. Atiyah, V.G. Drinfeld, N.J. Hitchin and Yu.I. Manin: Phys. Lett. **A65** (1978) 185
81. M.F. Atiyah and N.J. Hitchin: Phys. Lett. **A107** (1985) 21
82. M.F. Atiyah and N.J. Hitchin: Phil. Trans. R. Soc. Lond., **A315** (1985) 459
83. M.F. Atiyah: *Magnetic Monopoles in Hyperbolic Space*. In *Michael Atiyah: Collected Works 5 v* (Clarendon Press, Oxford 1988) vol 5 pp 579–611
84. J. Baacke: Z. Phys. **C53** (1992) 399
85. F.A. Bais and J.R. Primack: Phys. Rev. **D13** (1976) 819
86. F.A. Bais: Phys. Lett. **B64** (1976) 465
87. F.A. Bais: Phys. Rev. **D18** (1978) 1206
88. F.A. Bais and H.A. Weldon: Phys. Rev. Lett. **41** (1978) 601
89. F.A. Bais and B.J. Schroers: Nucl. Phys. **B512** (1998) 250
90. F.A. Bais and B.J. Schroers: Nucl. Phys. **B535** (1998) 197
91. D. Bak and C. Lee: Nucl. Phys. **B403** (1993) 315
92. D. Bak and C. Lee: Nucl. Phys. **B424** (1994) 124
93. D. Bak and C. Lee: Phys. Rev. **D57** (1998) 5239
94. D. Bak and C. Lee: Phys. Lett., **B468** (1999) 76
95. D. Bak, K. Hashimoto, B. Lee, H. Min and N. Sasakura: Phys. Rev. **D60** (1999) 046005
96. D. Bak, C. Lee, K. Lee and P. Yi: Phys. Rev. **D61** (2000) 025001
97. D. Bak, K. Lee and P. Yi: Phys. Rev. **D61** (2000) 045003
98. M. Baker, J. Ball and F. Zachariasen: Phys. Rev. **D37** (1988) 1036
99. M. Baker, J. Ball and F. Zachariasen: Phys. Rep. **209** (1991) 73
100. M. Baker, J. Ball and F. Zachariasen: Phys. Rev. **D44** (1991) 3949
101. M. Baker, J. Ball and F. Zachariasen: Phys. Rev. **D47** (1993) 3021
102. M. Baker, J. Ball and F. Zachariasen: Phys. Rev. **D51** (1995) 1968
103. M. Baker et al.: Phys. Rev. **D54** (1996) 2829
104. M. Baker, N. Brambilla, H.G. Dosch and A. Vairo: Phys. Rev. **D58** (1998) 034010
105. A.P. Balachandran, G. Marmo, B.-S. Skagerstam and A. Stern: Nucl. Phys. **B162** (1980) 385
106. A.P. Balachandran, et al.: Phys. Rev. Lett. **50** (1983) 1553
107. A.P. Balachandran and J. Schechter: Phys. Rev. Lett. **51** (1983) 1418

108. A.P. Balachandran and J. Schechter: Phys. Rev. **D29** (1984) 1184
109. G.S. Bali, V. Bornyakov, M. Müller-Preussker and K. Schilling: Phys. Rev. **D54** (1996) 2863
110. P. Banderet: Helv. Phys. Acta **19** (1946) 503
111. T. Banks, R. Myerson and J. Kogut: Nucl. Phys. **B129** (1977) 493
112. T. Banks and A. Casher: Nucl. Phys. **B169** (1980) 103
113. A.O. Barut: Phys. Rev. **D3** (1971) 1747
114. A.O. Barut and H. Becker: Nuovo Cim. **A19** (1974) 309
115. A.O. Barut: Phys. Rev. **D3** (1969) 1747
116. A.O. Barut and G. Bornzin: J. Math. Phys. **11** (1971) 896
117. A.O. Barut, Ya.M. Shnir and E.A. Tolkachev: Journal of Phys. **A26** (1993) L101
118. O. Bergman: Nucl. Phys. **B525** (1998) 104
119. O. Bergman and B. Kol: Nucl. Phys. **B536** (1998) 149
120. W. Bernreuther and M. Suzuki: Rev. Mod. Phys. **63** (1991) 313
121. K.J. Biebl and J. Wolf: Nucl. Phys. **B279** (1987) 571
122. R. Bielawski: Comm. Math. Phys. **199** (1998) 297
123. A. Bilal: *Duality in $N = 2$ SUSY $SU(2)$ Yang–Mills Theory: a Pedagogical Introduction to the Work of Seiberg and Witten.* In *Cargese 1996, Quantum Fields and Quantum Space Time* ed by G. 't Hooft et al.. NATO ASI **B 364** (1996) 21; hep-th/9601007.
124. A. Bilal: *Introduction to Supersymmetry*, Neuchâtel preprint NEIP-01-001 (2001); hep-th/0101055
125. A. Blaer, N. Christ and J.F. Tang: Phys. Rev. Lett. **47** (1981) 1364
126. A. Blaer, N. Christ and J.F. Tang: Phys. Rev. **D25** (1982) 2128
127. E.B. Bogomol'nyi: Sov. J. Nucl. Phys. **24** (1976) 449
128. S. Boguslavskyi: *Electron Paths in Electromagnetic Fields*, Moscow, 1929 (in Russian); *Selected works on physics* (Fizmatgiz, Moscow 1961)
129. S.K. Bose: Jour. of Phys. **A18** (1985) 1289
130. S.K. Bose: Jour. of Phys. **G12** (1986) 1135
131. D.G. Boulware et al.: Phys. Rev. **D14** (1976) 2708
132. M.C. Bowman: Phys. Rev. **D32** (1985) 1569
133. N. Brambilla and A. Vairo: Phys. Rev. **D55** (1997) 3974
134. R. Brandt and J. Primack: Phys. Rev. **D15** (1977) 1175
135. R.A. Brandt and F. Neri: Phys. Rev. **D18** (1978) 2080
136. R.A. Brandt, F. Neri and D.Zwanziger: Phys. Rev. Lett. **40** (1978) 147
137. R.A. Brandt, F. Neri and D.Zwanziger: Phys. Rev. **D19** (1979) 1153
138. R.A. Brandt and F. Neri: Nucl. Phys. **B161** (1979) 253
139. Y. Brihaye, B. Hartmann and J. Kunz: Phys. Lett. **B441** (1998) 77
140. Y. Brihaye, B. Hartmann, J. Kunz and N. Tell: Phys. Rev. **D60** (1999) 104016
141. Y. Brihaye, B. Hartmann and J. Kunz: Phys. Rev. **D62** (2000) 044008
142. Y. Brihaye and B. Piette: Phys. Rev. **D64** (2001) 084010
143. Y. Brihaye, V.A. Rubakov, D.H. Tchrakian and F. Zimmerschied: Theor. Math. Phys. **128** (2001) 1140
144. Y. Brihaye, D.Y. Grigoriev, V.A. Rubakov and D.H. Tchrakian: Phys. Rev. D **67** (2003) 034004
145. L.S. Brown, R.D. Carlitz and C. Lee: Phys. Rev. **D16** (1977) 417
146. L.S. Brown and D.B. Creamer: Phys. Rev. **D18** (1978) 3695

147. S.A. Brown, H. Panagopoulos and M.K. Prasad: Phys. Rev. **D26** (1982) 854
148. N. Cabibbo and E. Ferrari: Nuovo Cim., **23** (1962) 1147
149. C.G. Callan and D.J. Gross: Nucl. Phys. **B93** (1975) 29
150. C.G. Callan, R.F. Dashen and D.J. Gross: Phys. Rev. **D17** (1978) 2717
151. C.G. Callan: Phys. Rev. **D25** (1982) 2141
152. C.G. Callan: Phys. Rev. **D26** (1982) 2058
153. C.G. Callan: Nucl. Phys. **B212** (1983) 391
154. C.G. Callan and E. Witten: Nucl. Phys. **B239** (1984) 161
155. C.J. Callias: Phys. Rev. **D16** (1977) 3068
156. C.J. Callias: Comm. Math. Phys. **62** (1978) 213
157. G. Calucci, R. Jengo and M.T. Vallon: Nucl. Phys. **B197** (1982) 93
158. G. Calucci and R. Jengo: Nucl. Phys. **B223** (1983) 501
159. G. Calucci, R. Jengo and M.T. Vallon: Nucl. Phys. **B211** (1983) 77
160. J. Cardy and E. Rabinovici: Nucl. Phys. **B205** (1982) 1
161. A. Chakrabarti: Ann. Inst. H. Poincaré **23** (1975) 235
162. G. Chalmers: *Multi-Monopole Moduli Space for SU(N) Gauge Groups*, SUNY Stony Brook preprint ITP-SB-96-12 (1996); hep-th/9605182
163. M.N. Chernodub and F.V. Gubarev: JETP Lett. **62** (1995) 100
164. M.N. Chernodub, M.I. Polikarpov and V.I. Veselov: Phys. Lett. **B342** (1995) 303
165. M.N. Chernodub and M. Polikarpov: *Abelian Projections and Monopoles*. In *Cambridge 1997, Confinement, Duality and Nonperturbative Aspects of QCD* ed P. Van Baal, NATO ASI **B 368** (1998) 387; hep-th/9710205.
166. N. Christ, A. Guth and E.J. Weinberg: Nucl. Phys. **B114** (1976) 174
167. N. Christ and R. Jackiw: Phys. Lett. **B91** (1980) 228
168. S.Y. Chu: Phys. Rev. **D7** (1973) 853
169. S. Coleman and J. Mandula: Phys. Rev. **159** (1967) 1251
170. S. Coleman and E. Weinberg: Phys. Rev. **D7** (1973) 1888
171. S. Coleman, S. Parke, A. Neveu and C. Sommerfield: Phys. Rev. **D15** (1977) 544
172. S. Coleman: Phys. Rev. **D11** (1975) 2088
173. S. Coleman: Phys. Rev. **D15** (1977) 2929
174. E. D. Commins et al.: Phys. Rev. **D50** (1994) 2960
175. S. Connell: *The Dynamics of the SU(3) (1,1) Magnetic Monopoles*, PhD Thesis, University of South Australia, 1994
176. C. Corben and J. Schwinger: Phys. Rev. **58** (1940) 953
177. E. Corrigan, D. Olive, D. Fairlie and J. Nugts: Nucl. Phys. **B106** (1976) 475
178. E. Corrigan and D.Olive: Nucl. Phys. **B110** (1976) 237
179. E. Corrigan and D. Fairlie: Phys. Lett. **67** (1977) 69
180. E. Corrigan, D. Fairlie P. Goddard and R.G. Yates: Comm. Math. Phys. **58** (1978) 223
181. E. Corrigan and P. Goddard: Comm. Math. Phys. **80** (1981) 575
182. P. Cox and A. Yildiz: Phys. Rev. **D18** (1978) 1211
183. N.S. Craigie, W. Nahm and V.A. Rubakov: Nucl. Phys. **B241** (1984) 274
184. G. Cvetič and T.M. Yan: Nucl. Phys. **B299** (1988) 587
185. A.S. Dancer: Nonlinearity **5** (1992) 1355
186. A.S. Dancer: Comm. Math. Phys. **158** (1993) 545

187. A.S. Dancer and R.A. Leese: Proc. Roy. Soc. **440** (1993) 421
188. A.S. Dancer and R.A. Leese: Phys. Lett. **B390** (1997) 252
189. S. Dawson and A.N. Schellekens: Phys. Rev. **D27** (1983) 2119
190. S. Dawson and A.N. Schellekens: Phys. Rev. **D28** (1983) 3125
191. W. Deans: Nucl. Phys. **B197** (1982) 307
192. L. Del Debbio, A. Di Giakomo, G. Pafitti and P. Pier: Phys. Lett. **B355** (1995) 255
193. T. Dereli, J.H. Swank and L.J. Swank: Phys. Rev. **D11** (1975) 3541
194. T. Dereli, J.H. Swank and L.J. Swank: Phys. Rev. **D12** (1975) 1096
195. D. Diakonov and V. Petrov: Nucl. Phys. **B245** (1984) 259; ibid. **B272** (1986) 457
196. D. Diakonov: *Chiral Symmetry Breaking by Instantons.* In *Varenna 1995, Selected Topics in Nonperturbative OCD* ed by A. Di Giacomo and D. Diakonov (The Netherlands, IOS 1996) pp 397-432; hep-ph/9602375
197. D. Diakonov: *Chiral Quark-Soliton Model.* In *Peniscola 1997, Advanced School on Non-perturbative Quantum Field Physics* ed by M. Asorey and A. Dobado (World Scientific, Singapore 1998) pp 1-55; hep-ph/9802298.
198. K. Dietz and Th. Filk: Nucl. Phys. **B164** (1980) 536
199. R. Dijkgraaf: *Les Houches lectures on fields, strings and duality.* In *Les Houches 1995, Quantum symmetries* ed by A. Connes et al. (North-Holland, Amsterdam 1998) pp 3-147; hep-th/9703136
200. P.A.M. Dirac: Proc. Roy. Soc. **A133** (1931) 60
201. P.A.M. Dirac: Phys. Rev. **74** (1948) 817
202. P.A.M. Dirac: Canad. J. Math. **2** (1950) 129
203. C. Dokos and T. Tomaras: Phys. Rev. **D21** (1980) 2940
204. S.K. Donaldson: Comm. Math. Phys. **96** (1984) 387
205. S.K. Donaldson and P.B. Kronheimer: *The Geometry of Four-Manifolds* (Clarendon Press, Oxford 1990)
206. T. Eguchi and A.J. Hanson: Ann. Phys. (N.Y.) **120** (1979) 267
207. T. Eguchi, P. Gilkey and A. Hanson: Phys. Rep. **66** (1980) 213
208. U. Ellwanger: Nucl. Phys. **B560** (1999) 587
209. J. Ellis, D.V. Nanopoulos and K.A. Olive: Phys. Lett. **B116** (1982) 127
210. F. Englert and P. Windey: Phys. Rev. **D14** (1976) 2728
211. Z.F. Ezawa and A. Iwazaki: Z. Phys. **C 20** (1983) 335
212. L.D. Faddeev and V.E. Korepin: Phys. Rep. **C42** (1978) 1
213. L. Feher: Acta Phys. Pol. **B15** (1984) 919
214. M. Fierz: Helv. Phys. Acta **17** (1944) 27
215. R. Flume, L. O'Raifeartaigh and I. Sachs: *Brief Resume of Seiberg–Witten Theory*, Dublin preprint DIAS-STP-96-22 (1996); hep-th/9611118
216. A. Font et al.: Phys. Lett. **B249** (1990) 35
217. P. Forgács, Z. Horváth and L. Palla: Phys. Rev. Lett. **45** (1980) 505
218. P. Forgács, Z. Horváth and L. Palla: Ann. Phys. **136** (1981) 371
219. P. Forgács, Z. Horváth and L. Palla: Phys. Lett. **99B** (1981) 232
220. P. Forgács, Z. Horváth and L. Palla: *Physicist's Techniques for Multi-Monopole Solutions.* In *Monopoles in Quantum Field Theory* ed by N.S. Craigie, P. Goddard and W. Nahm (World Scientific, Singapore 1982) pp 21–57
221. P. Forgács, N. Obadia and S. Reuillon: *Numerical and Asymptotic Analysis of the 't Hooft–Polyakov Magnetic Monopole*, Tours preprint LMPT/YMU (2004); hep-th/0412057

222. E. Fradkin and L. Susskind: Phys. Rev. **D17** (1978) 2637
223. P.H. Frampton: *'t Hooft Monopoles and Singular Gauge Transformations*, preprint UCLA/76/TEP/12 (1976) unpublished
224. C. Fraser and T.J. Hollowod: Phys. Lett. **B402** (1997) 106
225. A. Frenkel and P. Hraskó: Ann. Phys. (N.Y.) **105** (1977) 288
226. J. Fröhlich and P.A. Marchetti: Europhys. Lett. **2** (1986) 933
227. J. Fröhlich and P.A. Marchetti: Comm. Math. Phys. **112** (1987) 343
228. K. Fujii, S. Otsuki and F. Toyoda: Progr. Theor. Phys. **81** (1979) 462
229. N. Ganoulis, P. Goddard and D. Olive: Nucl. Phys. **B205** [**FS5**] (1982) 601
230. J.P. Gauntlett: Nucl. Phys. **B411** (1994) 443
231. J.P. Gauntlett and D.A. Lowe: Nucl. Phys. **B472** (1996) 194
232. J.P. Gauntlett, C. Kim, J. Park and P. Yi: Phys. Rev. **D61** (2000) 125012
233. J.P. Gauntlett, C. Kim, K. Lee and P. Yi: Phys. Rev. **D63** (2001) 065020
234. H. Georgi and S.L. Glashow: Phys. Rev. **D6** (1972) 2977
235. J.L. Gervais and B. Sakita: Phys. Rev. **D11** (1975) 2943
236. J.L. Gervais, A. Jevicki and B. Sakita: Phys. Rev. **D12** (1975) 1038
237. G.W. Gibbons and C.N. Pope: Comm. Math. Phys. **66** (1979) 267
238. G.W. Gibbons and N.S. Manton: Nucl. Phys. **B274** (1986) 183
239. G.W. Gibbons and N.S. Manton: Phys. Lett. **B356** (1995) 32
240. C.P. Ginsparg: Nucl. Phys. **B170** (1980) 388;
    T. Appelquist and R.Pisarski: Phys. Rev. **D38** (1981) 2305;
    N.P. Landsman: Nucl. Phys. **B438** (1989) 498
241. P. Goddard, J. Nuyts and D. Olive: Nucl. Phys. **B125** (1977) 1
242. C. Goebel and M. Thomaz: Phys. Rev. **D30** (1984) 823
243. J.N. Goldberg, P.S. Jang, S.Y. Park and K. Wali: Phys. Rev. **D18**, 542 (1978)
244. A.S. Goldhaber: Phys. Rev. **140** (1965) B1407
245. A.S. Goldhaber: Phys. Rev. **D16** (1976) 1122
246. A.S. Goldhaber: Phys. Rev. **D16** (1977) 1815
247. M. Goodband: *Gauge Boson Monopole two Particle Bound States and Duality*. Sussex preprint SUSX-TH-96-019 (1996); hep-th/9612123, unpublished
248. V.N. Gribov: Physica Scripta **T15** (1987) 164;
    Eur. Phys. J. **C10** (1999) 71; hep-ph/980724;
    Eur. Phys. J. **C10** (1999) 91; hep-ph/9902279
249. D.Y. Grigoriev, P.M. Sutcliffe and D.H. Tchrakian: Phys. Lett. **B540** (2002) 146
250. B. Grossmann: Phys. Rev. Lett. **50** (1983) 464
251. A. Guth and E.J. Weinberg: Phys. Rev. **D14** (1976) 1660
252. R. Haag, J. Lopuszański and M. Sohnius: Nucl. Phys. **B88** (1975) 257
253. M.B. Halpern: Phys. Rev. **D12** (1975) 1684
254. Haris-Chandra: Phys. Rev. **74** (1948) 883
255. A. Hart and M. Teper: Phys. Rev. **D55** (1995) 3756
256. A. Hart and M. Teper: Phys. Lett. **B371** (1996) 261
257. A. Hart and M. Teper: Phys. Rev. **D58** (1998) 014504
258. B. Hartmann, B. Kleihaus and J. Kunz: Phys. Rev. Lett. **86** (2001) 1422
259. B. Hartmann, B. Kleihaus and J. Kunz: Phys. Rev. **D65** (2002) 024027
260. J.A. Harvey and A. Strominger: Comm. Math. Phys. **151** (1993) 221

261. P. Hasenfratz and G. 't Hooft: Phys. Rev. Lett. **36** (1976) 1119
262. P. Hasenfratz and D.A. Ross: Nucl. Phys., **B108** (1976) 482; Phys. Lett. **B64** (1976) 78
263. O. Heaviside: Phil. Trans. Roy. Soc. **183** (1893) 423
264. W. Heisenberg and H. Euler: Z. Phys. **98** (1936) 714;
V. Weisskopf: Mat. Fys. Medd. Dan. Vid. Selsk. **XIV** (1936) 6
265. S. Hioki et al.: Phys. Lett. **B272** (1991) 326
266. N.J. Hitchin: Comm. Math. Phys. **83** (1982) 579
267. N.J. Hitchin: Comm. Math. Phys. **89** (1983) 145
268. N.J. Hitchin, N.S. Manton and M.K. Murray: Nonlinearity **8** (1995) 661
269. E. D'Hoker and L. Vinet: Phys. Lett. **B137** (1984) 72
270. G. 't Hooft: Nucl. Phys. **B79** (1974) 276
271. G. 't Hooft: *Gauge Fields with Unified Weak, Electromagnetic, and Strong Interactions.* In *High Energy Physics*, Proceedings of the EPS International Conference, Palermo 1975, ed by A. Zichichi (Editrice Compositori, Bologna 1976) v 2 pp 1225-1249
272. G. 't Hooft: Phys. Rev. Lett. **37** (1976) 8
273. G. 't Hooft: Nucl. Phys. **B105** (1976) 538
274. G. 't Hooft: Nucl. Phys. **B138** (1978) 1
275. G. 't Hooft: Nucl. Phys. **B190** (1981) 455
276. G. 't Hooft: Physica Scripta **25** (1982) 133
277. G. 't Hooft: *Monopoles, Instantons and Confinement*, Lectures given at the 5th We-Heraeus Doktorandenschule Saalburg: Grundlagen und Neue Methoden der Theoretischen Physik, Saalburg, Germany, September 1999; hep-th/0010225
278. H. Hopf: Math. Ann. **104** (1930-31) 637
279. Z. Horváth and L. Palla: Phys. Rev. **D14** (1976) 1711
280. Y. Hosotani: Phys. Lett. **B69** (1997) 499
281. A. Hosoya and K. Kikkawa: Nucl. Phys. **B101** (1975) 271
282. C.J. Houghton and P.M. Sutcliffe: Nucl. Phys. **B464** (1996) 59
283. C.J. Houghton and P.M. Sutcliffe: Comm. Math. Phys. **180** (1996) 342
284. C.J. Houghton and P.M. Sutcliffe: Nonlinearity **9** (1996) 385
285. C.J. Houghton: *Multimonopoles* PhD thesis, University of Cambridge (1997)
286. C.J. Houghton, N.S. Manton and P.M. Sutcliffe: Nucl. Phys. **B510** (1998) 507
287. C.J. Houghton and K. Lee: Phys. Rev. **D61** (2000) 106001
288. P. Houston and L. O'Raifeartaigh: Phys. Lett. **B94** (1980) 153
289. P. Houston and L. O'Raifeartaigh: Z. Phys. **C8** (1981) 175
290. K. Huang and D.R. Stump: Phys. Rev. Lett. **37** (1976) 545; Phys. Rev. **D15** (1977) 3660
291. C.A. Hurst: Ann. Phys. **50** (1968) 52
292. J. Hurtubise: Comm. Math. Phys. **92** (1983) 195
293. J. Hurtubise: Comm. Math. Phys. **100** (1985) 191
294. Th. Ioannidou and P. Sutcliffe: Phys. Rev. **D60** (1999) 105009
295. P. Irwin: Phys. Rev. **D56** (1997) 5200
296. P. Irwin: *Radiation from SU(3) monopole scattering*, Montréal preprint UDEM-GPP-TH-2000-67 (2000); hep-th/0004054 unpublished
297. K. Isler, C. Schmidt and C.A. Trugenberger: Nucl. Phys. **B294** (1987) 925

298. R. Jackiw, and C. Rebbi: Phys. Rev. **D13** (1976) 3398; Phys. Rev. Lett. **36** (1976) 1116
299. R. Jackiw: Rev. Mod. Phys. **49** (1977) 681
300. R. Jackiw: Ann. Phys. (N.Y.) **129** (1980) 183
301. R. Jackiw and N. Manton: Ann. Phys. (N.Y.) **127** (1980) 257
302. J.P. Jacobs et al.: Phys. Rev. **A52** (1995) 3521
303. S. Jarvis: J. Reine Angew. Math. **524** (2000) 17
304. B. Julia and A. Zee: Phys. Rev. **D11** (1975) 2227
305. P. Jordan: Annalen der Physik **5** (1938) 66
306. Y. Kazama, C.N. Yang and A.S. Goldhaber: Phys. Rev. **D15** (1977) 2287
307. Y. Kazama and C.N. Yang: Phys. Rev. **D15** (1977) 2300
308. Y. Kazama: Phys. Rev., **D16** (1977) 3078
309. Y. Kazama: Progr. Theor. Phys. **70** (1983) 1166
310. Y. Kazama and A. Sen: Nucl. Phys. **B247** (1984) 190
311. S.V. Ketov: Fortsch. Phys. **45** (1997) 237
312. V.G. Kiselev and K.G. Selivanov: Phys. Lett. **B213** (1988) 165
313. V.G. Kiselev: Phys. Lett. **B249** (1990) 269
314. V.G. Kiselev and Ya.M. Shnir: Phys. Rev. **D57** (1997) 5174
315. T.W. Kirkman and C.K. Zachos: Phys. Rev. **D24** (1981) 999
316. D.A. Kirzhnits: Sov. Phys. JETP **71** (1990) 427
317. D.A. Kirzhnits: Sov. Phys. Usp. **30** (1987) 575
318. D.A. Kirzhnits and V.V. Losyakov: JETP Lett. **42** (1985) 279
319. H. Kleinert and W. Miller: Phys. Rev. **D38** (1988) 1239
320. B. Kleihaus, J. Kunz and A. Sood: Phys. Rev. **D54** (1996) 5070
321. B. Kleihaus and J. Kunz: Phys. Rev. Lett. **78** (1997) 2527
322. B. Kleihaus and J. Kunz: Phys. Rev. **D57** (1998) 6138
323. B. Kleihaus, J. Kunz, A. Sood and M. Wirschins: Phys. Rev. **D58** (1998) 084006
324. B. Kleihaus, J. Kunz and D.H. Tchrakian: Mod. Phys. Lett. **A13** (1998) 2523
325. B. Kleihaus: Phys. Rev. **D59** (1999) 125001
326. B. Kleihaus, D.H. Tchrakian and F. Zimmerschied: J. Math. Phys. **41** (2000) 816
327. B. Kleihaus and J. Kunz: Phys. Rev. **D61** (2000) 025003
328. B. Kleihaus, J. Kunz and Ya. Shnir: Phys. Lett. **B570**, (2003) 237
329. B. Kleihaus, J. Kunz and Ya. Shnir: Phys. Rev. **D68** (2003) 101701
330. B. Kleihaus, J. Kunz and Ya. Shnir: Phys. Rev. **D70** (2004) 065010
331. I. Kogan and A. Kovner: Phys. Rev. **D51** (1995) 1948
332. R.V. Konoplich: Nucl. Phys. **B323** (1989) 660
333. K. Konishi, *Who Confines Quarks? On Non-Abelian Monopoles and Dynamics of Confinement*. In *Nagoya 2002, Strong Coupling Gauge Theories and Effective Field Theories* ed M. Harada et al. (World Scientific, Singapore 2003) pp 34–52; hep-th/0304157
334. K. Konishi: Acta Phys. Polon. **B34** (2003) 3129
335. R. Auzzi, S. Bolognesi, J. Evslin, K. Konishi and Hitoshi Murayama: Nucl. Phys. **B701** (2004) 207
336. S.G. Kovalevich, P. Osland, Ya.M. Shnir and E.A. Tolkachev: *The Effective Lagrangian of QED with a Magnetic Charge*, ICTP preprint IC/188/95, Trieste 1995; hep-th/9601133

337. S.G. Kovalevich, P. Osland, Ya.M. Shnir and E.A. Tolkachev: Phys. Rev. **D55** (1997) 5807
338. H.A. Kramers and G.H. Wannier: Phys. Rev. **60** (1941) 252
339. A.S. Kronfeld, G. Schierholz and U.-J. Wiese: Nucl. Phys. **B293** (1987) 461
340. A.S. Kronfeld, M.S. Laursen, G. Schierholz and U.-J. Wiese: Phys. Lett. **B198** (1987) 516
341. J. Kunz and D. Masak: Phys. Lett. **B196** (1987) 513
342. J.S. Langer: Ann. of Phys. (N.Y.) **41** (1967) 108; **54** (1969) 258
343. J.R. Lapidus and J.L. Pietenpol: Am. J. of Phys. **28** (1960) 17
344. K. Lee, E.J. Weinberg and Piljin Yi: Phys. Rev. **D54** (1996) 6351
345. K. Lee, E.J. Weinberg and Piljin Yi: Phys. Lett. **B376** (1996) 97
346. K. Lee, E.J. Weinberg and Piljin Yi: Phys. Rev. **D54** (1996) 1633
347. W. Lerche, *Lecture on N=2 Supersymmetric Gauge Theory* in *Les Houches 1995, Quantum symmetries*, ed by A. Connes et al. (Amsterdam, North-Holland 1998) pp 613–640
348. A.N. Lesnov and M.V. Saveliev: Lett. Math. Phys. **3** (1979) 489
349. A.N. Lesnov and M.V. Saveliev: Comm. Math. Phys. **74** (1980) 111
350. S.B. Libby: Nucl. Phys. **B113** (1976) 50
351. H. Lipkin, W.I. Weisberger and M. Peshkin: Ann. Phys. (N.Y.) **53** (1969) 203
352. Changhai Lu: Phys. Rev. **D58** (1998) 125010
353. K. Lee and P. Yi: Phys. Rev. **D58** (1998) 066005
354. Changhai Lu: Phys. Rev. **D58** (1998) 125010
355. E. Lubkin: Ann. Phys. **23** (1963) 233
356. D. Lynden-Bell and M. Nouri-Zonoz: Rev. of Mod. Phys. **70** (1998) 427
357. J.D. Lykken, *Introduction to Superymmetry.* In *Theoretical Advanced Study Institute in Elementary Particle Physics (TASI 96): Fields, Strings, and Duality* ed by C. Efthimiou and B. Greene (World Scientific, Singapore 1997) pp 85–153; hep-th/9612114
358. J. Madore: Phys. Rep. **75** (1981) 127
359. S. Maedan and T. Suzuki: Prog. Theor. Phys. **81** (1989) 229
360. W.V.R. Malkus: Phys. Rev. **83** (1951) 899
361. S. Mandelstam: Ann. Phys. (N.Y.) **19** (1961) 1
362. S. Mandelstam: Phys. Lett. **53** (1975) 476
363. S. Mandelstam: Phys. Rev. **D11** (1975) 3026
364. S. Mandelstam: Phys. Rep. **23** (1976) 245
365. S. Mandelstam: *The Possible Role of Monopoles in the Confinement Mechanism.* In *Monopoles in Quantum Field theory*, ed by N.S. Craigie, P. Goddard and W. Nahm (World Scientific, Singapore, 1982) pp 289–313
366. N.S. Manton: Nucl. Phys. **B126** (1977) 525
367. N.S. Manton: Nucl. Phys. **B135** (1978) 319
368. N.S. Manton: Phys. Lett. **B110** (1982) 54
369. N.S. Manton: Phys. Lett. **B154** (1985) 397
370. N.S. Manton: Phys. Lett. **B198** (1987) 226
371. N.S. Manton: Phys. Rev. Lett. **60** (1988) 1916
372. N.S. Manton and T.M. Samols: Phys. Lett. **B215** (1988) 559
373. N.S. Manton and B.J. Schroers: Ann. of Phys. **225** (1993) 290
374. W.J. Marciano and H. Pagels: Phys. Rev. **D12** (1975) 1093

375. W.J. Marciano and I.J Muzinich: Phys. Rev. **D28** (1983) 973
376. W.J. Marciano and I.J Muzinich: Phys. Rev. Lett. **50** (1983) 1035
377. S.F. Margruder: Phys. Rev. **D17** (1978) 3257
378. A. Marshakov: *Seiberg–Witten Theory and the Integrable Systems* (World Scientific, Singapore 1999)
379. B. Martemyanov, S. Molodtsov, Yu. Simonov and A. Veselov: JETP Lett. **62** (1995) 695
380. Yu.S. Mavrutchev: Izv. Vuz. USSR, Physics **9**, (1970) 129
381. C.W. Misner and A.H. Taub: Sov. Phys. JETP **28** (1969) 122
382. T.F. Mitchel and J.A. Burns: J. Math. Phys. **9** (1968) 2016
383. M.I. Monastyrski and A.M. Perelomov: Pis'ma JETP **21** (1975) 94
384. C. Montonen and D. Olive: Phys. Lett. **B72** (1977) 117
385. E. Mottola: Phys. Lett. **B79** (1978) 242; Phys. Rev., **D19** (1979) 3170
386. M.K. Murray: Comm. Math. Phys. **196** (1984) 539
387. M.K. Murray: Comm. Math. Phys. **125** (1989) 661.
388. G. Nadeau: Am. J. of Phys. **28** (1960) 566.
389. W. Nahm: Phys. Lett. **B79** (1978) 426
390. W. Nahm: Phys. Lett. **B85** (1979) 373
391. W. Nahm: Phys. Lett. **B90** (1980) 413
392. W. Nahm: *All Self-Dual Monopoles for Arbitrary Gauge Groups*, CERN preprint, TH-3172 (1981); *The Algebraic Geometry of Multimonopoles*, Bonn preprint HE-82-30 (1982)
393. W. Nahm: *The Construction of All Self-Dual Monopoles by ADHM Method*. In *Monopoles in Quantum Field Theory* ed by N.S. Craigie, P. Goddard and W. Nahm (World Scientific, Singapore 1982) pp 87–94
394. H. Nakajima: *Monopoles and Nahm's equations*. In *Einstein Metrics and Yang–Mills Connections: Proceedings of the 27th Taniguchi International Symposium* ed by T. Mabuchi and S. Mukai (Marcel Dekker, New York 1993) pp 193–211
395. Y. Nambu: Phys. Rep. **C23** (1976) 250
396. P.N. Nelson: Phys. Rev. Lett. **50** (1983) 939
397. P.N. Nelson and A. Manohar: Phys. Rev. Lett. **50** (1983) 943
398. P.N. Nelson and S. Coleman: Nucl. Phys. **B237** (1984) 1
399. H. Nielsen and P. Olesen: Nucl. Phys. **B61** (1973) 45
400. H. Nielsen and B.Schroer: Nucl. Phys. **B120** (1977) 62
401. L.I. Nicolaescu: *Notes on Seiberg–Witten Theory* AMS Graduate Studies in Mathematics V 28 (2000)
402. V. Novikov, M.A. Shifman, A. Vainstein and V.I. Zakharov: Nucl. Phys. **B 229** (1983) 381, 407
403. K. Olaussen, H. Olsen, P. Osland and I. Overbo: Nucl. Phys. **B228** (1983) 567
404. K. Olaussen, H. Olsen, I. Overbo and P. Osland: Phys. Rev. Lett. **52** (1984) 325
405. K. Olaussen, H. Olsen, P. Osland and I. Overbo: Nucl. Phys. **B267** (1986) 1; ibid 25
406. M. Oleszczuk and E. Werner: Phys. Rev. **D35** (1987) 3225
407. H. Olsen, P. Osland and T.T. Wu: Phys. Rev. **D42** (1990) 665
408. H. Olsen and P. Osland: Phys. Rev. **D42** (1990) 690
409. H. Osborn: Phys. Lett. **B83** (1979) 321

410. H. Osborn: *Semiclassical Methods for Quantising Monopole Field Configurations*. In *Monopoles in Quantum Field Theory* ed by N.S. Craigie, P. Goddard and W. Nahm (World Scientific, Singapore 1982) pp 193–227
411. P. Osland and T.T. Wu: Nucl. Phys. **B247** (1984) 421
412. P. Osland and T.T. Wu: Nucl. Phys. **B247** (1984) 450
413. P. Osland and T.T. Wu: Nucl. Phys. **B256** (1985) 13
414. P. Osland and T.T. Wu: Nucl. Phys. **B256** (1985) 32
415. P. Osland, C.L. Schultz and T.T. Wu: Nucl. Phys. **B256** (1985) 449
416. P. Osland and T.T. Wu: Nucl. Phys. **B261** (1985) 687
417. N.K. Pak and R. Percacci: Nucl. Phys. **B188** (1981) 355
418. N.K. Pak, C.J. Panagiotakopoulos and Q. Shafi: *Magnetic Monopoles and Baryon Decay*, ICTP Preprint IC/82/174 Trieste (1982)
419. C.J. Panagiotakopoulos: J. Phys. **A16** (1983) 133
420. H. Panagopoulos: Phys. Rev. **D28** (1983) 380
421. V. Paturyan and T. Tchrakian: J. Math. Phys. **45** (2004) 302
422. M. Peshkin: Ann. Phys. (N.Y.) **66** (1971) 542
423. M.E. Peskin: Ann. Phys. (N.Y.) **113** (1978) 122
424. M.E. Peskin: *Duality in Supersymmetric Yang–Mills Theory*, SLAC preprint PUB-7393 (1997); hep-th/9702094.
425. H. Poincaré: Compt. Rend. Acad. Sci. **123** (1896) 530
426. J. Polchinski: Nucl. Phys. **B1242** (1984) 345
427. L. Polley and U.J. Wiese: Nucl. Phys. **B356** (1991) 629
428. A.M. Polyakov: Pis'ma JETP **20** (1974) 430
429. A.M. Polyakov: Phys. Lett. **B59** (1975) 82
430. A.M. Polyakov: Nucl. Phys. **B120** (1977) 429
431. M.K. Prasad and C.M. Sommerfield: Phys. Rev. Lett. **35** (1975) 760
432. M.K. Prasad: Physica **D1** (1980) 167;
M.K. Prasad and P. Rossi: Phys. Rev. **D24** (1981) 2182
433. M.K. Prasad and P. Rossi: Phys. Rev. Lett. **46** (1981) 806
434. M.K. Prasad: Comm. Math. Phys. **80** (1981) 137
435. Y. Qi: Phys. Lett. **B176** (1986) 115
436. A. Rabl: Phys. Rev. **179** (1969) 1363
437. L. O'Raifeartaigh, S.Y. Park and K. Wali: Phys. Rev. **D20**, (1979) 1941
438. N.F. Ramsey: Phys. Rev. **109** (1958) 225
439. F. Rohrlich: Phys. Rev. **150** (1966) 1104
440. C. Rebbi and P. Rossi: Phys. Rev. **D22** (1980) 2010
441. A. Ritz, M. Shifman, A. Vainshtein and M. Voloshin: Phys. Rev. **D63** (2001) 065018
442. P. Rossi: Nucl. Phys. **B127** (1977) 518
443. P. Rossi: Nucl. Phys. **B149** (1979) 170
444. P. Rossi: Phys. Rep. **86** (1982) 317
445. V.A. Rubakov: Pis'ma JETP **33** (1981) 658
446. V.A. Rubakov: Nucl. Phys. **B203** (1982) 311
447. V.A. Rubakov and M.S. Serebryakov: Nucl. Phys. **B218** (1983) 240
448. B. Rüber: *Eine axialsymmetrische magnetische Dipollösung der Yang–Mills–Higgs Gleichungen*. MS Thesis, University of Bonn (1985)
449. A. De Rújula: Nucl. Phys. **B435** (1995) 257
450. K. Rzazewski: Acta. Phys. Pol. **B2** (1971) 707
451. M.N. Saha: Ind. J. Phys. **10** (1936) 145; Phys. Rev. **75** (1949) 1968

452. T. Schäfer and E. Shuryak: Rev. Mod. Phys. **70** (1998) 323
453. J. Schechter: Phys. Rev. **D14** (1976) 534
454. A.N. Schellekens: Phys. Rev. **D29** (1984) 2378
455. A.N. Schellekens: Nucl. Phys. **B246** (1984) 494
456. K. Schilling, G.S. Bali and C. Schlichter: Nucl. Phys. Proc. Suppl. **73** (1999) 638
457. J. Schwinger: Phys. Rev. Lett. **3** (1959) 296
458. J. Schwinger: Phys. Rev. **127** (1962) 324; ibid **130** (1963) 800
459. J. Schwinger: Phys. Rev. **144** (1966) 1087; ibid **151** (1966) 1048
460. J. Schwinger: Phys. Rev. **173** (1968) 1536.
461. J. Schwinger: Science **165** (1969) 757; ibid **166** (1969) 690
462. J. Schwinger et al.: Ann. of Phys. (N.Y.) **101** (1976) 451
463. S. Shabanov: Mod. Phys. Lett. **A11** (1996) 1081
464. H. Suganuma, H. Ichie, A. Tanaka and K. Amemiya: Progr. Theor. Phys. Suppl. **131** (1998) 559
465. H. Shiba and T. Suzuki: Phys. Lett., **B333** (1994) 461
466. Ya. Shnir, E.A. Tolkachev and L.M. Tomilchik: Int. J. Mod. Phys. **A7** (1992) 3747
467. Ya. Shnir: Physica Scripta **69** (2004) 15
468. N. Seiberg: Phys. Lett. **B206** (1988) 75
469. N. Seiberg and E. Witten: Nucl. Phys. **B426** (1994) 19; ibid **B430** (1994) 485
470. A. Sen: Phys. Rev. **D28** (1983) 876
471. A. Sen: Phys. Rev. Lett, **52** (1984) 1755
472. A. Sen: Phys. Lett. **B329** (1994) 217
473. M. Shifman: Prog. Part. Nucl. Phys. **39** (1997) 1
474. M. Shifman and A. Yung: Phys. Rev. **D70** (2004) 045004
475. Yu. Simonov: Phys. Usp. **39** (1996) 313
476. Yu. Simonov: Sov. J. Yad. Phys. **42** (1985) 557
477. Yu. Simonov: *Dyons in QCD: Confinement and Chiral Symmetry breaking*. In *Varenna 1995, Selected Topics in Nonperturbative OCD* ed by A. Di Giacomo and D. Diakonov (The Netherlands, IOS 1996) pp 339–364; hep-ph/9509403.
478. T.H.R. Skyrme: Nucl. Phys. **31** (1962) 556
479. J. Smit and A. van der Sijs: Nucl. Phys. **B355** (1991) 603
480. M. Sohnius: Phys. Rep. **128** (1985) 39
481. H. Sonoda: Phys. Lett. **B143** (1984) 142
482. J.D. Stack, S.D. Neiman and R.J. Wensley: Phys. Rev. **D50** (1994) 3399
483. M. Stone: Phys. Lett. **B67** (1976) 186
484. T. Suzuki et al.: Nucl. Phys. Proc. Suppl. **B26** (1992) 441
485. T. Suzuki et al.: *The Dual Meissner Effect and Abelian Magnetic Displacement Currents*, preprint Kanazawa 04-15 (2004); hep-lat/0410001
486. I.E. Tamm: Z. Phys. **71** (1931) 141
487. J.F. Tang: Phys. Rev. **D26** (1982) 510
488. C.H. Taubes: Comm. Math. Phys. **81** (1981) 299
489. C.H. Taubes: Comm. Math. Phys. **86** (1982) 257
490. C.H. Taubes: Comm. Math. Phys. **91** (1983) 235
491. C.H. Taubes: Comm. Math. Phys. **95** (1984) 345

492. V.I. Tereshenkov, E.A. Tolkachev and L.M. Tomilchik: *Lienard–Wiechert Potentials and Nonstatic Monopole Type Solutions*, Minsk preprint, IPAS No 553 (1989) unpublished
493. R.V. Tevikyan: Sov. JETP **50** (1966) 911
494. E.A. Tolkachev, L.M. Tomilchik and Ya.M. Shnir: Sov. J. Nucl. Phys. **38** (1983) 320
495. E.A. Tolkachev, L.M. Tomilchik and Ya.M. Shnir: J. Phys. **G14** (1988) 1
496. E.A. Tolkachev and Ya.M. Shnir: Sov. J. Nucl. Phys. **55** (1992) 1596
497. E.A. Tolkachev and L.M. Tomilchik: Soviet Topics of History of Science and Technics **4** (1990) 71
498. L.M. Tomilchik: Sov. JETP **17** (1963) 111
499. E. Tomboulis and G. Woo: Nucl. Phys., **B107** (1976) 221
500. J.J. Tompson: Philos. Mag. **8** (1904) 331
501. D. Tong: Phys. Lett. **B460** (1999) 295
502. Yu.S. Typkin, V.A. Fateev and A.S. Schwarz: Pis'ma JETP **21** (1975) 91
503. S.N. Vergeles and S.B. Khokhlachev: Sov. J. Nucl. Phys **26** (1977) 883
504. J. Villain: J. Phys.(Paris) **36** (1975) 581
505. M.B. Voloshin, I.Yu. Kobzarev and L.B. Okun: Sov. J. Nucl. Phys. **20** (1975) 644
506. R.S. Ward: Phys. Lett. **A61** (1977) 81
507. R.S. Ward: Comm. Math. Phys. **79** (1981) 317
508. R.S. Ward: Phys. Lett. **B102** (1981) 136
509. R.S. Ward: Phys. Lett. **B107** (1981) 281
510. R.S. Ward: Comm. Math. Phys. **86** (1982) 437
511. R.S. Ward: Phys. Lett. **B158** (1985) 424
512. S. Weinberg: Phys. Rev. **138** (1965) B488
513. E.J. Weinberg: Phys. Rev. **D20** (1979) 936
514. E.J. Weinberg: Nucl. Phys. **B167** (1980) 500
515. E.J. Weinberg: Nucl. Phys. **B203** (1982) 445
516. E.J. Weinberg: Phys. Rev. **D49** (1994) 1080
517. E.J. Weinberg and Piljin Yi: Phys. Rev. **D58** (1998) 046001
518. E.J. Weinberg: *Massive Monopoles and Massless Monopole Clouds*. In *Proceedings of International Workshop on Mathematical and Physical Aspects of Nonlinear Field Theories* ed by Dong-Ho Chae and Sung-Ki Kim (RIM-GARC Lecture Notes Series No. 42, Seoul); hep-th/9908097
519. E.J. Weinberg: *Massive and Massless Monopoles and Duality*, Columbia preprint CU-TP-946 (1999); hep-th/9908095
520. G. Wentzel: Suppl. Progr. of Theor. Phys. **37–38** (1966) 163
521. K.G. Wilson: Phys. Rev. **D10** (1974) 2445
522. E. Witten: Phys. Rev. Lett. **38** (1977) 121;
    R. Jackiw, C. Nohl and C. Rebbi: Phys. Rev. **D15** (1977) 1642
523. E. Witten and D. Olive: Phys. Lett., **B78** (1978) 97
524. E. Witten: Phys. Lett. **B86** (1979) 283
525. S.R. Wadia: Phys. Rev. **D15** (1977) 3615
526. D. Wilkinson and F.A. Bais: Phys. Rev. **D19** (1979) 2410
527. A. Wipf: J. Phys. **A18**, (1985) 2379
528. T.T. Wu and C.N. Yang: In *Properties of Matter under Unusual Conditions*, eds H. Mark and S. Fernbach (Interscience, New York 1969) 349

529. T.T. Wu and C.N. Yang: Phys. Rev. **D12** (1975) 3845
530. T.T. Wu and C.N. Yang: Nucl. Phys. **B107** (1976) 365
531. T.T. Wu and C.N. Yang: Phys. Rev. **D16** (1977) 1018
532. T.T. Wu: Nucl. Phys. **B222** (1983) 411
533. H. Yamagishi: Phys. Rev. **D27** (1983) 2383
534. H. Yamagishi: Phys. Rev. **D28** (1983) 977
535. H. Yamagishi: Phys. Rev. **D32** (1985) 1576
536. H. Yamagishi: Phys. Rev. **D32** (1985) 2113
537. T.M. Yan: Phys. Rev. **150** (1966) 1349; ibid **155** (1967) 1423
538. T.M. Yan: Phys. Rev. **D28** (1983) 1496
539. C.N. Yang: Phys. Rev. Lett. **38** (1977) 1377
540. T. Yoneya: Nucl. Phys. **B 232** (1984) 356
541. K. Young: Nuovo Cim. **B35** (1976) 195
542. M.N. Chernodub, F.V. Gubarev, M.I. Polikarpov and V.I. Zakharov: Nucl. Phys. **B592** (2001) 107
543. K. Zarembo: Nucl. Phys. **B463** (1996) 73
544. Jian-zu Zhang: Europhys. Lett., **10** (1989) 639.
545. Jian-zu Zhang: J. of Phys. **G16** (1990) 195
546. Jian-zu Zhang: Phys. Rev. **D41** (1990) 1280; Erratum-ibid. **D44** (1991) 2618
547. B. Zumino: *Recent Developments in the Theory of Magnetically Charged Particles*. In *Strong and Weak Interactions – Present Problems* ed by A. Zichichi (Academic Press, New York 1966) 711
548. D. Zwanziger: Phys. Rev. **176** (1968) 1489
549. D. Zwanziger: Phys. Rev. **176** (1968) 1480
550. D. Zwanziger: Phys. Rev. **D3** (1971) 880

Index

C-conjugation   361
$SU(5)$ unification theory   397
α-plane   205
β-plane   205
θ-angle   59
θ-term   171
θ-vacuum   378
$\mathcal{R}$-symmetry   419
't Hooft tensor   347
1/4-BPS states   457

Kähler manifold   89

Abrikosov-Nielsen-Olesen string   325
Affine transformation   99
Almost complex structure   84
Almost complex structures   232
Angular momentum generalized   5, 20, 43, 200, 245
Atiyah–Hitchin metric   234
Atiyah–Ward matrix   209
Atiyah-Ward patching matrix   208
Atlas   71

Background gauge   162
Baker-Campbell-Hausdorff formula   421
Banderet potential   19
Baryon number   401
Berezin integration   423
Bianchi identities   145
BPS equations   157
BPS limit   152
Brouwer degree   150, 170
Bundle Instanton   101
Bundle Principal   97
Bundle Trivial   96

Characteristic Classes   99

Characteristic Polynomial   100
Charge lattice   63
Charge matrix   279
Charge quantization condition   28, 61, 150
Chern class   169
Chern form   100
Chiral condensate   321
Chiral massive N=1 SUSY multiplet   416
Chiral symmetry   320
Chiral symmetry breaking   320
Christoffel symbols   86
Clifford vacuum   416
Coleman–Mandula theorem   411
Coleman–Weinberg effect   470
Coleman-Weinberg potential   251
Collective coordinates   162
Compact QED   331
Confinement phase   471
Connection   98
Coulomb phase   470
Crossover   322
Current electric   186
Current topological   169

Dancer space   315
Debye mass   350
DeRham cohomology group   100
Diffeomorphism   77
Differential form   82
Differential forms   84
Differentiation Exterior   85
Dilaton   197
Dilute monopole gas   351
Diogen atom   37
Dirac potential   13
Dirac string   14

Dirac's veto 112
Dual current 23
Dual group 284
Dual lattice 335
dual Meissner effect 325
Dual prepotential 479
Dual root lattice 284
Dual transformation 112
Duality Electrodynamics 22, 24
Duality transformation 91
Dynamical supersymmetry 45
Dyon 6

Effective action 467
Electrodynamics Two-potential formulations 113
Elliptic curve 75
Euler–Poinsot equations 237
Euler-Poinsot equations 221
Extended Supersymmetry 412

Fiber bundle 94
First order constraint 257
Flat direction 230, 449
Flux tube 326
Fock–Schwinger formalism 123
Form Closed 85
Form Exact 85
Form Harmonic 87
Free magnetic phase 471
Free phase 471
Fubini-Study metric 204
Function Holomorphic 72

Gauge Abelian 148, 174
Gauge massless N=1 SUSY multiplet 418
Gauge massless N=2 SUSY multiplet 418
Gauge Zero mode 162
Gauge zero mode 163
Gauss law 161, 255, 257
Gauss-Bonnet formula 101
Georgi-Glashow model 144
Gluon condensate 321
Graded Lie algebra 412
Grassmann variables 44
Gribov copies 340

Haag-Lopuszański-Sohnius theorem 412
Haar measure 330
Higgs phase 471
Higgs vacuum 145
Hitchin equation 215
Hodge ∗-operation 91
Hodge star operation 85
Holomorphic charge 300
Holonomy group Local 86
Holonorphic charges 279
Homeomorphism 77
Homotopy 78
Homotopy class 79
Homotopy classes 167
Homotopy group 80, 168
Hooft tensor 149
Hopf fibration 105
hyper-Kähler manifold 89
Hyperkähler manifold 232

Index of the elliptic operator 369
Instanton 347
Instanton chain 159
Instanton liquid model 321
Isometry 86, 201

Jackiw-Rebbi ansatz 364, 374
Jacobi polynomial 503
Jacobi polynomials 35
Josephson effect 357
Julia–Zee correspondence 156

Kähler manifolds 204
Kähler potential 432, 440
Kaluza-Klein model 201
Killing vectors 502
Kleihaus-Kunz ansatz 180
Kleihaus-Kunz axially symmetric ansatz 181

Lax equations 223
Legendre transformation 310
Level crossing 379
Lienard–Wiechert potential 199
Line directed 203
Lorentz Group 408
Lubkin theorem 176

Möbius strip 95

Index 531

Magnetic charge 152
Magnetic cloud 302, 463
Magnetic current 149
Magnetic dipole 184
Magnetic mass 350
Magnetic mirror 50
Magnetic mirror effect 4
Magnetic orbit 300
Majorana spinor 409
Manifold 70
Manifold Almost complex 84
Manifold Complex 71
Manifold differentiable 71
Manifold Linearly connected 80
Manifold Riemannian 81
Manifold Simply connected 80
Marginal stability 456
Mass matrix 278
Massless N=2 SUSY hypermultiplet 418
Maurer–Cartan equations 502
Maximal embedding 287
Maximal symmetry breaking 278
Metastable vacuum decay 272
Metric 86
Metric Hermitian 88
Mini-twistor space 215
Minimal symmetry breaking 278
Modular group 64
Modular transformations 76
Moduli 467
Moduli space $SU(3)$ monopole 312
Moduli space metric 230
Moduli space N=2 SUSY Yang-Mills theory 467
Moduli space of monopole 163
Monodromy 208, 477, 485
Monopole catalysis 390
Monopole Core 153
Monopole dominance 348
Monopole harmonics 248
Montonen-Olive conjecture 241
Montonen-Olive duality 280

N=1 anti-chiral superfield 425
N=1 chiral superfield 424
N=1 rigid superspace 421
N=1 vector superfield 425
N=2 chiral superfield 439

N=2 dual chiral superfield 478
N=2 prepotential 440
N=2 supercurrent 440
Nahm equations 219
Non-linear sigma-model 401, 432, 440

O'Raifeartaigh theorem 411
Orbifold 468

Pauli-Lubanski vector 410
Poincare algebra 408
Poincaré–Hopf index 150
Poisson sum 335
Polyakov line 330
Pontryagin index 169, 378
Potential Wu-Yang Abelian 142
primary constraints 264
Projection 94
Projective complex space 204
Projective dual space 207
Projective plane 203
Projective quaternionic space 205
Pullback 87

Quenved QCD 322

R-symmetry 414
Rational map approach 225
Riemann–Hilbert problem 208
Root diagrams 283

Schwinger model 381
Schwinger potential 18
Secondary BPS equation 445
Section 98
Short $N=2$ SUSY multiplets 419
Simple roots 282
Skyrme model 400
Skyrmed monopoles 401
Spectral curve 213, 216, 223
Spherical harmonics generalized 35
Spin–statistics theorem 31
Spin-1 operator 245
Stereographic projection 73
Stokes theorem 87
String tension 322
Strong confinement 320
Structure group 95
Superfield 420

Superpotential   430
Superspace   420
Supersymmetric field strength   427
Symmetry dynamical   21

Tangent bundle   94
Taub–NUT metric   201
Taub-NUT metric   234, 311
Topological space   70
Twistor space   204
Two-torus   75

Vector field   81
Vector magnetic charge   288
Vector massive N=1 SUSY multiplet   417
Vector spherical harmonics   248
Vertex operator   391
Villain action   332
Vortex rings   191

Watson–Sommerfeld formula   130
Weak confinement   319
Wedge product   84
Weierstrass function   75
Wess-Zumino gauge   426
Weyl reflection   282
Wigner functions   503
Wilson loop   322, 329
Winding number   79, 169
Witten effect   59, 263
Wu–Yang potential   152
Wu-Yang non-Abelian potential   175

Zamolodchikov metric   473
Zero mode   51, 58
Zero mode fermionic   365
Zero modes translational   256
Zero translational modes   228

Printing: Krips bv, Meppel
Binding: Stürtz, Würzburg